T0362256

TUBULAR STRUCTURES XI

BALKEMA – Proceedings and Monographs
in Engineering, Water and Earth Sciences

PROCEEDINGS OF THE 11TH INTERNATIONAL SYMPOSIUM AND IIW INTERNATIONAL CONFERENCE ON TUBULAR STRUCTURES, QUÉBEC CITY, CANADA, 31 AUGUST–2 SEPTEMBER 2006

Tubular Structures XI

Editors

J.A. PACKER & S. WILLIBALD
Department of Civil Engineering, University of Toronto, Canada

Taylor & Francis Group

LONDON / LEIDEN / NEW YORK / PHILADELPHIA / SINGAPORE

COVER PHOTOGRAPHS:
Front: Ontario College of Art and Design, Toronto, Canada.
        Photo courtesy of interiorimages.ca
Back: Laboratory Experiment and Finite Element Model of a
        Tubular Connection, Department of Civil Engineering,
        University of Toronto, Canada.

Published by: Taylor & Francis/Balkema
               P.O. Box 447, 2300 AK Leiden, The Netherlands
               e-mail: Pub.NL@tandf.co.uk
               www.taylorandfrancis.co.uk/engineering, www.crcpress.com

ISBN 10: 0-415-40280-8   ISBN 13: 978-0-415-40280-4

Printed in Great Britain

Tubular Structures XI – Packer & Willibald (eds)
© 2006 Taylor & Francis Group, London, ISBN 0-415-40280-8

Table of Contents

Parallel session 2: Static strength of joints

Parallel session 3: Static strength of members/frames

Parallel session 4: Static strength of joints

ISTS Kurobane Lecture

Plenary session B: Code development & applications

Parallel session 5: Composite construction

Parallel session 6: Fire

Special session: CIDECT President's Student Awards

Research Award Competition

Design Award Competition

Parallel session 7: Composite construction

Parallel session 8: Cast steel

Parallel session 9: Static strength of joints

Parallel session 10: Fatigue & fracture

Plenary session C: Seismic

Tubular Structures XI – Packer & Willibald (eds)
© *2006 Taylor & Francis Group, London, ISBN 0-415-40280-8*

Preface

These Proceedings contain the papers presented at the *11th International Symposium and IIW International Conference on Tubular Structures*, held in Québec City, Canada, from August 31 to September 2, 2006. The 10 previous symposia, held between 1984 and 2003, are described in the "Publications of the previous symposia on tubular structures". This series of symposia or conferences began when the International Institute of Welding (IIW) decided to hold its 1984 International Conference on the topic of Tubular Structures, as part of the IIW Annual Assembly that year in Boston. So successful was the meeting that the tubular structures community decided to hold further conferences independently of IIW, but immediately prior or after the IIW Annual Assembly and in the same country. This practice continued from 1986 to 1996, with the exception of ISTS6 in 1994. The eighth, ninth and tenth symposia, from 1998 to 2003, had no relationship to the IIW Annual Assembly timing or venue. This 11th Symposium, or ISTS11, has again returned under the umbrella of the IIW Annual Assembly, from whence it originated, as the topic for the 2006 IIW International Conference. Throughout its 22-year history the frequency, location and technical content of all the symposia has been determined by the IIW Subcommission XV-E on Welded Tubular Structures.

This 11th Symposium marks the first time since the inaugural conference that the meeting has been held in North America. This has been made possible by the generous financial support of a number of industry associations: IIW, Comité International pour le Développement et l'Étude de la Construction Tubulaire (CIDECT), Steel Tube Institute of North America, American Institute of Steel Construction (AISC) and the International Tube Association. In addition several corporate sponsors provided key funding: Atlas Tube Inc., IPSCO Inc., Welded Tube of Canada, Novamerican Steel Inc., Russel Metals and AD Fire Protection Systems. In-kind support was also provided by the Canadian Institute of Steel Construction (CISC) and the University of Toronto. From modest beginnings this Symposium has now grown to become the principal showcase for manufactured tubing and the prime international forum for discussion of research, developments and applications in this field.

This volume contains 83 peer-reviewed papers, which have all been reviewed by two international experts in the field. Of these, 75 have resulted from an initial "Call for Papers", which produced a record number of submitted Abstracts for this symposium series. Two further papers were invited Keynote Addresses: the Houdremont Lecture (selected by IIW) given by Professor Jeffrey Packer of Canada, and the Kurobane Lecture (selected by the ISTS International Programme Committee) given by Professor Peter Marshall of the U.S.A. The "Houdremont Lecture" (or the "Portevin Lecture" in odd years) is the IIW International Conference Keynote Address which has been incorporated into the Opening Session of the IIW International Conference since 1964. The "Kurobane Lecture" is the International Symposium on Tubular Structures Keynote Address which was inaugurated at ISTS8 in 1998. Besides financial support, CIDECT has traditionally maintained a close interest in the ISTS and the CIDECT Annual Meeting has also been held in Québec City in 2006 in conjunction with ISTS11, as was also done in 2003 with ISTS10. Starting with ISTS9 (in 2001) and repeated at ISTS10 (in 2003), CIDECT has sponsored and administered a "CIDECT President's Student Awards" competition. In this, university students world-wide have been invited to submit entries pertaining to their tubular structures research and the three students judged to be the finalists then completed full papers and were invited to make presentations at the ISTS. For ISTS11, CIDECT has expanded this to two competitions: a Research Award Competition and a Design Award

Competition. A judging panel established by CIDECT has resulted in the three finalists from each of these competitions submitting papers and hence six Student Awards papers are also included in these Proceedings. The two winners of these competitions are announced at ISTS11, after the publication of this book. For a limited time after ISTS11, further details on the Symposium are available at: www.ists11.org

The information provided in this publication is the sole responsibility of the individual authors. It does not reflect the opinion of the editors, supporting associations, organizations or sponsors, and they are not responsible for any use that might be made of information appearing in this publication. Anyone making use of the contents of this book assumes all liability arising from such use.

The Editors hope that the contemporary applications, case studies, concepts, insights, overviews, research summaries, analyses and product developments described in this book provide some inspiration to architects, developers, contractors, engineers and fabricators to build ever more innovative and competitive tubular structures. This archival volume of the current "state of the art" will also serve as excellent reference material to academics, researchers, trade associations and manufacturers of hollow sections in the future.

Jeffrey A. Packer & Silke Willibald
Editors
University of Toronto, Canada
May 2006

Publications of the previous symposia on tubular structures

M. A. Jaurrieta, A. Alonso & J.A. Chica (Eds.) 2003. *Tubular Structures X*, 10th International Symposium on Tubular Structures, Madrid, Spain, 2003. Rotterdam: A.A. Balkema Publishers.

R. Puthli & S. Herion (Eds.) 2001. *Tubular Structures IX*, 9th International Symposium on Tubular Structures, Düsseldorf, Germany, 2001. Rotterdam: A.A. Balkema Publishers.

Y.S. Choo & G.J. van der Vegte (Eds.) 1998. *Tubular Structures VIII*, 8th International Symposium on Tubular Structures, Singapore, 1998. Rotterdam: A.A. Balkema Publishers.

J. Farkas & K. Jármai (Eds.) 1996. *Tubular Structures VII*, 7th International Symposium on Tubular Structures, Miskolc, Hungary, 1996. Rotterdam: A.A. Balkema Publishers.

P. Grundy, A. Holgate & B. Wong (Eds.) 1994. *Tubular Structures VI*, 6th International Symposium on Tubular Structures, Melbourne, Australia, 1994. Rotterdam: A.A. Balkema Publishers.

M.G. Coutie & G. Davies (Eds.) 1993. *Tubular Structures V*, 5th International Symposium on Tubular Structures, Nottingham, United Kingdom, 1993. London/Glasgow/New York/Tokyo/Melbourne/Madras: E & FN Spon.

J. Wardenier & E. Panjeh Shahi (Eds.) 1991. *Tubular Structures*, 4th International Symposium on Tubular Structures, Delft, The Netherlands, 1991. Delft: Delft University Press.

E. Niemi & P. Mäkeläinen (Eds.) 1990. *Tubular Structures*, 3rd International Symposium on Tubular Structures, Lappeenranta, Finland, 1989. Essex: Elsevier Science Publishers Ltd.

Y. Kurobane & Y. Makino (Eds.) 1987. *Safety Criteria in Design of Tubular Structures*, 2nd International Symposium on Tubular Structures, Tokyo, Japan, 1986. Tokyo: Architectural Institute of Japan, IIW.

International Institute of Welding 1984. *Welding of Tubular Structures/Soudage des Structures Tubulaires*, 1st International Symposium on Tubular Structures, Boston, USA, 1984. Oxford/New York/Toronto/Sydney/Paris/Frankfurt: Pergamon Press.

Tubular Structures XI – Packer & Willibald (eds)
© 2006 Taylor & Francis Group, London, ISBN 0-415-40280-8

Organisation

This volume contains the Proceedings of the **11th International Symposium on Tubular Structures – ISTS11** held in Québec City, Canada, from August 31 to September 2, 2006. ISTS11 has been organised by the International Institute of Welding (IIW) Subcommission XV-E, Comité International pour le Développement et l'Étude de la Construction Tubulaire (CIDECT) and the University of Toronto, Canada.

INTERNATIONAL PROGRAMME COMMITTEE

J. Wardenier, *Committee Chair*
Delft University of Technology, The Netherlands

Y.S. Choo
National University of Singapore, Singapore

R. Ilvonen, *Chair of CIDECT Technical Commission*
Rautaruukki Oyj, Hämeenlinna, Finland

Y. Kurobane
Kumamoto University, Kumamoto, Japan

M.M.K. Lee
Southampton University, UK

M. Lefranc
Force Technology Norway, Sandvika, Norway

S. Maddox
The Welding Institute, Cambridge, UK

P.W. Marshall
MHP Systems Engineering, Houston, USA

A.C. Nussbaumer
École Polytechnique Fédérale de Lausanne, Switzerland

J.A. Packer
University of Toronto, Canada

R.S. Puthli
University of Karlsruhe, Germany

G.J. van der Vegte
Kumamoto University, Japan/Delft University of Technology, The Netherlands

X.-L. Zhao, *Chair of IIW Subcommission XV-E*
Monash University, Melbourne, Australia

LOCAL ORGANISING COMMITTEE

J.A. Packer, *Chair*
University of Toronto, Toronto, Canada

S. Willibald, *Secretary*
University of Toronto, Toronto, Canada

G.S. Frater
Canadian Steel Construction Council, Toronto, Canada

S. Herth
Continental Bridge, Alexandria, Minnesota, USA

G. Martinez-Saucedo
University of Toronto, Toronto, Canada

F.J. Palmer
Steel Tube Institute of North America, Pittsburgh, USA

Acknowledgements

The Organising Committee wish to express their sincere gratitude for the financial assistance from the following associations: IIW (International Institute of Welding), CIDECT (Comité International pour le Développement et l'Étude de la Construction Tubulaire), the Steel Tube Institute of North America, AISC (American Institute of Steel Construction), the International Tube Association, as well as all the corporate sponsors: Atlas Tube Inc., Russel Metals, AD Fire Protection Systems, IPSCO, Welded Tube of Canada and Novamerican Steel Inc.

The technical assistance of the IIW Subcommission XV-E is gratefully acknowledged. We are also thankful to the International Programme Committee as well as the members of the Local Organising Committee. Finally, the editors (who also served as reviewers) want to acknowledge the kind assistance of the following reviewers:

M.A. Bradford	P.W. Marshall
Y.S. Choo	F. Mashiri
G.S. Frater	A.C. Nussbaumer
S. Herion	R.S. Puthli
V. Kodur	D.R. Sherman
Y. Kurobane	R. Tremblay
M.M.K. Lee	G.J. van der Vegte
M. Lefranc	J. Wardenier
S. Maddox	T.J. Wilkinson
Y. Makino	X.-L. Zhao

Jeffrey A. Packer & Silke Willibald
University of Toronto, Canada

IIW Houdremont Lecture

Tubular Structures XI – Packer & Willibald (eds)
© 2006 Taylor & Francis Group, London, ISBN 0-415-40280-8

Tubular brace member connections in braced steel frames

J.A. Packer
Department of Civil Engineering, University of Toronto, Canada

ABSTRACT: Diagonal bracings are extremely popular elements for lateral load resistance in steel-framed buildings. In turn, the most common shape used for bracing members is the hollow structural section. While the design of such members is straight-forward, the design of gusset-plate connections at the member ends is controversial and the fabrication of these connections can be expensive. This paper reviews the current "state-of-the-art" for the design of such connections, under both static and seismic loading conditions, and for fabricated and cast connections.

1 INTRODUCTION

The total global output of welded tubes, which represent the manufacturing process used for most of the world's structural tubing, has been approximately constant – despite some fluctuations – over the last 10 years: 40.1 million metric tons in 1995 and 41.1 million metric tons in 2004 (IISI 2005). In this same period, however, the world production of crude steel has increased by 41%, from 752 million metric tons in 1995 to 1,058 million metric tons in 2004. Thus, in 2004 welded tubes represent about 4% of the total steel market, but a very important component of the structural steel sector. While some countries have decreased welded tube output in the last decade (e.g. U.S.A.), there has been a huge increase in production in China (by 245% over the period 1995–2004). National production statistics, for the 10 leading countries, are shown in Figure 1 (IISI 2005). These figures do not

include other (less-common) types of hollow sections (e.g. seamless tubes and fabricated sections). While not all of these tonnages will be used for structural purposes, the data is indicative of local consumption and export levels.

In steel structures the most common applications for welded tubes are as columns, in trusses and as lateral bracing members, where the structural engineer can take advantage of excellent properties in compression and the architect can utilize aesthetic qualities in exposed steelwork. Simply-connected steel frames are typically laterally-braced with diagonal members as shown in Figure 2. The ends of the Hollow Structural Section (HSS) bracings are then usually connected to the steel frame via gusset plates, as shown in Figure 3. The design of the bracings, as compression or tension members, is performed in accordance with applicable national or regional structural steel specifications. For low-rise structures with lateral loads governed by static

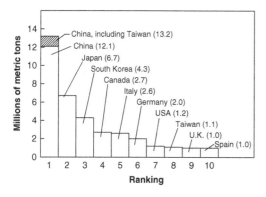

Figure 1. The 10 leading producers of welded tubes, by country, for 2004 (IISI 2005, Table 29).

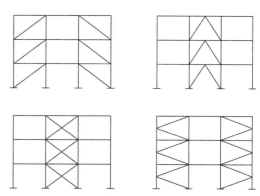

Figure 2. Typical configurations of concentrically-braced steel frames using hollow sections as bracings.

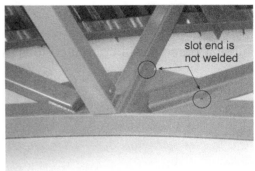

Figure 4. Square HSS diagonals with slotted ends, in a roof truss.

Figure 3. Statically-loaded steel frame, braced with diagonal hollow sections.

(wind) loading, bracing member selection will often be controlled by maximum permitted member slenderness limits. (For example, in Canada $(KL/r)_{max} = 200$ in compression and, generally, 300 in tension (CSA 2001)). In structures with lateral load design governed by seismic actions, bracing member selection will be further restricted by limits on the slenderness of the member cross-section. For example, for moderately ductile concentrically braced frames in Canada, where moderate amounts of energy are dissipated through yielding of bracing members with $(KL/r) \leq 100$, the flat width-to-thickness ratio of square and rectangular HSS must be $\leq 330/\sqrt{F_y}$ and the diameter-to-thickness ratio of circular HSS must be $\leq 10,000/F_y$. These cross-section slenderness limits, in which the yield stress F_y is expressed in MPa or N/mm^2, are considerably lower than the normal Class 1 limits (CSA 2001). In current U.S. provisions for "special" and "ordinary" concentrically braced frames, these cross-section slenderness limits are even more restrictive: $286/\sqrt{F_y}$ for square/rectangular HSS and $8,800/F_y$ for circular HSS (AISC 2005a).

2 GUSSET PLATE CONNECTIONS TO THE ENDS OF HOLLOW SECTIONS – STATIC LOADING

Single plates are often inserted into the slotted ends of a round or square HSS, concentric to the axis of the HSS member. This is done both in roof trusses (typically to avoid round-to-round HSS tube profiling associated with directly-welded members, as shown in Figure 4) and in diagonal bracing members in braced frames (Figure 3). This inserted plate is frequently then connected to a single gusset plate, usually by bolting. In such situations a bending moment is induced in the joint by the eccentricity between the plates which must be considered. Under compression loads the plates need to be proportioned as beam-columns, and assuming that both ends of the connection can sway laterally relative to each other. This is frequently overlooked, leading to periodic structural failures, but the American HSS Connections Manual (AISC 1997, Chapter 6) is not guilty of this omission and gives a reasonable and simple design method. Alternatively, the single gusset plate attached to the building frame can be stiffened, typically by adding another transverse plate along at least one edge of the gusset, thereby giving the gusset attached to the building frame a T-shape in cross-section.

With regard to the performance of the HSS in such connections, load is only transmitted initially to a portion of the HSS cross-section, thereby creating a shear lag effect which may result in a lower HSS capacity in both compression and tension. For tension loading on the HSS member, the effective area (A_e) is determined by the net area (A_n) multiplied by a shear lag factor, U. For the latter, the most recent specification version is given by AISC (2005b). These U factors have been revised by AISC from the previous specification (AISC 2000), where U had an upper limit of 0.9. Based on the work of Cheng & Kulak (2000) the U factor can now be taken as 1.0 for connections to circular HSS with a sufficiently-long inserted plate and weld length (L_w). Table 1 shows the current AISC U factors for circular HSS compared to those from other Canadian codes/guides, and Figure 5 illustrates the geometric parameters used. For the shear lag effect, Eurocode 3 (CEN 2005) only addresses bolted connections for angles connected by one leg and other un-symmetrically connected tension members.

North American specifications have gone through many revisions (Geschwindner 2004) concerning the design methods for the limit state of tensile fracture affected by shear lag. Table 1 illustrates the two main prevailing methods: based on the connection

Table 1. Shear lag design provisions for circular and elliptical hollow sections.

Specification or design guide	Effective net area	Shear lag coefficients		Range of validity
AISC (2005b):		$U = 1 - \dfrac{\bar{x}}{L_w}$	for $1.3D > L_w \geq D$	$L_w \geq D$
Specification for Structural Steel Buildings		$U = 1.0$	for $L_w \geq 1.3D$ (for circular HSS)	
CSA (1994): Limit States Design of Steel Structures		$U = 1.0$	for $L_w/w \geq 2.0$	$L_w \geq w$
		$U = 0.87$	for $2.0 > L_w/w \geq 1.5$	
		$U = 0.75$	for $1.5 > L_w/w \geq 1.0$	
CSA (2001): Limit States Design of Steel Structures	$A_e = A_n \cdot U$	$U = 1.0$	for $L_w/w \geq 2.0$	no restrictions
		$U = 0.5 + 0.25 \, L_w/w$	for $2.0 > L_w/w \geq 1.0$	
		$U = 0.75 \, L_w/w$	for $L_w/w < 1.0$	
Packer and Henderson (1997): Hollow Structural Section Connections and Trusses – A Design Guide		$U = 1.0$	for $L_w/w \geq 2.0$	shear lag not critical for $L_w < 0.6\,w$
		$U = 0.87$	for $2.0 > L_w/w \geq 1.5$	
		$U = 0.75$	for $1.5 > L_w/w \geq 1.0$	
		$U = 0.62$	for $1.0 > L_w/w \geq 0.6$	

$T_r = \phi \, A_e \, F_u$ (AISC (2005b) Specification, $\phi = 0.75$) or $T_r = 0.85 \, \phi \, A_e \, F_u$ (CSA (2001) Specification, $\phi = 0.9$), where T_r = factored tensile resistance, F_u = ultimate tensile stress and ϕ = resistance factor.

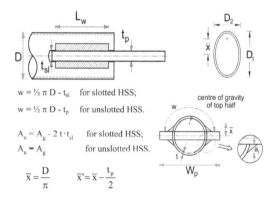

$w = \frac{1}{2} \pi D - t_{sl}$ for slotted HSS;

$w = \frac{1}{2} \pi D - t_p$ for unslotted HSS.

$A_n = A_g - 2\,t \cdot t_{sl}$ for slotted HSS;

$A_n = A_g$ for unslotted HSS.

$\bar{x} = \dfrac{D}{\pi}$ $\bar{x}' = \bar{x} - \dfrac{t_p}{2}$

Figure 5. Important geometric parameters influencing connection design.

eccentricity (AISC) or based on the distance between the welds (CSA). In this table it can be seen that the Packer and Henderson (1997) approach is just a modification of the CSA (1994) method. Note that the resistance factor of $\phi = 0.75$ for AISC (2005b) is approximately the same as $(0.9)(0.85) = 0.765$ for CSA (2001). The other tensile limit state for these connections is "block shear" (or tear-out) and the current North American and European design provisions are given in Table 2. As can be seen, all use a design model based on the summation of the resistance of the part in tension (where all use the net area in tension multiplied by the ultimate tensile stress) and the resistance of the part in shear. The latter can be calculated based on the net/gross area in shear multiplied

by the shear yield stress/shear ultimate stress, depending on the specification. At present the American and Canadian specifications use a common design model but quite different resistance factors. (The Canadian resistance factor is currently under review).

A study of both concentric gusset plate-to-slotted tube and slotted gusset plate-to-tube connections, under both static tensile and compression member loadings, using both round and elliptical HSS, has been underway at the University of Toronto since 2002. The connection fabrication details investigated, which include both end return welds and connections leaving the slot end un-welded, are shown in Figure 6. The latter fabrication detail is extremely popular in North-America because it provides maximum erection tolerance if the plate-to-HSS joint is fillet welded on site. Sometimes the slot end is roughly cut, in practice, as shown in Figure 7, but this has negligible influence on the connection capacity in quasi-static loading as cracking (leading to ultimate failure) initiates at the end of the weld, not at the end of the slot (see Figure 7). For dynamic loading situations, however, it is recommended that the slot end be drilled or cut accurately. Complete details of the experimental testing program can be found elsewhere (Willibald et al. 2006) but examples of the two classic failure modes are shown in Figure 8.

The experimental program by Willibald et al. (2006) concluded that the block shear design model (Table 2), although based on limited correlations, was suitable, particularly if predictions were calculated using a theoretical fracture path excluding the welds. Further proposals have been recently made to improve the

Table 2. Block shear (tear-out) design provisions.

Specification or design guide	Block shear strength
AISC (2005b): Specification for Structural Steel Buildings	$T_r + V_r = \phi\, U_{bs}\, A_{nt}\, F_u + 0.6\, \phi\, A_{gv}\, F_y \le \phi\, U_{bs}\, A_{nt}\, F_u + 0.6\, \phi\, A_{nv}\, F_u$ with $\phi = 0.75$ and $U_{bs} = 1$
CSA (2001): Limit States Design of Steel Structures	$T_r + V_r = \phi\, A_{nt}\, F_u + 0.6\, \phi\, A_{gv}\, F_y \le \phi\, A_{nt}\, F_u + 0.6\, \phi\, A_{nv}\, F_u$ with $\phi = 0.9$
Eurocode (CEN 2005): Design of Steel Structures – General Rules – Part 1–8: Design of Joints [a]	$T_r + V_r = \dfrac{1}{\gamma_{M2}}\, A_{nt}\, F_u + \dfrac{1}{\gamma_{M0}}\, \dfrac{1}{\sqrt{3}} A_{nv}\, F_y$ $\gamma_{M0} = 1.0$ and $\gamma_{M2} = 1.25$

[a] Design rule for bolted connections differs slightly.
T_r = factored tensile resistance, V_r = factored shear resistance, A_{nt} = net area in tension, A_{nv} = net area in shear, A_{gv} = gross area in shear and F_y = yield tensile stress.

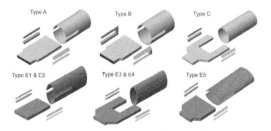

Figure 6. Fabricated connection details investigated.

Figure 7. Example of an inserted plate with the HSS slot end roughly cut and left open.

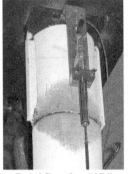
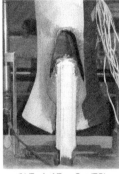

(a) Typical Circumferential Failure (CF) of the HSS, induced by Shear Lag

(b) Typical Tear-Out (TO) Failure along the Weld

Figure 8. Failure modes for gusset plate-to-HSS connections in tension.

general block shear model in Table 2 (Franchuk et al. 2004, Driver et al. 2006) by adjusting the shear resistance term. It should be noted, however, that their recommendations are based only on bolted connection data. These experiments by Willibald et al. (2006) also confirmed that both the AISC (2005b) and CSA (2001) shear lag factors (Table 1) were excessively conservative, as has been noted by other researchers.

The better shear lag factor method was that by AISC, but Willibald et al. (2006) suggested that the existing formulation could be much improved by reducing the connection eccentricity $\bar{x}$ term – used to calculate U – to $\bar{x}'$, as shown in Figure 5. This essentially accounts for the thickness of the gusset plate, which is often substantial relative to the tube size. Interestingly, a very similar conclusion has just been reached by Dowswell & Barber (2005) for slotted rectangular HSS connections, whereby they propose an "exact" $\bar{x}$ term calculated by using a distance from the edge of the gusset plate to the outside wall of the HSS. Dowswell & Barber (2005) verify their proposal by showing improved accuracy relative to published test data by others.

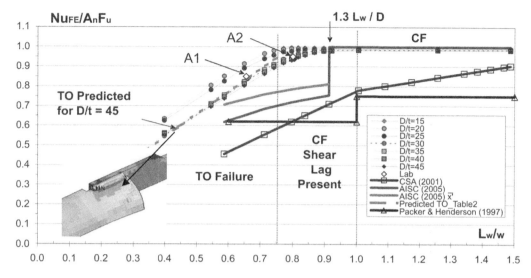

Figure 9. Results of parametric FE analysis and experiments (A1, A2) for connection type A (see Fig. 6) [Tension loading; circular HSS with the slot end not filled: a very popular bracing member detail in practice.] N_{uFE} = connection ultimate strength by FE analysis.

Following experimental research on the connection types shown in Figure 6, an extensive detailed numerical study followed on the same connections using non-linear Finite Element (FE) Analysis (Martinez-Saucedo et al. 2005). A full parameter study expanded the total experimental and numerical database to over 700 connections (Martinez-Saucedo et al. 2006). The FE models revealed a gradual transition between the failure modes of block shear/tear-out (TO) and circumferential tension fracture (CF), with the latter sometimes influenced by the shear lag phenomenon (see Figure 9). A continual monotonic increase in the connection capacity was achieved as the weld length increased. The transition point between these failure modes depended on factors such as: the connection type, the weld length, the tube diameter-to-thickness ratio and the connection eccentricity, $\bar{x}$ (the latter having a strong influence for elliptical HSS). This gradual transition between the failure modes is in contrast to the behaviour given by design models in current specifications, since these specifications do not consider a gradual change between these limit states. Thus, a more unified and less conservative design model for slotted gusset plate HSS connections can be expected in the near future. Figure 9 also confirms that a value of U = 1.0 (hence 100% of A_nF_u) for circular HSS with $L_w/D \geq 1.3$ (AISC 2005b) is indeed correct, and for all practical tube diameter-to-thickness (D/t) ratios. However, the conservative connection capacity predictions by over-estimating the severity of the shear lag effect at $L_w/D \leq 1.3$ are very apparent.

3 GUSSET PLATE CONNECTIONS TO THE ENDS OF HOLLOW SECTIONS – SEISMIC LOADING

If the results in Figure 9 are re-plotted in terms of $N_{uFE}/A_g F_y$, where A_g is the tube gross area, then it can be shown that long plate insertion lengths can achieve tension capacities very close to $A_g F_y$, even for this connection type with an open slot end. However, in tension-loaded energy-dissipating braces the connection will be required to resist an even greater load of $A_g R_y F_y$, where R_y is a material over-strength factor to account for the probable yield stress in the HSS bracing. This value of R_y is specified as 1.1 in Canada (CSA 2001), and 1.4 (for A500 Grades B and C (ASTM 2003)) or 1.6 (for A53 (ASTM 2002)) in the U.S. (AISC 2005a). The Canadian value is too low, based on personal laboratory testing experience, and a realistic value for the mean expected yield strength-to-specified minimum yield strength ratio is around 1.3, for CSA-grade HSS (CSA 2004). Tremblay (2002) reported a mean over-strength yield value of 1.29 for rectangular HSS surveyed, and Goggins et al. (2005) have reported a mean over-strength yield value of 1.49 for rectangular HSS (Europe), but the latter was the result of specifying low grade 235 MPa steel. The high U.S. values were determined by a survey of mill test reports by Liu (2003) and are not surprising because, in a market like North America with several different steel grades and production standards, manufacturers will produce to the highest standard and work to

7

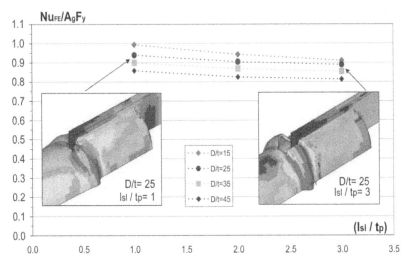

Figure 10. Results of parametric FE analysis for connection type A (see Fig. 6). [Compression loading; circular HSS with the slot end not filled: a very popular bracing member detail in practice.] N_{uFE} = connection ultimate strength by FE analysis, l_{sl} = length of open slot and t_p = plate thickness.

a "one product fits all" approach. (For example, in her survey Liu's ASTM A500 data all pertained to Grade B tubing, whereas manufacturers will knowingly produce to meet the higher Grade C strengths). AISC, however, has now introduced another material over-strength factor, R_t, to account for the expected tensile ultimate strength relative to the specified minimum tensile strength (AISC 2005a) with these values being 1.3 for ASTM A500 Grades B and C and 1.2 for ASTM A53. This R_t factor is applied to fracture limit states in designated yielding members – such as bracings in concentrically braced frames where circumferential fracture (CF) is a design criterion. Thus, applying capacity design principles to preclude nonductile modes of failure within a designated yielding member (bracing) and setting the resistance factors $\phi = 1.0$, one obtains the following, to avoid circumferential fracture of the HSS at the gusset plate (refer to the equations below Table 1):

AISC (2005a):
$\phi R_t F_u A_e \geq \phi R_y F_y A_g$, hence for ASTM A500 HSS and setting $F_y \leq 0.85 F_u$, $A_e \geq 0.92 A_g$
CSA (2001):
$(0.85 F_u A_e) R_y \geq R_y F_y A_g$, hence for CSA HSS and setting $F_y \leq 0.85 F_u$, $A_e \geq 1.00 A_g$

In the case of AISC (2005a) the handling of ϕ factors in the above inequality is still under debate. By retaining ϕ factors, and using a lower ϕ factor for the fracture criterion (the left hand side of the inequality), the required $A_e > A_g$. A good discussion of ϕ, R_t and R_y factor selection is given by Haddad and Tremblay (2006).

From the above, one can see that the required minimum effective net area – after consideration of shear lag and application of the U factor – is near, or above, the gross area of the HSS bracing.

In compression, type A connections (see Figures 6 and 10) can be shown to achieve capacities that also approach $A_g F_y$, provided the length of the open slot is kept short (in the order of the plate thickness) and the tube is relatively stocky (see Fig. 10). However, despite the achievement of high compression load capacity this is accompanied by considerable plastic deformation in the tube at the connection, which is likely to also render the connection unsuitable for use in energy-dissipating brace members.

Fabricated end connections to tubular braces, in concentrically braced frames, hence have great difficulty meeting connection design requirements under typical seismic loading situations. Reinforcement of the connection is then the usual route. It is difficult to fabricate cover plates for round HSS members so square HSS with flat sides have become the preferred section, resulting in costly reinforced connections as shown in Figure 11. Moreover, recent research on the performance of HSS bracings under seismic loading *still* concentrates on square/rectangular hollow sections (Goggins et al. 2005; Elghazouli et al. 2005; Tremblay 2002). A drawback of using cold-formed, North American square/rectangular HSS is that they have low ductility in the corners and are prone to fracture in the corners after local buckling during low-cycle fatigue.

A clear improvement is to use cold-formed circular hollow sections, which do not have corners, and to

Figure 11. Fabricated square HSS gusset connection for seismic application. (Photo courtesy of Pierre Gignac, Canam Group, Canada).

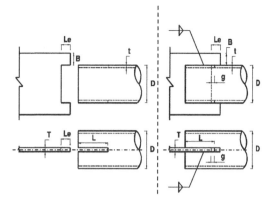

Figure 12. Fabricated connection detail using an over-slotted circular HSS but with $A_n = A_g$ at the weld termination.

end of the gusset plate which corresponds to a tube cross-section where the gross area applies.

attempt to avoid reinforcement. Yang & Mahin (2005) recently performed six tests on slotted square HSS and slotted circular HSS under seismic loading and highlighted the improved performance of the circular member, which was "much more resistant to local buckling". Additionally, the use of ASTM A53 Grade B (ASTM 2002) pipe, which is readily available in the U.S. but not Canada, provides a suitably low nominal F_y/F_u ratio of 0.58, which makes the connection much more resistant to fracture at the critical net section and a real design option without reinforcement. ASTM A53 Grade B can be compared to the popular ASTM A500 square HSS Grade C which has a nominal F_y/F_u ratio of 0.81. North American-produced square/rectangular HSS are also known to have poor impact resistance properties since, unlike their European cold-formed counterparts, they are normally produced with no impact rating (Kosteski et al. 2005). Regardless of the section shape and steel grade chosen for energy-dissipative bracings, it is clearly preferable to specify a maximum permissible material strength on engineering drawings, as per Eurocode 8 (CEN 2004).

The use of fabricated, slotted circular HSS gusset plate connections, without reinforcement, is hence being further explored at the University of Toronto. Fabrication with the slot end un-welded (i.e. without an end return weld) is a very popular practice in North America, so special details are being investigated which still permit this concept yet provide a net area (A_n) equal to the gross area (A_g) at the critical cross-section, such as shown in Figure 12. As can be seen in Figure 12, a small gap is still provided at the tube end for fit-up, but the weld terminates at the

4 CAST STEEL CONNECTIONS – SEISMIC APPLICATIONS

Cast steel joints have enjoyed a renaissance in Europe in conjunction with tubular steel construction, mainly as truss-type nodes in dynamically-loaded pedestrian, highway and railway bridges where fabricated nodes would have been fatigue-critical. Another popular application has been in tree-like tubular roof structures where the smooth lines of a cast node have great architectural appeal. Cast steel connectors to tubular braces under severe seismic load conditions have not been used to date, but cast steel connections represent a solution to the design dilemma of fabricated bracing member connections and these can be specially shaped to provide material where it is particularly needed. Types currently under investigation at the University of Toronto, which are designed to remain elastic under the full seismic loading regime, are shown in Figure 13. These connectors are being made using a sand-casting process, which involves some post-casting machining and hole drilling finishing operations. By mass-producing cast end connectors, to suit popular circular HSS bracing member sizes, an economic and aesthetic solution can be reached that still allows the use of regular HSS members and avoids the use of alternatives like buckling-restrained braces, which require pre-qualification by testing and a high level of quality assurance (AISC 2005a). Cast end connectors thus represent another exciting development in the evolution of tubular steel construction. Current work in Canada on cast connectors to tubular members is summarized elsewhere by De Oliveira et al. (2006).

Further research on cast steel nodes, mainly oriented to wide flange beam-to-column moment

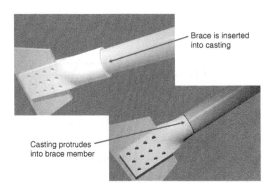

Brace is inserted into casting

Casting protrudes into brace member

Figure 13. Cast steel connections to tubular braces for seismic load applications.

connections and primarily for seismic applications, is also underway at present at the University of Arizona (Sumer et al. 2004, Fleischman et al. 2006). Such cast modular components avoid the use of diaphragms and special detailing requirements, which are currently advocated for moment connections to HSS columns under seismic loading (Kurobane et al. 2004). It should be noted that this latter publication (CIDECT Design Guide No. 9) does not cover connections to tubular braces under seismic loading.

Another innovative connection solution for wide flange beam-to-HSS columns has been launched by California-based ConXtech Inc., termed the SMRSF. With this, a pre-engineered collar connection is fitted around 102 mm or 203 mm square HSS columns and bolted together on site, resulting in very fast construction times. Although it uses machined components that are shop-welded in place, rather than cast components, this connection is also pre-qualified for use as a fully-restrained, Special Moment Resistant Frame connection under the latest FEMA and AISC seismic provisions. Novel connection solutions such as these herald a potential paradigm shift in HSS construction technology.

ACKNOWLEDGEMENTS

The major contributions of Dr. Silke Willibald, Mr. Gilberto Martinez-Saucedo and Mr. J. Carlos de Oliveira to research on this topic at the University of Toronto are gratefully acknowledged, as is the collaboration of Professors Constantin Christopoulos, Robert Tremblay and Xiao-Ling Zhao.

REFERENCES

AISC. 1997. *Hollow structural sections connections manual.* Chicago: American Institute of Steel Construction/Steel Tube Institute of North America/American Iron and Steel Institute.

AISC. 2000. *Load and resistance factor design specification for the design of steel hollow structural sections.* Chicago: American Institute of Steel Construction.

AISC. 2005a. Seismic provisions for structural steel buildings, *ANSI/AISC 341-05* and *ANSI/AISC 341s1-05.* Chicago: American Institute of Steel Construction.

AISC. 2005b. Specification for structural steel buildings, *ANSI/AISC 360-05.* Chicago: American Institute of Steel Construction.

ASTM. 2002. Standard specification for pipe, steel, black and hot-dipped, zinc-coated, welded and seamless, *ASTM A53/A53M-02.* West Conshohocken: ASTM International.

ASTM. 2003. Standard specification for cold-formed welded and seamless carbon steel structural tubing in rounds and shapes, *ASTM A500-03a.* West Conshohocken: ASTM International.

CEN. 2004. Eurocode 8: Design of structures for earthquake resistance – part 1: general rules, seismic actions and rules for buildings, *EN 1998-1: 2004(E).* Brussels: European Committee for Standardisation.

CEN. 2005. Eurocode 3: Design of steel structures – general rules – part 1–8: design of joints, *EN 1993-1-8: 2005(E).* Brussels: European Committee for Standardisation.

Cheng, J.J.R. & Kulak, G.L. 2000. Gusset plate connection to round HSS tension members. *Engineering Journal,* AISC 37(4): 133–139.

CSA. 1994. Limit states design of steel structures, *CAN/CSA-S16.1-94.* Toronto: Canadian Standards Association.

CSA. 2001. Limit states design of steel structures, *CAN/CSA-S16-01.* Toronto: Canadian Standards Association.

CSA. 2004. General requirements for rolled or welded structural quality steel/structural quality steel, *CAN/CSA-G40.20-04/G40.21-04.* Toronto: Canadian Standards Association.

De Oliveira, J.C., Willibald, S., Packer, J.A., Christopoulos, C. & Verhey, T. 2006. Cast steel nodes in tubular construction – Canadian experience. *Proc. 11th. Intern. Symp. on Tubular Structures.*

Dowswell, B. & Barber, S. 2005. Shear lag in rectangular hollow structural sections tension members: comparison of design equations to test data. *Practice Periodical on Structural Design and Construction,* ASCE 10(3): 195–199.

Driver, R.G., Grondin, G.Y. & Kulak, G.L. Unified block shear equation for achieving consistent reliability. *Journal of Constructional Steel Research* 62: 210–222.

Elghazouli, A.Y., Broderick, B.M., Goggins, J., Mouzakis, H., Carydis, P., Bouwkamp, J. & Plumier, A. 2005. Shake table testing of tubular steel bracing members. *Structures & Buildings,* ICE 158(SB4): 229–241.

Fleischman, R.B., Sumer, A., Pan, Y. & Palmer, N.J. 2006. Cast modular components for steel construction. *Proc. North American Steel Construction Conf.:* Session M5.

Franchuk, C.R., Driver, R.G. & Grondin, G.Y. 2004. Reliability analysis of block shear capacity of coped steel beams. *Journal of Structural Engineering,* ASCE 130(12): 1904–1912.

Geschwindner, L.F. 2004. Evolution of shear lag and block shear provisions in the AISC specification. *Proc. ECCS/AISC Workshop, Connections in Steel Structures V:* 473–482.

Goggins, J.M., Broderick, B.M., Elghazouli, A.Y. & Lucas, A.S. 2005. Experimental cyclic response of cold-formed hollow steel bracing members. *Engineering Structures* 27: 977–989.

Haddad, M. & Tremblay, R. 2006. Influence of connection design on the inelastic seismic response of HSS steel bracing members. *Proc. 11th. Intern. Symp. on Tubular Structures*.

IISI. 2005. *Steel statistical yearbook*. Brussels: International Iron and Steel Institute.

Kosteski, N., Packer, J.A. & Puthli, R.S. 2005. Notch toughness of internationally produced hollow structural sections. *Journal of Structural Engineering*, ASCE 131(2): 279–286.

Kurobane, Y., Packer, J.A., Wardenier, J. & Yeomans, N. 2004. *Design guide for structural hollow section column connections. CIDECT Design Guide No. 9*. Köln: CIDECT and TÜV-Verlag.

Liu, J. 2003. Examination of expected yield and tensile strength ratios, *Draft Report + Draft Addendum Report to AISC*. West Lafayette: Purdue University.

Martinez-Saucedo, G., Packer, J.A., Willibald, S. & Zhao, X.-L. 2005. Finite element modelling of gusset plate-to-tube slotted connections. *Proc. 33rd. CSCE Annual General Conf.*: paper no. GC-115.

Martinez-Saucedo, G., Packer, J.A., Willibald, S. & Zhao, X.-L. 2006. Finite element analysis of slotted end tubular connections. *Proc. 11th. Intern. Symp. on Tubular Structures*.

Packer, J.A. & Henderson, J.E. 1997. *Hollow structural section connections and trusses – A design guide*, 2nd. ed. Toronto: Canadian Institute of Steel Construction.

Sumer, A., Pan, Y., Wan, G. & Fleischman, R.B. 2004. Development of modular connections for steel special moment frames. *Proc. 13th. World Conf. on Earthquake Engineering*: Paper 2862.

Tremblay, R. Inelastic seismic response of steel bracing members. *Journal of Constructional Steel Research* 58: 665–701.

Willibald, S., Packer, J.A. & Martinez-Saucedo, G. 2006. Behaviour of gusset plate to round and elliptical hollow structural section end connections. *Canadian Journal of Civil Engineering* 33:

Yang, F. & Mahin, S. 2005. Limiting net section fracture in slotted tube braces, *Steel Tips – Structural Steel Education Council*. Berkeley: University of California.

Plenary session A: Applications & case studies

Tubular Structures XI – Packer & Willibald (eds)
© 2006 Taylor & Francis Group, London, ISBN 0-415-40280-8

Tubular structures for liquid design architecture

M. Eekhout
Octatube Space Structures, Delft, The Netherlands
Delft University of Technology, The Netherlands

S. Niderehe
Octatube Space Structures, Delft, The Netherlands

ABSTRACT: This paper deals with two applications of the technology of tubular structures, as it has been developed over the recent decades for applications in architecture thanks to the fundamental work of civil engineering scientists. Recent architecture is greatly influenced by 3D modelling computer programs which leads to 3D liquid design architecture. In structural engineering terms the design of structures will be governed by liquid design whims of architects, that do not conform to any of the efficient schemes developed in recent years, like space frames, shell structures, tensile or compression structures. Rather the form is not structurally logical. The form is simply given and has to be realized. This realization will require a mix of many primary structural systems. The two examples will illustrate this combined approach: the steel structure for the central body of the Great Hall of the Rabin Center in Tel Aviv (Israel) and the Glass House of Malmö (Sweden).

1 INTRODUCTION

The influence of 3D design in structures is greatly powered by new 2D computer design programs which can determine architectural building forms in almost every imaginable shape. As the main driving force of architects is sculptural, the realization of these dreams has become the task of structural designers, without very much feedback. Of course there are always easier ways to reach a stable structure, but usually the architect wants to see his liquid form realized. It is exactly at this point where the journey of the structural engineer starts. The path is more difficult than it has ever been. The result: an improbable structure, surprising and unbelievable. Nevertheless, after a demanding discovery route for the structural engineer only the result matters.

2 RABIN CENTER, TEL AVIV, ISRAEL

2.1 Introduction

In November 2002 we received tender drawings of a design by architect Moshe Safdie from Boston, USA as a part of the Yitzhak Rabin Center in Tel Aviv. The design of the building was an elaboration and extension of a former auxiliary electricity plant near a university campus, in order to become a memorial building for the late prime minister Yitzhak Rabin who was murdered in 1995, on November 4th. He was seen as a peace maker and was awarded the Nobel price for Peace (1994). His activities led to the so-called "Oslo peace talks". The tender we received provided for two building parts: the "Great Hall" and the "Library". These two big rooms both have large glass façades facing south towards the valley below. Both hall designs have remarkable and plastically designed roofs that resemble dove wings as a tribute to Rabin. Moshe Safdie is well known since he designed the 'Habitat' of Montreal as a part of the World Exhibition of 1967 when he was a 27 year old architect (Kohn et al. 1996).

2.2 Tender drawings

We had worked for Safdie before on the glass cone of the Samson Center in Jerusalem, overlooking a valley adjacent to the old city near the Jaffa Gate. In our eyes he is an almost prophetic designer who designs beautiful interior spaces. But he is also a perfectionist who supervises all of his delegates in his projects all over the world. He is gifted with a desire for ruthless accuracy in the engineering stage. The Samson glass dome, overlooking Jerusalem with its golden colour in the afternoon, is used for marriage feasts and other celebrations and is a great success. Safdie was very satisfied with our alternative design proposals and with the realized accuracy.

The complicated liquid design roofs of the Rabin Center contained in the tender were analyzed by ARUP

Figure 1. Exterior of the Samson Center, Jerusalem.

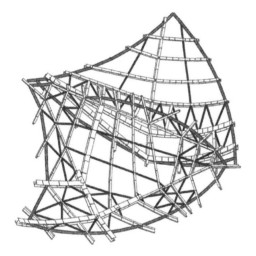

Figure 2. Tender drawing of "the Library" made by ARUP.

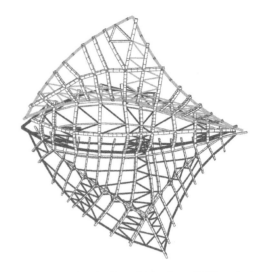

Figure 3. Tender drawing of "the Great Hall" made by ARUP.

New York to be made of a system of arbitrary open steel profiles with a layer of concrete on top. The specification left the roof cladding up to the contractors. On top of this the architect requested a seamless solution in the roof.

For two months we did not give the tender drawings and the thick specification much notice. We thought it was better to leave this project to one of our newly appeared competitors. The "seamless" requirement would make any prefabricated system very difficult and the success would depend entirely on local labor and supervision, which we do not like as a producer of industrial and prefabricated systems. However, the client and his building manager kept on reminding us of the tender date. They even postponed it for one month. It seemed difficult for them to find a trustworthy answer, according to their message. Finally it annoyed us that we would not be able to find a proper technical solution for a seamless fluid roof. This was too much of an intellectual challenge for us as technical designers and inventors.

2.3 Alternative idea

In a few brainstorms we came to the following basic idea: make the roofs as giant surfboards of foam with stressed GRP skins on both sides. The size of the roofs, subdivided into 5 different roof wings was a maximum of 30×30 m. Each wing had a maximum length of 30 m and a width of 15 to 20 m. It was a self-secluded form in itself – the very realization meant a possible step forward in larger sized architectural objects. It was the technology and the resulting technical product that was fascinating.

In a month we organized three successive brainstorms on the product idea, the structural concept and the logistics & pricing. These meetings were open discussions in which all concerned parties had to make up their mind whether they were able to live up to their part of the job and to estimate the financial complications. We noticed that all members of the team were a bit afraid of the adventure, which became apparent when the first cost estimations were added up. We decided to work out and price our stressed sandwich skin alternative as well as the original tender specification of the steel structure with a non-described, free covering as a variation. The deadweight of the steel structure was estimated by ARUP, so a price for the original with cladding variation was easy to make. The cladding we proposed for the original tender design was derived from the mega-sandwich idea, but now in a thinner scale version of 50 to 80 mm thickness, as it only needed to span the space between the steel structure elements (max. 3 m).

The budget calculations came out on a level of 2.5 million Euro for the original design with a thin 80 mm thick GRP sandwich cladding instead of concrete. The

Figure 4.   Two construction types: a steel structure of circular hollow sections (left) and the structural sandwich structure (right).

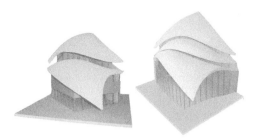

Figure 5.   Models made for the tender of 'The Library' and "The Great Hall".

alternative design with the full load bearing stressed skin sandwich would add up to more than 4 million Euro, largely due to the high estimates of the production of the polyester parts. We argued that the maximum extra costs could not exceed one million Euro, resulting in a total price of 3.5 million Euro. We were sure that any architect would fall in love with the alternative idea of the self-supporting stressed skin sandwich. This was what we faxed to Israel, just in time before the tender closing date, accompanied by a letter explaining the two quoted systems.

2.4   *Amazing solution*

Only two days after the tender closed we received a telephone call from the local representative architect Avi Halberstadt, speaking on behalf of Moshe Safdie. He gave us the compliment that the architect saw the alternative proposal as *"an amazing solution"*. Apparently they had not thought about this concept themselves and this was the most satisfactory compliment we ever received. For us, whether we were awarded the contract or not, intellectually we had already won. Halberstadt invited Eekhout to come over to Tel Aviv for a meeting, so he could present his ideas to the building commission. At the presentation he showed the polystyrene models that Haiko Dragstra had machined in a demountable model scale of 1 to 40. The model showed that the corner details in the design had not yet been accurately designed and that the overall stability was not satisfactory. The design needed

a lot of engineering work. The building commission went into a separate meeting.

2.5   *Extremely innovative but too expensive*

After one hour of fierce discussions, the outcome was that the tender original with the GRP covering was practically on the average tender price level. On the other hand they noted that the alternative proposal was indeed very attractive from the viewpoint of its extremely innovative design and construction, but was priced one million Euro over budget. We responded that this was known from the date we sent in the quotation. Knowing the intellectual value of the alternative proposal, it would have been stupid to sell it at a lower price than the tender proposals. Usually technical alternatives are more efficient solutions for the contractors and tend to be lower in price than the original. A more expensive alternative is rare and hence extraordinary. Since most of the tender results in Israel showed that Octatube was the most expensive party, it was and still is, a sport to get the order. Starting with the highest price and the best technology, may end with a contract at a compromised price. We also loose projects as our competitors can copy our technology after one completed project and execute this without the necessary research and without the higher Dutch labour costs of Octatube. But in the case of the wings, the alternative idea was to become a technical world novelty and Moshe Safdie understood this. He stated that the idea was unbelievable and never done before to his knowledge. If someone could make it work in his opinion, it was to be Octatube. He remembered our collaboration for the Samson Center two years before and he had made some phone calls to other Israeli architects to check our reputation. We had the complete support of the architect. He was our product hero, protecting us with an umbrella. The response of the chairman of the building committee was to come up with different logistics for the GRP sandwich proposal in a manner that the price level could be lowered to 2.5 million Euro. He suggested that it might be possible to transfer the foam machining and the GRP production to Israel in order to reduce costs for shipment and labour at the same time. This was the message we took home 29 April 2003. The committee wanted to speed up their decision, so we had to come up with an alternative idea within one or two weeks.

2.6   *Rethinking*

Back in Delft we discussed the consequences with our engineers and the external team members. The plan was born in the airplane from Israel to the Netherlands. If we could build the GRP sandwich roofs, it would be a hit on the world market. We were prepared to transfer more labour to Israel in order to reduce costs and talk to new Israeli partners if our current partners would let us down in order to realize this proposal.

Figure 6. Three prototypes (left: stressed membrane; centre: to be locally produced; right: prefab sandwich).

First we could try to decompose the big wings into transportable components, which we could assemble on-site on a jig and finish the broken GRP layers and give the shells a final top-layer or top-coat. We could have machined the polystyrene blocks and we could also set up an Israeli GRP plant in Tel Aviv on the building site. The most likely position to assemble a wing would be in a vertical position. More or less like stacking bricks, we would assemble the wall from placing the polystyrene blocks on top of each other. This way both polystyrene skins could be treated simultaneously and we could control shrinking of the foam. The subcontractor was not experienced in estimating larger productions than mock-ups. The bottom price of our subcontractor for composite productions in Israel did not give much hope either. At the same time the usual squeezing of tender prices came about, which forced us to land on another price level altogether. We decided upon a steel space frame with a locally made sandwich panel system on top, forgetting the world novelty of the stressed skin sandwich, just to stay in the race. Based on this price and on our abilities Moshe Safdie was convinced that Octatube could do the best job. We had to deliver further design development and prototypes of the construction of the Great Hall, assuming that the details of the Library would follow those of the Great Hall. Just for intellectual reasons we added the prototype of the stressed skin sandwich.

2.7 *Production and installation*

The largest challenge for Octatube proved to be the *central body*: the central part of the Great Hall. Due to the large forces from the upper and lower roof wings amongst others, that this part of the roof has to cope with, a steel tubular structure was the only solution to make this span possible. This resulted in a complex structure of tubular steel, later to be fitted with thin GRP panels. According to the 3D computer data, various tubes were to be rolled in both 3D (two directions) and 2D (one direction). Because accurate 3D rolling is a rather complex procedure, the 2D rolled tubes possessed a greater accuracy. The 3D tubes, mainly situated along the length of the central body, at best approached the desired shape, therefore had to be connected to the accurately shaped 2D tubes. As if the 3D shape of the central body alone was not complicated enough! Nonetheless, after a lot of thinking and welding an impressive "artwork" came into being.

Figure 7. Renderings of the roofs for the Great Hall with special view of the *central body*.

Figure 8. Two more views of the *central body*, the small cubes (image right) mark the position of the steel *inserts* to connect the GRP shells to the steel of the columns and the *central body* (right).

Figure 9. Test assembly of the *central body* (spanning 30 m) in Lelystad at Holland Composites.

At Holland Composites the entire central body was assembled in order to fit the panels. After every panel was fitted, the structure was disassembled and transported to Tel Aviv. In Israel it was to be assembled in two parts, hoisted on its position and only later would these two parts be connected.

After the production of the roof parts for the lower wing of the Library, discrepancies between the theoretical drawings and the practical distortions and tolerances from shrinking of the polyester resin in the vacuum bags were measured. Tolerances because of warping of the negative moulds resulted in unforeseen deformations of the produced GRP components. These components together had to form the ruthless smooth surface of the complete wing in the end. All aspects were approached in an engineering manner: measuring, analyzing problems and deducing solutions. Improvisations bring the Law of Murphy nearer. Analytical engineering in the best traditions of the TU Delft made the initial amazing, improbable design solution finale a reality. The resulting design is a combination of structural design with architectural flavour,

Figure 10. GRP roof segment assembled on a temporary steel structure at Holland Composites in March 2005.

incorporating the technologies from aeronautics, ship building, industrial design and geodetic surveying and poses an example of multiple innovation of technology, thanks to the involvement of co-makers Octatube International, Holland Composites Industries and Solico Engineering.

2.8 *Assembly*

Due to the experimental character of the production process and unfamiliarity with the consequences of vacuum deformation, we decided to perform a test-assembly or pre-assemblage on the premises of Holland Composites in Lelystad. The fitting took place on a positive steel frame, the shell would therefore be curved upward. When a technician fell, he would not fall in the shell, falling instead off the shell. Subsequently we would gently turn the shell over with a mobile crane, by means of three temporary hoisting fixtures in the shell. From the pre-assembly we could also draw conclusions regarding the theoretical versus the practical measurements of the individual segments. All the segments were produced on individual foam moulds and they all had their own shrinkage and shrink-direction. Yet together, these segments had to form the unforgiving smooth surface desired by the client and architect. It was exciting to see if the total fitting of the individual deformed segments would still form a smooth surface when the entire shell was assembled. In order to acquire this smooth surface we indeed needed a solid frame with clamps in order to force the segments into the desired position. In general the segments proved to be somewhat smaller than intended. When filling up the seams during assembly a bigger seam meant more fibre (due to the required ratio between fibre and resin) thus causing a larger weight of the shell.

The connections between the individual segments can be divided into connections in the length and connections in the width of the segments. On the side of the segments a groove has been made of 220 mm width and

Figure 11. Hoisting of one of the roof shell wings.

Figure 12. During installation of the roofs the *central body* was temporarily supported by scaffolding.

15 mm depth. In this groove a prefabricated reinforcement of 200 mm width and 10 mm depth (of high density glass fibre meshes that have been vacuum injected with resin) was placed. This reinforcement was glued and clamped by bolts. After the segments of the two wings in Lelystad were fitted on the steel frame, the frame was dismantled and shipped in special containers to Tel Aviv. The build-up in Tel Aviv had to take place on the south side of a tall wall of the building. The segments were assembled inversely, measured, touched-up and finished with the structural reinforcement meshes and filler. Next, the shells were turned over and identically finished on the other side. After hoisting onto the Library, the shell was positioned on a steel sub-structure, which in turn rests on a concrete wall with a much larger tolerance difference. Until the end, theoretical drawings remained the decisive factor. In all phases of engineering, production, assembly up until the hoisting and positioning, theoretical drawings were always present. Building parts were simultaneously produced in locations all over the world. In this project the steel was manufactured in Delft, the glass in Luxembourg and Belgium, the polyester segments in Lelystad and the concrete in Tel Aviv.

Figure 13. The Yitzhak Rabin Center in November 2005.

2.9 *Combination of factors*

The design is the result of a combination of architectural design, building technical design, structural design, material design, with major influences from aeronautical and yacht design, from mechanical engineering (machining moulds) and industrial design, composite production design, assisted by the newest techniques of geodetic surveying to accurately measure the 3D forms in any stage of production, assembly and installation, all smeared by standard computer software like Maya and Rhinoceros and static analysis programs, including Computational Fluid Dynamics. In all of the successive steps of the process many of the used technologies have applications of theoretical developments done in more fundamental design or science areas. The bold design proposals from this design process challenge the more fundamental partners in the design process to come up with new answers. This process was an illustration of the "Delft Silicone Valley-effect" that the TU Delft as a whole has, with its range of faculties, on the world novelty of the end result. The co-makers Octatube, Holland Composites and Solico joined forces and developed this very experimental project with high economic risks, but with great endeavour and the eternal optimism of designers envisaging a new future in the thrill of this experimental design and build process.

3 GREEN HOUSE, MALMÖ, SWEDEN

Not only is the shape of the Green House in Malmö special, the technical aspects of the project are also very interesting. The control of the complex geometry as well as the structural design were very carefully approached to ensure that both were up to the high standard a project like this was asking for. The goal set by landscape architect Monika Gora was to develop a structure with as much transparency as possible. She asked Ian Liddell of Buro Happold of Bath (UK) to support her design in a structural scheme suitable for tender. A major consideration was the realization of the glass house with flat glass quadrangular panels. This model was sent for tender on the international market. Some Swedish companies as well as Octatube tendered.

Figure 14. Transparent model of the Green House describing the final geometry.

Although the price of Octatube would not have been the lowest, it was supported by great experience in the field of tubular structures and glass structures. Also juggling with 3D geometries is fairly common business in the engineering department of Octatube. The project was awarded to Octatube. At the start of the redesign phase after contracting, Monika Gora provided a small physical surface model of the Green House and a 3D CAD drawing that Octatube could work with. As the 3D drawing was not nearly accurate enough to develop a working basis for the further engineering process, a new computer model had to be developed. This was done using sophisticated design software (like Maya and Autodesk Mechanical Desktop). Rendered visualizations helped to finalize the shape.

Early in the design process the decision was made to use only flat glass panels for the outside surface. Although Octatube had ample experience with twisted and bent glass panels, the use of warm-bent glass panels was ruled out, as the cost for cladding the complete building with free-formed panels would be astronomical. Cold-twisted panels (panels pressed into shape on site) might have been an option but were also ruled out, as their twisted surface shape is difficult to control. Cold-twisted panels normally form a hyperplane (known from tent constructions), which would contradict the chosen shape of the Green House completely. The challenge therefore was to develop a geometry for the glass panels that would follow the original shape as close as possible, while keeping all the glass panels flat.

The challenge could have been solved easily if the surface was divided into triangles. This easy solution was abandoned, as triangles would mean a lot more glass divisions with extra silicone seams and glass nodes, which would lessen the aspired transparency of the surface. Instead, a natural growth algorithm was used to develop a surface consisting of flat panels with four corners, starting at the highest point of the Green House and working onward and downward to both

Figure 15. Test assembly of the steel structure in Delft.

Figure 16. Close-up of the glass panels and the supporting steel structure during installation.

ends of the building. A couple of different divisions of the surface were worked out, adjusting the parameters for the generation of the 4 corner panels each time to arrive at a solution which had logical divisions and endings at all places. By using a natural growth algorithm it was made sure that all panels were perfectly flat, while every corner point of every panel still rested perfectly on the originally chosen surface!

The glass panels are each made from two laminated, heat-strengthened glass panes with each having 8 mm thickness. To get an as transparent façade as possible, all glass panels were made from a special, extra white (low-iron) glass. Also the glass for the doors and the louver systems at the bottom are made from extra white glass. The glass material of these panels has a very low level of iron, reducing the normal, slight green tint of the glass to almost zero. The high glass thickness was necessary, as the site location of the Green House is exposed to high wind loadings, as it is situated directly on the shore of the Öresund between Sweden and Denmark.

For the glass node, a special clamped connection in the seam between the glass panels was chosen. This connection consists of stainless steel plates, both inside and outside of the glass, with a plastic spacer in between. Four clamped glass nodes are fixed to one spider with four arms. Each of the spiders is laser cut from a steel plate and bent into the necessary shape. Due to the complex geometry of the surface none of the spiders is the same. A central bolt to the steel structure inside the Green House then fixes the spider. The clamped connection gives the possibility to use the glass surface to stiffen the whole structure, as each glass panel can transmit forces in its plane through the plastic spacer to its neighbouring panel. This would have been impossible with a drilled connection due to the high stresses around the holes in the glass. The glass surface is therefore used as a continuous shell.

The structural design of the tubular steel structure of the Green House started with the already developed divisions of the glass surface as a boundary condition. The structure had to follow the glass seams, as the spiders had to be fixed to it. Different possibilities were discussed:

1. delta trusses as columns,
2. delta trusses for the backbone,
3. plane trusses for the columns,
4. a single layered space frame for the whole surface.

In the end a very simple and clear structure of a double tubular spine and single tubes as legs was chosen. Along the ridge of the Green House (the backbone) a double beam from CHS profiles ø168 mm was designed. As columns behind each glass seam, CHS profiles ø159 are used. Overall this choice leads to a very clean appearance of the Green House, with a very high transparency of the façade. To minimize the steel structure, the glass surface was used as a shell to reduce overall deflections. The connections between the different steel parts are made by sleeve connections, to achieve a continuous appearance of the CHS profiles. All connections of the structure are designed to be moment-resisting, to reduce the profile dimensions as much as possible.

For production and assembly the highest accuracy had to be administered. Due to the choice to keep all parts as small as possible almost all adjustment space was left out. The glass panels were fitted into the glass nodes with a tolerance of ±1 mm. This tolerance between the glass and the plastic spacer of the glass nodes was then filled up by silicone strips, as otherwise the glass surface could not work as a shell and reduce deflections of the whole structure. The glass seams between the glass panels were made weatherproof by using structural silicon sealant.

To make sure that all parts fitted together and all connection points for the spiders were fixed at the right

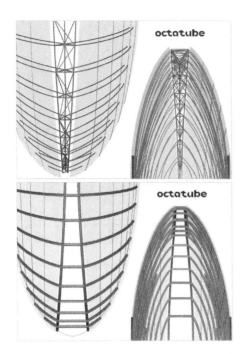

Figure 17. Tubular scheme 2 (top) and 4 (bottom).

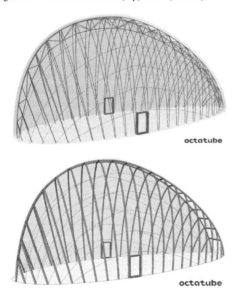

Figure 18. Tubular scheme 2 (top) and 4 (bottom).

place, pointing in the right direction the whole steel structure was set up in Delft before being galvanized and coated. After all dimensions were checked, the parts were galvanized and then powder coated to achieve the best possible surface finish.

Transport to Sweden and assembly on site went very smoothly in the late summer of 2005. Last finishing

Figure 19. The Green House just after completion.

touches as well as the technical installations were done in the winter. The outside landscape gardening was finished in November, the inside in January 2006. The official opening of the Green House was in February 2006.

4 CONCLUSION

Realizing two structures in liquid design form has taught the engineers that the combined virtues of different structural schemes have to be employed, and that insight into tubular structures has become ever so complicated. By combining the structural action of shells and space frames, and even for structures with bent elements, the resulting structures could be made. The engineering required a great many bent-tube pieces in very different bending radii. The accuracy required for producing structures assembled from bent elements of different radius is well known. Recognizing mismatches and dealing with them in order to arrive at a structure with a high tolerance, dictated by the use of frameless glazing that cannot allow larger tolerances than 1 to 2 mm at very seam, requires very careful production and frequent 3D surveying during production and assembly stages. The result, however, shows a great optical logic.

REFERENCE

Kohn, W., Rowe, P., Rybczynski, W., Goldberger, P. & Sorkin, M. 1996. *Moshe Safdie*, Academy Editions, London.

Tubular Structures XI – Packer & Willibald (eds)
© 2006 Taylor & Francis Group, London, ISBN 0-415-40280-8

Tubular steel roof for Spencer Street railway station in Melbourne, Australia

P. Skene
Leightons Contractors Pty Ltd

R. DiBlase
Winward Structures

ABSTRACT: With its striking visual appearance, the redeveloped Spencer Street (Southern Cross) Station in Melbourne, Australia is now a distinctive landmark and a world-class transport facility, and opened in early 2006. Through a sequence of staged deliveries, the station remained fully operational throughout the construction phase. The new station brings together the best elements of a modern, international-style station, creating a user-friendly facility that will be comfortable, convenient and safe for passengers. This paper describes the design and construction of the unique and innovative domed tubular roof structure, comprised of spine trusses along the platforms and long span arches across multiple railway lines.

1 BACKGROUND

Spencer Street Station is the key transport hub for both the Melbourne metropolitan and the Victorian regional and interstate rail network. Upgrading of the old station (including its renaming to Southern Cross Station) has included revamping of commuter traffic flows from an underground basement access to a modern open-air transport interchange facility where the platforms are accessed via a newly constructed elevated concourse. The track network and platform layouts are predominantly unchanged by the redevelopment; however, new infrastructure has been built for ticketing, administration and retailing facilities. A feature of the new station is the total enclosure of the station, comprising an entire city block, with a spectacular wave-form roof enclosed on all sides by a glass façade.

The project is a Public Private Partnership between the Victorian Government and the Civic Nexus consortium. The Civic Nexus consortium comprises a maintenance and retail operator with the design and construction being undertaken by Leighton Contractors. The design consultants for the project are as follows:

– Project Architect – Grimshaw-Jackson Joint Venture
– Structural Engineering – Winward Structures
– Mechanical and Electrical Engineering – Lincolne Scott
– Track and Signalling – GHD
– Durability and Rail Engineering – Maunsell
– Wind Tunnel Testing – MEL Consulting
– Environmental Engineering – Advanced Environmental Concepts
– Shop Detailing (Roof) – Precision Design Australia.

2 PRELIMINARY DESIGN DEVELOPMENT

The main feature of the station is its large wave-form roof. Occupying a complete city block and covering 36,000 square metres, it is not only an architectural icon of Melbourne, but performs an important functional requirement for the station. The concept for the roof developed by Grimshaw Jackson was to not only provide an architectural feature but also to provide a form that would allow gases from both diesel locomotives and in some instances steam locomotives to be exhausted without the use of costly mechanical fans or extraction systems. Grimshaw's past experience with railway stations in both Europe and the UK indicated that covered stations with large barrel vaulted structures tended to trap diesel fumes – hence the development of the domed or individual mogul-shaped roof which allows the fumes to rise to isolated points which can then be extracted from the station.

Computer modelling in the form of computational fluid dynamics (CFD) confirmed that this could be achieved. The roof moguls at Spencer Street reach a height of up to 24 metres above the platform levels at their highest point, dropping to a minimum height of 6 metres in some places. The roof is also required to be elevated at both the northern and southern ends to clear the existing Collins Street and Bourke Street bridges.

In order to assist the roof ventilation under prevailing cross winds, the moguls are offset across grid lines. This also reduces the risk of fumes being sucked back into the station environs under a cross-wind through an adjacent mogul. At the apex of each mogul or dome is a louvre cap which enables fumes and/or gases to be extracted, but which must also operate to prevent wind-driven rain from being blown into the station.

The roof structure essentially comprises a series of structural steel tubes forming a two-way net system. The original roof design was cut back to make way for a future commercial development on the western side of the station above the existing rail lines. This structure is an elevated slab which will support a future 10 storey building above. Introduction of this structure to the project after the initial design was commenced created a problem for the main station roof, as the largest internal bay of the roof now became an end bay. This resulted in higher stresses in the main roof rafters, requiring increased wall thicknesses in a substantial amount of members. A number of options were investigated, with the final solution adopted being to "lean" the roof on to the substantially stiffer elevated deck. This avoided having to use heavy wall pipe but meant that temporary works in the form of a series of arch ties could not be removed until the elevated deck had been completed.

3 ADOPTED DESIGN

The roof is set out on a geometric grid, with a series of five grid lines running down the centre of every alternate platform in a north–south direction. The platforms are set at a 12 degree skew to the adjacent Spencer Street. Grid lines labelled A to E run in the east–west direction. A total of 29 column supports for the roof are located at the intersection of all grid lines. Grid lines numbered 1 to 5 run in the north–south direction and are the central axis for large triangulated trusses referred to as spine trusses. The 12 degree skew has resulted in a complex intersection of trusses at the corner of Spencer Street and Collins Street. A series of curved roof rafters, referred to as primary arches, run in the east–west direction and are spaced equally at 4 metre centres. These arches are not true arches as such, as their geometry has a reverse curve as part of the arch shape. Again, these intersect with the spine trusses at a 12 degree skew in plan. A series of short secondary members between adjacent arches help form the complex two-way system. Columns at 40 metre centres along the spine trusses support the truss in its Y-shaped form. Columns are cantilevered out of the ground from a rigid pile cap and piled foundations.

The structure described above forms the main substructure, or the bottom of the roof's three layers. The second layer above the structure is a ceiling consisting of a series of triangulated ceiling panels. Over 7000 ceiling panels of similar design but each with their own unique geometry provide both an architectural ceiling and an insulating barrier for the station roof as well as providing a work platform to enable construction of the third and final layer of roof sheeting above. The ceiling itself also contains an important diagonal structural brace which is connected to the main supporting steel substructure. The architect intended that this diagonal bracing member should be hidden from view, although it is an essential element of the roof structure. An important feature of the ceiling panel design is that it allows air to flow and naturally ventilate the station environs. This was achieved by maintaining nominal gaps between panels to allow air to track through the gaps into the ceiling space, and rise to the top of each mogul before being vented out through the louvre cap.

The third and final layer is a weatherproof skin in the form of a metal deck to the station roof. The chosen product had to have the ability to be manufactured and installed to suit the continually changing geometry of the wave roof. It also needed to have a 25 year warranty. The choice of roof sheeting was a key issue in the final design solution. A standing seam aluminium tray deck system, stainless steel shingles and zinc roof sheeting were the main alternatives considered. A number of early concepts using GRC panels, tension fabrics and sandwich panels were also considered, but dismissed as being either too difficult to weatherproof, not able to meet the warranty requirements or unable to meet the defined geometry of a flexible roof structure capable of accommodating thermal movements, settlement of the structure under self weight and fabrication/erection tolerances.

Other key features of the design include a siphonic drainage system. The steep roof sheeting drains to gutters located each side of the spine truss. Outlet discharge points are located at each column, where a sump gives way to a siphonic outlet discharging into the column arm down to the base of the column, piped along the platform into an abandoned subway tunnel, and then discharging to a common surge pit on the western side of the station.

A clear skylight system is located above each grid line, running north–south along the centre of alternate platforms. The skylight sits above each line of spine trusses. A number of solutions for the skylight were investigated. A solution suggested by the architect was the use of ETFE cushions. This is a tensioned fabric technology, with each cushion consisting of a twin layer fabric structure of ETFE (basically clear Teflon) in the shape of a cushion. A continual air supply is pumped into the cushion to maintain both shape and strength to the finished product. Grimshaw had successfully used this technology in a number of projects in the United Kingdom – most notably at the Eden Project in Cornwall, which is an environmentally controlled biosphere constructed entirely of EFTE cushions over a steel framework. Alternatives such as traditional glazing were also investigated for the skylights. However, the logistics of installing large sheets of glass in-situ on the roof together with the continuously changing geometry of the trusses would have made glazing a difficult option. The roof itself is a flexible structure, both in its final configuration and during the various stages of construction. This could have potentially created additional problems for glazing and sealants at construction completion. The flexible nature of ETFE ensures that movements of the roof will not impede the performance of the skylights.

Some statistics for the roof are:

– Approximately 3200 tonnes of structural steel
– Roof area of 36,000 square metres of which 7000 square metres are skylights
– 29 supporting columns
– 207 primary arches
– 1370 secondary members and 1600 diagonal braces
– 7050 ceiling panels.

4 STRUCTURAL DESIGN

The roof structure consists of structural steel tube, with the predominant size being 356 mm OD circular hollow section (CHS). This size includes the main boom members of the spine trusses and all primary arch members (curved roof rafters). Wall thicknesses vary from 6.4 mm to 12.7 mm; however, some heavier wall tube up to 23.8 mm was used in some of the flatter arches. Secondary and diagonal bracing members in the roof proper consist of 168 mm diameter CHS of varying wall thicknesses between 4.8 mm and 25.0 mm.

The main support for the roof is a series of triangulated spine trusses running in a north–south direction along the centre of the platforms. The inverted triangle section is 8 metres in width at the top, with a depth varying between 4 metres at the columns and 2 metres at mid-span. These trusses also form an undulating profile arching between column supports along the platform. The spine trusses consist of 356 mm diameter members for the main chords and cross member of the triangle, with 273 mm diameter vertical members and 168 mm diameter plan bracing.

The roof structure is unique in that the primary arches spanning between spine trusses span one way in the construction condition. Once all the lacing members (i.e. secondary and diagonal members) are installed, the roof mogul begins to form a two-way system in the final condition. The roof mogul acts as a shell transferring the bulk of the roof loads back to the four columns in the corners of each dome, thus substantially reducing the amount of vertical load or dead load to the spine trusses on the grid lines. This was a chief architectural requirement, as the architect wished to see only a two-way net system/structure from beneath. The trusses therefore became a Vierendeel truss, with no diagonal bracing in elevation but diagonal bracing in plan to achieve the lateral restraint required by the out-of-balance forces provided by a flat arch opposing a steep arch on either side of the truss.

The choice of typical key connection types was critical early in the design process. These included the following connections:

Primary arch to spine truss connection. A bolted halving joint was chosen as it provided a seat for the primary arch to land, enabling quick bolting and securing of the arch.

Primary arches up to 40 metres in total length were generally broken into three sections for transport. This was not just due to its length, but also because the varying radii of most arches produced an over-width transport problem. Depending on the location of these joints, some varied from a full moment connection to pure compression, depending on the load in the arch and the point of contraflexure (virtually all arches contained a reverse curve in geometry). To maintain a consistent approach it was decided to butt weld all these joints on site. This also removed the difficulty where a connection of a secondary/diagonal member at a node point clashed with a primary arch connection, had these been bolted together. This decision was made during the early design development phase, before the geometry of the roof was fixed and the nodal points defined.

Spine truss to spine truss main chord or boom member connections between adjacent truss lengths. A spliced angle plate/cruciform connection was chosen to allow some construction tolerance as trusses were erected over each column first with an infill section placed between them. The installation of the infill spine truss was one of the more complex lifts on site and needed to be bolted and secured in the limited time slot available during the night occupation. The connection was therefore detailed to allow a rapid bolting and securing of the truss.

Connections of vertical, diagonal and main chord members in each truss were fully welded. The use of similar tube sizes avoided issues with punching shear failure modes in the majority of cases. However in some connections, in particular in and around the column where a diagonal brace was installed, heavy wall pipe and/or stiffener plates were installed as part of the design and fabrication process.

Primary arch connection to diagonal roof bracing. An innovative connection detail was developed for this connection. As described previously, the diagonal brace sits above the main structure hidden within the ceiling panels. The diagonal bracing is an important element in transferring load from the roof shell back to the columns. Some of the diagonal roof members in and around the columns carry substantial loads. This connection therefore induces an eccentric load on the primary arch. Due to this substantial load and the varying orientation of cleat plate connections for the diagonal braces, it was decided to locate individual cleats around a circular stub upstand welded on top of the arch. This stub also provided for a further upstand to pick up a roof purlin at a higher level again.

5 FABRICATION

Structural steel tube was purchased from OneSteel, with some of the heavier wall pipe being sourced from overseas. On delivery of the tube, it was transported to a fabricator in South Australia to be cold-rolled into its correct shape and radius. Once this was completed the members were then transported to various fabricators in Victoria, Tasmania and New South Wales for the next stage of fabrication.

The columns were made in two parts: the bottom section was a circular tapered tube, and the top section was a fabricated box structure forming the Y-shaped arms. The fabrication process was made difficult by the 12 degree skew of the column arm – because the cross section of the arm was not rectangular but effectively a rhombus. The fitting of the internal services including the siphonic downpipes and an electrical conduit also made the fabrication process difficult. Once on site, the bottom section was installed, the top part of the "Y" inserted via a spigot connection into the bottom pedestal section, services connections were made and then the bottom pedestal section was filled with concrete, pumped from the bottom.

The primary arches were set out in a workshop to their completed final shape. The shop drawings were used to then set out all connection points and all cleat connections for secondary and diagonal members were fabricated in the shop, including the end connections for bolting to the spine trusses. Arches were then cut into three sections for transport, painted and then sent to site with set-out marks for welding on site.

The fabrication of the spine trusses was a more complex process. The triangulated trusses were fabricated in an inverted position. The shop detailer was able to provide additional set-out information to enable the fabricator to set up the truss in purpose-made jigs. Tube elements were cut with ends profiled to match the complex connection points around the main boom members. Some node points on the main boom member had five different tubes connecting at the same point. The structural designer also had to nominate a priority for members coming into that joint, together

with the specified welding type and extent. Set-out of the halving joint stub for connection to the primary arches was critical, as the connection relied on the full mating of matching plates in the field prior to bolting. Taking into account the 12 degree skew and the continually changing pitch of the arch and the undulating spine truss, this was no mean feat. Once the set-up was done, the trusses were fully welded, lifted and rotated into their upright position and the steelwork for the gutter beams was fixed. Finally, the truss was painted, installed into a transport frame and loaded on to a barge for its trip to the Port of Melbourne, for its final delivery to site via road transport at night.

Connection of the spine truss to the Y-shaped column arm was the one of the most complex connections. The concept, both architecturally and structurally, was to provide a three-point pin connection, one at the "crutch" of the Y and the top two at the column "ears". However, lining up or matching three pins in this configuration within 2 mm oversized holes was almost impossible owing to fabrication tolerances, deformation of items during both transport and erection, and thermal movement. It was therefore decided to provide an architectural pin connection only. This involved large steel billets being shaped and fitted to the spine trusses and then welded to the column arm on site. This required a detailed survey of both the spine truss at the end of fabrication and the as-installed column arm to confirm whether any clashes or excessive gaps were likely to occur prior to erection of the element.

Secondary and diagonal members were simpler, being straight members with bolted cleat connections at both ends. However, the logistics of tracking these members through the fabrication, painting, expediting and final erection processes was a massive task as all elements were of varying length and had their own unique position on the roof. Similarly, the ceiling panels were relatively straightforward to fabricate but each had its own unique geometry resulting in its unique position on the roof.

6 CONSTRUCTION

One of the key motives behind building a complete roof covering the entire Spencer Street Station site was to come up with a concept that was both architecturally acceptable and capable of being constructed over an operating railway station. Up to 60,000 commuters use the station per day, with 700 metropolitan trains serving six platforms and 240 interstate and country trains serving eight regional platforms.

This was a key issue that had to be taken into account throughout the entire design development process, from early concepts through to the final detailed design. An important part of the process was to engage the shop detailer early in the design process, to not only assist and provide input during the design development, but also to save significant time in the fabrication process. It reduced the amount of unnecessary and repetitive architectural drawing by making the shop detailer responsible for the roof geometry and all dimensional control. This also assisted greatly in the off-site fabrication process and the on-site survey installation by being able to interrogate any part of the theoretical model of the roof. The shop detailer therefore became an integral part of the design team with the architect and the structural consultant.

The choice of structural connection types and location of construction splices was another key factor in the design stage. Connections had to be located at convenient locations from both a design and installation point of view. Connection types had to allow for easy installation given the time constraints of constructing over live railway lines in limited blocks of time, as well as achieving the architectural and structural capacity desired by the roof designers. Bolted connections being the preferred choice.

The station constraints can be split into two main areas. Approximately one third of the roof was built over the electrified metropolitan lines, with the balance over the non-electrified country and regional platforms. The particular challenge over the electrified lines was that opportunities for installation were only available between the last train at night and the first train the next morning. Allowance within this period also had to be made for the overhead power to be isolated. This therefore further restricted the available window of opportunity to approximately two and a half hours each night to install all elements of the roof over the electrified lines.

To minimise the number of lifts over the metropolitan lines, it was therefore decided to construct large roof modules on the western side of the site, in the median strip of Wurundjeri Way, and to lift these large units into position with a 600 tonne crawler crane. The main spine trusses that supported the roof modules and ran along the length of the platforms were installed with this same crane. These large roof modules consisted of 4 primary arches with associated secondary and diagonal structural members and ceiling panels assembled as a complete unit (minus the roof sheeting) in the median strip ready for lifting.

The spine trusses were fabricated off-site in Geelong, Tasmania and New South Wales and towed to Port Melbourne on large barges. They were then transported at night to the site from the South Wharf, approximately 5 kilometres away, and lifted directly into their final positions. Each truss section was fabricated in 20 metre lengths. With 40 metres between column supports, a truss section was installed firstly over each column and then a 20 metre long infill truss was dropped into position and bolted. A similar process was adopted for installation of the remaining spine trusses over the entire project, but the full-time occupation of other areas of the station, such as over the regional platforms, enabled smaller cranes to be used.

Owing to the full-time occupation of regional platforms being attained in stages, it was more opportune to install the roof segments between trusses piece by piece. This meant the primary arch or main roof rafter (spanning from truss to truss) was installed first, followed by the series of secondary members and diagonal members, followed by the ceiling panels. The ceiling panel installation ensured that a working deck was in place to enable the roof sheeting to be installed during normal working hours, without the danger of dropping materials on to the public or workers below.

A series of safety nets were also installed immediately beneath the top of the trusses where the ETFE skylights were to be installed. Again, these nets enabled installation of the skylights to proceed during the day.

7 INNOVATION

The constraints of the site dictated that the main spine trusses had to run centrally down the platform centres on every other platform. This set the dimension/span between grid lines in the east–west direction, whereas along the length of the platform a nominal grid spacing of columns at 40 metres (and therefore main roof arches at 4 metre centres) was adopted in the north/south direction. The complex intersection at the corner of Spencer Street and Collins Street where the two spine trusses entwine presented its own construction challenges. This intersection was fabricated in one piece in the workshop off site, strategically cut to enable its transport to site, then reassembled on the ground beneath the supporting columns. The sections were then re-welded into one piece and lifted into its final position as a complete unit. The resulting effect is a remarkable feature of the main entrance to the station.

A full-scale prototype of an 8 metre by 8 metre section of the roof was constructed early in the design development stage. This was intended to not only provide a real-life part of the roof to be reviewed for its architectural aspects, but also to confirm concepts developed for the connection of steel members to steel members, and the connection of ceiling panels and roof sheet purlins to the main structure. Standard connections had to be adaptable for the full range of geometries encountered by intersecting members over the entire roof. The prototype also confirmed a connection detail for the triangulated ceiling panels that cover the entire roof. There are over 7000 individual triangulated ceiling panels, each with its own unique geometry, however connection details were required that would be applicable over the entire structure. From this prototype, more testing was done to replicate the performance and load capacity of the ceiling panels.

Although the ceiling panel was a relatively straightforward design, the connection details were unique. The connections had to allow for not only structural adequacy, but also for tolerances associated with steelwork fabrication and erection out of position. The connection also had to allow for movement as the structure was progressively loaded during construction and again at completion under in-service thermal and wind loading. A unique pin/barrel connection was developed which allowed rotation to suit the geometry of the roof, and lateral movement with an adjustable pin for construction tolerances.

A significant number of temporary works were required to stabilise the roof during the construction and to avoid excessive member sizes in the permanent structure to accommodate temporary construction loads. Temporary propping had to be minimised to avoid track pits and commuter traffic on platforms. Temporary works included:

- An arch tie "bow-string" which connected at the base of all primary arches (roof rafters). These were designed to take out the out-of balance lateral forces on the spine trusses prior to the adjacent bay being installed.
- A temporary spine truss prop located midway between columns to provide support for the spine truss until the roof modules were fully completed in adjacent areas of the roof. As described previously, the spine trusses do not have sufficient vertical capacity to carry the weight of the roof. At the completion of construction, the shell action of the completed roof structure transfers loads diagonally back to the columns.
- An adjustable jig stand was installed in the median strip of Wurundjeri Way to assemble the large roof modules for installation over the metropolitan lines. This jig had to replicate the changing geometry and the connection details of the spine trusses to which the roof modules would be landed. Again, this jig had to be uniquely developed for this roof. The jig was set at the spacing of the spine trusses where the roof modules were to be placed. The jig allowed for vertical adjustment to match the individual locations where the roof module would land in its final position. The roof modules were erected in the jig, but effectively floated in the jig, relying on the arch tie connected at the feet of each arch to maintain the horizontal spread of the arch. This ensured that once the roof module was removed from the jig on lifting, it did not spring apart or contract on lifting, thus maintaining its correct shape (spread on connections) which ensured a correct fit in its final position.

The two systems considered for the roof cladding were zinc and aluminium. Both met the long-term durability requirements for the station, and both were capable of being profiled to the unusually shaped roof. However, the support structure required by each system was totally different. Zinc sheeting, being a soft material, requires a fully supported surface in the form of a curved plywood surface. The ceiling panels for zinc sheeting would have therefore required a curved top surface. This could have been achieved by either curving the top surface of each individual ceiling panel (without visible steps at the joint of each panel) or by building the curved plywood surface in-situ after the ceiling panels were installed. The aluminium solution was a standing seam tray deck system, with traditional purlin supports at nominal spacings. Both these systems were trialled on the 8 metre by 8 metre prototype before the aluminium solution was selected.

The aluminium roof sheeting system adopted for the roof is a proprietary system called Kalzip. This system, supplied by Corus Bausysteme from Europe, was chosen for both its long-term durability and its flexibility in both design and installation to accommodate the continually changing shape of the Spencer Street wave roof in three directions. The support system of purlins was a series of circular steel purlins that snaked their way continuously along the entire length of the roof. More than 24 kilometres of uniquely shaped purlins in nominal 8 to 9 metre lengths were designed, drawn, fabricated and installed on the roof. The circular purlin was ideal for this roofing system as the clip for fixing the roof sheeting to the steelwork had to be preset for location and a correct angle in two directions. This entailed a complex survey set-out to orientate and place the clips prior to landing the roof sheeting.

The roof sheets were individually made. The width of each sheet between standing seams varies across the roof, as does the vertical curve, somewhat like the effect of peeling an orange. The profiling equipment for producing roof sheets was set up in an off-site facility. Flat coil was imported, then cut to length and shape, the standing seam was produced, before a taper was introduced and then finally a vertical curve. Each sheet was required to go through this four stage process. Sheets were then delivered to site and installed on the clips in their own unique position. A similar process was required for cladding the louvre caps, also in Kalzip. The system creates a unique finish for the outer skin of the station roof.

8   PHOTOGRAPHS

This section shows some of the main features of the truss roof and its connections.

Figure 1. Erecting the spine truss.

Figure 2. Insertion of an arch section.

Figure 3.   Station in operation during construction.

Figure 4.   Multiplanar tubular truss connection.

Figure 5.   Tree columns supporting the roof.

Figure 6.   Completed fully clad roof.

Tubular Structures XI – Packer & Willibald (eds)
© 2006 Taylor & Francis Group, London, ISBN 0-415-40280-8

João Havelange Olympic Stadium – Tubular arches for suspended roof

F.C. D'Alambert

Projeto Alpha Engenharia de Estruturas, São Paulo, Brazil

ABSTRACT: This paper presents the basic concepts for the development of a Tubular Roof Structure for the João Havelange Olympic Stadium, which will be home of the 2007 Pan American Games in the city of Rio de Janeiro, Brazil, where the main design criteria were are safety and economy.

1 INTRODUCTION

1.1 *Pan American Games – 2007*

The city of Rio de Janeiro is known all over the world for the natural beauty of its beaches and mountains, together with hot climate and friendly people.

In 2007 Rio will host The Pan American Games. It is high the interest on this event since it is the first sport event of intercontinental interest that will take place in Rio de Janeiro second to the World Cup of soccer of 1950 which too place in another stadium, The Maracanã.

The roof project of the Olympic Stadium was developed with this objective of: bring the architectural and structural modernity to the world of sports.

As the budget to implement the necessary infrastructure for this event was small, this led to a detailed study of the project that had to be adapted to the available funds.

1.2 *Sequence*

This paper will be presented with emphasis on the following topics:

- Structural conception
- Construction methods
- Architectural and structural originality
- Respect to the environment
- Structural behavior – deflections and slenderness
- Better use of the materials and aesthetics.

2 STRUCTURAL CONCEPTION

2.1 *General data*

- Sport stadium with 45 000 capacity, meant to be the home of the Pan American Games – 2007 in the city of Rio de Janeiro, Brazil.

4) ISOMÉTRICA 01

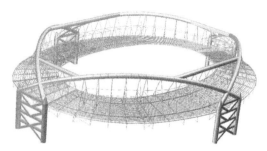

Figure 1. First version of the Arches.

- Structural Analysis: Computer Program: STRAP 12 Version
- Wind Tunnel: RWDI – Guelph city, Ontario, Canadá

2.2 *Historical*

The ideal geometric definition was one of the largest challenges in the development of the arches project. A first version was soon be discarded (Figure 1), because require to use tubes with very large diameters and therefore the structure's weight would be out of the usual parameters of weight per square meter.

A second version was developed with the use of arches composed of trusses of tubes with 500 mm diameter, however this version could not be adopted because of its serious geometric problems like the intersection of the arches (Figure 2).

2.3 *Final version*

- Finally the Structural/Architectural team decided, for lack of local qualified suppliers to adopt the following structural design
- The roof of the stadium would be composed of 42 trusses 50 meters long, 20 meters distant to each

Figure 2.   Second version of the Arches.

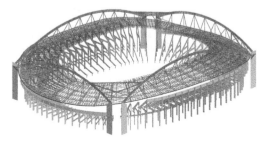

Figure 3.   General view.

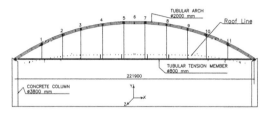

Figure 4.   Arch No 1.

other and distributed in a radial pattern forming in plan a roof in the space of a ring with approximately 40 000 m².

This ring is supported by trusses, 4 great tubular arches.

ARCH No 1 – Formed by a 2000 mm diameter tube of variable thickness, with 221 meters span and 34 m high.

It Will be supported by 2 tubular concrete columns with 42 meters high, and 4,40 meters diameter.

ARCH No 2 – Formed by a 2000 mm diameter tube and variable thickness, with 163 meters span and 29 m high.

It Will be supported by 2 tubular concrete columns with 42 meters high, and 4,40 meters diameter.

2.4   Structural behavior

The Structure can be defined by four steel arched trusses with tubular sloped suspension hangers. The Tubular arches are supported by 8 massive concrete columns at the four corners of the Stadium. Beneath the arches are the egg shaped roof, suspended by steel hangers, 2 for each truss.

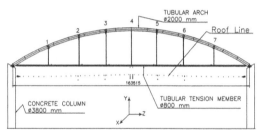

Figure 5.   Arch No 2.

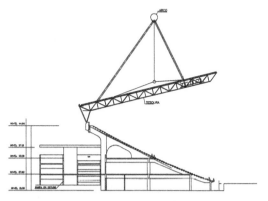

Figure 6.   Typical section.

2.5   Loads

The Roof's project for Olympic Stadium took into account the proper distribution to the dead loads, live loads (0,25 KN/m²), equipment, etc, thermal effect of the variation (+/−15° Celsius), 16 cases of wind effects, according to the results of the wind tests done by RWDI Engineering and Scientists (www.rwdi.com) in Canada.

They obtained $50 \times 4 = 200$ combinations which allowed static and dynamic analysis of the structure for the 1st and 2nd order.

3   CONSTRUCTIVE METHODS

3.1   Fabrication shop

The tubes will be formed and welded in a fabrication shop, where they will be dimensionally inspected and weld quality controlled.

After approved by the quality control inspector, they will be shipped to the construction site in 11 meter long segments.

3.2   Field works

At the site, the tube segments will be connected by welding, forming polygonal shapes 33 meters long.

The roof trusses will be raised to temporary supports until the final closure of the arch.

At the end of the erection of each arch and respective suspension hangers, connecting the roof, they will be loosened of the temporary supports. Control of deformations will be done in all stages of erection.

As soon as the structure has been stabilized, the roof tiles will be placed.

4 ARCHITECTURAL AND STRUCTURAL ORIGINALITY

Theatre of dreams, stage of emotions, these are the principles used in the development of the architectural project for the João Havelange Olympic Stadium, this monument to the sport will be prepared not only the sporting events of the Pan American Games, but also for official Games of soccer according to the regulations of FIFA (International Federation of Soccer).

The tubular arches will be reference for those who enter the city. In Brazil similar structures do not exist. The impact of the vision of the group of arches has for objective to give structural solidity to the notion and "I throw".

5 STRUCTURAL BEHAVIOR – DEFLECTIONS, SLENDER

5.1 Structural model

Figure 7. Structural model.

5729 nodes –
11 343 beams –
58 supports –
39 properties –

5.2 Local buckling

The instability modes for an axially compressed cylinder are mainly overall column buckling and local wall buckling.

In our case, the ratio of cylinder diameter to wall thickness, D/t is high, then the tubes can buckle in either as inelastic or an elastic shell buckling mode.

Elastic local buckling:
For the column buckling

$$\sigma xc = \frac{\pi^2 E}{(L/r)^2}$$

where:
L = is the length between ring stiffeners (20 m)

Inelastic Shell Buckling (Galambos 2004):
The inelastic buckling Stress of cylindrical shells is usually obtained in one of two ways.

Either the elastic formula is used with an effective modulus in place of the elastic modulus, or empirical relations are developed for specific classes of material.

The former approach is applicable only when the material stress-strain curves varies smoothly.

The slenderness parameters are usually either D/t or a nondimensional local buckling parameter, α, where:

$$\alpha = \frac{E/\sigma y}{D/t}$$

E = Modulus of elasticity
σy = Yield Strength (300 MPa)
D = External diameter of tube (mm)
t = Wall thickness (mm)

The structure of the arches was designed according API Code (API) which is recognized code worldwide.

The API Code (API) resulted as the best fit to a series of tests of fabricated cylinders with several strength levels, having the yield strength as part of the slenderness parameter in addition to α, where:

$$\text{For } \alpha < \frac{E/\sigma y}{60}$$

$$\sigma xc / \sigma y = 1{,}64 - 0{,}23/(\sigma y\, \alpha/E)^{0{,}25}$$

$$\text{For } \alpha > \frac{E/\sigma y}{60}$$

$$\sigma xc / \sigma y = 1$$

5.3 Local buckling – bending (Galambos 2004)

The buckling behavior of cylinders in bending differs from that of axially compressed cylinders in that bent cylinders have a stress gradient which is not present in axially cylinders and the cross section tends to ovalize.

The buckling behavior of cylinders in bending can be reasonably represented by a linear expression in terms of α for critical axial stresses.

For $\alpha > 14$ Mu/Mp = 1,0
For $\alpha < 14$ Mu/Mp = 0,775 + 0,016 α

Table 1. Critical stress in axially loaded steel shell.

D (mm)	t (mm)	α	$\sigma xc/\sigma y$
2000	19	6,65	0,903
2000	22	7,70	0,930
2000	25	8,75	0,952
800	19	16,63	1,000

Table 2. Critical moment capacity.

D (mm)	t (mm)	α	Mu/Mp
2000	19	6,65	0,880
2000	22	7,70	0,900
2000	25	8,75	0,920
800	19	16,63	1,000

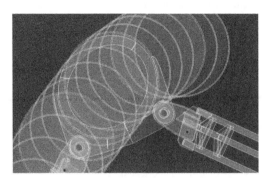

Figure 8. Design of the suspension ring.

5.4 Imperfections

For our design, imperfections are considered by specifying permissible out-of-roundness, using the API tolerances (Galambos 2004).

5.5 Theoretical analysis

For checking the design, models were developed using the method of the finite elements, for determination of the working stresses.

5.6 Global Behavior and deflections

The study of the deformations was accomplished in a meticulous way to allow a control of the serviceability limit state, tables were created containing the theoretical elastic and inelastic, 1st and 2nd order deformations in the vertical plans and horizontal to the arches.

The extensive data in the tables will be considered in the detailing drawings and in the production of the structural parts.

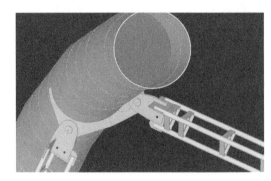

Figure 9. Detail of a suspension ring – Image.

Table 3. Theoretical vertical deflections – Arch 1.

	VERTICAL (mm)				
	41	5	6	7	8
DEAD LOAD	157	274	361	274	157
LIVE LOAD	70	95	123	95	70
TEMP +	−250	−277	294	−277	−250
TEMP −	250	277	294	277	250
WIND 1a	22	23	21	18	14
WIND 1b	26	44	50	41	39
WIND 2a	−12	−22	−26	−29	−27
WIND 2b	−35	−52	−65	−67	−57
WIND 3a	−33	−41	−48	−46	−41
WIND 3b	44	62	66	65	54
WIND 3c	62	96	142	114	111
WIND 4a	71	35	1	−30	−60
WIND 4b	76	70	40	−98	−115
WIND 4c	33	−30	−64	−85	−100
WIND 5	44	23	1	−25	−68
WIND 6	13	2	−12	−43	−46

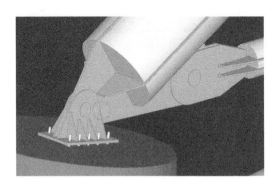

Figure 10. Detail of structural supports.

6 BETTER USE OF THE MATERIALS AND AESTHETICS

The use of circular tubes in the composition of arches is important, because through excellent structural

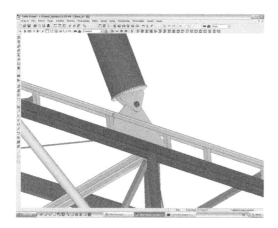

Figure 11. Detail connection for tension member x-suspension truss.

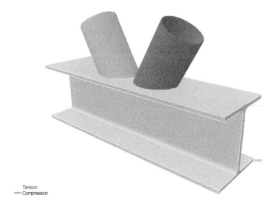

Tension
— Compression

Figure 12. Detail connection for diagonal-chord x trusses.

characteristics they provide a modern and beautiful architectural aspect. In spite of Brazil not having tradition in major tubular structures, this solution was adopted for reasons of economy and beauty.

The connection details are individually identified allowing a correct structural assembly, easy maintenance and careful architectural finishing.

The choice of tubular steel structures considered easiness of handling the parts, speed of erection, less labor and ease interface with other materials.

7 ACTUAL STAGE OF CONSTRUCTION

Actually the construction is with the structure of concrete ready and manufacturing the pieces in steel, the assembly is foreseen to begin in May with end foreseen for December of 2006.

Figure 13. Image during the construction.

Table 4. Comparison between steel roof structures.

Stadium	Weight (ton)	Area (m^2)	W/A (kg/m^2)
Porto	4 050	36 000	112,5
Boavista	670	8 880	75,45
Aveiro	2 160	20 870	103,49
Coimbra	1 000	12 930	77,33
Leiria	2 600	20 700	125,6
Sporting	3 120	24 850	125,55
Faro	1 350	9 400	143,63
Benfica	4 620	29 380	157,25
EOJH-Brazil	3 700	39 000	94,87

8 CONCLUSION

After the sequence adopted is shown for the project of the João Havelange Olympic Stadium roof steel structure, we concluded that the analysis followed globally accepted concepts. However partly due to the favorable loading condition considered in Brazil (non-existence of snow, hurricanes or earthquakes) and mainly for the appropriate use of tubular structures, we achieved our objective of designing a light and economical structure.

Examples of similar structures built recently in Portugal are shown below.

REFERENCES

American Petroleum Institute-Recommended practice for Planning, Designing and Constructing Fixed Offshore Platforms, API RECOMMENDED PRACTICE 2A-LRFD.[a]
Galambos, 2004. Guide for Stability Design Criteria for Metal Structures – T.V.
J. A. Packer , J.E. Henderson, 1997. Hollow Structural Section – Connections and Trusses – A Design Guide.

Tubular Structures XI – Packer & Willibald (eds)
© 2006 Taylor & Francis Group, London, ISBN 0-415-40280-8

Roof of Santa Caterina market in Barcelona

J.M. Velasco
Amatria Ingeniería, Barcelona, Spain

ABSTRACT: Santa Caterina market roof cover is a structure made of steel and wood. It is composed of a set of different wooden arches with straight and curved axis, supported by tubular steel beams of various cross-sections. These beams also hang from a set of three circular arches by means of a set of vertical hangers. The wooden arches are either tri-pinned or two-pinned. The steel arches are tied by cables which control the vertical deformations.

1 INTRODUCTION

The Institute of Barcelona Markets, controlled by the City council, is in a process of remodeling old markets around the city. Santa Caterina market is located in the centre of an old neighborhood that carries the same name.

The winning architect of the project, Enric Miralles, along with his wife Benedetta Tagliabue, conceived the roof as a curved and light surface that floated on a set of cables. Its outer surface is formed by ceramic mosaics that find their roots in the "trencadis" of Catalan architect Antoni Gaudí.

The initial idea materialized while the structure started to be conceived and the calculations were initiated (end of 1997). It was finally fixed as a set of wooden arches supported by steel beams, with these supported by three metallic arches through a set of hangers. The structure was completed in 2004, and the market opened to the public in May 2005. All the steel structure is made only of tubular sections and steel plates.

2 GENERAL DESCRIPTION OF THE STRUCTURE

The support at level zero consists of 13 columns:

Steel columns: Four curved tubular section columns curved are located in the main facade while two columns, the lower part made of concrete and the upper part of steel tubes, are located at the rear end of the building.

Concrete columns and beams: Two pre-stressed concrete beams are located on the sides of the building. One of 69 m long supported by 4 concrete columns, and another one of 30 m supported by 3 concrete columns of section 0.90 × 0.90 m.

V shaped beams: Taking advantage of the V shape of the zone at the lower edges of the roof, 6 steel

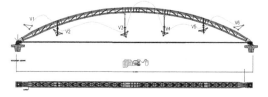

Figure 1. General cross-section of roof through the tubular steel arches.

beams were made to run the facade to the opposite end. They are distributed in the following way: The two outer beams are parallel to the market and the concrete beams, and the four other steel beams change in direction and bend to form a pair of rhombuses. Two beams begin from each of the central columns at the front. One of the two columns at the back of the structure has three beams and the other two.

Props: These are the set of bars that hold the roof uniting the far steel beams with the lateral concrete beams.

Wooden arches: The space between the steel beams is completed with wooden arches. The highest arch is tri-pinned and the lower arches are two-pinned.

Steel arches: Three arches have been designed in order to hold the four central beams by means of 12 hangers. The support points of each arch are tied at their base. The tension of these cables holds the arches firmly in place. These arches are unique in that they go in and out of the roof (see Figures 15 and 16).

3 ANALYSIS

The structure was analyzed by means of a three dimensional finite element model, using bar and shell elements. Twenty-six different load combinations were

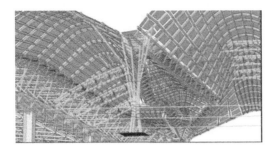

Figure 2. Numerical model.

Figure 3. The tubular arches and beams.

Figure 4. Arches supports.

taken into account: the wind load being the most disadvantageous in the analysis of the structure. The different thermal coefficients of wood and steel produced considerable stresses at the interface. The analysis of lateral buckling of the steel arches was difficult to achieve because horizontal loads, originating from roof movements, destabilize the arches if they do not have enough vertical load. Eurocodes have been used for loading hypotheses and analysis.

The stress distribution in the roof cover greatly depends on the way and the order that each of its components is placed during construction. If the steel beams are installed first, and are then released to start receiving loads, the wooden arches and the braces of the arches will only support loads near to them. If the cover is installed entirely and then released to start receiving loads, it will work as a whole set and it will behave as a big composite beam of steel and wood, where the braces and the cover (wood) work mainly in compression, while the steel beams will be working in tension.

4 THE CENTRAL STEEL ARCHES

4.1 The arches

The three arches are 42.78 m long, with a height of 6.02 m from the base of the arch (1/7 ratio), and a radius along its axis of 41.44 m (approximately 60° open). The section of the arch is composed of three tubes 219.1 mm in diameter and 25 mm thick, arranged in a shape like an equilateral triangle, with its base located at the top (see Figure 4).

The tubes form Howe-type trusses, in which the inclined members (tubes of 80 mm in diameter and 12 mm thick) are under compression and the vertical members, which are plates of 15 mm thickness, are under tension.

To prevent lateral buckling, calculations were made using classical formulae and non-linear analysis.

4.2 Supports and cables

The supports of the arches are "pin" type which allow rotation. One of them is a support guided in the plane of the arch and the other one is restrained from translation. This way, the in-plane deformation of the arches does not affect the concrete beams horizontally.

A pair of tendons each one formed by 19 cables, of 15.24 mm diameter, resist the horizontal thrusts. The cable tension of 3000 kN (about, in service) controls vertical deflections of the arches, therefore, the beam's deflections are controlled through hangers.

4.3 The bracing

The bracing between the arches has been solved by bridging members. These members (tubes) go from the centre of one triangle to the other, and when they reach the arch, the bar divides into two props, one directed to one upper tube and the other to the lower one (see Figure 6).

4.4 The Hangers

From the steel arches, 12 hangers descend (four per arch) and connect the arch to the steel beams, and work as a set of vertical elastic supports (see Figure 5). The

Figure 5. Connection of the hangers to the steel beams.

Figure 6. Local view of the bracing.

position of the hangers differs from one arch to another. Also, each one has a different length.

5 THE WOODEN ARCHES

5.1 General description

The curved shape of the cover is achieved by wooden arches. This wood is structural, and its cross-section

Figure 7. General view of the bracing.

Figure 8. General perspective under construction.

Figure 9. Steel beams.

is 200×400 mm as a flat beam. The arches are all different in length and shape. They are comprised, in general, of two straight sides and one curved part above flat aprons and one curved higher part. The arches are

41

Figure 10.   Steel beams.

Figure 11.   Tubular steel beams. South facade.

Figure 12.   North facade.

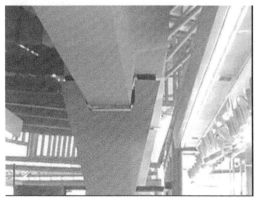

Figure 13.   Concrete beams and columns.

articulated at the base for the purpose of not transmitting moments between the wood parts and the steel beams.

5.2   Three-pinned arches

The more inclined arches are three-pinned for the main purpose of releasing forces due to internal movement of the wood. Also, for construction, it was necessary to have two independent pieces because of its length. The span varies between 6 m and 18 m and the height is between 3.80 m and 4.70 m from the base.

5.3   Two-pinned arches

The reduced arches are two-pinned to enhance their stiffness, because they form part of the central rhombuses that work like a great horizontal beam.

The maximum span is 10.94 m with a height of 1.69 m.

Figure 14.   Concrete cantilever, tubular steel props and beam, and wooden arches.

Figure 15.   General view of the Santa Caterina roof during construction.

6   THE STEEL BEAMS

The span is about 50 m but as the beams are curved their real length is 56 to 90 m. The cross-section of the beams is triangular. It is formed by three tubes of 219.1 mm diameter and of 25 mm thickness. The three tubes are joined by smaller pipes of 80 mm diameter × 12 mm thick, and triangular steel plates 20 mm wide. Every plate is different in shape. Buckling of the plates has been studied using non-linear analysis.

7   TUBULAR STEEL COLUMNS

These columns are shaped as bunches of independent steel tubes. For every load combination each tube works in a different way. Therefore the box needs to be

Figure 16.   General view of the Santa Caterina roof.

studied carefully. The buckling has been studied using non-linear analysis. The geometry is complex and the connections have been studied in-depth.

8   CONCRETE BEAMS AND COLUMNS

Seven concrete square columns support two concrete beams (four columns for one beam and three columns for the other) that are perpendicular to the main entrance. The concrete beams support the big steel arches and the props then support the outer steel beams.

The concrete beams are T-shaped. Horizontal loads are basically supported by the flanges of the beams and the vertical loads by the whole beams.

The beams are post-tensioned and are 2.50 m wide × 1.20 m high. The flanges are 0.30 m wide and the webs are 0.90 m deep. The concrete beams are supported at the flanges, on roller supports. At each column there are two vertical supports to the flanges and two horizontal supports to the web of the beam. The torsional movement of the beam is prevented by the two vertical supports to the flanges with a span of 1.65 m. The lateral movement is prevented by the

vertical placed supports. The longitudinal movement is free in each column except for one in each beam that is prevented from translation.

9   MATERIALS USED

The type of steel used is S355. The concrete used is C45/55 according to the Eurocode. The wood sections are type MC-30, (30 MPa).

10   CONCLUSIONS

– Tubular steel sections can be combined with wood to form suitable, with low weight, achieving complicated shapes and aesthetic forms.
– Wood and steel can work together but special care must be taken in the thermal analysis because of the different thermal coefficients of the materials used.
– The lateral buckling analysis of the arches is critical and should take into account horizontal loads perpendicular to the arches due to roof cover movements.

Tubular Structures XI – Packer & Willibald (eds)
© 2006 Taylor & Francis Group, London, ISBN 0-415-40280-8

Advantages of using tubular profiles for telecommunication structures

M.G. Nielsen & U. Støttrup-Andersen
Ramboll Denmark, Denmark

ABSTRACT: The telecommunication industry is one of the fastest growing industries; consequently the telecommunication industry focuses on the towers for supporting the antennas. The focus is particularly on the total costs as well as production and erection time of the towers. However, aesthetics become more and more important since building permits become more difficult to achieve. Steel lattice towers are primarily produced of tubular or angular profiles. Compared to lattice towers of angular profiles the towers of tubular profiles have the advantages of lower wind resistance, increased buckling capacity and more aesthetic appearance. On the other hand, towers of angular profiles are easier to produce and demand less skilled people in the shop. Finally, angular profiles are less expensive per kilogram than tubular profiles. In practice the towers of tubular profiles are up to twice as expensive per kilogram as towers produced of angular profiles. However, towers of tubular profiles can compete with towers of angular profiles, but this calls for intelligent design and innovative solutions by the tower designer. This paper gives an overview of advantages and disadvantages of using tubular profiles compared to angular profiles as well as some practical examples on innovative connection details to reduce production costs. Furthermore some aspects regarding maintenance and problems regarding ice are mentioned. Finally, the difference between use of hot-rolled, cold-formed and welded profiles is discussed.

1 INTRODUCTION

The need for towers to support telecommunication antennas is increasing rapidly at the moment. The need for towers in Europe is big at the moment with the UMTS mobile communication system. Thus the need for towers is even bigger in the developing countries. This can be seen from the GSM world coverage (See figure 1).

The telecommunication industry focuses on the total cost and delivery time for the towers. This includes the manufacturing of the tower itself, but also foundation and erection, that all should be taken into account.

Figure 1. GSM world coverage (www.gsmworld.com).

Traditionally lattice towers have been produced of angular profiles, circular tubes or solid round bars.

In the very beginning, more than 100 years ago, the first steel lattice towers for telecommunications were produced of flat-sided profiles like the angular profiles since it was easy to produce and easy to assemble. However, some 50 years ago the first lattice towers were produced of tubular profiles and solid round bars in order to reduce the wind load and save material.

Nowadays towers are in most cases produced of tubular profiles in the northern part of Europe. In the UK and America the majority of the towers is however produced of angular profiles. The choice of structure is controlled by the options according to the national codes, manufacturing process but also traditions and innovations within the design.

In the following the angular profiles are considered as 90° angle profiles.

2 TOWERS FOR TELECOMMUNICATIONS

Towers for telecommunications are designed to withstand the wind load on antennas, cables, ladders etc. and on the structure itself. In some regions the towers are furthermore designed to withstand ice load and the combination of wind and ice load.

Since the towers carries antennas and often parabolas for microwave links the stiffness criteria is set up for the towers in order to be able to use the network under severe weather conditions. The stiffness criteria are often the design driver of the towers, especially when they carry parabolas and the height of the tower is more than 40 m.

The arrangement of the antennas is an important parameter in the design of the towers. The antennas are often arranged in one of the following two configurations:

- Road configuration covering two directions using two antenna directions.
- Normal configuration covering all directions using three antenna directions.

A triangular cross section enables the attachment of the antennas directly to the legs.

3 DESIGN BASIS

The basis of the design rules the design of the structures since the most cost efficient solutions are chosen.

Comparing the towers of angular profiles versus tubular profiles the following design parameters are important:

- Wind load
- Ice load
- Buckling capacity

3.1 Wind load

Apart from the wind load on the antennas, cables and other ancillaries, the lattice structure itself contributes significant to the wind resistance of the tower. The wind resistance of the flat-sided profiles such as angular profiles is larger than for the circular profiles. Consequently is the demand for the strength and the stiffness of the sections of the tower and the foundation dependent on the type of members.

The wind resistance of the lattice sections is dependent on various parameters: e.g. type of cross section, solidity ratio and type of members. The wind resistance is larger for square cross sections than for triangular cross sections. The drag coefficients for lattice bracing is decreasing for increasing solidity ratio in situations were the solidity ratio is moderate. Finally is the wind resistance for flat-sided profiles often up to 50 % larger that the circular profiles. For circular profiles the wind resistance is furthermore dependant on the wind speed – if the flow is supercritical of subcritical.

Figure 2 illustrates the drag coefficient dependent on the solidity ratio, type of cross section and profile. The values are based upon data from wind tunnel tests and are given in EC 3:Part 3-1 1997. For circular profiles the drag coefficient is dependent on the Reynolds

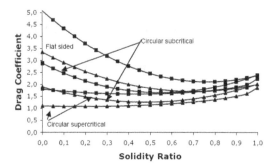

Overall Drag Coefficients for Masts

Figure 2.   Drag coefficients for lattice triangular and square cross section.

number (proportional to the wind speed and the diameter) since the wind generates some turbulence around the cylinder which decreases the wind drag for larger circular profiles. Some codes like the old American TIA/EIA-222-F 1996 do not take this reduction of the wind load for circular profiles into account.

3.2 Ice load

In some regions heavy ice load occurs on the structure and the dimensioning load can be the weight of the ice or the combination of ice load and wind load.

The weight per meter of the ice on a profile is dependent on the free surface area and since all the surfaces on angular profiles are exposed to ice load; the amount of ice on an angular profile is more than for the tubes. See figure 3.

The special considerations concerning ice load is further described in Nielsen, M.G. & Nielsen, S.O. 1998.

3.3 Buckling capacity

The design of the members in the bracing of lattice towers is normally controlled by their buckling capacity. Important parameters for the buckling capacity are radius of inertia, buckling length, eccentricity and the buckling curve.

When comparing a circular tube to a single angular profile with identical width and area of cross sections, the radius of inertia of a circular tube will typical be 10% larger than the radius of inertia about the strong axis of the angular profile and 70% larger compared to the radius about the weak axis. This result in a significant lower buckling capacity of the single angular profiles for the same distance between the bracing.

Furthermore are the diagonals for the sections with single angular profiles often eccentric loaded, which results in even lower buckling capacity.

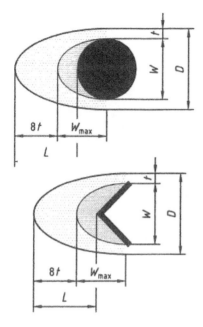

Figure 3. Ice accretion model for rime on circular and angular profiles (ISO 12492 2001).

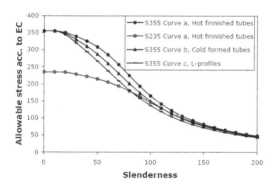

Figure 4. Buckling curves according to EC 3:Part 1-1 1992. The red curve is for tubes (hot finished) the light blue is for cold finished tubes, while the blue is for angular profiles and round bars. The green curve is hot finished tubes in steel quality S235.

The buckling curve according to EC 3:Part 1-1 1992 gives less critical stresses for angular profiles than the buckling curve used for hot rolled or even cold formed circular tubes. These results gives approx. 20% higher buckling capacity for the hot rolled circular profiles compared to the angular profiles for a typical slenderness of 60–120. The old American standard, TIA/EIA-222-F 1996, does not separate the buckling curves. Consequently it is more difficult for the towers with circular tubes to compete with the towers

Figure 5. Standard towers of tubular (left) and angular profiles (right).

of angular profiles when the American standard is followed.

In order to meet the requirements laid down in the codes the design of towers of angular profiles demands more bracings and more members than for towers of circular profiles. This makes the towers of angular profiles more complicated to erect.

4 COMPARISON OF DIFFERENT STRUCTURAL DESIGNS

The production costs per kilo steel for the towers of angular profiles are relatively low since the angular profiles are cheep and the production of the joints is straightforward. Normally no welding is needed. The members can therefore be manufactured by automatic machinery in the industrialized countries or with less heavy machinery in developing countries.

However, the lattice towers of angular profiles are quite heavy and the number of bracing members is relative high. The lattice towers of circular tubes compete in this particular respect.

Figure 6 shows a number of different lattice towers of tubular profiles. Towers of circular tubes both as leg and diagonals members are often build using pattern no 1. However this pattern has some limitations if the tower is tall and has a strict rotation criterion. Since

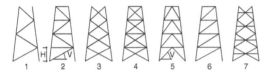

Figure 6. Typical lattice patterns (All diagonals can take compression as well as tension except pattern 4 that has special slender tension-only diagonal members).

Table 1. Comparison of different designs of lattice towers.

Type	I	II	III	IV	V
Cross section	Square	Square	Trian.	Trian.	Trian.
Leg members	Angle	Tube	Tube	Tube	Tube
Bracing configuration	Cross	Cross	Cross	V-bra.	V-bra.
Diagonal members*	Angle	S.tube	S.tube	S.tube	C.tube
Weight of steel (t)	3.6	2.9	2.7	2.4	2.0
Overturning moment, foundation (kNm)	810	670	590	530	420

*C.tube: Circular tube, S.tube: Square tube.

the most cost efficient solution in this case requires towers with a large face width. In this case the pattern no 3 has an advantage since it reduces the buckling length of the diagonals.

Table 1 from Støttrup-Andersen, U. 2000 shows the weight of steel and the overturning moment on the foundation for a typical 40 m lattice tower for mobile communications using five different layouts of the structure, where each design is optimized based on the different conditions.

The table shows that towers produced of circular tubes (Type V) are superior to angle profile constructions (Type I) with respect to weight of steel and overturning moment with potential savings of approximately 45%.

Furthermore, the tower of angular profiles (Type I) is more visible than the triangular tower of circular tubes (Type V) since the solidity of the tower is greater and the tower will look more massive with more and wider bracings. This is a disadvantage for the tower of angular profiles since aesthetics become more and more important in order to get the building permits.

5 DETAILS

In order to reduce the delivery costs it is not only important to reduce the costs of the raw material but also the costs of manufacturing. For the towers produced of angular profiles the costs of the manufacturing are rather low since normally the joints consists

Figure 7. Traditional joint between diagonal and legs, both circular tubes (ONE, Austria).

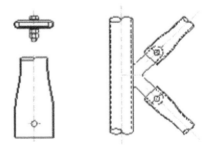

Figure 8. Special design of the joint without using welding on the diagonals.

of bolts and plates and no welding is included. The joints for lattice sections of circular towers are traditionally more complicated and time consuming. As an example hereof the bolted joint between the leg and diagonals is mentioned. This has traditionally been rather complicated with gusset plates welding etc. as shown in Figure 7 from a series of standard towers to Connect-Austria. However the towers were very competitive and more than 800 of these towers were delivered within a short period.

5.1 *Diagonals of circular tubes without welding*

Ramboll has introduced joints were the amount of welding is reduced: the ends of the circular tubes are squeezed together on a plate in order to have two plane faces in the ends of the diagonal.

The principle of the joint is shown in figure 8.

Figure 9. New joint between diagonal and leg, both circular tubes. BASE, Belgium.

Figure 11. Eccentric joint between diagonal and leg, PANNON, Hungary.

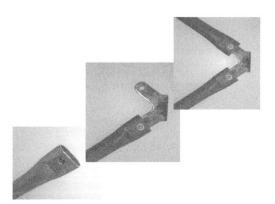

Figure 10. Principles of the assembly of the diagonals.

Figure 12. Joint in the intersection between cross bracing, PANNON, Hungary.

The diagonal is accordingly bolted to a special designed gusset plate welded to the legs as shown in figure 9.

The principle of the joint is more illustrative shown in figure 10. The gusset plate is prepared in a way that is possible to exchange the diagonals.

5.2 *Diagonals of square tubes without welding*

Another type of joint is used between the cross bracing of square tubes and circular leg members as shown in figure 11 and figure 12. Here no welding is used for the square tubes; only drilling and a saw cut under an angle of approximate 45° is needed. Eventhough this method brings eccentricity into the diagonals it has some clear advantages compared to angular profiles.

6 ERECTION OF TOWERS

The towers of tubes consist typically of less number of elements than the corresponding towers of angular profiles. The tower of angular profiles often with a square cross section and a large number of secondary bracing, whereas the towers of tubular members often has a triangular cross section and do not need secondary bracing. This implies that the number of elements in the tower of angular profiles often has the double or three times as many elements.

This makes the logistics more complicated and the erection more time consuming for the tower of angular profiles compared with the tower of tubular profiles.

However the towers of angular profiles have the advantage during transportation that the needed space is significantly lower for the tower of angular profiles compared with the tower of tubular profiles – however this does not always imply lower costs.

Figure 13. Erection of a tower of tubular elements in Pakistan using simple methods.

7 SPECIAL PROBLEMS FOR TUBULAR PROFILES

In some designs the designers have chosen not to protect the inner surface of the tubes but just closed the profiles e.g. with welded flanges etc. In case of weld porosities air will come into the tube from the exterior and water will condense in the tube. Apart from the corrosion problem the water can break the tube wall when it freezes and expands. This has been seen quite often and consequently all tubular structures should be equipped with holes for draining. It is especially important to drain the leg members of circular tubes in the bottom at the foundation, since here is a risk to collect water from the leg members in all the sections above. The hole for draining should as a minimum be Ø15 mm in order not to be blocked by leaves, dirt etc.

The inner surfaces of tubular profiles have to be protected against corrosion, but the inner surface is not as exposed to corrosion as the outside even if it is open in both ends. Hot dip galvanizing will be the best solution as protection against corrosion, since it is unrealistic to protect the inner surface by painting.

When the above mentioned special precautions are taken into account, experience has shown that lattice towers of circular tubes are maintenance free and can fully compete with towers of angular profiles.

8 CONCLUSIONS

The design of the structures is governed by the basis of the design rules since the most cost efficient solutions are chosen.

Lattice towers of tubular profiles have many advantages compared to towers of angular profiles, since the wind resistance is lower, the weight of steel is less and the overturning moment is less. These advantages can fully counterbalance the less expensive angular profiles; especially if the designer focus on the minimizing the labor cost for the manufacturing of the joints.

However some codes like the old American does not take due account to the decreased wind load and increased buckling capacity of the circular profiles compared to angular profiles. This has challenged the designer to make even better designs in order to be able to compete with the lattice towers of angular profiles. The new American code has diminished these advantages.

Our experiences show that lattice towers of tubular profiles are very competitive with towers of angular profiles.

REFERENCES

ISO 12492 2001. Atmospheric icing of structures, International Standard

EC 3:Part 1-1 1992. Eurocode 3: Design of steel structures – Part 1-1: General rules and rules for buildings; ENV 1993-1-1:1992.

EC 3:Part 3-1 1997. Eurocode 3: Design of steel structures – Part 3-1: Towers and masts; ENV 1993-3-1:1997.

Nielsen, M.G. & Nielsen, S.O. 1998, Telecommunication Structures in Arctic Regions, POLARTECH, Nuuk, Greenland

Nielsen, M.G 2001, Advantages of using tubular profiles for telecommunication structures, NSCC, Helsinki, Finland

Nielsen, M.G. & Støttrup-Andersen, U. 2005. Design of guyed masts, 2nd Latin American Symposium on Tension Structures, Caracas, Venezuela

Packer J.A. & Henderson J.E. 1997, Hollow Structural Section: Connections and Trusses – A Design Guide, 2. edition

Støttrup-Andersen, U. 2000. Reducing the costs of Masts and Towers, Nordic Mast Seminar 2000, Stockholm, Sweden

Støttrup-Andersen, U. 2005. Masts and towers for the UMTS network in Sweden, Eurosteel 2005, Maastricht, The Netherlands

TIA/EIA-222-F 1996. Structural standards for Steel Antenna Towers and Antenna Supporting Structures, Telecommunications Industry Association, USA.

TIA/EIA-222-G 2005. Structural standards for Steel Antenna Towers and Antenna Supporting Structures, Telecommunications Industry Association, USA.

Tubular Structures XI – Packer & Willibald (eds)
© *2006 Taylor & Francis Group, London, ISBN 0-415-40280-8*

Survey of support structures for offshore wind turbines

A.M. van Wingerde & D.R.V. van Delft
Knowledge Centre WMC, Wieringerwerf, The Netherlands

J.A. Packer
University of Toronto, Department of Civil Engineering, Toronto, Canada

L.G.J. Janssen
Energy Research Centre of the Netherlands, Unit Wind Energy, Petten, The Netherlands

ABSTRACT: This paper aims to provide an overview of the state-of-the-art in offshore wind turbines, including an introduction to general design principles and the foundations. For use in shallow water, gravity foundations or buckets (caissons) are currently used, whereas monopiles are being applied for medium depth water. It is foreseen that future larger wind turbines (say 10 MW nominal power output, 160 m rotor diameter) in deeper water will be based on a tripod foundation. For very deep waters, such as around Japan, floating wind turbines are also being considered.

1   INTRODUCTION

1.1   *Wind energy*

Wind energy is quickly growing to the point where it is becoming a measurable contribution in the "war on global warming". Shell expects that by the year 2050 one third of all required energy will come from sustainable energy (Shell 2001).

After the first boost in the eighties in California, the focus of attention shifted to Europe, where the need for a more durable form of energy was felt early on. Here, research projects and positive tax climates resulted in the predominant position of Europe in the wind energy sector today and a thriving industry with a market share of about 90%. The "old" EU (15 countries, excluding the new members in East Europe) aim to have over 20% of the electricity production from renewable sources as early as the year 2010 (European Parliament 2001). A large part of that production will have to come from (on- and offshore) wind energy.

It is only recently that countries and companies outside of Europe are beginning to realise the potential. But nowadays an amazingly broad number of countries, including even countries like Iran, are participating. This can be seen in Figure 1, which shows wind turbines placed in a valley which acts as a giant wind funnel. Also in the USA (especially General Electric), India and China have started programs, companies and research projects to kick-start their wind industry, often with an amazing degree of aplomb.

Figure 1.   Wind turbines in Iran.

1.2   *Offshore wind energy*

Offshore wind has been used by mankind as a prime energy source for a very long time – from the first sailing boats onwards. In the case of wind turbines, the drive towards the sea has two major causes. Firstly, the need to minimize negative effects, such as the impact on the horizon and, in the case of close proximity, shade and sometimes noise. This is especially the case in densely populated areas, where the need for electricity is largest. Secondly, the availability and quality (lack of disturbance) of offshore wind is better than on most onshore locations.

Since wind is not constantly available and wind turbines are stopped for maintenance and due to failures, the actual power output of a wind turbine is only a part of the rated nominal output (the electricity production under optimal conditions). In fact the actual power generation offshore is about 1/3 of the nominal power output. However, this compares quite favourably to onshore wind electricity production, where the actual output is about 1/4 of the nominal power output.

As a result, an important part of the future wind energy which Europe is aiming at can be expected to be generated offshore. This poses a new major opportunity for the offshore industry worldwide.

The current applications are for wind turbines with a diameter of about 90 to over 120 m and a rated nominal power output of 2.5 to 5 MW and over, per wind turbine. The current turbines are typically positioned in shallow (up to 5 m) to medium (5–20 m) water depth, at distances of 5 to 10 km from the shore.

It is worth noting that, unlike oil production platforms, wind turbines can be placed anywhere that sufficient wind exists and there is no necessity for placement in deep waters.

Although it would seem that the possibilities for placing wind parks are plentiful, in practice the available area is considerably diminished by shipping routes, fishing grounds, military zones and ecologically vulnerable areas (such as the Waddenzee in the north of The Netherlands), to the point where some wind parks will be placed in less favourable areas which are further from the shore (cabling), in deeper water (foundation and support structure) or at less favourable wind locations (lower electricity production).

1.3 Cost of offshore wind energy

The realisation of offshore wind parks is inherently more expensive than for onshore conditions for a number of reasons:

– Long cables are often necessary to connect the turbines to the onshore electricity grid. The cost of electricity cables can be 20 to 30% of the total costs of offshore wind energy and thus cabling is an important factor in the positioning of wind turbines.
– Installation and maintenance have to be carried out at sea which is quite expensive.
– In rougher offshore areas, such as the North Sea, larger maintenance campaigns are only possible for 6–7 months per year. Larger maintenance problems developing outside this period can result in several months of lost electricity production.
– The foundation and support, discussed in detail in Part 6 takes up also about 15% of the overall cost.

Altogether, the contribution to the electricity costs of the foundation, cabling, installation and operation & maintenance are about as high as that of the tower and turbine. These additional costs are the major reason for the higher cost of offshore wind energy, outweighing the effect of the better availability of the wind.

2 THE DESIGN PROCESS OF AN OFFSHORE WIND TURBINE

General structural design documents contain a number of basic elements. Slightly adapted towards wind turbines, these could be summed up as:

- A description of the project, containing aspects such as locations and other aspects affecting the design:
  – The overall capacity of the wind park
  – The general location
  – Number and position of wind turbines
  – Identification of nearby fishing grounds or ports, etc.
- An identification of design codes and standards to be used. In the case of wind turbines, a lot of codes deal with the load side of the design
  – Wind
  – Waves
  – Current
  – Control modes, such as for instance grid failure.

A few examples of relevant standards for the design of wind turbines are:

- IEC 2006
- GL 2004
- DNV 2004

The design process can be schematised as follows:

- Concept study, in which the main design parameters like diameter, installed power, control strategy, foundation type, etc. are used to define the basic design loads
- Pre-design phase in which the design is checked more carefully, now including the effects of wave loads, etc.
- Detailed design phase: all parts are designed in detail. Checks are run for both static (ultimate) and fatigue loads, the design is adapted and rechecked until a satisfactory design has been reached

3 WIND LOADING

Obviously, wind loading tends to be a fairly major issue wherever wind turbines are placed. For the tower and the support structure, the main wind loading on a wind turbine can be rather easily characterized by:

$$F = \frac{1}{2} \cdot \rho \cdot v^2 \cdot \pi \cdot r^2 \cdot C_t \qquad (1)$$

Where :
ρ = density of air: 1.225 kg/m³
v = wind speed in m/s
r = the radius of the rotor in m
C_t = drag coefficient
The upper limit value of C_t can be taken as 1.0 in certain circumstances, depending on the mode of the

Figure 2.   Wind turbines in Middelgrunden, Denmark.

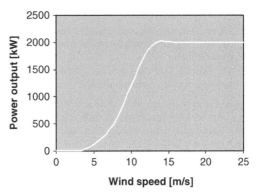

Figure 3.   Power output of a 2 MW offshore turbine.

wind turbine. This means that the force exerted by the wind on a wind turbine is essentially equal to that on a massive disc with the same diameter as the wind turbine rotors, even though wind turbines are anything but massive, as is obvious from Figure 2.

The reason for this surprisingly high coefficient is that wind turbines extract about 50% of the kinetic energy of the wind and therefore pose more of a barrier for the airflow than the "empty" cross section of the rotor would suggest.

The moment on the base of the tower is simply the force times the height of the hub above the water plus the water depth.

For instance, for a wind turbine with a radius of 60 m and a wind speed of 13 m/s (about 6 Beaufort) the force acting on the rotor becomes:

$$½ \cdot 1.225 \cdot 13^2 \cdot \pi \cdot 60^2 \cdot 1.0 = 1.17\,\text{MN}.$$

The bending moment at seabed level is then:

(water depth + hub height) $\cdot$ Axial force =
$(45 + 85) \cdot 1.17 = 178\,\text{MNm}.$

Obviously, significantly higher wind speeds do occur, but in that case the blades of the wind turbines would be pitched so as to catch less wind in order to control the power output. For modern pitch-controlled wind turbines with slender blades, the factor C_t drops to levels that make these high wind speeds result in about the same wind forces on the tower.

4   POWER GENERATION

In contrast to most structures, wind is not just another load case but is very much the raison d'être for the whole industry. A typical offshore wind turbine does not generate any energy at wind speeds below 4 m/s. Between 4 and 13 m/s, the turbine gears up towards the nominal output which is reached at around 13 m/s. At higher wind speeds, the output stays at this nominal output level, save for extreme wind speeds at which times the turbines will be halted. Some prototypes of modern turbines are not even turned off during storms anymore.

As an example of a typical power curve, the Bonus 2 MW turbine, used in Middelgrunden, the turbine itself shown in Figure 2, has a power curve as shown in Figure 3 (Sørensen et al. 2002).

The twenty 2 MW turbines of the Middelgrunden wind farm delivered, since their installation, the electricity consumption of about 25,000 families.

5   WAVE LOADING

Wave loading can be much more detrimental to the fatigue life of the steel structure than the wind loading, thanks in part to the fatigue behaviour of steel which is considered insensitive to the many cycles in the area above 10^8 cycles, where a lot of the wind-induced fatigue damage occurs for fibre-reinforced plastics. The impact of waves relative to wind is strongly dependent on the water depth: for shallow waters (up to 10 m) or sheltered locations, the wave loading tends to be less dominant, but for larger depths wave loads tend to outweigh wind loads.

The calculation of the impact of wave loading is compounded by the fact that the Eigen frequency of a large monopile system is close to the second order wave frequency. This topic is quite beyond the scope of this paper. It is, however, important to notice that the total fatigue load of a wind turbine structure cannot be determined by superimposing the wind and wave loads. Since the wind turbine essentially acts as a giant gyroscope, the exitation of the top of the tower and the effects of the wave loads on the structure are much mitigated, and the calculated fatigue life might be double for a wind turbine in operation compared to a parked turbine (Tempel 2000).

6   SUPPORT STRUCTURES

Although the designs for wind turbine support structures are derived from the equivalent structures for offshore oil platforms, a number of differences occur:

(1) There exists a strong interaction between the foundation compliance and the reaction of the turbine to wind forces.
(2) The fatigue load does not just come from wave actions, but also from wind loads.

Interestingly enough, the wind loads help to reduce the effect of wave loads, so the fatigue load on wind turbines is significantly smaller than superposition of wind and wave loads would have yielded.

As stated before, the support and foundation of offshore wind turbines is a major contribution to the cost of offshore wind energy and therefore an important design parameter. A number of support structures are currently in use for offshore wind turbines which will be discussed in this section.

Remarks regarding the (economic) feasibility of certain types of support structures should be taken as an indication only since, depending on the particular issues of a given project and the views of the person in charge, different opinions may result: no consensus on the best support type for a given wind turbine is available in this young industry.

6.1 Gravity foundation

For shallow waters, particularly when ice floes are around, a steel or concrete gravity foundation often provides a simple, economic and technically sufficient solution. For instance for Middelgrunden, this solution was employed.

The principle is the same as that of a garden umbrella: a heavy block with a wide base is employed to counter the overturning moment of wind and waves. In this case, the foundation was strengthened at the water line, to counter ice floes, which occur in that region. The tower is simply mounted on the foundation.

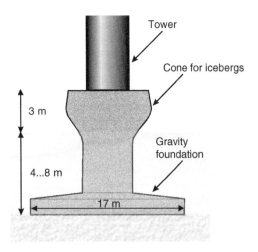

Figure 4. Gravity foundation: Middelgrunden.

Steel structures have the advantage of a lower weight when empty, making for easier transportation and often providing a more economic solution than concrete. Although this type of foundation is usable for all types of seabed, preparation of the seabed is necessary, involving work by divers which is both slow and expensive. In order to prevent undermining of the foundation by erosion, a layer of rocks typically needs to be placed around the base, adding to the cost of the foundation.

This type of foundation was employed in Frederikshavn in Sweden. The bucket is placed on the bottom and the water is subsequently sucked out, in order to get the sand out from underneath the skirt of the bucket, essentially driving the bucket into the seabed.

6.2 Bucket foundation

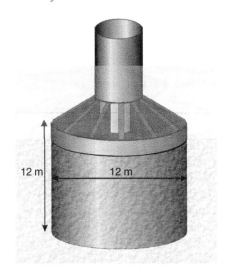

Figure 5. Bucket foundation: Frederikshavn.

6.3 Monopile foundation

In a sense, the monopile foundation is the simplest solution available: the tower is simply elongated, and the lower part is hammered, drilled or vibrated into the seabed, much like the poles of a garden fence. Unlike gravity or bucket foundations, no preparations of the seabed are necessary. A disadvantage of this foundation is that it requires heavy-duty piling equipment.

The method is not suitable for rocky types of soil. Although it is possible to blast a large rock which is blocking the drilling with explosives, this is obviously a slow and expensive process. In order to facilitate the mounting of the tower, it is possible to use a profile with a smaller diameter but a larger wall thickness in this area, as shown in Figure 6.

6.4 Tripod foundation

For water depths of 20 metres, the tripod may offer the most economical option. The required preparations for the seabed are minimal, and the piles required for anchoring the foundation are only about one metre in diameter, making the piling a lot easier. Also scouring, erosion of material around the pile, which is dependent on the diameter of the pile, is less of a problem than for monopiles.

A disadvantage of the tripod in shallow waters is that ships will be hindered in their approach to the tower, making this foundation suitable for deeper waters only. A further optimization, simplifying the steel structure, can be reached by using the tower itself as the third point of the tripod, as shown in Figure 8. This form is believed to yield a more economic design. It should be realised that, unless a figure specifically refers to an application, the dimensions are indicative only.

Instead of piles, also buckets can be used to anchor the tripod to the seabed.

6.5 Lattice structure

For deeper waters, where waves become the dominant loading, a lattice structure, like the one employed for the DOWNVInD turbine, shown in Figure 9 might prove to be the best solution. The DOWNVInD project is an EU-sponsored demonstration project, consisting of two 5 MW wind turbine generators in deepwater near the Beatrice Alpha oil production platform in the Moray Firth offshore of north-east Scotland. The diagonals are 580 mm diameter and have a wall thickness of 10 to 16 mm, whereas the chords are 870 mm diameter, with a wall thickness of about 50 mm.

6.6 Floating wind turbines

In even deeper waters, such as around Japan, there may be possibilities for floating and moored wind turbines. The high costs of the mooring, as well as the rather complicated dynamic behaviour of these structures,

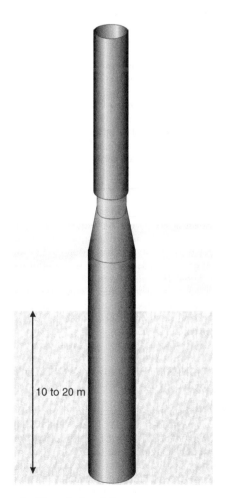

Figure 6. Monopile foundation.

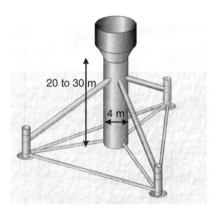

Figure 7. Traditional tripod foundation.

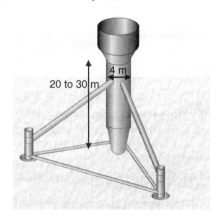

Figure 8. Optimized tripod foundation.

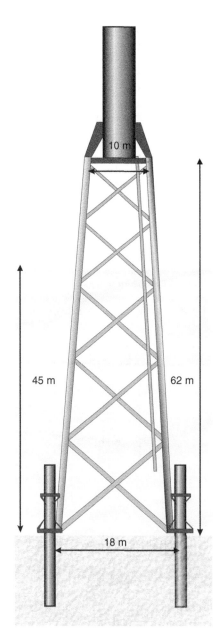

Figure 9.   Lattice structure for deeper waters: DOWNVInD.

Figure 10.   Claymore oil jacket – 30 times this amount of steel is needed for the 1 GW Beatrice offshore wind park. Source: talisman Energy.

For instance, the 1 GW Wind farm planned adjacent to the Beatrice offshore platform in the United Kingdom, for which the DOWNVInD turbines are the prototypes, will require about 250,000 tonnes of steel, equivalent to 30 times that of the Claymore jacket, an oil jacket of 25 by 45 by 135 m, shown in Figure 10. The old Beatrice oil platform will house the site transmission equipment and serve as an operations & maintenance base for the wind park.

It is obvious that many structural issues can be taken directly from existing offshore and tubular structures design codes and recommendations. Indeed, an important certification bureau like Det Norske Veritas (DNV 2004) is, for example, using Efthymiou and Lloyd's SCF equations for tubular connections.

However, older design guidelines like EC 3, with its outdated classification rules, are still recommended. Thus, the IIW (International Institute of Welding) Subcommission XV-E, which has extensive expertise on the topic, could take up a role in the development of appropriate standards. Such a group could show the benefits of modern design methods, as outlined in the CIDECT and IIW guidelines, to a receptive crowd, since the wind industry is much more sharply driven by the economy of design, as opposed to the more traditional oil industry where, for instance, timing tends to overrule construction costs.

make this solution seem rather extreme at this time, although some preliminary investigations are being carried out by International Energy Agency.

7   OPPORTUNITIES FOR STEEL DESIGNERS

It will be clear that the upcoming wave of offshore wind parks poses tremendous opportunities for the construction, wind energy and offshore industries.

8   S-N LINES TO BE USED FOR STEEL DESIGN RECOMMENDATIONS

For tubular steel structures, the following S-N lines are recommended by the IIW for use with the hot spot stress method; see Figure 11 (Wingerde et al. 1997).

Of particular significance for this application is the right-hand side of the figure: it matters quite a bit whether the fatigue limit occurs (the S-N lines become horizontal) at 10^8 cycles or at another level. As can be seen in Figure 11, thin-walled steel structures made out

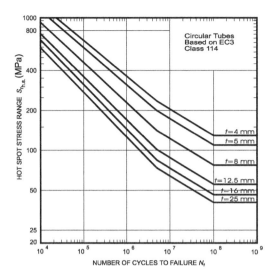

Figure 11. S-N lines of CIDECT/IIW design recommendations for tubular structures.

of circular hollow sections (CHS) have a fatigue limit of about 128 MPa, whereas for thicker steel structures with a wall thickness of 25 mm the fatigue limit is at 46 MPa.

These limits are meant for connections of tubular structures and there is reason to believe that in the case of less severe stress concentration factors, such as the welds in the various segments of a tower, the thickness effect is smaller. It is important to realize that the S-N lines given here are valid only in cases where corrosion is effectively prohibited, using for instance cathodic protection.

9 CONCLUSIONS

In this paper, an introduction to the role of support structures for wind turbines has been presented. As the market for offshore wind turbines is rapidly developing, major opportunities exist in research and contracting in the coming decennia.

ACKNOWLEDGEMENTS

The generous help and contributions of Reinout Prins & Bas Kuilman of Genius Vos BV, Peter Coutts of AMEC, Herman Snel of Energy Centre of The Netherlands, Jan van der Tempel of the Delft University of Technology and of Prof. Martin Kühn of the University of Stuttgart are gratefully acknowledged.

REFERENCES

DNV 2004, *DNV-OS-J101, Design of Offshore Wind Turbine Structures*, Oslo: Det Norske Veritas.
European Parliament & Council 2001, *Directive 2001/77/EC of the European Parliament and of the Council*, Brussels: European Parliament and Council.
GL 2004, *Guideline for the Certification of Wind Turbines*, Hamburg: Germanischer Lloyd (2004).
IEC 2006, *Wind Turbines – part 3: Design recommendations for offshore wind turbines*, Geneva: International Electrotechnical Commission 61400-3, 88/257/CD (draft).
Shell 2001, *Energy Needs, Choices and Possibilities, Scenarios to 2050*, London: Shell International.
Sørensen, H.C., Hansen, L.K. & Larsen, J.H. 2002, Middelgrunden 40 MW offshore wind farm Denmark- lessons learned. *Proc. Offshore Wind Conference*, Copenhagen, 26–28 October 2005.
Tempel, J. van der 2006. *Design of Support Structures for Offshore Wind Turbines*. Enschede: FEBO Druk B.V. (Ph.D. Thesis).
Tempel, J. van der 2000. *Lifetime fatigue of an offshore wind turbine support structure*. Delft: Delft University of Technology (Master thesis).
Wingerde, A.M. van, Packer, J.A. & Wardenier, J. 1997, IIW Fatigue rules for tubular joints. In S.J. Maddox and M. Prager (ed.), *Performance of Dynamically Loaded Welded Structures, Proc. IIW intern. conf., San Francisco, 14–15 July 1997*. New York; Welding Research Council.

Parallel session 1: Fatigue & fracture

Tubular Structures XI – Packer & Willibald (eds)
© *2006 Taylor & Francis Group, London, ISBN 0-415-40280-8*

Fatigue of bridge joints using welded tubes or cast steel node solutions

A. Nussbaumer & S.C. Haldimann-Sturm
Swiss Federal Institute of Technology Lausanne (EPFL), ICOM – Steel structures Laboratory, Switzerland

A. Schumacher
Swiss Federal Laboratories for Materials Testing and Research (EMPA), Structural Engineering Research Laboratory, Dübendorf, Switzerland

ABSTRACT: In the design of recently constructed steel-concrete composite bridges using hollow section trusses for the main load carrying structure, the fatigue verification of the tubular truss joints has been a main issue. Recent research on the fatigue behaviour of such joints has focussed on circular hollow section (CHS) K-joints with low diameter-to-thickness ratios – a geometric characteristic typical to tubular bridge trusses. Analytical and experimental research was carried out and joints with both directly welded tubes and cast steel nodes were studied. This paper presents the main results of these studies and shows comparisons between welded and cast steel solutions. The key issues for the design and fabrication of both types of nodes are reviewed and recommendations for the design and fabrication of tubular bridge structures are made.

1 INTRODUCTION

Tubular joints represent the critical point in the design of a bridge structure with circular hollow section (CHS) members. Presently, three main methods of joint fabrication can be found in existing bridges. A conventional possibility consists of brace-to-chord connections using gusset plates. This solution will not be dealt with in this paper. The second possibility is the directly welded joint, where the braces are cut to fit and welded to the continuous chord. Cast steel nodes offer a third alternative whereby castings are employed to provide a smooth transition between the brace and chord members, which are welded to the casting stubs.

The second and third mentioned solutions have been extensively studied by the offshore industry (Marschall 1992). However, the relatively new concept of steel-concrete composite CHS truss bridges presents the designer with new challenges, in particular with respect to the fatigue design of the joints. In comparison to offshore structures, differences in member sizes, tube slenderness, fabrication techniques, etc. make the direct application of the current offshore knowledge difficult and demonstrate that the current behaviour models for welded joints and cast nodes subjected to fatigue are incomplete.

For the designer, an important question is whether to choose welded joints or cast nodes in order to insure better fatigue strength for a given project. The lack of existing fatigue design rules has been the underlying motivator for a series of studies on the fatigue behaviour of directly welded joints and of cast steel nodes in tubular truss bridges carried out at the Steel Structures Laboratory (ICOM-EPFL) in Lausanne, Switzerland. The experimental studies are fully described by Schumacher (Schumacher 2003a, Schumacher & Nussbaumer 2006) and Haldimann-Sturm (Sturm et al. 2003 and Haldimann-Sturm 2005). In this paper, the comparison between both solutions is discussed. In the case of welded CHS K-joints, the determination of the stress concentration factors (SCF), synthesised in the form of a modified stress concentration factor (referred to as $SCF_{total}.$), is explained. In the case of cast steel nodes, the importance of a balanced design between the various potential crack initiation sites in a joint is shown.

2 EXPERIMENTAL STUDIES

2.1 General

In order to include all aspects influencing the fatigue strength, fatigue tests on large-scale specimens have been carried out. The specimens consisted of K-joints in a planar truss girder (Figure 1). The material used for the truss members was steel grade S 355J2H according to EN 10210-1:1994 and EN 10210-5:1997. The material used for the castings was steel grade GS20Mn5V according to EN 17182. Regarding the allowable casting defects, the quality of the castings was defined according to DIN 1690, Part 2. The test specimen (shaded in Figure 1) was bolted in place by means

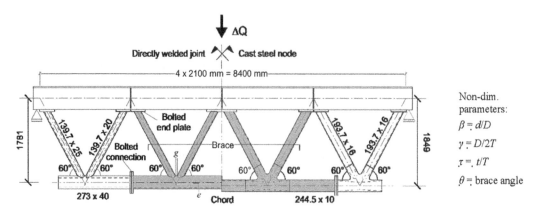

Figure 1. Test specimen, only half represented with directly welded joints (left) and cast steel nodes (right).

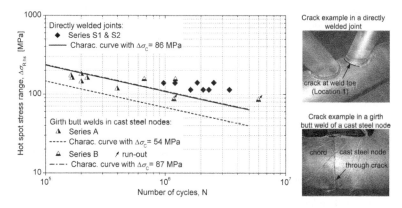

Figure 2. Hot-spot S-N results for tests on welded joints and cast nodes (girth butt welds).

of end plates. The top chord I-girder and the two outer K-joints acted as a load frame to the two inner K-joints. Both the welded joints and cast nodes were tested in the same way. For comparison purposes, Figure 1 shows half of a truss girder with welded joints (on the left side) and the other half with cast steel nodes (on the right side). The constant amplitude fatigue loading, with a load ratio of 0.1, was applied at midspan through hydraulic actuators. Joint failure was defined as through-cracking of the tube or cast node wall.

For the directly welded joints, Schumacher (Schumacher 2003a) carried out a total of 4 test series with 2 test specimens per series. Each test series examines a particular parameter: dimension, fabrication method or weld improvement. The non-dimensional parameters (β, γ, τ, θ, defined in Figure 1) of the specimens were chosen to reflect actual CHS truss bridge parameters. Except for the weld-improved specimens, twelve out of the sixteen welded joints cracked at the same location and in the same manner. Cracks initiated in the chord gap region at the tension brace weld toe, referred to as Location 1 (Figure 2). This corresponded with the location of the highest measured hot-spot

stress. Crack propagation occurred through the depth of the chord as well as along the weld toe.

For the cast steel nodes, two series (A and B) with 2 resp. 3 test specimens per series were carried out. The tests were carried out to investigate the fatigue strength of the cast steel nodes as well as the fatigue strength of the girth butt welds between the CHS members and the casting stubs. Several parameters were studied: cast steel quality, weld type at stub connections and stiffness parameters. The last parameter includes changes in the CHS thickness, thickness ratios and stub length. This serves the purpose of studying the influence of the stiffness difference between cast steel nodes and CHS members on the global behaviour as well as on the stress distribution near the butt welds and in the cast steel nodes. Except for two specimens of series B, where no cracking occurred and testing was stopped after a certain number of cycles, all cracks initiated in the girth butt welds between the casting stubs and the CHS members with the highest measured stress (Figure 2). That is, no fatigue cracking due to casting defects were detected in the cast steel nodes. In additional destructive and non-destructive testing on cast steel nodes, no non-allowable casting defects, which could have been an initiation site of fatigue cracking, could be found. It could be concluded that in these tests, the girth butt welds were much more susceptible to fatigue loading.

2.2 Comparison of results

Since tests on girders with welded joints and cast steel nodes were carried out on a truss girder with similar geometry and loading, it is interesting to compare the results. When mentioning the test results of cast steel nodes, it is the test results of the girth butt welds of the cast nodes to which reference is being made, since no fatigue cracking was observed in the nodes themselves. In other words, the fatigue strength of the girders with cast nodes was, in fact, determined by the fatigue strength of the girth butt welds.

The comparison is made on the basis of hot spot stress range values $\Delta\sigma_{R,hs}$. Therefore, the hot spot stresses in the girth butt welds of the cast steel nodes are obtained by multiplying the nominal stresses in the chord with a stress concentration factor according to DNV recommendations (DNV 2001) to take into account the eccentricity in the weld due to differences in thickness and diameter. The hot spot stresses in the directly welded joints were extrapolated from the strain measurements near the weld toe of the tension brace.

In Figure 2, test results of series S1 and S2 (with identical chord thickness) for the directly welded nodes are represented by the black rhombi. The corresponding characteristic S-N curve was found based on IIW rules. The triangular points show the test results for the girth butt welds without and with backing bars (series A and B, resp.). The corresponding characteristic S-N curve has been established by subtracting two standard deviations of log(N) from the mean curve. A standard deviation of 0.2 typical for welded girders was used because only few test results are available to represent a significant sample of test data. It can be seen that the results for the welded joints come to a fatigue strength at $2 \cdot 10^6$ cycles of $\Delta\sigma_C = 86$ MPa which is similar to $\Delta\sigma_C = 87$ MPa for the girth butt welds with backing bars. There is, however, a significant difference in the fatigue behaviour between the two types of tests: in the welded joints, cracking is directed by high stress concentrations at the weld toes and in the cast steel node joints cracking is caused by weld root defects in the girth butt welds at the casting stubs. Furthermore, the fatigue strength of $\Delta\sigma_C = 54$ MPa of the girth butt welds *without* backing bars is much lower. More testing on girth butt welds between casting stubs and CHS members is currently being carried out by the Versuchsanstalt für Stahl, Holz und Steine at the Technische Universität Karlsruhe, Germany.

As shown in Figure 2, the fat. strength of the welded joints at $2 \cdot 10^6$ cycles is 86 MPa. In comparison, is it known from literature that the fat. strength at $2 \cdot 10^6$ cycles of the cast steel nodes alone lies between 100 and 165 MPa (Haldimann-Sturm 2005), depending mainly upon casting quality. Although the strength of the cast nodes is considerably higher than the welded joints, this was found to be irrelevant in the present tests, since this higher capacity could not be mobilized as failure was governed by the connections between the casting stubs and the CHS members. In the present study, any type of comparison (nominal stress, hot-spot, etc.), if properly done, shows that both solutions have about the same fatigue strength. As discussed in Section 3.2, it is possible to find an optimum in the design by relaxing the quality requirements on the cast nodes, thus reducing their cost. This way, a balanced and economical design between the various potential crack initiation sites in a joint is obtained.

2.3 Pros and cons

The root cracking of the girth butt welds for cast steel nodes can be seen as a disadvantage in comparison to the directly welded joints. Two main reasons can be given. First, it is impossible to perform non destructive testing (NDT) or to apply a post weld treatment to the girth butt welds, because the root is generally not accessible. During a bridge inspection, only through-thickness cracks can be visually detected. Second, there are at least four such welds in each node (in the case of a K node). They are the most critical detail and diminish the fatigue strength of the whole structure if not properly designed.

Looking at a truss girder with cast nodes, two girth butt welds are necessary at each node in the chord to ensure continuity, which means that weld quality must be of a high and constant level to guarantee fatigue reliability. Also, with the high number of girth butt welds, pre-assembly and assembly sequences must be well studied in order to properly manage weld shrinkage and to respect tolerances. The advantage is that all weld preparations are simple, on straight edges and with constant bevel angle. This is, however, not the case for on-site assembly. The last tube cannot be placed in the assembled truss if the backing rings, necessary to ensure good weld quality, are already welded. This problem can be solved by adding cut-outs. In this way, the backing rings are movable during the assembly of the tube. Once the tube is put in place, the backing rings are pushed to the right position for welding. An example of these movable backing rings is shown on Figure 3 (Raoul 2005).

In the case of welded truss girders, the chord continuity is an advantage, especially for the tension chord, when speaking about fatigue reliability. There are no girth butt welds except at CHS section changes, at assembly joints between two tube members and at site joints. However, this advantage disappears

Figure 3. Movable backing ring for on-site assembly.

partially for large bridges with spans over 80 m, since the maximum deliverable tube length (for diameters over 500 mm and thicknesses over 25 mm) is less than 10 m. In these cases, the number of butt welds will approach the one of the cast steel node solution.

There are, on the other hand, also several advantages of cast nodes, for example, their ability to manage any geometry and number of incoming members. For welded joints, there are limitations due to minimum required angles between incoming CHS members, gaps and eccentricities. Also, the smooth geometry of cast nodes leads to low SCFs and it is possible to integrate special features such as parts of bearings or connectors to the concrete deck directly into the castings. As a result, cast nodes are particularly suited for nodes near or at supports, where the bending moment is negative and where there are usually more incoming tubes at the nodes.

3 NUMERICAL STUDIES AND DESIGN METHODS

3.1 Tubular welded joints

The concept of the hot spot stress (Marshall 1992, Zhao et al. 2000 and Zhao & Packer 2000b) was used in the interpretation of test results. It has been proven to be the best solution to properly account for the complexity in the stress distribution in tubular joints and to quantify the observed fatigue behaviour. Moreover, hot spot stress values can also be determined by FEM calculations using a validated standard procedure as explained in Schumacher (2003a). The software programs I-DEAS® and ABAQUS® were used to develop FE models of the truss girder and of the K-joints. The welds were also included in the models.

A large parametric study with over 200 FE models was performed on a range of welded joints to examine the effects of geometry and load. In particular,

the study of low γ, between 4 and 12, values typical to bridges, differentiates the present parametric study from previous, comparable studies. The study led to the proposal of SCFs (Schumacher 2003a) for ranges of non-dimensional parameters (β, γ, τ, θ) not yet available. Furthermore, the intuitive fatigue design method summarized below was proposed.

The concept of the hot spot stress gives the total hot spot stress $\sigma_{hs,i}$, at a joint Location i in the node, due to a combination of nominal member stresses (both in the brace *and* the chord). It is usually expressed as the summation of the contribution from each individual load case (noted LC):

$$\sigma_{hs,i} = \sum_{LC=1}^{n} \sigma_{LC} \cdot SCF_{i,LC} \qquad (1)$$

Where n is the total number of individual load cases necessary to represent a real load condition. This equation can be rearranged by expressing the percentage of each stress component, for example σ_{ax_br}, as a percentage P_{ax_br} of the total nominal stress σ_{total}. In a more general way, this results in the following:

$$\sigma_{hs,i} = \sigma_{total} \cdot \sum_{LC=1}^{n} \frac{P_{LC}}{100} \cdot SCF_{i,LC} = \sigma_{total} \cdot SCF_{total,i} \qquad (2)$$

$\sigma_{hs,i}$ hot spot stress at joint Location i

σ_{total} total nominal stress affecting the joint, $\sigma_{total} = \sigma_{nom\_ch} + \sigma_{nom\_br}$. The nominal stresses in chord σ_{nom_ch} resp. in brace σ_{nom_br} correspond to the superposition of axial and bending stresses.

P_{LC} percentage of member stress, due to load case LC, with respect to total stress

$SCF_{i,LC}$ stress concentration factor at joint Location i, for load case LC

$SCF_{total,i}$ total stress concentration factor at joint Location i

The total stress concentration factor can be represented as shown in Figure 4, for given geometry parameters and a given stress partition in the members. A given ratio of stresses due to axial load and in-plane bending in the brace and chord member was assumed as recommended in current design guidelines (Zhao 2000b).

The total stress concentration factor is given as a function of the percentage of member stresses, which are expressed as the brace stress ratio:

$$\frac{\sigma_{nom\_br}}{\sigma_{total}} = \frac{\sigma_{nom\_br}}{\sigma_{nom\_br} + \sigma_{nom\_ch}} \qquad (3)$$

σ_{nom_br} total nominal stress in the tension brace

σ_{nom_ch} total nominal stress in the chord.

The advantages of this representation are that all potential crack locations can be plotted on one graph. For design, only the envelope can be given (Locations

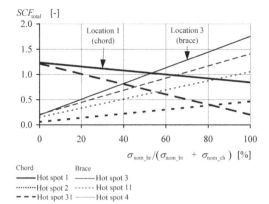

SCF_{total} [-]

Chord   Brace
——— Hot spot 1   ——— Hot spot 3
········· Hot spot 2   ------ Hot spot 11
– – – Hot spot 31   ——— Hot spot 4

Figure 4. SCF_{total} calculated for non-dimensional parameters $\beta = 0.5$, $\gamma = 4$, $\tau = 0.3$ and $\theta = 60°$.

Table 1. Parameter values and stress partitions covered.

θ [°]	β [–]	γ [–]	τ [–]	Stress partition in members [%]	
				Brace $\dfrac{\Delta\sigma_{ax}}{\Delta\sigma_{bend}}$	Chord $\dfrac{\Delta\sigma_{ax}}{\Delta\sigma_{bend}}$
45	0.50–0.70	4.0–30.0	0.30–0.70	75/25	65/35
60	0.50–0.70	4.0–30.0	0.30–0.70	75/25	65/35
45	0.50–0.70	4.0–30.0	0.30–0.70	100/0	100/0
60	0.50–0.70	4.0–30.0	0.30–0.70	100/0	100/0
45	0.50–0.70	4.0–30.0	0.30–0.70	0/100	0/100
60	0.50–0.70	4.0–30.0	0.30–0.70	0/100	0/100

1 and 3 in Figure 4), since these locations govern. Note that the denominator in Equation (3) implies a scalar superposition of member stresses; the fact that the stresses act in different directions is accounted for by individual SCFs. For usual stress partition values, the designer has at his disposal a set of graphs from which can be extracted the proper SCF_{total}.

Table 1 summarizes the range of parameters for the set of graphs that were developed for practical applications (Schumacher 2003b). From Figure 4, it can also be seen that the SCF_{total} always stays below a value of two. This figure does not show an isolated result; relatively low SCF values were found throughout the study. It was concluded that there is a strong tendency for SCF values to decrease with decreasing γ values. A low γ value, inferior to seven as seen with existing bridges, is therefore a desirable geometric characteristic. This should be reflected more in design specifications, which currently fix a recommended minimum stress concentration factor of $SCF = 2.0$, a value very penalizing if applied to bridges with low γ values. For bridge applications, the minimum value can be taken as $SCF = 1.0$. However, one reason for a recommended minimum value of 2.0 comes from the concern of possible root cracking if too low values are used for weld toe fatigue design. If the recommended minimum SCF is lowered, it must be done in conjunction with the requirement of full penetration for all welds in CHS joints.

3.2 Cast steel nodes with girth butt welds

The fatigue test results have shown that cast nodes have a far better fatigue strength, but that this capacity cannot be mobilized as failure is governed by the connections between the casting stubs and the CHS members. A numerical study has been carried out to quantify allowable initial sizes of defects in cast nodes in order to provide a balanced design between the various potential crack initiation sites, especially between the girth butt welds and the cast steel node.

To find the internal forces under fatigue and ultimate load acting on a cast steel node in a tubular bridge structure, a typical steel-concrete composite bridge was defined based on the properties of existing tubular bridge structures described in Sturm et al. (2003) and Velselcic et al. (2003). The study is limited to the two most critical nodes at different locations along the bridge with varying internal load configurations acting on the casting stubs. At midspan, the axial brace forces are much lower than in a node near an intermediate support. For this truss configuration, a cast node shape with minimal stub length has been chosen because secondary bending moments acting at the casting stub ends reach their maximum for this case.

The software BEASY® was used to develop a boundary element (BE) model of the cast steel node to simulate propagation of cracks initiating from a casting defect (Figure 5). Due to the longitudinal symmetry, only half of the cast steel node was modelled. The out-of-plane displacements were constrained in the plane of symmetry (not shown in Figure 5). The casting defects at different locations i in the node were modelled as two-dimensional cracks, representing all types of casting defects. This is a very conservative assumption as the fatigue behaviour of a two-dimensional crack is much more critical than that of casting defects like gas holes, slack inclusion or shrinkages. Figure 5 shows the 9 different locations in the cast nodes where the allowable initial sizes $a_{0,i}$ of casting defects are quantified. At Location 7, an inner crack has been assumed; at all other locations, more critical surface cracks were introduced in the cast node model.

The difference of the stress intensity factor $\Delta K(a)$ is needed for crack propagation calculations. When moving the fatigue load (for this bridge example according to the Swiss design code, SIA 261 (2003)) across the bridge, it is not obvious which load position induces the minimum and the maximum stress intensity factor (SIF) at a defect in the node. For this reason, the SIF influence lines of an identical crack at the 9 locations

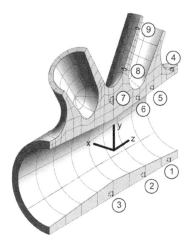

Figure 5. Different locations i of casting defects.

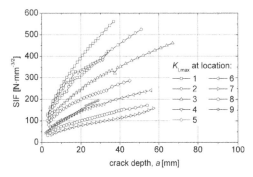

Figure 6. Examples of SIF results at different crack locations in the node at midspan.

in the node were calculated considering the internal forces which are induced by the moving fatigue load. Knowing the influence line of the SIF and the crack propagation plane, the two load positions defining $\Delta K_I(a)$ for mode I could be determined.

Under fatigue loading, crack propagation at Location i of the node was simulated using the software BEASY®. The results are given at intermediate steps in terms of crack configuration, crack depth a and the appropriate maximum and minimum stress intensity factors $K_{I,max}$ and $K_{I,min}$ (Figure 6 shows $K_{I,max}$). The final crack depth a_f was limited to a minimum of 90% of the wall thickness and the critical crack size in the case of brittle fracture, which was found using the failure assessment diagram (FAD) (Milne 1986). For this purpose, the stress intensity factor under ultimate static loading $K_{I,ULS}$ was needed. The results showed that brittle failure of a node containing cracks does not occur. A crack depth equal to 90% of the wall thickness at the crack location was therefore chosen as the failure criterion, assuming that a through-thickness crack is unacceptable.

Once the final crack size a_f was known, the initial allowable crack size a_0 could be back-calculated using the Paris-Erdogan equation. As the BEASY® output for crack propagation simulations consists of discrete stress intensity factor values (Figure 6), the Paris-Erdogan equation can only be solved discretely:

$$N_{tot} = \sum_{k=0}^{n-1} \Delta N_k = \frac{1}{C} \cdot \left[\frac{a_1 - a_0}{\Delta K_{I,0}^m} + \frac{a_2 - a_1}{\Delta K_{I,1}^m} + \ldots + \frac{a_n - a_{n-1}}{\Delta K_{I,n}^m} \right] \quad (4)$$

N_{tot} service life in terms of the number of loading cycles

C crack propagation parameter, $C = 2 \cdot 10^{-13}$ $(mm/cycle) \cdot (N \cdot mm^{-3/2})^{-m}$

m crack propagation parameter, $m = 3$

a_n final crack size a_f

According to (Blair 1995) the crack propagation parameters for cast steel correspond to those of ferritic-perlitic steel. For the back-calculation, the following assumptions were made:

– A service life of 70 years according to (SIA 261 2003); the fatigue loads were adapted to correspond to a service life of $2 \cdot 10^6$ cycles.
– The cracks behaved as long cracks (and not as short cracks) and with no initiation period.
– Deterministic calculations were carried out.
– For the brittle failure verification, the fracture toughness under quasi-static loading was used.

The resulting allowable initial defect sizes for the typical steel-concrete composite bridge are very large. They vary between approximately 28 and 88% of the wall thickness at the defect's location. The smallest size is found in the casting stub at Location 1, the biggest at Location 7 between the two braces.

A further step of the numerical study consists in the analysis of the SIF values with the aim to generalize and simplify the procedure to determine the allowable initial crack sizes. This simplified procedure should enable an engineer to estimate the allowable initial crack size (and so the defect size) without numerical crack propagation simulations. The SIF results obtained for the typical truss bridge (Figure 6) show that this simplification could be realised by expressing the stress intensity factor $K_I(a)$ in function of a constant correction factor Y:

$$K_I(a) = Y \cdot \sigma_1 \sqrt{\pi \cdot a} \quad (5)$$

$K_I(a)$ stress intensity factor for mode I at crack depth a

Y constant correction factor

σ_1 first principal stress at defect location

a crack depth

The constant correction factor can be obtained by normalising the SIF with the principal stress at Location i. Figure 7 shows the normalised SIF as a function

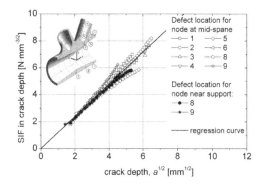

Figure 7. Normalised SIF at different crack locations in nodes at midspan and near the support.

Table 2. Constant correction factor at crack depth and length.

	Y_a	Y_c
Surface crack	0.67	0.77
Inner crack	0.62	0.64

of the square root of the depth $\sqrt{a}$. The constant correction factor value can be extracted from the slope of the regression curve, also shown in the Figure 7.

Table 2 summarises the values of the constant correction factor in the crack depth (Y_a) and in crack length (Y_c). In general, the correction factor is not constant, but depends for example on the position of the crack in the node and on the crack shape. This simplification shows that it can be assumed constant in cast steel nodes. The factors for an inner crack are nearly equal to 0.64. This is typical for a two dimensional circular inner crack. As seen in Table 2, Y_c is higher than Y_a, indicating that the surface crack propagates faster on the surface than in the depth.

An error on the correction factor has a big influence on allowable initial defect size, since the correction factor in Equation (5) is inserted into Equation (4) with an exponent of $m = 3$. Comparison of the allowable initial defect sizes obtained from the numerical SIF and from the constant correction factors shows good agreement ($\pm 20\%$ deviation) and confirms the validity of the constant correction factor approach.

This allows a simplification of the aforementioned procedure to quantify the initial allowable defect sizes in cast steel nodes used for the typical steel-concrete composite bridge. The simplified steps are summarised in Figure 8:

– Determination of the critical crack size in the case of brittle fracture using the failure assessment diagram with a constant correction factor Y_c to calculate the stress intensity factor $K_{I,ULS}^c$ (a) for ultimate static load.

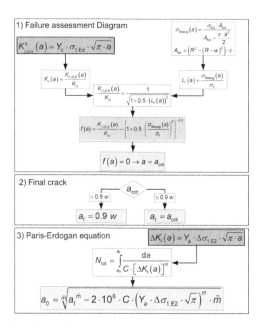

Figure 8. Calculation steps of the simplified procedure.

– Determination of the final crack size as the minimum of 90% of the wall thickness and the critical crack size.
– Back-calculation using the Paris-Erdogan equation with a constant correction factor Y_a to calculate the difference in the stress intensity factor ΔK_I (a) for the fatigue load.

The simplified procedure can now be implemented as an algorithm. This has been done as an user-friendly function for Microsoft Excel® written in Visual Basic for Applications (VBA). With the aid of this algorithm, the allowable initial defect sizes in any cast steel node can be calculated without any crack propagation simulations. The first principal stress under ultimate static and fatigue load at the critical defect location in the cast node is needed as input to the algorithm. The stress can be calculated using a FE or a BE model of the node.

The algorithm was also found suitable for a parametric study in order to determine the influence of the utilisation ratio under ultimate static and fatigue loads, the fracture toughness and the yield strength of the cast steel as well as the node dimensions on the defect size. The algorithm and the results of the parametric study can be found in Haldimann-Sturm (2005).

4 SUMMARY AND CONCLUSIONS

From the interpretation and comparison of the test results between directly welded and cast steel nodes the main points are recalled:

- The welded joints were found to have a fatigue strength of $\Delta\sigma_C = 86$ MPa, which is similar to the fatigue strength of $\Delta\sigma_C = 87$ MPa for the girth butt welds with backing bars.
- The hot spot fatigue strength curve for girth butt welds between casting stubs and CHS members corresponds to $\Delta\sigma_C = 54$ MPa *without* backing bars and $\Delta\sigma_C = 87$ MPa *with* backing bars. When computing the stress range, the stress concentration factor (SCF) due to differences in thickness and diameter of the tubular members and casting stubs must be taken into account.
- A significant difference in the fatigue behaviour between welded joints and cast nodes was observed as cracking in the welded joints is directed by high stress concentration at weld toes and in the cast nodes it is caused mainly by weld root defects in the girth butt welds at the casting stubs.
- Cast steel nodes are particularly suited for nodes near or at supports. In the tension chord at midspan, directly welded joints are the better solution regarding fatigue reliability.

From the numerical studies on welded nodes the following recommendations are given:

- The hot-spot stress method has been found to be the only current, widely accepted fatigue design method for welded tubular joints. For the determination of SCFs, existing design guide formulas cannot be extrapolated to low γ values, such as $\gamma \leq 12$, typical to bridges. For these cases, new SCFs were derived from FE analysis. The results show a strong tendency for SCFs to decrease with decreasing γ.
- The recommended minimum SCF at critical joint locations in current design specifications, $SCF = 2.0$, is highly penalizing if applied to welded CHS bridge K-joints. When $\gamma \leq 12$, the recommended minimum SCF can be taken as $SCF = 1.0$ for CHS K-joints.

From the numerical studies on cast steel nodes the following recommendations are given:

- A procedure to quantify the allowable initial casting defect sizes to ensure a balanced design between the various potential crack initiation sites in a cast steel node has been developed for a typical steel-concrete composite bridge. The defect sizes vary between approximately 28 and 88% of the wall thickness at the location of the casting defect.
- The procedure can be simplified by expressing the SIF with a constant correction factor.

ACKNOWLEDGEMENT

The research carried out is part of the project P591, supervised by the Versuchsanstalt für Stahl, Holz und Steine at the Technische Universität Karlsruhe, which is supported financially and with academic advice by the Forschungsvereinigung Stahlanwendung e. V. (FOSTA), Düsseldorf, within the scope of the Stiftung Stahlanwendungsforschung, Essen. The material presented herein also contains results from previous research sponsored by the Swiss Federal Roads Authority (FEDRO) and the Swiss National Science Foundation (SNF).

REFERENCES

Blair, M., Stevens, T. L., Steel castings handbook, Steel founders' Society of America and ASM International, 1995.

DNV, Fatigue strength analysis of offshore steel structures. Recommended practice RP-C203. Det Norske Veritas, 2001.

Haldimann-Sturm, S.C., Ermüdungsverhalten von Stahlgussknoten in Brücken aus Stahlhohlprofilen, EPFL Thesis No. 3274, Lausanne, 2005.

Marshall, P.W., Design of welded tubular connections, Basis and use of AWS provisions, Elsevier Science Publishers, Amsterdam, 1992.

Milne, I., Ainsworth, R.A., Dowlling, A.R., Stewart, A.T., Assessment of the integrity of structures containing defects, R/H/R6 – Revision 3, Central Electricity Generating Board, Mai 1986.

Raoul, J., Picture taken during construction of the St-Kilian bridge on the A73 highway, near Schleusingen, Germany, 2005.

Schumacher, A., Nussbaumer, A., Experimental study on the fatigue behaviour of welded tubular K-joints for bridges. Engineering Structures, Elsevier Science, 28(5), pp. 745–755.

Schumacher, A., Fatigue behaviour of welded circular hollow section joints in bridges. EPFL Thesis No. 2727, Lausanne, 2003a.

Schumacher, A., Sturm, S., Walbridge, S., Nussbaumer, A., Hirt, M.A., Fatigue design of bridges with welded circular hollow sections ICOM report n° 489E, ICOM/EPFL, Lausanne, 2003b.

SIA 261, Einwirkungen auf Tragwerke, Schweizerischer Ingenieur- und Architektenverein, 2003.

Sturm, S., Nussbaumer, A., Hirt, M. A., Fatigue behaviour of cast steel nodes in bridge structures, Proceed. of the 10th Int. Symp. on Tubular Structures, pp. 357–364, Tubular Structures X, A.A. Balkema Publishers, Madrid, 2003.

Veselcic, M., Herion, S., Puthli, R., Cast steel in tubular bridges – new applications and technologies, Proceed. of the 10th Int. Symp. on Tubular Structures, pp. 135–142, Tubular Structures X, A.A. Balkema Publishers, Madrid, 2003.

Zhao, X. L. et al., Design guide for circular and rectangular hollow section joints under fatigue loading, CIDECT, No. 8, TÜV-Verlag Rheinland, Köln, 2000.

Zhao, X. L., Packer, J. A., Fatigue design procedure for welded hollow section joints, Doc. XIII-1804-99, XV-1035-99, IIW, Cambridge, Abington, 2000.

Tubular Structures XI – Packer & Willibald (eds)
© 2006 Taylor & Francis Group, London, ISBN 0-415-40280-8

Probabilistic fatigue analysis of a post-weld treated tubular bridge

S. Walbridge & A. Nussbaumer

Swiss Federal Institute of Technology Lausanne, ICOM – Steel Structures Laboratory, Lausanne, Switzerland

ABSTRACT: In this paper, several probabilistic models for evaluating the fatigue performance of post-weld treated tubular bridge structures are briefly described. The application of the models is then demonstrated on a full-scale example bridge. Under realistic traffic loading conditions, it is shown that a significant increase in fatigue performance can be obtained with post-weld treatment. In comparing several treatment strategies, a partial treatment strategy is presented, which results in the same fatigue performance improvement as full treatment with a fraction of the treatment effort. One potential concern with the use of post-weld treatment methods such as needle peening is that they may simply shift the critical crack site to an untreatable location such as the weld root. To address this concern, a deterministic verification of this fatigue detail is presented. Shortcomings of this verification are then discussed, and future research needs highlighted.

1 INTRODUCTION

Bridges consisting of steel tubes welded together to form truss girders represent an industry trend that has received much recent attention from practicing engineers, code writing authorities, and researchers. The merits of these structures are cited in a number of references. One of their often cited weaknesses, and thus an area meriting further study, is the relatively poor fatigue performance of their joints.

Two ways of improving this fatigue performance have been considered in recent studies: 1) replacing the directly welded joints with cast steel nodes, and 2) improving the performance of the fatigue-critical welds by post-weld treatment (PWT).

The use of residual stress-based PWT methods such as needle or hammer peening to improve the fatigue performance of tubular bridge joints has recently been investigated in a laboratory test-based study (Schumacher 2003) and a subsequent probabilistic fracture mechanics-based study (Walbridge et al. 2003, Walbridge 2005). This paper presents a number of key findings from this second study.

Specifically, probabilistic models developed in this second study are used to analyze a full-scale tubular bridge structure under realistic loading conditions. In Section 2 of this paper the probabilistic models used for this analysis are briefly described. Following this, an example bridge structure is described and the results of a deterministic, code-based fatigue verification of this structure are presented. A probabilistic analysis of the example structure is then performed. With the results of this analysis, the deterministic verification

procedure is validated and the benefit of post-weld treatment by needle peening is evaluated. Finally, in order to assess the implications of fatigue cracking at the weld root, a deterministic verification of this fatigue detail is presented. Shortcomings of this verification are then discussed, and future research needs highlighted.

2 PROBABILISTIC MODEL OVERVIEW

The probabilistic models employed herein allow the probabilities of fatigue failure of single potential crack sites (i.e. *hot-spots*) in tubular joints and entire tubular truss structures with multiple potential crack sites to be determined at various points during the life of the structure. The details of these models are presented in Appendix I. The basis for the single site model is a linear elastic fracture mechanics-based approach. The model has been developed for applications involving non-over lapping single circular hollow section (CHS) K-joints (see Fig. 1) such as those common to planar Warren trusses.

The basis for the multiple site model employed herein is a systems reliability-based approach, wherein the entire tubular joint or truss is considered as a simple series system (see also Appendix I).

For the probabilistic analysis of each potential crack site, the required input includes the following parameters: the initial defect depth, a_0, the initial defect shape, $(a/c)_0$, the critical crack depth, a_c, the crack propagation parameters: C, m, and ΔK_{th}, the various parameters describing the joint geometry (see Fig. 1),

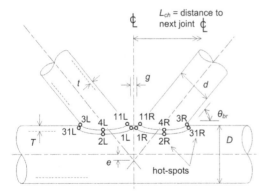

Figure 1. Non-overlapping single K-joint.

the weld angle and footprint length, θ_w and L_w, the hot-spot stress, σ_{hs}, due to the dead load, and a range-mean histogram of the hot-spot stresses due to the passage of vehicles over the bridge.

Among these parameters, a_c, m, and the joint geometry parameters are treated deterministically, while the parameters a_0, $(a/c)_0$, C, and ΔK_{th} are described by statistical variables. Several additional statistical variables are introduced in the form of factors to consider uncertainties associated with: the dead load and traffic induced hot-spot stresses, the weld angle and footprint length, and a number of other parameters contained in the crack propagation model including the magnification and correction factors, Mk and Y, stress concentration factor, SCF, and degree of bending, DOB (see Appendix I).

Along with these deterministic parameters and statistical variables, additional parameters are introduced to describe the residual stresses along the anticipated crack path due to: (1) the welding process, and (2) the subsequent post-weld treatment application. These parameters are also discussed in greater detail in Appendix I. Herein, it is assumed that treatments such as needle peening have the primary effect of introducing compressive residual stresses near the treated surface, which tend to reduce crack propagation rates at smaller crack depths.

3 DESCRIPTION OF BRIDGE STRUCTURE

The characteristics of the example bridge structure studied herein were established based on a survey of existing tubular truss bridges. These characteristics can be summarized as follows (see Fig. 2):

- highway bridge with 3×40 m spans supporting two opposing lanes of traffic,
- cross section consisting of an 11 m wide concrete slab sitting on two planar tubular trusses (full composite action assumed), and

- Warren truss configuration assumed with single CHS K-joints ($\theta_{br} = 55°$, 12 truss bays per span).

The bridge was designed based on a deterministic code-based fatigue verification carried out in accordance with the requirements of the Swiss SIA Codes (2003), including the following assumptions:

- Planned service life = 70 years.
- 40 tonne legal truck weight limit.
- Principal road traffic $\rightarrow$ $5 \cdot 10^5$ trucks/dir./year.

The design truck used in the SIA (2003) code-based fatigue verification consists of two 270 kN axle loads spaced 1.2 m apart. The verification consists of evaluating the following relationship:

$$\Delta\sigma_{E2} \leq \frac{\Delta\sigma_{c,t}}{\gamma_{Mf}} \qquad (1)$$

where $\Delta\sigma_{E2}$ is the equivalent design hot-spot stress range at $2 \cdot 10^6$ cycles, calculated as follows:

$$\Delta\sigma_{E2} = \lambda_l \cdot \Delta\sigma\left(Q_{fat}\right) \qquad (2)$$

where $\Delta\sigma(Q_{fat})$ is the hot-spot stress range, $\Delta\sigma_{hs}$, at the location of interest due to a single passage of the design truck and λ_l is a damage equivalence factor ($\lambda_l = 1.42$ for a 40 m bridge span on a principal road). In Equation 1, $\Delta\sigma_{c,t}$ is the fatigue strength corresponding with $2 \cdot 10^6$ applied stress cycles. According to Schumacher (2003), $\Delta\sigma_{c,20} = 86$ MPa should be used for the hot-spot stress-based design of tubular bridge joints, with a reference wall thickness of 20 mm, and the following size effect correction:

$$\frac{\Delta\sigma_{c,t}}{\Delta\sigma_{c,20}} = \left(\frac{20}{T \text{ or } t}\right)^{0.25} \qquad (3)$$

In Equation 1, γ_{Mf} takes on a value between 1.0 and 1.35 depending on the ease with which fatigue damage may be detected/repaired and the consequence of fatigue failure. For each joint along the bottom chord of the interior span, this verification was carried out for each of the hot-spots in Figure 1.

Under a given set of loads, the hot-spot stress, σ_{hs}, at a given potential crack site or hot-spot can be determined using the following expression:

$$\begin{aligned} \sigma_{hs} = {}&\sigma_{ax\_br} \cdot SCF_{ax\_br} + \sigma_{ax\_ch} \cdot SCF_{ax\_ch} \\ &+ \sigma_{ipb1\_br} \cdot SCF_{ipb1\_br} + \sigma_{ipb2\_br} \cdot SCF_{ipb2\_br} \\ &+ \sigma_{ipb\_ch} \cdot SCF_{ipb\_ch} \end{aligned} \qquad (4)$$

where σ_{ax_br} is the nominal member stress due to balanced axial brace load case; and SCF_{ax_br} is the corresponding stress concentration factor, etc. The five

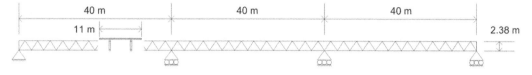

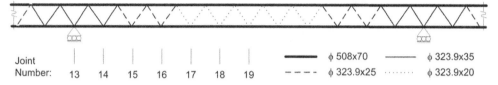

| Joint | | | | | | | | ● φ508x70 ──── φ323.9x35 |
| Number: | 13 | 14 | 15 | 16 | 17 | 18 | 19 | – – – – φ323.9x25 ·········· φ323.9x20 |

Figure 2. Example tubular bridge structure.

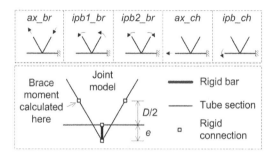

Figure 3. Joint model and nominal member load cases.

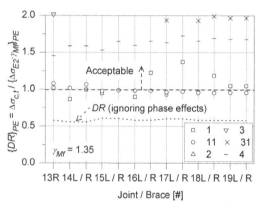

Figure 4. Design ratios, $\{DR\}_{PE}$, considering phase effects.

load cases implicated in Equation 4 are shown in Figure 3. The associated nominal member stresses can be determined by structural analysis.

In the current study, the *SCF*s in Equation 4 were determined using tables for CHS K-joints from (Schumacher 2003). These tables require as input the hot-spot location and the following parameters: $\beta (= d/D)$, $\gamma (= 0.5 \cdot D/T)$, $\tau (= t/T)$, and θ_{br}.

To determine the nominal member stresses, a simplified structural analysis was carried out wherein no interaction or load sharing between the two trusses was considered (i.e. each truss was assumed to support half of the dead load and half of the traffic load). The slab was assumed to be 300 mm thick over each truss, thinning to 250 mm at the edges and centre of the deck. Reinforcement ratios of 1% and 1.5% were assumed at the mid-span and over the support respectively. A modulus of elasticity ratio ($E_{steel}/E_{concrete}$) of 10 was assumed. The slab was assumed to be cracked over the supports, meaning that only the reinforcement was assumed to contribute to the top chord stiffness in these areas.

At each hot-spot location, the equivalent design hot-spot stress range, $\Delta\sigma_{E2}$, was calculated by taking the nominal member stress ranges for each of the five load

cases in Figure 3, multiplying them by the appropriate *SCF*s, and summing the results to get the total hot-spot stress range, $\Delta\sigma(Q_{fat})$, as recommended in (Zhao et al. 2000). In employing this approach, phase effects were conservatively ignored, i.e. it was effectively assumed that the stress peaks for each load case occur at the same truck position.

Using the verification approach described above, the member sizes in Figure 2 were selected, resulting in a bridge truss with joints that were more-or-less uniformly under-designed for fatigue, regardless of the value for γ_{Mf} assumed. Considering phase effects, i.e. by first calculating the hot-spot stress for each truck position using Equation 4 and then determining the hot-spot stress range using the resulting hot-spot stress influence line, was seen to improve the situation somewhat, as seen in Figure 4.

In this figure the results are presented of the deterministic verification considering phase effects (assuming $\gamma_{Mf} = 1.35$) for each hot-spot on each brace member of each joint along the bottom chord of the interior span (numbered as in Fig. 2). An envelope of the design ratio, DR, ignoring phase effects for the worst hot-spot on each brace member is included for comparison purposes.

From this figure, it can be deduced that the example structure can be made to pass the deterministic verification if phase effects are considered and a lower value for γ_{Mf} is permitted, such as 1.0. If use of the higher γ_{Mf} value is deemed necessary, then the example structure almost passes the verification if phase effects are considered, but fails by a considerable margin if these effects are ignored.

Also important to note in this figure is that the performance of each of the joints appears to be largely determined by Sites 1L, 11L, 1R, and 11R.

4 PROBABILISTIC ANALYSIS

Following the code-based fatigue design of the example bridge structure, a probabilistic analysis was conducted of each hot-spot (untreated) on each of the seven joints numbered in Figure 2.

To perform these analyses, the single site probabilistic model described in Appendix I was employed. As discussed in (Walbridge 2005), the required hot-spot stresses were determined in the same way as in the code-based verification considering phase effects. The required weld geometry parameters were determined at each hot-spot using parametric equations based on the AWS Code (2000).

In order to simulate realistic loading conditions, a traffic model was needed. The model eventually adopted consists of three truck types, with weight distributions as shown in Figure 5. The contribution of each truck type to the total traffic volume is also indicated in this figure. This model is based on weigh scale measurements taken on the main highway between Bern and Zurich, Switzerland (Kunz & Hirt 1991), modified to consider the new 40 tonne Swiss legal truck weight limit as discussed in (Walbridge 2005). In using this model, the truck weights were multiplied by a dynamic factor of 1.3.

The assumed values for the statistical variables required to apply the single site probabilistic model are summarized in Table 1. In this table the 'VAR' variables are typically factors by which the input parameters described by the subscripts are multiplied. The applied stresses due to the traffic and dead loads, for example, are multiplied by the variables: $VAR_{traffic}$ and VAR_{dead}. Regarding these variables, it should be pointed out that the mean values for Variables 2, 12, and 13 were determined by calibration using the test results from (Schumacher 2003). With these values, the model was seen to closely predict the mean and scatter of these test results.

To solve the resulting probabilistic fracture mechanics problem, a FORTRAN 90 subroutine employing

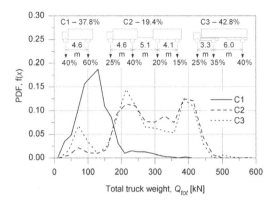

Figure 5. Traffic model used in probabilistic analysis.

Table 1. Statistical variables.

Variable	i	μ_x	σ_x	Dist.	Units
a_0	1	0.2	0.045	LN	mm
$(a/c)_0$	2	0.5	0.16	LN	–
$VAR_{traffic}$	3	1.0	0.15	N	–
VAR_{dead}	4	1.0	0.10	N	–
VAR_{DOB}	5	1.0	0.08	N	–
VAR_{SCF}	6	1.0	0.04	LN	–
VAR_{Mk}	7	1.0	0.05	LN	–
VAR_{Lw}	8	1.0	0.10	N	–
$VAR_{\theta w}$	9	1.0	0.10	N	–
VAR_{weld}	10	1.0	0.25	N	–
VAR_{pwt}	11	0.5	0.10	N	–
$LN(C)$	12	-28.80	0.55	N	LN((mm/cycle)· $(N/mm^{-3/2})^m$)
ΔK_{th}	13	100.0	15.0	LN	MPa$\sqrt{}$mm
a_c	–	$0.5 \cdot T$	–	det.	mm
f_y	–	355	–	det.	MPa
m	–	3.0	–	det.	–

the *Monte Carlo simulation* (MCS) solution method with a crude importance sampling scheme was used to determine the probabilities of failure, p_f, for the various hot-spots corresponding with the passage of a given number of trucks, Tr.

With the resulting p_f vs. Tr data, a reliability envelope was calculated for the interior span of one truss on the example bridge structure (see Fig. 6). This envelope takes into account the uncertainty in the true level of correlation between the probabilities of failure of the various hot-spots, as discussed in Appendix I. To produce this envelope, several approaches were considered to account for the top chord joints in the example structure. For illustrative purposes, the fatigue lives of the hot-spots in these joints are assumed herein to be sufficiently high that the fatigue reliability of the entire truss can be taken as roughly equal to that of the bottom chord.

Following the analysis of the untreated structure, a second series of calculations was performed with each

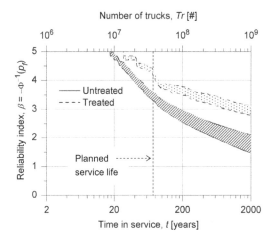

Figure 6. β vs. Tr envelopes for example bridge structure.

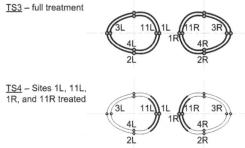

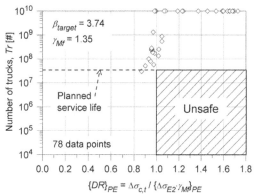

Figure 7. Full (TS3) and partial (TS4) treatment strategies.

Figure 8. Verification of code-based design.

potential crack site treated by needle peening. With the results of these calculations, a second reliability envelope was produced for the treated structure (see also Fig. 6). Comparing the envelopes for the untreated and treated structures, the potential benefit of the treatment could be determined.

Looking at Figure 6, it can be seen that the benefit of post-weld treatment was significant for the studied example bridge structure. In order to quantify this benefit, the fatigue lives of the untreated and treated bridges can be compared at a given target reliability index, β_{target}. In (Walbridge 2005), for example, comparisons are made at $\beta_{target} = 3.74$. This is the target index suggested in the Eurocode EN 1990 (2002) for a structure with a design life of 70 years, a low level of redundancy, and limited possibility for inspection/repair. If this target index is used to compare the envelopes in Figure 6, the benefit of treatment can be seen to be a 98 to 267% fatigue life improvement, depending on whether the results obtained with the upper or the lower bound series system models are used for the comparison.

Using the analytical approach described above a number of studies where carried out to determine the effects of variations in a number of the treatment parameters, as discussed in (Walbridge 2005). One useful result of these studies was the finding that the same treatment benefit could be achieved with a partial treatment strategy (TS4 in this reference), wherein only Sites 1 and 11 on the left and right braces are treated, as illustrated in Figure 7.

This finding can be explained by examining the deterministic design ratios in Figure 4. Apparently post-weld treatment by needle peening is sufficient to increase the fatigue lives of Sites 1 and 11 to a certain extent, but this improvement is not so great as to make these sites less critical than the next most critical untreated site with treatment strategy TS4.

Another potentially useful result of the probabilistic calculations is that the single site results for the untreated bridge can be used to perform a verification of the deterministic, code-based design. The results of such a verification are summarized in Figure 8. In this figure, the design ratio considering phase effects, $\{DR\}_{PE}$, and assuming a low level of redundancy and limited possibility for inspection/repair ($\gamma_{Mf} = 1.35$) is plotted verses the calculated fatigue life corresponding with $\beta_{target} = 3.74$.

Looking at this figure, it can be concluded that according to the probabilistic model, the code-based design gives safe results for single potential crack sites, even when phase effects are considered. In other words, there is no case where the code-based design deems a potential crack site to be adequate, while the probabilistic model suggests it is not.

A number of the sites in Figure 8 are seen to perform much better than might be expected simply by looking at the deterministic design ratio. The main reason for this is that the code-based verification does not consider the beneficial effects of compressive stresses at the various potential crack site imposed by the dead loads. These stresses have a positive effect on the

73

fatigue lives of these sites similar to that induced by the post-weld treatment.

As discussed in (Walbridge 2005), this verification is conservative in that the failure criterion for the probabilistic calculation is crack growth to a critical depth of $a_c = 0.5 \cdot T$ (or t). In fact, tubular structures are known to possess a certain reserve capacity beyond crack growth to this depth. According to van Wingerde et al. (1997), for example, the time to total joint failure is on average 1.49 times as long as the time to through thickness cracking. In addition, a constant, but conservative DOB has been assumed for all of the load cases in Figure 3.

One potentially unconservative assumption made by the probabilistic model is that the crack tip loading mode is essentially the same for all of the load cases in Figure 3 (i.e. primarily *opening* or *Mode I* loading). It is thought that the error due to this assumption should be small, although further study of the effect of the true crack tip loading mode may be of value if phase effects are to be routinely considered in the code-based design. The results of this further study are not, however, expected to have serious implications on the calculated benefit of post-weld treatment as seen, for example, in Figure 6.

5 WELD ROOT VERIFICATION

One potential concern with the use of post-weld treatment methods is the increased possibility, with treatment, that the eventual failure of the joint will result from fatigue cracking at the weld root. Cracking at this location is generally considered to be much less desirable, as there is no possibility in this case for early detection by visual inspection.

One possible fatigue assessment approach for the weld root, inspired by the hot-spot stress approach commonly used for the weld toe, is discussed in (Health and Safety Executive 1999). This reference provides a set of parametric equations for a factor, R_{SCF}, which relates the maximum SCF at the weld root to that at the weld toe, i.e.:

$$R_{SCF} = \frac{SCF_{root,max}}{SCF_{toe,max}} \qquad (5)$$

These equations require as input the following parameters: $\beta, \gamma, \tau, \zeta\,(=g/D)$, and θ_{br}. Equations are provided for two of the brace load cases (*ax_br* and *ipb2_br*) in Figure 3. As is often the case with research conducted primarily for the offshore industry, the validity range for the γ parameter is much higher ($12 \le \gamma \le 30$) than the range for this parameter common to tubular bridge structures. If this range is ignored, however, these equations can be applied to the example bridge to determine the possible implications of weld root cracking. The

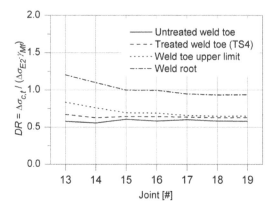

Figure 9. Deterministic weld root verification.

results of such an application are summarized in Figure 9.

To generate the design ratios for the weld root in this figure, it was assumed that R_{SCF} was the same for both brace bending load cases in Figure 3. Furthermore, it was assumed that the chord load cases would not affect R_{SCF}. On this basis, a hot-spot stress range for the weld root was calculated. Phase effects were ignored in this calculation because it was necessary to separate the effects of the various load cases to apply these equations. The applied stress range was then compared to a fatigue strength at $2 \cdot 10^6$ applied stress cycles, $\Delta\sigma_{c,t}$, of 90.6 MPa for the weld root, as proposed in (Health and Safety Executive 1999), corrected for size using Equation 3, assuming (conservatively) that the weld root crack will occur in the thicker chord wall.

Looking at Figure 9, it can be concluded that cracks will most likely always initiate from the weld toe in the untreated bridge. In this figure, a theoretical design ratio is presented for the bridge treated using strategy TS4. As the deterministic design ratios for the treated sites could not be calculated, these sites were simply not considered in the determination of the design ratios for the joints, effectively modelling the case of *perfect* treatment. Comparing the data for the treated weld toe and the untreated root, it can be seen that the post-weld treatment should not result in a shift in the critical crack site from the weld toe to the weld root if this partial treatment strategy is employed.

To make sure of this, it is recommended that the weld root be designed with a fatigue life 100 to 200% greater than that of the weld toe (Health and Safety Executive 1999). This can be equated to ensuring that the design ratio of the root is 1.25 to 1.44 that of the toe if an S-N curve slope of $m = 3.0$ is assumed. A fourth curve in Figure 9 shows the maximum allowable design ratio for the weld toe using the more severe of these two limits. As can be seen in this figure, the benefit of treatment using strategy TS4 can still be deemed permissible on this basis.

6 CONCLUSIONS

The analyses presented herein demonstrate the potential of post-weld treatment methods such as needle peening for increasing the fatigue lives of tubular bridge structures. Herein, it is also shown that the same treatment benefit can be achieved for tubular bridges similar to the studied example bridge structure with a partial treatment strategy that would likely be much less costly than full treatment.

Finally, a deterministic verification of the weld root is presented. Although this verification shows that the adopted partial treatment strategy will not cause the critical crack location to shift to the weld root, further work is clearly needed to develop parametric equations for the R_{SCF} factor that are adapted to tubular joints with geometries typical of bridge structures. In addition, there may be value in modifying the weld root verification procedure so that phase effects can be more easily considered.

ACKNOWLEDGEMENTS

The work presented herein was supported by the Swiss Federal Highway Administration (OFROU Project No. AGB2002/011) and the Swiss National Research Foundation (SNF Grant 200020-101521).

REFERENCES

Albrecht, P. & Yamada, K. 1977. Rapid Calculation of Stress Intensity Factors. *Journal of Structural Engineering* 103(ST2):377–389.

American Welding Society 2000. *ANSI/AWS D1.1-2000 Structural Welding Code – Steel*. U.S.A.

Bowness, D. & Lee, M.M.K. 1999. Weld Toe Magnification Factors for Semi-Elliptical Cracks in T-Butt Joints. Offshore Technology Report – OTO 199 014, Health and Safety Executive (HSE), U.K.

Bremen, U. 1989. Amélioration du comportement à la fatigue d'assemblages soudés : étude et modélisation de l'effet de contraintes résiduelles. EPFL Thesis No. 787, Lausanne.

Connolly, M.P., Hellier, A.K., Dover, W.D., & Sutomo, J. 1990. A parametric study of the ratio of bending to membrane stress in tubular Y- and T-joints. *International Journal of Fatigue* 1:3–11.

Dubois, V. 1994. Fatigue de détails soudés traités sous sollicitations d'amplitude variable. EPFL Thesis No. 1260, Lausanne.

European Committee for Standardization 2002. *EN1990 – Basis of structural design*. Brussels.

Health and Safety Executive 1999. Fatigue Life Implications for Design and Inspection for Single Sided Welds at Tubular Joints. HSE Report No. OTO 99/022, U.K.

Kunz, P. & Hirt M.A. 1991. Grundlagen und Annahmen für den Nachweis der Ermüdungsfestigkeit in den Tragwerksnormen des SIA. Doc. D 076, Swiss Society of Engineers and Architects (SIA), Zurich.

Manteghi, S. & Maddox, S.J. 2004. Methods for fatigue life improvement of welded joints in medium and high strength steels. Doc. XIII-2006-04, International Institute of Welding (IIW).

Newman, J.C. & Raju, I.S. 1981. An Empirical Stress-Intensity Factor Equation for the Surface Crack. *Engineering Fracture Mechanics* 15(1–2):185–192.

Romeijn, A., Karamanos, S.A., & Wardenier, J. 1997. Effects of joint flexibility on the fatigue design of welded tubular lattice structures. 7th International Offshore and Polar Engineering Conference.

Schumacher, A. 2003. Fatigue behaviour of welded circular hollow section joints in bridges. EPFL Thesis No. 2727, Lausanne (http://icom.epfl.ch).

Stacey, A., Barthelemy, J.-Y., Leggatt, R.H., & Ainsworth, R.A. 2000. Incorporation of Residual Stresses into the SINTAP Defect Assessment Procedure. *Engineering Fracture Mechanics* 67:573–611.

Stephens, R.I., Fatemi, A., Stephens, R.R., & Fuchs, H.O. 2001. *Metal Fatigue in Engineering (2nd Ed.)*. John Wiley & Sons.

Swiss Society of Engineers and Architects (SIA) 2003. *SIA 260/261/263:2003 (Basis for structural design/Actions on structures/Steel structures)*. Zurich.

van Wingerde, A.M., van Delft, D.R.V., Wardenier, J., & Packer, J.A. 1997. Scale Effects on the Fatigue Behaviour of Tubular Structures. WRC Proceedings, International Institute of Welding 123–135.

Walbridge, S. 2005. A probabilistic study of fatigue in postweld treated tubular bridge structures. EPFL Thesis No. 3330, Lausanne (http://icom.epfl.ch).

Walbridge, S., Nussbaumer, A., & Hirt, M.A. 2003. Fatigue Behaviour of Improved Tubular Bridge Joints. Proceedings of the 10th International Symposium on Tubular Structures, Madrid.

Zhao, X.-L., Herion, S., Packer, J. A., et al. 2001. *Guide No. 8 – Design guide for circular and rectangular hollow section joints under fatigue loading*. Cologne: CIDECT.

APPENDIX I – PROBABILISTIC MODELS

Single site model

The limit state function, G(**z**), for the probabilistic single site model employed herein is founded on the Paris-Erdogan crack growth law, modified to consider crack closure effects and a threshold stress intensity factor (SIF) range, ΔK_{th}, and integrated over a crack depth range, a_0 to a_c. Specifically:

$$G(\mathbf{z}) = N_c - N = \int_{a_0}^{a_c} \frac{da}{C \cdot \left(\Delta K_{e\!f\!f}^{\ m} - \Delta K_{th}^{\ m}\right)} - N \qquad (A1)$$

where:

$$\Delta K_{e\!f\!f} = \mathrm{MAX}\left(K_{app,max} - K_{op}, 0\right)$$
$$- \mathrm{MAX}\left(K_{app,min} - K_{op}, 0\right) \qquad (A2)$$

where $K_{app,max}$ and $K_{app,min}$ = maximum and minimum SIFs due to the applied load; and K_{op} = applied SIF at which crack tip opens upon loading.

Herein, K_{op} is calculated as follows:

$$K_{op} = -\left(K_{res} + K_{pl}\right) \qquad (A3)$$

where K_{res} = SIF due to residual stress distribution along anticipated crack path; and K_{pl} = crack closure SIF. Herein, K_{pl} is calculated using the following empirical expression from (Bremen 1989):

$$K_{pl} = -\mathrm{MIN}\left(0.2/\left(1 - R_{e\!f\!f}\right), 0.28\right)$$
$$\cdot \left(K_{app,max} + K_{res}\right) \qquad (A4)$$

where $R_{e\!f\!f}$ = effective stress ratio. Specifically:

$$R_{e\!f\!f} = \frac{K_{app,min} + K_{res}}{K_{app,max} + K_{res}} \qquad (A5)$$

K_{res} is calculated at each crack depth increment using the approach proposed by Albrecht & Yamada (1977). The assumed stress distributions needed for this approach are based on those proposed by Stacey et al. (2000) and Bremen (1989) for the welding and post-weld treatment (PWT) residual stresses.

The assumed residual stress distribution due to the welding process is as follows:

$$\sigma_{weld}(b) = f_y \cdot \begin{pmatrix} 0.62 + 2.33 \cdot (b/T) \\ -24.13 \cdot (b/T)^2 \\ +42.49 \cdot (b/T)^3 \\ -21.09 \cdot (b/T)^4 \end{pmatrix} \cdot VAR_{weld} \qquad (A6)$$

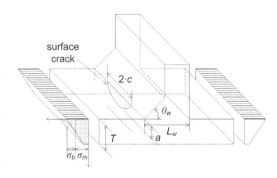

Figure A1.　Crack at weld toe of T-butt joint.

where b = depth below surface; and T = wall thickness of cracked member. In this expression, VAR_{weld} is a statistical variable introduced to consider the uncertainties associated with this stress distribution.

The assumed PWT residual stress distribution is:

$$\sigma_{pwt}(b) = -f_y \cdot \left(VAR_{pwt}\right) \text{ if } b \le 0.1 \cdot d_p$$
$$= f_y \cdot \begin{pmatrix} (b/d_p) \cdot (5/6) - \\ (1/12 + VAR_{pwt}) \end{pmatrix} \text{ if } b > 0.1 \cdot d_p \qquad (A7)$$

where d_p = imprint diameter of peening tool (1.5 mm for needle peening). To determine the combined (welding plus treatment) stress distribution, $\sigma_{res}(b)$, the parameter PWT is introduced, such that:

$$\sigma_{res}(b) = \mathrm{MIN}\left(\sigma_{pwt}(b), \sigma_{weld}(b)\right) \text{ if } PWT = 1$$
$$= \sigma_{weld}(b) \text{ if } PWT = 0 \qquad (A8)$$

To solve Equation A2, K_{app} is determined using the following expression from (Bowness & Lee 1999):

$$K_{app} = \begin{pmatrix} Mk_m \cdot Y_m \cdot (1 - DOB) \\ + Mk_b \cdot Y_b \cdot DOB \end{pmatrix} \cdot \sigma_{hs,app} \cdot \sqrt{\pi \cdot a} \qquad (A9)$$

where $\sigma_{hs,app}$ = applied hot-spot stress; Mk_m, Mk_b, Y_m, and Y_b = magnification and correction factors for the bending (σ_b) and membrane (σ_m) stress cases; and DOB = degree of bending ($= \sigma_b/(\sigma_b + \sigma_m)$).

In using this approach, it is essentially assumed that a weld toe crack anywhere on a tubular joint behaves in the same way as a similar crack in a T-butt joint such as the one in Figure A1.

Herein, Mk_m and Mk_b are solved using parametric equations from (Bowness & Lee 1999). These equations require as input the following parameters: a/T, a/c, L_w/T, and θ_w. Y_m and Y_b are solved using parametric equations from (Newman and Raju 1981). These equations require as input the following parameters: a/T and a/c. The DOB is solved using parametric equations from (Connolly et al. 1990), making assumptions for the missing load cases, and ignoring the limits on γ. These equations require as input the hot-spot location along with the following parameters: α $(= 2 \cdot L_{ch}/D)$, β, γ, τ, and θ_{br}.

A one-dimensional crack propagation model is employed herein with the aspect ratio, a/c, varied according to a predefined crack shape evolution function wherein the initial aspect ratio, $(a/c)_0$, may vary, but this ratio then evolves smoothly, converging on a fixed value of 0.2 at $b/T = 0.25$.

Multiple site model

In order to determine the probabilities of failure of structures comprised of tubular K-joints with multiple potential crack sites, lower and upper bound reliability models for series systems are employed herein. Specifically, it is first assumed that each K-joint in the structure can be modelled as a series system with 16 constituent elements, corresponding with each of the hot-spots identified in Figure 1 (Note: Sites 2L, 4L, 2R, and 4R each occur twice).

The lower bound model assumes that the probabilities of failure of the individual hot-spots are fully independent. On this basis the probability of failure of the joint can be written as follows:

$$p_{f,joint} = 1 - \left(1 - p_{f,1L}\right) \cdot \left(1 - p_{f,11L}\right) \cdot \ldots \cdot \left(1 - p_{f,4R}\right) \quad \text{(A11)}$$

where $p_{f,joint}$ = probability of joint failure; $p_{f,1L}$ = probability of failure of Site 1L, etc. To determine the probabilities of failure of tubular structures with multiple joints, a similar approach is employed.

The upper bound reliability model assumes full correlation of the probabilities of failure at each potential crack site, and takes the following form:

$$p_{f,joint} = \text{MAX}\left(p_{f,1L}, p_{f,11L}, \ldots, p_{f,4R}\right) \quad \text{(A12)}$$

Additional considerations for analysis under realistic loading conditions

As the probabilistic single site model is presented in Section 8.1, it can only be used for analysis under constant amplitude (CA) loading conditions.

In order to analyze structures under realistic, variable amplitude loading conditions, modifications to the adopted fracture mechanics model are therefore required. A number of candidate models for predicting crack growth rates under variable amplitude (VA) loading conditions are summarized in (Stephens et al. 2001). Among the simplest of these methods are the *Equivalent constant amplitude stress range* and *Equivalent block loading* approaches. Dubois (1994) and Manteghi and Maddox (2004) both found that fatigue life predictions made using the first approach are often highly unconservative for post-weld treated details. For this reason, the second approach is adopted herein.

According to this approach, the VA stress spectrum is divided into a number of CA stress range blocks. At each crack depth, a, the crack closure SIF due to the maximum stress for each block is calculated. The rate of crack growth is then determined herein by calculating the damage due to each CA stress range block, assuming a crack closure SIF for all blocks equal to the largest crack closure SIF caused by any one block at that crack depth.

Tubular Structures XI – Packer & Willibald (eds)
© *2006 Taylor & Francis Group, London, ISBN 0-415-40280-8*

Notch stress factor of welded thick-walled tubular dragline joints by effective notch stress method

N.L. Pang & X.-L. Zhao
Department of Civil Engineering, Monash University, Clayton, Australia

F.R. Mashiri & P. Dayawansa
Maintenance Technology Institute, Department of Mechanical Engineering, Monash University, Australia

J.W.H. Price
Mechanical Engineering Department, Monash University, Clayton, Australia

ABSTRACT: In Australia, draglines are used extensively in the coal mining industry. The tubular booms of the dragline are susceptible to fatigue cracking due to the large number of cycles they are subjected to during operation. Fatigue cracks may occur at either weld toe or weld root of the dragline joints. This paper acts as the first attempt to use the effective notch stress (ENS) method to determine the notch stress factors (NSFs) of the dragline joints. Only simplified 2D finite element analysis (FEA) was carried out at this stage. It was found that the NSF at the weld toe is higher than the NSF at the weld root for LC(I). For LC(II), the NSF at the weld root is in general higher than the NSF at the weld toe. The NSF results obtained are also used to compare with those obtained by using ENS method in the literature.

1 INTRODUCTION

Dragline is large mining equipment with a boom length of about 100 m, used to lift a total bucket weight of approximately 180–210 tonnes. The boom of a dragline machine may be made up of welded, large thick-walled circular hollow section (CHS). These connections consist of a main chord member with 3- to 5-lacing members, all welded to the chord at one point forming complex overlapped joint. Multi-planar double overlapping CHS NN-joint consisting of 4-lacing members is the most common type of joint in the dragline boom (Pang et al. 2005). Constant monitoring of dragline structures has shown that fatigue cracks are prevalent in the dragline CHS connections and hence making the draglines susceptible to fatigue failure.

Large variations of stress around the thick-walled, double overlapping CHS NN-joints was reported to occur in the region of the intersection on the main chord, between the braces, especially at the areas where the bracing members terminate (Dayawansa et al. 2003). The repair history data for dragline boom structures shows the cracking in joints typically initiates on the surface of the main chord in the footprint of a bracing weld root and grows into the main chord. Adequate penetration may become difficult to achieve as the wall

thickness increases and hence resulting in the susceptibility of welded joints to weld root failure, instead of conventional weld toe failure.

The most frequent type of cracking occurs in the main chords, perpendicular to the axis of the chord. These cracks generally start from the outer surface of the chord at the root of a lacing weld and grow into the wall of the chord, eventually penetrating through the wall (Dayawansa et al. 2003). The location of cracking is embedded underneath the lacing, and hence cannot be detected using visual inspection methods. These cracks are not normally detected until they have propagated through the thickness.

In the past decades, both experimental and finite element methods have been used to establish the design of welded tubular structures using the hot-spot stress (HSS) method. However, the hot-spot stress method determines the stress concentration factor (SCF) at the weld toe, not the weld root (Zhao et al. 2005).

This paper introduces the effective notch stress (ENS) method which can deal with both the weld toe and weld root failure by determining the notch stress factor (NSF) at both the weld toe and weld root. To avoid confusion, the terminology of NSF is used in ENS method, instead of SCF, which is traditionally used by HSS method. Simplified 2D FE model was adopted as the first attempt. Both the mesh convergence study and the parametric study were carried out. The relationship between the NSF at weld root and NSF at weld toe is established. Results are also compared with those reported in literatures using ENS method.

2 EFFECTIVE NOTCH STRESS METHOD

2.1 The method

ENS method is a relatively new method which refers to the stresses directly at the notch, at which the fatigue crack initiation is expected, i.e. either at the weld root or the weld toe. The ENS is the maximum principal stress at the root of a notch, assuming linear-elastic material behaviour, and obtained by finite element analysis (FEA). The geometry of the notch cannot be modeled directly by FEA because of the irregularity of the weld. However, this can be overcome by the introduction of a fictitious effective notch radius as shown in Figure 1. The effective notch radius is introduced such that the tip of the radius touches the root of the real notch. As the effective notch radius is an idealization, the ENS cannot be measured directly in the welded component. For structural steel, the effective notch radius is taken as 1 mm (IIW 1996).

As recommended by IIW, this method is restricted to welded joints which are expected to fail from either the weld toe or weld root. Other causes of fatigue failure, e.g. from surface roughness or embedded defects, are not covered. In addition, this method is not applicable where considerable stress components parallel to the weld or parallel to the root gap exist. This method is also limited to a minimum wall thickness of 5 mm.

This method has been applied with success to the fatigue damage assessment of a main load-carrying structure of the wind energy converter, a T-joint with permanent backing bar and a few cruciform joints (Hobbacher 2002). Most recently, this method has been used to investigate the fatigue crack on the weld of steel deck plate and trough rib for the bridge structures (Suganuma et al. 2005).

2.2 The fictitious notch radius

The concept of fictitious notch radius is based on Neuber's micro-support theory. Morgenstern et al. (2004) outlined this method, which assumes

$$r_f = r_r + \rho^* \cdot s \tag{1}$$

where
r_f is the fictitious notch radius,
r_r is the real radius,
ρ^* is the micro-support constant and
s is the multiaxiality constant

As suggested by Neuber, $\rho^* = 0.4$ mm and $s = 2.5$ in accordance to von Mises, for structural steel. In the

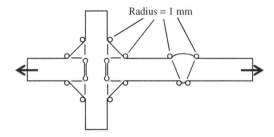

Figure 1. Effective notch stress model (IIW 1996).

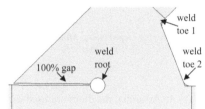

(a) Location of weld root and weld toes where the effective notch stresses are determined.

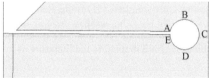

(b) Weld root region where effective notch stresses are determined.

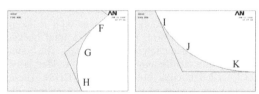

(c) Weld toe 1 region where effective notch stress is determined.

(d) Weld toe 2 region where effective notch stress is determined.

Figure 2. NSF determination by effective notch method.

worst case of a crack, where the real radius is 0 mm, Equation (1) will give $r_f = 1.0$ mm.

2.3 NSF definition

The definition of the NSF for the effective notch method is different from the commonly used fatigue resistance method, i.e. the HSS method. In HSS method, the HSS has to be determined at the weld toe position, located away from the stress field region influenced by the local weld toe geometry. Hence, the location from which the stresses have to be extrapolated is depending on the dimensions of the joint and the position around the intersection. However, for the ENS method, the ENS is determined at the surface of the fictitious effective notch radius of 1 mm. Figure 2(a) shows the region where the ENS is determined for a typical Y-joint with 100% gap. The ENS for the weld root is the maximum notch stress along the line ABCDE as shown in Figure 2(b). The ENS for the weld toes is the maximum notch stress along the line FGH for weld toe 1 or along the line IJK for weld toe 2. Figure 2(c) and Figure 2(d) shows the region where ENS is determined for the weld toe 1 and weld toe 2 respectively.

The NSF for the weld root or weld toe is the ratio between the respective ENS at the weld root or weld toe, and the nominal stress in the member.

In fatigue resistance assessment using HSS method, SCF is used to calculate the HSS range S_{rhs}. Then, the fatigue strength of the joint can be determined from the S_{rhs}-N curves. However, for the effective notch method, the fatigue class (FAT) of the parent material is fixed to the value of $\Delta\sigma = 225$ MPa, i.e. the effective notch stress range at 2×10^6 cycles at a survival probability of 95%, which is equal to mean minus two standard deviations (IIW 1996). The fatigue class is calculated by dividing the FAT 225 of notch stress by the NSF of the notch in consideration.

3 SIMPLIFICATION OF FINITE ELEMENT MODEL

3.1 Laboratory specimens

Test of four dragline joints, with different main chord and bracings' wall thicknesses, will be carried out in the Civil Engineering Laboratory at Monash University. The dimensions for the four joint specimens are given in Table 1. These joints are selected after careful study of all the joints of typical dragline (model 1370W) so that the widest validity range can be covered in the test. The length of the main chord, L_0, and brace L_1, is 2.7 m and 1.0 m respectively. These lengths are chosen such that the strains to be measured will not be influenced by the end conditions. The measurements have to be taken place at about two to two and half times the diameter from any end conditions (van Wingerde 1992). Figure 3 shows the photo of specimen 2 set-up with end plates.

The validity range for the specimens is as follows:

- brace-to-chord diameter ratio, $0.41 \leq \beta \leq 0.80$
- chord radius-to-thickness ratio, $16 \leq 2\gamma \leq 21.33$
- brace-to-chord thickness ratio, $0.28 \leq \tau \leq 1.0$
- in-plane angle of braces with respect to the chord axis, $\theta = 44.30°$
- out-of-plane angle of braces, $\varphi_{1,4} = 33.69°$ for member 1 and 4 and $\varphi_{2,3} = 24.96°$ for member 2 and 3

Table 1. Dimensions and parameters of laboratory joints.

Specimen	d_0 (mm)	t_0 (mm)	d_1 (mm)	t_1 (mm)	d_2 (mm)	t_2 (mm)
1	406.4	25.4	323.8	19.0	219.1	8.2
2	406.4	25.4	219.1	8.2	168.3	7.1
3	406.4	19.0	323.8	19.0	219.1	8.2
4	406.4	19.0	219.1	8.2	168.3	7.1

where, d_0 and d_i is the diameter of the main chord and the bracing respectively; t_0 and t_i is the thickness of the main chord and the bracing respectively; and i is the positive integer.

Figure 3. Specimen 2 with the end plates.

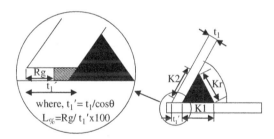

where, $t_1' = t_1/\cos\theta$
$L_\% = Rg/t_1' \times 100$

Figure 4. Definition of the weld toe profile and the depth of the weld root gap.

3.2 Weld profile measurement

The silicon imprint technique was used to determine the weld profile of the specimen 2. Figure 4 shows the definition of the weld toe profile and the depth of the weld root gap, Rg. A 8" feeler gauge was used to measure the depth of the weld root gap.

The result of the weld profile and the depth of the root gap is shown in Figure 5.

3.3 Model simplification

The 3D multiplanar double overlapping CHS NN-joint of the laboratory specimens can be simplified to

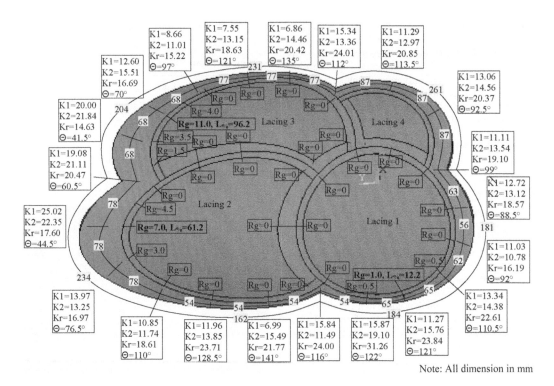

Note: All dimension in mm

Figure 5. Weld profile and depth of root gap.

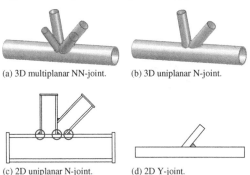

(a) 3D multiplanar NN-joint.   (b) 3D uniplanar N-joint.

(c) 2D uniplanar N-joint.   (d) 2D Y-joint.

Figure 6. Specimen model simplifications process.

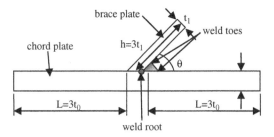

Figure 7. Nomenclatures of the joint.

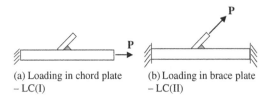

(a) Loading in chord plate – LC(I)   (b) Loading in brace plate – LC(II)

Figure 8. Schematic diagram of the load cases and the boundary conditions.

3D uniplanar overlapping N-joint. The 3D uniplanar overlapping N-joint can further be simplified to 2D uniplanar joint. Finally the 2D uniplanar joint could be simplified to 2D Y-joint. The simplification sequence is shown in Figure 6.

The simplified 2D model for the laboratory specimens were modeled and analysed by using the finite element programme ANSYS 10.0 (SAS IP Inc. 2005). Each individual joint is modeled with length of 3 times the thickness of the plate to minimize, if not eradicate, the boundary condition effect on the joint. Figure 7 shows the nomenclatures of the joint used in the FEA of dragline joint.

Two different load cases were used in the FEA. The schematic diagram of the two load cases and the boundary conditions used for the FEA of the joint is shown in Figure 8. In both load cases, unit load of 1N was applied to the FE model. The first load case, LC(I),

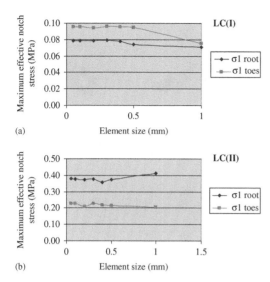

(a) Element size (mm)

(b) Element size (mm)

Figure 9. Element size versus effective notch stress at weld root and weld toes for (a) LC(I); (b) LC(II).

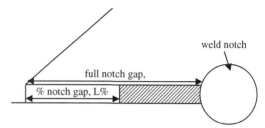

Figure 10. Notch gap definition.

was the loading in the chord plate and the second load case, LC(II), was the loading in the brace plate.

4 MESH CONVERGENCE STUDY

The most common type of dragline joint was used to carry out the mesh convergence study. This joint has the main chord thickness of 25.4 mm, bracing thickness of 8.2 mm, in-plane angle of brace with respect to the chord axis, θ, of 44.3° and 50% notch gap, which will be explained in Section 5. Seven different element sizes, ranging from 0.05 mm to 1.0 mm, were used to mesh the area of high stress concentration, i.e. at both the weld root and weld toes of the model. This study was carried out to find the effect of mesh size on the FE analysis of this particular type of joint and to verify the use of mesh size of r/20, as recommended by Morgenstern et al. (2004).

4.1 Results of mesh convergence study

Figure 9(a) shows the plot of the element size versus the effective notch stress at the weld root and the weld toes for LC(I). Figure 9(b) shows the plot of the element size versus the effective notch stress at the weld root and the weld toes for LC(II).

4.2 Remarks for mesh convergence study

For both LC(I) and LC(II), the ENS start converging at element size of 0.4 mm, though the element size of 0.1 mm has a closest ENS values in comparison to 0.05 mm element size. The element size of 0.3 mm has

a relatively smaller percentage difference of ENS values in comparison with other element sizes apart from element size of 0.1 mm. The percentage difference of ENS values for element size of 0.3 mm in comparison to 0.05 mm element size is less than 1% for both the weld root and the weld toes. Hence, a maximum element size of 0.3 mm may be used to mesh this particular type of joint. This recommendation will reduce both the analysis time and file size for each joint model especially for 3D model.

5 NOTCH GAP STUDY

The most common type of dragline joint used in the mesh convergence study was used to carry out the notch gap study too. This joint has the same geometry as the model used in the mesh convergence study except the notch gap. Figure 10 shows the definition of notch gap for the dragline joint.

5.1 Results of notch gap study

Figure 11(a) shows the plot of the notch gap percentage versus the NSF at the weld root and the weld toes for LC(I). Figure 11(b) shows the plot of the notch gap percentage versus the NSF at the weld root and the weld toes for LC(II). The parameters used are the main chord thickness of 25.4 mm, the bracing thickness of 8.2 mm and the in-plane angle of brace with respect to the chord axis, θ, of 44.3°.

5.2 Remarks for notch gap study

As the notch gap percentage increases, the NSF at the weld toes decreases marginally for both LC(I) and LC(II). Conversely, the NSF at the weld root increases as the notch gap percentage increases for both LC(I) and LC(II). The rate of NSF increment is significant when the notch gap percentage is more than 75.

For LC(I), the NSF at weld root is smaller than the NSF at weld toes if the notch gap percentage is less than 75. For LC(II), the NSF at weld root is larger than the NSF at weld toes if the notch gap percentage is greater than 30. Therefore, it is critical to determine

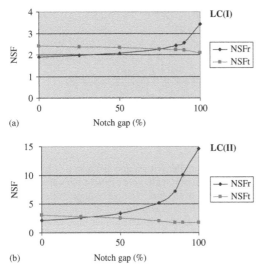

(a)

(b)

Figure 11. Notch gap percentage versus NSF at weld root and weld toes for (a) LC(I); (b) LC(II).

the notch gap percentage of the dragline line joint since the NSF varies considerably with it.

For the dragline joint, the welding of the bracings on to the main chord is by a combination of grove weld and fillet weld (Bucyrus International 1999). In this paper, the notch gap percentage is set as 50 since any gap larger than this may not occur in practice.

6 PARAMETRIC STUDY

6.1 Parameters

Three parameters used in this study were the main chord thickness, t_0, bracing thickness, t_1, and the in-plane angle of brace with respect to the chord axis, θ. All the three parameters of the joint are shown in both Figure 7 and Figure 10.

A typical joint geometry with the mesh at angle $\theta = 90°$, $\theta = 44.3°$ and notch gap of 50% is shown in Figure 12(a) and Figure 12(b) respectively. Element size of r/20, as recommended by Morgenstern et al. (2004) is used in this study.

6.2 Results of the parametric study

Figure 13(a) and Figure 13(b) show the plot of the main chord thickness, t_0, versus the NSF at weld root and weld toes for LC(I) and LC(II) respectively. The brace thickness, t_1, is 8.2 mm and the angle θ is 90°, while t_0 varies for the models.

Figure 14(a) and Figure 14(b) show the plot of the bracing thickness, t_1, versus the NSF at weld root and weld toes for LC(I) and LC(II) respectively. The main

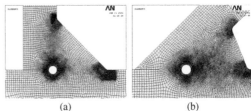

Figure 12. Typical joint geometry and meshing near the joint with (a) $\theta = 90°$; (b) 44.3°.

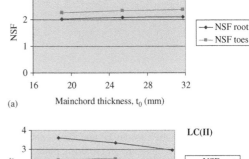

(a)

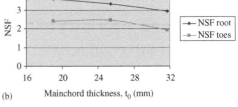

(b)

Figure 13. Main chord thickness versus NSF at weld root and weld toes for (a) LC(I); (b) LC(II).

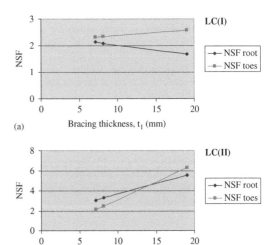

(a)

(b)

Figure 14. Bracing thickness versus NSF at weld root and weld toes for (a) LC(I); (b) LC(II).

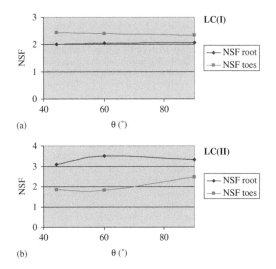

(a)

(b)

Figure 15. In-plane angle of brace with respect to the chord axis versus NSF at weld root and weld toes for (a) LC(I); (b) LC(II).

chord thickness, t_0, is 25.4 mm and the angle θ is 90°, while t_1 varies for the models. Figure 15(a) and Figure 15(b) shows the plot of the in-plane angle of brace with respect to the chord axis, θ, versus the NSF at weld root and weld toes for LC(I) and LC(II) respectively. The main chord thickness, t_0, is 25.4 mm and the brace thickness, t_1, is 8.2 mm, while the angle is varies for the models.

6.3 Remarks for the parametric study

The NSF at the weld toes is higher than the NSF at the weld root for LC(I). For LC(II), the NSF at the weld root is higher than the NSF at the weld toes except when the bracing thickness is greater than 14 mm. The NSF ratio of the weld root and weld toes is ranging from 0.65 to 0.89 for LC(I). However, this ratio is ranging from 0.88 to 1.92 for LC(II).

As the thickness of the main chord increases, the chord radius-to-thickness ratio (d_0/t_0), 2γ, decreases. The NSF at the weld root and weld toes seems to remain constant for LC(I) but decreases for LC(II).

As thickness of the bracing increases, the brace-to-chord thickness ratio (t_1/t_0), τ, increases. Consequently, the NSF at the weld toes increases especially for LC(II). The NSF at the weld root increases for LC(II) but decreases for LC(I). The rate of NSF increment for weld toes is higher than the weld root for LC(II).

As the in-plane angle of brace with respect to the chord axis, θ, increases, the NSF at the weld root and weld toes seems to remain constant for LC(I). For

Table 2. NSFs comparison.

Reference	Type of joint	NSF at root	NSF at toe
Hobbacher 2002	2D, half Y-weld with double fillets, unwelded out of centre gap of 2 mm.	1.60	2.55
IIW 2004	1. 2D, cruciform joint with double fillets weld and centre gap.	5.63	4.80
	2. 3D, rectangular hollow section, fillet welded on to a rigid plate.	9.13	Not reported
This paper	2D, Y-joint with combination of grove and fillet weld.	1.68–5.55	1.83–6.31

Table 3. Fatigue classes of the respective lacing numbers for the laboratory specimen 2 under different load cases.

Load case	Lacing number	Notch gap (%)	NSF for weld root	FAT
I	1	12.2	1.94	116
	2	61.2	2.15	105
	3	96.2	3.10	73
II	1	12.2	2.34	96
	2	61.2	4.20	54
	3	96.2	13.00	17

LC(II), the NSF at the weld root and the weld toes increases marginally as θ increases.

The results obtained are compared with the NSFs obtained by using ENS method for several welded joints reported in the literature as shown in Table 2. The fatigue classes of the respective lacing numbers of laboratory specimen 2 under different load cases are shown in Table 3.

7 CONCLUSIONS

Through this study, the following are observed:

- The element size of 0.3 mm, instead of r/20 (0.05 mm) as recommended by Morgenstern et al. (2004), at highly stressed area, e.g. the weld roots and weld toes, can give accurate results.
- The NSF at the weld toes remains constant with changes in notch gap. The NSF at the weld root changes significantly for notch gap value greater than 75%.
- The NSF at the weld toes is higher than the NSF at the weld root for LC(I). For LC(II), the NSF at the

85

weld root is in general higher than the NSF at the weld toes.

- The NSF at both the weld root and the weld toes is not significantly affected by t_0 and θ for both LC(I) and LC(II). However, the NSF is affected by t_1 especially for LC(II).
- Depending on the notch gap at weld root for the laboratory specimen 2, different lacing member will have different NSF for weld root and hence different FAT.

8 FUTURE RESEARCH

This paper is the first attempt to use the ENS method on the simplified 2D FE model of the dragline joint in order to determine the NSFs of the dragline structure. 3D modeling of the dragline joint is scheduled to be carried out to confirm the accuracy of the results.

ACKNOWLEDGEMENT

The authors wish to thank the supporters of this research, the Australian Research Council (ARC), Monash University, the dragline manufacturer, Bucyrus International (BE), and the dragline operator, BHP Billiton Mitsubishi Alliance (BMA). The authors also like to thank Professor Adolf F. Hobbacher, Professor Wolfgang Fricke and Professor Cetin Morris Sonsino for their help and valuable comments in the effective notch stress method.

REFERENCES

Bucyrus International 1999: Tubular Boom Manual. Bucyrus International, Manual No. 10033, December.

Dayawansa, D., Kerezsi, B., Chitty, G., Price, J. & Bartosiewicz, H. 2003. Chronic Problem Areas, Repair Quality and Improved Repairs for Dragline Structurals. BHP-Monash Maintenance Technology Institute Report.

Hobbacher, A.F. 2002. Application of the Effective Notch Stress Method for Fatigue Assessment of Welded Joints, Proceedings of the IIW Fatigue Seminar, IIW Commission XIII: pp. 19–29.

IIW (The International Institute of Welding) 1996. Fatigue design of welded joints and components. Author: A. Hobbacher, XIII-1539-96/XV-845-96, Abington Publishing, Abington, Cambridge, England.

IIW (The International Institute of Welding) 2004. Round-robin notch stress analysis. IIW interim meeting of commission XIII/WG3, Osaka, Japan, July 11, 2004.

Morgenstern, C., Sonsino, C.M., Hobbacher, A. & Sorbo, F. 2004. Fatigue design of aluminium welded joints by the local stress concept with the fictitious notch radius of $r_f = 1$ mm. International Institute of Welding Commission XIII, IIW Doc. XIII-2009-04, Osaka, Japan, July 12–14, 2004.

Pang, N.L., Zhao, X.-L., Mashiri, F.R., Dayawansa, P. & Price, J.W.H. 2005. Fatigue design of welded thick-walled tubular dragline joints. Australian Structural Engineering Conference 2005, Newcastle, Australia, 11–14 September.

SAS IP Inc. 2005. ANSYS 10.0 Users Manual, Swanston Analysis Systems (SAS) Inc, USA.

Suganuma, H. & Miki, C. 2005. Investigation of fatigue crack on the weld of deck plate and trough rib with effective notch stress method. International Institute of Welding Commission XIII, IIW Doc. XIII-2084-05.

van Wingerde, A.M. 1992. The fatigue behaviour of T and X joints made of square hollow sections. Heron, Volume 37, No. 2.

Zhao, X.-L., Herion, S., Packer, J.A., Puthli, R.S., Sedlacek, G., Wardenier, J., Weynand, K., Wingerde, A.M. van. & Yeomans, N.F. 2000. Design Guide for Circular and Rectangular Hollow Section Welded Joints under Fatigue Loading (8). CIDECT (Ed.) and TÜV Verlag, Cologne, Germany.

Tubular Structures XI – Packer & Willibald (eds)
© *2006 Taylor & Francis Group, London, ISBN 0-415-40280-8*

Effect of thickness and joint type on fatigue performance of welded thin-walled tube-to-tube T-joints

F.R. Mashiri & X.-L. Zhao
Department of Civil Engineering, Monash University, Clayton, Australia

ABSTRACT: This paper summarizes the proposed fatigue design curves for three different thin-walled T-joints following the testing of 99 specimens in the welded condition. The specimens are made up of cold-formed high-strength thin-walled (t < 4 mm) circular hollow sections (CHS) and square hollow sections (SHS). Fatigue tests were carried out under constant amplitude stress range on thin CHS-CHS, CHS-SHS and SHS-SHS T-joints, for the load "in-plane bending in the brace". Fatigue failure occurred in chord members of thicknesses 3 mm, 3 mm and 3.2 mm for SHS-SHS, CHS-SHS and CHS-CHS T-joints respectively. Different options are recommended for the design of the T-joints based on the deterministic method and least-squares method of statistical analysis.

1 INTRODUCTION

Thin-walled cold-formed hollow sections are widely used in the road transport and agricultural industry. They are used for the construction of undercarriages and support systems for road and agricultural trailers, swing ploughs, graders and haymakers. Typical sections used in these structures have a thickness less than 4 mm. These structures are subjected to fatigue loading under service. A number of fatigue failures have been observed in these kinds of structures. Most of the current design codes such as Australian Standard, AS4100-1998 (SAA 1998), Eurocode 3 (EC3 2003), American Welding Society (AWS 2004) and the CIDECT Design Guide No. 8 (Zhao et al 2000) have fatigue design recommendations for steel hollow sections of thickness greater than or including 4 mm only. Recent publications and those prior to them also show that research on fatigue of tubular nodal joints has been limited to tubes of thicknesses greater than 4 mm (van Wingerde et al 1997, Packer and Henderson 1997, Wardenier 1982). There is therefore a need to undertake research into the fatigue strength of thin-walled cold-formed hollow sections with thicknesses below 4 mm.

This paper summarizes the proposed fatigue design curves for three different T-joints following the testing of 99 specimens in the welded condition. These tests were carried out under the CIDECT Projects 7T and 7U (Mashiri et al 2001a, Mashiri and Zhao 2005). Fatigue tests were carried out under constant amplitude stress range on thin CHS-CHS, CHS-SHS and SHS-SHS T-joints, for the load "in-plane bending in

the brace". The following specimens were tested to failure for each connection type: (a) 23 Thin CHS-CHS T-joints, (b) 18 Thin CHS-SHS T-joints and (c) 58 Thin SHS-SHS T-joints. The specimens are made up of grade C350LO and Grade C450LO which comply to AS1163-1991 (SAA 1991). Grade C350LO has a specified minimum yield stress and minimum ultimate tensile strength of 350 MPa and 430 MPa respectively. On the hand, Grade C450LO has a specified minimum yield stress and minimum ultimate tensile strength of 450 MPa and 500 MPa respectively. The specimens were joined using a pre-qualified welding procedure for the gas metal arc welding method complying with AS1554.1 and 5 (SAI 2004a, 2004b).

The validity range of the thin SHS-SHS, CHS-CHS and CHS-SHS T-joints is given in terms of the non-dimensional parameters of the specimens. For the thin SHS-SHS, CHS-SHS and CHS-CHS T-joints the width ratio, $\beta(= b_1/b_0)$, the wall thickness ratio, $\tau(= t_1/t_0)$, the chord slenderness ratio, $\gamma(= b_0/2t_0)$ and the relative chord length $\alpha(= 2L/b_0)$ are given in Table 1.

Different options are recommended for the design of thin CHS-CHS, CHS-SHS and SHS-SHS T-joints as follows:

1. Option 1: Using the deterministic method and adopting the existing fatigue design curves from IIW (2000) and CIDECT Design Guide No. 8 (Zhao et al 2000).
2. Option 2: Using the least-squares method of statistical analysis to determine design $S_{r.hs}$-N curves with a natural slope of the fatigue test data for each

Table 1. Summarised results of welded thin-walled tube-to-tube T-joints.

Type of joint	Parameter range	Thickness effect on fatigue life	Effect of joint type on SCFs
SHS-SHS T-joint	$0.35 \leq \beta \leq 0.71$ $23 \leq 2\gamma \leq 33$ $0.5 \leq \tau \leq 1.0$ $15.3 \leq \alpha \leq 16.4$	The current trend (the thinner the tube, the higher the $S_{r.hs}$-N curve) does not apply to tubes with a thickness less than 4 mm.	In general, SHS-SHS T-joints give the highest SCFs whereas CHS-CHS T-joints give the lowest SCFs. SCFs for CHS-SHS T-joints are in the same order as those for SHS-SHS T-joints except for $\beta < 0.4$ where smaller SCFs are found for CHS-SHS T-joints.
CHS-SHS T-joint	$0.34 \leq \beta \leq 0.64$ $25 \leq 2\gamma \leq 33$ $0.67 \leq \tau \leq 0.97$ $15.3 \leq \alpha \leq 16.4$		
CHS-CHS T-joint	$0.33 \leq \beta \leq 0.63$ $24 \leq 2\gamma \leq 32$ $0.63 \leq \tau \leq 1.0$ $15.0 \leq \alpha \leq 16.1$		

of the 3 different types of specimens namely CHS-CHS, CHS-SHS and SHS-SHS T-joints.

3. Option 3: Using the least-squares method of statistical analysis to determine *one* design $S_{r.hs}$-N curve with a natural slope. The fatigue test data for all the 3 different types of specimens namely CHS-CHS, CHS-SHS and SHS-SHS T-joints are analysed as one sample.

A summary of the results and observations based on fatigue test data and experimental SCFs is given in Tables 1 and 2.

2 EFFECT OF TUBE WALL THICKNESS ON FATIGUE STRENGTH

The thickness effect in welded tubular joints can be seen in the current fatigue design guidelines for the hot spot stress method IIW (2000) and Zhao et al (2000), for CHS tubular joints with thicknesses ranging from 4 mm to 50 mm inclusive and for SHS tubular joints with thicknesses ranging from 4 mm to 16 mm inclusive. The design $S_{r.hs}$-N curves in IIW (2000) and CIDECT Design Guide No. 8 (Zhao et al 2000), show that fatigue strength increases as the tube-wall thickness of the member failing, under fatigue loading decreases.

The fatigue test data of the thin CHS-CHS, CHS-SHS and SHS-SHS T-joints tested in this investigation is plotted in Figure 1. The critical tube wall thicknesses of the 3 types of specimens tested are 3 mm, 3 mm and 3.2 mm for SHS-SHS, CHS-SHS and CHS-CHS T-joints respectively. The critical tube wall thicknesses correspond to the chord wall thickness since fatigue crack initiation and propagation is predominant in the chord member, where the highest SCFs occur.

If the fatigue strength of the tested specimens follows that of the existing fatigue design $S_{r.hs}$-N curves,

Table 2. Summarised results of welded thin-walled tube-to-tube T-joints.

Type of joint	Option 1: Adopt IIW design curves	Option 2: Different curves for each type of connection	Option 3: One curve for all types of connections tested
SHS-SHS T-joint (h_1, t_1, b_1, Fillet Weld, SHS, b_0, t_0, h_0, L)	Use IIW SCFs for SHS-SHS T-joints under IPB. Adopt IIW $t = 8$ mm curve (see Figure 3 and Table 3).	Use IIW SCFs for SHS-SHS T-joints under IPB. Adopt curve for thin SHS-SHS T-joints based on natural slope (see Figure 4 and Table 4).	Use IIW SCFs for SHS-SHS T-joints under IPB. Adopt curve for thin tubular joints based on natural slope and test data for all specimens (see Figure 5 and Table 5).
CHS-SHS T-joint (d_1, t_1, Fillet Weld, CHS, SHS, b_0, t_0, h_0, L)	Use IIW SCFs for SHS-SHS T-joints under IPB. Adopt IIW $t = 8$ mm curve (see Figure 3 and Table 3).	Use IIW SCFs for SHS-SHS T-joints under IPB. Adopt curve for thin CHS-SHS T-joints based on natural slope (see Figure 4 and Table 4).	Use IIW SCFs for SHS-SHS T-joints under IPB. Adopt curve for thin tubular joints based on natural slope and test data for all specimens (see Figure 5 and Table 5).
CHS-CHS T-joint (d_1, t_1, Fillet Weld, CHS, d_0, t_0, L)	Use IIW SCFs for CHS-CHS T-joints under IPB. Adopt IIW $t = 12$ mm curve. (see Figure 3 and Table 3). If $S_{r.hs}$-N data below 300,000 cycles is ignored (i.e. for high cycle fatigue), adopt IIW $t = 8$ mm Curve	Use IIW SCFs for CHS-CHS T-joints under IPB. Adopt curve for thin CHS-CHS T-joints based on natural slope (see Figure 4 and Table 4).	Use IIW SCFs for CHS-CHS T-joints under IPB. Adopt curve for thin tubular joints based on natural slope and test data for all specimens (see Figure 5 and Table 5).

it would be expected that the fatigue test data for a critical tube wall thickness of 3 mm, should lie on or above the derived IIW curve for $t = 3$ mm. The $S_{r.hs}$-N data in Figure 1 is derived using stress concentration factors from existing parametric equations in IIW (2000) and CIDECT Design Guide No. 8 (Zhao et al 2000).

Figure 1 shows that although most of the fatigue data points lie above the derived IIW curve for $t = 3$ mm some of the $S_{r.hs}$-N data points fall below the IIW $t = 3$ mm curve. This shows that the fatigue strength for welded tubular joints with a critical tube wall thickness of 3 mm have a lower than expected fatigue strength. The lower than expected fatigue strength for welded thin-walled ($t < 4$ mm) joints is attributed to the greater negative impact of weld defects such as undercut on the fatigue crack propagation life of these thin-walled joints as reported by Mashiri et al (2001b). A similar phenomenon was also observed in the thin CHS-plate and SHS-plate T-joints as discussed in Mashiri and Zhao (2005b).

3 EXPERIMENTAL SCFS IN THIN TUBULAR T-JOINTS

Experimental SCFs for welded thin-walled CHS-CHS, CHS-SHS and SHS-SHS T-joints, with different non-dimensional parameters, were determined from measured strains and are summarised in Figure 2. Figure 2 shows the experimental SCFs, at weld toes in the chord, for the different specimens plotted against the β value. It can be noted that the thin CHS-CHS T-joints have the lowest SCFs. The thin CHS-CHS T-joints, by virtue of their lower SCFs are therefore likely to have a better fatigue performance compared to thin CHS-SHS and SHS-SHS T-joints when the S-N data of

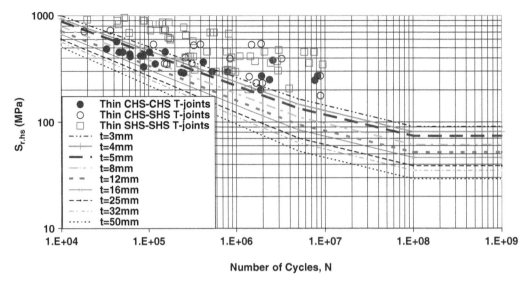

Figure 1.   Effect of tube wall thickness on fatigue life in thin tubular nodal joints.

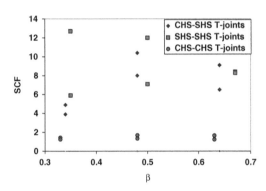

Figure 2.   Comparison of measured SCFs from thin CHS-SHS, SHS-SHS and CHS-CHS T-joints.

the specimens is expressed in terms of nominal stress range (in the classification method). Figure 2 shows that for larger β values, the SCFs for thin CHS-SHS T-joints fall within the same range as those for the thin SHS-SHS T-joints. For smaller β values however, there is a pronounced difference between the SCFs for thin CHS-SHS and thin SHS-SHS T-joints. As an approximation it may be assumed that for design purposes the SCFs for thin-walled CHS-SHS and SHS-SHS T-joints are equal. This assumption will yield conservative estimates of fatigue life for thin CHS-SHS T-joints. Note that this assumption deviates from the generally accepted criterion that the magnitudes of the SCFs for CHS-SHS T-joints, lies between that of the SHS-SHS and CHS-CHS T-joints as shown by Gandhi and Berge

(1998) and Bian and Lim (2003) in their investigation on thicker walled (t > 4 mm) CHS-SHS T-joints.

4   FATIGUE DESIGN CURVES FOR THIN TUBULAR T-JOINTS (IPB)

Current fatigue design guidelines for tubular nodal joints such as IIW (2000) and CIDECT Design Guide No. 8 (Zhao et al 2000), give parametric equations for both SHS-SHS and CHS-CHS T-joints. As a design tool these parametric equations can be adopted for thin CHS-CHS and SHS-SHS T-joints. There are however no parametric equations for SCFs for CHS-SHS T-joints in current fatigue design guidelines such as IIW (2000) and CIDECT Design Guide No. 8 (Zhao et al 2000). The trend in Figure 2 shows that the SCFs for thin CHS-SHS T-joints can conservatively be assumed to be equal to those for thin SHS-SHS T-joints. Therefore, for design, the parametric equation SCFs for SHS-SHS T-joints in current fatigue design guidelines can be used for estimating the SCFs in thin CHS-SHS T-joints.

The $S_{r.hs}$-N data of the thin CHS-CHS, SHS-SHS and CHS-SHS T-joints shown in Figure 3 has therefore been obtained by multiplying the nominal stress ranges by the parametric equation SCFs from IIW (2000). The parametric equation SCFs for SHS-SHS T-joints were used to convert the S-N data of both the thin SHS-SHS and CHS-SHS T-joints in the classification method to $S_{r.hs}$-N data in the hot spot stress method. On the other hand, the parametric equation SCFs for the CHS-CHS T-joints, were used to convert the nominal

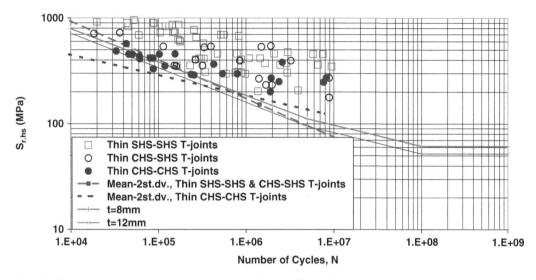

Figure 3. Derived and adopted curves for thin tubular nodal joints (Option 1).

stress ranges of the thin CHS-CHS T-joints into hot spot stress ranges.

Different options can be adopted for the design of thin CHS-CHS, CHS-SHS and SHS-SHS T-joints as explained in the introduction and detailed in the next sections.

4.1 Option 1, adopting design curves from existing standards

Using the least-squares method of statistical analysis, the mean-minus-two-standard-deviation curve with a natural slope for a combined data set of thin SHS-SHS and CHS-SHS T-joints was obtained and is shown in Figure 3. For design purposes, it can also be seen that using the deterministic method all of the $S_{r.hs}$-N data for thin CHS-SHS and SHS-SHS T-joints lie above the IIW t = 8 mm design curve. This represents a very high probability of survival if the IIW t = 8 mm design curve is used for the design of thin CHS-SHS and SHS-SHS T-joints considering the scatter obtained from a total of 76 specimens. Figure 3 shows that compared to the mean-minus-two-standard-deviation curve of the $S_{r.hs}$-N data (for thin CHS-SHS and SHS-SHS T-joints), the IIW t = 8 mm curve is conservative at relatively higher stress ranges and non-conservative at relatively lower stress ranges. Adopting the IIW t = 8 mm $S_{r.hs}$-N design curve is reasonable taking into account the better fatigue performance that is seen in the test results at lower stress ranges as illustrated in Figure 3.

The $S_{r.hs}$-N data of the 23 thin CHS-CHS T-joints was also analysed to determine the mean-minus-two-standard-deviations curve with a natural slope also shown in Figure 3. Using the deterministic method,

it can also be seen that from the scatter of the $S_{r.hs}$-N data, 22 out of 23 data points lie on or above the IIW t = 12 mm design curve, representing a 95.77% chance of survival of the thin CHS-CHS T-joints considering the scatter in the present data. Compared to the mean-minus-two-standard-deviations curve of $S_{r.hs}$-N data, the IIW t = 12 mm curve is non-conservative at relatively higher stress ranges and conservative at relatively lower stress ranges.

For the purposes of design, it is logical to recommend the use of existing parametric equations for determining SCFs and to recommend the use of existing design $S_{r.hs}$-N curves for determining fatigue life at given hot spot stress ranges. The fatigue design of both thin SHS-SHS and thin CHS-SHS T-joints under in-plane bending (IPB) is therefore recommended to be as follows:

(i) Use of existing parametric equations from IIW (2000) and CIDECT Design Guide No. 8 (Zhao et al 2000) for SHS-SHS T-joints under in-plane bending for determining SCFs of thin SHS-SHS and CHS-SHS T-joints and

(ii) Use the IIW t = 8 mm curve for the subsequent fatigue design of thin SHS-SHS and thin CHS-SHS T-joints.

The fatigue design of thin CHS-CHS T-joints under in-plane bending (IPB) is therefore recommended to be as follows:

(i) Use of existing parametric equations from IIW (2000) and CIDECT Design Guide No. 8 (Zhao et al 2000) for CHS-CHS T-joints under in-plane bending for determining SCFs of thin CHS-CHS T-joints and

Table 3. Equations for fatigue design curves adopted from IIW (2000).

Joint	Equation	Range of number cycles	Hot spot stress range at 2 mil. cycles	Constant amplitude fatigue limit	Cut-off limit
Thin CHS-CHS T-joints	**IIW t = 12 mm Curve** $\log(N) = 12.763 - 3.0690 \log(S_{rhs})$ $\log(N) = 16.578 - 5 \log(S_{rhs})$	$10^3 \leq N \leq 5 \times 10^6$ $5 \times 10^6 \leq N \leq 10^8$	128	95	52
Thin CHS-SHS & SHS-SHS T-joints	**IIW t = 8 mm Curve** $\log(N) = 13.191 - 3.1719 \log(S_{rhs})$ $\log(N) = 16.932 - 5 \log(S_{rhs})$	$10^3 \leq N \leq 5 \times 10^6$ $5 \times 10^6 \leq N \leq 10^8$	149	111	61

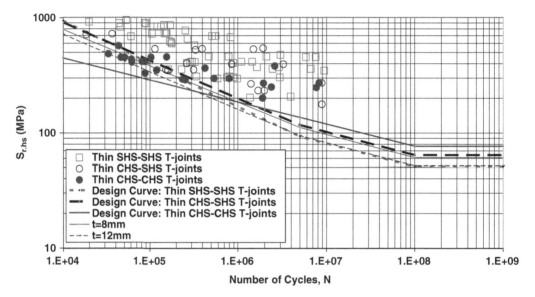

Figure 4. Design curves based on natural slope, different design curves for each type of connection (Option 2).

(ii) Use the IIW t = 12 mm curve for the subsequent fatigue design of thin CHS-CHS T-joints.

The equations for the fatigue strength curves for thin CHS-CHS, CHS-SHS and SHS-SHS T-joints adopted from the existing fatigue design guidelines are shown in Table 3. Table 3 also shows the corresponding constant amplitude fatigue limit, cut-off limit and the hot spot stress range at 2 million cycles. The constant amplitude fatigue limit is the stress range of a given fatigue design curve when the number of cycles is equal to 5 million cycles. The cut-off limit is the stress range of a given fatigue design curve when the number of cycles is equal to 100 million cycles.

4.2 Option 2, design curves for each connection type based on natural slope

Using the least-squares method of statistical analysis, design $S_{r.hs}$-N curves were obtained from the experimental fatigue data for each of the 3 different types of connections namely thin CHS-CHS, CHS-SHS and SHS-SHS T-joints. The experimental fatigue data was used to obtain the fatigue strength curves based on the natural slope for the constant amplitude loading region lying between 10^3 and 5×10^6 cycles. For the variable amplitude loading region between 5×10^6 and 10^8 cycles, a negative slope of 1:5, that is similar to that in IIW (2000) and CIDECT Design Guide No. 8 (Zhao et al 2000) was adopted. The fatigue design curves based on the natural slope of the experimental data for each of the thin CHS-CHS, CHS-SHS and SHS-SHS T-joints are shown in Figure 4. Note that because the natural slope of the thin CHS-CHS is very close to 5, the same slope has been adopted for both constant amplitude and the variable amplitude regions.

The corresponding equations for the fatigue strength curves based on the natural slope of the $S_{r.hs}$-N

Table 4. Equations for fatigue design curves for each type of connection derived from statistical analysis using the natural slope of experimental fatigue data.

Joint	Equation	Range of number cycles	Hot spot stress range at 2 mil. cycles	Constant amplitude fatigue limit	cut-off limit
Thin CHS-CHS T-joints	$\log(N) = 15.8672 - 5.2324 \log(S_{rhs})$	$10^3 \leq N \leq 10^8$	162	136	77
Thin CHS-SHS T-joints	$\log(N) = 12.9757 - 3.0343 \log(S_{rhs})$ $\log(N) = 17.04 - 5 \log(S_{rhs})$	$10^3 \leq N \leq 5 \times 10^6$ $5 \times 10^6 \leq N \leq 10^8$	158	117	64
Thin SHS-SHS T-joints	$\log(N) = 11.9875 - 2.6871 \log(S_{rhs})$ $\log(N) = 16.54 - 5 \log(S_{rhs})$	$10^3 \leq N \leq 5 \times 10^6$ $5 \times 10^6 \leq N \leq 10^8$	131	93	51

data for the 3 different welded thin-walled connections are given in Table 4. Table 4 also gives the corresponding constant amplitude fatigue limit, cut-off limit and hot spot stress range at 2 million cycles for the design curves given in Figure 4.

The recommended design procedure using the curves based on natural slope for each of the 3 different welded thin-walled connections is as follows:

(i) Use of existing parametric equations from IIW (2000) and CIDECT Design Guide No. 8 (Zhao et al 2000) for CHS-CHS T-joints under in-plane bending for determining SCFs of thin CHS-CHS T-joints. Similarly use of existing parametric equations from IIW (2000) and CIDECT Design Guide No. 8 (Zhao et al 2000) for SHS-SHS T-joints under in-plane bending for determining SCFs of thin SHS-SHS and thin CHS-SHS T-joints and

(ii) Use of the design curves based on equations in Table 4 and given in Figure 4.

Figure 4 also shows the IIW t = 8 mm and t = 12 mm design curves adopted in Option 1 for comparison with the design curves proposed for Option 2.

For thin SHS-SHS T-joints, the design curve based on natural slope gives a non-conservative estimate of fatigue life at relatively high stress ranges and a conservative estimate of fatigue life at relatively lower stress ranges compared to the IIW t = 8 mm curve.

For thin CHS-SHS T-joints, the design curve based on natural slope lies above the IIW t = 8 mm curve. This means that the design curve based on natural slope gives non-conservative estimates of fatigue life for both constant amplitude and variable amplitude regions compared to the IIW t = 8 mm curve.

Compared to the IIW t = 12 mm curve, the design curve for thin CHS-CHS T-joints based on natural slope gives a conservative estimate of fatigue life at relatively high stress ranges and a non-conservative estimate of fatigue life at relatively low stress ranges.

4.3 Option 3, design curve for all connection types based on natural slope

By considering the experimental fatigue data from the 3 different types of connections tested as one sample, one design curve was obtained based on the natural slope of the data. The least squares method of statistical analysis was used to obtain the design $S_{r.hs}$-N curve in the constant amplitude loading region lying between 10^3 and 5×10^6 cycles. A curve with a negative slope of 1:5 was adopted for the variable amplitude loading region between 5×10^6 and 10^8 cycles. Figure 5 shows the design curve obtained from this analysis. The equation for the fatigue strength curve in Figure 5 is shown in Table 5. The constant amplitude fatigue limit, cut-off limit and the hot spot stress range at 2 million cycles for the design curve shown in Figure 5 is also given in Table 5.

Figure 5 also shows the IIW t = 8 mm and t = 12 mm design curves for comparison with the design curve proposed for Option 3.

Figure 5 shows that the design curve based on natural slope and taking into account all test data as one sample for statistical analysis, gives conservative estimates of fatigue life compared to both IIW t = 8 mm and t = 12 mm design curves.

5 CONCLUSIONS

The fatigue performance of thin tubular T-joints made up of circular hollow sections (CHS) and square hollow sections (SHS) was analysed. The following observations were made for thin tubular T-joints:

(a) The fatigue life for thin SHS-SHS, CHS-CHS and CHS-SHS T-joints were found to be below the expected design $S_{r.hs}$-N curves from the current trends in IIW (2000) and CIDECT Design Guide No. 8. The lower than expected fatigue strength can be attributed to the greater negative impact of weld toe defects such as undercut on fatigue crack propagation life of welded thin-walled joints.

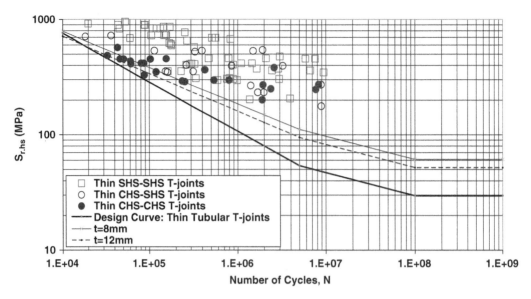

Figure 5.   Design curves based on natural slope, one design curve for all 3 types of connections (Option 3).

Table 5.   Equations for fatigue design curve, for all 3 types of connections, derived from statistical analysis using the natural slope of all experimental fatigue data.

Joint	Equation	Range of number cycles	Hot spot stress range at 2 mil. cycles	constant amplitude fatigue limit	cut-off limit
Thin CHS-CHS, CHS-SHS & SHS-SHS T-joints	$\log(N) = 10.7695 - 2.3496 \log(S_{rhs})$ $\log(N) = 15.36 - 5 \log(S_{rhs})$	$10^3 \leq N \leq 5 \times 10^6$ $5 \times 10^6 \leq N \leq 10^8$	80	54	30

(b) Experimental SCFs for thin CHS-CHS joints were found to be lower that those for thin SHS-SHS and CHS-SHS T-joints. The SCFs for thin CHS-SHS T-joints were found to be of similar magnitude to those of thin SHS-SHS T-joints except at lower β values where the SCFs for thin SHS-SHS T-joints were found to be relatively larger than the SCFs for thin CHS-SHS T-joints.

(c) When expressed in terms of the hot spot stress method, the thin CHS-CHS T-joints exhibit a lower than expected fatigue strength compared to the thin CHS-SHS and SHS-SHS T-joints. This is despite the fact that in the classification method the thin CHS-CHS T-joints have a better fatigue life compared to the thin CHS-SHS and SHS-SHS T-joints and that they have critical tube wall thicknesses of 3.2 mm, 3 mm and 3 mm respectively which are almost the same. This is contrary to the expected fatigue strength of nodal tubular joints where the critical tube wall thickness is the same. In the hot spot stress method, different tubular nodal joints with the same critical tube wall thickness are designed using one design $S_{r.hs}$-N curve for that particular tube wall thickness. The different behaviour witnessed in welded thin-walled ($t < 4$ mm) tubular T-joints can be attributed to the difficulty associated with welding of the different connection types and the greater negative impact of weld defects resulting from the welding on fatigue life. This phenomenon results in the normal trend of thickness effect in the hot spot stress method not being followed in welded thin-walled joints.

(d) The thin CHS-SHS and SHS-SHS T-joints can be designed using parametric equation SCFs for SHS-SHS T-joints in IIW (2000) and CIDECT Design Guide No. 8 and adopting the IIW $t = 8$ mm design $S_{r.hs}$-N curve. The thin CHS-CHS T-joints can be designed using parametric equation SCFs for CHS-CHS T-joints in IIW (2000) and CIDECT Design Guide No. 8 and adopting the IIW $t = 12$ mm design $S_{r.hs}$-N curve. Fatigue $S_{r.hs}$-N curves based on natural slope can also be adopted for the design of welded thin-walled tubular T-joints.

REFERENCES

AWS 2004, Structural Welding Code-Steel, AWS D1.1/D1.1M:2004, 19th edition, American Welding Society, Miami, U.S.A

Bian L.C. and Lim J.K. 2003, "Fatigue Strength and Stress Concentration Factors of CHS-to-RHS T-Joints", Journal of Constructional Steel Research, Vol. 59, pp. 561–663.

EC3 2003, "Eurocode 3: Design of Steel Structures – Part 1.9: Fatigue Strength of Steel Structures", prEn 1993-1-9: 2003(E), European Committee for Standardisation, Brussels, Belgium

Gandhi P. and Berge S. 1998, "Fatigue Behaviour of T-joints: Square Chords and Circular Braces", Journal of Structural Engineering, ASCE, Vol. 124, No. 4, pp. 399–404

IIW 2000, Fatigue Design Procedures for Welded Hollow Section Joints, IIW Doc. XIII-1804-99, Recommendations for IIW Subcommission XV-E, Eds. X.L. Zhao and J.A. Packer, Abington Publishing, Cambridge, UK

Mashiri F.R. and Zhao X.L. 2005a, "Fatigue of Cold-Formed CHS Joints (t < 4 mm)" CIDECT Project No. 7U, Revised Final Report No. 7U-5/05, Department of Civil Engineering, Monash University, Australia, August 2005, 153 pages

Mashiri F.R. and Zhao X.L. 2005b, "Effect of Thickness and Joint Type on Fatigue Performance of Welded Thin-Walled Tube-Plate T-Joints", Proceedings of The 1st International Conference on Advances in Experimental Structural Engineering (AESE 2005), Eds.: Itoh Y. and Aoki T. 19–21 July 2005, Nagoya, Japan, pp. 877–884

Mashiri F.R., Zhao X.L., Grundy P. and Tong L. 2001a, "Fatigue Analysis of Welded Connections in High Strength Cold-Formed SHS (t < 4 mm)" CIDECT Project No. 7T, Final Report No. 7T-8-01, Department of Civil Engineering, Monash University, Australia, August 2001, 92 pages

Mashiri F.R., Zhao X.L. and Grundy P. 2001b, "Effect of weld profile and undercut on fatigue crack propagation life of thin-walled cruciform joint" Thin-Walled Structures, Vol. 39, Issue 3, Elsevier Science Ltd, March 2001, pp. 261–285

Packer J.A. and Henderson J.E. 1997, "Hollow Structural Section Connections and Trusses – A Design Guide", 2nd Edition, Canadian Institute of Steel Construction, Ontario, Canada

SAA 1991, Structural Steel Hollow Sections, Australian Standard AS1163-1991, Standards Association of Australia, Sydney, Australia

SAA 1998, Steel Structures, Australian Standard AS 4100-1998, Standards Association of Australia, Sydney

SAI 2004a, Structural steel welding Part 1: Welding of steel structures, AS/NZS1554.1: 2000, Standards Australia International, Sydney, Australia

SAI 2004c, Structural steel welding Part 5: Welding of steel structures subject to high levels of fatigue loading, AS/NZS1554.1: 2004, Standards Australia International

van Wingerde A.M., van Delft D.R.V., Wardenier J. and Packer J.A. 1997, Scale Effects on the Fatigue Behaviour of Tubular Structures, IIW International Conference on Performance of Dynamically Loaded Welded Structures, July 14–15, San Francisco, CA, pp. 123–135

Wardenier, J. 1982, Hollow Section Joints, Delft University Press, Delft, The Netherlands

Zhao X.L., Herion S., Packer J.A., Puthli R., Sedlacek G., Wardenier J., Weynand K., van Wingerde A., and Yeomans N. 2000, "Design Guide for Circular and Rectangular Hollow Section Joints under Fatigue Loading", Verlag TÜV Rheinland, Köln, Germany

Tubular Structures XI – Packer & Willibald (eds)
© *2006 Taylor & Francis Group, London, ISBN 0-415-40280-8*

Welded thin-walled RHS-to-RHS cross-beams under cyclic bending

F.R. Mashiri & X.-L. Zhao
Department of Civil Engineering, Monash University, Clayton, Australia

ABSTRACT: Galvanized sections are suitable for use in the road transportation industry, the agricultural industry and for recreational structures. The structural systems in which galvanized sections can be used include chassis boxes, roof frames, base frames and drawbars among others. These structural systems are subjected to cyclic loading in service. The structural systems are therefore prone to fatigue failure. In Australia, galvanized sections, commonly known as DuraGal, are thin-walled with wall thicknesses less than 4 mm. There are currently no fatigue design rules for sections of thicknesses less than 4 mm. The connections under investigation namely cross-beam connections, are not covered in current fatigue design guidelines. This paper reports on fatigue tests of cross-beam connections made up of galvanized rectangular hollow sections (RHS). The fatigue test data in the present investigation is compared to data from previously reported work on galvanized cross-beam connections. Stress concentration factors have been determined in a typical connection to verify the crack pattern observed in these connections.

1 INTRODUCTION

Galvanized sections such as angle, channel and rectangular hollow sections are suitable for use in the road transportation industry for the manufacture of structural systems such as chassis boxes, roof frames, base frames and drawbars. A significant feature of galvanized sections, commonly known as DuraGal in Australia, is the fact that they are galvanized in-line during the forming process. The in-line galvanising has been found to enhance the strength properties of the sections (an increase of up to 30%) (Zhao and Hancock 1996, Zhao and Mahendran 1998). To allow the use of galvanized sections in the road transportation industry, agricultural equipment and recreation structures fatigue design recommendations need to be developed. The fatigue design recommendations are based on the testing of prototype connections. Most of the galvanized sections manufactured in Australia by OneSteel Market Mills have a thickness less than 4 mm. Fatigue failure is a major concern when these sections are welded and subjected to repeated loading.

Existing research on fatigue of cold-formed steel hollow sections has mainly been applied to sections with a thickness larger than 4 mm (Zhao et al 2000, IIW 2000). Recent research projects have been carried out at Monash University to investigate the fatigue strength of welded thin cold-formed square hollow sections (SHS) and circular hollow sections (CHS) (Mashiri et al 2002a, 2002b, 2002c, 2004a, 2004b). Typical types of connections investigated are SHS-to-plate, SHS-to-SHS, CHS-to-Plate, CHS-to-CHS

and CHS-to-SHS T-joints as used in trusses and communication towers. Limited research has been performed in the USA on fatigue of welded connections in cold-formed sheeting steel (Klippstein 1980, 1981, 1985; LaBoube and Yu 1999). Several fatigue S-N curves are now included in AS4600-2005 (SAA 2005) for welded thin cold-formed connections based on these research. However the above mentioned research does not cover the problems to be investigated in this project because of different cross-sections, different weld details and different loading conditions.

The major activity in this research involves the fatigue testing of galvanized cross-beam welded connections to obtain S-N data that can be used in determining the design class in the classification method.

A schematic diagram showing the connection layout of thin-walled RHS-RHS cross-beam connection tested in this investigation is shown in Figure 1(a). Figure 1(b) shows a photograph of a typical RHS-RHS cross-beam connection whose fatigue behaviour is reported in this paper. Figure 1(c) gives details of the orientation of sections in the RHS-RHS cross-beam connections. Table 1 shows the RHS-RHS cross-beam connections studied in this paper. For each connection, Table 1 shows the section sizes used in the manufacture of the RHS-RHS connections and the corresponding non-dimensional parameters. The mean measured yield stress ($f_{y,m}$) values for each of the sections used are also shown in Table 1. The measured yield stresses show that the RHS and SHS tubes are of grade C350LO or C450LO. The RHS and SHS tubes comply to AS1163-1991 (SAA 1991).

This paper describes the material properties, welding procedure qualification, fatigue tests, S-N curves and stress concentration factors for galvanized RHS-RHS cross-beam welded connections.

The gas metal arc welding method was used in joining the RHS bottom member to the RHS section top member shown in Figure 1. Category SP (structural purpose) welds complying with both AS1554.1-2004

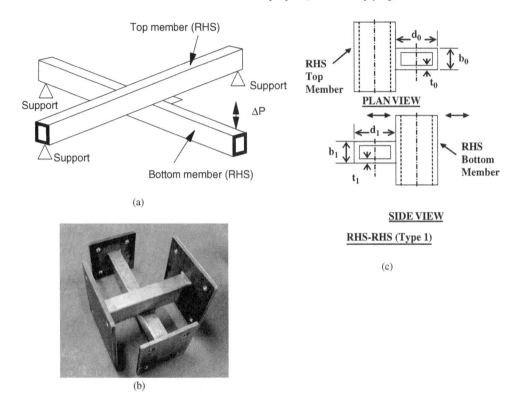

(a)

(b)

PLAN VIEW

SIDE VIEW

RHS-RHS (Type 1)

(c)

Figure 1. (a) Connection layout for galvanized cross-beam connection (b) RHS-RHS connection (c) Orientation of sections in RHS-RHS cross-beam connections.

Table 1. Fatigue tests for RHS-RHS cross-beams welded on all four faces.

Connection Number	Bottom Member $d_0 \times b_0 \times t_0$	Mean Yield Stress $f_{y,m}$ MPa	Top Member $d_1 \times b_1 \times t_1$	Mean Yield Stress $f_{y,m}$ MPa	$\dfrac{b_0}{t_0}$	$\dfrac{t_0}{t_1}$	$\dfrac{d_0}{b_0}$	Nominal Stress Range (Sr-nom) MPa	Number of Cycles to Failure	Failure Mode
S1R1L1A	$50 \times 50 \times 3$SHS	475	$75 \times 50 \times 3$RHS	459	16.7	1	1	149.7	871688	BMC[#]
S1R1L2A	$50 \times 50 \times 3$SHS	475	$75 \times 50 \times 3$RHS	459	16.7	1	1	149.7	461734	BMC
S2R1L1A	$50 \times 50 \times 1.6$SHS	402	$75 \times 50 \times 3$RHS	459	31.3	0.5	1	93.5	2304714	BMC
S2R1L2A	$50 \times 50 \times 1.6$SHS	402	$75 \times 50 \times 3$RHS	459	31.3	0.5	1	124.6	211506	BMC
S4R1L1A	$35 \times 35 \times 3$SHS	533	$75 \times 50 \times 3$RHS	459	11.7	1	1	214.4	156087	BMC
S4R1L2A	$35 \times 35 \times 3$SHS	533	$75 \times 50 \times 3$RHS	459	11.7	1	1	193.0	179735	BMC
S5R1L1A	$35 \times 35 \times 1.6$SHS	493	$75 \times 50 \times 3$RHS	459	21.9	0.5	1	135.0	1316243	BMC
R2R1L1A*	$50 \times 25 \times 3$RHS	519	$75 \times 50 \times 3$RHS	459	8.3	1	2	65.2	7738790	No Cracks
R2R1L1Aa	$50 \times 25 \times 3$RHS	519	$75 \times 50 \times 3$RHS	459	8.3	1	2	130.5	235219	BMC
R3R1L1A	$50 \times 25 \times 1.6$RHS	398	$75 \times 50 \times 3$RHS	459	15.6	0.5	2	103.8	675844	BMC

*Runout.
BMC: Bottom Member Cracking.

(SAA 2004a) and AS1554.5-2004 (SAA 2004b) were used. In order to verify compliance of the welding procedure to Australian Standards, AS1554 Part 1 and 5, hardness tests were performed on a representative joint as reported in Mashiri and Zhao (2005a). The hardness tests included comparison tests between parent metal and weld metal as well as hardness tests for weld-heat-affected zones (SAA 2004a,b). The hardness tests complied with Australian Standards AS1554.1 and AS1554.5 (SAA 2004a, 2004b). The Vickers Hardness tests showed that the maximum value of the heat-affected zone (HAZ) was less than 350HV10. When the hardness value of the HAZ is less than 350HV10, there is little danger of HAZ cracking (WTIA 1998). Cracking of the HAZ in structures subjected to cyclic loading, increases their susceptibility to fatigue failure and should therefore be avoided.

Fatigue tests were carried out in a multiple fatigue test rig to obtain S-N data for establishing the fatigue design class for the cross-beam connections. Constant amplitude cyclic bending moment was applied to the bottom member as shown in Figure 1(b). The cyclic loading was applied at a stress ratio of 0.1. the Stress concentration factors were derived through the determination of hot spot stresses, experimentally from strain measurements and numerically through three-dimensional finite element modelling using STRAND 7 (Strand7 Pty Ltd 2004).

2 FATIGUE TESTS AND FAILURE MODES

An RHS-RHS cross-beam connection is shown in Figure 1(b). The setup of the cross-beam tests for cyclic bending load is shown in Figure 2(a). The test rigs used for the cyclic bending tests of RHS-RHS cross-beam connections are part of the multiple fatigue test rig described by Mashiri et al (2002a).

The RHS-RHS cross-beam connections consist of an RHS top member and an RHS bottom member, see Figures 1(a) and (b). Cyclic bending is applied through the RHS bottom member as shown in Figures 1(a) and 2(a). Crack initiation was observed to occur in the corner region at the weld toes in the bottom member as shown in Figure 2(b). If crack initiation only started at one corner, propagation of the first corner crack occurred to a certain extent before a crack initiated at the weld toes of the second corner in the bottom member signifying a redistribution of stresses from the cracked corner to the un-cracked corner. Cracks from the two corners then propagated until the full width of the bottom RHS member was cracked. The subsequent crack growth after the width of the bottom member is fully cracked, is away from the weld toes and occurs in the parent metal. The cracking in the parent metal as shown in Figure 2(b) confirms the fact that the applied bending creates stresses acting parallel to the longitudinal axis of the bottom member in these RHS-RHS cross beam connections. The stresses parallel to the longitudinal axis of the RHS bottom member act as the principal stress contributing to the growth of the cracks in the parent material under Mode I, opening of the cracks. This failure is similar to the fatigue failure under cyclic bending observed in RHS-Angle and RHS-Channel cross-beam connections reported by Mashiri and Zhao (2005b).

Apart from the non-dimensional parameters, yield stresses and sections sizes, Table 1 also shows the nominal stress range (S_{r-nom}) applied to the specimens, the corresponding number of cycles to failure and the failure mode. The nominal stress range is determined at the weld toes of the bottom member where

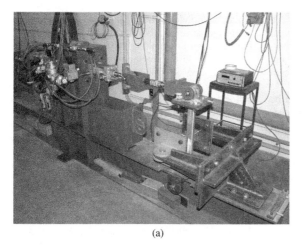

(a)   (b)

Figure 2. (a) Test setup for cross-beam fatigue tests (b) Failure in RHS-RHS cross-beams, S5R1L1A (Note: The bottom member is oriented vertically in these pictures).

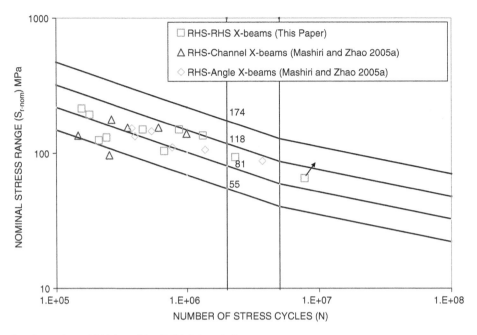

Figure 3. Comparison of S-N data with AS4600 fatigue design curves.

cracking occurs and is due to the maximum and minimum applied bending stresses defining the constant amplitude stress cycles.

Figure 3 shows the fatigue test data from the 10 RHS-RHS cross-beam connections. The experimental S-N data is compared to the existing design S-N curves in the Australian Standard AS4600 (SAA 2005). Figure 3 shows that most of the S-N data from the RHS-RHS cross-beam connections lies above the Class 55 design S-N curve. Also shown in Figure 3 is the experimental S-N data of RHS-Channel and RHS-Angle cross-beam connections from previous research (Mashiri and Zhao 2005a, 2005b). The experimental S-N data of the RHS-Channel and RHS-Angle cross-beam connections lie within the same scatter band as the fatigue test data of the RHS-RHS cross-beam connections.

Using the deterministic method, Class 55 can be adopted for designing RHS-RHS cross-beam connections. Since more tests of RHS-Channel and RHS-Angle cross-beam connections are still to be tested, future work will also determine the fatigue design S-N curves for the cross-beam connections using the least-squares method of statistical analysis on the completed set of test data.

3 STRESS CONCENTRATION FACTORS

Stress concentration factors were determined in specimen, S1R1L2A whose properties are shown in Table 1.

Stress distributions, the resultant hot spot stresses and hence SCFs were determined in order to understand the cracking pattern in typical RHS-RHS cross-beam connections. Stress distributions were determined both experimentally and numerically.

Static loads were applied and for each load level strain measurements were recorded from strip strain gauges located at the hot spot locations. The test set up used for the determination of stress distribution at hot spot locations is shown in Figure 4(a). Figure 4(b) shows the location of the strip strain gauges at the welded interface of the top and bottom member. The strain gauge on the bottom member was located at the transition line making the end of the outside corner radius and the beginning of the flat face of the square hollow section member. This transition line marks the point where a notch exists because of the sudden change in shape making it a potential "hot spot" location. The strain gauge in the top member was located in line with the line of transition in the bottom member. The hot spot locations were chosen following the crack initiation observed in the fatigue tests. Crack initiation in RHS-RHS cross beam connections was observed to occur at weld toes in the bottom RHS member. The cracks occurred in the corner region of the RHS member as shown in Figure 2(b).

The strip strain gauges that were used were of type KFG-1-120-D9-11N10C2, made by Kyowa, Japan. The strip strain gauges were used to measure stress distribution at the weld toes. The stress distribution was used to determine hot spot stresses at the toe of the

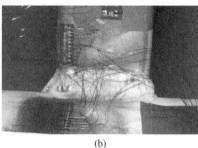

(a)   (b)

Figure 4. (a) Test set-up for strain distribution at hot spot locations, (b) Location of strain gauges used in the measurement of strain distribution (Note: The bottom member is oriented vertically in these pictures).

weld. Within one strip strain gauge, there are five strain gauges, 2 mm apart, that can measure stresses to establish a relationship between stress and distance from the toe of the weld, and hence determine the hot spot stresses. The gauge factor of all the strip gauges is 2.06. The first gauge in the strip strain gauge was located at a distance of 4 mm, the minimum distance from the weld toe, of the first gauge used in the extrapolation for hot spot stresses, recommended for wall thicknesses less than 10 mm (van Wingerde 1992, IIW 2000, Zhao et al 2000).

The stress perpendicular to the toe of the weld can be determined from the true strain obtained from the strain gauge measurements using equation 1 (van Wingerde 1992). Equation 1 assumes that the Poisson's ratio, v is 0.3. It also assumes that due to the stiffening effect of the weld the strains parallel and normal to the weld toe are negligible at locations close to the weld toe within the extrapolation region for the hot spot stresses. Therefore, the stress perpendicular to the toe of the weld can be estimated as:

$$\sigma_x = \frac{E}{1-v^2}(\varepsilon_x) = \frac{E}{0.91}(\varepsilon_x) = 1.1E\varepsilon_x \qquad (1)$$

Stress distribution at the weld toes in the bottom RHS member were obtained from the strain measurements and corresponding hot spot stresses obtained through linear extrapolation of stresses falling within region of extrapolation recommended by CIDECT Design Guide No. 8 (Zhao et al 2000) and IIW (2000).

Apart from the measurement of stress distribution at the weld toes using strip strain gauges, three dimensional finite element analyses were also carried out to determine hot spot stresses in the RHS-RHS cross-beam connections. The software, STRAND7 was used as the modeling tool. All specimens were modeled using solid elements for the bottom and top members. The end plates for each member in the connection were modeled using shell elements and a pin support. The pin support only allowed rotation but no translation along the major neutral axis of each member. The load was applied using a point load on a 74 mm beam element. This was done to simulate the height at which the load was applied to the specimen in the test rig.

A plot of the stress distribution at weld toes in the bottom RHS member and the top RHS section member at the hot spot locations is shown in Figure 5. The stress distribution was used in the determination of hot spot stresses at weld toes in both the bottom RHS and top RHS members at different levels of applied bending load. The resultant hot spot stresses from linear extrapolation of the stresses within the IIW (2000) recommended extrapolation region were used in the determination of stress concentration factors. It can be seen from Figure 5 that the hot spot stresses at weld toes in the bottom RHS member are significantly larger than the hot spot stresses at weld toes in the top RHS section member, also see, Table 1. This is in agreement with the crack initiation and propagation pattern that was observed in the RHS-RHS cross-beam connections and reported in Section 2, where all the cracks initiated and propagated in the bottom RHS member.

The stress concentration factors in the bottom RHS member were determined as the ratio between the hot spot stress at the weld toe derived from stress distribution within the extrapolation region divided by the nominal stress at the weld toe in the bottom RHS member due to the applied load. The resultant stress concentration factors for the RHS-RHS cross-beam

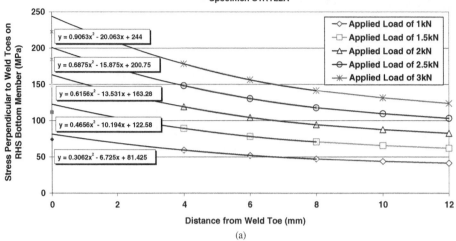

Stress Perpendicular to Weld Toe vs Distance from Weld toe on SHS Bottom Member: Specimen S1R1L2A

(a)

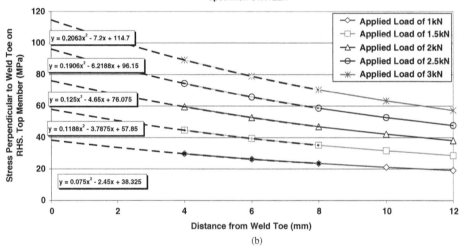

Stress Perpendicular to Weld Toe vs Distance from Weld Toe on RHS Top Member: Specimen S1R1L2A

(b)

Figure 5.   Stress distribution and hot spot stresses from FE analysis at weld toes in (a) the bottom RHS member and (b) the top RHS section member.

connections from the experimental investigation and the numerical investigation using STRAND 7 are shown in Table 2. The mean ratio of experimental SCFs to numerical SCFs is 0.8 and 1.07 for the bottom chord member and the top member respectively as shown in Table 3. This shows that the 3D finite element models adopted in this investigation can reliably be used to estimate the stress concentration factors in RHS-RHS cross-beam connections. The SCFs in RHS-RHS cross-beam connections are generally lower than SCFs in square hollow section nodal joints which have recommended minimum SCFs of 2.0.

4   CONCLUSIONS

(a) Under cyclic bending loading, crack initiation in RHS-RHS cross-beam connections was observed to occur in the corner region, at the weld toes in the bottom RHS member. Subsequent crack growth in the RHS-RHS cross-beam connections resulted in the cracks growing at the weld toes along the entire width of the RHS bottom member and then away from the weld toes into the parent metal.

(b) The S-N data from the RHS-RHS cross-beam connections all lie on or above the Australian Standard

Table 2. Determination of SCFs for S1R1L2A.

		Experimental SCFs				Numerical SCFs			
Load (kN)	Nominal Stress (MPa)	Hot Spot Stress Bottom Member	SCF Bottom Member	Hot Spot Stress Top Member	SCF Top Member	Hot Spot Stress bottom Member	SCF Bottom Member	Hot Spot Stress Top Member	SCF Top Member
1	41.59	66	1.58	46.86	1.13	81.245	1.95	38.325	0.92
1.5	62.39	96.94	1.55	56.92	0.93	122.58	1.96	57.85	0.93
2	83.18	122.32	1.47	74.36	0.89	163.28	1.96	76.075	0.91
2.5	103.98	155.54	1.50	98.78	0.95	200.75	1.93	96.15	0.92
3	124.78	177.98	1.43	126.72	1.01	244.00	1.96	114.7	0.92
	Average SCFs		1.5		0.98		1.95		0.92

Table 3. Comparison of Stress Concentration Factors (SCFs) in RHS-RHS cross-beam connection, S1R1L2A at weld toes in the bottom and top RHS members from experimental and numerical investigation.

Stress Concentration Factor (SCF) at weld toes in RHS Bottom Member			Stress Concentration Factor (SCF) at weld toes in RHS Top Member		
Experimental SCFs	Numerical SCFs	Ratio Experimental to Numerical SCFs	Experimental SCFs	Numerical SCFs	Ratio Experimental to Numerical SCFs
1.5	1.95	0.8	0.98	0.92	1.07

AS4600-2005 Class 55 fatigue design curve and is comparable to the data from RHS-Angle and RHS-Channel cross-beam connections. However tests are still continuing and design recommendations will be determined through statistical analysis of the full test data and also through the deterministic method.

(c) Stress concentrations in a typical RHS-RHS cross-beam connection was studied using experimental and numerical methods to gain an understanding of the observed crack pattern in RHS-RHS cross-beam connections. These studies showed that the magnitude of hot spot stresses and hence the SCFs in the bottom RHS member were significantly larger than that in the top RHS member. The higher stress concentrations in the bottom RHS member is manifested in the crack initiation and subsequent propagation that is observed at weld toes in the bottom RHS member of the RHS-RHS cross-beam connections.

ACKNOWLEDGEMENTS

The authors would like to thank the following laboratory staff in the Department of Civil Engineering at Monash University for their assistance in this project: Mr. Graham Rundle, Mr. Roy Goswell, Mr. Len Doddrell, Mr. Geoff Doddrell and Mr. Roger Doulis. Thanks also to Mr. Don McCarthy for taking the photographs shown in this paper. Thanks to Mr. Roy Goswell for his enthusiasm and for his insight and improvisation in the continuously changing support systems used for the fatigue tests. The authors also wish to that the following 4th year students for their supporting roles, Mr. Nick Kantourogiannis for his work on fatigue of RHS-Angle cross-beam connections in 2004 and Mr. James Aroin for continuing the work in this project in 2005. This Project is sponsored by OneSteel Market Mills and a Monash University SMURF2 fund. A special thanks to Mr. Hayden Dagg of OneSteel Market Mills for his continuous support and help.

REFERENCES

IIW 2000: Fatigue Design Procedures for Welded Hollow Section Joints, IIW Doc. XIII-1804-99, IIW Doc. XV-1035-99, Recommendations for IIW Subcommission XV-E, Edited by X.L. Zhao and J.A. Packer, Abington Publishing, Cambridge, UK

Klippstein, K.H. (1980), "Fatigue Behaviour of Sheet-Steel Fabrication Details", Fifth Specialty Conference on Cold-Formed Steel Structure, St. Louis, MO, 681–701

Klippstein, K.H. (1981), "Fatigue Behaviour of Steel-Sheet Fabrication Details", SAE Technical Paper Series 810436, Int. Congress and Exposition, Society of Automotive Engineers, Inc., Detroit, MI

Klippstein, K.H. (1985), "Fatigue of Fabricated Steel-Sheet Details – Phase II", SAE Technical Paper Series 850366, Int. Congress and Exposition, Society of Automotive Engineers, Inc., Detroit, MI

LaBoube, R.A. and Yu, W.W. (1999), "Design of Cold-Formed Steel Structural Members and Connections for

Cyclic Loading (Fatigue)", Cold-formed Steel Series 99-1, Univ. of Missouri-Rolla, MO

Mashiri F.R., Zhao X.L. and Grundy P. 2002a, "Fatigue Tests and Design of Thin Cold-Formed Square Hollow Section-to-Plate T-Connections under In-Plane Bending" Journal of Structural Engineering, ASCE, Vol. 128, No. 1, January 2002, pp. 22–31

Mashiri F.R., Zhao X.L., Grundy P. and Tong L. 2002b, "Fatigue Design of Very Thin-Walled SHS-to-plate Joints under In-Plane Bending", Thin-Walled Structures, Vol. 40, Issue. 2, Elsevier Science Ltd, February 2002, pp. 125–151

Mashiri F.R., Zhao X.L. and Grundy P. 2002c, "Fatigue Tests and Design of Welded T-Connections in Thin Cold-Formed Square Hollow Sections under In-Plane Bending" Journal of Structural Engineering, ASCE, Vol. 128, No. 11, November 2002, pp. 1413–1422

Mashiri F.R., Zhao X.L. and Grundy P. 2004a, "Stress Concentration Factors and Fatigue Failure of Welded T-Connections in Circular Hollow Sections under In-Plane Bending", International Journal of Structural Stability and Dynamics, World Scientific, September 2004, Vol 4, No. 3, pp. 403–422

Mashiri F.R., Zhao X.L. and Grundy P. 2004b, "Stress Concentration Factors and Fatigue Behaviour of Welded Thin-Walled CHS-SHS T-Joints under In-Plane Bending", Engineering Structures, Vol 26, No. 13, Elsevier Science Ltd, 2004, pp. 1861–1875

Mashiri F.R. and Zhao X.L. 2005a, "Fatigue Behaviour of Welded Thin-Walled RHS-to-Angle Cross-Beams under In-Plane Bending", Structural Engineering-Preserving and Building into the Future, Australian Structural Engineering Conference (ASEC2005), Editors: Stewart M.G. and Dockrill B, 11–14 September 2005, Newcastle, Australia, 10 pp. CDROM. ISBN: 1 877040 37 1

Mashiri F.R. and Zhao X.L. 2005b, "Fatigue Behaviour of DuraGal Cross-Beam Welded Connections", OneSteel Market Mills Project, Interim Report No. 04/05, Department of Civil Engineering, Monash University, Clayton, Australia, 63pp

SAA 1991: Structural Steel Hollow Sections, Australian Standard AS1163-1991, Standards Association of Australia, Sydney, Australia

SAA 2004a: Structural Steel Welding, Part 1: Welding of Steel Structures, Australian Standard AS/NZS 1554.1-2004, Standards Association of Australia, Sydney, Australia

SAA 2004b: Structural Steel Welding, Part 5: Welding of Steel Structures subject to High Levels of Fatigue Loading, Australian Standard AS/NZS 1554.5-2004, Standards Association of Australia, Sydney, Australia

SAA 2005, Cold-Formed Steel Structures (Draft), Australian Standard AS 4600-2005, Standards Association of Australia, Sydney, Australia

Strand 7 Pty Ltd 2004, "Introduction to Strand7 finite element analysis system", Sydney, Australia

van Wingerde A.M. 1992: The fatigue behaviour of T- and X-joints made of SHS, Heron, Vol. 37, No. 2, pp. 1–180

WTIA 1998, "Commentary on the Structural Steel Welding Standard AS/NZS 1554", TN 11-98, WTIA Technical Note No. 11, Published jointly by the Welding Technology Institute of Australia and the Australian Institute of Steel Construction, NSW, Australia

Zhao, X.L. and Hancock, G.J. (1996), "Welded Connections in Thin Cold-Formed RHS", In: Connections in Steel Structures III, Eds. Bjorhovde, et al, Oxford: Pergamon, 89–98

Zhao, X.L. and Mahendran, M. (1998), "Recent Innovation of Cold-Formed Tubular Sections", J. Construct. Steel Research, 46 (1–3)

Zhao, X.L., Herion, S., Packer, J.A., Puthli, R.S., Sedlacek, G., Wardenier, J., Weynand, K., van Wingerde, A.M. and Yeomans, N. F. (2000), Design Guide for Circular and Rectangular Hollow Section Welded Joints under Fatigue Loading, Verlag TÜV Rheinland, Cologne, Germany

Tubular Structures XI – Packer & Willibald (eds)
© 2006 Taylor & Francis Group, London, ISBN 0-415-40280-8

Effects of hydrogen removal heat treatment on residual stresses in high strength structural steel welds

P. Wongpanya & Th. Boellinghaus
Federal Institute for Materials Research and Testing (BAM), Berlin, Germany

G. Lothongkum
Department of Metallurgical Engineering, Chulalongkorn University, Bangkok, Thailand

ABSTRACT: High strength structural steels with yield strengths of up to 1100 MPa are increasingly applied to various industrial branches for weight optimization. Recent failure cases make evident that special attention has to be paid to hydrogen assisted cold cracking avoidance in welds, if such materials are used for tubular structures, as for instance for penstocks in hydropower plants. In order to gain more insight into the cold cracking resistance of high strength steel welds, Instrumented Restraint Cracking (IRC) Tests have been performed with a S 1100 steel. As a particular item, the efficiency of various hydrogen removal heat treatments (HRHT) is currently discussed versus softening effects by annealing and an increase of the stresses and strains in the welds. For this reason, the effects of pre- and postheating on the stress-strain distribution in the root weld of a 60° V-butt joint have been investigated by finite element analyses validated by the IRC tests. Maximum transverse residual stresses at the fusion line beneath the top surface of the root weld increased with increasing restraint intensity. It turned out that preheating causes higher stresses and strains in the root weld than postheating at the same temperature and time. As a particular item, time-strain-fracture diagrams represent a very helpful tool to combine experimental and simulation results for a firm assessment of the cold cracking resistance of specific welds.

1 INTRODUCTION

Fine grained high strength structural steels gain increasing importance in welding of modern tubular steel constructions, in particular for pipelines transporting less aggressive media. But, application of such materials with yield strengths of up to 1100 MPa at new constructions has been accompanied by failure cases which occur mostly and fortunately less spectacularly during fabrication, but they are usually associated with extensive costs due to fabrication delays and extensive repairs. Frequently, the origin of such failures has been identified as hydrogen assisted cracking, predominantly hydrogen assisted cold cracking during fabrication welding. One reason for these failures is a lack of basic knowledge about the hydrogen effects on steel microstructures. Two very important facts have to be reminded. The first is that hydrogen reduces predominantly the ductility of steel microstructures instead of its technical strength, as schematically illustrated in Fig. 1. This means that a decrease of the tensile strength of a steel microstructure to the level of the yield strength or the 0.2% proof stress already represents a significant ductility loss. An additional reason is that the susceptibility for hydrogen assisted cracking

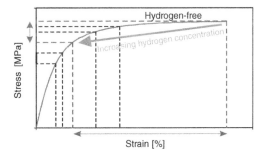

Figure 1. Schematic illustration of the effect of hydrogen on the technical stress and strain parameters of steel microstructures (Zimmer, Seeger & Boellinghaus 2005).

of high strength steel microstructures increases with the strength level of the base material. This is shown by Fig. 2, where the hydrogen concentration has been considered as a critical value for such a non tolerable ductility loss has been assigned to the strength level of the base material. While such a limit has not been identified for base materials with a yield strength of up to 690 MPa, it well exists at for the high strength steels, like the S1100QL. In the different weld microstructures hydrogen can actually decrease the

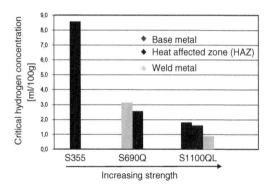

Figure 2. Critical hydrogen concentration indicating a ductility loss by coincidence of the tensile strength with the 0.2% proof stress dependent on the strength level of structural steels (Zimmer, Seeger & Boellinghaus 2005).

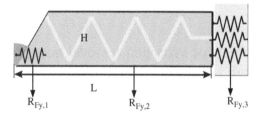

Figure 3. Schematic illustration of the three design parts contributing to the total restraint intensity $R_{Fy,total}$.

ductility to very low values, if the strength level of the base material exceeds 700 MPa. For the heat affected zone, the limiting hydrogen concentration ranges at a value of 2 ml/100 g for the materials S690Q and the S1100QL, while the HAZ of the steel S355 still maintains the ductility up to hydrogen concentrations of about 8 ml/100 g. It can only be emphasized that such results are still not considered in welding procedure specifications for cold cracking avoidance. For instance, the guidelines for preheating and hydrogen removal heat treatments (DIN EN 1011) have been developed for steels with yield strengths of up to 690 MPa and it is not certain that they can be applied to materials with higher yield strengths. Only a limited number of attempts have been undertaken, to evaluate welding and heat treatment procedures for cold cracking avoidance in welded structures of steels with yield strengths above 700 MPa (Heinemann 2001, Meyer 2003 & Wegmann 2003). But, it still remains in question which effects such hydrogen removal heat treatment might have on the stress-strain level in the structure after cooling of the joints. As an additional point, welds in structural components, in particular, if they are stiffened for weight reduction and to save material, are usually highly restrained. In this context, it has to be considered that tubular structures are per se much stiffer than plate constructions. As a contribution for save application of structural steels in tubular components in the future, the present study has been focussed on the effects of hydrogen removal heat treatments (HRHT) on the residual stresses and strains in S1100QL steel butt joints. In order to elucidate the basic effects of the heat treatment procedures on the residual stresses and to investigate the most sensitive part of the welds, the investigations were carried out by finite element simulations of a cross section of a circumferential pipeline root weld at various restraint levels transverse to the welding direction which were validated by respective IRC Tests.

2 EXPERIMENTAL AND NUMERICAL PROCEDURES

2.1 Experimental procedure

For the evaluation of the cold cracking susceptibility of weld joints many different tests exist (Viyanit & Boellinghaus 2004). In contrast to other test procedures, the IRC-test allows to study the time dependent stress state during and after welding at realistic shrinkage restraint levels (Boellinghaus et al. 1999, Hoffmeister 1986) which contribute significantly to the residual stresses and strains in welds at a real construction. The shrinkage restraint transverse to the weld is determined ahead at the real component by the intensity of restraint

$$R_{F_y} = \frac{F_y}{2\Delta y \cdot l_w} \qquad (1)$$

where F_y = the reaction force transverse to the welding direction, l_w = the total length of the weld bead, $2\Delta y$ = the transverse root gap displacement in both directions. As shown in Fig. 3, the weld edge preparation, $R_{Fy,1}$, the plate or pipe dimensions, $R_{Fy,2}$ and the surrounding construction, $R_{Fy,3}$, contribute individually to the intensity of restraint and thus, these three parts of welded structure have to be evaluated to calculate the total restraint intensity by

$$\frac{1}{R_{F_y,total}} = \frac{1}{R_{F_y,1}} + \frac{1}{R_{F_y,2}} + \frac{1}{R_{F_y,3}} \qquad (2)$$

which is then transferred to the IRC Test set up by adjustment of the assembly parts (Fig. 4).

The IRC Test plates with a thickness of H = 20 mm and a restraint length of L = 100 mm, were welded with a root pass measuring with a height of h_w = 6.5 mm by Gas Metal Arc Welding (GMAW) process with the parameters presented in Table 1. Figure 5 shows the dimensional details of the weld and, in particular, the position of the thermocouples and the strain gage used for comparison with the results of subsequent numerical simulations. The time histories of the temperatures and strains at those locations were registered by a respective data acquisition system.

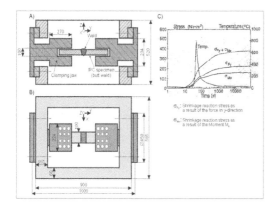

Figure 4. Schematic illustration of the IRC-Test principle with exemplary dimensions: A) View for butt weld and fillet weld, B) Top view, C) IRC-Test diagram (Viyanit & Boellinghaus 2004).

Table 1. GMAW welding parameters.

Parameters	Unit	Value
Voltage	(V)	31.5
Current	(A)	280.0
Heat input	(kJ/cm)	9.0
Speed of wire	(cm/min)	9.6

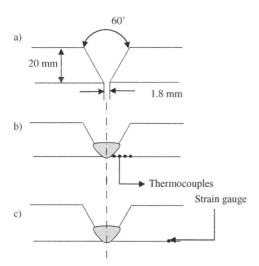

Figure 5. Weld Details, a) Groove preparation and dimensions, b) Root pass and location of Ni-Cr-Ni thermocouples (2.9, 4.9, 6.9 and 8.9 mm from the weld centerline), c) Location of strain gage (20.9 mm from weld centerline).

2.2 Numerical procedure

Two-dimensional nonlinear transient heat flow calculations and coupled thermo-mechanical analyses with

Table 2. Chemical composition of base and filler materials.

	C	Si	Mn	Cr	Mo	Ni
S 1100	0.1700	0.2710	0.8540	0.4590	0.4510	1.8800
Union X96	0.1206	0.7800	1.8600	0.4600	0.5300	2.3600

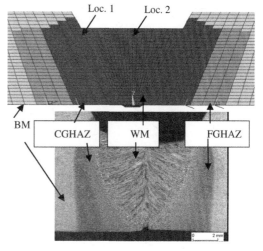

Figure 6. Four different material properties in each weldment.

the commercial software ANSYS were performed to calculate the temperature distribution and the resulting stresses and strains in the root weld.

Oriented by the cross sectional dimensions of the real welds, a two dimensional finite element model of the root joint has been created, as shown in Figure 6. The dimensions were those of the actual IRC Tests with a restraint length of $L = 55$ mm, but for additional variation of the external shrinkage restraint produced by the surrounding construction, $R_{Fy,3}$, respective spring elements have been assigned to both ends of the model. Both sides of the symmetric welds were represented in the model for better illustrations of the results. However, to decrease CPU-time and storage capacity, a finer mesh density was selected for the weld metal and the HAZ.

Since the calculated temperatures are used as loads for the subsequent thermal-mechanical analyses, the thermo-physical material properties such as density, heat convection coefficient, enthalpy and thermal conductivity will significantly affect the finally calculated stresses and strains. They have thus been selected very carefully from respective diagrams (Richter 1973) for a steel with similar composition than the relatively new S1100QL type (Figure 7) and have been applied to all microstructures of the joint, i. e. the weld metal, the HAZ and the base material. In order to simulate the

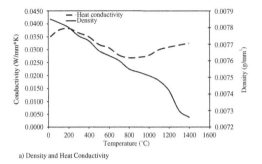

a) Density and Heat Conductivity

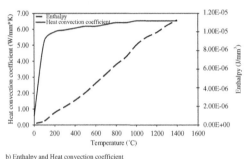

b) Enthalpy and Heat convection coefficient

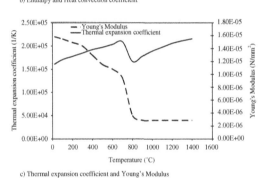

c) Thermal expansion coefficient and Young's Modulus

Figure 7. Thermo-physical properties selected for the thermal analyses (Richter 1973).

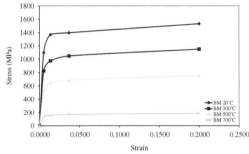

a) Modified value for base metal

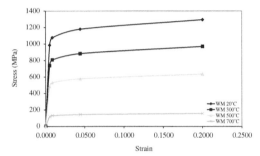

b) Modified value for weld metal

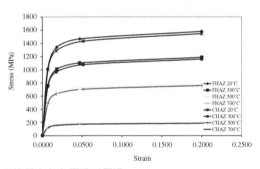

c) Modified value for FHAZ and CHAZ

Figure 8. Thermo-mechanical properties.

heat dissipation into the clamping jaws of the IRC Test or similarly, into a surrounding construction, a constant temperature has been assigned to both ends of the model. Heat convection has been modeled at all free surfaces of the joint. As described previously (Boellinghaus & Hoffmeister 1995) and as confirmed by respective pre-analyses, heat loss by radiation can be neglected for solid materials in contrast to liquids and has thus been ignored in the modeling procedure.

In contrast to the physical properties, the thermo-mechanical properties vary significantly with the weld microstructures (Figure 8). The weld metal and the heat affected zone have a different stress-strain behavior than that of the base material and the various stress-strain relationships used for modeling are shown in Figure 8a–8c. Phase transformation of the weld metal and the heat affected zone has been considered by

interpolation between the various temperature dependent stress-strain curves (Figure 8) and by the temperature dependent thermal elongation coefficient. For coupling, the same number and width of the time steps were chosen for the thermal and the structural analyses.

Heat treatment of welds is usually carried out using electric heating pads and two frequently applied procedures have been investigated, i. e. preheating for 30 minutes and postheating between 30 and 120 minutes with an onset during the first 400 seconds after welding. The temperature used for heat treatment procedure ranged between 50°C and 300°C. Heating with such pads has been simulated by application of a constant temperature on the surface of the model as shown in Figure 9.

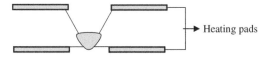

Figure 9. Heating pad position for heat treatment process.

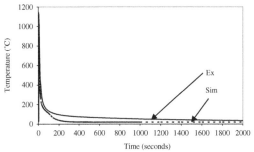

a) Thermocouple at 2.9 mm

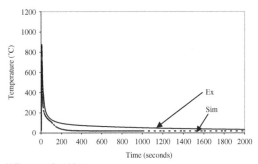

b) Thermocouple at 4.9 mm

Figure 10. Measured and simulated temperature histories.

3 RESULTS AND DISCUSSION

3.1 Validation of numerical simulations by experimental results

As shown by Figure 10, the experimentally and numerically obtained temperature histories at the surface of the IRC-test plates at the positions indicated in Figure 5 are fairly good consistent. At temperatures below 100°C, a slightly higher calculated cooling rate can be observed which may be related to the boundary condition, i. e. the constant temperature at both ends of the model. However, in this temperature range no phase transformation takes place and it can thus be anticipated that the faster cooling rate does not significantly influence the thermo-mechanical behavior.

Figure 11 represents a comparison of the time dependent plot of the strain transverse to the welding direction at the location shown in Figure 5.

Obviously, the calculated strains range at the same level than the measured values. However, the numerically calculated values increase faster and up to slightly

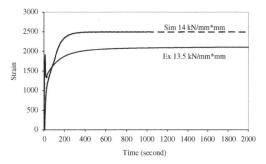

Figure 11. Comparison of the thermal strain of IRC-test weld with the corresponding finite element simulation.

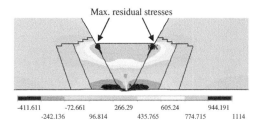

Figure 12. Transverse residual stress distribution in MPa after cooling down to room temperature at a restraint intensity of $R_{Fy} = 14.91$ kN/(mm·mm).

higher levels than in the experiments which has to be attributed to the faster cooling rates below 100°C in the calculation. From this point of view and also considering potential heat treatment, the calculations have thus to be regarded as conservative compared to the experimental data. By consideration of different thermo-mechanical properties for the various weld microstructures in the finite element analyses the consistency between experimental and numerical results might be approved further.

3.2 Effect of restraint intensity on residual stresses

Figure 12 shows exemplarily the residual stress distribution transverse to the welding direction after cooling down to room temperature for the restraint intensity of $R_{Fy,total} = 14.91$ kN/(mm·mm). Due to the asymmetrically welded root and respective bending of the total joint, the stresses decrease gradually from positive (tension) values from the top of the weld towards the middle and the bottom into negative values (compression).

This also becomes evident by the plots of the residual stresses transverse to the welding direction alongside the top, the middle and the bottom of the weld assigned to Figure 13.

At the top of the weld the values show the typical M-shaped curve. With increasing distance from the

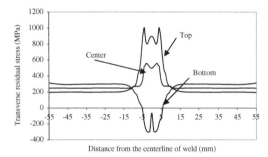

Figure 13. Residual stresses transverse to the welding direction at the top, middle and bottom of the weld after cooling down to room temperature at a total restraint intensity of $R_{Fy} = 15\,kN/(mm \cdot mm)$.

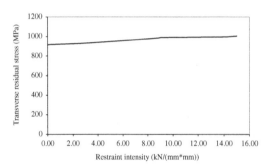

Figure 14. Maximum transverse residual stresses in the weld metal adjacent to the fusion line (Loc. 1) dependent on the intensity of restraint R_{Fy}.

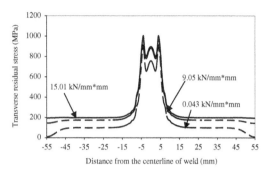

Figure 15. Distribution of as-weld transverse residual stresses for 0.043, 9.05 and 15 kN/(mm·mm).

of $R_{Fy,3} = 15.01\,kN/(mm \cdot mm)$ the stress values range at about 1070 MPa in the weld metal adjacent to the fusion line (Figure 11) with respect to practical welding such conditions mean that the tubulars are clamped rigidly at a distance of 55 mm from the weld metal centerline at a total restraint intensity of about $R_{Fy,total} = 15\,kN/(mm \cdot mm)$. By the plot of the residual stresses alongside the topside of the weld in both directions shown in Figure 15, it becomes obvious that with increasing the external shrinkage restraint up to the conditions of rigid clamping of the tubulars at a position of 55 mm will increase the residual stresses by about 100 MPa all over the joint. This effect is even more pronounced in the weld metal, where an increase of about 200 MPa has been registered.

weld centerline the transverse residual stresses first slightly decrease then establish a sharp tensile peak of about 1000 MPa near the fusion line and gradually decrease again to a constant value of about 180 MPa at the distance of 15 mm away from weld centerline. In the middle of the weld metal maximum stresses of about half the yield strength of the base material are reached. The most susceptible region for hydrogen assisted cold crack initiation is thus located at the top in the weld metal near the fusion line and the transition to the HAZ (Figure 12).

The effect of the restraint intensity on the residual stresses transverse to the welding direction becomes obvious by the plot in Figure 14 showing the maximum residual stresses at the location identified by Figure 12 for various shrinkage restraints. As to be expected for butt joints, the weld residual stresses increase linearly with the shrinkage restraint. If no external shrinkage restraint ($R_{Fy,3} \Rightarrow 0$) is present the total restraint intensity is reduced to fairly low values of about $R_{Fy,total} = 0.04\,kN/(mm \cdot mm)$. This represent nearly free shrinkage and the maximum transverse residual stresses range at about 940 MPa. In contrast, at the maximum external restraint intensity values

3.3 Effect of preheating on residual stresses

Figure 16 shows the maximum transverse residual stresses at the location as marked in Figure 12 dependent on the intensity of restraint and for various preheating temperatures. As to be expected by the additional temperature field before welding, the transverse residual stresses are significantly increased by preheating. By the three dimensional plot in Figure 17 it becomes clear that increasing preheating temperatures might cause a higher stress increase than an increase of the shrinkage restraint, at least at this root weld of a 60°-V butt joint. It is also remarkable that the transverse stresses are more increased at preheating temperatures of up to 150°C than at higher temperature ranges between 150°C and 300°C. Preheating of the S1100QL steel at such high temperatures is usually avoided to prevent annealing effects and a loss in strength. But, for practical welding applications these results show that preheating at lower temperatures also produces a significant stress increase in such joints.

As shown previously for an offshore steel of the S355 type (Boellinghaus & Hoffmeister 1995), preheating at such temperatures appears as not sufficient for hydrogen removal and only shifts the highest

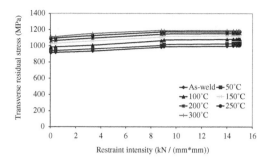

Figure 16. Maximum transverse residual stresses dependent on the restraint intensity for various preheating temperatures.

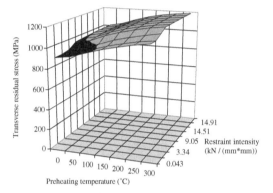

Figure 17. Stress-Restraint-Heating (SRH) diagram for 30 min preheating of the 60°-V S1100QL tubular joint with a wall thickness of 20 mm.

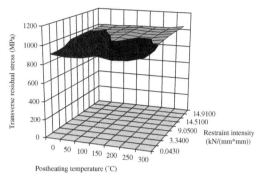

Figure 18. Stress-Restraint-Heating (SRH) diagram 30 min for postheating of the 60°-V S1100QL tubular joint with a wall thickness of 20 mm.

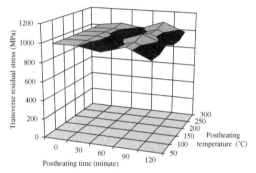

Figure 19. Transverse residual stresses after postheating of the 60°-V S1100QL tubular joint with a wall thickness of 20 mm at various temperatures and times.

concentration levels towards the fusion line. Even preheating at only 50°C frequently applied in arctic regions to remove moisture on the materials before welding increases the transverse residual stresses in this type of joint by about 125 MPa.

3.4 Effect of postheating on residual stresses

Postheating in this publication is considered as typical hydrogen removal heat treatment (HRHT) and is not considered in terms of stress relief which would cause detrimental strength reduction and microstructure changes in the investigated S1100QL steel. As shown by comparison of the diagrams in Figure 17 and Figure 18, the increase of the transverse residual stresses by postheating at the same temperature and time is much lower than by preheating, in particular at higher restraint intensities. Also, the residual stresses significantly increase only, if the postheating temperature is increased from 250°C to 300°C (Figure 18 and Figure 19). As one reason for such different effects of pre- and postheating, it has to be considered that the tubulars have already been joined when the heating is

applied. This means that postheating is acting into both directions at the joined weld, i. e. compression during the heating process and tension after relief of the pads. In contrast, compression during the onset of preheating cannot act on the joint before welding and thus, preheating causes only an additional temperature field producing additional thermal stresses and strains in tension after completion of the joint. It has to be mentioned that, according to Figure 19, transverse residual stresses increase also with both, postheating temperature and time. But, the onset of postheating has been assumed at 400 s after welding at which the joint has cooled down almost to room temperature (Figure 7). It can thus be anticipated that the transverse stresses can additionally be reduced by starting postheating at higher temperatures during the cooling cycle. Summarizing this, postheating appears generally as much more beneficial regarding the stress-strain build up in the investigated joint and as compared to preheating. Postheating thus also contributes to minimization of the cold cracking risk of tubular joints from a mechanical aspect and not only by reduction of the hydrogen concentration at critical locations.

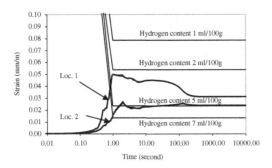

Figure 20. TSF diagram of numerical determination of equivalent strain in critical region of IRC-Test specimen.

3.5 Assessment of hydrogen assisted cold cracking by Time-Strain-Fracture (TSF) Diagrams

Although hydrogen assisted cold cracking is generally influenced by the local microstructure, the mechanical loads and hydrogen concentration, the cracking risk can simply be evaluated from the mechanical aspect, if the critical strains for the respective hydrogen concentration for the specific microstructure are known. A helpful tool for such assessment represent Time-Strain-Fracture (TSF) Diagrams. As exemplarily shown by the TSF Diagram in Figure 20 for the non-heat treated S1100 joint at the highest restraint, the local strains at various crack critical locations (Figure 6) increase gradually to a value of about $\varepsilon_{local} = 0.05$ mm/m while the critical strains for different hydrogen amounts decrease. As can be seen in the diagram, the local strains exceed the critical strains for a hydrogen concentration in the weld metal of $HD = 5$ ml/100 g and $HD = 7$ which is indicating cracking and corresponds to experimentally observed failure of the respective specimens in the IRC Test.

4 CONCLUSIONS AND PERSPECTIVES

The following conclusions can be drawn from the present state of modeling the effects of restraint and heat treatment on the thermo-mechanical stress-strain behavior of the root weld at a 60° V circumferential butt joint with a wall thickness of 20 mm:

- For the selected filler materials the highest local stresses and strains have been identified in the weld metal adjacent to the fusion line, indicating also the most cold crack susceptible region. Further investigations have to show, if the crack critical region is shifted even more into the weld metal at overmatching conditions.
- Residual stresses in the root weld of the 60°-V butt joint increase linearly with the restraint intensity which has been varied between nearly free shrinkage

and rigid clamping. For practical welding applications this means that rigid clamping of the parts for circumferential welding has to be avoided wherever possible.

- Preheating causes significantly higher stresses in the root weld than welding without heat treatment already at temperatures below 150°C. Preheating has thus to be avoided, in particular, if the capabilities for sufficient hydrogen removal remain in question.
- Postheating causes much lower stresses in the root weld than preheating and thus appears as much more effective for cold cracking avoidance by such mechanical aspects, but also by prolonging the time for hydrogen removal after welding.
- Time-Strain-Fracture (TSF) Diagrams provide the data to assess a potential failure risk of welded components by hydrogen assisted cold cracking already from the mechanical aspect, if the hydrogen related mechanical properties in terms of the true fracture strain are known for the particular microstructure in crack susceptible regions.

REFERENCES

Boellinghaus, Th. & Hoffmeister, H. 1995, Finite element calculations of pre- and postheating procedures for sufficient hydrogen removal in butt joints In H. Cerjak (ed.), Mathematical Modelling of Weld Phenomena 3, IOM, London: 276–756.

Boellinghaus, Th., Hoffmeister, H. & Littich, M. 1999. Application of the IRC-Test for assessment of reaction stresses in tubular joints with respect to hydrogen assisted cracking. Welding in the world, 43 (2): 27–35.

DIN EN 1011 – Guidelines for welding of metallic materials, Part I.

Heinemann, H. 2001. Metall-Aktivgasschweißen hochfester, vergüteter Feinkornbaustähle, Jahrbuch der Schweißtechnik: 50–57.

Hoffmeister, H. 1986. Concept and procedure description of the IRC Test for assessing hydrogen assisted weld cracking, Steel Research, 57 (7): 345–347.

Meyer, B. & al. 2003. Entwicklung neuer und Optimierung vorhandener MSG-Fülldrähte für das Schweißen hochfester Feinkornstähle, DVS-Bericht 225: 443–451.

Richter, F. 1973. Die wichtigsten physikalischen Eigenschaften von 52 Eisenwerkstoffen, Heft 8, Verlag Stahleisen, Düsseldorf.

Viyanit, E. & Boellinghaus, Th. 2004. Cold Cracking Tests, IIW-Doc. No. II-A-111-04.

Wegmann, H. & Gerster P. 2003. Schweißtechnische Verarbeitung und Anwendung hochfester Baustähle im Nutzfahrzeugbau, DVS-Bericht 225: 429–435.

Zimmer, P., Boellinghaus, Th. & Kannengiesser, Th. 2004. Effects of Hydrogen on Weld Microstructure Mechanical Properties of High Strength Structural Steels, IIW Doc. No. II-A-139-04.

Zimmer, P., Seeger, D.M. & Boellinghaus, Th. 2005. Hydrogen Permeation and Related Material Properties of High Strength Structural Steels, Proc. High Strength Steels for Hydropower Plants, Graz, 5–6 July 2005: paper 17.

Parallel session 2: Static strength of joints

Tubular Structures XI – Packer & Willibald (eds)
© 2006 Taylor & Francis Group, London, ISBN 0-415-40280-8

The ultimate strength of axially loaded CHS uniplanar T-joints subjected to axial chord load

G.J. van der Vegte[1,2] & Y. Makino[1]

[1]*Kumamoto University, Kumamoto, Japan;* [2]*Delft University of Technology, Delft, The Netherlands*

ABSTRACT: In recent years, extensive numerical research was conducted into the chord stress effect of CHS uniplanar joints. Among the three basic types of geometries (K-, T- and X-joints), the effect of chord load on the ultimate strength is most complicated for a uniplanar T-joint under axial brace load since its strength is governed by a combination of (local) joint failure and failure due to (global) chord in-plane bending moments. The presence of additional axial chord load further aggravates the complexity of load transfer. As a follow-up on the investigation by Van der Vegte & Makino (2005), the current study identifies the effects of both compressive and tensile axial chord pre-load on the ultimate strength of axially loaded T-joint configurations for a wide range of the brace-to-chord diameter ratio β and chord diameter-to-chord wall thickness ratio 2γ. Based on the available FE data, a new strength formulation is established for T-joints under axial chord pre-load, describing the interaction between axial brace load and in-plane bending chord moments for the values of the chord load considered.

1 INTRODUCTION

During the past five years, in the framework of a CIDECT programme, multiple numerical studies were conducted into the chord stress effects of circular hollow section (CHS) joints. The rationale for this interest can be found in the different approaches used for CHS and rectangular hollow section (RHS) joints to account for the influence of chord stress on the ultimate strength. For CHS joints, the chord stress function is based on the chord pre-stress, while for RHS joints the maximum chord stress is used. As it is widely accepted that the maximum chord stress is the most appropriate parameter, the effect of chord stress of CHS joints has been re-analyzed in order to establish a chord stress function consistent with the approach adopted for RHS joints.

In order to generate more evidence on the strength of tubular joints under chord load, extensive numerical analyses were conducted on uniplanar K- and X-joints with the chord subjected to either axial load or in-plane bending moments. For example, finite element (FE) research presented by Pecknold et al. (2001) and Van der Vegte et al. (2002) focused on the chord stress effect of uniplanar gap K-joints. More recently, Van der Vegte et al. (2005) studied the interaction between chord pre-stress and chord boundary conditions of axially loaded gap K-joints. Whereas the aforementioned

studies consider thin-walled joints, Choo et al. (2003) investigated the influence of tensile and compressive chord pre-stress on the strength of thick-walled X-joints for various types of brace load.

For the third type of the basic uniplanar joint configurations, the T-joint, the evaluation of the chord stress effect is not as straightforward as for K- and X-joints. Unlike the X-joint configuration, where axial brace loads do not cause equilibrium induced chord loads, for uniplanar T-joints, axial brace loads will unavoidably lead to chord in-plane bending moments affecting the joint strength. As presented by Van der Vegte (1995), the strength of uniplanar T-joints is governed by a combination of local (joint) failure and failure due to overall chord bending. In order to exclude the chord bending effect and to derive the "local strength" of a T-joint, in-plane bending moments are applied to both chord ends proportional to the brace load, so that the global chord moment in the chord cross sections between the brace remain zero throughout the loading history.

In 2005, Van der Vegte & Makino conducted a more rigorous study as compared to 1995 into the effect of chord length, boundary conditions and chord end conditions on the strength of uniplanar T-joints under brace compression. Sixteen axially loaded T-joint configurations (four β- and four 2γ-values) were analysed numerically for five values of the chord length parameter α. The approach of applying compensating moments to the chord end with the purpose to eliminate equilibrium induced chord bending moments was confirmed to provide ultimate strength values for T-joints not affected by the chord length parameter α.

The current study further evaluates the ultimate strength of the sixteen above-mentioned configurations under axial chord pre-load. Both compressive and tensile chord pre-load are considered. Material- and geometric non-linear FE analyses are conducted with the general-purpose package ABAQUS/Standard (2003). Based on the FE data generated, a new strength formulation is established for T-joints under axial chord pre-load, presenting the interaction between axial brace load and chord bending moments for the values of the chord pre-load considered. The chord in-plane bending moments include both the equilibrium induced moments and the in-plane pre-moments applied to the chord end.

Recently, improved strength formulations for tubular joints were also proposed by API (2003) in the framework of an extensive project to update the API RP2A (1993). A relatively small part of the research programme was devoted to axially loaded uniplanar T-joints. The strength formulations developed for T-joints, adopting the aforementioned moment-free baseline approach to account for the effect of chord stress, will be included in the 22nd edition of API RP2A.

Since the current study is part of a larger project into the effect of chord stress on CHS joints, not only involving T-joints but also K- and X-joints, the set of ultimate strength equations and chord stress functions derived for T-joints is in close relationship with the formulations for K- and X-joints. An overview of the results as well as the similarities between the strength equations for the various types of CHS joints will be presented in the near future. Furthermore, at the moment a re-evaluation of the chord stress formulations for RHS joints is in progress (Liu & Wardenier, 2006). The format of the new formulations is identical to that proposed in the current study.

2 RESEARCH PROGRAMME

The configuration of uniplanar T-joints and the definition of the geometric parameters are illustrated in Figure 1. The geometric parameters of the T-joints considered are summarized in Table 1. For all joints, the chord diameter $d_0 = 406.4$ mm. The chord length parameter α is taken as 20.

Each of the sixteen combinations of β and 2γ has been analysed for the following nine values of the chord pre-load ratios $N_{0,p}/A_0 f_{y0}$: +0.9, +0.8, +0.6, +0.3, 0.0, −0.3, −0.6, −0.8 and −0.9, where positive (negative) values refer to tension (compression). In this

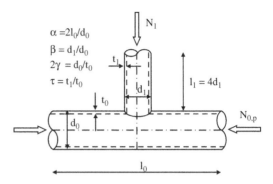

Figure 1. Configuration of a CHS uniplanar T-joint.

Table 1. Geometric parameters investigated for uniplanar T-joints under brace compression.

$d_0 = 4406.4$ mm			
2γ			
25.4	36.9	50.8	63.5
$\beta = 0.25$ T1	T2	T3	T4
$\beta = 0.48$ T5	T6	T7	T8
$\beta = 0.73$ T9	T10	T11	T12
$\beta = 0.98$ T13	T14	T15	T16

study, the chord is subjected to pre-load i.e. the chord load has been applied prior to loading of the brace. An alternative approach, considering the chord- and brace load to be proportional throughout the loading history, has not been explored herein.

Table 2 shows the T-joint configurations subjected to a combination of chord pre-load $N_{0,p}$ and externally applied in-plane bending moment $M_{0,p}$. Each β-value in Table 2 represents the four 2γ values mentioned in Table 1. The T-joints presented in the centre column of Table 2 (indicating zero pre-moment) were analysed by Van der Vegte & Makino (2005). The pre-moment ratios different from 0.0 are used to further "fill" and verify the interaction contours displaying local joint strength against chord in-plane bending moment. No analyses are made for chord pre-tensioning in combination with external chord moments.

The joints with $\beta = 0.98$ (T13–T16) have been subjected to additional analyses. After having been pre-loaded, these joints are analysed for chord bending moment caused by shear force applied to the chord end. In these simulations, where the shear forces applied to the chord end are reacted at the chord centre, the brace remains un-loaded. These analyses, which basically determine the in-plane bending strength of a chord with an unloaded brace attached to it, allow the interaction contours to be extended to the horizontal axis.

The steel grade used for the tubular members is S355 with $f_y = 355$ N/mm^2 and $f_u = 510$ N/mm^2.

Table 2. Overview of chord pre-moment ratios $M_{0,p}/M_{pl,0}$ applied to uniplanar T-joints.

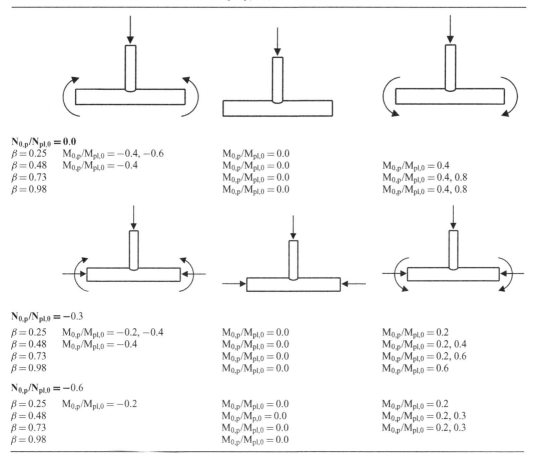

$N_{0,p}/N_{pl,0} = 0.0$

$\beta = 0.25$	$M_{0,p}/M_{pl,0} = -0.4, -0.6$	$M_{0,p}/M_{pl,0} = 0.0$	
$\beta = 0.48$	$M_{0,p}/M_{pl,0} = -0.4$	$M_{0,p}/M_{pl,0} = 0.0$	$M_{0,p}/M_{pl,0} = 0.4$
$\beta = 0.73$		$M_{0,p}/M_{pl,0} = 0.0$	$M_{0,p}/M_{pl,0} = 0.4, 0.8$
$\beta = 0.98$		$M_{0,p}/M_{pl,0} = 0.0$	$M_{0,p}/M_{pl,0} = 0.4, 0.8$

$N_{0,p}/N_{pl,0} = -0.3$

$\beta = 0.25$ $M_{0,p}/M_{pl,0} = -0.2, -0.4$ $M_{0,p}/M_{pl,0} = 0.0$ $M_{0,p}/M_{pl,0} = 0.2$
$\beta = 0.48$ $M_{0,p}/M_{pl,0} = -0.4$ $M_{0,p}/M_{pl,0} = 0.0$ $M_{0,p}/M_{pl,0} = 0.2, 0.4$
$\beta = 0.73$ $M_{0,p}/M_{pl,0} = 0.0$ $M_{0,p}/M_{pl,0} = 0.2, 0.6$
$\beta = 0.98$ $M_{0,p}/M_{pl,0} = 0.0$ $M_{0,p}/M_{pl,0} = 0.6$

$N_{0,p}/N_{pl,0} = -0.6$

$\beta = 0.25$ $M_{0,p}/M_{pl,0} = -0.2$ $M_{0,p}/M_{pl,0} = 0.0$ $M_{0,p}/M_{pl,0} = 0.2$
$\beta = 0.48$ $M_{0,p}/M_{p,0} = 0.0$ $M_{0,p}/M_{pl,0} = 0.2, 0.3$
$\beta = 0.73$ $M_{0,p}/M_{pl,0} = 0.0$ $M_{0,p}/M_{pl,0} = 0.2, 0.3$
$\beta = 0.98$ $M_{0,p}/M_{pl,0} = 0.0$

Remark: chord in-plane bending moments are assumed as negative (positive) when causing compressive (tensile) stress at the brace footprint.

3 FE MODELLING

The numerical analyses were carried out with the finite element package ABAQUS/Standard (2003). Due to symmetry in geometry and loading, only one quarter of each joint has been modelled, whereby the appropriate boundary conditions have been applied to the nodes in the various planes of symmetry.

The joints are modelled using eight-noded thick shell elements employing reduced integration (ABAQUS element S8R). Seven integration points through the shell thickness are applied. For all joints except $\beta = 0.98$, the geometry of the welds at the brace-chord intersection is modelled using shell elements. Previous research (Van der Vegte 1995) revealed that the use of shell elements to simulate the welds provides accurate predictions of the load-displacement response of axially loaded uniplanar and multiplanar T- and X-joints. The dimensions of the welds in the numerical model are in accordance with the specifications recommended by the AWS (1992).

Figure 2 shows the FE mesh modelled for the joints with $\beta = 0.48$. The FE meshes generated for the three other β values have a similar element density.

Various approaches can be identified to restrain the chord and apply chord axial load. In the approach adopted in this study, rigid beams are attached to the chord end. Chord pre-load is modelled through a concentrated force applied to the reference node located in the centre of the cross section at the chord end. Boundary conditions necessary to react the brace load are applied to the reference node as well. Because of the rigid behaviour of the beams, the chord-end cross section cannot deform, but is able to rotate. The reference node at the chord end is free to translate in horizontal direction.

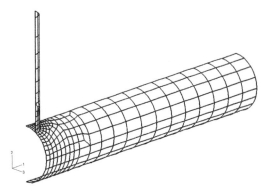

Figure 2. FE mesh generated for T-joints with $\beta = 0.48$.

4 RESULTS AND OBSERVATIONS

4.1 Definition of ultimate load

For each T-joint under axial brace load, a load-ovalisation curve has been derived, where ovalisation is defined as the vertical displacement of the crown point minus the vertical displacement of the chord bottom surface at mid-span.

In preliminary analyses, it was found that for all T-joints under zero- or compressive chord pre-load the load-ovalisation curves have a distinct peak. However, in some cases the ovalisation at peak load may exceed Lu's deformation limit of $0.03d_0$ (Lu et al. 1994). Strictly adhering to Lu's deformation limit by ignoring the peak load and assuming the "artificially" defined load at the deformation limit as ultimate load, may cause disruptions in the interaction contours. In order to avoid such "kinks", for this set of T-joints the peak loads are taken as ultimate load, even if the ovalisation at peak load exceeds the deformation limit. In general, the differences between the peak load and the load at the deformation limit of $0.03d_0$ are rather small i.e. in the order of a few percent.

For T-joints under tensile chord pre-load, not all of the load-ovalisation curves have a clear top. Hence, for the joints under tensile chord load, ultimate load is defined as the force on the brace first exceeding one of the following three failure criteria:

- peak load in the load-ovalisation diagram.
- for the joints which curves do not show a peak at ovalisation levels less than $0.03d_0$, the brace load at an ovalisation of $0.03d_0$ is taken as the "ultimate" load.
- early termination of the numerical analysis, indicating member failure.

Since the current analyses do not include tensile loads applied to the braces, no fracture failure criterion was considered.

4.2 Failure of the T-joints

For uniplanar T-joints under axial brace load, failure is caused by a combination of (local) chord face plastification and (global) chord in-plane bending failure. The contribution of each failure mode depends on the geometric parameters β and 2γ and the chord preload ratio $N_{0,p}/A_0 f_{y0}$.

For T-joints under tensile chord pre-load with small β values, chord face plastification clearly dominates failure, whereby the in-plane bending moments at failure load are relatively small and not sufficient to cause cross sectional yielding of the chord member. T-joints with large β values, subjected to tensile chord loading, however, primarily fail as a result of chord bending accompanied with pronounced strain hardening. Especially for thick-walled joints with $\beta = 0.98$ subjected to large tensile chord pre-loads, severe plasticity due to strain hardening is observed at the bottom half of the chord.

On the other hand, for compressive chord pre-load, in addition to the failure modes of chord plastification and chord bending failure, buckling or significant plastic deformation may occur in chord cross sections beside the brace adjacent to the crown point. The extent of these chord wall instabilities depends on β, 2γ and the chord pre-load ratio. While for joints with small β values, no buckling is observed, for thin-walled, large β T-joints, chord wall buckling takes place even for very low values of the chord pre-load.

4.3 Interaction contours

Figure 3 displays the interaction contours derived from the FE results obtained for uniplanar T-joints for each β value subjected to a chord pre-load of $N_{0,p}/A_0 f_{y0} = -0.3$. In interaction contours, the contribution of local joint failure (displayed on the vertical axis) and overall chord bending failure (plotted on the horizontal axis) can be easily quantified. As mentioned before, the approach of applying compensating moments to the chord end with the purpose of eliminating equilibrium induced chord bending moments provides ultimate strength values for T-joints not affected by the chord length parameter α. Furthermore, a chord length of $10d_0$ ($\alpha = 20$) appeared to be sufficiently long to exclude the influence on the ultimate strength caused by chord end conditions. Hence, the vertical axis of Figure 3 depicts the strength of each T-joint divided by the strength of the corresponding T-joint ($\alpha = 20$) without chord pre-load but with compensating chord end moments. As the in-plane bending chord moments cause compressive stresses at the brace-chord intersection, the bending moments on the horizontal axis are assumed as negative.

For circular hollow sections, the interaction between axial load and bending moment applied to a member can be approximated by the expression of

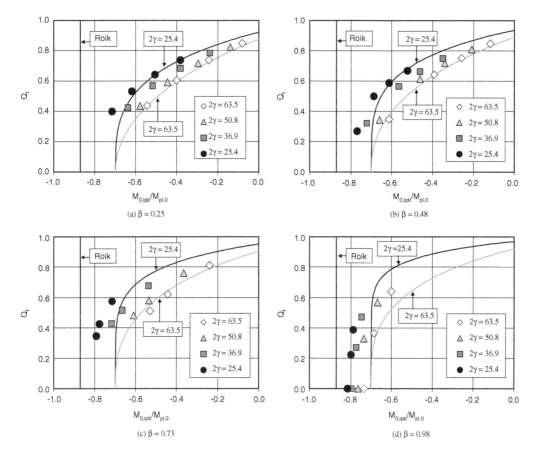

Figure 3. Interaction contours for $N_{0,p}/A_0 f_{y0} = -0.3$.

Roik (Wardenier 1982). In case of chord pre-loaded T-joints, Roik's equation is as follows:

$$\left(\frac{N_{0,p}}{N_{pl,0}}\right)^{1.7} + \frac{M_{0,ipb}}{M_{pl,0}} = 1.0 \qquad (1)$$

where $N_{0,p}$ = chord pre-load; $N_{pl,0}$ = chord squash load; $M_{0,ipb}$ = in-plane bending moment in chord; and $M_{pl,0}$ = full plastic moment capacity of chord.

The vertical line drawn in Figures 3a–d displays the maximum allowable in-plane bending chord moment according to Equation 1, corresponding to a pre-load ratio of $N_{0,p}/A_0 f_{y0} = -0.3$.

The following observations can be made from Figure 3:

– In line with the behaviour of K- and X-joints, compressive chord pre-load has a detrimental effect on the ultimate strength of axially loaded T-joints. The reduction is strength is proportional to the applied chord load.
– For increasing β values, chord in-plane bending moments at failure become larger, while at the same time, the contribution of local failure diminishes. Hence, for increasing β values, the data points tend to move to the left side of the interaction contours closer to the horizontal axis, as shown in Figure 3d.
– Smaller values of 2γ lead to relatively larger strength ratios Q_f. A similar trend was also found for axially loaded uniplanar X- and K-joints subjected to compressive chord loading.

The following observations become clear after examining all contours, i.e. not only those for $N_{0,p}/A_0 f_{y0} = -0.3$:

– Tensile chord pre-load significantly strengthens the T-joints in comparison to compressive chord pre-load. For tensile chord pre-load, the deteriorating effect of chord buckling is not observed, while chord strengthening due to membrane action becomes active.
– For T-joints under compressive chord pre-load, all data points fall within the area limited by Roik's formula. Even for the selected T-joints with $\beta = 0.98$ loaded by shear force applied to the chord

119

end only (without brace load), the in-plane bending strength will not reach Roik's limit as shown in Figure 3d. The combination of compressive chord pre-load and in-plane bending moment causes the chord to "buckle" prematurely, even for the relatively thick walled chords with $2\gamma = 25.4$. For T-joints subjected to zero chord pre-load, it is found that for low 2γ values, the moment ratio $M_{0,ipb}/M_{pl,0}$ may exceed -1.0, primarily due to strain hardening. For joints under tensile chord pre-load, membrane action causes the chord to strengthen significantly. As a result, in many cases, the data points may be located beyond Roik's limit, also for the thin walled joints with $2\gamma = 63.5$.

– Another point of interest is the question whether equilibrium induced in-plane bending moments have a similar effect on the ultimate strength as externally applied chord end moments. An answer to this question can be obtained by examining the chord stress contour for T-joints with $\beta = 0.48$ and $N_{0,p}/N_{pl,0} = 0.0$, shown in Figure 4. This graph indicates that for the joints with $2\gamma = 63.5$, no difference in chord stress effect can be observed between the "ordinary" T-joint (i.e. the T-joint without externally applied in-plane bending moments) and the data where in-plane bending pre-moments are applied to the chord end, as indicated in Table 2. Similar results are obtained for T-joints with different β and 2γ values or subjected to different chord pre-stress ratios.

5 ULTIMATE STRENGTH FORMULATION

5.1 General formulation

The general format of the capacity equation to describe the ultimate strength of tubular joints is as follows:

$$N_{1,u} = \frac{f_{y0} t_0^2}{\sin \theta} Q_u Q_f \tag{2}$$

The strength factor Q_u expresses the strength of uniplanar T-joints with compensating bending moments applied to the chord end. Based on the complex formulation derived from the analytical "ring model" approach, Van der Vegte & Makino (2005) adopted the following expression for the strength factor Q_u:

$$Q_u = \frac{A}{1 - B\beta} \gamma^{C\beta + D\beta^2} \tag{3}$$

The coefficients A, B, C and D were determined by regression analyses. Table 3 summarizes the values derived for axially loaded T-joints with compensating chord end moments. Figure 5 displays the available FE data points as a function of β as well as the predictions for $2\gamma = 25.4$ and 63.5 obtained with Equation 3.

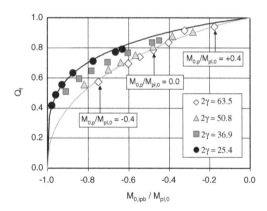

Figure 4. Interaction contour for $\beta = 0.48$ and $N_{0,p}/A_0 f_{y0} = 0.0$.

In general, various formulations such as linear equations or parabolic functions can be used to describe the influence of chord stress. In order to derive functions that are applicable to multiple joint configurations of circular and rectangular hollow sections and load types, the following equation is selected as basis for the chord stress function Q_f:

$$Q_f = (1.0 - |n|^{F1})^{(G+H\beta+K\gamma)} \tag{4}$$

with n defined as:

$$n = \frac{N_{0,p}}{N_{pl,0}} + \left(\frac{M_{0,ipb}}{M_{pl,0}}\right)^{F2} \tag{5}$$

5.2 Chord stress function Q_f for compressive chord pre-load

For T-joints subjected to compressive- or zero chord pre-load, a total of 251 T-joint analyses without compensating chord end moments were made, summarized in Table 4. The 95 analyses for zero chord pre-load not only include the analyses by Van der Vegte & Makino (2005), but also 15 T-joint analyses made by Van der Vegte (1995).

It is noticed that with the proposed format, the chord stress function becomes very steep when the formulation approaches the horizontal axis. Data points in this region (i.e. joints subjected to large in-plane bending moments) may exhibit large errors in percentage terms, having a strong effect on the final equation. Hence, for compressive and zero chord pre-load, two sets of data are considered:

– The largest set (206 out of the 251 available data points) includes all data that allow the regression analyses to proceed. Data causing the regression analyses to abort, due to mathematical errors, are excluded. For example, for $N_{0,p}/N_{pl,0} = 0.0$ the proposed chord stress function is not able to

Table 3. Results of the regression analyses for uniplanar T-joints.

General format: $\dfrac{N_{1,u}\sin\theta}{f_{y,0}t_0^2} = Q_u Q_f$

Reference strength function: $Q_u = \dfrac{A}{1-B\beta}\gamma^{C\beta+D\beta^2}$

			Q_u			
No. of data	Mean	CoV (%)	A	B	C	D
33	1.000	4.4	3.9	0.8	0.85	−0.55

Chord stress function: $Q_f = (1.0 - |n|^{F1})^{(G+H\beta+K_\gamma)}$ with $n = \dfrac{N_{0,p}}{N_{pl,0}} + (\dfrac{M_{0,ipb}}{M_{pl,0}})^{F2}$

	No. of data	Mean	CoV (%)	F1	F2	G	H	K
Compressive and zero chord pre-load								
1 – exact	141	1.001	4.7	1.14	0.939	0.213	−0.224	0.0090
2 – exact	206	0.988	18.0	1.42	1.000	0.219	−0.270	0.0132
3 – simplified	141	0.996	5.4	1.2	1.0	0.24	−0.26	0.01
4 – simplified	206	0.965	18.4					
Tensile chord pre-load								
Insufficient data. Use chord stress function for X-joints subjected to tensile chord pre-load								
5 - simplified				1.7	2.0	0.22	0.0	0.0

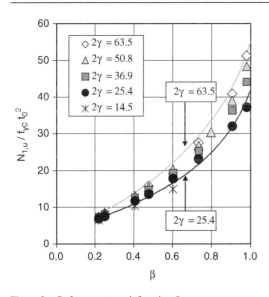

Figure 5. Reference strength function Q_u.

produce a value if $M_{0,ipb}/M_{pl,0} < -1.0$. For the other pre-stress ratios, $M_{0,ipb}/M_{pl,0}$ should be limited in a similar way.

– A second set includes the relevant data points only. Data points where the chord stress contours become steep are excluded. Furthermore, for thin-walled T-joints with large β values and $\alpha = 12$ or 16, chord end conditions may enhance the strength of the joints. As this effect is limited to these few configurations, these data are removed from the regression analyses as well. Hence, 141 data points remain.

Rows 1 and 2 of Table 3 show the results of the regression analyses for both sets of data. The CoV of the regression analyses for the 206 data points is much worse than for the 141 data points. As mentioned, this is not caused by a possible poor formulation but primarily due to the steep decline of the chord stress function for large chord stress ratios, resulting in large deviations in terms of percentage between the actual and predicted values.

As observed from the two sets of regression results, the regression coefficients can vary within a certain range. Simplifying the overall equation, by fixing the regression coefficients F2 = 1.0 and K = 0.01, and re-analysing F1, G and H using the 141 relevant data points, gives the results shown in rows 3 and 4 of Table 3. The CoV's are well in line with the values obtained for the exact solution. Figure 3 displays the chord stress function Q_f for $2\gamma = 25.4$ and 63.5 as well as all FE data points derived for $N_{0,p}/A_0 f_{y0} = -0.3$. The contour lines drawn in Figure 4 represent the chord stress function Q_f for $2\gamma = 25.4$ and 63.5 for $N_{0,p}/A_0 f_{y0} = 0.0$.

5.3 Chord stress function Q_f for tensile chord pre-load

For tensile chord pre-load, a total of 64 T-joint analyses are made, summarized in Table 4. No analyses are

Table 4. Database generated for T-joints without compensating chord end moments pre-loaded by chord axial load and in-plane bending moment.

Brace load	Chord pre-load	Chord pre-load ratio $N_{0,p}/N_{pl,0}$	α	No. of analyses
Axial compression	Zero pre-load	0.0	12, 16, 18, 20, 24, 28	95
Axial compression	Axial compression	$-0.9, -0.8, -0.6, -0.3$	20	64
Axial compression	Axial compression + In-plane bending moment	See Table 2	20	92
Axial compression	Axial tension	$+0.3, +0.6, +0.8, +0.9$	20	64

made for chord pre-tensioning in combination with externally applied chord moments.

As mentioned before, for circular hollow sections, the interaction between axial load and bending moment applied to the chord member can be approximated by the expression of Roik (Equation 1). Only 13 of the 64 data generated for T-joints fall within the contour defined by Roik's expression, indicating that the 51 remaining data points refer to member failure, which are to be excluded from the regression analyses. The number of valid data is considered to be insufficient to develop a chord stress function for T-joints under tensile chord pre-load. However, it is observed that the chord stress function for X-joints subjected to tensile pre-load (Van der Vegte, unpubl.) gives conservative predictions for the 13 T-joint data. It is recommended to use this chord stress function for T-joints under tensile chord pre-load as well.

6 SUMMARY AND CONCLUSIONS

Finite element analyses are conducted on uniplanar T-joints under brace compression after having been pre-loaded by axial chord load (compression or tension). Sixteen combinations of the geometric parameters β and 2γ are analysed for nine values of the chord pre-load ratio $N_{0,p}/A_0 f_{y0}$. In addition, for selected pre-load ratios, external in-plane bending moments are applied to the chord end. Based on the results, the following conclusions can be drawn:

- Compressive chord pre-load has a detrimental effect on the ultimate strength of axially loaded T-joints, in line with the behaviour found for K- and X-joints. The reduction is strength is proportional to the applied chord load. Tensile chord pre-load significantly strengthens the T-joints in comparison to compressive chord pre-load.
- For T-joints under zero-or compressive pre-load, in-plane chord bending moments at failure load may not reach the limit strength obtained with Roik's interaction formula. For the majority of the T-joints under tensile load, the brace load at failure causes the chord bending moment to exceed Roik's limit, indicating member failure.

- For axially loaded T-joints, equilibrium induced in-plane bending moments or externally applied in-plane bending moments have the same effect on the ultimate strength of the joints.
- The reference strength function Q_u and chord stress function Q_f summarized in Table 3 are proposed to describe the ultimate strength of uniplanar T-joints subjected to axial chord pre-load. For T-joints under tensile chord pre-load, it is recommended to use the chord stress function derived for X-joints.
- The formats of the reference strength- and chord stress functions can also be used as a basis for the ultimate strength equations of uniplanar K- and X-joints subjected to chord pre-load. The results of the strength analyses on K- and X-joints will be published in the near future.

REFERENCES

ABAQUS/Standard 2003. Version 6.4, Hibbitt, Karlsson and Sorensen, Inc, USA.
American Petroleum Institute 1993. Recommended practice for planning designing and constructing fixed offshore platforms – Working stress design, API RP2A-WSD, 20th Edition.
American Petroleum Institute 2003. Proposed updates to tubular joint static strength provisions in API RP2A 21st Edition.
American Welding Society 1992. Structural Welding Code, AWS D1.1-92.
Choo, Y.S., Qian, X.D., Liew, J.Y.R. & Wardenier, J. 2003. Static strength of thick-walled CHS X-joints – Part II. Effect of chord stresses, *Journal of Constructional Steel Research*, Vol. 59, No. 10, pp. 1229–1250.
Liu, D.K. & Wardenier, J. 2006. New chord stress functions for RHS gap K-joints, *to be presented at the 11th International Symposium on Tubular Structures*, Quebec City, Canada.
Lu, L.H., Winkel, G.D. de, Yu, Y. & Wardenier, J. 1994. Deformation limit for the ultimate strength of hollow section joints, *Proc. of the 6th International Symposium on Tubular Structures*, Melbourne, Australia, pp. 341–347.
Pecknold, D.A., Park, J.B. & Koppenhoefer, K.C. 2001. Ultimate strength of gap K tubular joints with chord preloads, *Proc. of the 20th International Conference on Offshore Mechanics and Arctic Engineering*, Rio de Janeiro, Brazil.

Vegte, G.J. van der 1995. The static strength of uniplanar and multiplanar tubular T- and X-joints, *Doctoral Dissertation*, Delft University of Technology, Delft, the Netherlands, Delft University Press, ISBN 90-407-1081-3.

Vegte, G.J. van der, Makino, Y. & Wardenier, J. 2002. The effect of chord pre-load on the static strength of uniplanar tubular K-joints, *Proc. of the 12th International Offshore and Polar Engineering Conference*, Kitakyushu, Japan, Vol. IV, pp 1–10.

Vegte, G.J. van der, Makino, Y. & Wardenier, J. 2005. The influence of boundary conditions on the chord load effect for CHS gap K-joints, *Proc. of the ECCS-AISC Workshop: Connections in Steel Structures V*, Amsterdam, the Netherlands, pp. 433–443.

Vegte, G.J. van der & Makino, Y. 2005. Ultimate strength formulation for axially loaded CHS uniplanar T-joints, *Proc. of the 15th International Offshore and Polar Engineering Conference*, Seoul, Korea, Vol. IV, pp. 279–286.

Wardenier, J. 1982. Hollow section joints, *Doctoral Dissertation*, Delft University of Technology, Delft, the Netherlands, Delft University Press, ISBN 90-6275-084-2.

Tubular Structures XI – Packer & Willibald (eds)
© 2006 Taylor & Francis Group, London, ISBN 0-415-40280-8

Experimental study on overlapped CHS K-joints with hidden seam unwelded

X.Z. Zhao, Y.Y. Chen, Y. Chen & G.N. Wang
Tongji University, Shanghai, China

L.X. Xu, R.Q. Zhang & B.C. Tang
Shanghai Baoye Construction Corp. LTD, Shanghai, China

ABSTRACT: Overlapped Circular-Hollow-Section K-joints are widely used in large span structures in China. In practice, the hidden toe of this type of joints is normally left unwelded due to the difficulty resulted from the fabrication sequence. It is thus necessary to investigate how the unwelded hidden seam may affect the behaviour of overlapped CHS K-joints. This paper reports seven static tests on uniplanar overlapped CHS K-joints with hidden seam unwelded under axial loading. Among these 7 specimens, parameters including the overlapped ratio, the brace-to-chord diameter ratio, the brace-to-chord thickness ratio, the brace inclination, the throat thickness of fillet welds, the unbalanced loading at the joint as well as the loading hierarchy reversal were varied to study their effect on the joint behaviour. In order to study the effect of the hidden seam on the behaviour of the connection, three additional specimens with the hidden seam welded were also tested for comparison purposes. Results of these tests show that the welding situation of the hidden seam has some effect on the stress distribution and failure mechanism, but the static ultimate capacity of the overlapped CHS K-joints is not affected significantly given that the through brace is under compression. When the brace-to-chord thickness ratio is smaller, the local failure is the main failure mechanism observed. Comparison of the joint capacities obtained from the tests with strength prediction from current design codes was presented.

1 INTRODUCTION

In the past decade, Circular Hollow Sections (CHS) have been widely used in large span structures in China. This largely owes to the excellent properties of the CHS shape with regard to loading in torsion, compression and bending in all directions, the attractive shape in appearance and the rapid development of special CHS end preparation machine. To guide designers to use CHS properly, considerable tests have been carried out on uniplanar joints in order to investigate the behaviour of the joints and to establish formulae for static strength calculation. However, little substantial information on design, fabrication, assembly and erection of overlapped K-joints is available from current international database for tubular connections. For partially overlapped CHS K-joints, it is a common practice to mount the CHS members on an assembly rig and tack welds them in fabrication workshops. Final welding is then carried out in a following separated operation. This sequence makes it impossible to weld the hidden toe of overlapped K-joints. In tubular structures, this type of joints is widely used. It is thus

necessary to investigate what parameters and how they affect the behaviour of overlapped CHS K-joints with unwelded hidden seam.

2 DESCRIPTION OF OVERLAPPED CHS K-JOINT

Overlapped CHS K-joints are one of the fundamental joint configurations used in truss structures. The configuration and notation describing the geometry of overlapped K-joints are shown in Figure 1. The through brace is welded entirely to the chord, whereas the overlapping (lap) brace is welded to both the chord and the through brace. In an overlapped joint, all or part of the load is transferred directly between the two braces through their common weld, which may lead to a reduction in the chord wall thickness required.

Compared with gap K-joints, more factors, including loading hierarchy reversal (i.e. tension or compression in the through or lap brace) and the presence or absence of the hidden weld, need to be considered for overlapped K-joints. Dexter (Dexter & Lee, 1999a)

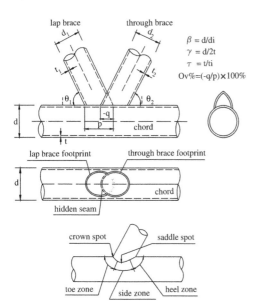

Figure 1. Configuration of overlapped CHS K-joint.

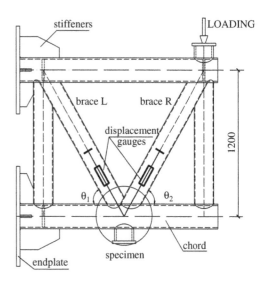

Figure 2. Overlapped CHS K-joint specimen.

listed the areas requiring further in-depth investigation for an overlapped K-joint:

(1) The effects of varying the joint geometry, including the parameter β and γ, which are generally accepted to affect the strength of typical joints, and the additional parameter τ.
(2) The effects of the relative material strengths of the braces and the chord.
(3) The effects of loading hierarchy.
(4) The effects of the presence or absence of the hidden seam.

The experiments and analysis presented in this paper concentrated on item(1), (3) and (4) given above.

3 TESTS OF OVERLAPPED K-JOINTS

3.1 Test specimens

In total twelve specimens of uniplanar overlapped CHS K-joints were tested. Nine of them are absence of the hidden seam; and for comparison purpose, the other three are presence of the hidden seam to study the effect of hidden seam on the behaviour of the connection. Among the nine specimens with unwelded hidden seam, 7 were subjected to static axial loading, whereas cyclic loading were applied to the other two, which are not in the scope of this study.

A typical CHS overlapped K-joint specimen is shown in Figure 2. To simulate the loading pattern that may occur within a K-joint in the framework, the specimen was designed to be part of a cantilever truss.

Because the chord and the brace axes for all specimens meet at one point, there is no eccentricity in joints. To minimize the effects of the secondary moments in the joints, the chord/brace length is designed to be at least seven times of the chord/brace diameter. End profiles of braces were prepared by automatic flame cutting. Weld details were in accordance with the Technical Specification for Welding of Steel Structures of Building JGJ81-2002 (China); but the throat thickness of the fillet weld was varied for different specimens. The details of all the specimens are given in Table 1. The material properties of the chord and braces are listed in Table 2.

As shown in Table 1, specimens SJ1 ~ SJ3, specimens SJ4 ~ SJ5, and SJ6 ~ SJ12 have the same size of chord/braces, respectively. The three specimens which have the hidden seam welded are SJ1, SJ4 and SJ6. Except specimens SJ3 and SJ8 which were subjected to cyclic loading, the others were tested under static loading. Specimen SJ9 had unbalanced load applied, the fillet weld in SJ10 had a smaller throat thickness, while SJ11 had a different overlap ratio and SJ12 a different loading hierarchy reversal.

3.2 Test setup

The specimens were tested using the truss-system facility housed in the State Key Laboratory for Disaster Reduction in Civil Engineering (Tongji University, China), as shown in Figure 3. The overlapped CHS-K joint specimen is part of a cantilever truss, which was fixed to a long truss beam through its end plates. The vertical load was applied to the end of the cantilever truss by a 320 ton hydraulic jack with a spreader beam. As for specimen SJ9, the unbalanced load at the K-joint was applied by a 100 ton hydraulic jack directly seated on the solid test floor.

126

Table 1. Details of test specimens.

No.	d × t	$d_i × t_i$	θ_i	Ov(%)	β	γ	τ	Throat thickness	Condition of hidden seam	Loading hierarchy
SJ1	$\phi 203 × 12$	$\phi 159 × 10$	60	36	0.783	8.458	0.833	$1.5t_i$	Welded	TBIC*
SJ2	$\phi 203 × 12$	$\phi 159 × 10$	60	36	0.783	8.458	0.833	$1.5t_i$	Unwelded	TBIC
SJ3	$\phi 203 × 12$	$\phi 159 × 10$	60	36	0.783	8.458	0.833	$1.5t_i$	Unwelded	TBIC
SJ4	$\phi 203 × 12$	$\phi 168 × 8$	60	40	0.828	8.458	0.667	$1.5t_i$	Welded	TBIC
SJ5	$\phi 203 × 12$	$\phi 168 × 8$	60	40	0.828	8.458	0.667	$1.5t_i$	Unwelded	TBIC
SJ6	$\phi 203 × 12$	$\phi 168 × 6$	60	40	0.828	8.458	0.500	$1.5t_i$	Welded	TBIC
SJ7	$\phi 203 × 12$	$\phi 168 × 6$	60	40	0.828	8.458	0.500	$1.5t_i$	Unwelded	TBIC
SJ8	$\phi 203 × 12$	$\phi 168 × 6$	60	40	0.828	8.458	0.500	$1.5t_i$	Unwelded	TBIC
SJ9	$\phi 203 × 12$	$\phi 168 × 6$	60	40	0.828	8.458	0.500	$1.5t_i$	Unwelded	TBIC
SJ10	$\phi 203 × 12$	$\phi 168 × 6$	60	40	0.828	8.458	0.500	$1.2t_i$	Unwelded	TBIC
SJ11	$\phi 203 × 12$	$\phi 168 × 6$	50	22	0.828	8.458	0.500	$1.5t_i$	Unwelded	TBIC
SJ12	$\phi 203 × 12$	$\phi 168 × 6$	60	40	0.828	8.458	0.500	$1.5t_i$	Unwelded	TBIT**

* TBIC = through-brace-in-compression; ** TBIT = through-brace-in-tension.

Table 2. Results of the tensile coupon tests.

Specimen Nominal size	Yield strength (N/mm^2)	Ultimate strength (N/mm^2)	Elongation (%)	Grade of Steel
Chord ($\phi 203 × 12$)	277.3	476.9	34.5	Q235B
Brace ($\phi 159 × 10$)	327.3	465.0	25.7	Q235B
Brace ($\phi 168 × 8$)	310.5	458.4	29.5	Q235B
Brace ($\phi 168 × 6$)	278.0	473.5	27.2	Q235B

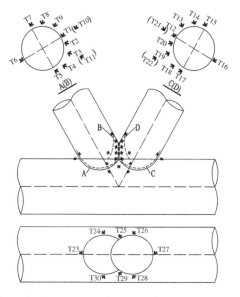

Figure 4. Arrangement of rosette strain gauges.

The deformations of the chord wall under axial brace loadings were measured using four wire displacement gauges (D1~D4) attached to bolts welded at the intersection of axes of the chord/braces and at the central line of each brace (as shown in Figure 2). Another five displacement gauges were placed vertically and horizontally along the chord to obtain the vertical displacement of the end of the cantilever truss and the out-of-plane deflections of the truss.

Figure 3. Test setup.

3.3 Measurement program

In total 30 rosette strain gauges were applied at the intersections of chord and braces, as shown in Figure 4. In order to obtain the actual loads applied in the chord and the braces, 12 strain gauges were attached on the chord/braces surfaces along the length of each member.

4 TEST RESULTS OF OVERLAPPED K-JOINTS

4.1 Brace load-displacement relationships

Figure 5 shows typical brace load-displacement plots. The load, N, is the internal force in the braces

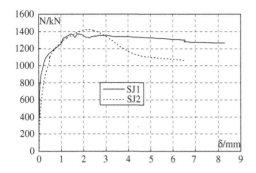

Figure 5a. Tensile brace load-deformation plot of SJ1 & SJ2.

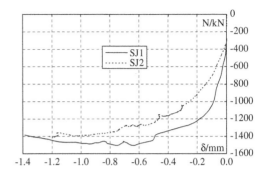

Figure 5b. Compressive brace load-deformation plot of SJ1 & SJ2.

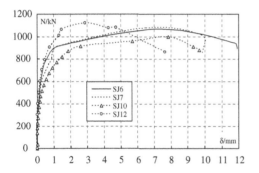

Figure 5c. Tensile brace load-deformation plot of SJ6, SJ7, SJ10 & SJ12.

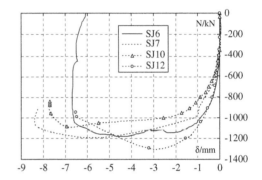

Figure 5d. Compressive brace load-deformation plot of SJ6, SJ7, SJ10 & & SJ2.

calculated from a semi-rigid analytical model which have been verified by the measured values in the braces; the displacement is the average deformation of the chord wall under the respective through/lap brace loading in the direction of the original brace axis, calculated from the measured value of the wire displacement gauges deducting the elastic deformation of the braces. Tensile force in the brace is nominated to be positive and compressive force negative.

4.2 Strain distribution

Strain distribution near the welds between braces and chord are shown in Figure 6 in the form of the equivalent strain versus the rosette strain gauge number shown in Figure 4. The strain distribution of the specimen SJ6 with hidden seam welded is shown by the solid line and SJ7 with hidden seam unwelded by the dashed line.

It can be seen that: (1) While the internal forces of the braces were about one-third of the design resistance of the overlapped K-joint specified by Eurocode3, the intersection point (IP) of the through brace, lap brace and the chord, reached the yield stress of the steel; and the equivalent strain at the IP, obtained from the rosette stain stuck on the outer surface of the tube, remains to

be the largest during the whole loading process. (2) The strain gradient at the crown spots of the braces does not vary significantly (T1 & T10, T12 & T21). However, the longer the distance away from the crown spot, the steeper the gradient. (3) The strain at the saddle spot is larger than that at the crown spot. (4) Plasticity first appeared at IP, then the intersection line between the through brace and the lap brace; and finally the chord near the intersection point. This implies that part of the load in the braces is transferred directly between the two braces through their common weld. (5) Strain in the brace of the specimen SJ7 is larger than that of specimen SJ6 with hidden seam welded.

4.3 Failure modes

Various failure modes were observed in the tests: (1) local buckling of the compressive brace near the common weld, and the necking of the tensile brace (2) cracks, starting from the IP then propagated along the welds. Figure 7a shows an example of the second type of failure, where specimen SJ1 & SJ2 with relative larger τ cracked at the IP just after the peak load, the cracks then propagated in the whole welds and the lap brace was finally broken off from the joint. For specimen SJ4 with smaller τ, local buckling occurred

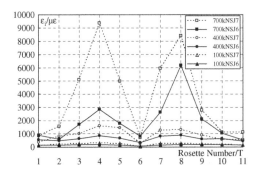

Figure 6a.　Strain distribution of lap brace of SJ6 & SJ7.

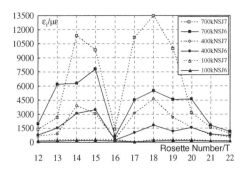

Figure 6b.　Strain distribution of through brace of SJ6 & SJ7.

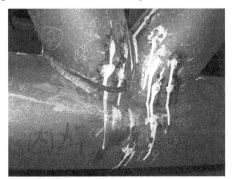

Figure 7a.　Failure mode of SJ1.

Figure 7b.　Failure mode of SJ14.

Figure 7c.　Failure mode of SJ5.

Figure 7d.　Failure mode of SJ12.

in the through brace before cracks appeared at the IP, as shown in Figure 7b. It should be noted that local buckling is not a member failure mode, instead, the buckling was caused by geometrical stress concentration. Upon the happening of the local buckling failure, the panel at the IP was forced out-of the joint, as shown in Figure 7c. For specimen SJ12 with through-brace-in-tension, a large local buckling had developed in the compressive lap brace before the cracks along the whole weld between the through brace and the chord occurred suddenly; moreover, the cracks extended to the lap brace as shown in Figure 7d.

4.4　Ultimate capacity of overlapped K-joints

Three potential failure criteria were considered and checked for each joint to get the ultimate capacity: (1) the peak load on the brace load-displacement plot; (2) deformation reaches its limit, i.e. 3% of the chord diameter; (3) visually observable crack initiates. The ultimate capacity of joints is defined as the lowest load obtained from (1), (2) and (3). The ultimate capacity of each specimen, for both the through brace and the lap brace, is listed in Table 3.

Table 3. Measured and predicted specimen ultimate capacity.

| No. of specimen | Test ultimate capacity/kN | | Ratio of prediction to test ultimate capacity | | |
	Through brace	Lap brace	GB50017	Euro-code3	AWS
SJ1	−1507	1373	0.68	0.67	0.79
SJ2	−1407	1409	0.66	0.66	0.78
SJ4	−1330	1209	0.85	0.81	0.95
SJ5	−1193	1060	0.95	0.90	1.06
SJ6	−1181	1043	0.97	0.92	1.08
SJ7	−1189	1071	0.95	0.91	1.07
SJ9	−1355	1268	0.87	0.84	1.02
SJ10	−1069	976	1.06	1.02	1.20
SJ11	−1259	1306	0.89	0.84	0.99
SJ12	1121	−1290	1.00	0.97	1.18

4.5 Comparison with formulae predictions

A comparison of the measured ultimate capacity for each joint with the formulae predictions from GB50017 (2003), Eurocode (1992) and AWS (2004) are shown in Table 3. The measured dimensions and material properties were used for calculating the ultimate capacities by these code predictions with no safety factor considered. Note that (1) for specimen SJ1 & SJ2 with larger τ, the ultimate capacity of the joint increased dramatically due to the effective load transfer between the two braces through the common weld, resulting a larger measured joint capacity than code predictions; (2) for specimen SJ6 to SJ12 with smaller τ, the measured joint capacity agrees well with the code predictions. However, the joint tested failed by local buckling while the current code prediction is based on chord face failure.

5 COMPARISON BETWEEN TESTS AND FEM ANALYSIS

Elasto-plastic large displacement analysis was carried out on 10 joint specimens tested using the general purpose FE package ANSYS to obtain the stress distribution and load-carrying capability of the joints.

5.1 Finite element model

Element and mesh: 3-D 8-node solid element (Brick45), which also supports large displacement, was used in this analysis. As the joint was symmetric about the centre line of the tubes, only half of the joint was modeled. Figure 8a shows the final mesh adopted. Finer mesh was applied to the chord-braces joining area where the stress distribution is very complex and is of main interest of this study. For the specimens having welded hidden seam, the chord and the through brace were connected as a whole by the weld.

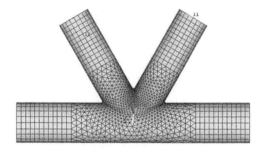

Figure 8a. Element and meshes.

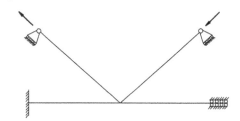

Figure 8b. Boundary conditions and loading mode.

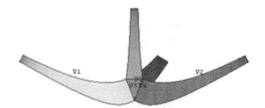

Figure 8c. Weld modeling.

Material property: The yield strength, elastic modulus of the steel as well as the target modulus obtained from the tensile coupon tests were used in the FE analysis. Bi-linear kinematic strain hardening model was adopted. Self-weight was ignored as it has little effect on the joint behaviour compared with other external loads applied.

Boundary conditions: The left end of the chord was fully fixed. As the right end was subjected to moment and deformed axially, roller support was applied. The braces were subjected to mainly axial force, and in addition, axial displacement was allowed, so that pin support was adopted, as shown in Figure 8b. The length of all the chord and braces measured from their joining point was three times of the tube diameter in order to eliminate the local effect of the applied load.

Weld modeling: As shown in Figure 8c, the weld detail was also included in the FE model generated in this paper to take into account the effect of the weld on the joint behaviour. The dimensions of the weld were the same as that measured from the experiment. The material properties of the weld were assumed to be the same as the parent metal. Residual stresses induced

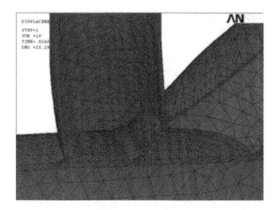

Figure 9. Failure mode by numerical simulation.

Table 4. Comparison of test joint capacity with FE.

No. of specimen	Test results kN	FE results kN	Error %
SJ1	1373	1321	−3.8
SJ2	1407	1330	−5.6
SJ4	1209	1119	−7.3
SJ5	1060	1025	−3.1
SJ6	1043	1032	−1.0
SJ7	1071	981	−8.4
SJ9	1268	1135	−10.3
SJ10	976	939	−3.7
SJ11	1259	1166	−6.9
SJ12	1121	1050	−6.3

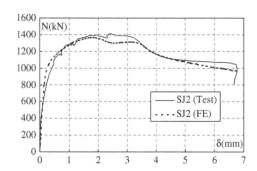

Figure 10a. Comparison of lap brace load-displacement curve.

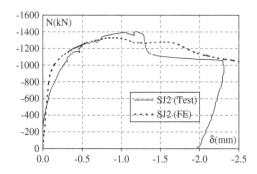

Figure 10b. Comparison of through brace load-displacement curve.

from the welding process were not considered in this analysis.

5.2 FE results and comparison with test results

Figure 9 illustrates the failure mode of brace local buckling and necking of the tensile brace obtained from FE analysis. The comparisons of the brace load-displacement plot and the ultimate capacities obtained from both the test and FE are shown in Figure 10 and

Table 4, respectively. Good agreement is achieved not only in the ultimate loads but also in the shape of the curve and the failure modes. The accuracy of the finite element model generated is thus validated.

6 DISCUSSION

Based on the validated FE model, parametric study was performed on 81 models. Together with the test results on 10 specimens, the influence of the geometry, the hidden weld absence, the loading hierarchy and throat thickness of welds can be studied.

6.1 Influence of geometry

Parameter β and θ affect the behaviour of the overlapped joint in the similar way as to gap K-joint. However, both the test and FE analysis results show that parameter γ, which dominates the ultimate capacity of gap K-joint, has less effect on the ultimate capacity of overlapped K-joint. On the contrary, the brace-to-chord thickness ratio τ, (1) for a relatively small $\tau(<0.4)$, joints are found to fail by brace member failure (BMF) or brace local buckling (BLB), and the overlap of braces has no contribution to the joint capacity; (2) for an intermediate τ, failure modes of brace local buckling (BLB) or brace local buckling and chord plasticity (BLB-CP) have been identified, and the ultimate capacity of overlapped joints is generally larger than that of gap joints; (3) for a large $\tau(\geq 1.0)$, only the BLB-CP failure mode is found, and the ultimate capacity of overlapped joints increases dramatically compared with gap joints.

6.2 Influence of hidden weld absence

Current China design code stipulates that all brace intersections in overlapped K-joints should be fully welded, whereas Eurocode 3 allows the hidden part of the connection left unwelded. Test results of comparative specimens SJ1 (presence of hidden weld, PHW)

131

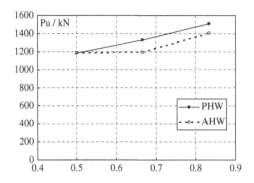

Figure 11. Joint capacities with presence/absence of hidden weld.

and SJ2 (absence of hidden weld, AHW), SJ4 (PHW) and SJ5 (AHW), SJ6 (PHW) and SJ7 (AHW), showed that the ultimate capacity of overlapped K-joints is not generally affected by leaving the hidden weld out, as shown in Figure 11, though the failure mode of each comparative tests were different. CIDECT design guide (Dutta, 1998) suggests that if the vertical components of load in the two braces are different by more than 20%, the hidden weld has to be made. Results of specimen SJ9 agree with this statement.

6.3 Influence of loading hierarchy reversal

CIDECT design guide (Dutta, 1998) stipulates that in the case of the diameters and wall thickness of the two braces being of the same order, the tensile member should be welded to the chord first (TBIT) and the compression member shall partially cover the tensile one by welding. But the test and FE results do not agree this well, especially under the condition of leaving the hidden weld out. Although the ultimate capacity of specimen SJ12 (TBIT, and AHW) is larger than that of specimen SJ7 (TBIC, and AHW), the SJ7-like joints are still preferred in practice due to its relatively larger plasticity, compared with the sudden failure as the SJ12-like joint. Further FE results indicate that (1) for PHW joints, while the overlap ratio is small or intermediate, the capacity of TBIT joint is slightly higher than that of TBIC joint as shown by the test result, whereas for relatively large overlap ratio, the TBIT joint is poorer than the TBIC joint; (2) for AHW joints, the ultimate capacity of TBIT joint decreases dramatically, which is thus not recommended to be used in practice.

6.4 Influence of throat thickness of welds

The design resistance of the weld should not normally be less than the design tension resistance of the cross-section of the member to prevent the premature joint failure. But as shown in Figure 6, the strain distribution across the weld is highly non-uniform, with the peak value being three times higher than that indicated on the basis of the nominal geometry. Thus the throat thickness should be enlarged to prevent "unzipping" or progressive failure of the welds and ensure the redistribution of the uneven stress through local yielding and reach their design capacity. Specimen SJ10 with less throat thickness failed prematurely. For overlapped K-joint, a relative larger throat thickness, basically at least one and a half times of the thickness of the thinner brace member is recommended.

7 CONCLUSIONS

Experimental and FE studies on overlapped CHS K-joint with hidden seam unwelded were carried out and presented in this paper. The study concentrated on the effect of varying the joint geometry, the loading hierarchy as well as the presence or absence of the hidden weld on the behaviour of the joints. Results show that the welding situation of the hidden seam has some effect on the stress distribution and failure mechanism, but the static ultimate capacity of the overlapped CHS K-joints is not affected significantly given that the through brace is under compression.

Results obtained from the generated FE model are found to have good agreement with the experimental measurements. Based on the validated FE model, a more detailed parametric study and design recommendations could be carried out and will be presented in the future.

REFERENCES

AWS D1.1/D1.1 M: 2004. *Structural Welding Code – Steel (19th Edition)*. American Welding Society.

BS EN 1993-1-8: 2005. *Eurocode 3: Design of steel structures – Part1-8: Design of joints*. British Standards Institution.

Dexter, E. & Lee, M. 1999a. Static Strength of Axially Loaded Tubular K-Joints I: Behaviour. *Journal of Structural Engineering, ASCE*, 125(2): 194–201.

Dexter, E. & Lee, M. 1999b. Static Strength of Axially Loaded Tubular K-Joints II: Ultimate Capacity. *Journal of Structural Engineering, ASCE*, 125(2): 202–210.

Dutta, D., Wardenier, J., Yeomans, N., Sakae, K., Bucak, O. & Packer, J. 1998. *Design Guide for Fabrication, Assembly and Erection of Hollow Section Structures*. CIDECT.

Gazzola, F., Lee, M. & Dexter, E. 2000. Design Equations for Overlap Tubular K-Joints under Axial Loading. *Journal of Structural Engineering, ASCE*, 126(7): 798–808.

GB50017: 2003. *Code for Design of Steel Structures*. National Standards of P.R.China.

Makino, Y. & Kurobane, Y. 1994. Tests on CHS KK-Joints under Anti-symmetrical Loads. In R. Puthli & S. Herion (ed.), *Proceedings of the Sixth International Symposium on Tubular Structures, Australia, 14–16 December 1994*. Rotterdam: Balkema.

Wardenier, J. 2002. *Hollow Sections in Structural Applications*. CIDECT.

Tubular Structures XI – Packer & Willibald (eds)
© 2006 Taylor & Francis Group, London, ISBN 0-415-40280-8

Test results for RHS K-joints with 50% and 100% overlap

T. Sopha & S.P. Chiew
Nanyang Technological University, Singapore

J. Wardenier
Delft University of Technology, The Netherlands

ABSTRACT: This paper details the experimental investigation and testing of three (3) large-scale overlap square hollow section K-joint specimens with the following joint parameters: $0.6 \leq \beta \leq 0.75$; $30 \leq 2\gamma \leq 35$ and $O_v = 50\%$ and 100%. The experimental results showed for these overlap joints a chord member yield capacity failure due to axial load and bending moment at the joint location. In one specific case for the 100% overlap joint, this failure mode was combined with an overlapped brace ultimate shear failure. Based on this study and earlier reported numerical works it is recommended to improve the design of the overlapped RHS K-joints by checking also against failures due to the overlapped brace shear and local chord yield criterions in addition to the current overlapping brace effective width criterion.

1 INTRODUCTION

The CIDECT programme 5BN deals with a detailed investigation of chord failures of rectangular hollow section (RHS) K-joints with 50% and 100% overlap. The total programme consists of 2 parts; an experimental part carried out at Nanyang Technological University, Singapore (Sopha et. al., 2005) and an analytical and numerical part carried out at Delft University of Technology, The Netherlands. The experimental programme basically consists of three (3) large-scale overlap K-joint specimens with $\theta = 45°$ made from square hollow sections with different β and τ ratios, and is reported in detail in this paper. The same chord sections are used in all 3 test specimens and only the brace sections, the degree of overlap and chord preload are varied.

2 TEST PROGRAMME

The dimensions of the 3 specimens are summarized in Table 1 and its respective loading schemes are shown in Figure 1.

3 TEST FRAME

The configuration of the main test frame to be used in the testing of the overlap RHS K-joint specimen is depicted schematically in Figure 2.

4 SPECIMEN DETAILS

The specifications and criteria for all test specimen fabrication are as follows:

- Structural square hollow sections are all S355 J2H to BS EN10210.
- All welds are full penetration butt welds unless otherwise stated.
- All welding procedures are to comply with AWS D1.1-2004 or equivalent approved standards.
- Welding electrodes of E70X type conforming to AWS D1.1-2004 or equivalent approved standards.
- Welders are qualified to 6GR standard under AWS D1.1 or equivalent.
- Bolt M30 Grade 8.8.
- All fillet weld leg sizes are 8 mm.

Table 1. Dimensions of the overlap RHS K-joint specimens.

Joint Ref	Braces $h_i \times b_i \times t_i$	Chord $h_0 \times b_0 \times t_0$	$\theta_{1,2}$	β	τ	γ	$O_v\%$	Remark
S1	$150 \times 150 \times 6$	$200 \times 200 \times 6.3$	45°	0.75	0.95	15.88	100	No chord preload
S2	$120 \times 120 \times 5$	$200 \times 200 \times 6.3$	45°	0.6	0.79	15.88	100	Comp. chord preload
S3	$120 \times 120 \times 5$	$200 \times 200 \times 6.3$	45°	0.6	0.79	15.88	50	Comp. chord preload

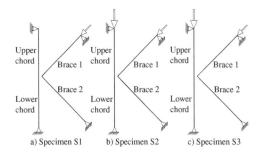

a) Specimen S1   b) Specimen S2   c) Specimen S3

Figure 1.   Loading scheme.

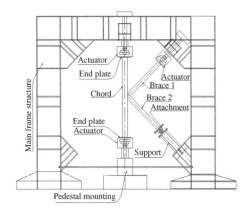

Figure 2.   View of the test frame with overlap K-joint specimen.

The measured dimensions of overlap RHS K-joint test Specimens S1, S2 and S3 are shown in Table 2 and Figure 3.

5   MATERIAL PROPERTIES

The material properties of the specimens are obtained from material coupon tests. The tensile coupon test results of S1, S2 and S3 are summarized in Table 3.

6   INSTRUMENTATION

The type of instrumentation used is dictated by the following considerations:

- Measurement of local stresses, and
- Measurements of applied loads and deformations throughout the test duration.

(i) Strain Gauges
Linear strain gauges of BFLA-5-3-5LT type and 5mm-gauge length are used. The positions of these strain gauges are shown in Figure 4.

(ii) Displacement Gauges
Linear Variable Displacement Transducers (LVDTs) of RDP type and 50 mm gauge lengths are used to monitor displacements at different positions of the specimens. The positions of all these LVDTs are shown in Figure 5.

Table 2.   Measured geometrical parameters of specimens.

Ref.	Braces $h_i \times b_i \times t_i$	Chord $h_0 \times b_0 \times t_0$	$\theta_{1,2}$	β	τ	γ	$O_v(\%)$	Remark
S1	$150 \times 150 \times 5.75$	$200 \times 200 \times 6.1$	45°	0.75	0.77	16.47	100	No Chord Preload
S2	$120 \times 120 \times 5$	$200 \times 200 \times 6$	45°	0.6	0.83	16.67	100	Chord Preload
S3	$120 \times 120 \times 4.96$	$200 \times 200 \times 6.09$	45°	0.6	0.81	16.42	50	Chord Preload

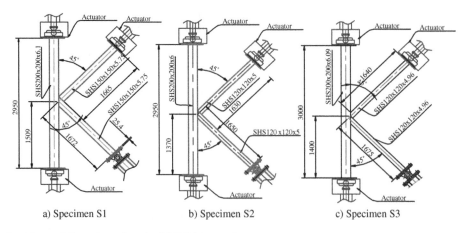

a) Specimen S1   b) Specimen S2   c) Specimen S3

Figure 3.   Measured dimensions of overlap RHS K-joint specimens.

134

(iii) Load and Displacement cells

Load cells are available at the ends of brace 1, upper chord and lower chord. However, at the end of brace 2 there is no load cell available. The positions of load and displacement cells are shown in Figure 5.

7 TEST SET-UP

The following steps are observed during the testing of the specimens:

- Full care is taken to ensure good alignment as shown in Figure 6 for Specimen S1. All bolts are tightened to full load.
- All strain gauges are wired up and their validity is checked using the data logger. The gauges are initialized and their readings are monitored.
- Displacement gauges are fixed in the desired positions by using scaffolding as support; these gauges are initialized and their readings are monitored. These displacement gauges are used to measure the axial deformations of compression brace and the displacement of chord in three sides.

Table 3. Tensile coupon test results.

Joint Ref.	Member	Yield strength N/mm²	Ultimate tensile strength N/mm²	Modulus of elasticity kN/mm²
S1	Chord	405	530.0	213.0
	Brace	411	517.0	207.5
S2	Chord	405	505.5	204.0
	Brace	391	502.5	202.0
S3	Chord	406	524.0	212.0
	Bracc	379	515.0	210.0

- Small trial pre-loads are applied to ensure that the equipments are functioning properly; load cells and strain gauges readings, displacement cells and LVDT readings are compared at each stage.

A typical test set-up is shown in Figure 6 for Specimen S1.

8 TEST PROCEDURE

All the specimens are tested according to their loading schemes indicated in Figure 1.

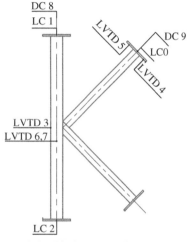

DC = Displacement cell

LC = Load cell

LVTD = Linear VariableDisplacement Transducer

Figure 5. Load and displacement cell locations.

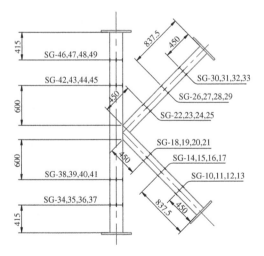

Figure 4. Strain gauge locations.

Figure 6. Test set up for specimen S1.

9 FAILURE MODES

9.1 *Failure mode of specimen S1*

The test specimen is loaded axially beyond the ultimate load. The close-up views of the chord yield failure of Specimen S1 are shown in Figure 7.

9.2 *Failure mode of specimen S2*

The test specimen is also loaded axially beyond the ultimate load. The close-up views of the chord yield failure of Specimen 2 are shown in Figure 8.

9.3 *Failure mode of specimen S3*

The test specimen is also loaded beyond the ultimate load. The close-up views of the chord yield failure of Specimen 3 are shown in Figure 9.

10 EXPERIMENTAL RESULTS

10.1 *Load-displacement curves*

The typical load-displacement curves for Specimens S1 and S3 are given in Figure 10. The load which is based on the strain gauge readings in brace 1 is plotted against the average displacement of transducers located at the end plate of brace 1.

(a) Left Side View

(b) Right Side View

Figure 8. Side view of the joint after failure of specimen S2.

(a) Left Side View

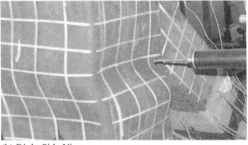

(b) Right Side View

Figure 7. Side view of the joint after failure of specimen S1.

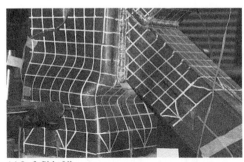

(a) Left Side View

(b) Right Side View

Figure 9. Side view of the joint after failure of specimen S3.

10.2 Check equilibrium of forces

To ascertain whether the test results are correct or not, the equilibrium of forces are checked and it has been shown that the force equilibriums are always maintained.

11 COMPARISON WITH EQUATIONS

11.1 Member check

The interaction of the chord axial load and chord bending moment is checked using the member non-linear interaction formula (1) according to Roik and Wagenknecht (1977), see Wardenier (1982):

$$\left(\frac{N}{N_p}\right)^{1.5} + \left(\frac{M}{M_p}\right) = 1 \tag{1}$$

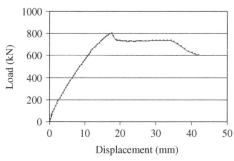

(a) Load-displacement curve of S1

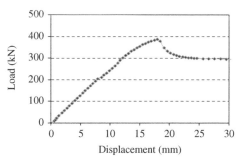

(b) Load-displacement curve of S2

Figure 10. Load-displacement curves.

Further, a linear interaction formula (2) is used, which was found in the numerical work to limit the chord capacity at the joint location.

$$\left(\frac{N}{N_p}\right) + \left(\frac{M}{M_p}\right) = 1 \tag{2}$$

The axial load and moment components that caused the joint to fail in the various modes are indicated in Tables 4, 5 and 6 for Specimens S1, S2 and S3 respectively.

11.2 Brace effective width

The predicted failure loads using EN1993-1-8 (2005) overlapping brace effective width criterion are tabulated in Table 7 where it can be seen that these strengths cannot be reached due to chord member interaction failure.

11.3 Shear check

Specimens S1 and S2 with 100% joint overlap are checked for shear failure using the equations proposed by Wardenier and Choo (2005).

- Check Shear for S1

$$V_u = \frac{N_i \cos \theta_i + N_j \cos \theta_j}{\dfrac{f_{uj}}{\sqrt{3}} \cdot \dfrac{(2h_j + b_j + b_e) \cdot t_j}{\sin \theta_j}}$$

$$= \frac{800.2 \cos 45° + 613.6 \cos 45°}{\dfrac{0.411}{\sqrt{3}} \cdot \dfrac{(2 \times 150 + 150 + 48) \cdot 5.75}{\sin 45°}} = 1.04$$

- Check Shear for S2

$$V_u = \frac{N_i \cos \theta_i + N_j \cos \theta_j}{\dfrac{f_{uj}}{\sqrt{3}} \cdot \dfrac{(2h_j + b_j + b_e) \cdot t_j}{\sin \theta_j}}$$

$$= \frac{649.4 \cos 45° + 598.6 \cos 45°}{\dfrac{0.391}{\sqrt{3}} \cdot \dfrac{(2 \times 120 + 120 + 45) \cdot 5.0}{\sin 45°}} = 1.3$$

Table 4. Member capacities of specimen S1.

Members	σ_e (N/mm²)	N(kN)	N_p(kN)	M(kN-m)	M_p(kN-m)	Eq (2)	Eq (1)
Upper chord	405	5.6	1916.1	38.54	139.3	0.24 < 1	0.12 < 1
Lower chord	405	985.0	1916.1	121.37	139.3	1.26	1.16
Brace 1	411	800.2	1363.6	8.36	73.8	0.7 < 1	0.62 < 1
Brace2	411	613.6	1363.6	37.97	73.8	0.95 < 1	0.8 < 1

Table 5. Member capacities of specimen S2.

Members	σ_e (N/mm^2)	N(kN)	N$_p$(kN)	M(kN-m)	M$_p$(kN-m)	Eq (2)	Eq (1)
Upper chord	405	56.2	1871.1	52.57	137.2	0.40 < 1	0.27 < 1
Lower chord	405	923.3	1871.1	84.38	137.2	1.10	0.98
Brace 1	391	649.4	887.6	12.3	38.8	1.05	0.91 < 1
Brace2	391	598.6	887.6	4.1	38.8	0.78 < 1	0.71 < 1

Table 6. Member capacities of specimen S3.

Members	σ_e (N/mm^2)	N(kN)	N$_p$(kN)	M(kN-m)	M$_p$(kN-m)	Eq (2)	Eq (1)
Upper chord	406	652.3	1917.8	51.8	148.3	0.69 < 1	0.55 < 1
Lower chord	406	1113.7	1917.8	91.4	148.3	1.20	1.06
Brace 1	379	385.6	865.8	8.1	40.6	0.65	0.54 < 1
Brace2	379	276.3	865.8	10.6	40.6	0.58 < 1	0.71 < 1

Table 7. Comparison of failure loads using brace effective width criterion.

Joint Ref.	Load failure (kN)	Brace effective width criterion (kN)	Diff (%)
S1	800.2	1128.3	29.1
S2	649.4	789.75	18.0
S3	385.6	592.0	35.0

12 CONCLUSIONS

For the range of joint parameters studied in this paper, the critical failure mode is chord member yield failure due to the interaction between axial compression and bending moment at the joint. This was in fact the governing failure criterion for Specimens S1 and S3. For Specimen 2 the chord member yield capacity was reached together with the overlapped brace ultimate shear capacity.

REFERENCES

Wardenier J. and Choo Y.S. 2005. *Some Developments in Tubular Joint Research*. Proc. of the 4*th* International Conference on Advances in Steel Structures, Vol. I, Shanghai, PR China.

EN1993-1-8: 2005. *Eurocode 3: Design of steel structures Part 1-8: Design of Joints*, CEN.

Roik K. and Wagenknecht G. 1977. *Traglastdiagramme zur Bemessung von Druckstäben mit doppelsymmetrischem Querschnitt aus Baustahl, Institut für Konstruktiven Ingenieurbau-Ruhr* Universität Bochum, Heft 27.

Sopha T., Chiew S.P. and Wardenier J. 2005. *Design Recommendations for RHS K-Joints with 50% and 100% overlap: Test Results.* CIDECT Report 5BN-3/05, Part 3, Nanyang Technological University, Singapore.

Wardenier J. 1982. *Hollow Section Joints*, Delft University Press, The Netherlands.

Tubular Structures XI – Packer & Willibald (eds)
© *2006 Taylor & Francis Group, London, ISBN 0-415-40280-8*

First practical implementation of the component method for joints in tubular construction

K. Weynand & E. Busse
PSP Technologien GmbH, Aachen, Germany

J.-P. Jaspart
Department M&S, Liège University, Belgium

ABSTRACT: The extension of the component method, recommended in Eurocode 3 and 4 for steel and composite joints between H or I profiles, to the design of steel joints in tubular construction has been already discussed in a paper presented at the 9th ISTS in Düsseldorf in 2001. In the meantime further investigations have been undertaken in order to develop rules for specific components. Some have been presented at the 10th ISTS in Madrid in 2003; others have been derived in various CIDECT research projects. Furthermore existing rules published in the CIDECT Design Guides and in EN 1993 Part 1.8 have been "converted" into a component format. As a result of these works, this paper presents a survey on the application of the component method to "tubular" joints and a first practical implementation of the component method by referring to so-called "design sheets" particularly useful for the day-to-day design.

1 INTRODUCTION

In a paper presented at the Ninth International Symposium on Tubular Structures (Jaspart & Weynand 2001), the extension of the component method, as recommended in Eurocode 3 and 4 for steel and composite joints between H or I profiles, to the design of steel joints in tubular construction was discussed. First related design rules for bolted beam-to-column joints between RHS columns and H or I beams were also briefly described. At the end of the paper, the following conclusions were drawn:

– the successful application of the component method to joints with RHS columns appears as promising even if further investigations are required;
– the CIDECT committee is supporting further developments;
– a wider implementation of this method would lead to simplifications in view of the standardisation and hence, it would help to facilitate the daily work of designers.

In the meantime several further investigations have been undertaken on the scientific level in order to develop more rules for components specific to joints in tubular construction. Some have been for instance presented during the tenth Symposium on Tubular Structures (Bortolotti & al 2003, Pietrapertosa & Jaspart 2003). Others have been derived in the framework of different CIDECT research projects; they are

all referenced in (Jaspart & al 2005). Furthermore, existing rules published in the CIDECT Design Guides and in EN 1993 Part 1.8 have been "converted" into a component format.

As a result of these recent works, a survey on the application of the component method for the design of joints used in tubular construction is first presented in this paper. Then a full design procedure for one specific one is detailed, as an example, by referring to so-called "practical design sheets" particularly useful for the day-to-day design.

2 THE COMPONENT METHOD AS A BASIS FOR THE UNIFIED APPROACH

The component method is a three step procedure which requires successively:

– to identify the constitutive individual components of the joint;
– to evaluate the stiffness/resistance/ductility properties of all these components by using appropriate design formulae;
– to combine or "assemble" these individual components so as to derive the stiffness/resistance/ductility properties of the complete joint.

The properties of joints to be evaluated in practice strongly depend on the type of global frame analysis and design process which is followed by the designer; for instance:

- for an elastic analysis combined with an elastic verification of the member sections and joints, the stiffness and the elastic resistance of the joints should be derived;
- for an elastic analysis combined with a plastic verification of the most heavily loaded member section or joint, the stiffness and the plastic resistance are required;
- for a rigid-plastic analysis, only the plastic resistance and the ductility of the joints will have to be evaluated.

This approach which is recommended in Eurocode 3 and 4, respectively for steel and composite joints between I or H sections, is very comprehensive. As already said, the objective of the authors was to extend it to joints involving other types of member cross-sections and, in particular, hollow sections. Practically speaking, this requires:

- to extend the list of available components to those met in joints between hollow sections;
- to verify that the available "assembly" procedures (which are based on general principles like equilibrium, compatibility of displacements, ...) are general enough to be considered as independent of the actual nature of the constitutive components and are therefore still relevant.

It has again to be stated that this approach is in full agreement with all the principles and rules stated in Eurocode 3, and especially in its Part 1–8.

3 FIELD OF APPLICATION

Basically the field of application of the "component based" unified design approach is rather wide. The only limitations to its use may be expressed as follows:

- design rules for the evaluation of the stiffness/resistance/ductility properties would not be available for some or all the constitutive individual components;
- these rules would have a limited range of application;
- an appropriate assembly procedure would not be available.

In practice, the main limitation may come from the lack of information on a specific component (unknown or with a limited field of application). But as stated in Eurocode 3, any validated component design rule which would be found in the literature or result from specific investigations by the user (experiments, numerical simulations, ...) may be adopted and combined with those provided by the code. From that point of view, all the results of past, ongoing and future research projects (referring or not to the component approach) are and will remain adequate and useful pieces of information.

In Table 1 the reader will find a schematic view of the general field of application of the here-promoted unified design procedure for joints. And in Figure 1 few examples of joints to which the unified design approach applies are shown.

A wide range of possible joint configurations may in fact nowadays be covered, as detailed knowledge is available for 37 hereafter-listed components.

1 Web panel in shear
2 I or H section web in transverse compression
3 I or H section web in transverse tension
4 I or H section flange in out of plane bending
5 End-plate in out of plane bending

Table 1. Field of application of the unified design approach.

Parameters	Values, range or field
Joint configurations	In-plane joints in trusses: T-joints, X joints, ... In-plane joints in frames: beam-to-column joints, beam splices, column splices, column bases Others
Loading situations	Axial forces (tension or compression) Shear forces Bending moments Any combination of these forces
Cross-sections of the connected members	Hollow sections: RHS, SHS, CHS, EHS Hot-rolled or built-up I or H sections Others (L or U sections, ...)
Connectors	Bolted connections (with preloaded or non-preloaded bolts, injection bolts, studs, anchors, ...) Welded connections Combination of welding and bolting
Connection elements	End-plates: partial-depth, flush or extended Cleats: web or flange cleats Fin plates Inserted plates Splices Diaphragms Contact plates
Stiffening elements	Transverse and diagonal web stiffeners (I or H sections) Web plates Haunches Backing plates
Steel grades for members and connection elements	S235 to S460

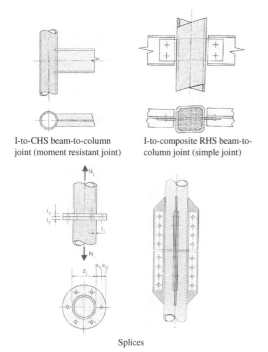

I-to-CHS beam-to-column joint (moment resistant joint)

I-to-composite RHS beam-to-column joint (simple joint)

Splices

Figure 1. Examples of joint configurations covered by the unified design procedure.

6　Flange cleat in out of plane bending
7　I or H section flange and web in compression
8　I or H section web in longitudinal tension
9　Plate in tension or compression
10　Bolts (studs) in tension
11　Bolts (studs) in shear
12　Plate in bearing(plate in general, beam flange or web, column flange or face, end-plate, cleat or base plate)
13　Concrete in compression including grout
14　Base plate in bending under compression
15　Base plate in bending under tension
16　Anchor bolts in tension
17　Anchor bolts in shear
18　Welds
19　Haunch
20　Longitudinal steel reinforcement in tension
21　Steel contact plate in compression
22　Partial depth end plate or fin plate in shear: Gross section
23　Partial depth end plate or fin plate in shear: Net section
24　Partial depth end plate or fin plate in shear: shear block
25　Partial depth end plate or fin plate in in-plane bending
26　Web in shear in partial depth end plate connection
27　Buckling of the fin plate

28　Web in shear in fin plate connection: net section
29　Web in shear in fin plate connection: shear block
30　CHS in transverse compression or tension: chord face failure
31　CHS in transverse compression or tension: Punching shear failure
32　RHS in transverse compression or tension: chord face failure
33　RHS in transverse compression or tension: brace failure
34　RHS in transverse compression or tension: chord side wall crushing
35　RHS in transverse compression or tension: punching shear failure
36　Inserted Plate in shear in CHS/RHS
37　Diaphragm in tension or compression

In (Jaspart & al 2005), detailed rules are provided for the evaluation of the stiffness and resistance properties of all these 37 components. Additionally assembly procedures are given that cover the usual loading situations met in steel construction.

"To assemble the components" means to express the fact that the forces acting on the whole joint distribute amongst the constitutive components in such a way that:

– the internal forces in the components are in equilibrium with the external forces applied to the joint;
– the resistance of a component is nowhere exceeded;
– the deformation capacity of a component is nowhere exceeded.

As far as the resistance of the whole joint to external forces is concerned, the fulfilment of these three rules is enough to ensure that the evaluated design resistance is smaller than the actual joint resistance.

For stiffness calculation, the elastic distribution of internal forces in the joint is requested to fulfil one more condition:

– the compatibility of displacements amongst the constitutive components.

4　DESIGN SHEET FOR PRACTICAL APPLICATION

4.1　General

The component based unified design approach is a quite powerful tool for the evaluation of the stiffness and/or resistance properties of structural steel and composite joints under several loading situations. Amongst its advantages the user will express the ability, through a unique concept:

– to study all the joints of the considered structure, whatever their configurations and the member cross-sections (open or hollow);

- to characterise joints made of other materials or of a combination of different materials;
- to follow with the same approach the evolution of the properties of the composite joints, for instance, during the erection time (bare steel joints before and during concreting, composite joints after concrete drying);
- to take easily into consideration all the loading situations resulting from the several structural load cases.

For people who are acquainted to the use of the component approach, the advantages that it brings in comparison to more traditional approaches will much more than compensate the sometimes heavier requested calculation time. That is why, in parallel to the development of the component method, various practical design tools have to be developed and disseminated throughout the construction community. For steel and composite joints connecting members with open sections, a huge work has been achieved which led to the publication of design tables of standardised joints, simple design sheets, etc. Dedicated software are also available on the market.

For tubular construction or "mixed open/hollow sections" joints, such an effort will have to be achieved too. Hereafter, a simple calculation design sheet is presented just as an example for a specific joint configuration (moment resistant I beam-to-RHS column joint with flush end plate and one row of studs in tension).

4.2 Joint layout, notations

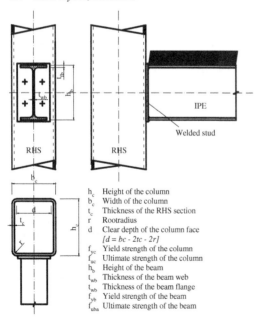

Figure 2. Notations for column and beam.

4.3 Design moment resistance

4.3.1 Beam flange in compression

$$F_{Rd1} = M_{b.Rd} / (h_b - t_{fb})$$

with:
$M_{b,Rd}$ is the design moment resistance, depending on the class of the cross section (M_{pl}, M_{el} or M_{eff})

4.3.2 Studs in tension

$$F_{Rd2} = n \times \min[F_{t,Rd}; B_{p,Rd}]$$

with:
n is the number of studs in the tension zone of the joint

$$F_{t,Rd} = \frac{k_2\, f_{ub}\, A_s}{\gamma_{M2}}$$

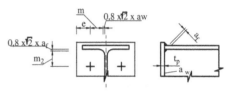

a_f Throat thickness of the weld adjacent to the beam flange
a_w Throat thickness of the weld adjacent to the beam web
e Edge to bolt row distance
m Distance as indicated
m_2 Distance as indicated
t_p Thickness of the end plate
f_{yp} Yield strength of the end plate
f_{up} Ultimate strength of the end plate

Figure 3. Notations for end plate.

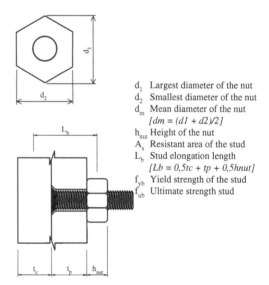

d_1 Largest diameter of the nut
d_2 Smallest diameter of the nut
d_m Mean diameter of the nut
    $[dm = (d1 + d2)/2]$
h_{nut} Height of the nut
A_s Resistant area of the stud
L_b Stud elongation length
    $[Lb = 0,5tc + tp + 0,5hnut]$
f_{yb} Yield strength of the stud
f_{ub} Ultimate strength stud

Figure 4. Notations for studs.

h_c Height of the column
b_c Width of the column
t_c Thickness of the RHS section
r Rootradius
d Clear depth of the column face
    $[d = bc - 2tc - 2r]$
f_{yc} Yield strength of the column
f_{uc} Ultimate strength of the column
h_b Height of the beam
t_{wb} Thickness of the beam web
t_{wb} Thickness of the beam flange
f_{yb} Yield strength of the beam
f_{uba} Ultimate strength of the beam

where:

$k_2 = 0,63$ for countersunk bolt,
$k_2 = 0,9$ otherwise
$B_{p,Rd} = 0,6\pi d_m t_p f_{up}/\gamma_{M2}$

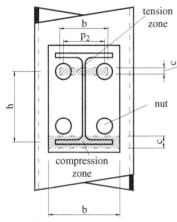

Figure 5. Notations for studs.

Tension zone:
- p_2 Horizontal distance between the studs
- b Width of the tension zone $[b = p2 + 0,9dm]$
- c Height of the tension zone $[c = 0,9dm]$

Compression zone:
- b Width of the compression zone
  [For the determination of b in the compression zone, a dispersion of the load under 45° from the edges of the welds can be assumed. b is limited by the dimensions of the end plate]
- c Height of the compression zone
  [For the determination of c in the compression zone a dispersion of the load under 45° from the edges of the welds can be assumed. b is limited by the dimensions of the end plate]
- h Distance between the centre of the tension zone and the compression zone

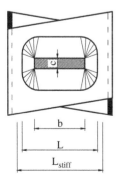

L   Width of the yield area
    $[L = bc - 2tc - 1,5r]$
L_{stiff} Equivalent yield length for stiffness determination
    $[Lstiff = d + r = bc - 2tc + r]$

Figure 6. Notations for the yield area (tension or compression).

4.3.3 End-Plate in bending

$F_{Rd,3} = min[\, F_{T,Rd,1}\,;\,F_{T,Rd,2}]$

with:

$$F_{T.Rd,1} = \frac{(8n - 2e_w)M_{pl,1,Rd}}{2m\,n - e_w(m + n)}$$

where:

$n = min[e;1,25 \times m]$

$e_w = \dfrac{d_1}{4}$

$M_{pl,1,Rd} = 0,25\sum l_{eff,1} t_p^2 f_{yp}/\gamma_{M0}$

and:

$\sum l_{eff,1} = l_{eff,1}$ (only 1 row)
$\quad\quad\quad = min[2\pi m; \alpha \times m]$

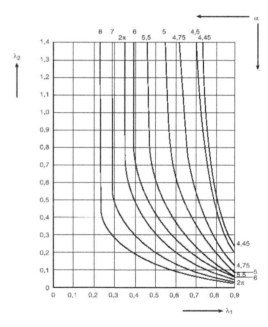

α is based on the pre-evaluation of:
where:

$\lambda_1 = \dfrac{m}{m+e}$ and $\lambda_2 = \dfrac{m_2}{m+e}$

$$F_{T.Rd,2} = \frac{2\,M_{pl,2,Rd} + n\sum F_{t,Rd=}}{m + n}$$

where:

$M_{pl,2,Rd} = 0,25\sum l_{eff,2} t_p^2 f_{yp}/\gamma_{M0}$

and:

$\sum l_{eff,2} = l_{eff,2}(only\,1row) = \alpha \times m$

4.3.4 Beam web in tension

$$F_{Rd.4} = b_{eff,t,wb} t_{wb} f_{yb} / \gamma_{M0}$$

with

$$b_{eff,t,wb} = l_{eff,1}$$

4.3.5 RHS in transverse compression and tension: Chord face failure

$$F_{Rd,5\,and\,6} = \min[F_{pl,loc}; F_{pl,glob}]$$

for determination of $F_{pl,loc}$:

if b> $b_m = L\left[1 - 0,82\dfrac{t_c^2}{c^2}\left(1 + \sqrt{1 + 2,8\dfrac{c^2}{t_c L}}\right)^2\right]$

$$\Rightarrow F_{pl,loc} = \beta 4 m_{pl,Rd}\left[\dfrac{\pi\sqrt{L(a+x)} + 2c}{a+x} + \dfrac{1,5cx + x^2}{\sqrt{3}t_c(a+x)}\right]$$

where:

$\beta = 1$ $\qquad$ if $\dfrac{b+c}{L} \geq 0,5$

$\beta = 0,7 + 0,6\dfrac{b+c}{L}$ $\quad$ if $\dfrac{b+c}{L} \leq 0,5$

$m_{pl,Rd} = \dfrac{1}{4}t_c^2 f_{yc} / \gamma_{M0}$

$a = L - b$

$$x = -a + \sqrt{a^2 - 1,5ac + \dfrac{\sqrt{3}t_c}{2}\left[\pi\sqrt{L(a+x_0)} + 4c\right]}$$

and:

$$x_0 = L\left[\left(\dfrac{t_c}{L}\right)^{\frac{2}{3}} + 0,23\dfrac{c}{L}\left(\dfrac{t_c}{L}\right)^{\frac{1}{3}}\right]\left(\dfrac{b - b_m}{L - b_m}\right)$$

if b $\leq b_m = L\left[1 - 0,82\dfrac{t_c^2}{c^2}\left(1 + \sqrt{1 + 2,8\dfrac{c_2}{t_c L}}\right)^2\right]$

$$\Rightarrow F_{pl,loc} = \beta\dfrac{4\pi m_{pl,Rd}}{1 - \dfrac{b}{L}}\left(\sqrt{1 - \dfrac{b}{L}} + \dfrac{2c}{\pi L}\right)$$

for determination of $F_{pl,glob}$:

$$F_{pl,glob} = \dfrac{F_{pl,loc}}{2} + m_{pl,Rd}\left(\dfrac{2b}{h} + \pi + 2\rho\right)$$

where:

$\rho = 1$ $\qquad$ if $0,7 \leq \dfrac{h}{L - b} \leq 1$

$\rho = \dfrac{h}{L - b}$ $\quad$ if $1 \leq \dfrac{h}{L - b} \leq 10$

if $h/L - b$ smaller than 0,7 or greater than 10, the formula for $F_{pl,glob}$ is not valid.

4.3.6 RHS in transverse tension: Punching shear failure

$$F_{Rd,7} = \min[F_{punch,nc}; F_{punch,cp}]$$

with

$$F_{punch,nc} = 2(b + c)V_{pl,Rd}$$

where:

$$V_{pl,Rd} = \dfrac{t_c f_{yc}}{\sqrt{3}\gamma_{M0}}$$

$$F_{punch,cp} = n\pi d_m V_{pl,Rd}$$

4.3.7 RHS in transverse compression: Punching shear failure

$$F_{Rd,8} = F_{punch,nc}$$

with

$$F_{punch,nc} = 2(b + c)V_{pl,Rd}$$

where:

$$V_{pl,Rd} = \dfrac{t_c f_{yc}}{\sqrt{3}\gamma_{M0}}$$

4.3.8 Resistance of the weakest component

$$F_{Rd} = \min_{i=1}^{8} F_{Rd,i}$$

4.3.9 Plastic design moment resistance

$$M_{Rd} = F_{Rd} \times h$$

4.3.10 Elastic moment resistance

$$M_e = \dfrac{2}{3}M_{Rd}$$

4.4 Initial joint stiffness

4.4.1 Beam flange in compression

$$k_1 = \infty$$

4.4.2 Studs in tension

$$k_2 = 1,6\dfrac{A_s}{L_b}$$

4.4.3 End-plate in bending

$$k_3 = \dfrac{0,9l_{eff} t_p^3}{m^3}$$

4.4.4 Beam web in tension

$$k_4 = \infty$$

144

4.4.5 RHS in transverse compression and tension: Chord face failure

$$k_{5\,and\,6} = \frac{t_c^3}{14,4\beta L_{stiff}^2}\left(\frac{L_{stiff}^2}{bt_c}\right)^{1,25} \frac{\dfrac{c}{L_{stiff}} + \left(1 - \dfrac{b}{L_{stiff}}\right)\tan\theta}{\left(1 - \dfrac{b}{L_{stiff}}\right)^3 + \dfrac{10,4\left(k_1 - k_2\dfrac{b}{L_{stiff}}\right)}{\left(\dfrac{L_{stiff}}{t_c}\right)^2}}$$

with:

$$\theta = 35 - 10\frac{b}{L_{stiff}} \quad \text{if} \quad \frac{b}{L_{stiff}} < 0,7$$

$$\theta = 49 - 30\frac{b}{L_{stiff}} \quad \text{if} \quad \frac{b}{L_{stiff}} \geq 0,7$$

$k_1 = 1,5$
$k_2 = 1,6$

the formula for k_{5and6} is only valid if these requirements are fulfilled:

$$10 \leq L_{stiff} / t_c \leq 50$$

$$0,08 \leq b / L_{stiff} \leq 0,75$$

$$0,05 \leq c / L_{stiff} \leq 0,20$$

4.4.6 RHS in transverse tension: Punching shear failure

$$k_7 = \infty$$

4.4.7 RHS in transverse compression: Punching shear failure

$$k_8 = \infty$$

4.4.8 Initial stiffness

$$S_{joint,init} = Eh^2 / \sum_{i=1}^{8} 1/k_i$$

5 CONCLUSIONS

The component approach is nowadays considered as the general procedure for the evaluation of the design properties of structural joints made of a single construction material or of a combination of different materials and subjected to various loading situations (moments, shear and/or axial forces, ...) and loading conditions (static, seismic, fire, ...). It is explicitly referred to in Eurocodes 3 and 4, respectively for steel and steel-concrete joints between I or H profiles. Since few years, the authors investigate, with the help of CIDECT, its extension to steel joints between members made of tubular profiles or of tubular and open profiles. The steps to cross are the following ones:

– derivation of design formulae for still-unknown components;

– validation of design procedures for whole joints by comparisons with tests and numerical simulations;
– proposal of practical guidelines for the day-to-day application of the component method.

Progress is regularly made on the two first steps through appropriate scientific works; it consists either in the reformatting of existing design formulae or the derivation of new ones. A review of the today available material is proposed in (Jaspart & al 2005).

In the present paper a first practical design tool, with a so-called "design sheet" format, is suggested. Its application is rather simple and a reliable programming (EXCEL sheet for instance) may be easily contemplated. In order to simplify further the work of the designer, tables of standardised joints may even be produced in which the properties of the joints, evaluated by means of the "design sheets" are directly listed (moment resistance, rotational stiffness, failure mode, ...).

Such design tools (software, design sheets, tables of standardised joints) are now widely available and used in Europe for joints between members with I and H sections and the authors hope that the work presented in this paper will be the trigger of a similar evolution process for the tubular construction.

REFERENCES

Bortolotti, Jaspart, Pietrapertosa, Nicaud, Petitjean & Grimault 2003. Testing and modelling of welded joints between elliptical hollow sections. *Proc. of the 10th International Symposium on Tubular Structures, 18–20 September 2003, Madrid, Spain, pp. 259–266.*

CEN 2005. EN 1993-1-1 "Eurocode 3 Design of Steel Structures – Part 1.1: General rules", Brussels.

CEN 2005. EN 1993-1-8: "Eurocode 3: Design of Steel Structures – Part 1.8: Design of joints" Brussels.

CEN 2005. EN 1994-1-1: "Design of composite steel and concrete structures – Part 1.1: General rules and rules for buildings" Brussels.

Jaspart, J.-P. & Weynand, K. 2001. Extension of the component method to joints in tubular construction. In Puthli, R. & Herion, S. (ed.), *Proc. of the 9th Intern. Symposium on Tubular Structures, Düsseldorf 3–5 April 2001.* Rotterdam.

Jaspart, J.P., Pietrapertosa, C., Weynand, K., Busse, E. & Klinkhammer R. 2005. Development of a full consistent design approach for bolted and welded joints in building frames and trusses between steel members made of hollow and/or open sections – Application of the component method. CIDECT research project 5BP, Aachen, Liège.

Pietrapertosa & Jaspart 2003. Study of the behaviour of welded joints between elliptical hollow sections. *Proceedings of the 10th International Symposium on Tubular Structures, 18–20 September 2003, Madrid, Spain, pp. 601–608.*

Weynand, K & Jaspart, J.-P. 2002. Application of the component method to joints between hollow and open sections. CIDECT research project 5BM. Aachen & Liège.

Tubular Structures XI – Packer & Willibald (eds)
© 2006 Taylor & Francis Group, London, ISBN 0-415-40280-8

Local buckling behaviour of cold-formed RHS made from S1100

G. Sedlacek, B. Völling, D. Pak & M. Feldmann
Institute for Steel and Lightweight Construction, Aachen University, Germany

ABSTRACT: New structural steels, such as (liquid) quenched and tempered (QT) steels, with extra high strength, good toughness and weldability have been developed in Europe in recent years and are included in Eurocode 3 Part 1–12 for strengths up to S700. However, extra high strength steels (EHSS) with a yield strength of $1100\,N/mm^2$ so far have not been taken into account in the present version of Eurocode 3 due to insufficient knowledge on their buckling behaviour and fatigue strength and lack of associated experimental data. In this paper results of experimental and numerical investigations concerning the local buckling behaviour of rectangular hollow sections (RHS) made of EHSS (S1100) are summarised. The investigations have been carried out within the scope of an European research project. Results have been used to check the applicability of existing design rules for local buckling to EHSS. As a consequence Eurocode 3 Part 1.12 could be expanded to cover stability rules for such steels.

1 INTRODUCTION

The limited knowledge on the stability behaviour and fatigue strength of extra high strength steels (EHSS) has been a major obstacle for their application in the structural field. E.g. in (EC 3 Part 1.1 2005) and (EC 3 Part 1.12 2006) EHSS with a yield strength between $700–1100\,N/mm^2$ are disregarded due to a lack of knowledge and insufficient experimental data.

Within the scope of the ECSC Project 4553 "Efficient lifting equipment with extra high strength steel" (Lifthigh 2005) investigations were made to expand the limits for structural steels available for crane manufacturers as well as for other applications. One focus was drawn on the local (plate) buckling behaviour of EHSS specimens subjected to direct stresses. Therefore experimental and numerical investigations have been carried out on cold formed, rectangular and circular hollow sections of EHSS of grade S1100. The sections have been tested in bending as this is a typical loading situation for the applications in the crane industry. The major aims of these investigations were

- to check the applicability of existing design rules for local buckling, which have been established for ordinary steel grades (e.g. (EC 3 Part 1.5 2006)), to EHSS,
- to improve existing design rules for local buckling to make them applicable to EHSS if necessary,
- to determine the influence of geometric imperfections on the global buckling behaviour of EHSS,
- to determine the influence of residual stresses on the global buckling behaviour of EHSS.

2 ENGINEERING APPROACHES

2.1 General

Thin walled plane elements subjected to in-plane loading are sensitive against buckling due to compression stress and/or shear stress. Buckling is influenced by many parameters such as:

- the component's geometry
- the material properties
- the boundary- and bearing-conditions
- imperfections
- residual stresses

Various buckling theories can be derived the complexity and accuracy of which differ depending on the degree of consideration of the various parameters and their interaction. They range between extremes as highly idealised bifurcation theories and complete non-linear ultimate loading theory as applied in the FEM simulations described in Chapter 4.

There are also different engineering approaches that may be applied to estimate the resistance of plates and shells vulnerable to buckling. They all are based on the use of slenderness dependent buckling curves:

$$R_{d,i} \leq \chi(\bar{\lambda}_i)\frac{R_{k,i}}{\gamma_M} \qquad (1)$$

with

$$\bar{\lambda}_i = \sqrt{\frac{R_{k,i}}{R_{crit,i}}} \qquad (2)$$

where

- i describes the buckling phenomenon,
- $R_{d,i}$ is the design value of resistance (e.g. N_{Rd}, M_{Rd}, f_{yd}),
- $R_{k,i}$ is the characteristic value of resistance (e.g. N_{Rk}, M_{Rk}, f_{yk}),
- $R_{crit,i}$ is the critical buckling load,
- λ_i is the slenderness,
- χ_i is the reduction factor dependent on the slenderness and buckling phenomenon i and
- γ_M is the partial safety factor.

The major difference between the different procedures described hereafter is the way how to determine the slenderness and to define the (slenderness dependent magnitude of the) reduction factor.

2.2 Plate buckling according to EN 1993-1-5

In (EC 3 Part 1.5 2006) two different methods are described to calculate the resistance of plated members without longitudinal stiffeners to direct stresses:

1. Method of effective^p areas (the subscript "p" indicates effects of plate buckling) of plate elements in compression for calculating class 4 cross sectional data (A_{eff}, I_{eff}, W_{eff}) to be used for cross sectional verifications.
2. Method using gross areas of plate elements in compression for calculating cross sectional data, however with stress limitations for the various plate elements.

Following method 1) the effective^p area A_{eff} of plate elements in compression (Fig. 1) is obtained by reducing the gross sectional area A using the reduction factor χ for plate buckling of a single plate panel

$$\chi = \frac{\overline{\lambda}_p - 0.055 \times (3+\psi)}{\overline{\lambda}_p^2} \leq 1 \tag{3}$$

with

$$\overline{\lambda}_p = \sqrt{\frac{f_y}{\sigma_{crit}}} \tag{4}$$

where ψ is the stress ratio ($\psi = 1$ for uniform compression) and σ_{crit} is the elastic critical buckling stress (for determination of σ_{crit} see Annex A of (EC 3 Part 1.5 2006)).

According to method 2) stated in (EC 3 Part 1.5 2006) the stress limits for stiffened or unstiffened plates i subjected to combined stresses $\sigma_{x,Ed}$, $\sigma_{y,Ed}$, τ_{Ed} shall be determined in accordance with the following equation

$$f_{yd,red,i} = \chi_j \times \frac{f_{yk}}{\gamma_M}, \quad j = x, z, v \tag{5}$$

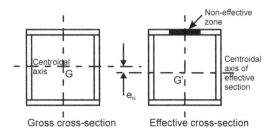

Figure 1. Effective cross section – bending moment (EC 3 Part 1.5 2006).

where χ_x, χ_z are the reduction factors for x-direction and z-direction respectively as stated in Equation 3, χ_v is the reduction factor for shear as stated in clause 5.2 of (EC 3 Part 1.5 2006).

In contrast to method 1) the slenderness $\overline{\lambda}_p$; for the calculation of the reduction factors χ_x, χ_z, χ_v shall be taken from

$$\overline{\lambda}_p = \sqrt{\frac{\alpha_{ult}}{\alpha_{crit}}} \tag{6}$$

where α_{ult} is the minimum load amplifier of the design loads to reach the characteristic value of resistance of the most critical point of the plate and α_{crit} is the minimum load amplifier of the design loads to reach the elastic critical resistance of the plate under the compression stress field. For the calculation of α_{crit} finite element analysis or any suitable literature such as (Klöppel & Möller 1968, Klöppel & Scheer 1960) may be used.

The concept of Equation 6 may be applied to the full stress field comprising $\sigma_{x,Ed}$, $\sigma_{y,Ed}$, τ_{Ed} leading to a global slenderness $\overline{\lambda}_p$; and requiring a single buckling curve. It may also be used for the individual components of the stress field $\sigma_{x,Ed}$, $\sigma_{y,Ed}$, τ_{Ed} leading to a slenderness and a resistance for each component according to Equation 5. In this case an interaction formula for the stress components $\sigma_{x,Ed}$, $\sigma_{y,Ed}$, τ_{Ed} is necessary to achieve safety:

$$\left(\frac{\sigma_{x,Ed}}{f_{yd,red,x}}\right)^2 + \left(\frac{\sigma_{z,Ed}}{f_{yd,red,z}}\right)^2 - \left(\frac{\sigma_{x,Ed}}{f_{yd,red,x}}\right) \times \left(\frac{\sigma_{z,Ed}}{f_{yd,red,z}}\right)$$
$$+ 3\left(\frac{\tau_{Ed}}{f_{yd,red,v}}\right)^2 \leq 1 \tag{7}$$

If only a single stress component $\sigma_{x,Ed}$ has to be considered Equation 7 can be simplified to:

$$\sigma_{x,Ed} \leq f_{yd,red,x} = \chi_x \times \frac{f_{yk}}{\gamma_M} \tag{8}$$

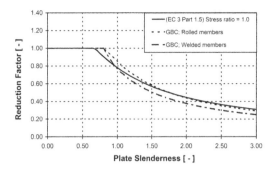

Figure 2. Reduction factor for plate buckling according to (EC 3 Part 1.5 2006) and general buckling curves (GBC) according to (Müller 2003).

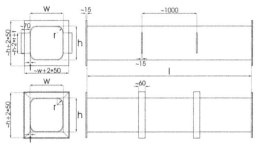

Figure 3. Test specimen without/with collar.

2.3 Plate buckling using global slenderness

The concept of global slenderness is applicable to calculate the resistance of plated members in one step, but it may also be applied to shells, columns and even complex systems such as frames etc. where different buckling phenomena may interact.

For the calculation of α_{ult} and α_{crit} in Equation 6 finite element analysis can be used. For a single plated element with only a single stress component the global slenderness is equivalent to the plate slenderness $\bar{\lambda}_p$; as defined in Equation 4.

The verification with a global slenderness may be carried out corresponding to Equation 1 using a single buckling curve to determine the reduction factor χ. The type of the single buckling curve only depends on the buckling phenomenon considered, i.e. column buckling, lateral-torsional buckling, local buckling or shell buckling.

For local plate buckling the reduction factor χ_s is defined by the Generalized Buckling Curve (GBC) as follows

$$\chi_s = \frac{1}{\Phi_s + \sqrt{\Phi_s^2 - \bar{\lambda}_s}} \qquad (9)$$

with

$$\Phi_s = \frac{1}{2}\left(1 + \alpha_s\left(\bar{\lambda}_s - \bar{\lambda}_{s0}\right) + \bar{\lambda}_s\right) \qquad (10)$$

The format of this curve is similar to the one used as "European buckling curve" (Ayrton Perry format) The parameters $\bar{\lambda}_s$; 0 and α_s of this buckling curve have been determined from tests and take the following values:

$\bar{\lambda}_s$; 0 = 0.8, α_s = 0.13 (for rolled members) and α_s = 0.34 (for welded members).

This generalised buckling curve is presented together with the buckling curve as defined in (EC 3 Part 1.5 2006) (see Equation 3) in Figure 2.

Table 1. Parameters of the cold formed hollow sections (measured values).

specimen	quant. [-]	wc** [-]	l mm	h* mm	w* mm	t mm
A	1	no	2999	205.0	200.7	8.05
B	2	no	2998	208.5	205.5	6.70
C	1	no	2996	257.5	257.0	6.70
D	2	no	2996	307.5	308.0	6.70
E	1	yes	2997	208.0	206.0	6.70
F	1	yes	2998	308.5	306.5	6.70

* outer dimensions.
** welded collar.

3 EXPERIMENTAL INVESTIGATIONS

3.1 General

Within the project "Lifthigh" buckling tests have been performed on cold formed rectangular hollow sections made from EHSS which have been selected such that a relatively wide range of different slendernesses is covered (see Equation 4).

The specimens consist of two identical half-tubes, which have been bent from plane plates (cold-formed) and subsequently been welded together along the axis neutral to bending. At both ends vertical end plates with a thickness of 15 mm were welded to the tube. The load introduction points have been detailed in two different ways (with and without collar) to investigate the influence of the detailing on the ultimate resistance (Fig. 3).

Considering the specimens without a collar (A–D in Table 1), the loading is applied on four stiffeners. These stiffeners also have a thickness of 15 mm and are welded to the web of the tube. In case of specimen E and F these stiffeners are replaced by a revolving collar made up of U-profiles with a thickness of 15 mm welded to the tube.

To investigate the applicability of existing design rules for plate buckling, reduction factors have been calculated from the test results and subsequently been

Table 2. Tensile test/RFDA.

t_{nom} (mm)	$R_{p0.2}$ (MPa)	E (GPa)
8.0	1199	195.56
6.5	1227	195.64
4.0	1233	not measured

Figure 4. Test set-up.

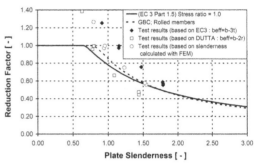

Figure 5. Comparison of reduction factors obtained from test results with plate buckling rules.

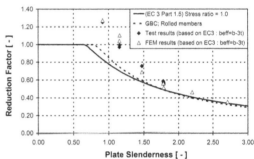

Figure 6. Test vs. FEM results for rectangular hollow sections.

compared with the reduction curves displayed in Figure 2.

To this end tensile tests and measurements of the modulus of elasticity have been conducted. The modulus of elasticity has been determined by "Resonant Frequency and Damping Analyser (RFDA)". The results of the measurements are as follows:

The values stated in Table 2 have been used for further evaluation of the buckling tests and for FEM calculations conducted subsequently (compare Chapter 4).

3.2 Results

The sections have been tested in four-point bending (Fig. 4). Besides the usual geometric values for dimensions also geometric imperfections have been measured. However, the results of the measurements have shown that the geometric imperfections are small enough to be neglected (geometric imperfections <0.5 mm). All significant parameters of the tested specimens are summarized in Table 1.

The slendernesses have been calculated according to (EC 3 Part 1.5 2006) and as suggested in (Dutta 1999).

The difference between these two approaches is the different plate width b_{eff} to be applied for the

flat parts between the radii. Results show that the approach as proposed by (Dutta 1999) may yield to non-conservative results.

If the slenderness ratio is calculated according to (EC 3 Part 1.5 2006) (Fig. 5) both the corresponding reduction curve in EC 3 Part 1.5 as well as the generalised buckling curve (GBC) (which yield to even better results) appear to be appropriate for the investigated specimens. If the slenderness is calculated by finite element methods results are further improved (Fig. 6).

Furthermore it was found that the influence of the different type of load introduction (specimens with and without welded collar) is negligible.

4 NUMERICAL INVESTIGATIONS

4.1 General

Mechanical (based on real stress – strain behaviour obtained from tension tests) and geometrical nonlinear finite element calculations have been carried out in order

– to calculate the critical buckling stress for rectangular hollow sections tested in bending and

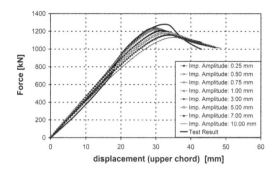

Figure 7. Influence of imperfection amplitude on load-deflection curve.

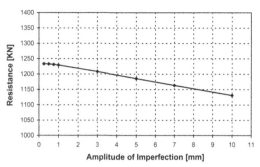

Figure 8. Ultimate resistance of rectangular hollow sections.

– to conduct parameter studies concerning geometric and structural imperfections.

The finite element models of the cold formed specimens tested in bending have been generated and subsequently verified by test results (Fig. 6). All models showed a very good correlation with tests.

To expand the range of slendernesses considered in the tests three more rectangular hollow sections with a higher slenderness of the compression chord have been generated and calculated by FEM assuming the same material properties as for the specimens tested.

4.2 Results of parametric study

4.2.1 Geometric imperfections

The geometric imperfection was applied in terms of the mode shape of the first buckling mode. Subsequently the influence of the amplitude of the applied imperfection on the ultimate resistance of the specimen has been investigated.

As expected the ultimate resistance of the specimens with rectangular hollow sections tested in bending is not significantly influenced by geometric imperfections as long as amplitudes are small (i.e. <1 mm). Measured imperfection amplitudes of the test specimens have been even smaller than 0.5 mm. For large imperfections with amplitudes in the order of l/300 (=10 mm) the ultimate resistance is decreased by 8% (Figs 7, 8).

4.2.2 Structural imperfections

Within the scope of the "Lifthigh" – project a general study has been carried out to determine also the influence of structural imperfections (residual stresses) caused by cold-forming on the ultimate resistance of cold-formed rectangular hollow sections from EHSS (Lifthigh 2005). Therefore measurements have been carried out and evaluated. The results provided a basis for the input parameters for parametric studies which have been carried out by FEM-calculations.

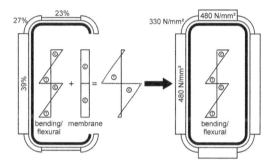

Figure 9. Distribution of residual stresses (Schafer & Peköz 1998).

For the FEM-calculations a residual stress pattern as shown in Figure 9 was applied and for the parametric studies the amplitude of the residual stresses was varied.

In Figure 10 results of the parametric study are shown for the example of specimen D. It can be seen that the amplitude of the residual stresses applied to the section has only a small influence on the ultimate resistance.

This can be explained by the fact that the residual stress patterns applied in the calculations have no membrane part, which is in good accordance with residual stress measurements (Lifthigh 2005, Schafer & Peköz 1998). For welded sections the reduction of the ultimate capacity due to residual stresses would be much higher due to the high membrane part of the residual stresses (Lifthigh 2005).

5 CONCLUSIONS

The investigations yield to the following conclusions:

1. The Winter formula for plate buckling as well as the generalized buckling curve (GBC) (all in EC 3 Part 1.5) are generally applicable to calculate the ultimate resistance. However

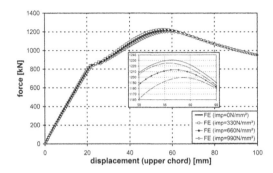

Figure 10. Influence of structural imperfections – results of a parametric study for test specimen D.

- results are conservative if the slenderness is calculated as recommended in (EC 3 Part 1.5 2006) (effective width is assumed to be smaller than in reality),
- results are non-conservative if calculated according to (Dutta 1999) (effective width is assumed to be larger than in reality),
- a very good correlation is achieved if the slenderness is calculated with FEM (global slenderness Method), the effective width is considered correctly.

2. The ultimate resistance of the specimens (which have been tested in bending) is only slightly influenced by geometric imperfections. This is due to the fact that the bending deformation itself already yields to geometric deviations to the initial shape so that an additional imperfection of small amplitude does not effect the result significantly (it should be mentioned that the imperfection due to the bending deformation has the shape of the first eigenmode (of an eigenfrequency analysis) whereas the relevant imperfection that was applied in FEM-calculations was affine to the first buckling mode (collapse-affine imperfection). However, the application of a collapse-affine imperfection generally yields to conservative results).

3. Residual stresses that are induced by cold-forming do not significantly effect the ultimate resistance. This is basically due to the type of the residual stresses that are distributed over the plate's thickness only. There is almost no membrane part of the residual stress. However, if a significant membrane part would be present (e.g. due to welding) the resistance would be decreased.

REFERENCES

Dutta, D. 1999. *Hohlprofil – Konstruktionen*. Berlin: Ernst & Sohn Verlag
EC 3 Part 1.1. Standard. Eurocode 3 – Design of steel structures, Part 1.1. 2005. *General rules and rules for buildings (EN 1993-1-1:2005)*
EC 3 Part 1.5. Standard. Eurocode 3 – Design of steel structures, Part 1.5. 2006. *Plated structural elements (EN 1993-1-5:2006)*
EC 3 Part 1.6. Standard. Eurocode 3 – Design of steel structures, Part 1.6. 2004. *General rules: Strength and stability of shell structures (ENV 1993-1-6:2004)*
EC 3 Part 1.12. Standard. Eurocode 3 – Design of steel structures, Part 1.12 in prep. *Additional rules for the extension of EN 1993 up to steel grades S 700 (prEN 1993-1-12:20xx)*
Klöppel, K. & Möller, K. 1968. *Beulwerte ausgesteifter Rechteckplatten, Band II*. Berlin: Ernst & Sohn Verlag
Klöppel, K. & Scheer, J. 1960. *Beulwerte ausgesteifter Rechteckplatten*. Berlin: Ernst & Sohn Verlag
Lifthigh 2005. *Final Report: Efficient lifting equipment with extra high strength steel; ECSC P 4553*. not published yet
Müller, C. 2003. *Zum Nachweis ebener Tragwerke aus Stahl gegen seitliches Ausweichen, Schriftenreihe Stahlbau – RWTH Aachen; Heft47*. Aachen: Shaker Verlag
Schafer, B.W. & Peköz, T. 1998. Computational modelling of cold-formed steel: characterizing geometric imperfections and residual stresses. *Journal of Constructional Steel Research* 47 : 193–210
Timoshenko, S.P. & Woinowsky-Krieger, S. 1959. *Theory of Plates and Shells*. New York: McGraw-Hill Book Company, Inc.

Tubular Structures XI – Packer & Willibald (eds)
© 2006 Taylor & Francis Group, London, ISBN 0-415-40280-8

Study on effect of curved chords on behavior of welded circular hollow section joints

L.W. Tong, B. Wang & Y.Y. Chen
Dept. of Building Engineering, College of Civil Engineering, Tongji Univ., Shanghai, P. R. China

Z. Chen
Dept. of Civil Engineering, National Univ. of Singapore, Singapore

ABSTRACT: Welded trusses made of circular hollow sections can be commonly seen in large span roof structures of buildings. The curved chord members are used sometimes, which causes structural engineers to answer the question how to check the strength of welded truss joints. In the present paper, nine model static tests on circular hollow section joints with curved and straight chord members were carried out. Behavior in failure modes, load-deformation curves, stress distributions and load capacities is compared between joints with curved and straight chord members. All test results are discussed and show that the curved circular chord members do not exert more significant influence upon the behavior of joints, in comparison with the straight chord members. The conclusion is further confirmed through the finite element analysis of a series of joint types. Both the experimental and numerical results indicated that the strength of the joints with curved chords at a wide range of curvature radiuses used in practical engineering can be checked like the joints with straight chords based on the current design specifications.

1 INTRODUCTION

Circular hollow sections, with their excellent structural and mechanical properties, are widely used in structural engineering. Welded tubular joints are commonly adopted as the main component of construction elements when using such circular hollow section. Intensive researches concerning the static behavior of welded tubular joints have been carried out since 1950s. Many experimental and empirical formulas have been developed for most types of circular joints (Hamed et al. 2002, Wardenier et al. 2002, Lee & Wilmshurst 1997).

In China, welded trusses made of circular hollow sections can be commonly seen in large span roof structures of buildings such as airport terminals, stadiums, gymnasiums and exhibitions centers. In these structures, curved circular hollow sections have been increasingly used in order to make the needs of more attractive architectural appearance. The needs cause structural engineers to answer the question how to check the strength of welded tubular joints with curved chord members. However, none of the above formulas or any current design codes in the world provide any specific guidance for the static strength of tubular joints with curved chords. Most engineers just treat them as tubular joints with straight chords, while it is still unknown whether it is ok or not that the tubular joints with curved chords can be considered those with straight chords to calculate their strength.

In this paper, to compare the static behavior between joints with curved and straight chord members but the same joint types, nine model static tests on circular hollow section joints with curved and straight chord members were carried out, the background of which is Shanghai Qizhong Tennis Court. The configurations of joints experimented were concentrated on multiplanar TT, KK and KTT joints and planar K joints. These specimens were then modeled by means of the finite-element (FE) method using ANSYS software package. The results obtained from such numerical analysis were compared with those obtained from the tests. Good agreement was obtained. Further researches were made for the multiplanar TT joints.

2 EXPERIMENTAL INVESTIGATION

2.1 *Specimen details*

Throughout this work, three multiplanar TT-joints (two with curved chords and another one with straight chord), two multiplanar KK-joints (one with curved chord and the other one with straight chord), two

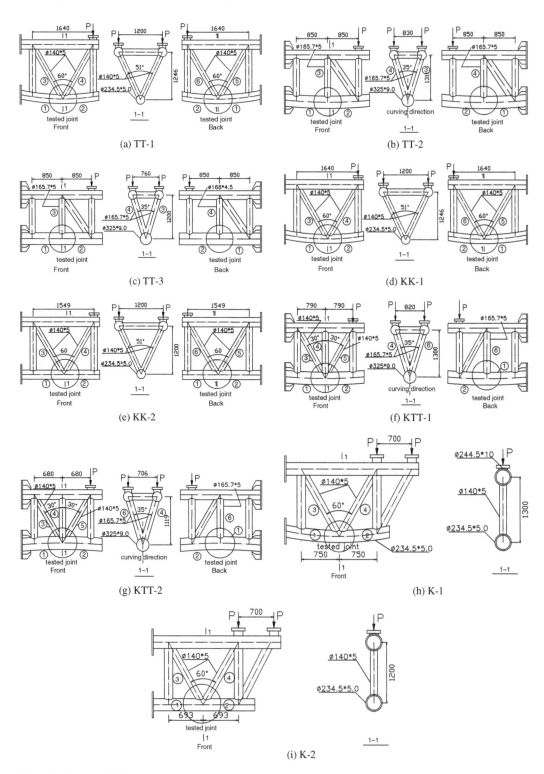

Figure 1. Specimen details.

Table 1. Geometrical parameters of specimen.

Specimens	D_0 (mm)	t_0 (mm)	β	γ	τ
TT-1 ∼ TT-3	325.0	9.0	0.510	18.056	0.556
KK-1/KK-2	234.5	5.0	0.597	46.900	1.000
KT-M	234.5	5.0	0.597	46.900	1.000
KTT-1/KTT-2	325.0	9.0	0.473	18.056	0.556
KT-P	325.0	9.0	0.431	18.056	0.556
K-1/K-2	234.5	5.0	0.597	46.900	1.000

Table 2. Material properties.

Tube size	Yield strength, f_y (N/mm^2)	Ultimate strength, f_u (N/mm^2)
Ø140.0 × 5.0	438	557
Ø165.7 × 5.0	398	538
Ø234.5 × 5.0	369	560
Ø325.0 × 9.0	387	572

Figure 2. Overview of test arrangement.

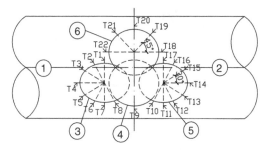

Figure 3. General arrangement of strain gauges.

Figure 4. General arrangement of transducer.

multiplanar KTT (both with curved chord) and two planar K-joints(one with curved chord and the other one with straight chord) were used for tests. The tests of all types of joints were performed through a series of multiplanar or planar trusses with a few panels. The configurations and the detailed dimensions as well as geometric properties of the specimen are shown in Figure 1 and Table 1. The curvature radius of all curved chord is 6000 mm.

Noting that for curved chord, there are two possible curving directions. In the present paper, if the center of curvature of a curved chord and braces lie on the same side, like TT-2, KK-1, KTT-1 and K-1, the curved chord is called "downward curved chord". If the center of curvature of a curved chord and braces lie on opposite sides, like TT-1, KK-2 and KTT-2, the curved chord is called "upward curved chord".

The chord and braces of each specimen are fabricated from seamless hot finished, carbon steel pipe. Table 2 summarizes the measured yield stress for each of the chord and brace materials.

These dimensions and material properties were also used for the finite element input file where similar models were created and loaded.

2.2 Test set-up

Figure 2 shows the general arrangement for the specimen tests (Chen Yiyi et al. 2003). One end of each specimen was fixed to a triangular truss through a connecting flat plate and the other end was free. The loading was applied to the free end of the specimen.

For each specimen, the instrumentations include strain gauges around the crown and saddle points to measure the strain distributions at the hot spots and transducers to measure the displacements at selected points. Figure 3 shows the general arrangement of the strain gauges. Figure 4 shows the general arrangement of the transducers. Although not visible in Figure 4, three transducers were placed under the chord at the two ends and the center of the chord, monitoring the vertical displacements of the whole specimen.

Load-displacement relationships were recorded for each test carried out. The critical load is the maximum load which the joint can withstand. After reaching that load, the test was bought to an end.

2.3 Test results

2.3.1 Failure modes

Figures 5 and 6 show the failure modes of specimen joints. From the experimental observation, there

Figure 5. Failure mode one.

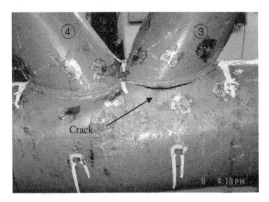

Figure 6. Failure mode two.

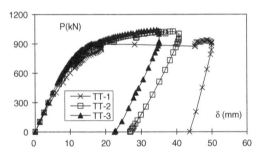

Figure 7. Global load-displacement curves of TT-1 ~ TT-3.

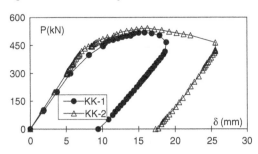

Figure 8. Global load-displacement curves of KK-1 & KK-2.

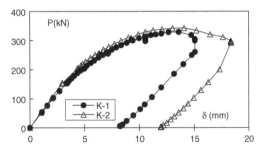

Figure 9. Global load-displacement curves of K-1 & K-2.

are two main failure modes: (1) brace local buckling; (2) combination of weld toe crack and chord side wall yielding.

The failure modes of all specimens with chord type of Ø325 × 9, including TT-1 ~ TT-3, KTT-1 ~ KTT-2, belong to the first failure mode. Those of the others with chord type of Ø234.5 × 5, including KK-1 ~ KK-2, K-1~K-2, belong to the second failure mode. Noting that for all specimens, the brace sizes are quite similar, one can see that the thickness of chord wall has significant influence on the failure modes of tubular joints.

From the test results, it also can be seen that, specimens with the same joint types (TT, KK etc.) but different chord types (straight, upward curved and downward curved) always have the same failure mode. This suggests that the existence of curved chords would not lead to significant change of failure mode of tubular joints.

2.3.2 Global load-displacement curves

Figures 7 to 9 show some typical global load-displacement curves of specimens. From the three figures, it can be seen that for the tubular joints with same joint type, the global load-displacement curves

of joints with curved chords and those of joints with straight chords have good agreements, especially in the linear parts, which are almost the same.

From Figure 7, it also can be seen that the global load-displacement curves of specimen joints with upward curved chords (TT-1) and downward curved chord (TT-2) also approximate each other very well.

In addition, Figures 7 to 9 also show that, except for a little difference in the plastic deformation, not only the loads at which the non-linearity of the curves were initiated but also the peak loads for the joints with same joint types and different chord types are all quite similar.

2.3.3 Strain strength distributions of joints

The strain strength distributions around the braces and chords intersections of specimen joints are

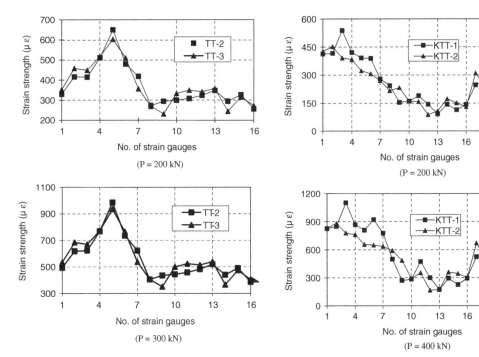

Figure 10. Strain strength distributions of TT-2 & TT-3.

complicated and very important for the understanding of the static behavior of joints. To investigate the properties of these strain strength distributions and their developments, strain gauges were arranged on the chords around braces as shown in figure 3. According to the stains measured at these points, corresponding strain strengths can be calculated, which can reflect the state of the development of plasticity at these locations. Figures 10 to 12 show some strain strength distributions of specimens at certain global loading (P) level.

From Figures 10 and 12, it can be seen that at the same loading level, the strain strength distributions along the brace and chord intersections of the specimen joints with downward curved chords (TT-2 and K-1) and those of the specimen joints with straight chords (TT-3 and K-2) are quite similar. From Figures 11, it can be seen that at the same loading level, the strain strength distribution along the brace and chord intersection of specimen joint with downward curved chord (KTT-1) and that of specimen joint with upward curved chord (KTT-2) also approximate each other quite well.

2.3.4 Ultimate capacity of joints

The experimental results drawn from this study, for the ultimate capacity of joints, were tabulated and presented for all specimens in Table 3.

The ultimate capacity given in Table 3 for each specimen point is the lowest brace axial force when the joint lost its loading capacity.

Figure 11. Strain strength distributions of KTT-1 & KTT-2.

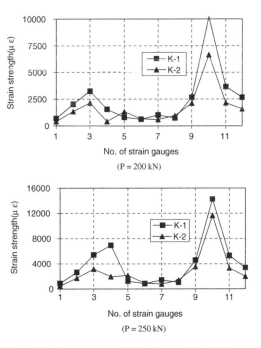

Figure 12. Strain strength distributions of K-1 & K-2.

Table 3. Experimental results for ultimate loads.

| Type | Straight chord | | Curved chord | | |
	No.	Ultimate capacity N_s (kN)	No.	Ultimate capacity N_c (kN)	N_c/N_s
TT	TT-3	625	TT-1	628	1.01
			TT-2	611	0.98
KK	KK-1	442	KK-2	425	0.96
KTT			KTT-1	759	/
			KTT-2	786	/
K	K-1	373	K-2	365	0.98

From Table 3, it can be seen that the ultimate capacity of specimen joints with the same joint types (TT, KK etc.), but different chord types (curved or straight) as well as different curving direction (upward or downward), do not show significant difference.

3 FINITE ELEMENT SIMULATION

3.1 *Finite element model*

For the numerical analysis, the FE package ANSYS is employed. The dimensions and material properties of the tubular members assumed in the FE analysis are accordance with the measured dimensions and material properties of the test specimens. Due to the complexity of specimens, only tested joint parts of the specimen trusses were modeled. The mesh was created using shell elements (four nodes SHELL 181 Element). A fine regular mesh was created around the intersection of the brace and chord. Figure 13 shows the typical finite-element models devised for the investigation of multiplanar TT-joint specimen and planar K-joint specimen. In this investigation, in-plane bending and out-of-plane bending loading were not included.

3.2 *Comparison between experimental and numerical results*

3.2.1 *Failure modes based on FE analysis*
Figures 14 and 15 show two typical deformation shapes of numerical models when the computations finished. From the two figures, it can be seen that the FE models have the same failure modes with the corresponding tested specimens. Noting that no weld crack happened for the FE model, it may be due to the fact that when simulated the specimen using ANSYS, detailed welded fillet was not modeled.

3.2.2 *Strain developments at points*
The computed strain strength developments with the increasing of global loading at saddle and crown points

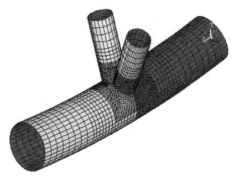

(a) TT-2 (with downward curved chord)

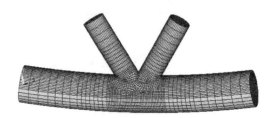

(b) K-1(with downward curved chord)

Figure 13. Typical FE models.

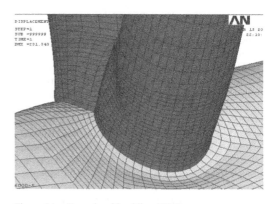

Figure 14. Brace local buckling (TT-2).

of FE model of joint TT-1 are plotted in Figure 16 together with corresponding test results. In this figure, the x-axis is corresponding to strain strength while the y-axis is corresponding to the global loading.

It can be seen from Figure 16 that the numerical and test results have very good agreements.

3.2.3 *Strain strength distribution at joints*
Figure 17 shows some comparisons of the stain strength distributions around the intersections of the braces and chords between the numerical and test results for the joint KK-2.

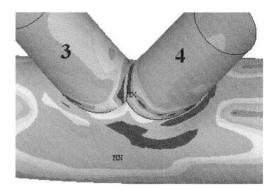

Figure 15. Chord side wall yielding (K-1).

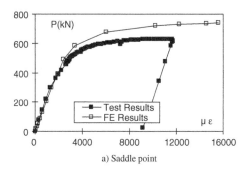

a) Saddle point

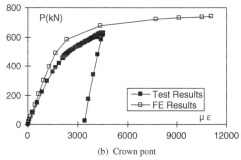

(b) Crown pont

Figure 16. Strain developments of TT-1.

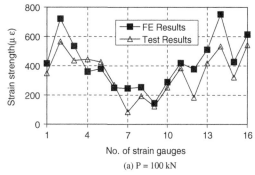

(a) P = 100 kN

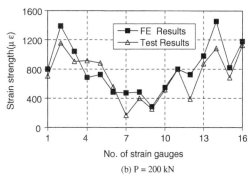

(b) P = 200 kN

Figure 17. Strain strength distributions for specimen KK-2.

Table 4. Comparison between test and numerical ultimate capacities.

Joints	$N_{u,test}$ (kN)	$N_{u,num}$ (kN)	$N_{u,num}/N_{u,test}$
TT-1	628	739	1.17
TT-2	611	626	1.02
TT-3	625	720	1.15
KK-1	442	354	0.80
KK-2	425	342	0.80
K-1	373	435	1.17
K-2	365	438	1.20

As it is shown in Figure 17, the computed strain strength distributions around the intersections of braces and chords and the corresponding test results match quite well.

3.2.4 *Ultimate capacity*

The computed ultimate capacities of specimen joints are summarized in Table 4.

As can be observed from Table 4, the numerical and experimental results match very well for the ultimate loads and the discrepancies are between −20% and 20%.

4 NUMERICAL PARAMETRIC STUDY ON TUBULAR JOINTS WITH CURVED CHORD

Because of the ability of FE models to produce results which closely match the static response found in the experiments, a parametric numerical study is set up to further investigate the static behavior of tubular joints with curved chord under brace compression. In the parametric study, the dimensions and material properties of TT-3 were used and two possible curving directions (upward and downward) were analyzed. All geometrical parameters remained unchanging while the curvature radius of the chord changed from 1.1 m to 36 m. A total of 58 TT-joints with curved chord

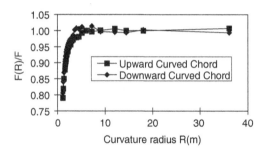

Figure 18. Ultimate load of TT- joints with curved chord.

(29 with upward curved chord and 29 with downward curved chord) were analyzed. A summary of the joints investigated is given in Figure 18, in which the x-axis is the curvature radius of chord and the y-axis is the ratio between the ultimate load P(R) of TT-joint with curved chord (R is the curvature radius of corresponding curved chord) and the ultimate load P of corresponding TT-joint with straight chord.

From Figure 18, it can be seen that, when larger than certain value, the value of curvature radius almost has no influence on the ultimate loads of TT-joints with curved chord. However, when the value of curvature radius is small enough, the ultimate load of TT-joint with curved chord shows an obvious drop. It also can be seen that in this investigation, the lowest ultimate load found for TT-joints with curved chord is only about 20% below that of TT-joint with straight chord. When that obvious drop in ultimate load appears, the curvature radius of curved chord is close to 3 m, which is much smaller than the curvature radius usually used in real structures. In addition, at the same curvature radius, the curving direction of the curved chord (upward or downward) does not show obvious influence on the ultimate load of joint.

Similar numerical analyses for welded tubular joints with other joint types also have been carried out. According to these computed results, it can be concluded that if the value (R) of the curvature radiuses of the curved chords is larger than 5m and the ratio between the curvature radiuses and the outer diameter of chords $(a = R/D_0)$ is between 12 and 110, the tubular joints with the curved chords can be treated as corresponding joints with the straight chords. As a result, their strength can be calculated using formulas drawn from corresponding joints with the straight chords.

5 CONCLUSIONS

A series of 9 tests on multiplanar and uniplanar joints with straight and curved chords have been conducted. FE analyses were conducted to simulate the static behaviors on these tested joints. The main variable studied is the curvature radius of curved chord. From the present results, the following conclusions can be drawn:

1. For all types of joints tested, the static behaviors of tubular joints with straight and curved chord, if having the same joint types, do not show obvious differences.
2. The thicknesses of chords have significant influence on the failure mode of tubular joints.
3. Reasonable agreement was obtained between the nonlinear finite element analysis results and the experimental results for all tests.
4. The existence of curved chord did have some influences on the ultimate capacity of tubular joint, but it is still safe to calculate their strength like those with straight chord in normal cases.
5. If the value of curvature radius of curved chord (R) is larger than 5 m and the ratio between the curvature radius and outer diameter of chord $(a = R/D_0)$ is between 12 and 110, the influences of curved chords on the ultimate capacity of tubular joints can be ignored.

REFERENCES

A.F. Hamed, Y.A. Khalid, B.B. Sahari & M.M. Hamdan. 2002. Finite Element and Experimental Analysis for the Effect of Elliptical Chord Shape on Tubular T-Joint Strength. Proc Instn Mech Engrs, Vol 215 Part E: 123–130.

Chen Y., S. Zuyan, Z. Hong, C. Yangji, C. Rongyi, G. Mosong. 2003. Experimental Research on Hysteretic Property of Unstiffened Space Tubular Joints. Journal of Building Structures, Vol 24, No 6:57–62.

J. Wardenier. 2002. Hollow Sections in Structural Applications. the Netherlands. Bouwen met Staal.

M.M.K. Lee & S.R. Wilmshurst. 1997. Strength of Multi-planar Tubular KK-Joints under Antisymmetrical Axial Loading. Journal of Structural Engineering, Vol 123, No 6:755–764.

Parallel session 3: Static strength of members/frames

Tubular Structures XI – Packer & Willibald (eds)
© *2006 Taylor & Francis Group, London, ISBN 0-415-40280-8*

Experimental and numerical studies of elliptical hollow sections under axial compression and bending

T.M. Chan & L. Gardner
Department of Civil and Environmental Engineering, Imperial College London, London, UK

ABSTRACT: A series of tests were performed on hot-rolled steel elliptical hollow section members. Nineteen stub column tests and six four-point bending tests about the minor axis were conducted. Measurements were taken of cross-section geometry, local and global initial geometric imperfections and material properties in tension. Results including full load-end shortening curves for the stub column tests and full moment-deflection histories for the beam tests are presented; these results have been used to calibrate the numerical models and for the development of a system of cross-section slenderness parameters and limits.

1 INTRODUCTION

The earliest examples of the use of elliptical hollow sections (EHS) in construction date back to the mid-nineteenth century, one of the most prominent being Brunel's Royal Albert Bridge in the UK, where the main structural skeleton is elliptical. However, historically such structural forms have been of a bespoke nature and it is only recently that EHS have been introduced to the construction industry as a standard hot-rolled product. Recent projects that have adopted hot-rolled EHS include the Coach Station at Heathrow Terminal 3 in London, UK and the Jarrold Department Store in Norwich, UK (Gardner & Ministro 2005).

Despite widespread interest in the structural application of EHS, there is currently a lack of structural performance data (experimental and numerical) and verified structural design rules. In developing design rules for connections between EHS, Bortolotti et al. (2003), Choo et al. (2003) and Pietrapertosa & Jaspart (2003) have expressed the need for research on EHS in a range of configurations. Thus, the primary objective of this paper is to present the results from a series of full-scale laboratory experiments and numerical studies in compression and bending. A companion paper utilises the results of the studies described herein to develop a system of section classification for EHS. The described research forms part of a wider study on the structural behaviour of elliptical hollow sections currently underway at Imperial College London.

2 EXPERIMENTAL STUDIES

A series of precise full-scale laboratory tests on EHS (grade S355), manufactured by Corus Tubes (Corus 2004), was performed at Imperial College London. The test programme comprised tensile coupon tests, stub column tests and bending tests.

2.1 Tensile coupon tests

The basic stress-strain behaviour of the material for each of the tested section sizes were determined through tensile coupon tests carried out in accordance with EN 10002-1 (2001).

Two parallel coupons, each with the nominal dimensions of 360×30 mm, were machined longitudinally along the centreline of the minor axis (the flattest portions) of each of the tested cross-sections. The tensile tests were performed using an Amsler 350 kN hydraulic testing machine. To prevent slippage of coupons in the jaws of the testing machine, holes were drilled and reamed 20 mm from each end of the coupons for pins to be inserted.

Linear electrical strain gauges were affixed at the midpoint of each side of the tensile coupons. A series of overlapping proportional gauge lengths was marked onto the surface of the coupons to determine the elongation parameters with an average value of elongation at fracture of 32%. Load, strain, displacement and input voltage were all recorded using the data acquisition equipment DATASCAN and logged using the DALITE computer package. All data were recorded at one second intervals. Mean measured dimensions and the key results from the nineteen tensile coupon tests are reported in Tables 1 and 2. A typical stress-strain curve is depicted in Figure 1.

2.2 Stub column tests

Stub column tests were conducted to develop the relationship between cross-section slenderness,

Table 1. Mean measured dimensions of tensile coupons.

Specimen	Width b (mm)	Thickness t (mm)
150 × 75 × 6.3-TC1	29.97	6.30
150 × 75 × 6.3-TC2	29.91	6.35
150 × 75 × 8.0-TC1	29.97	8.3
150 × 75 × 8.0-TC2	29.93	8.35
300 × 150 × 8.0-TC1	29.90	7.63
300 × 150 × 8.0-TC2	29.95	7.67
300 × 150 × 8.0-TC3	29.93	7.79
400 × 200 × 8.0-TC1	29.60	7.53
400 × 200 × 8.0-TC2	29.60	7.62
400 × 200 × 10.0-TC1	30.05	9.52
400 × 200 × 10.0-TC2	30.07	9.54
400 × 200 × 12.5-TC1	29.25	12.03
400 × 200 × 12.5-TC2	29.26	11.98
400 × 200 × 14.0-TC1	29.49	14.34
400 × 200 × 14.0-TC2	29.55	14.33
400 × 200 × 16.0-TC1	29.98	15.13
400 × 200 × 16.0-TC2	30.00	15.33
500 × 250 × 8.0-TC1	29.72	7.59
500 × 250 × 8.0-TC2	29.57	7.56

Table 2. Key results from tensile coupon tests.

Specimen	Young's Modulus E (N/mm^2)	Yield stress σ_y (N/mm^2)	Ultimate tensile stress σ_u (N/mm^2)
150 × 75 × 6.3-TC1	212100	406	517
150 × 75 × 6.3-TC2	221100	415	541
150 × 75 × 8.0-TC1	209500	369	502
150 × 75 × 8.0-TC2	216700	386	518
300 × 150 × 8.0-TC1	217700	415	536
300 × 150 × 8.0-TC2	209600	419	537
300 × 150 × 8.0-TC3	215100	408	524
400 × 200 × 8.0-TC1	222000	434	559
400 × 200 × 8.0-TC2	221200	424	541
400 × 200 × 10.0-TC1	191800	396	527
400 × 200 × 10.0-TC2	202400	406	540
400 × 200 × 12.5-TC1	215200	388	525
400 × 200 × 12.5-TC2	215000	402	544
400 × 200 × 14.0-TC1	220100	387	533
400 × 200 × 14.0-TC2	220100	408	535
400 × 200 × 16.0-TC1	221200	377	531
400 × 200 × 16.0-TC2	221300	380	519
500 × 250 × 8.0-TC1	219900	409	532
500 × 250 × 8.0-TC2	227700	417	540

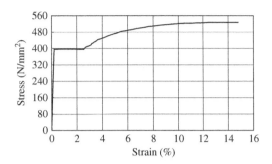

Figure 1. 400 × 200 × 10-TC1 stress–strain curve.

Figure 2. Testing arrangements for stub column.

shortening of the stub columns. Four linear electrical resistance strain gauges were affixed to each specimen at mid-height, and at a distance of four times the material thickness from the major axis. The strain gauges were initially used for alignment purposes. The testing arrangement is shown in Figure 2. The end platens were fixed flat and parallel. Load, strain, displacement, and input voltage were all recorded using the data acquisition equipment DATASCAN and logged using the DALITE and DSLOG computer package. All data were recorded at one second intervals.

A typical failure mode for the elliptical hollow section stub columns is shown in Figure 3.

The geometry of an elliptical hollow section is depicted in Figure 4 and the mean measured dimensions and key results from the nineteen stub column tests are summarised in Tables 3 to 5. In Table 5, the ultimate test load F_u has been normalised by the squash load $F_y = A\sigma_y$.

Figures 5 and 6 show the normalised load versus end-shortening curves for the twelve most recently performed stub column tests (on the 400 × 200 and

deformation capacity and normalised load-carrying capacity for elliptical hollow sections under uniform axial compression. A total of nineteen stub column tests were performed. Full load-end shortening curves were recorded, including into the post-ultimate range. The nominal length of the stub column was two times the largest outer diameter of the cross-section. Four LVDTs located between the parallel end platens of the testing machine were used to determine the end

Figure 3. Typical failure mode of stub columns.

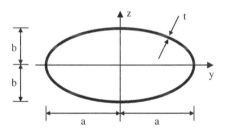

Figure 4. Geometry of elliptical hollow section.

500×250 section sizes). Similar curves for the seven earlier tests (on the 150×75 and 300×150 section sizes) have been reported by Gardner (2005). In these Figures, the test load F has been normalised by the squash load F_y and plotted against end-shortening δ which has been normalised by the stub column length L. Each stub column test was repeated. The curves in Figure 5 are from the first series of tests (SC1), whilst those in Figure 6 are from the second series (SC2).

2.3 Bending tests

A full scale four-point minor axis bending experimental study of the bending behaviour of EHS has

Table 3. Mean measured dimensions of stub column specimens.

Specimen	Major axis outer diameter 2a (mm)	Minor axis outer diameter 2b (mm)
$150 \times 75 \times 6.3$-SC1	149.83	74.87
$150 \times 75 \times 6.3$-SC2	150.24	75.16
$150 \times 75 \times 8.0$-SC1	150.11	75.10
$150 \times 75 \times 8.0$-SC2	149.17	75.07
$300 \times 150 \times 8.0$- SC1	299.67	149.99
$300 \times 150 \times 8.0$-SC2	300.04	149.79
$300 \times 150 \times 8.0$-SC3	301.64	148.90
$400 \times 200 \times 8.0$-SC1	395.73	207.36
$400 \times 200 \times 8.0$-SC2	399.41	202.70
$400 \times 200 \times 10.0$-SC1	394.55	209.49
$400 \times 200 \times 10.0$-SC2	396.20	207.16
$400 \times 200 \times 12.5$-SC1	402.20	200.41
$400 \times 200 \times 12.5$-SC2	402.31	199.47
$400 \times 200 \times 14.0$-SC1	400.51	199.47
$400 \times 200 \times 14.0$-SC2	399.59	201.91
$400 \times 200 \times 16.0$-SC1	403.45	201.18
$400 \times 200 \times 16.0$-SC2	403.43	200.63
$500 \times 250 \times 8.0$-SC1	492.35	261.20
$500 \times 250 \times 8.0$-SC2	488.42	259.09

Table 4. Mean measured dimensions of stub column specimens.

Specimen	Thickness t (mm)	Area A (mm^2)	Length L (mm)
$150 \times 75 \times 6.3$-SC1	6.52	2236	451.3
$150 \times 75 \times 6.3$-SC2	6.34	2184	298.5
$150 \times 75 \times 8.0$-SC1	8.66	2920	302.6
$150 \times 75 \times 8.0$-SC2	8.51	2859	297.2
$300 \times 150 \times 8.0$-SC1	7.95	5578	598.7
$300 \times 150 \times 8.0$-SC2	7.97	5595	599.4
$300 \times 150 \times 8.0$-C3	7.80	5493	600.1
$400 \times 200 \times 8.0$-SC1	7.63	7228	799.5
$400 \times 200 \times 8.0$-SC2	7.65	7250	799.4
$400 \times 200 \times 10.0$-SC1	9.60	9044	799.4
$400 \times 200 \times 10.0$-SC2	9.56	9004	799.8
$400 \times 200 \times 12.5$-SC1	12.01	11249	799.9
$400 \times 200 \times 12.5$-SC2	12.07	11285	799.5
$400 \times 200 \times 14.0$-SC1	14.42	13341	799.7
$400 \times 200 \times 14.0$-SC2	14.43	13369	800.0
$400 \times 200 \times 16.0$-SC1	15.35	14267	799.6
$400 \times 200 \times 16.0$-SC2	15.37	14273	799.7
$500 \times 250 \times 8.0$-SC1	7.59	9019	1000.0
$500 \times 250 \times 8.0$-SC2	7.63	8992	999.8

also been conducted. The four-point bending arrangement (Figures 7 and 8) provides a uniform moment region between loading points and allows study of the pure moment-curvature behaviour of EHS. The tested beams were loaded at the third points along the span using two 50 T Instron hydraulic actuators with displacement control through an Instron control cabinet

Table 5. Summary of results from stub column tests.

Specimen	Ultimate load F_u (kN)	End shortening at F_u (mm)	F_u/F_y
150 × 75 × 6.3-SC1	931	13.7	1.03
150 × 75 × 6.3-SC2	952	10.5	1.05
150 × 75 × 8.0-SC1	1367	18.1	1.27
150 × 75 × 8.0-SC2	1435	18.8	1.30
300 × 150 × 8.0-SC1	2777	1.6	1.20
300 × 150 × 8.0-SC2	2792	1.7	1.19
300 × 150 × 8.0-SC3	2574	–	1.15
400 × 200 × 8.0-SC1	2961	1.7	0.95
400 × 200 × 8.0-SC2	3081	2.2	0.99
400 × 200 × 10.0-SC1	3521	2.6	0.97
400 × 200 × 10.0-SC2	3693	1.7	1.02
400 × 200 × 12.5-SC1	4727	2.4	1.06
400 × 200 × 12.5-SC2	4623	7.4	1.04
400 × 200 × 14.0-SC1	5610	18.7	1.06
400 × 200 × 14.0-SC2	5610	19.1	1.06
400 × 200 × 16.0-SC1	6310	24.7	1.17
400 × 200 × 16.0-SC2	6159	21.4	1.14
500 × 250 × 8.0-SC1	3684	2.5	0.99
500 × 250 × 8.0-SC2	3546	2.8	0.96

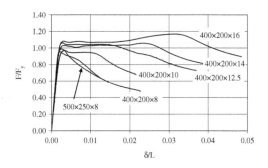

Figure 5. Normalised load-end shortening curves for stub columns (series 1).

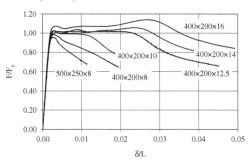

Figure 6. Normalised load-end shortening curves for stub columns (series 2).

at a rate between 1 mm/min and 3 mm/min. Plates were welded at the loading points and supports to prevent localised failure. Three LVDTs were located along the underside of the specimens, between the loading

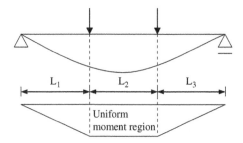

Figure 7. Schematic illustration of four-point bending arrangement.

Figure 8. Four-point bending arrangement.

Table 6. Measured dimensions for four-point bending arrangement.

Specimen	L_1 (mm)	L_2 (mm)	L_3 (mm)
400 × 200 × 8.0-B1	1495	1495	1494
400 × 200 × 10.0-B1	1495	1497	1493
400 × 200 × 12.5-B1	1494	1499	1499
400 × 200 × 14.0-B1	1493	1497	1496
400 × 200 × 16.0-B1	1495	1500	1495
500 × 250 × 8.0-B1	1494	1498	1496

points, to determine the average curvature. Two additional LVDTs were positioned at each end of the beam to measure end rotation. Two linear electrical resistance strain gauges were affixed to the extreme tensile and compressive fibres of the section at a distance of 150 mm from the mid-span of the beam. Load, strain, displacement, and input voltage were all recorded using the data acquisition equipment DATASCAN and logged using the DSLOG computer package. All data were recorded at one second intervals.

Mean measured dimensions and the key results from the bending tests are reported in Tables 6 to 8. The ultimate test moment M_u is normalised by the elastic moment resistance M_{el} and the plastic moment

Table 7. Mean measured cross-sectional dimensions for bending specimens.

Specimen	Major axis outer diameter 2a (mm)	Minor axis outer diameter 2b (mm)	Thickness t (mm)
400 × 200 × 8.0-B1	396.09	207.63	7.75
400 × 200 × 10.0-B1	396.06	207.54	9.65
400 × 200 × 12.5-B1	401.54	201.01	12.13
400 × 200 × 14.0-B1	400.32	200.04	14.48
400 × 200 × 16.0-B1	403.16	201.08	15.63
500 × 250 × 8.0-B1	495.34	255.85	7.78

Table 8. Summary of results of four-point bending tests.

Specimen	Ultimate moment M_u (kNm)	M_u/M_{el}	M_u/M_{pl}
400 × 200 × 8.0-B1	186	1.09	0.86
400 × 200 × 10.0-B1	232	1.19	0.94
400 × 200 × 12.5-B1	288	1.27	0.99
400 × 200 × 14.0-B1	343	1.31	1.01
400 × 200 × 16.0-B1	331	1.23	0.94
500 × 250 × 8.0-B1	291	1.12	0.89

resistance M_{pl} in Table 8. Full experimental moment-deflections curves are compared with those obtained numerically in section 3 of this paper.

3 NUMERICAL SIMULATIONS

A numerical modelling study, using the finite element (FE) package ABAQUS (2005), was carried out in parallel with the experimental programme. The initial aim of the programme was to replicate the experimental compression and bending behaviour numerically, before performing parametric studies. The elements chosen for the FE models were 4-noded, reduced integration shell elements, designated as S4R in the ABAQUS element library, and suitable for thin or thick shell applications (ABAQUS 2005). Convergence studies were conducted to decide upon an appropriate mesh density, with the aim of achieving suitably accurate results whilst minimising computational time. Satisfactory results were obtained using uniform mesh densities throughout the models.

3.1 Stub columns

The stub column tests were modelled using the measured dimensions of the test specimens, measured material stress-strain data from the corresponding tensile tests and geometric imperfections of the form of

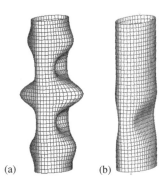

(a)   (b)

Figure 9. (a) Lowest eigenmode for 400 × 200 × 14-SC1 stub column, (b) Failure mode of 400 × 200 × 14-SC1 stub column.

Table 9. Comparison of stub column test results with FE results for varying imperfection amplitude w_0.

Specimen	FE F_u/Test F_u		
	$w_0 = t/10$	$w_0 = t/100$	$w_0 = t/500$
400 × 200 × 8.0-SC1	1.03	1.05	1.04
400 × 200 × 8.0-SC2	0.99	1.01	1.01
400 × 200 × 10.0-SC1	1.01	1.03	1.03
400 × 200 × 10.0-SC2	0.96	0.98	0.98
400 × 200 × 12.5-SC1	0.93	0.95	0.95
400 × 200 × 12.5-SC2	0.95	0.97	0.97
400 × 200 × 14.0-SC1	0.96	0.97	0.96
400 × 200 × 14.0-SC2	0.94	0.98	1.01
400 × 200 × 16.0-SC1	0.86	0.90	0.92
400 × 200 × 16.0-SC2	0.87	0.92	0.94
500 × 250 × 8.0-SC1	0.99	1.01	1.01
500 × 250 × 8.0-SC2	0.94	0.97	0.97

the lowest elastic eigenmode (Figure 9a). Sensitivity to variation in imperfection amplitude w_0 was assessed by considering three values of w_0 – t/10, t/100 and t/500, where t is the material thickness. Material non-linearity was incorporated by means of a multi-linear stress-strain model. Boundary conditions were applied to model fixed ends. This was achieved by restraining all displacements and rotations at the base of the stub columns, and all degrees of freedom except vertical displacement at the loaded end of the stub columns. Constraint equations were used to ensure that the loaded end of the stub columns remained in a horizontal plane. The modified Riks method (ABAQUS 2005) was employed to solve the geometrically and materially non-linear stub column models, which enabled the unloading behaviour to be traced. A typical failure mode is shown in Figure 9b. Results of the numerical simulations are tabulated in Table 9. In the Table, the ratio between the FE ultimate load and the experimental ultimate load are shown and compared at different imperfection amplitudes.

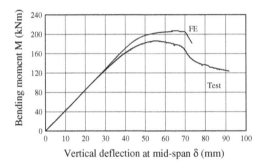

Figure 10. Moment-deflection curve for 400 × 200 × 8.0-B1.

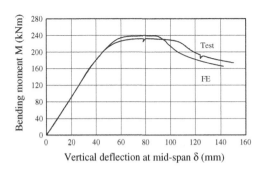

Figure 11. Moment-deflection curve for 400 × 200 × 10.0-B1.

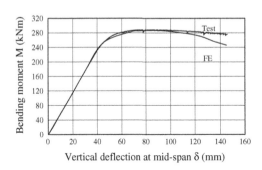

Figure 12. Moment-deflection curve for 400 × 200 × 12.5-B1.

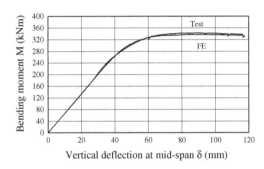

Figure 13. Moment-deflection curve for 400 × 200 × 14.0-B1.

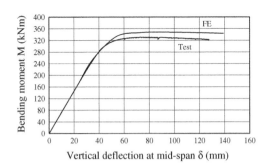

Figure 14. Moment-deflection curve for 400 × 200 × 16.0-B1.

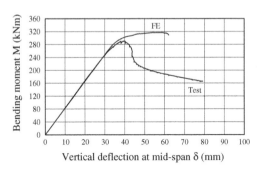

Figure 15. Moment-deflection curve for 500 × 250 × 8.0-B1.

3.2 Bending

For the case of bending, a similar modelling technique to that described for the stub columns was employed. For the replication of tests, measured geometry and measured material properties were specified. Boundary conditions were applied to model the simple supports of the testing arrangement. The test and FE moment versus mid-span vertical deflection curves are shown in Figures 10 to 15. The displayed FE models contain local imperfections of amplitude w_0 equal to $t/10$, where t is the material thickness.

Studies of geometric imperfections revealed that the FE models were insensitive to variation in global imperfection amplitude, thus, a global imperfection amplitude or geometrical 'out-of-straightness' of $L/500$ (prEN 10210-2 2003) was employed throughout the study. For the local imperfection, three different amplitudes w_0 equal to $t/10$, $t/100$ and $t/500$, where t is the material thickness, were considered. Results of the numerical simulations are summarised in Table 10. A deformed FE model is shown in Figure 16.

The results of Figures 10 to 15 and Table 10 generally reveal good agreement between test and FE

168

Table 10. Comparison of beam test results with FE results for varying imperfection amplitude w_0.

Specimen	FE M_u/Test M_u		
	$w_0 = t/10$	$w_0 = t/100$	$w_0 = t/500$
$400 \times 200 \times 8.0$-B1	1.10	1.14	1.14
$400 \times 200 \times 10.0$-B1	1.03	1.05	1.05
$400 \times 200 \times 12.5$-B1	0.99	1.01	1.01
$400 \times 200 \times 14.0$-B1	0.98	1.00	1.00
$400 \times 200 \times 16.0$-B1	1.05	1.07	1.07
$500 \times 250 \times 8.0$-B1	1.09	1.09	1.09

Figure 16. Deformed FE model of $400 \times 200 \times 14$-B1 beam.

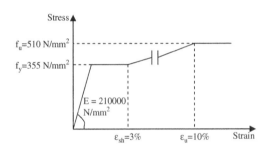

Figure 17. Multi-linear stress–strain relationship.

behaviour. Disparities in the region of the peak moment, particularly apparent in Figure 15, are currently being investigated. Further development of the FE models is underway.

Having verified the general ability of the FE models to replicate test behaviour, a series of parametric studies were conducted. For the parametric studies, Young's modulus E, Poisson's ratio v, yield strength f_y and ultimate strength f_u were determined with reference to EN 1993-1-1 (2005) for grade S355 steel, whilst the plastic plateau and the strain hardening regime were defined from the average results of tensile coupon tests. The strain-hardening strain ε_{sh} was taken as 3% and the strain corresponding to the maximum stress ε_u was taken as 10%, as shown in Figure 17. A local imperfection amplitude w_0 of $t/10$ was used throughout the parametric studies.

Results of the parametric study are presented and utilised for the validation of proposed slenderness parameters and a system of cross-section classification limits for elliptical hollow sections in the companion paper in this Conference (Gardner & Chan 2006).

4 CONCLUSIONS

Results from nineteen tensile coupon tests, nineteen stub column tests and six four-point bending tests have been presented. The tensile tests performed on material cut from finished hot-rolled steel elliptical hollow section members have been presented. Geometric properties and the key findings from the stub column tests and minor axis bending tests have been reported, including full load-end shortening and moment-deflection curves. Numerical models, created using the non-linear FE package ABAQUS, were verified against the test results. Following good agreement between tests and numerical results, parametric studies were performed to extend the range of structural performance data. Results of the described study have been utilised in the development of a system of slenderness parameters and cross-section classification limits for hot-rolled structural steel elliptical hollow sections. The described work forms part of wider study on the structural behaviour of elliptical hollow sections currently underway at Imperial College London.

ACKNOWLEDGEMENTS

The authors are grateful to the Dorothy Hodgkin Postgraduate Award Scheme for the project funding, and would like to thank Corus for the supply of test specimens and for funding contributions, Eddie Hole and Andrew Orton (Corus Tubes) for their technical input and Ron Millward and Alan Roberts (Imperial College London) for their assistance in the laboratory works.

REFERENCES

ABAQUS 2005. *ABAQUS/Standard User's Manual Volumes I-III and ABAQUS CAE Manual. Version 6.5.* Pawtucket, USA, Hibbitt, Karlsson & Sorensen, Inc.

Bortolotti, E., Jaspart, J.P., Pietrapertosa, C., Nicaud, G., Petitjean, P.D. & Grimault, J.P. 2003. Testing and modelling of welded joints between elliptical hollow sections. *Proceedings of the 10th International Symposium on Tubular Structures.* September 2003, Madrid.

Choo, Y.S., Liang, J.X. & Lim, L.V. 2003. Static strength of elliptical hollow section X-joint under brace compression. *Proceedings of the 10th International Symposium on Tubular Structures.* September 2003, Madrid.

Corus 2004. *Celsius 355® Ovals – Sizes and Resistances Eurocode Version*, Corus Tubes – Structural & Conveyance Business.

EN 10002-1 2001. *Metallic materials – Tensile testing – Part 1: Method of test at ambient temperature*, CEN.

EN 1993-1-1 2005. *Eurocode 3: Design of steel structures – Part 1-1: General rules and rules for buildings*, CEN.

Gardner, L. 2005. Structural behaviour of oval hollow sections. *International Journal of Advanced Steel Construction*. **1(2)**, 26–50.

Gardner, L. & Chan, T.M. 2006. Cross-section classification of elliptical hollow sections, *Proceedings of the 11th International Symposium on Tubular Structures*, August 2006, Québec.

Gardner, L. & Ministro, A. 2005. Structural steel oval hollow sections. *The Structural Engineer*. **83(21)**. 32–36.

Pietrapertosa, C. & Jaspart, J.P. 2003. Study of the behaviour of welded joints composed of elliptical hollow sections. *Proceedings of the 10th International Symposium on Tubular Structures*. September 2003, Madrid.

prEN 10210-2 2003. *Hot finished structural hollow sections of non-alloy and fine grain steels – Part 2: Tolerances, dimensions and sectional properties*. CEN.

Tubular Structures XI – Packer & Willibald (eds)
© 2006 Taylor & Francis Group, London, ISBN 0-415-40280-8

Cross-section classification of elliptical hollow sections

L. Gardner & T.M. Chan

Department of Civil and Environmental Engineering, Imperial College London, London, UK

ABSTRACT: Cross-section classification is a fundamental aspect of structural metallic design. This paper proposes slenderness parameters and a system of cross-section classification limits for elliptical hollow sections, developed on the basis of laboratory tests and numerical simulations. Four classes of cross-sections, namely Class 1 to 4 have been defined with limiting slenderness values. For the special case of elliptical hollow sections with an aspect ratio of unity, consistency with the slenderness limits for circular hollow sections in Eurocode 3 has been achieved. The proposed system of cross-section classification underpins the development of further design guidance for elliptical hollow sections.

1 INTRODUCTION

Elliptical hollow sections (EHS) combine the merits of traditional circular hollow sections (CHS) and sections with different major and minor axis properties such as I sections. The smooth streamlined shape is not only architecturally appealing but also favourable for reducing wind resistance. Currently, there is a lack of structural performance data and verified structural design guidance for EHS. A tentative proposal on cross-section classification for EHS was made by Corus (2004). The companion paper in this Conference (Chan & Gardner 2006) presents the results of experimental and numerical studies on EHS in compression and bending. This paper proposes slenderness parameters and a system of cross-section classification limits for EHS in compression and bending. More details are reported by Gardner & Chan (submitted). The present research forms part of a wider study on the structural response of elliptical hollow sections currently underway at Imperial College London.

2 CROSS-SECTION CLASSIFICATION

The majority of structural steel design codes including Eurocode 3, place cross-sections into one of four behavioural classes based upon their susceptibility to local buckling. Class 1 cross-sections are capable of reaching and maintaining their full plastic moment in bending (and may therefore be used in plastic design). Sufficient deformation capacity or rotation capacity has to be demonstrated in this behavioural class. Class 2 cross-sections are also capable of reaching their full plastic moment in bending but have somewhat

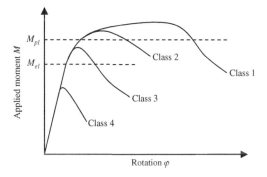

Figure 1. Four behavioural classes of cross-section.

lower deformation capacity. In Class 3 cross-sections, local buckling prevents attainment of the full plastic moment and the bending moment resistance is limited to the yield (elastic) moment. For Class 4 cross-sections, local buckling occurs in the elastic range and bending resistance is determined on the basis of an effective cross-section defined by the width-to-thickness (or diameter-to-thickness) ratios of the constituent elements. The moment-rotation characteristics of the four classes are summarized in Figure 1.

In this paper, Class 1 to 3 slenderness limits are determined on the basis of tests and numerical models in three-point and four-point bending configurations. Limits are established for bending about both the major and minor axes, and validated against the CHS limits in EN 1993-1-1 (2005). Compression tests have been used to confirm the applicability of the Class 3 limit on the basis of whether or not the yield load is reached. Development of a method for determination of effective section properties for Class 4 sections is underway.

3 ROTATION CAPACITY

In plastic design, members must be capable of forming plastic hinges which allow rotation whilst sustaining the plastic moment resistance until a collapse mechanism is formed. The total rotation of the first plastic hinge to form in a collapse mechanism defines the required rotation capacity. Class 1 cross-sections must have sufficient rotation capacity to meet this requirement.

Rotation capacity can be determined by two commonly adopted methods. One is evaluated from the moment-curvature relationship and the other is based on the moment-rotation curve. The former method has been widely utilised in the literature (Korol & Hudoba 1972, Hasan & Hancock 1988, 1989, Wilkinson & Hancock 1998, Jiao & Zhao 2004) to determine the rotation capacity of structural hollow sections in a four-point bending arrangement. The rotation capacity R of a plastic hinge based on the moment-curvature relationship is defined by Equation 1.

$$R = \frac{\kappa_{rot}}{\kappa_{pl}} - 1 \qquad (1)$$

where κ_{pl} is evaluated as M_{pl}/EI, where M_{pl} is the plastic moment resistance, E is Young's modulus and I is the second moment of area, and κ_{rot} is the limiting curvature at which the moment resistance drops back below M_{pl} (Figure 2).

Similarly, the definition of rotation capacity based on the moment-rotation relationship (Figure 3) has been commonly used in the literature (Stranghöner et al. 1994, Rondal et al. 1995, Sedlacek et al. 1995, Sedlacek & Feldmann 1995, Gioncu, et al. 1996, Vagas & Rangelov 2001), and is given by Equation 2.

$$R = \frac{\varphi_{rot}}{\varphi_{pl}} - 1 \qquad (2)$$

where φ_{pl} is the elastic component of rotation upon reaching M_{pl} and φ_{rot} is the limiting rotation at which the moment resistance falls back below M_{pl}.

In the current study, four-point bending is the principal testing arrangement employed; this enables study of the cross-section behaviour under uniform moment with negligible influence from shear and axial force. Therefore, the former definition of rotation capacity from the moment-curvature relationship is adopted for evaluating the four-point bending test results and numerical simulations, whilst the moment-rotation relationship definition is used to evaluate the three-point bending numerical simulation results.

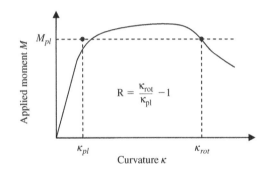

Figure 2. Definition of rotation capacity (moment-curvature).

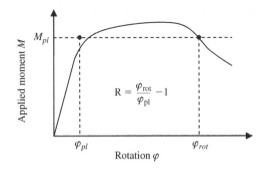

Figure 3. Definition of rotation capacity (moment-rotation).

A number of studies have been conducted to determine the required level of rotation capacity to allow sufficient moment redistribution in plastic design. By considering plastic collapse mechanisms in a variety of frames and multi-span beams, required values of rotation capacity R of 3 (Yura et al. 1978, Stranghöner et al. 1994, Rondal et al. 1995, Sedlacek et al. 1995, Sedlacek & Feldmann 1995, Elchalakani et al. 2001, Jiao & Zhao 2004) and 4 (Korol & Hudoba 1972, Hasan & Hancock 1988, 1989, Wilkinson & Hancock 1998, Elchalakani et al. 2001, Jiao & Zhao 2004) have been proposed. A rotation capacity R of 3 was adopted in the development of the current European (EN 1993-1-1 2005) and North American (AISC 2005a, b) steel design codes. Likewise, a value of rotation capacity of 3 has been assumed for the development of the Class 1 classification limits in this paper.

4 DEFINITION OF CROSS-SECTION SLENDERNESS

The elastic critical buckling stress σ_{cr} of a uniformly compressed oval shell may be closely approximated by substituting the expression for the maximum radius of curvature r_{max} into the classical buckling stress

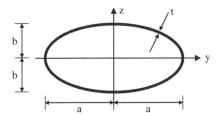

Figure 4. Geometry of elliptical hollow section.

of a circular cylinder (Kemper 1962), as given by Equation 3.

$$\sigma_{cr} = \frac{E}{\sqrt{3(1-\nu^2)}(r_{max}/t)} \qquad (3)$$

where E is the Young's modulus, ν is Poisson's ratio and t is the thickness of the shell.

This assumes that buckling initiates at the point of maximum radius of curvature and ignores the restraining effect of the surrounding material of lower radius of curvature. This approximation provides a lower bound solution to the critical buckling stress of an oval section. For an aspect ratio a/b, where a and b are the major and minor axis radii, respectively, as shown in Figure 4 of less than 5, the resulting critical buckling stress is within 5% of the exact solution (Tennyson et al. 1971).

For an elliptical section, the maximum radius of curvature occurs at the ends of the minor axis, and may be shown to be equal to a^2/b. Thus, the elastic critical buckling stress for an elliptical cylinder may be approximated by Equation 4.

$$\sigma_{cr} = \frac{E}{\sqrt{3(1-\nu^2)}(a^2/bt)} \qquad (4)$$

Note that for the case where $a = b$, Equation 4 reverts exactly to the elastic critical buckling stress of a circular cylinder, whilst for high a/b ratios the critical buckling stress approaches that predicted by the classical buckling expression for a flat plate.

With reference to Equation 4, it is therefore proposed that under compression and bending about the minor axis, the cross-section slenderness of an elliptical cross-section is defined by Equation 5.

$$\frac{D_e}{t\varepsilon^2} = 2\frac{(a^2/b)}{t\varepsilon^2} \qquad (5)$$

where D_e is the effective diameter and $\varepsilon^2 = 235/f_y$ to allow for a range of yield strengths.

For bending about the major axis, buckling would initiate in general neither at the point of maximum radius of curvature (located at the neutral axis of the cross-section with negligible bending stress) nor at

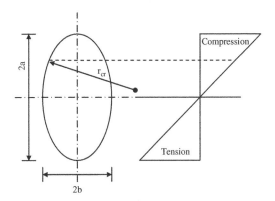

Figure 5a. Location of critical radius of curvature in major axis bending (elastic).

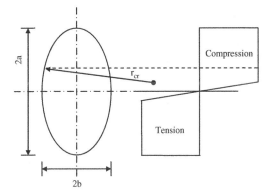

Figure 5b. Location of critical radius of curvature in major axis bending (elasto-plastic).

the extreme of the major axis (where the maximum compressive stress occurs). Gerard & Becker (1957) suggested that in this bending situation, determination of the elastic buckling stress involves the location of a point of critical curvature. This critical radius of curvature r_{cr} was calculated by optimizing the function composed of the varying curvature expression and the elastic bending stress distribution and was found to be equal to $0.65a^2/b$ (Figure 5a). For an aspect ratio a/b of less than 1.155 where the section is approaching circular, Gerard & Becker (1957) observed that buckling would occur at the extreme of the major axis and r_{cr} would be equal to a.

The approximate location of the critical radius of curvature was confirmed with elastic FE models by examining the location where buckling initiated. It was also observed that where buckling occurred in the inelastic range, the location of the critical radii of curvature shifted towards the neutral axis. This phenomenon could be explained with reference to Figure 5 and observing that in the elasto-plastic case (Figure 5b), the compressive stress distribution is more severe towards the neutral axis of the cross-section where the radius of curvature is greatest.

For a general slenderness parameter in major axis bending for the cross-section classification of EHS, it is proposed to utilize the findings of Gerard & Becker (1957) based on an elastic stress distribution. The proposed slenderness parameters are therefore given by Equations 6 and 7.

$$\frac{D_e}{t\varepsilon^2} = 1.3 \frac{(a^2/b)}{t\varepsilon^2} \qquad \text{for a/b} > 1.155 \qquad (6)$$

$$\frac{D_e}{t\varepsilon^2} = \frac{2a}{t\varepsilon^2} \qquad \text{for a/b} < 1.155 \qquad (7)$$

Note that for the special case of an EHS with an aspect ratio of unity, the cross-section slenderness defined by Equation 7 reverts to that for CHS in Eurocode 3.

5 CROSS-SECTION CLASSIFICATION

On the basis of the aforementioned slenderness parameters, experimental and numerical results of EHS in the companion paper (Chan & Gardner 2006) are plotted against cross-section slenderness as shown in Figures 6 to 12. Figure 6 summarises the behaviour of EHS in compression, whilst bending behaviour is depicted in Figures 7 to 12.

To compare EHS with CHS, experimental results of the flexural behaviour of hot-rolled steel CHS (Sherman 1986, Sedlacek et al. 1995) and cold-formed stainless steel CHS (SCI 2000) are also included in Figures 7–12. It should be stated that the experimental research from Sherman (1986) formed the basis for the system of section classification of the current North American (AISC 2005a) steel design code; the research work from Sedlacek et al. (1995) underpins the European (EN 1993-1-1 2005) limits and the data from SCI (2000) has been used for the development of structural design guidance on the use of stainless steel in construction (prEN 1993-1-4 2004).

Results are considered in more detail in the following sections, though it may generally be observed that, on the basis of the proposed slenderness parameters, EHS and CHS data follow similar trends. This suggests that the CHS limits in EN 1993-1-1 (2005) may also be safely applied to EHS (adopting the proposed measures of slenderness); though the data reveals that relaxation of the limits for both section types may be appropriate. It is worth noting that the current classification limits given in EN 1993-1-1 (2005) and AISC (2005a) are clearly sensitive to the range of data upon which they were developed; this will be discussed further in the following sections.

5.1 Compression

Axial compression represents one of the fundamental loading arrangements for structural members. For

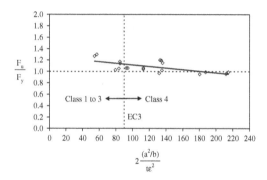

Figure 6. F_u/F_y versus cross-section slenderness.

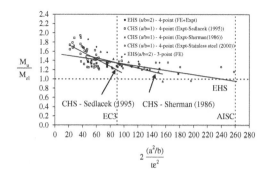

Figure 7. M_u/M_{el} versus cross-section slenderness (minor axis bending).

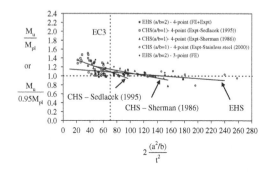

Figure 8. M_u/M_{pl} or $M_u/0.95\,M_{pl}$ versus cross-section slenderness (minor axis bending).

cross-section classification under pure compression, of primary concern is the occurrence of local buckling in the elastic material range (i.e. below the yield stress). Cross-sections that reach the yield load are considered Class 1–3, whilst those where local buckling of the slender constituent elements prevents attainment of the yield load are Class 4. Local buckling is accounted for in Class 4 sections through the effective area concept.

Results of stub columns tests conducted at Imperial College London are reported in Chan & Gardner

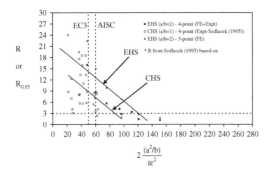

Figure 9. Rotation capacity versus cross-section slenderness (minor axis bending).

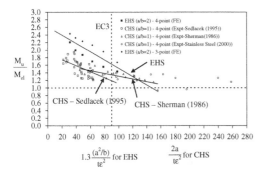

Figure 10. M_u/M_{el} versus cross-section slenderness (major axis bending).

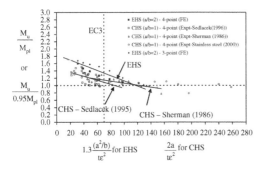

Figure 11. M_u/M_{pl} versus cross-section slenderness (major axis bending).

(2006). The relationship between F_u/F_y and cross-section slenderness $2(a^2/b)/t\varepsilon^2$ is plotted in Figure 6. A value of F_u/F_y greater than unity represents meeting of the Class 1-3 requirements, whilst a value less than unity indicates a Class 4 section where local buckling prevents the yield load from being reached. Figure 6 exhibits the anticipated trend of reducing values of F_u/F_y with increasing slenderness, and indicates that the Class 3 slenderness limit of 90 from EN 1993-1-1 (2005) for CHS may be safely adopted.

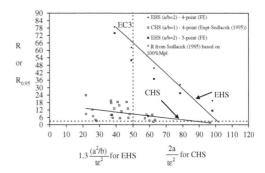

Figure 12. Rotation capacity versus cross-section slenderness (major axis bending).

5.2 Bending

For cross-section classification in bending, distinction is made between cross-sections depending on their rotation capacity and their ability to reach the plastic and elastic moment resistance. The following sub-sections compare test and numerical results with the four cross-section class requirements under two loading configurations: four-point bending (where the member is subjected to uniform moment) and three-point bending (where the member is subjected to a moment gradient).

The behaviour of beams under uniform bending differs from that under a moment gradient (Galambos 1968, Gioncu et al. 1996): Under uniform bending, bending moments remains constant (along a moment plateau) until the average outer fibre strain reaches the strain hardening strain ε_{sh} along the entire uniform moment length. Only then may the bending moment rise above M_{pl}. Many tests (Lay & Galambos 1965, Sedlacek et al. 1995, Gioncu et al. 1996) and the numerical simulations have demonstrated that the moment plateau occurs below M_{pl} and thus the definition of rotation capacity (Equation 1) is not appropriate. Conversely, for beam under a moment gradient, the plastic hinge is localized and strain-hardening occurs as soon as M_{pl} is reached. The moment will continue to increase until the yielded length of the compression flange is equal to the full local buckling wavelength (Galambos 1968).

To take account of this phenomenon, researchers (Lay & Galambos 1965, Sedlacek et al. 1995, Gioncu et al. 1996) have suggested that for beam with uniform moment, the rotation capacity should be determined at a reduced plastic moment $0.95 M_{pl}$. Thus, rotation capacity $R_{0.95}$ is defined by Equation 8.

$$R_{0.95} = \frac{\kappa_{rot, 0.95}}{\kappa_{pl, 0.95}} - 1 \qquad (8)$$

This definition is used throughout this paper for beams in the four-point bending arrangement.

5.2.1 Minor axis bending

Figure 7 shows the relationship between minor axis moment resistance and the proposed cross-section slenderness. In this figure, the ultimate test and FE moment M_u has been normalised by the elastic moment resistance M_{el}, and plotted against cross-section slenderness $2(a^2/b)/t\varepsilon^2$. A value of M_u/M_{el} greater than unity represents meeting of the Class 3 requirements, whilst a value less than unity indicates a Class 4 section where local buckling prevents the yield moment being reached.

A linear regression fit through all experimental and numerical EHS results suggests that the Eurocode limit of 90 representing the boundary between Class 3 and 4 cross-sections may be safely adopted. However, the two additional regression fits in Figure 7 also demonstrate that the classification limits in EN 1993-1-1 (2005) and AISC (2005a) are sensitive to the range of data upon which they were developed. Sedlacek et al. (1995) tested stocky sections where the results led to the stricter Eurocode Class 3 limit, whilst Sherman (1986) tested sections with a wider range of slenderness and derived the less strict value for the Class 3 limit that has been adopted in AISC (2005a). Similar observations were made by SCI (2000) for the development of Class 3 section classification limits for stainless steel CHS.

The ultimate moments attained in the tests and numerical analyses have also been normalised against the plastic moment resistance (M_{pl} for three-point bending and $0.95\,M_{pl}$ for four-point bending) and plotted against cross-section slenderness $2(a^2/b)/t\varepsilon^2$ in Figure 8. A value of $M_u/0.95\,M_{pl}$ or M_u/M_{pl} greater than unity represents meeting of the Class 2 requirements, while a value less than unity indicates a Class 3 or 4 section where local buckling prevents attainment of the full plastic moment. Similarly, a linear regression fit is plotted through the data points indicating that the EN 1993-1-1 (2005) Class 2 limit may be safely adopted.

Both Class 1 and Class 2 cross-sections are capable of reaching their plastic bending moment resistance ($0.95\,M_{pl}$ for four-point bending and M_{pl} for three-point bending). Distinction between these two classes is made on the basis of rotation capacity R. Figure 9 plots rotation capacity (as defined by Equations 1, 2 and 8, as appropriate) against cross-section slenderness.

As discussed earlier, a rotation capacity R of 3 is required for a Class 1 cross-section. From the displayed regression analysis a Class 1 classification limit of 50 from EN 1993-1-1 (2005) may be safely adopted.

5.2.2 Major axis bending

Results of numerical simulations and experiments in four-point and three-point bending about the major axis have been plotted in Figures 10 to 12 based on the proposed slenderness parameters of Equations 6 and 7. The results show similar trends to the case of minor axis bending, indicating that the Eurocode limits applied to CHS may be safely adopted for EHS, using the proposed slenderness parameters. Figure 12 shows that although the EHS exhibit greater rotation capacity at low slenderness, results converge towards those for CHS at the required rotation capacity of 3. It may be seen from the results in compression, minor axis bending and major axis bending that, using the proposed slenderness parameters for EHS, the Eurocode classification limits for CHS may be safely adopted.

Further analysis of the results indicates that slenderness limits for CHS and EHS (bending about either axis) may be relaxed to $70\varepsilon^2$ for Class 1 cross-sections, $100\varepsilon^2$ for Class 2 cross-sections and $150\varepsilon^2$ for Class 3 cross-sections.

5.2.3 Combined compression and bending

For cross-section classification under combined compression and bending, designers can initially check the cross-section against the most severe loading case of pure compression. If the classification is Class 1, then there is no benefit to be gained from checking against the actual stress distribution. Similarly, if plastic design is not being utilized, there would be no benefit in re-classifying a Class 2 cross-section under the actual stress distribution. Under combined compression and minor axis bending, clearly buckling will initiate in the region of the maximum radius of curvature, similar to the case of pure compression and pure minor axis bending. Hence, for this case, the same slenderness parameters and classification limits are recommended. Under combined compression and major axis bending, the critical radius of curvature will shift towards the centroidal axis. Conservatively, classification may be carried out assuming pure compression, though development of a method for determination of the critical radius of curvature and the corresponding slenderness parameter and limit is currently underway.

6 CONCLUSIONS

Cross-section classification is a fundamental aspect of structural metallic design. This paper has proposed slenderness parameters and a system of cross-section classification limits for elliptical hollow sections in compression, bending about both principal axes and combined compression plus bending. Compatibility with CHS limits has been achieved. The results demonstrate that, using the proposed measures of slenderness for EHS, the Eurocode 3 section classification limits for CHS may be safely adopted, though the results indicate that the classification limits for both CHS and EHS could be relaxed, and proposals for improved limits have also been made. The developed classification system underpins the development of further structural design guidance for elliptical hollow sections.

ACKNOWLEDGEMENTS

The authors are grateful to the Dorothy Hodgkin Postgraduate Award Scheme for the project funding, and would like to thank Corus for the supply of test specimens and for funding contributions, Eddie Hole and Andrew Orton (Corus Tubes) for their technical input and Ron Millward and Alan Roberts (Imperial College London) for their assistance in the laboratory works.

REFERENCES

AISC 2005a. *Specification for Structural Steel Buildings.* American Institute of Steel Construction, Chicago.

AISC 2005b. *Commentary on the Specification for Structural Steel Buildings.* American Institute of Steel Construction, Chicago.

Chan, T.M. & Gardner, L. 2006. Experimental and numerical studies of elliptical hollow sections under axial compression and bending. *Proceedings of the 11th International Symposium on Tubular Structures,* August 2006, Québec.

Corus 2004. *Celsius 355® Ovals – sizes and resistances eurocode version.* Corus Tubes – Structural & Conveyance Business.

Elchalakani, M., Zhao, X.L. & Grzebieta, R.H. 2001. Plastic slenderness limits for cold-formed circular hollow sections. *Australian J. Str. Eng.* 3(3): 127–141.

EN 1993-1-1 2005. *Eurocode 3: Design of steel structures – Part 1-1: General rules and rules for buildings.* CEN.

Galambos, T.V. 1968. Deformation and Energy Absorption Capacity of Steel Structures in the Inelastic Range. *AISI.* Bulletin No.8.

Gardner, L. & Chan, T.M. submitted. Cross-section classification of Elliptical Hollow Sections. *Steel and Composite Structures.*

Gerard, G. & Becker, A. 1957. *Handbook of structural stability: Part III – buckling of curved plated and shells.* NACA, Washington.

Gioncu, V., Tirca, L. & Petcu, D. 1996. Rotation capacity of rectangular hollow section beams. *Proceedings of the 7th International Symposium on Tubular Structures.* August 1996, Budapest.

Hasan, S.W. & Hancock, G.J. 1988. *Plastic bending tests of cold-formed rectangular hollow sections.* School of Civil and Mining Engineering, Centre for Advanced Structural Engineering, The University of Sydney, Sydney.

Hasan, S.W. & Hancock, G.J. 1989. Plastic bending tests of cold-formed rectangular hollow sections. *Steel Construction (Sydney, Australia)* 23(4): 2–19.

Jiao, H. & Zhao, X.L. 2004. Section slenderness limits of very high strength circular steel tubes in bending. *Thin-Walled Structures* 42(9): 1257–1271.

Kempner, J. 1962. Some results on buckling and postbuckling of cylindrical shells. *Collected papers on Instability Shell Structures.* NASA TN D-1510: 173–186.

Korol, R.M. & Hudoba, J. 1972. Plastic behaviour of hollow structural sections. *J. Str. Div., Proc. for the ASCE* 98(5): 1007–1022.

Lay, M.G. & Galambos, T.V. 1965. Inelastic steel beams under uniform moment. *J. Str. Div., Proc. for the ASCE* 91(ST6): 67–93.

prEN 1993-1-4 2004. *Eurocode 3 – Design of steel structures – Part 1-4: General rules – Supplementary rules for stainless steels.* CEN.

Rondal, J., Boeraeve, P., Sedlacek, G., Stranghöner, N. & Langenberg, P. 1995. *Rotation Capacity of Hollow Beam Sections.* CIDECT.

SCI 2000. *Report to ECSC – Development of the use of stainless steel in construction: Structural design of stainless steel circular hollow sections.* Steel Construction Institute, UK.

Sedlacek, G., Dahl, W., Rondal, J., Boreaeve, Ph., Stranghöner, N. & Kalinowski, B. 1995. *Investigation of the rotation behaviour of hollow section beams.* ECSC Convention 7210/SA/119.

Sedlacek, G. & Feldmann, M. 1995. *Background document 5.09 for chapter 5 of Eurocode 3 Part 1.1 – The b/t ratios controlling the applicability of analysis models in Eurocode 3, Part 1.1.* Aachen.

Sherman, D.R. 1986. Inelastic flexural buckling of cylinders. *Proceedings of the Steel Structures: Recent Research Advances and Their Applications to Design.* September 1986, Budva.

Stranghöner, N., Sedlacek, G. & Boeraeve, P. 1994. Rotation requirement and rotation capacity of rectangular, square and circular hollow sections beams. *Proceedings of the 6th International Symposium on Tubular Structures.* September 1994, Rotterdam.

Tennyson, R.C., Booton, M. & Caswell, R.D. 1971. Buckling of imperfect elliptical cylindrical shells under axial compression. *AIAA* 9(2): 250–255.

Vagas, I. & Rangelov, N. 2001. Classification of girders with I- or Box- cross-sections. *Int. J. Steel Structures* 1(3): 153–165.

Wilkinson, T. & Hancock, G.J. 1998. Compact or class 1 limits for rectangular hollow sections in bending. *Proceedings of the 8th International Symposium on Tubular Structures.* September, 1998, Singapore.

Yura, J.A., Galambos, T.V. & Ravindra, M.K. 1978. The bending resistance of steel beams. *J. Str. Div., Proc. for the ASCE* 104(9): 1355–1370.

Tubular Structures XI – Packer & Willibald (eds)
© 2006 Taylor & Francis Group, London, ISBN 0-415-40280-8

Finite element analysis of structural steel elliptical hollow sections in pure compression

Y. Zhu & T. Wilkinson
School of Civil Engineering, The University of Sydney, Australia

ABSTRACT: This paper presents a finite element investigation of the local buckling behaviour of the structural steel elliptical hollow section (EHS) in compression. The theoretical elastic buckling load of an EHS is similar to that of a Circular Hollow Section (CHS) except that the diameter term, D, is replaced by D_1^2/D_2, representing the major and minor diameters of the ellipse. The overall aim is to examine whether an "equivalent CHS" can be used to model the local buckling of EHS when considering imperfections and non-linear material properties as well as in pure elastic circumstance, through four steps of analysis. The results are benchmarked against experimental results. It was found that the use of an equivalent CHS was a reasonably good predictor of capacity of slender sections and the deformation capacity of compact sections. However, further benchmarking against experimental results is recommended.

1 BACKGROUND

The elliptical hollow section (EHS) is a new shape of high strength, hot-rolled steel sections being used in structural building applications. Compared with circular sections, elliptical sections not only show unique architectural effects but also offer the structural advantages of sections with differing major and minor axis properties.

However, despite strong interest in their usage, a lack of fundamental test data and verified structural design guidance is inhibiting uptake. Currently, few test results on EHS exist and there is a limited understanding of their properties and structural performance.

Past research has been performed on Circular Hollow Section regarding the local buckling behaviour. Equation 1 gives the expression to calculate the elastic buckling stress of a CHS subject to pure compression.

$$\sigma_{cr} = \frac{2Et}{D\sqrt{3(1-v^2)}} \tag{1}$$

where E is the material Young's modulus, t is the thickness of the circular cross section, v is Poisson's ratio and D is the diameter. From this equation, it can be seen that the ratio of D/t is the key parameter affecting local buckling of CHS.

Early test results on elliptical sections of very slender, non-structural, proportions that exhibit high imperfection sensitivity and sharp unloading were conducted by the aeronautical industry (Hutchinson,

1968). In the 1950s and 1960s, the elastic critical buckling and post-buckling behaviour of elliptical shells first received attention from the aeronautical industry with the principal investigations conducted in the USA. From these initial studies, formulae to predict the elastic critical buckling and post-buckling response of EHS under axial loading were derived (Marguerre, 1951 and Kempner, 1962). The elastic critical buckling stress for an elliptical cross section (whose geometry is defined by the equation of an ellipse) subjected to pure compression is given by Equation 2.

$$\sigma_{cr} = \frac{Et}{\left(\dfrac{A^2}{B}\right)\sqrt{3(1-v^2)}} \tag{2}$$

where A and B are the major and minor radii, respectively, as shown in Figure 1.

The form of Equation 2 is very similar to that of Equation 1. By changing both formulas slightly and replace $2A$ and $2B$ as D_1 and D_2 which refer to the length of major and minor diameter of an ellipse respectively, it can be concluded that an EHS can be considered equivalent to a CHS whose diameter is $D = D_1^2/D_2$. Indeed for the case where A equals B, Equation 2 reverts exactly to Equation 1.

This paper will further investigate the relationship between EHS and CHS through finite element analysis, benchmarked against unpublished test results from The University of Toronto (2005).

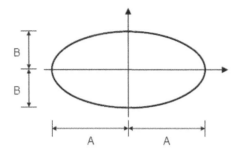

Figure 1. Geometry of elliptical sections.

Figure 2a. ABAQUS mesh of EHS columns.

2 FINITE ELEMENT ANALYSIS

2.1 *Element type*

Finite element analyses were performed using ABAQUS to simulate the local buckling behaviour of the stub columns of EHS and CHS. ABAQUS includes general-purpose shell elements as well as elements that are valid for thick and thin shell problems. The element type S4R was used in this report. S4R is a general-purpose, finite-membrane-strain, reduced integration shell element. The ratio of the length to width of the element is about 1:1. For the whole column, different mesh densities were adopted. In the transverse direction, a higher mesh density was used in the tighter corners of the higher aspect ratio ellipses. In the longitudinal direction, the mesh density was kept consistence. The mesh density for ABAQUS models of EHS and CHS sections is shown in Figure 2(a) and (b).

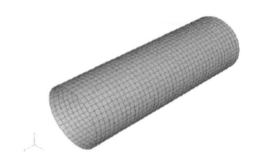

Figure 2b. ABAQUS mesh of CHS columns.

2.2 *Material behaviour*

Most materials of engineering interest initially respond elastically. If the load exceeds some limit, some part of the deformation will remain when the load is removed. Plasticity theories model the material's mechanical response as it undergoes such nonrecoverable deformation in a ductile fashion. Most of the plasticity models in ABAQUS are "incremental" theories in which the mechanical strain rate is decomposed into an elastic part and a plastic part.

This project considered different material properties in different analysis stages as detailed in the following chapters.

2.3 *Boundary conditions*

For each of the two ends, two different types of boundary conditions were used to simulate the test situation in the column tests. The ends were divided into a fixed end and a movable end, which was the loaded end.

Initially, load was exerted directly on the movable end with an even stress distribution. Secondly, for the purpose of comparing with the first method and to

eliminate the concern of the top edge being damaged for such thin sheet steel, a displaced rigid body was used. The top end of the column was linked to the rigid body. The results obtained from the two methods were not found to be significantly different, so to simulate the behaviour of the stub columns either method can be used.

2.4 *Geometric imperfection*

For ABAQUS, the approach to define an imperfection in this paper involved two analysis runs with the same model definition: (a) In the first analysis run, an eigenvalue buckling analysis was performed on the "perfect" structure to establish probable collapse modes and to verify that the mesh discretizes those modes accurately. (b) In the second analysis run, an imperfection in the geometry was introduced by adding these buckling modes to the "perfect" geometry. (c) Finally, a geometrically nonlinear load displacement analysis of the structure was performed containing the imperfection using the Riks method. In this way the Riks method could be used to perform post-buckling analyses of "stiff" structures that show linear behaviour prior to buckling, if perfect. By performing a load-displacement analysis, other important nonlinear effects, such as material inelasticity or contact, can be included.

3 FEA PROCEDURES OF EHS AND CHS COLUMNS

3.1 Pure elastic buckling

The first stage was pure elastic buckling. The analysis was performed on EHS with a range of aspect ratios of major and minor axis from 1:1 (CHS) to 3:1 to examine the transitional behaviour.

3.1.1 FEA model

Nominally a CHS with diameter of 400 mm and length of 1200 mm was chosen for analysis. Various degrees of aspect ratio of the ellipse ($D_1:D_2$) were used: $D_1/D_2 = 1.00, 1.25, 1.50, 1.75, 2.00, 2.25, 2.50, 2.75$, and 3.00. Meanwhile, different slenderness values were achieved with D/t ranging from 20 to 120 by varying t. Thus, 9 different groups with 21 different models in each group were set up. Totally 189 models were simulated using ABAQUS.

The material properties of the models were assumed as pure elastic, which means no plastic data other than $E = 200 \times 10^9$ Pa, $\nu = 0.3$ as elastic data were input.

3.1.2 Analysis method

Eigenvalue buckling analysis was generally used to estimate the critical bucking loads of stiff structures. An incremental loading pattern was defined in *BUCKLE step. A general eigenvalue buckling analysis can provide useful estimates of collapse mode shapes and calculate the buckling stress as well. The equation for calculation of buckling stress can be written as Equation 3.

$$\sigma_{cr} = E \cdot \varepsilon = E \cdot \lambda \cdot \frac{\Delta l}{l} \qquad (3)$$

where E is the material Young's modulus, λ is the eigenvalue obtained from the results of FEA, Δl is the initial displacement at the movable end input in the boundary conditions in ABAQUS, l is the length of the column.

3.1.3 Results

The nine groups of EHS column models were simulated using ABAQUS. With the results of eigenvalues obtained, the buckling stress of each model was calculated. On the other hand, knowing the section properties of each model, the buckling stress can be calculated from Equation 2. Some results from ABAQUS simulation and Equation 2 were shown in Figure 3 and Table 1.

3.1.4 Discussion

The results obtained from ABAQUS as well as Equation 2 illustrate that the change of section properties obviously changes the local buckling stress of the EHS columns. Table 1 shows that for CHS ($D_1 : D_2 = 1.00$), ABAQUS results and Equation 2

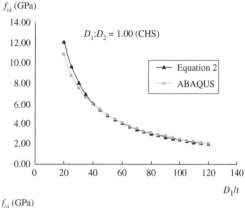

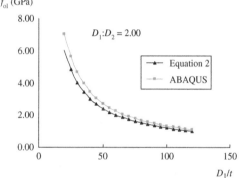

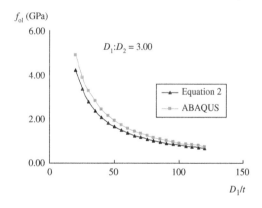

Figure 3. Some Results of Elastic Buckling Stress from ABAQUS and Equation 2.

results match closely, within 1%. However, for higher aspect ratios ($D_1 : D_2 > 1.50$), the elastic buckling load calculated from Equation 2 is consistently about 10% less than that from ABAQUS.

3.2 Elastic buckling with geometric imperfection

This stage investigated the behaviour of an EHS column and that of an equivalent CHS, assuming their

Table 1. Difference between ABAQUS results and calculated results from Equation 2.

D_1/D_2	$f_{ol,equation2}/f_{ol,ABAQUS}$	
	Mean value	Standard deviation
1.00	1.01	0.04
1.25	0.94	0.01
1.50	0.92	0.01
1.75	0.91	0.02
2.00	0.90	0.02
2.25	0.89	0.02
2.50	0.89	0.02
2.75	0.88	0.02
3.00	0.88	0.02

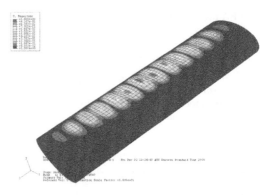

Figure 4. Buckling shape of EHS $(D_1:D_2 = 2.00)$ with imperfection in eigenvalue buckling analysis.

material properties are elastic with some geometric imperfection as well. Some models of CHS were established in a similar way of EHS.

3.2.1 FEA model

$D_1:D_2 = 2.00$, $D_1 = 400$ mm was selected as a typical section properties of EHS for research. According to $D = D_1^2/D_2$ which is the predicted equation to relate an EHS to an equivalent EHS concluded in Section 2, CHS with $D = 800$ mm is expected to behave similarly as EHS $(D_1:D_2 = 2.00, D_1 = 400$ mm). So CHS models of this cross-section dimension were set up in the same way of EHS except the lengths of CHS were extended to 2400 mm to adapt the same L/D_1 ratio so as to guarantee the occurrence of local buckling. Furthermore, each group had three models of different thickness including $t = 1$ mm, 5 mm and 10 mm.

3.2.2 Analysis method

Finite element analyses were carried out both on EHS models and CHS models based on the method discussed in Section 2.4. Three different imperfections which are $t/10$, $t/50$ and $t/100$ were input to observe how geometric imperfection affected the buckling stress of EHS and CHS. Figure 4 presents the buckling shape of an EHS caused by geometric imperfection using eigenvalue buckling analysis.

3.2.3 Results

Table 2 shows the results of the elastic buckling with geometric imperfection, where f_{crit} is the value of elastic buckling stress with no imperfection, and f_{ol} is the value of elastic buckling stress with certain imperfection. After getting these two groups of data, the percentage of f_{ol}/f_{crit} can be calculated and made for comparison.

3.2.4 Discussion

From the results, it is observed that for all the models, high imperfection is quite likely to induce low buckling stress. Table 2 shows how the elastic buckling load is reduced from the perfect specimen to a specimen with initial geometric imperfections. For sections with the same aspect ratio, the reduction in buckling stress f_{ol} is similar for the same relative imperfection size (eg. $t/100$) regardless of the slenderness of the section.

In addition, the reduction in buckling stress is greater for a CHS compared to the equivalent EHS. In other words, the elastic local buckling stress of CHS is more sensitive to imperfections than the corresponding EHS. Hence, this illustrates that the behaviour of the equivalent CHS based on the equation $D = D_1^2/D_2$ does not exactly match that of the EHS. However, they behave similarly since the differences are averagely within 10%.

3.3 Inelastic buckling

The material properties of real structural products are practically non-linear. This section investigates the effect of using close to real material properties in the analysis. For slender sections, comparison of buckling loads is significant. For compact sections, comparison of the deformation capacity prior to buckling is important.

3.3.1 FEA model

A series of EHS models were set up with a range of aspect ratios from 1.25:1 to 3:1 followed by a series of the equivalent CHS models with a range of diameter from 500 mm to 1200 mm. Plastic material properties were added to all the models and the data input is shown in Table 3.

3.3.2 Analysis method

Generally, analysis were conducted on all the models of EHS and the equivalent CHS with the range of thickness changing from 30 mm to 1 mm and it is summarized that different D_1/t ratio changes the buckling behaviour of both EHS and CHS. Figure 5

Table 2. ABAQUS Results of comparison of elastic buckling with geometric imperfection.

Imperfection	f_{ol} (MPa)	f_{ol}/f_{crit} (%)	f_{ol} (MPa)	f_{ol}/f_{crit} (%)
EHS ($D_1 = 400$ mm $= 2.00D_2$), $t = 10$ mm, $f_{crit} = 3445$ MPa			CHS ($D = 800$ mm), $t = 10$ mm, $f_{crit} = 3286$ MPa	
$t/100$	3139	91.1	2714	82.6
$t/50$	2966	86.1	2672	81.3
$t/10$	2283	66.3	1953	59.4
EHS ($D_1 = 400$ mm $= 2.00D_2$), $t = 5$ mm, $f_{crit} = 1668$ MPa			CHS ($D = 800$ mm), $t = 5$ mm, $f_{crit} = 1667$ MPa	
$t/100$	1557	93.3	1390	83.4
$t/50$	1469	88.1	1316	78.9
$t/10$	1113	66.7	1026	61.5
EHS ($D_1 = 400$ mm $= 2.00D_2$), $t = 1$ mm, $f_{crit} = 325$ MPa			CHS ($D = 800$ mm), $t = 1$ mm, $f_{crit} = 375$ MPa	
$t/100$	300	92.3	329	87.7
$t/50$	285	87.7	286	76.3
$t/10$	222	68.3	248	66.1

Table 3. Plastic data using in fea models.

Stress (MPa)	Strain
400	0.00
400	0.03
460	0.05
500	0.08

shows the $P/A_g f_y$-Displacement diagram of an EHS ($D_1:D_2 = 2.00$, $D_1 = 400$ mm), where P is the compressive load, A_g is the gross area of the cross section and F_y is the yield stress of the structural material. From this typical example, it is clearly observed that the compact ones reach yield stress prior to buckle, while the slender ones buckle before reaching yield stress.

Two parts of analysis methods were tried due to the different buckling behaviours of compact models and slender models. As illustrated in Figure 4, for compact sections, ΔL, which indicates the displacement at which an EHS or a CHS reaches ultimate load P_u, was examined. The ratio of $\Delta L/L$ gives the structural ductility of EHS and CHS models because it explains the deformation during the stage from yielding to buckling. However, for slender sections, since they buckle prior to yield, the ratio of $P/A_g f_y$ is less than 1 and it shows the percentage of the inelastic buckling stress divided by the yield stress. These two ratios of both EHS and the equivalent CHS deserved to investigate in this section.

3.3.3 Results

Table 4 and Table 5 shows the results of comparison of the $\Delta L/L$ ratios for compact EHS and the equivalent CHS and that of $P/A_g f_y$ ratios for slender ones, respectively.

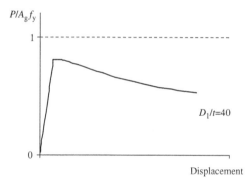

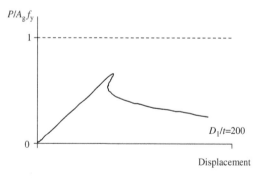

Figure 5. ABAQUS $P/A_g f_y$–Displacement Curves of EHS and CHS with Different D_1/t ratio.

3.3.4 Discussion

The results above show the difference of the $\Delta L/L$ ratios for the thick models are within 7%, and that of the $P/A_g F_y$ ratios for thin ones are within 3%, on average.

Table 4. ABAQUS results of comparison of the ratio of $\Delta L/L$ for compact models.

EHS			CHS			
D_1/D_2	t	$\Delta L/L$ (%)	D	t	$\Delta L/L$ (%)	Difference (%)
1.25	18	1.52	500	18	1.42	6.58
1.50	18	1.41	600	18	1.34	4.96
1.75	18	0.81	700	18	0.76	6.17
2.00	20	1.30	800	20	1.22	6.15
2.25	25	1.98	900	25	1.88	5.05
2.50	25	1.93	1000	25	1.87	3.11
2.75	25	1.89	1100	25	1.77	6.35
3.00	25	1.56	1200	25	1.44	7.69

Table 5. ABAQUS results of comparison of the ratio of P/A_gF_y for slender models.

EHS			CHS			Difference
D_1/D_2	t	P_u/A_gF_y	D	t	P_u/A_gF_y	(%)
1.25	0.8	0.892	500	0.8	0.887	0.56
1.50	0.8	0.795	600	0.8	0.786	1.13
1.75	1.0	0.851	700	1.0	0.831	2.35
2.00	1.0	0.750	800	1.0	0.730	2.67
2.25	1.0	0.666	900	1.0	0.676	1.50
2.50	1.2	0.722	1000	1.2	0.725	0.42
2.75	1.2	0.656	1100	1.2	0.671	2.29
3.00	1.5	0.758	1200	1.5	0.755	0.40

Hence, for thin sections which buckle elastically, the equivalent CHS gives a buckling stress very close to the EHS (within 3%). For thick sections, which exhibit a plastic plateau prior to yielding, the strains at buckling (or deformation capacity) for an EHS and the equivalent CHS are very similar (within 7%).

All EHS show slightly higher ductility than the equivalent CHS. Nevertheless, these results indicate that the equation $D = D_1^2/D_2$ provides reasonable agreement for the inelastic local buckling behaviours of EHS and CHS.

3.4 Simulation of real tests

Researchers at The University of Toronto (2005) performed some column tests of steel elliptical hollow sections under compression. Deformed shapes are shown in Figure 6. In this stage, ABAQUS was used to simulate the local buckling behaviour of the specimens of the real tests.

3.4.1 FEA model

The section dimensions and material properties of the test specimen were shown in Table 6.

The FEA model was established according to the actual section dimensions and material properties. The plastic data using in ABAQUS is shown in Table 7.

Figure 6. Deformed shape of specimen from tests.

Table 6. Cross-section dimensions and material properties of the test specimen.

Dimensions

Height (mm)	Width (mm)	Thickness (mm)	Area (mm²)
221.2	110.9	5.94	3054

Tube material properties

Coupon	E (MPa)	f_y (MPa)	f_u (MPa)	ε (%)
1	231	420.6	527.4	34.55
2	209	431.5	537.8	33.96
3	211	420.9	529.0	34.84
4	211	412.5	526.5	35.43
Average	215	421.4	530.2	34.70

Table 7. Plastic data of ABAQUS models.

Stress (MPa)	Strain
420	0.000
422	0.015
424	0.018
428	0.029
439	0.034
462	0.045
479	0.056
500	0.078
511	0.097
516	0.108
520	0.123
525	0.133
529	0.175

3.4.2 Analysis method

Finite element analyses were carried out on the model of the real tests, in the method outlined in Section 2.4. Since the real tests did not include the measurement of real geometric imperfection of the specimen, different varieties of imperfection data were attempted in FEA and a couple of load-displacement curves were obtained respectively.

3.4.3 Results

The results of ultimate load P_u from tests as well as ABAQUS are shown in Table 8. Various geometric

Table 8.   Results from real tests and ABAQUS.

		P_u 1392	P_{ABAQUS}/P_{test} (%)
ABAQUS	0	1314	94.40
results	$t/100$	1314	94.40
with	$t/50$	1312	94.25
different	$t/20$	1302	93.53
imperfection	$t/10$	1286	92.39

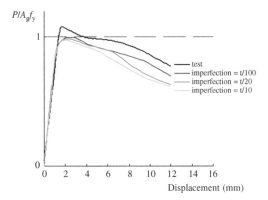

Figure 7.   ABAQUS load–displacement curves of test models.

imperfections were applied. The differences between the ABAQUS results and the test results were obtained by the ratio of P_{ABAQUS}/P_{test}.

Figure 7 plots the load–displacement curve obtained from the real specimen models as well as other three curves from the results of three different imperfection data which are $t/100$, $t/20$ and $t/10$ respectively. The buckling shapes of the models from ABAQUS are shown in Figure 8 & 9.

3.4.4   Discussion

Firstly, the FEA simulations tend to underestimate the experimental results. Even with zero imperfection, there is an underestimate of just over 5%. Possible reasons for this may include a slight discrepancy in the cross-section area when modeling a hollow section with a shell element. In addition, the small yield stress spikes of the real specimen as shown in Figure 10 were not included in the FEA material properties.

Since the experimental section were quite stocky ($D_1/t = 37$) and buckled after yielding, they are not so sensitive to geometric imperfections. Hence, it can be seen that the ABAQUS buckling load for various imperfection sizes (from $t/100$ to $t/10$) are very similar as yielding dominates the behaviour. The post buckling load shedding curves are similar in all cases.

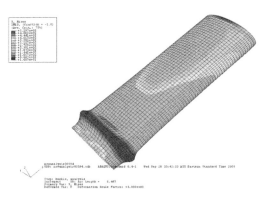

Figure 8.   ABAQUS local deformed shape of test model with imperfection $= t/100$.

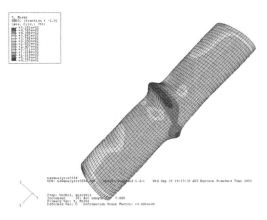

Figure 9.   ABAQUS local deformed shape of test model with imperfection $= t/20$.

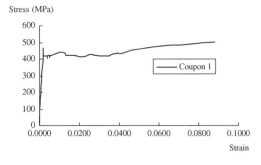

Figure 10.   Strain–stress curve of coupon 1 of real specimen.

4   SUMMARY

This report has provided an introduction to the behaviour of structural steel elliptical hollow sections and described a series of finite element analysis on the columns of EHS and the predicted equivalent CHS

185

using ABAQUS. Several stages of numerical simulations were conducted to get an overall understanding of local buckling behaviour of EHS and CHS.

The use of the finite element program ABAQUS for simulating the behaviour of the EHS and CHS columns was successful since the ABAQUS results were generally in good agreement with experimental values. The ABAQUS analyses gave reasonable and reliable results to be collected and compared with the calculation results of the predicted theory on EHS, through the stages of elastic, inelastic and real material properties.

Firstly, ABAQUS predictions of buckling stress of perfect EHS were approximately 10% higher than calculated results of equation

$$\sigma_{cr} = \frac{Et}{(\frac{A^2}{B})\sqrt{3(1 - v^2)}}$$

(Marguerre, 1951 and Kempner, 1962). However, the prediction of the equivalent CHS compared to an EHS using $D = D_1^2/D_2$ gave reasonably close approximation of buckling strain (ductility) once inelastic material properties were considered. This suggests that an equivalent CHS approach might be suitable to classify EHS as compact or slender with regard to stub column strength.

The ABAQUS analysis was compared to a single test results on a compact EHS. Some further comparison to a wider range of sections is required before this equivalent slenderness approach could be confirmed.

REFERENCES

ABAQUS (2003). ABAQUS/Standard User's Manual Volumes I–III and ABAQUS CAE Manual. Version 6.4. Hibbitt, Karlsson & Sorensen, Inc. Pawtucket, USA.

Marguerre, K. (1951). Stability of cylindrical shells of variable curvature. NACA TM 1302.

Hutchinson, J.W. (1968). Buckling and initial post-buckling behaviour of oval cylindrical shells under axial compression. Transactions of the American Society of Mechanical Engineers, Journal of Applied Mechanics. March 1968, pp 66–72.

Kempner, J. (1962). Some results on buckling and post-buckling of cylindrical shells. Collected papers on Instability of Shell Structures. NASA TN D-1510, pp. 173–186.

Gardner, L. and Ministro, A (2004). Testing and numerical modelling of structural steel oval hollow Sections. Department of Civil and Environmental Engineering Structures Section. Imperial college, UK.

Eckhardt, C. (2004). Classification of oval hollow sections. School of Civil Engineering Report. University of Southampton.

The University of Toronto (2005), Unpublished test results of elliptical hollow section (EHS) stub columns in pure compression.

Riks E., "An incremental approach to the solution of snapping and buckling problems", Int. J. Solids Structures, Vol. 15, 529–551, 1979.

Tubular Structures XI – Packer & Willibald (eds)
© *2006 Taylor & Francis Group, London, ISBN 0-415-40280-8*

Behaviour of hollow flange channel sections under concentrated loads

T. Wilkinson & Y. Zhu
School of Civil Engineering, The University of Sydney, Australia

D. Yang
Siemens Logistics & Assembly Systems, Belrose, Australia

ABSTRACT: A new range of hollow flange channel sections presents some unique failure modes and design challenges for engineers. The single slender web and hollow flanges have unique bearing failure mechanisms, and the forming process produces varying material properties throughout the section. This paper reports bearing and compression test results. The results show that current design procedures are conservative and that new proposals are required to model the behaviour less conservatively.

1 HOLLOW FLANGE SECTIONS

1.1 Introduction

Smorgon Steel Tube Mills have used their patented dual welding technology to manufacture a new section – the Hollow Flange Channel (HFC) as shown in Figure 1. This section is being marketed as the LSB, or LiteSteel Beam.

Hollow flange sections are designed to take advantage of properties of hot-rolled sections – in which area is concentrated away from the neutral axis, and the torsional stiffness of hollow sections.

In the mid 1990s, a slightly different cross-sectional shape was manufactured – the Hollow Flange Beam (HFB). The HFB had some unique failure modes such as flexural distortional buckling and bearing capacity failure. Research was required to investigate these failure modes before these sections could be used efficiently and safely. This research included analytical, experimental and numerical studies (Hancock et al (1994), Sully et al (1994), Pi and Trahair (1997), Avery et al (2000)).

During the cold-forming process, the flat web receives very little additional cold work, compared to the flanges. As a result the nominal yield stress of the flange is $f_{yf} = 430\,\text{MPa}$, while for the web it is $f_{yw} = 380\,\text{MPa}$.

1.2 Key structural behaviour issues

Prior to production, the likely behaviour of LSB compared to existing cold-formed steel design rules was considered. Several key areas were identified in which it was possible that new design models would need to

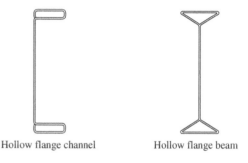

| Hollow flange channel | Hollow flange beam |

Figure 1. Cross-sectional shapes.

be considered – these included connections, compression capacity, lateral distortional buckling and direct bearing.

This paper details investigations into two of the key areas – stub column behaviour and bearing capacity.

2 STUB COLUMN TESTS

2.1 Behavioural issues

It was identified above that the webs and flanges would have different yield stresses. Compression design for sections with unequal yield stresses is potentially complicated, since the effective width formulation used in calculations is based on the assumption of reaching the yield stress. Considering a fully effective section, in order for the flanges to reach their yield stress of 450 MPa, the webs will not only experience their own yield stress of 380 MPa, but they will experience additional strain past the yield strain, so that the total strain is $450/E$ which is as if it had a yield stress of 450 MPa.

Average stress in plate

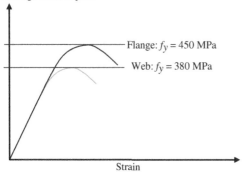

Flange: $f_y = 450$ MPa

Web: $f_y = 380$ MPa

Strain

Figure 2. Stress strain behaviour.

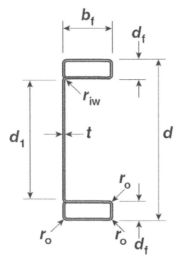

Figure 3. Section dimension definitions.

This is because all parts of the cross section must have equal strain.

Figure 2 illustrates the possible compression behaviour of the flange and the web in compression. It is possible that the web strength will start to reduce with increased strain after it has reached yield but before the flange has reached yield. There is some doubt as to whether it is appropriate to assess the web and flange strengths separately, using their respective yield stresses, and then adding the components together. There has been little research into this type of behaviour.

2.2 Preparation of stub column specimens

Two stubs columns in each of the thirteen size ranges were tested. The nominal dimensions are outlined in Figure 3 and Table 1. The ends were milled flat and parallel to achieve accurate seating in the testing

Table 1. Nominal section dimensions of specimens (mm).

Specimen designation	d	d_f	b_f	t	r_o	r_{iw}	d_1
300 × 75 × 3.0	300	25	75	3.0	6	3	244
300 × 75 × 2.5	300	25	75	2.5	5	3	244
300 × 60 × 2.0	300	20	60	2.0	4	3	254
250 × 75 × 3.0	250	25	75	2.0	6	3	194
250 × 75 × 2.5	250	25	75	2.5	5	3	194
250 × 60 × 2.0	250	20	60	2.0	4	3	204
200 × 60 × 2.5	200	20	60	2.5	5	3	154
200 × 60 × 2.0	200	20	60	2.0	4	3	154
200 × 45 × 1.6	200	15	45	1.6	3.2	3	164
150 × 45 × 2.0	150	15	45	2.0	4	3	114
150 × 45 × 1.6	150	15	45	1.6	3.2	3	114
125 × 45 × 2.0	125	15	45	2.0	4	3	89
125 × 45 × 1.6	125	15	45	1.6	3.2	3	89

Table 2. Measured section dimension of specimens (mm).

Specimen designation	d	d_f	b_f	t	r_o	r_{iw}	d_1
300 × 75 × 3.0 (A)	302.5	25.5	75.3	2.88	6.8	3	244
300 × 75 × 3.0 (B)	302.3	25.5	75.2	2.87	6.5	3	244
300 × 75 × 2.5 (A)	303.8	25.7	75.2	2.51	5.8	3	244
300 × 75 × 2.5 (B)	303.8	25.8	75.5	2.51	5.8	3	244
300 × 60 × 2.0 (A)	302.3	20.8	59.8	1.95	4.6	3	254
300 × 60 × 2.0 (B)	302.3	20.8	59.5	1.94	4.6	3	254
250 × 75 × 3.0 (A)	250.3	25.2	75.8	2.82	6.3	3	194
250 × 75 × 3.0 (B)	250.3	25.2	75.2	2.81	6.9	3	194
250 × 75 × 2.5 (A)	250.3	25.8	75.7	2.50	7.1	3	194
250 × 75 × 2.5 (B)	250.3	25.5	75.3	2.50	6.5	3	194
250 × 60 × 2.0 (A)	252.3	20.3	59.5	1.93	4.3	3	204
250 × 60 × 2.0 (B)	252.0	20.2	59.7	1.94	4.4	3	204
200 × 60 × 2.5 (A)	200.8	19.5	60.0	2.51	4.3	3	154
200 × 60 × 2.5 (B)	200.8	19.3	60.0	2.52	4.3	3	154
200 × 60 × 2.0 (A)	200.0	20.2	59.7	1.93	3.7	3	154
200 × 60 × 2.0 (B)	200.0	20.0	60.0	1.93	3.9	3	154
200 × 45 × 1.6 (A)	200.8	15.5	44.7	1.60	3.3	3	164
200 × 45 × 1.6 (B)	201.0	15.3	44.5	1.59	3.5	3	164
150 × 45 × 2.0 (A)	152.5	15.2	45.2	1.97	4.0	3	114
150 × 45 × 2.0 (B)	151.3	15.0	45.0	1.96	4.4	3	114
150 × 45 × 1.6 (A)	150.0	15.3	44.8	1.59	3.4	3	114
150 × 45 × 1.6 (B)	150.0	15.0	44.7	1.58	3.1	3	114
125 × 45 × 2.0 (A)	125.8	15.3	44.8	1.92	3.8	3	89
125 × 45 × 2.0 (B)	126.0	15.2	45.0	1.92	3.5	3	89
125 × 45 × 1.6 (A)	126.0	15.3	45.3	1.56	3.3	3	89
125 × 45 × 1.6 (B)	126.0	15.2	44.7	1.55	3.1	3	89

machine. Seven groups of linear electrical resistance strain gauges were affixed to both surfaces of the webs of some selected specimens at mid-height, in order to investigate the behaviour of the web. The strain is the average of the two strain gauge readings. Measurements of geometry were taken. Regardless of the nominal section sizes, the actual section dimensions were precisely measured and recorded. Tables 1 & 2

188

Figure 4.   Testing arrangement for stub columns.

Table 3.   Results from stub column tests.

Designation	Group A (kN)	Group B (kN)
300 × 75 × 3.0	728	735
300 × 75 × 2.5	563	600
300 × 60 × 2.0	356	375
250 × 75 × 3.0	673	651
250 × 75 × 2.5	564	572
250 × 60 × 2.0	369	377
200 × 60 × 2.5	488	489
200 × 60 × 2.0	332	326
200 × 45 × 1.6	248	250
150 × 45 × 2.0	305	310
150 × 45 × 1.6	250	228
125 × 45 × 2.0	314	311
125 × 45 × 1.6	239	217

show the nominal and actual section dimensions of LSB stub column specimens respectively.

2.3   Test set-up

Testing of the 26 LSB stub columns was carried out in a self-contained 2000 kN DARTEC hydraulic testing machine. Both set-ups were load-controlled through a DARTEC control cabinet. The test arrangement is shown in Figure 4.

2.4   Loading rates

Loading rate for each test was set such that ultimate load would be reached after 5–10 minutes, and the test would be completed following an appropriate amount of unloading after 20–25 minutes.

2.5   Instrumentation and data acquisition

Three linear variable displacement transducers (LVDTs) were used to determine the end displacement of the stub columns. As mentioned in former sections, two strain gauges were affixed to each specimen. Load, stroke, strain, displacement were all recorded using the data acquisition equipment and the equivalent computer package.

2.6   Stub column results

A summary of the results showing the ultimate load for each test from the LSB stub column tests is presented in Table 3.

From the measured end shortening readings from the LVDTs and the load recordings from the computer package, the load-end displacement curves from the 26 stub column tests were plotted and some of typical diagrams are shown in Figures 5 to 7. Photographs of deformed test specimens are shown in Figure 8.

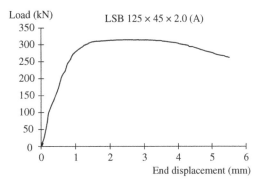

Figure 5.   Load-End displacement curves for LSB 125 × 45 × 2.0 (A) stub column.

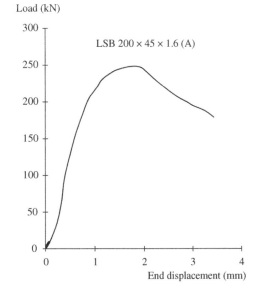

Figure 6.   Load-End displacement curves for LSB 200 × 45 × 1.6 (A) stub column.

189

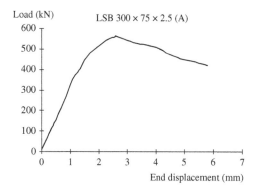

Figure 7. Load-End displacement curves for LSB 250 × 75 × 3.0 (A) stub column.

Figure 8. Deformations of tested specimens (300 × 75 × 2.5 (A) and 150 × 45 × 1.6 (A)).

2.7 Comparison of the capacities between test results and calculated results

Since the yield stresses of web and flange are different, no definite design standard is available for the capacity calculation. In this paper, four different methods of capacity calculation were attempted and they were compared with the tests results.

Coupon tests were carried out and actual yield stresses of flange and web were obtained. Table 4 shows the results of the actual yield stresses.

The four methods to calculate the capacity are stated as following based on the equation $P = A_e \times f_y$ where f_y refers to the actual yield stress of web or flange:

(1) Assume $f_{y,flange}$ for both flange and web through whole calculation.
(2) Assume $f_{y,web}$ for both flange and web through whole calculation.
(3) Assume $f_{y,flange}$ for flange and $f_{y,web}$ for web through whole calculation.
(4) Assume $f_{y,flange}$ for both flange and web when calculating effective section properties. Assume

Table 4. Results of the actual yield stresses from coupon tests.

Designation	Yield stress of web $f_{y,web}$ (MPa)	Yield stress of flange $f_{y,flange}$ (MPa)
300 × 75 × 3.0	425	580
300 × 75 × 2.5	410	520
300 × 60 × 2.0	420	575
250 × 75 × 3.0	400	540
250 × 75 × 2.5	380	510
250 × 60 × 2.0	400	515
200 × 60 × 2.5	400	530
200 × 60 × 2.0	380	520
200 × 45 × 1.6	380	450
150 × 45 × 2.0	410	540
150 × 45 × 1.6	380	560
125 × 45 × 2.0	415	520
125 × 45 × 1.6	380	550
Nominal	380	450

$f_{y,flange}$ for flange and $f_{y,web}$ for web when calculating design capacities.

The key point is to calculate effective section area A_e which can be expressed as $A_e = b_e \times t$. According to Clause 2.2.1.2 of AS/NZS 4600 (1996), effective width (b_e) for capacity calculations shall be determined from Equation 2.2.1.2(1) or Equation 2.2.1.2(2), as appropriate.

For $\lambda \le 0.673$: $b_e = b$ 2.2.1.2   (1)

For $\lambda > 0.673$: $b_e = \rho\, b$ . . . 2.2.1.2   (2)

where
b = flat width of element excluding radii

$$\rho = \text{effective width factor} = \frac{\left(1 - \dfrac{0.22}{\lambda}\right)}{\lambda}$$

The slenderness ratio (λ) shall be determined as follows:

$$\lambda = \left(\frac{1.052}{\sqrt{k}}\right)\left(\frac{b}{t}\right)\left(\sqrt{\frac{f^*}{E}}\right)$$

where
k = plate buckling coefficient
   = 4 for stiffened elements supported by a web on each longitudinal edge
t = thickness of the uniformly compressed stiffened elements
f^* = design stress in the compression element calculated on the basis of the effective design width
E = Young's modulus of elasticity (200×10^3 MPa).
The results of group A and B were shown in Tables 5, 6 & 7.

Table 5–7. Comparison of load capacities (M1, M2, M3, M4 refer to method 1), 2), 3), 4) respectively.).

Specimen designation	Design capacities (kN)				Ultimate load from tests (kN)	
	M 1	M 2	M 3	M 4	Group A	Group B
300 × 75 × 3.0	781	582	744	735	728	735
300 × 75 × 2.5	615	483	586	578	563	600
300 × 60 × 2.0	418	310	401	396	356	375
250 × 75 × 3.0	712	536	681	673	673	651
250 × 75 × 2.5	582	440	559	553	564	571
250 × 60 × 2.0	366	289	352	347	369	377
200 × 60 × 2.5	505	388	483	477	488	489
200 × 60 × 2.0	372	275	357	353	332	325
200 × 45 × 1.6	204	175	201	198	248	250
150 × 45 × 2.0	307	238	294	290	305	310
150 × 45 × 1.6	250	172	237	234	250	228
125 × 45 × 2.0	284	230	274	271	314	311
125 × 45 × 1.6	241	169	229	227	239	216

	$P_{calculation}/P_{test}$ (Group A)			
	M1	M2	M3	M4
300 × 75 × 3.0	1.07	0.80	1.02	1.01
300 × 75 × 2.5	1.09	0.86	1.04	1.03
300 × 60 × 2.0	1.17	0.87	1.13	1.11
250 × 75 × 3.0	1.06	0.80	1.01	1.00
250 × 75 × 2.5	1.03	0.78	0.99	0.98
250 × 60 × 2.0	0.99	0.78	0.95	0.94
200 × 60 × 2.5	1.04	0.80	0.99	0.98
200 × 60 × 2.0	1.12	0.83	1.07	1.06
200 × 45 × 1.6	0.82	0.70	0.81	0.80
150 × 45 × 2.0	1.01	0.78	0.96	0.95
150 × 45 × 1.6	1.00	0.69	0.95	0.94
125 × 45 × 2.0	0.90	0.73	0.87	0.86
125 × 45 × 1.6	1.01	0.71	0.96	0.95
Mean value	1.02	0.78	0.98	0.97
Standard deviation	0.090	0.057	0.082	0.081

	$P_{calculation}/P_{test}$ (Group B)			
	M1	M2	M3	M4
300 × 75 × 3.0	1.06	0.79	1.01	1.00
300 × 75 × 2.5	1.03	0.81	0.98	0.96
300 × 60 × 2.0	1.11	0.83	1.07	1.06
250 × 75 × 3.0	1.09	0.82	1.05	1.03
250 × 75 × 2.5	1.02	0.77	0.98	0.97
250 × 60 × 2.0	0.97	0.77	0.93	0.92
200 × 60 × 2.5	1.03	0.79	0.99	0.98
200 × 60 × 2.0	1.14	0.85	1.10	1.09
200 × 45 × 1.6	0.82	0.70	0.80	0.79
150 × 45 × 2.0	0.99	0.77	0.95	0.94
150 × 45 × 1.6	1.10	0.75	1.04	1.03
125 × 45 × 2.0	0.91	0.74	0.88	0.87
125 × 45 × 1.6	1.12	0.78	1.06	1.05
Mean value	1.03	0.78	0.99	0.98
Standard deviation	0.092	0.039	0.082	0.082

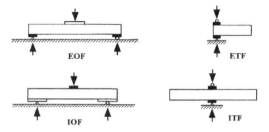

Figure 9. Definition of bearing cases.

2.8 Discussion

From Table 3, it can be seen that the tested ultimate loads of Group A and B were reasonably in good agreement despite that the maximum difference of two sets of them are approximately 8% and 9%. This may be caused by slight discrepancy of section dimensions and material yield stress of the same set.

From Table 4, it is observed that the tested ultimate loads of all sets of section sizes are about 20% higher than the calculated results using method 2, which is the method to calculate the load capacities in the current product manual of LSB. Hence, it is concluded that this method is conservative to apply when calculating the load capacities of LSB stub columns. In addition, the calculated results using method 3 and 4 are quite close to the tested ones, both within 3% averagely. This indicates a suggestion to use either of these two methods to calculate the load capacities of LSB stub columns. However, further reliability analysis is required before full conclusion could be confirmed.

3 END BEARING TESTS

3.1 Behavioural issues

The bearing capacity of hot-rolled I-sections has been well researched, but cold-formed sections have specific problems related to their rounded corners, and hollow flange sections can fail by "crushing" of the hollow flange as well as a web failure. The bearing capacity of hollow flange beams and plain channels has been investigated (Hancock et al (1994), Sully and Hancock (1994), Young and Hancock (2001)), however the rotational restraint provided to LSB webs from the hollow flanges was unknown. Hence it was appropriate to examine the bearing strength. The EOF (exterior bearing – one flange) case is considered in this paper.

3.2 Test method

A diagram and photographs of the test set-up are shown in Figures 10 & 11. The EOF bearing load was applied at the bottom flanges at two ends of the beam through

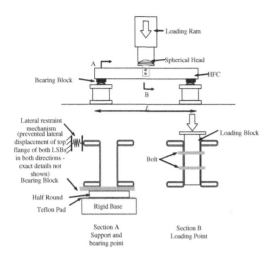

Figure 10. EOF test method.

Figure 11. EOF test method (bearing block hidden behind lateral support).

the bearing blocks, and three different stiff bearing lengths (25 mm, 50 mm and 100 mm) were selected.

3.3 Results

All specimens failed by bearing in a failure mode that involved web crippling only. A typical failed specimen is shown in Figure 12. From observation of the surface of the specimen, the buckling behaviour can be seen. The fall of the load resulted from the failure of the web crippling. As shown in Figure 12, the bottom flange did not have large rotation, which occurred in the plain channel test (Young & Hancock 2003). The large rotation only happened after the test reached the peak load.

Tables 8 and 9 give the test results (R_{max}) and compares them with the predictions of the current AS/NZS 4600 and the proposed changes.

Figure 12. Failed specimen after EOF test.

Table 8. Summary of EOF results.

Dimension	Bearing length (mm)	f_y (MPa)	R_{max} (kN)
125 × 45 × 15 × 1.6	50	465	17.5
125 × 45 × 15 × 1.9	50	398	25.3
125 × 45 × 15 × 1.9	25	398	20.6
125 × 45 × 15 × 1.6	25	465	12.9
150 × 45 × 15 × 2.4	50	364	36.0
150 × 45 × 15 × 1.9	50	377	22.3
150 × 45 × 15 × 1.6	50	348	15.1
150 × 45 × 15 × 1.9	25	377	20.3
150 × 45 × 15 × 2.4	25	364	29.0
200 × 60 × 20 × 2.4	50	503	38.0
200 × 60 × 20 × 1.9	50	352	20.0
200 × 60 × 20 × 2.4	100	503	45.8
200 × 60 × 20 × 1.9	100	352	24.3
250 × 60 × 20 × 2.0	100	404	24.3
250 × 60 × 20 × 2.0	50	404	21.4
250 × 75 × 25 × 2.5	100	406	38.8
250 × 75 × 25 × 2.5	50	406	33.4
250 × 75 × 25 × 3.0	100	397	55.3
250 × 75 × 25 × 3.0	50	397	50.3
300 × 60 × 20 × 2.0	100	371	20.2
300 × 60 × 20 × 2.0	50	371	18.9
300 × 75 × 25 × 2.5	100	430	36.9
300 × 75 × 25 × 2.5	50	430	33.8
300 × 75 × 25 × 3.0	100	426	46.5
300 × 75 × 25 × 3.0	50	426	42.6

The design rules for bearing (EOF type) in the current AS/NZS 4600 (1996) are as follows:

$$R_b = 10.0t^2 f_y C_3 C_4 C_\theta \{1-0.00183(d_1/t_w)\} \{1+0.01(l_b/t_w)\}$$

(Case 1 of Table 3.3.6(1) for one flange end loading)

Table 9. EOF result comparison.

Dimension	DR 03518		AS/NZS 4600	
	R_b (kN)	R_{max}/R_b	R_b (kN)	R_{max}/R_b
$125 \times 45 \times 15 \times 1.6$	12.4	1.40	12.7	1.38
$125 \times 45 \times 15 \times 1.9$	14.2	1.78	14.2	1.78
$125 \times 45 \times 15 \times 1.9$	12.7	1.62	13.0	1.59
$125 \times 45 \times 15 \times 1.6$	11.1	1.16	11.5	1.12
$150 \times 45 \times 15 \times 2.4$	20.9	1.73	20.3	1.77
$150 \times 45 \times 15 \times 1.9$	13.8	1.61	13.8	1.62
$150 \times 45 \times 15 \times 1.6$	9.2	1.64	9.3	1.62
$150 \times 45 \times 15 \times 1.9$	12.4	1.64	12.6	1.61
$150 \times 45 \times 15 \times 2.4$	18.7	1.55	18.8	1.54
$200 \times 60 \times 20 \times 2.4$	29.4	1.29	28.2	1.35
$200 \times 60 \times 20 \times 1.9$	12.5	1.60	12.2	1.63
$200 \times 60 \times 20 \times 2.4$	33.8	1.35	32.2	1.42
$200 \times 60 \times 20 \times 1.9$	14.5	1.68	14.3	1.70
$250 \times 60 \times 20 \times 2.0$	17.4	1.40	16.7	1.46
$250 \times 60 \times 20 \times 2.0$	15.3	1.40	14.6	1.47
$250 \times 75 \times 25 \times 2.5$	30.2	1.28	28.2	1.37
$250 \times 75 \times 25 \times 2.5$	26.4	1.27	24.7	1.35
$250 \times 75 \times 25 \times 3.0$	37.2	1.49	33.9	1.63
$250 \times 75 \times 25 \times 3.0$	32.8	1.53	30.2	1.66
$300 \times 60 \times 20 \times 2.0$	15.5	1.30	14.4	1.40
$300 \times 60 \times 20 \times 2.0$	13.7	1.38	12.6	1.50
$300 \times 75 \times 25 \times 2.5$	31.4	1.17	28.8	1.28
$300 \times 75 \times 25 \times 2.5$	27.8	1.21	25.7	1.32
$300 \times 75 \times 25 \times 3.0$	40.1	1.16	36.0	1.29
$300 \times 75 \times 25 \times 3.0$	35.4	1.20	32.1	1.33
Mean		*1.43*		*1.49*

where

$$C_3 = 1.33 - 0.33k$$
$$C_4 = 1.15 - 0.15r_i/t_w$$
$$C_\theta = 0.7 + 0.3(\theta/90)^2$$
$$k = f_y/228$$

The public draft for the new version of AS/NZS 4600, DR 03518 (2003), has a different formulation for bearing.

$$R_b = Ct_w^2 f_y \sin\theta \{1 - C_r\sqrt{(r_i/t_w)}\}\{1 + C_l\sqrt{(l_b/t_w)}\}\{1 - C_w\sqrt{(d_l/t_w)}\}$$

(Equation 3.3.6.1)
where (for EOF loading for stiffened flanges given in Table 3.3.6.1(B))
$C = 4$
$C_r = 0.14$
$C_l = 0.35$
$C_w = 0.02$

3.4 Discussion

The following observations can be made from the results of these tests:

All tests show capacities well above those predicted by either the current AS/NZS 4600 (mean overstrength 49%) or the draft DR 03518 (mean overstrength 43%).

For smaller web depths (125 mm and 150 mm), the predictions of AS/NZS 4600 and DR 03518 are similar. As the total section depth increase (200 mm, 250 mm, 300 mm), the DR 03518 is of the order of 10% higher than the AS/NZS 4600 prediction. In all cases, the test load exceeded the predictions of both specifications.

As shown in Figure 12, the large deformation occurred in the web and there was little evidence of flange crushing. This is a significant observation since it suggests that the flange crushing failure mode is not significant.

3.5 Development of a new design model

The bearing design equations in the current AS/NZS 4600 and the new draft (based on the AISI specification) are derived from parametric curve fitting of test results. Two approaches are currently being used to derive a less conservative formulation. The first is to derive new coefficients that can be substituted into the current equations. The second approach is to treat the web as an equivalent column compression member. The size of the "column" depends on the load dispersal mechanism, and the effective length is a function of the restraint conditions that are applied to the webs via the hollow flanges.

4 CONCLUSIONS

This paper has presented the results of initial investigations into the behaviour of hollow flange beams. The unique shape makes the section prone to different failure modes under direct bearing loads. In the exterior one flange loading situations the current design rules are conservative and new approaches are being taken to devise a less conservative set of equations. The flange and web have different yield stresses, which make it difficult to use the effective width approach to determine the stub column strength. it is concluded that AS/NZS 4600 is conservative to apply when calculating the load capacities of LSB stub columns.

REFERENCES

Avery, P., Mahendran, M. and Nasir, A., (2000), "Flexural Capacity of Hollow Flange Beams", Journal of Constructional Steel Research, Elsevier, Vol 53, No 2, pp 201–223.
CASE, (1993), "Hollow Flange Beams Under Combined Bending and Bearing", Investigation Report S958, Centre for Advanced Structural Engineering, School of Civil and Mining Engineering, The University of Sydney, Sydney, Australia.
Hancock, G. J., Sully, R. M., and Zhao, X-L., (1994), "Hollow Flange Beams and Rectangular Hollow Sections Under Combined Bending and Bearing", Tubular Structures VI, Proceedings of the 6th International Symposium on Tubular Structures, Melbourne, Australia, Balkema (publ), Grundy, Holgate and Wong (eds.), pp 47–54.

Standards Australia/Standards New Zealand (1996), Australian/New Zealand Standard AS/NZS 4600 Cold-Formed Steel Structures, Standards Australia/Standards New Zealand, Sydney, Australia.

Standards Australia/Standards New Zealand (2003), Australian/New Zealand Standard Public Draft DR 03518 Cold-Formed Steel Structures, Standards Australia/Standards New Zealand, Sydney, Australia.

Sully, R. M., and Hancock, G. J., (1994), "Hollow Flange Beams Under Combined Bending and Bearing", Proceedings, Australasian Structural Engineering Conference, Institute of Engineers, Australia, Sydney, Australia, September 1994, pp 1033–1038.

Yang, D. and Wilkinson, T., (2005), "LiteSteel Beams (LSB) Under Interior and End Bearing Forces", Research Report No R849, Department of Civil Engineering, The University of Sydney, September 2005.

Young, B. and Hancock, G. J., (2001), "Design of Cold-Formed Channels Subjected to Web Crippling", Journal of Structural Engineering, American Society of Civil Engineers, Vol. 127, No. 10, October, 2001, pp 1137–1144.

Young, B. and Hancock, G. J., (2003), "Cold-Formed Steel Channels Subjected to Concentrated Bearing Load", Journal of Structural Engineering, American Society of Civil Engineers, Vol. 129, No. 8, August, 2003, pp 1003–1010.

Tubular Structures XI – Packer & Willibald (eds)
© 2006 Taylor & Francis Group, London, ISBN 0-415-40280-8

Analytical modelling of non-uniform deformation in thin-walled orthotropic tubes under pure bending

M.A. Wadee
Department of Civil & Environmental Engineering, Imperial College London, UK.

M.K. Wadee
School of Engineering, Computer Science and Mathematics, University of Exeter, UK

A.P. Bassom
School of Mathematics and Statistics, The University of Western Australia, Australia

ABSTRACT: An analytical model employing both large displacement and Timoshenko beam theories is presented for tubes with circular cross-sections under pure bending. The resulting equations are solved numerically to simulate the inhomogeneous buckling deformation and the moment–curvature relationships for various tubes constructed from orthotropic materials. Not only is this an advancement on classical Brazier buckling theory, but a significant modal interaction is identified that severely reduces the moment capacity of tubes with particular levels of the shear modulus.

1 INTRODUCTION

The response to homogeneous bending of thin-walled tubular members with circular cross-sections has been well documented for nearly eighty years (Brazier 1927). The principal phenomenon associated with this is the so-called Brazier effect which manifests itself as an ovalization of the initially circular cross section during pure bending. However, most of the work in this field has tended to assume that the cross section ovalization is uniform along the length of the member; a good assumption for small deflections and bending to small curvatures, but practically this breaks down when deflections become large and engineer's bending theory becomes inadequate. The Brazier effect can lead to instability, with increasing curvature the material within the zone of compression can buckle locally. In terms of the relationship between moment M and curvature κ the gradient, which is initially constant, fades away and a limiting value of M is encountered when buckling occurs; the deformation seen in the tube becomes decidedly nonuniform, a regime of *negative stiffness* is encountered and kinks tend to form in many cases.

In this study, a new analytical model is presented that uses both large-displacement and shear deformable beam theories to account for the inhomogeneity in the physical response, as found in earlier work on sandwich structures (Hunt and Wadee 1998). The formulation is based on fundamental work by Reissner (1961), but is extended using a Fourier decomposition of the radial deformation of the cross-section. A total potential energy functional is developed and subsequently minimized using variational principles leading to a system of nonlinear ordinary differential equations describing the radial deflections in the cross-section along the length of the tube. These equations have been solved numerically to simulate the moment–curvature relationships for various tube geometries found in the literature and comparisons have been made with published experiments and earlier analytical and numerical models (Wadee et al. 2006). In the present study, an extension to the model is made to include tubes constructed with orthotropic materials (Tatting et al. 1996), such that in future tubes made from composite materials may be modelled. In these cases the deformations are more pronounced for certain material property configurations and kinking localization can be observed more readily such that the response is seen to be highly sensitive to initial defects.

2 MODEL DEVELOPMENT

Consider an initially straight hollow tube of length L with a circular cross-section of radius r and wall thickness t made from a homogeneous but two dimensionally orthotropic linear elastic material with Young's moduli E_x and E_y, Poisson's ratios ν_x and ν_y and the shear modulus G; with G being independent of

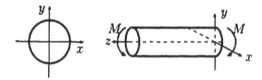

(a) Orientation of the undeformed tube and axes.

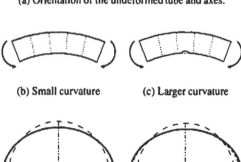

(b) Small curvature   **(c) Larger curvature**

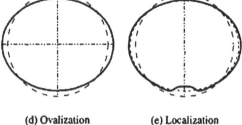

(d) Ovalization   **(e) Localization**

Figure 1. (a) Schematic layout of tube. (b)–(e): Representation of progressive ovalization and localization of a tube in pure bending.

the Young's moduli and the relationship between the Young's moduli and the Poisson's ratios being:

$$E_x v_y = E_y v_x. \tag{1}$$

Suppose also that a coordinate system is aligned so that the centreline of the tube is horizontal and coincides with the z-axis. Moreover, the remaining axes are rotated so that the x-direction is also horizontal and the tube is acted upon by a bending moment M about the x-axis—the configuration is summarized in Figure 1(a).

The remaining sub-parts of Figure 1 illustrate the plausible behaviour of a typical cross-section of the tube as it is subjected to increasing loading. In the unloaded state the tube is straight and the section is perfectly circular, but if a small moment is applied it attains a constant curvature and ovalizes (Figure 1(b), (d)). Notice that at this stage the deformation continues to have both vertical and horizontal axes of symmetry. However, once the load is sufficiently high, the tube becomes unstable and a breaking of symmetry (about the horizontal axis) occurs, see Figure 1(c),(e). The cross-section is now characterized by the presence of a kinking phenomenon but this is not expected to occur simultaneously at all sections along the tube. Rather, the kink forms over a relatively short length giving

rise to the desired localization of buckle pattern along the tube.

Taking modelling ideas from the analysis of sandwich panels (Hunt and Wadee 1998), where the structure is assumed to behave according to the bending theory of Timoshenko, makes it possible to formulate an analytical model for non-uniform ovalization for tubes under flexure. For a tube under constant end moment the vertical deflection of the neutral axis, $W(z)$, is a circular arc which here is approximated by a parabola and its associated angle of tilt of the initially vertical plane sections $\theta(z)$ are both given by:

$$W(z) = q_s z(z - L)/L,$$
$$\theta(z) = q_t(2z - L)/L, \tag{2}$$

where W is measured downwards and the quantities q_s and q_t denote the dimensionless amplitudes of the sway and tilt. The magnitude of tilt is readily identified with the radius of curvature ($R = 1/\kappa$) of the bent tube via

$$\kappa = \frac{d\theta}{dz} = \frac{2q_t}{L} = \frac{1}{R}. \tag{3}$$

For Euler–Bernoulli beams q_s and q_t are always equal; this is the upshot of requiring that the shear strain be zero (i.e. $dW/dz = \theta$). However, in Timoshenko's more general beam theory, this constraint is relaxed and q_s and q_t can be different; the adoption of this beam theory is necessary in order to allow for the potential variation in the shape of the cross-section along the tube.

By itself Timoshenko's theory is incapable of capturing ovalization but this is addressed by incorporating Reissner's model which includes the Brazier effect in a systematic manner. The formulation begins by summarizing the uniform phenomenon before embarking on modelling of the localization.

2.1 Homogeneous ovalization of the tube

It is first convenient to introduce the following elastic constants

$$C = Et, \quad D = \frac{Et^3}{12(1-v_x v_y)} \quad \text{and} \quad \Delta = \frac{r^4 C}{L^2 D}. \tag{4}$$

Additionally, a dimensionless measure of curvature is given as α where

$$\alpha \equiv 2q_t \sqrt{\Delta}. \tag{5}$$

Figure 2 summarizes some of the notation that will be referred to in the course of the development. Suppose that a point (x, y) lies on the original undeformed circular section; it is helpful to make reference to the angle φ denoting the angle made by the radius from the origin to (x, y) with the negative y-axis. When the section is ovalized the point originally at (x, y) is now supposed to reside at $(x + \zeta, y + \eta)$ and that the

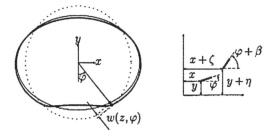

Figure 2. Definitions of coordinates and deformation of a section of the tube. The open circle on the perimeter shows the coordinate of the homogeneously deformed tube with position $(x + \zeta, y + \eta)$. Right: Displacement of an element (thick line) of length ds.

deviation of the tangent to the oval from its direction when the tube is undeformed is β; in all that follows β is assumed to be small.

The moment–curvature relationship for a tube undergoing homogeneous ovalization of the cross section can be written in dimensionless terms as

$$m \equiv \frac{M}{\pi r \sqrt{CD}} = \alpha - \frac{1}{8}\alpha^3 - \frac{1}{96}\alpha^5 + \dots \qquad (6)$$

after Reissner (1961) and is noted for future reference in the discussion of our results.

2.2 Deviation from Reissner's formulation

Reissner's approach accounts for the ovalization of the cross-section under increasing moment. As the section becomes flattened so the moment–curvature graph destiffens. Among the assumptions made in that analysis is that the uniform deformation of the tube under load is such that no stretching of the material occurs in the circumferential direction; that is, an element of the section of length ds remains unchanged during deformation. This requirement is not enforced and instead the tube is allowed to undergo an additional small radial displacement $w(z, \varphi)$ which in turn accounts for any z-dependent variation in the profile. It is this that enables the tube to exhibit axially varying deformation after an initial global ovalization phase.

According to linear bending theory the first-order axial strain in the z-direction is

$$\varepsilon_z = \frac{y + \eta}{R} = 2q_t(y + \eta)/L. \qquad (7)$$

In order to incorporate nonlinear strain effects the von Karman expression for strain is used (Timoshenko and Woinowsky-Krieger 1970)

$$\varepsilon = \varepsilon_z + \frac{\partial \tilde{u}}{\partial z} + \frac{1}{2}\left(\frac{\partial w}{\partial z}\right)^2. \qquad (8)$$

This quantity includes the influence of the in-plane deflection, $\tilde{u}$, which is assumed to vary linearly with

the vertical distance from the neutral axis of the section i.e. $\tilde{u}(\varphi, z) = u(z)y/r \equiv -u(z)\cos\varphi$. Note that if the section were to remain undeformed during loading then these terms would reduce to the standard expression for bending energy.

The strategy underpinning the derivation of the salient governing equations relies on variational techniques applied to an appropriate potential energy functional. So far it is seen that with the ovalization followed by localization the point originally at (x, y) moves to $(x + \zeta + w\sin\varphi, y + \eta - w\cos\varphi)$ with $w = w(z, \varphi)$. In order to proceed the appropriate form for the potential energy needs to be deduced in terms of these quantities and remark that contributions arise from five sources.

2.3 Bending energy

Under conditions of constant moment, the tube initially behaves like a beam with a general cross section and deflects into a circular arc. For small deflections, this can be approximated by a parabola, see (2). As the intention is to account for ovalization of the tube during this deformation, it is necessary to incorporate the incremental change in shape as it occurs. Hence the conventional term for second moment of area of an elemental piece of the section, $y^2 t$ ds, is replaced by the term $(y + \eta)^2 t$ ds. The strain energy of this component plus that due to additional z-dependent local deflection of the section is then

$$U_b = \frac{1}{2}Et\int_0^L \int_0^{2\pi}\left[\ddot{w}^2 + \ddot{w}^2\right](y + \eta)^2 \, \mathrm{d}s \, \mathrm{d}z, \qquad (9)$$

where a dot denotes differentiation with respect to z.

2.4 Membrane energy

The strain energy stored axially within the membrane of the wall of the hollow tube is given by the von Karman strain expression. Therefore this can be cast as

$$U_m = \frac{1}{2}Et\int_0^L \int_0^{2\pi}\varepsilon^2 \, \mathrm{d}s \, \mathrm{d}z$$

$$= \frac{1}{2}Et\int_0^L \int_0^{2\pi}\left[\varepsilon_z^2 + 2\varepsilon_z\left(\dot{\tilde{u}} + \frac{1}{2}\dot{w}^2\right) + \dot{\tilde{u}}^2 + \dot{\tilde{u}}\dot{w}^2 + \frac{1}{4}\dot{w}^4\right]\mathrm{d}s\,\mathrm{d}z$$

$$\qquad (10)$$

2.5 Shear strain energy

Shear strain is incorporated within the formulation and including the components arising from the incremental deformation w the following expression is obtained

$$\gamma_{yz} = \frac{\partial W}{\partial z} - \theta + \frac{\partial w}{\partial z} + \frac{\partial \tilde{u}}{\partial y}$$

$$= (q_s - q_t)\left(\frac{2z}{L} - 1\right) + \dot{w} + \frac{u}{r}. \qquad (11)$$

The contribution to the potential energy from shear strains is given by

$$U_s = \frac{1}{2}Gt\int\limits_0^L \int\limits_0^{2\pi} \gamma_{yz}^2 \, ds \, dz. \tag{12}$$

2.6 Circumferential bending energy

A small but important contribution to the energy arises due to the additional deformation w. The additional curvature is $\partial^2 w = \partial s^2$ leading to the term

$$U_{cb} = \int\limits_0^L \int\limits_0^{2\pi} \frac{1}{2}D\left(\frac{\partial^2 w}{\partial s^2}\right)^2 ds \, dz$$

$$= \frac{D}{2r^3}\int\limits_0^L \int\limits_0^{2\pi} \left(\frac{\partial^2 w}{\partial \varphi^2}\right) d\varphi \, dz. \tag{13}$$

In passing, notice that this contribution is analogous to the role played by the elastic foundation in the strut model—the focus of much study as an archetypal localized buckling problem. Its effect is to cause the response to become independent of the length of the structure as long as the boundaries are sufficiently far away from the position of significant deformation (Hunt and Wadee 1991).

2.7 Work done by load

The work done is the sum of the contributions from the internal and external moments. Each term is basically the moment multiplied by the local rotation, which is therefore

$$M\Theta = \int\limits_0^L \int\limits_0^{2\pi} Et\varepsilon_z \dot{\tilde{u}} \, ds \, dz + \int\limits_0^L M\left(\frac{\dot{u}}{r} + \frac{2q_t}{L}\right) dz. \tag{14}$$

3 POTENTIAL ENERGY FUNCTIONAL

The total potential energy for the structural system is given by the sum of the contributions of strain energy formulated above in (9)–(13) minus the work done by the load (14). It is also helpful to note that the requisite extremal problem would be much simplified if the double integrals could be reduced to single ones and this can be done with relatively little difficulty. There are added bonuses as well: when performing the s (or equivalently φ) integrals in (9)–(14) it is ensured that $w(z, \varphi)$ is necessarily periodic in φ by assuming a Fourier cosine series for $w(z, \varphi)$ so

$$w(z,\varphi) = \sum_{n=0}^{\infty} w_n(z)\cos n\varphi \tag{15}$$

of course it must ensured that $w(z, \varphi)$ is even-valued in φ for the deflection of the tube to be symmetric about the y-axis in the context of Figure 2.

At this stage the potential energy is given in terms of integrals of the functions $w_n(z)$; application of variational principles would lead to an infinite system of ODEs for these quantities. In practice one needs to truncate this system at some point and here an admittedly very severe form was adopted. For all the computations described below $w_n \equiv 0$ is set for $n \geq 3$. To illustrate the usefulness and power of the proposed model, the desire was to minimize the complexity of the underlying problem and some experimentation revealed that ignoring everything except for w_0, w_1 and w_2 was sufficient both to capture the ovalization and localization phenomena sought and provides good comparison with previous results (see §4). Of course a more refined picture could be constructed by inclusion of more terms (with a correspondingly more involved numerical task).

The functional form of the potential energy is now typified by $\int \Im(Q, \dot{Q}, \ddot{Q}) dz$ where Q is any of $w_0(z)$–$w_2(z)$ or $u(z)$. Such an integral is extremized if (Fox 1987)

$$\delta \int\limits_0^L \Im(Q, \dot{Q}, \ddot{Q}) \, dz$$

$$= \left[\frac{\partial \Im}{\partial \ddot{Q}} \partial \dot{Q} + \left\{\frac{\partial \Im}{\partial \dot{Q}} - \frac{d}{dz}\left(\frac{\partial \Im}{\partial \ddot{Q}}\right)\right\}\partial Q\right]_0^L$$

$$+ \int\limits_0^L \left\{\frac{d^2}{dz^2}\left(\frac{\partial \Im}{\partial \ddot{Q}}\right) - \frac{d}{dz}\left(\frac{\partial \Im}{\partial \dot{Q}}\right) + \frac{\partial \Im}{\partial Q}\right\}\partial Q \, dz$$

$$= 0. \tag{16}$$

With the assumption of simple supports at the ends of the tube the physical boundary conditions

$$w_n(0) = w_n(L) = 0, \quad \ddot{w}_n(0) = \ddot{w}_n(L) = 0, \tag{17}$$

eliminate most of the boundary terms in (16) when $Q \equiv w_n$, where $n = 0, 1, 2$. However, slightly more complicated boundary conditions arise for $Q \equiv u$:

$$\dot{u}(0) + \frac{1}{2}\dot{w}_1(0)[2\dot{w}_0(0) + \dot{w}_2(0)] = \frac{M}{\pi r^2 Et}$$

$$\dot{u}(L) + \frac{1}{2}\dot{w}_1(L)[2\dot{w}_0(L) + \dot{w}_2(L)] = \frac{M}{\pi r^2 Et}. \tag{18}$$

The second part of the minimizing process is to eliminate the integrand in (16). This leads to sets of ODEs for each of the variables $w_0(z)$, $w_1(z)$, $w_2(z)$ and $u(z)$ plus two integral constraints for the parameters q_s and q_t, respectively. The ODEs are all nonlinear: for w_n they are fourth order and for u it is second order. For the sake of brevity, the governing ODEs and integral constraints are not listed here but the corresponding set for an isotropic tube can be found in Wadee et al. (2006).

4 NUMERICAL RESULTS

The system of ODEs was solved using the numerical continuation package AUTO97 (Doedel et al. 1997). To conserve the inherent symmetry of the system the equations were solved over half the length ($z = [0, L/2]$) with the symmetry conditions at $z = L/2$:

$$\dot{w}_n(L/2) = \ddot{w}_n(L/2) = u = 0 \qquad (19)$$

which also minimize V, being imposed. Solutions were obtained from the initially undeformed and unloaded state ($q_s = q_t = w_n(z) = u(z) = M = 0$) and q_t was increased parametrically with AUTO97 solving the ODEs subject to the boundary conditions and integral constraints, giving numerical values of q_s and M that satisfied the equilibrium equations. The ability of the software to evaluate the location of bifurcation points is well-known and was employed to determine the limit points and the post-buckling behaviour discussed below.

4.1 Variation of shear modulus

In previous studies comparing isotropy to orthotropy for the core materials in sandwich struts (Da Silva and Hunt 1990; Wadee and Hunt 1998), the principal finding has been that the struts with orthotropic core are much susceptible to sensitivity to initial defects than their isotropic counterparts. This manifests itself in the secondary buckling that leads to localization. In the tube bending model, however, secondary buckling is not yet accounted for and is speculated to be connected with material, rather than geometric, nonlinearities. Therefore, a pilot study is initiated in the current study to illustrate the effect of orthotropy for the tube bending problem. A tube configuration studied earlier (Wadee et al. 2006) that had isotropic properties, is modified from that state by changing its shear modulus G. The initial tube was assumed to be made of steel with length $L = 250$ mm, radius $r = 10$ mm and wall thickness $t = 1.0$ mm. The Young's modulus of steel was assumed to be 205 kN/mm² with its Poisson's ratio being $v = 0.3$ and therefore using the isotropic relationship for the shear modulus G_i, where:

$$G_i = \frac{E}{2(1+v)}, \qquad (20)$$

and therefore $G_i = 78.85$ kN/mm². For the present study, the value of the shear modulus is changed such that the assumption of isotropy no longer holds. Two other tubes are tested with differing r/t ratios as these have been previously shown to have quite distinct behaviours (Axelrad and Emmerling 1984). It is worth noting at this point that the scale of the tube is on the small side, but that the absolute length plays only a

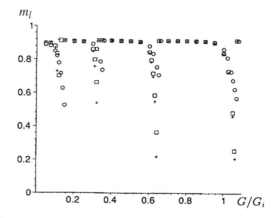

Figure 3. Scatter plot of nondimensional limit moments m_l against G/G_i. Results for three sets of tubes are shown: $r/t = 10$ (○), 33.3 (□) and 100(+).

minor role in the response; it is the ratios of r/t and L/r that are more important and these have been chosen to be practically realistic. Moreover, it should be noted that the Poisson's ratios v_x and v_y are left fixed at 0.3 and studies of these varying are left for the future.

Figure 3 shows results for the three tube configurations: $r/t = 10$, 30 and 100 respectively. The graph shows the nondimensional limit moment ml where this is the ratio of M_l (the actual limit moment) and M_R (Reissner's limit moment) plotted against the shear modulus G of the material used which is varied from the isotropic case G_i. It is clear from the graph that there are certain ratios of G/G_i that substantially reduce the limit moment around 0.15, 0.35, 0.65 and 1.05. In these regions of G/G_i not only is the limit moment reduced but the post-critical moment versus bending curvature equilibrium curve also has a much less stable profile with steep unloading and sometimes even snap-back (Lord et al. 1997). The graph would of course continue for higher values of the shear modulus G, but from a practical viewpoint the material is not likely to have a value of G much above the isotropic case and so the graph is truncated.

Two representative examples from the scatter plot of Figure 3 are presented in Figures 4 and 5, which show a tube with a substantially reduced limit moment and one without the reduction respectively. The moment–curvature curves are also presented in nondimensional form against Reissner's curve to emphasize the difference in the responses (Figure 6).

The response for the $G/G_i = 0.70$ is typical for all the cases where $m_l \approx 0.91$ for all r/t cases considered in the current study. Note also that the response shown for $G/G_i = 0.35$ is typical for all the cases where the limit moment is substantially below $m_l \approx 0.91$ and the total curvature is also much smaller. For the $G/G_i = 0.35$ case the deformed shape is much more

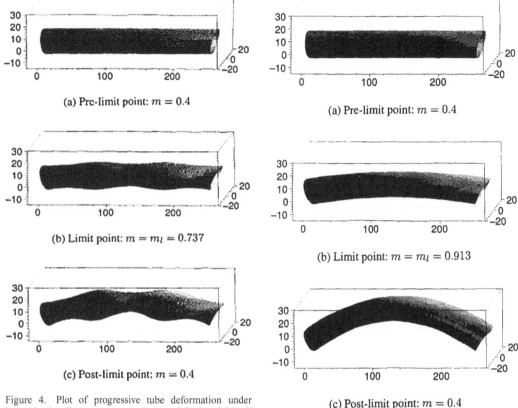

(a) Pre-limit point: $m = 0.4$

(b) Limit point: $m = m_l = 0.737$

(c) Post-limit point: $m = 0.4$

Figure 4. Plot of progressive tube deformation under increasing bending curvature: $t = 1$ mm and $G/G_i = 0.35$. All dimensions are in mm.

(a) Pre-limit point: $m = 0.4$

(b) Limit point: $m = m_l = 0.913$

(c) Post-limit point: $m = 0.4$

Figure 5. Plot of progressive tube deformation under increasing bending curvature: $t = 1$ mm and $G/G_i = 0.70$. All dimensions are in mm.

pronounced at the limit point and in the post-buckling range with kinking seen clearly. The non-uniform nature of the ovalization is also much easier to distinguish for this case if the cross-section profiles are compared from the ends to the mid-span as shown in Figure 7.

4.2 Discussion

The reasoning behind the response as seen in Figure 3 seems to lie with the interaction of buckling mode wavelengths. In the neighbourhood of these special values there is a definite change in the wavelengths of the buckle pattern components of the radial displacements w_n, see Figure 8, which shows the results for $r/t = 10$ and similar results exist for the other two cases presented above.

This is not a coincidence: what are seen in these special regions are certain modes interacting nonlinearly that further erode the already unstable post-critical response of the tube by reducing the limit point moment ml. An analogous phenomenon is found in the post-buckling of flat plates where instead of the

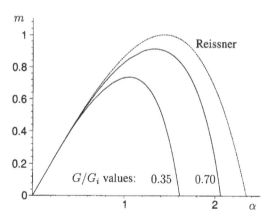

Figure 6. Nondimensional moment m versus curvature plots for the tubes of thickness 1 mm with $G/G_i = 0.35$ and 0.70. Reissner's curve (dashed) is also plotted as a benchmark for the assumption of uniform ovalization.

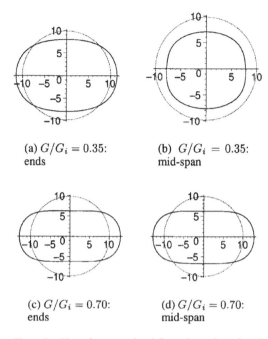

(a) $G/G_i = 0.35$: ends

(b) $G/G_i = 0.35$: mid-span

(c) $G/G_i = 0.70$: ends

(d) $G/G_i = 0.70$: mid-span

Figure 7. Plots of cross-section deformation at the ends and mid-span of the tubes with thickness $t = 1$ mm. The plots compare the cross-sections at $m = 0.4$ in the pre-limit and the post-limit points. Note that the tube with $G/G_i = 0.35$ has a much clearer nonuniform cross-section deformation when compared to the $G/G_i = 0.70$ case. All dimensions are in mm.

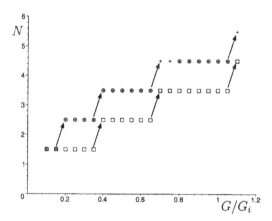

Figure 8. Plot showing the number of waves along the length N in the functions w_0 (o), w_1 ($\square$) and w_2 (+) versus G/G_i. Note that an increase in at least two of the wave numbers simultaneously, shown by the arrows, marks the boundary between the strong and weak mode interactions.

shear modulus being modified, the varying parameter could be the plate aspect ratio. At certain values of the aspect ratio a number of buckling modes with differing wavelengths could be triggered simultaneously and the usual stiff post-buckling response of a flat plate can be lost; this is usually associated with the phenomenon of mode jumping (Everall and Hunt 1999). In the present case, increasing G/G_i causes at first a steady and then rapid reduction from $m_l \approx 0.91$, the approximately uniform ovalization case that behaves similarly to Reissner's model, with a corresponding increase in the non-uniform nature of the cross section ovalization. This non-uniformity increases until at least two of the radial displacement functions w_n (where $n = 0$, 1, 2) increase their wave numbers simultaneously—when this happens m_l returns approximately to 0.91, the cross-section deformation becomes approximately uniform again. As G/G_i continues to increase the process repeats with ml then proceeding to decrease until the next increase in the wave number of at least two of the w_n functions.

The heightened interaction between the modes of buckling which reduces the limit moment is an important aspect of the post-critical response of orthotropic tubes. As the model has been compared favourably with published experimental results previously the current study should caution designers to the sensitivity of certain material property parameters such as the shear modulus and their effect on the load carrying capacity of the structure.

5 CONCLUDING REMARKS

A model for the bending of circular tubes accounting for the length dependent nature of the distortion of the cross-section from circular to an ovalized profile has been presented with an extension to account for orthotropic material properties. A pilot study varying the material shear modulus G has revealed that at certain ranges of values of G the interaction between modes of buckling reduces the limiting moment significantly at which the tube cannot sustain greater external loading. Moreover, it is found that the postcritical behaviour is also more severe and the deformations are less uniform and can lead to kinking. This has significant implications for the behaviour of tubes in practical circumstances as the type of structural response found in the present study is well-known to lead to a high degree of sensitivity to any initial defects that arise from manufacturing processes. Therefore designers of circular tubular structures should take note that accurate nonlinear modelling becomes increasingly important particularly when material and geometric behaviour of this structural form become more complex.

REFERENCES

Axelrad, E. L. and F. A. Emmerling (Eds.) 1984. *Flexible shells: Theory and applications*. New York: Springer.

Brazier, L. G. 1927. On the flexure of thin cylindrical shells and other "thin" structures. *Proc. R. Soc. A* 116, 104–114.

Da Silva, L. S. and G. W. Hunt 1990. Interactive buckling in sandwich structures with core orthotropy. *Mech. Struct. & Mach.* 18(1), 353–372.

Doedel, E. J., A. R. Champneys, T. F. Fairgrieve, Y. A. Kuznetsov, B. Sandstede, and X.-J. Wang 1997. *AUTO97*: Continuation and bifurcation software for ordinary differential equations. Technical report, Department of Computer Science, Concordia University, Montreal, Canada. (Available by FTP from ftp.cs.concordia.ca in /pub/doedel/auto).

Everall, P. R. and G. W. Hunt 1999. Arnold tongue predictions of secondary buckling in flat plates. *J. Mech. Phys. Solids* 47(10), 2187–2206.

Fox, C. 1987. *An introduction to the calculus of variations.* New York: Dover Publications, Inc.

Hunt, G. W. and M. A. Wadee 1998. Localization and mode interaction in sandwich structures. *Proc. R. Soc. A* 454, 1197–1216.

Hunt, G. W. and M. K. Wadee 1991. Comparative lagrangian formulations for localized buckling. *Proc. R. Soc. A* 434, 485–502.

Lord, G. J., A. R. Champneys, and G. W. Hunt 1997. Computation of localized post buckling in long axially compressed cylindrical shells. *Phil. Trans. R. Soc. Lond. A* 355, 2137–2150.

Reissner, E. 1961. On finite pure bending of cylindrical tubes. *Österr. Ing. Arch.* 15, 165–172.

Tatting, B. F., Z. Gürdal, and V. V. Vasiliev 1996. Nonlinear response of long orthotropic tubes under bending including the Brazier effect. *AIAA Journal* 34(9), 1934–1940.

Timoshenko, S. P. and S. Woinowsky-Krieger 1970. *Theory of plates and shells* (2nd ed.). Singapore: McGraw-Hill.

Wadee, M. A. and G. W. Hunt 1998. Interactively induced localized buckling in sandwich structures with core orthotropy. *Trans. ASME J. Appl. Mech.* 65(2), 523–528.

Wadee, M. K., M. A. Wadee, A. P. Bassom, and A. A. Aigner 2006. Longitudinally inhomogeneous deformation patterns in isotropic tubes under pure bending. *Proc. R. Soc. A* 462, 817–838.

Tubular Structures XI – Packer & Willibald (eds)
© 2006 Taylor & Francis Group, London, ISBN 0-415-40280-8

Three-dimensional second order analysis of scaffolds with semi-rigid connections

U. Prabhakaran & M.H.R. Godley
Department of Real Estate and Construction, Oxford Brookes University, Oxford, United Kingdom

R.G. Beale
Department of Mechanical Engineering, Oxford Brookes University, Oxford, United Kingdom

ABSTRACT: Proprietary scaffold structures are generally slender and normally fail by elastic instability. The elastic buckling load of a scaffold is affected by the stiffness of the connections. In proprietary scaffold structures these connections exhibit different behaviour under clock-wise and anti-clockwise rotations. The moment rotation curves are highly non-linear and frequently have very low stiffnesses including the possibility of connection looseness with different loading and unloading curves.

This paper describes a computer model developed to incorporate this non-linear joint behaviour and also includes the development of a three-dimensional stability function of the uprights including torsion. Good agreement has been achieved between theory and experiment. The results of the analysis show that the deflection behaviour of the scaffold is sensitive to the shape of the approximations used to model connection moment rotation curves.

1 INTRODUCTION

Steel scaffolds are extensively used to provide access and support to permanent works during different stages of their construction. These structures are generally slender and prone to fail by elastic instability. The elastic buckling load of a scaffold is strongly influenced by the stiffness of the connections, which exhibits semi-rigid deformation behaviour that can contribute substantially to the stability of the structure as well as to the distribution of member force.

Traditional analyses of framed structures have assumed that the connections between uprights and beams, or between uprights and bases, are either rigid or pinned. In the past, analyses of structures containing semi-rigid joints have been concerned primarily with hot-rolled sections (Chen and Lui, 1991). In these structures the joint rotations are usually small and therefore non-linear moment rotation curves for the connection are usually not considered. Joints in hot-rolled sections are made by means of bolts or welding and hence the stiffnesses of the connections are large. However in proprietary scaffolds the connections are often push fit joints and hence often exhibit different behaviour under clockwise and anti-clockwise rotations and frequently have very low stiffnesses including the possibility of rotational looseness at the connection. The looseness contributes to the overall

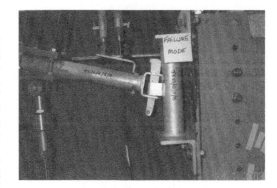

Figure 1. Typical standard-ledger connection under test.

deflection of the structure under loads (Godley and Beale, 2001). The joints in these structures are subjected to frequent loading and unloading for example due to wind pressure suction effect. It is observed that these joints often deform plastically at low loads and hence elastic unloading curves are not parallel to the initial moment rotation curves. Reloading occurs along the same line as the unloading line. Figures 1 and 2 show a typical proprietary scaffold ledger-standard connection under test and the moment rotation curve for the connection. To obtain the moment rotation curve for a ledger connection about an axis along the

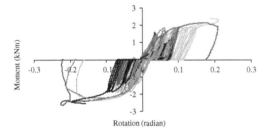

Figure 2. Typical standard-ledger moment curvature relation.

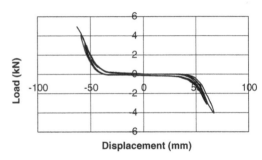

Figure 4. Typical load-displacement curve for standard ledger joint about an axis through the standard.

Figure 3. Experiment to determine rotational stiffness of connection about axis of standard.

standard a frame test has to be conducted (Godley and Beale, 1997). An example of a frame test is given in Figure 3 with the resulting load-displacement curve in Figure 4. It is to be noted that this connection has a looseness of ±50 mm about the mean. In addition, due to the slender nature of scaffold structure, the geometric non-linear interactions between the axial load and lateral deformations increase the complexity of the analysis.

For the purpose of design and analysis, the moment rotation curve is often assumed to be bilinear or multilinear and the same curve is used for both sagging and hogging moments. In this case, the predicted behaviour of the structure may be quite unrealistic compared to that of the actual structure. A more accurate method will be to carry out a second order analysis of frames considering the nonlinear behaviour of the joint connection.

Previous experimental and analytical research into scaffold structures (for example, Beale and Godley, 1995; Chan et al, 1995; Milojkovic et al, 1996; Peng et al, 1996, Huang et al, 2000 and Chan et al 2003) concentrated on modelling the joints as an elastic semi-rigid connection assuming linear behaviour with the same moment rotation stiffness used for clockwise and anti-clockwise rotations. However, little work has been reported on the effects of non-linear moment curvature relationships of these joints on scaffold performance.

Similar non-linear moment rotation curves have been published in relation to pallet rack structures (Agatino et al, 2001 and Bernuzzi and Castiglioni, 2001). These papers were primarily concerned with experimental testing of joints and obtaining empirical formulae. Abdel-Jaber et al (2005, 2006) conducted an experimental and numerical study on semi-rigid racking studied portal frames under combinations of side and axial loading. For the study the authors considered three different models to approximate the moment rotation curve of the semi-rigid beam end connector. A tri-linear loading curve in conjunction with a bi-linear unloading curve was shown to give good agreement with experiments conducted on portal frames free to sway. However the work was done using a MathCAD spread sheet and cannot be generalised.

This paper presents a generalised procedure for the second-order analysis of three-dimensional scaffold structures using stability functions. The analysis procedure incorporates the non-linear moment curvature behaviour of the semi-rigid connections.

2 THEORETICAL MODEL

For most of the analysis reported the second order elastic analysis has usually been performed by finite element modelling in which case a large number of elements is required to get the desired accuracy. The number of elements can be considerably reduced by using beam-column approach wherein the geometric

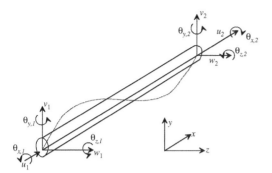

$$\begin{bmatrix} c_1 & 0 & 0 & 0 & 0 & 0 & -c_1 & 0 & 0 & 0 & 0 & 0 \\ 0 & c_2 & 0 & 0 & 0 & c_3 & 0 & -c_2 & 0 & 0 & 0 & c_3 \\ 0 & 0 & c_4 & 0 & -c_5 & 0 & 0 & 0 & -c_4 & 0 & c_5 & 0 \\ 0 & 0 & 0 & c_6 & 0 & 0 & 0 & 0 & 0 & -c_6 & 0 & 0 \\ 0 & 0 & -c_5 & 0 & c_7 & 0 & 0 & 0 & c_5 & 0 & -c_8 & 0 \\ 0 & c_3 & 0 & 0 & 0 & c_9 & 0 & -c_3 & 0 & 0 & 0 & c_{10} \\ -c_1 & 0 & 0 & 0 & 0 & 0 & c_1 & 0 & 0 & 0 & 0 & 0 \\ 0 & -c_2 & 0 & 0 & 0 & -c_3 & 0 & c_2 & 0 & 0 & 0 & -c_3 \\ 0 & 0 & -c_4 & 0 & c_5 & 0 & 0 & 0 & c_4 & 0 & c_5 & 0 \\ 0 & 0 & 0 & -c_6 & 0 & 0 & 0 & 0 & 0 & c_6 & 0 & 0 \\ 0 & 0 & -c_5 & 0 & c_8 & 0 & 0 & 0 & c_5 & 0 & c_7 & 0 \\ 0 & c_2 & 0 & 0 & 0 & c_{10} & 0 & -c_3 & 0 & 0 & 0 & c_9 \end{bmatrix} \quad (1)$$

Figure 5. Beam-column model.

The stiffness matrix of the beam-column element using stability functions is given in Equation 1 above where:

non-linear effect is taken into account in the analysis by the use of the stability functions.

In the present study this approach has been used for the second order analysis of frames. The programme was written using the C language.

$$c_1 = \frac{AE}{L}, c_2 = \frac{12EI_z}{L^3}\phi_1, c_3 = \frac{6EI_z}{L^2}\phi_2$$

$$c_4 = \frac{12EI_y}{L^3}\phi_3, c_5 = \frac{6EI_y}{L2}\phi_4, c_6 = \frac{GJ}{L}$$

$$c_7 = \frac{4EI_y}{L}\phi_7, c_8 = \frac{2EI_y}{L^2}\phi_8, c_9 = \frac{4EI_z}{L}\phi_5 \quad (2)$$

and $c_{10} = \dfrac{2EI_z}{L}\phi_6$

2.1 Model assumptions

The following assumptions are made:

- All members are prismatic and member imperfections are neglected.
- Material nonlinearity is not considered.
- All the members have linear elastic or second order elastic behaviour. The $P - \Delta$ effect is considered only for the standards. For the transoms and ledgers the affect of axial force is assumed to be too small to have an effect.
- The rotational deformation of connections is assumed to be concentrated at a point, which is the end of the beam. The affect of connection flexibility is accounted by modifying the stiffness matrix for beam.
- The joints exhibit non-linear moment rotation curve.
- The effect of eccentricity at joints is neglected.

Previous work by the authors (Godley and Beale, 2001, Milojkovic et al, 1996) has shown that failure of scaffold structures is predominantly elastic, material nonlinearity only occurring after elastic instability has taken place and that effects of member imperfection and joint eccentricity are negligible for structures involving more than 10 elements.

and $k_z = \sqrt{\dfrac{P}{EI_z}}$, $k_y = \sqrt{\dfrac{P}{EI_y}}$

For axial compressive force in standards

$$\phi_1 = \frac{(k_zL)^3 \sin k_zL}{12\phi_{cz}}, \qquad \phi_2 = \frac{(k_zL)^2(1-\cos k_zL)}{6\phi_{cz}}$$

$$\phi_3 = \frac{(k_yL)^3 \sin k_yL}{12\phi_{cy}}, \qquad \phi_4 = \frac{(k_yL)^2(1-\cos k_yL)}{6\phi_{cy}}$$

$$\phi_5 = \frac{k_zL(\sin k_zL - k_zL\cos k_zL)}{4\phi_{cz}}, \phi_6 = \frac{k_zL(k_zL - \sin k_zL)}{2\phi_{cz}} \quad (3)$$

$$\phi_7 = \frac{k_yL(\sin k_yL - k_yL\cos k_yL)}{4\phi_{cy}}, \phi_8 = \frac{k_yL(k_yL - \sin k_yL)}{2\phi_{cy}}$$

$$\phi_{cz} = 2 - 2\cos k_zL - k_zL\sin k_zL$$
$$\phi_{cy} = 2 - 2\cos k_yL - k_yL\sin k_yL$$

and for axial tensile force in standards

2.2 Derivation of stiffness matrix

For the analysis, two types of elements are used: (1) Beam-column elements using three-dimensional stability functions (Figure 5). (2) Semi-rigid beam elements modified to include end connection flexibility.

$$\phi_1 = \frac{(k_zL)^3 \sinh k_zL}{12\phi_{tz}}, \phi_2 = \frac{(k_zL)^2(\cosh k_zL - 1)}{6\phi_{tz}}$$

$$\phi_3 = \frac{(k_yL)^3 \sinh k_yL}{12\phi_{ty}}, \phi_4 = \frac{(k_yL)^2(\cosh k_yL - 1)}{6\phi_{ty}}$$

$$\phi_5 = \frac{k_zL(k_zL\cosh k_zL - \sinh k_zL)}{4\phi_{tz}}, \phi_6 = \frac{k_zL(\sinh k_zL - k_zL)}{2\phi_{tz}}$$

$$\phi_7 = \frac{k_yL(k_yL\cosh k_yL - \sinh k_yL)}{4\phi_{ty}}, \phi_6 = \frac{k_yL(\sinh k_yL - k_yL)}{2\phi_{ty}} \quad (4)$$

$$\phi_{tz} = 2 - 2\cosh k_zL + k_zL\sinh k_zL$$
$$\phi_{cy} = 2 - 2\cosh k_yL + k_yL\sinh k_yL$$

205

The influence of connection flexibility is accounted for by modifying the slope deflection equations for a beam element. The stiffness matrix of Equation 1 is used with in this case the coefficients given by:

$$c_1 = \frac{AE}{L}, c_2 = \left(s_{iiz} + 2s_{ijz} + s_{jjz}\right)\frac{EI_z}{L^3}, c_3 = \left(s_{iiz} + s_{ijz}\right)\frac{EI_z}{L^2}$$

$$c_4 = \left(s_{iiy} + 2s_{ijy} + s_{jjy}\right)\frac{EI_y}{L^3}, c_5 = \left(s_{iiy} + s_{ijy}\right)\frac{EI_y}{L^2}, c_6 = \frac{GJ}{L}$$

$$c_7 = s_{iiz}\frac{EI_z}{L}, c_8 = s_{ijz}\frac{EI_z}{L}, c_9 = s_{iiy}\frac{EI_y}{L}, c_{10} = s_{ijy}\frac{EI_y}{L} \quad (5)$$

where

$$s_{iiz} = \frac{\left(4 + 12EI_z / LR_{kzB}\right)}{R_z^*}, s_{jjz} = \frac{\left(4 + 12EI_z / LR_{kzA}\right)}{R_z^*}$$

$$s_{ijz} = s_{jiz} = \frac{2}{R_z}$$

$$s_{iiy} = \frac{\left(4 + 12EI_y / LR_{kyB}\right)}{R_y^*}, s_{jjy} = \frac{\left(4 + 12EI_y / LR_{kyA}\right)}{R_y^*}$$

$$s_{iiy} = s_{jiy} = \frac{2}{R_y} \quad (6)$$

and

$$R_z = \left(1 + \frac{4EI_z}{LR_{kzA}}\right)\left(1 + \frac{4EI_z}{LR_{kzB}}\right) - \left(\frac{EI_z}{L}\right)^2 \frac{4}{R_{kzA}R_{kzB}}$$

$$R_y = \left(1 + \frac{4EI_y}{LR_{kzA}}\right)\left(1 + \frac{4EI_y}{LR_{kzB}}\right) - \left(\frac{EI_y}{L}\right)^2 \frac{4}{R_{kzA}R_{kzB}}$$

where $R_{kzA}, R_{kzB}, R_{kyA}, R_{kyB}$ are beam left and right joint stiffness value about the z and y axes respectively, P the axial force within a standard, E is the modulus of elasticity of material, G the torsional modulus, I_z and I_y the second moments of area about the z and y axes, J the torsional constant, A is the cross-section area and L the length of the member.

2.3 Solution algorithm

The Newton-Raphson procedure is used for the non-linear analysis. The stiffness matrix is modified at the end of each iteration based on the updated geometry. The connection stiffness is evaluated from the joint moments of the previous iteration. Convergence is obtained when differences between the member forces in two subsequent cycles reaches a specified tolerance. A summary of the procedure is:

1. Evaluate the element stiffness matrix. To evaluate the stiffness matrix for the column member, initially the member axial force is set to zero. For subsequent iterations the element stiffness matrix is evaluated based on the axial force from the previous iteration. For a beam element, the connection stiffness is evaluated from the moment rotation

curve at the end each iteration. The moment rotation curve is expressed in terms of piecewise linear curve.

2. Assemble the structure stiffness matrix, K, fixed joint force vector, P_f if the member is subjected to external loads and the joint load vector, P.

3. Evaluate the displacement increment at nodes.

$$\left[K_i^{j-1}\right]\left[\Delta d_i^j\right] = \left[\Delta P_i^j\right] \quad (7)$$

$$\left[\Delta d_i^j\right] = \left[K_i^{j-1}\right]^{-1}\left[\Delta P_i^j\right] \quad (8)$$

where $\lfloor \Delta P_i^j \rfloor = \lambda(P - P_f)$ or $\lfloor \text{Re } s_i^{j-1} \rfloor$, i the load step, j the iteration count, λ the load factor and $\lfloor \text{Re } s_i^{j-1} \rfloor$ is the unbalanced force at the end of the $(j-1)$th iteration. For the first iteration at any load increment, $\lfloor \Delta P_i^j \rfloor = \lambda P$ and for subsequent iterations $\lfloor \Delta P_i^j \rfloor = \lfloor \text{Re } s_i^{j-1} \rfloor$.

4. Calculate the total displacement at each node.

$$d_i = d_{i-1} + \sum_{j=1}^{n}\left[\Delta d_i^j\right] \quad (9)$$

5. Update the geometry of the structure.

$$\left[x\right]_i = \left[x\right]_{i-1} + \sum_{j=1}^{n}\left[\Delta d_i^j\right] \quad (10)$$

6. Calculate the element displacements by extracting them from the global displacement vector and transforming to the local coordinate matrix.

$$\left[\Delta d_i^j\right]_e = T\left[\Delta d_i^j\right] \quad (11)$$

where T is the transformation matrix.

7. Compute member end forces in the local coordinate system R_e using the relation

$$\left[\Delta R_i^j\right]_e = k\left[\Delta d_i^j\right]_e + P_f \quad (12)$$

where k is the element stiffness matrix. For the first iteration of any load cycle, P_f is the fixed end force vector if any and zero for subsequent iterations.

8. Compute the member end forces in global coordinate system R using the relation

$$\left(\Delta R\right)_i^j = T^T\left(\Delta R_e\right)_i^j \quad (13)$$

9. Calculate the total force in each member in local and global coordinate systems using:

$$\left(\Delta R\right)_i^j = T^T\left(\Delta R_e\right)_i^j \quad (14)$$

$$\left(R\right)_i^j = \left(R_e\right)_{i-1} + \sum_{j=1}^{n}\left(\Delta R\right)_i^j \quad (15)$$

10. Compute the internal force vector Q_i^j by assembling the member forces in the global coordinate system for all the elements obtained from step 8.

11. Compute the residual forces.

$$\left[\operatorname{Re} s_i^{j+1}\right] = \left[\Delta P_i^j\right] - \left[Q_i^j\right] \qquad (16)$$

12. Compute the tolerance:

$$\text{TOL} = \sqrt{\sum_{m=1}^{NDOF}\left(\frac{\left[\operatorname{Re} s_i^{j+1}\right]}{Q_i^j}\right)^2 \Bigg/ NDOF} < 0.0001 \qquad (17)$$

13. Repeat step 1 to 9 until required convergence is achieved.
14. Apply next load increment and repeat step 1 to 13.

3 VALIDATION OF THEORETICAL MODEL

Two structures were analysed to check the validity of the new procedure. Initially, two-dimensional sway portal frames tested experimentally by Abdel-Jaber et al (2005, 2006) were considered. These frames were used because the joints are subjected to load reversal. Secondly, a three-dimensional scaffold model was compared with the finite element model constructed using the programme Lusas (Lusas, 2003).

3.1 Portal frame

Figure 6 shows a schematic of the portal frame. The beam had a cross-section area of 216 mm² and a second moment of area of $1.3 \cdot 10^5$ mm⁴. The uprights had a cross-section area of 280 mm² and a second moment of area of 0.7×10^5 mm⁴. The modulus of elasticity of the frame was taken to be 2.09×10^8 kN/m². The base of the portal frame was pinned. The ratio of side load, H, to axial load P was 2%.

The moment rotation curve of the joints of the portal frame is shown in Figure 7, which was obtained from a cyclic cantilever test. The details about the test can be found in Abdel-Jaber (2002). The moment rotation curve obtained from tests on different connections show quite a variation from one another; however the general behaviour remains the same. Hence for the present study, a typical moment rotation curve was considered. The moment rotation curve was divided into three segments, segment 1 being the moment curvature relationship when the joints are under the effect of hogging moment, segment 2 being the unloading curve and segment 3 is for the sagging moment.

Initially, the nonlinear moment rotation curve was approximated using a bi-linear curve in the hogging moment region (segment 1) and with two different linear curves for unloading (segment 2) and sagging moment regions (segment 3). Using this approximation, an analysis of the portal frame was carried out for a ratio of side load to axial load of 2%. Under such loading, initially both the joints in the portal frame

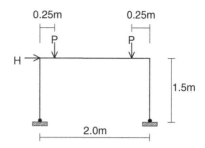

Figure 6. Schematic of portal frame test.

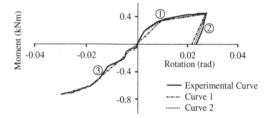

Figure 7. Moment rotation curve of joint.

loaded but eventually the left joint unloaded due to the $P - \Delta$ effect while the right joint continued to load. In the analysis, when joints were subjected to increasing causing a hogging moment, the connection stiffness was obtained from segment 1. When the left joint started to unload, the moment curvature relation followed segment 2 and when the joint started loading in the reverse direction, the moment curvature relation followed segment 3.

The analysis carried out using the above approximation yielded reasonably good results for the left joint. However for the right joint the behaviour predicted is stiffer than the actual test results. From the experimental results it can be seen that the left joint started unloading after reaching a maximum moment of approximately 0.036 kNm at a corresponding load of about 1 kN. The moment rotation behaviour of the joint was found to be highly non-linear between −0.12 and 0.6 kNm and various curves could be drawn with considerable variation in rotation stiffness. For unloading, segment 2 was used for calculating the stiffness value of the connection, however it was possible that the left joint might have elastically unloaded along a path similar to the loading curve (segment 1) as the moment was very low at the time of unloading. The behaviour of the right joint compared reasonably well with the test results below a load of 1 kN. Above 1 kN there is quite a variation which could be due the unloading of the left joint and the selection of stiffness value for calculating the moment and forces.

A further study was conducted to study the sensitivity of the approximation made for the moment rotation curve. In the second case, the moment rotation curve

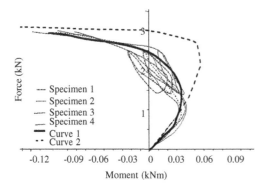

Figure 8. Relation between load and moment for left joint for H/P ratio of 2%.

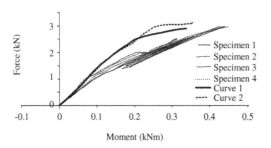

Figure 9. Relation between force and moment at right joint for H/P ratio 2%.

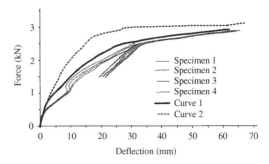

Figure 10. Load against displacement curve for H/P ratio 2%.

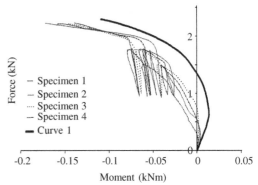

Figure 11. Relation between force and moment at left joint for H/P ratio of 4%.

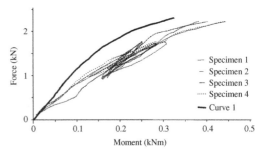

Figure 12. Relation between force and moment at right joint for H/P ratio of 4%.

(curve 2) was approximated using a tri-linear curve in the hogging moment region (segment 1). Two different linear curves with varied slopes from the first case were used for unloading (segment 2) and sagging moment regions (segment 3). The analysis of frame using this approximation shows a significant difference from the first case. This indicates that the analysis results are highly influenced by the approximations made for the moment rotation curve. Figures 8 and 9 show the load – moment behaviour of the left (windward) and right (leeward) joints for the two different approximations of the moment rotation curve. Figure 10 shows the load displacement relationship of the windward joint.

Note that there are four experimental curves shown in these figures labelled specimens 1–4. In the figures below individual experimental results are not significant. They are drawn to show the variation in moment, displacement, etc. The experiment was conducted using four portal frames tied together and allowed to sway (see Abdel-Jaber 2005, 2006). During the experiments the load was cycled and the experimental loops show the hysteresis in the system.

The analysis of portal frame was repeated with different value of the side ratio H/P of 4%. This analysis was mainly conducted to study the $P - \Delta$ effect. In this analysis, only the first approximation of moment-rotation curve was used.

Figures 11 and 12 show the load – moment behaviour of the left (windward) and right (leeward) joints. It is observed that as H/P increases the left joint starts unloading at very early stage due to the $P - \Delta$ effect. Figure 13 shows the load displacement relationship of the windward joint.

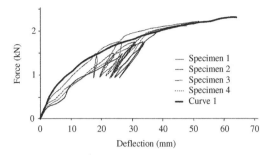

Figure 13. Load against displacement curve for H/P ratio of 4%.

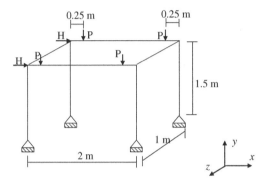

Figure 14. Three dimensional analytical frame.

3.2 Three dimensional model

In this section the results of the computer programme are compared with Lusas (Lusas, 2003) software to validate the non-linear analysis procedure of three dimensional (3D) scaffold models. Figure 14 shows the 3D scaffold model considered for the validation. The section and material properties of beams and columns are similar to that of the portal frame shown in Figure 6.

The ratio of side load to axial load, H/P considered was 2%. Since Lusas is not capable of handling non-linear moment rotation curve for the joints, a constant stiffness value for the joints was used. For comparison the joint stiffness about the axis perpendicular to the longitudinal axis (z axis) is taken as 60 and about the vertical axis (y axis) is taken as 0.5.

Figure 15 shows the load displacement curve obtained from Lusas and the computer programme. Figure 16 and Figure 17 show the load – moment behaviour of the left (windward) and right (leeward) joint of the beam. The results are in close agreement to that obtained from Lusas. A point worth noting here is while Lusas needs at least ten elements per beam to get the desired accuracy, this program requires only one element per beam as it uses the stability functions.

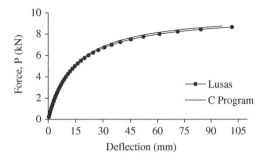

Figure 15. Load against displacement curve for 3D model.

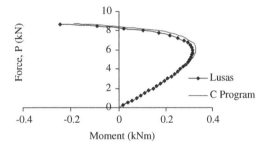

Figure 16. Relation between force and moment at left joint.

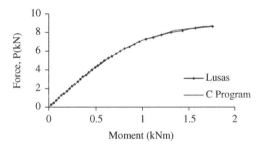

Figure 17. Relation between force and moment at right joint.

Hence there is both time and memory saving in the computation.

The 3D model was then analysed using the non-linear moment rotation curve. For the joint stiffness about z axis, first approximation of moment rotation curve shown in Figure 7 was used. The joint stiffness about y axis is taken as 0.5. Figure 18 shows the load displacement curve obtained from the analysis. Since there are no experimental results available for three dimensional analyses of the frame with non-linear moment rotation curve, therefore the results obtained from the computer program could not be compared. However the results obtained from the program are presented here. Nevertheless the programme is capable of handling a highly non-linear moment rotation

209

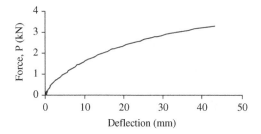

Figure 18. Load against displacement curve for 3D model with non-linear moment rotation curve.

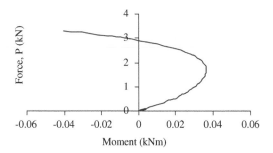

Figure 19. Relation between force and moment at left joint of 3D model with non-linear moment rotation curve.

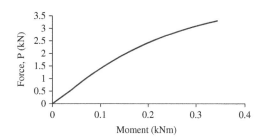

Figure 20. Relation between force and moment at right joint of 3D model with non-linear moment rotation curve.

curve for the connection joints. Figures 19 and 20 show the load – moment behaviour of the left (windward) and right (leeward) joints.

4 CONCLUSIONS

This paper has presented a computational algorithm for the second order analysis of three-dimensional scaffold and racking structures using stability functions. The procedure incorporates the non-linear behaviour of semi-rigid proprietary scaffold connections.

The load-displacement results predicted by the program compare well with available experimental data on structures including the effects of load reversal. The behaviour of a frame is shown to be particularly sensitive to changes in the moment rotation curve.

The results of three dimensional analyses are also presented. A good correlation with the Lusas is achieved. However the correlation of non-linear moment rotation curve of connection joint could not be done due to the unavailability of experimental data.

REFERENCES

Abdel-Jaber, M. The influence of semi-rigid connections on the behaviour of beam and column structural systems. PhD Thesis, Oxford Brookes University, UK, 2002
Abdel-Jaber, M., Beale, R.G. and Godley, M.H.R. 2005. Numerical study on semi-rigid racking frames under sway. Comp. & Struct. 83: 2463–2475
Abdel-Jaber, M., Beale, R.G. and Godley, M.H.R. 2006. A theoretical and experimental investigation of pallet rack structures under sway. J. Const. Steel Research. 62: 68–80
Agatino, M.R., Bernuzzi, C. & Castiglioni, C.A. 2001. Joints under cyclic reversal loading in steel storage pallet racks. Giornate Italiane Della Construzione in Acciaio. Venice, 10p
Beale, R. G. & Godley, M.H.R. 1995. The analysis of scaffold structures using LUSAS, Proc. of Lusas 95, Stratford, 9–16
Bernuzzi, C. & Castigliano, C.A. 2001. Experimental analysis on the cyclic behaviour of beam-to-column joints in steel storage pallet racks. Thin-Walled Struct. 39(10): 841–59
Chan, S.L., Zhou, Z.H., Chen, W.F., Peng, J.L & Pan, A.D. 1995. Stability analysis of semi-rigid steel scaffolding. Engineering Structures. 17(8): 568–574
Chen, W.F. & Lui, E.M. 1991. Stability Design of Steel Frames: CRC Press
Godley, M.H.R. & Beale, R.G. 1997. Sway stiffness of scaffold structures, The Structural Engineer. 75(1): 4–12
Godley, M.H.R. & Beale, R.G. 2001. Analysis of large proprietary access scaffold structures. Proc. Instn Civ. Engrs Structs & Bldgs. 146: 31–39
Huang, Y.L., Kao, Y.G. & Rosowsky, D. V. 2000. Load carrying capacities and failure modes of scaffold shoring systems, Part II: An analytical model and its closed form solution. Structural Engineering and Mechanics. 10(1): 67–79
Lusas, 2003. Lusas 13.4 User Manual. FEA Ltd.
Milojkovic B., Beale, R.G. & Godley, M.H.R. 1996. Modelling Scaffold Connections. Proc. of the Fourth ACME UK Annual Conference. Glasgow, 85–88
Peng, J.L., Pan, A.D., Rosowsky, D.V., Chen, W.F., Yen, T. and Chan, S.L. 1996. High clearance scaffold systems during construction-I. Structural modelling and modes of failure. Engineering Structures 18(3): 247–257

Tubular Structures XI – Packer & Willibald (eds)
© 2006 Taylor & Francis Group, London, ISBN 0-415-40280-8

Prevention of disproportionate collapse of tubular framed construction by performance based specification of connections

Y.C. Wang
University of Manchester, UK

A.H. Orton
Corus Tubes, UK

ABSTRACT: This paper presents a method to calculate the required tying force requirement for connections in steel framed structures, by considering catenary action in steel beams after removal of a column. The requirement for sufficient tying force resistance is to safeguard the structure against disproportionate collapse. To enable this calculation, it is necessary to have information on the strength and stiffness of the beam-column connections when subjected to an axial tensile load in the beam. For fin plate connections to rectangular/square tubular columns, this paper will present a simple method to obtain these parameters based on results of robustness tests on fin plate connections carried out by the Building Research Establishment in the UK. This paper suggests that the tying force approach can be made to work if an internal column is removed. In this case, the calculated tying force may be considerably higher than the nominal tying force requirement specified in the current British Standards. Fortunately, this paper demonstrates that owing to a number of favourable design conditions (over-design, unused bending moment resistance of partial strength connections, reduced accidental loading, higher connection fracture resistance than the nominal tying force requirement), the connections studied in this paper would still have sufficient tying force resistance to control disproportionate collapse, perhaps explaining why it is rare to observe disproportionate collapse in steel framed buildings. Nevertheless, this paper suggests that a combination of unfavourable design conditions may increase the risk of disproportionate collapse and that connection performance plays an important part in resisting disproportionate collapse of steel framed structures. For these buildings, this paper makes a contribution to developing a sensible calculation method for checking connection tying force resistance.

1 INTRODUCTION

Structural design for robustness, or disproportionate collapse, or progressive collapse, has become an important consideration for structural engineers. The current UK Building Regulations (BR 2005) and structural steel design code (BSI 2000) contain a number of design approaches to control disproportionate collapse for different types of buildings, including the tying force approach, the key element approach and the risk assessment approach. Depending on their risk to disproportionate collapse, buildings are classified into Class 1, Class 2A, Class 2B and Class 3. For Class 1 buildings such as houses not exceeding 4 storeys or agricultural buildings, there is no requirement. For Class 2A buildings, which include buildings such as hotels or offices not exceeding 4 storeys, effective horizontal ties should be provided, in which structural members should be tied together in the principal horizontal directions. The connections

should be able to resist a tying force equal to the beam reaction force under full factored design load at the connections, with a minimum of 75 kN. For Class 2B buildings which include buildings such as educational buildings greater than 1 storey but not exceeding 15 storeys, offices greater than 4 storeys but not exceeding 15 storeys, etc., the building should be effectively tied together both horizontally and vertically. Alternatively the building may be checked that upon notional removal of any member, the building is stable and the damaged area is limited; in the case the damaged area is greater, then the member notionally removed should be considered a key member and should be designed to resist the full accidental loading. For Class 3 buildings, which are buildings not covered by Class 2, a systematic risk assessment approach should be taken, in which consideration should be given to all the normal hazards that may be reasonably foreseen as well as any abnormal hazards. At present, there is neither an accepted procedure to establish abnormal hazards, nor accepted methods to assess robustness of the building to resist abnormal hazards. Whatever the abnormal hazard, a highly likely consequence is removal of columns. This is the situation assumed in this study and the objective of this paper is to evaluate the required connection tying force resistance so that removal of a column does not lead to disproportionate collapse. It should be noted that the tying force approach is also the principal method of controlling disproportionate collapse in the prescriptive method for Class 2 buildings. Although the results of this paper may be used to judge whether the required tying force resistance in the current prescriptive method is technically correct, this is not the purpose of this paper as the prescriptive method has been in existence for a considerable period of time for Class 2 buildings and the risk of disproportionate collapse has been considered to be acceptably low. In the existing approach, the notional tying force resistance should be seen as no more than a number that has been agreed by the code drafting committee so as to provide the structure with some degree of continuity. For Class 3 buildings where a more rigorous approach is required, if the tying force approach is to be used, then specifying a notional tying force resistance is no longer appropriate. However, should tying of structural members be used to control disproportionate collapse, then a much more accurate method of calculating the tying force requirement as well as assessing whether the structure is adequate should be provided. This is the objective of the present paper. Although the methodology of this paper can be applied to different types of structures, fin-plate connections to steel tubes will be used as examples.

2 MODEL OF ANALYSIS ON COLUMN REMOVAL

2.1 Assumptions and general behaviour

A realistic structure is a complex 3-dimensional assembly, with all components of the construction contributing to the behaviour and resistance of the structure under extreme (abnormal) loading. Being a first step in developing a rational method of quantifying structural robustness, this paper is restricted

to framed structures, in which the important contributions of floors and walls are not included. Also dynamic effects, which may be a result of explosion that caused the column to be removed in the first place or immediately after partial structural collapse, are not considered. Furthermore, because the design objective is to control disproportionate collapse of the structure, rather than to maintain usability of the structure by limiting structural deformations, large deflections are considered acceptable.

This paper will only deal with removal of an internal column (including internal edge columns). The tying force approach is not considered to be appropriate in the case of a corner column removal. The reason is that if a corner column is removed from a frame designed as simple construction, the floor supported by the corner column would simply fold around the line joining the two adjacent edge columns as shown in figure 1. One possible way of arresting disproportionate collapse is to design the floor slab to have sufficient bending resistance along the aforementioned fold line, as indicated in figure 1.

2.2 Analysis of tying force

Figure 2(a) shows the general loading condition of a beam segment on removal of a central column. From equilibrium condition:

$$F\delta + \left(M_h + M_s\right) = M_e + Vx \qquad (1)$$

where F is the tension (tying) force in the beam; M_h and M_s are the hogging and sagging bending moments

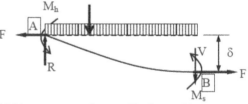

(a) A beam segment under general loading condition

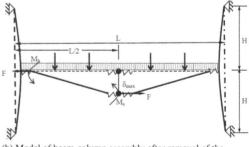

(b) Model of beam-column assembly after removal of the central column

Figure 2. Analysis model.

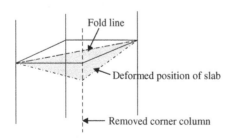

Figure 1. Possible behaviour on removal of a corner column.

at the two ends of the beam segment; M_e is the bending moment of the externally applied load, taken about the left end (point A) of the beam segment; V is the shear force at the right end (point B) of the beam.

For demonstration, assume the beam is symmetrically loaded as shown in figure 2(b), which may represent the case of removing the central column that originally supports two identical beams on each side. $V = 0$ and equation (1) may be re-written as:

$$WL/\gamma = F.\delta_{max} + M_c \tag{2}$$

where W is the total accidental load applied on the beam; L is the total beam span (the span of each of the original two beams $= L/2$); WL/γ gives the externally applied maximum bending moment at the position of the removed column about the left end of the beam, e.g. $\gamma = 4$ for central point load and $\gamma = 8$ for uniformly distributed load. δ_{max} is the maximum beam deflection at the position of the removed column; M_c ($= M_s + M_h$) is the total of the connection bending moment capacities at the beam end and at the position of the removed column.

A tying force (axial force) in the beam is generated because the axial contraction of the beam at large beam vertical deformations is restrained. If the equivalent stiffness of the restraint to the beam tying action is K, the tying force is:

$$F = K.\Delta L \tag{3}$$

where ΔL is the axial shortening of the beam.

The equivalent stiffness of the restraint may be calculated using the following equation:

$$\frac{1}{K} = \frac{2}{K_s} + \frac{4}{K_e} + \frac{1}{EA/L} \tag{4}$$

in which EA/L is the axial stiffness of the beam and K_e the stiffness of the beam to column connection. The coefficient 4 associated with K_e indicates that there are four connections to the beam as shown in figure 2(b). K_s is the stiffness of the surrounding structures (columns) and the coefficient 2 means that there are two surrounding structures, one on each side of the beam.

The axial shortening of the beam may be related to its maximum vertical deflection as:

$$\Delta L = \phi \frac{\delta_{max}^2}{L} \tag{5}$$

The coefficient ϕ depends on the beam deflection profile. If the profile is linear, $\phi = 2$. For parabolic beam deflection profile, $\phi = 8/3 = 2.67$. When the exact beam deflection profile is unknown, approximately, $\phi = 2.5$.

Substituting equations (3), (4) and (5) into equation (2) gives:

$$\left(\frac{1}{\gamma} - \frac{M_c}{WL}\right)WL = \phi \frac{\delta_{max}^2}{L}.K.\delta_{max} \tag{6}$$

which gives:

$$\delta_{max} = \left(\frac{1}{\xi\phi}\right)^{1/3}\left(\frac{W}{LK}\right)^{1/3} \tag{7}$$

where

$$\frac{1}{\xi} = \frac{1}{\gamma} - \frac{M_c}{WL}$$

The maximum tying force in the beam is:

$$F = \frac{(\phi)^{1/3}}{\xi^{2/3}}(LK)^{1/3}(W)^{2/3}$$

$$= \frac{4(\phi)^{1/3}}{\xi^{2/3}}\left(\frac{LK}{W}\right)^{1/3}\frac{W}{W_0}\left(\frac{W_0}{4}\right) \tag{8}$$

in which W_0 is the total factored load under normal design. Recognizing that L/2 is the original span of the beam on each side of the removed column, then $W_0/4$ is the nominal required tying force resistance specified in the current UK steel design code (BSI 2000), which is denoted by F_{rule}. Equation (8) is rewritten as:

$$F = \frac{4W}{W_0}\left(\frac{\phi LK}{W\xi^2}\right)^{1/3}F_{rule} \tag{9}$$

According to the UK steel design code (BSI 2000), under gravity load:

$$\frac{W}{W_0} = \frac{1.05(DL + LL/3)}{1.4DL + 1.6LL} = \frac{1.05 + 0.35LL/DL}{1.4 + 1.6LL/DL} \tag{10}$$

where DL is dead load and LL imposed load.

Usually, $LL/DL \gg 1$. From equation (10), the maximum value of W/W_0 is about 0.5. Assuming $\phi = 2.5$, then the maximum tying force is approximately:

$$F = \left(\frac{20LK}{W\xi^2}\right)^{1/3}F_{rule} \tag{11}$$

Since the axial stiffness of the beam is usually much greater than the stiffness of the connections and that of the surrounding structure, equation (4) may be simplified to:

$$\frac{1}{K} = \frac{2}{K_s} + \frac{4}{K_e} = \frac{4}{K_e}\left(1 + \frac{K_e}{2K_s}\right) \tag{12}$$

213

$$K = \frac{K_e}{4\left(1 + \dfrac{K_e}{2K_s}\right)} \tag{13}$$

Equation (11) becomes:

$$F = \left[\frac{4K_e}{(W/L)\xi^2}\frac{1}{1 + K_e/(2K_s)}\right]^{1/3} F_{rule} \tag{14}$$

2.3 *Examples of application*

Although the method presented in this paper is simplistic and a more comprehensive model should be developed to include many other aspects of structural behaviour under accidental loading, applications of the above method to some examples may be used to indicate: (1) whether the current nominal tying force resistance is sufficient; (2) whether steel framed structures have sufficient resistance to control disproportionate collapse; and (3) how to improve resistance to disproportionate collapse of steel framed structures.

The examples are based on the robustness tests on fin plate connections to rectangular and circular steel tubes, conducted by Jarrett and Grantham of the UK's Building Research Establishment (BRE) in 1993 (Jarrett and Grantham 1993). In these tests, I-section stubs ($305 \times 150 \times 25$ UB) are connected to RHS columns ($250 \times 250 \times 10$ RHS, $250 \times 150 \times 10$ RHS or $250 \times 150 \times 5$ RHS) via a $220 \times 100 \times 8$ mm fin plate welded to the tube and bolted to the web of the I-stub beam using 3 M20 Grade 8.8 bolts. Two test configurations (Tee specimens in which only one I-stub was attached to the tube and cruciform specimens in which one I-stub was attached on each side of the tube) were adopted and figure 3 shows a cruciform specimen.

In this example, it is assumed that the beam size in figure 2 is the same as that used in the BRE tests and the beam span/depth results of 15, 20 and 25 are considered, which give a beam span of 4.5 m, 6 m and 7.5 m respectively for the beam with $305 \times 150 \times 25$ UB section. The beam is loaded by a uniformly distributed load. The factored design load on the beam is calculated on the basis of simple construction and the beam achieving plastic bending moment capacity with the design yield strength of steel being 275 N/mm^2. The secant stiffness of the connections from the BRE tests (connection maximum load divided by the corresponding displacement) is used as K_e in equation (14). With regard to the stiffness of the surrounding structure, the maximum tying force is obtained by setting $K_s = \infty$. Table 1 presents results of the above analysis.

In table 1, the shaded cells indicate that the connections may not have sufficient fracture resistance to control disproportionate collapse under the assumed conditions.

From the results in table 1, the following conclusions may be drawn:

1 All the connections would have sufficient resistance to the nominal tying force requirement (F_{rule}). In

Figure 3.   Cruciform specimen of BRE test.

Table 1.   Ratio of tying forces.

BRE Test ID	Ratio of F/F_{rule} from eqn. (14) for beam span/depth			Ratio of test fracture load to F_{rule} for beam span/depth		
	15	20	25	15	20	25
1	3.02	3.66	4.24	3.67	4.90	6.12
1a	4.01	4.86	5.64	4.40	5.87	7.34
2	2.39	2.90	3.36	1.75	2.33	2.91
3	2.66	3.22	3.73	1.76	2.35	2.93
4	3.09	3.74	4.34	4.18	5.57	6.96
4a	3.80	4.60	5.34	3.73	4.98	6.22
5	2.72	3.29	3.82	2.62	3.49	4.37
6	2.83	3.42	3.97	2.82	3.77	4.71
7	3.14	3.91	4.42	5.30	7.07	8.84
8	3.58	4.34	5.03	5.09	6.78	8.48
9	2.07	2.51	2.91	1.97	2.63	3.29
10	2.53	3.06	3.56	2.07	2.76	3.45
11	2.92	3.53	4.10	3.53	4.71	5.88
12	2.66	3.23	3.74	1.99	2.65	3.31
13	2.53	3.06	3.55	1.93	2.67	3.21
14	3.13	3.79	4.40	4.02	5.36	6.70
14a	3.51	4.25	4.93	1.84	2.46	3.07
15	3.24	3.92	4.55	4.15	5.54	6.92

fact, the connection tying resistance is a few times the nominal tying force requirement.

2 However, the required tying force resistance calculated using the model in this paper is also much higher than the nominal tying force requirement. This indicates that the prescriptive method of specifying a nominal tying force requirement used for Class 2 buildings in the UK should not be used in the more formal risk assessment for robustness for Class 3 buildings.

Table 1 indicates that in a number of cases, the required tying force resistance calculated using the model in this paper may be higher than the fracture resistance of connections, as indicated by the shaded cells. However, it may be possible to take into account a number of conservatisms in real design situations to ensure that the connections do have sufficient tying force resistance. The following aspects may be considered:

1 Actual live load to dead load ratio (LL/DL). Results in table 1 are based on the assumption that the accidental loading is 50% of the ultimate factored load for normal design ($W/W_0 = 0.5$), approximately corresponding to $LL/DL = 1$. If $LL/DL \gg 1$, then $W/W_0 \ll 0.5$ and equation (10) may be used to obtain a proportionally reduced tying force requirement.
2 Partial bending moment resistance of the connections. To obtain the results in Table 1, the connection bending moment resistance is neglected, i.e. $M_c = 0$ in equation (6). Including a contribution from M_c would reduce the tying force requirement. In fact, if the connections have full bending resistance of the beam (this bending moment resistance is not utilized in normal design), then it can be demonstrated that the required tying force is zero. To demonstrate this, assume $M_s = M_h = M_p$, where M_p is the bending moment resistance of the beam. Then $M_c = 2M_p$. According to normal design, $(W_0/2)(L/2)/\gamma = M_p$, which gives

$$\frac{1}{\xi} = \frac{1}{\gamma} - \frac{M_c}{WL} = \frac{1}{\gamma}\left(1 - \frac{2M_p}{4(W/W_0)M_p}\right) \qquad (15)$$

If $W/W_0 < 0.5$, the value in the above equation is non-positive, indicating that the beam would still have sufficient bending moment resistance after removal of the central column so that $F = 0$.
3 Actually applied load. The results in table 1 are based on the assumption that the beam is fully loaded to its ultimate factored load in normal design. In many cases, the beam will not be fully loaded. Any actual reduction in the applied load will result in proportional reduction in the tying force requirement.

4 Flexibility of the surrounding structure. The calculated tying forces in table 1 are calculated by assuming infinite rigidity of the surrounding structure, i.e. $K_s = \infty$ in equation (13). In many cases, the surrounding structure would be much more flexible, i.e. $K_e/K_s \gg 1$. This could substantially reduce the value of F calculated using equation (14) compared to assuming infinite surrounding structure rigidity.
5 Real structural behaviour. As previously mentioned, the model in this paper is for framed structures without any contribution from the floor slabs or walls. Including floor slabs and walls in the model would reduce the required tying force.

Therefore, for most steel framed structures using tubular columns, the structure would have inherent resistance to disproportionate collapse after removal of a column. Nevertheless, for buildings with high risk of accidental loading (e.g. Class 3 buildings in the UK), checking the structure to have sufficient resistance for the nominal tying force requirement is not sufficient. The model presented in this paper may be used as part of the systematic risk assessment procedure.

3 CONNECTION BEHAVIOUR

3.1 A fracture model for fin plate connections

In order to use the above model, it is necessary to have information on the connection behaviour. Assuming that the connection force-deformation relationship is linear until fracture, it is necessary to obtain information on connection stiffness and fracture strength. This paper will present a simple model for fin plate connections to tubular columns.

The model is based on the assumption that the tubular wall is in pure membrane action. Connection fracture is due to the tubular steel reaching its fracture strain. It is assumed that although the tubular wall would be in biaxial loading in its plane, since the tubular wall is not restrained longitudinally, the longitudinal stress is small and is neglected. Therefore, the tubular wall may be assumed to be loaded only in the direction across the tubular width. Figure 4

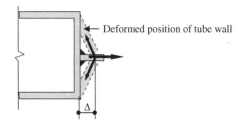

Figure 4. Assumed loading condition in tube at fracture.

shows the uniaxial loading condition of the tubular wall. The failure modes of the whole connection assembly include (a) beam web pull out; (b) fracture of the RHS around the weld; (c) bolt shear; (d) fin plate fracture; (e) RHS punching shear failure. The failure loads for modes (a), (c), (d) and (e) are well established and will not be discussed further.

The authors are conducting detailed research to derive accurate prediction of connection failure load under failure mode (b). This paper will present a simple analysis of this failure mode so as to identify the main influential factors. It should be pointed out that this failure mode is different from that considered in the current CIDECT design guide (Kurobane et al 2004), in which the connection failure strength is based on the connection deformation at a small percentage of the tube width. In this paper, the failure mode corresponds to ultimate fracture of the connection and the deformation limit has long been exceeded.

Refer to figure 4 which sketches a cross-section of the RHS at the connection. For the RHS fracture failure mode considered in this paper, it is assumed that the maximum tensile strain of the tubular wall at the position of the fin plate has reached fracture strain of the steel. However, away from the fin plate, the tubular wall will be much less stressed with much lower strains. Also since the tubular wall will be undergoing combined bending and membrane action, the average strain in the tube wall at the fin plate position will be lower than the fracture strain of steel. Therefore, if the RHS tube wall is assumed to be rigid so that the deformation is concentrated at the fin plate location, it is reasonable to assume that the average tensile strain and the average tensile stress at connection fracture are respectively half the fracture strain and stress of steel. Clearly this is a gross approximation and the stresses and strains in the RHS wall at fracture will depend on many factors such as the type of steel (cold-formed/hot-rolled, mild steel/high strength) and reduced ductility of steel due to welding. These factors will be considered in more detailed investigations.

To derive the general equations, assume the average strain and stress in the RHS are ε and f_ε respectively. Assuming the deformed profile of the RHS wall is linear, then the deformation of the RHS at centre is:

$$\Delta = \frac{d}{2}\left(\sqrt{(1+\varepsilon)^2 - 1}\right) = \frac{d}{2}\sqrt{2\varepsilon + \varepsilon^2} \quad (16)$$

where d is the width of the steel tube.

The fracture resistance of the connection may be obtained by using the principle of virtual work. The virtual work done by the tensile force in the connection is:

$$\delta W_{ext} = P.\delta\Delta = P.\frac{d}{2}\frac{(1+\varepsilon)}{\sqrt{2\varepsilon + \varepsilon^2}}\delta\varepsilon \quad (17)$$

The total tension force in the RHS is:

$$N = ht_{rhs}f_\varepsilon \quad (18)$$

where h is the height of the fin plate and t_{rhs} the RHS thickness.

The virtual work done by the tensile force in the RHS is:

$$\delta W_{int} = N.d.\delta\varepsilon = ht_{rhs}f_\varepsilon d.\delta\varepsilon \quad (19)$$

From principle of virtual work, $\delta W_{int} = \delta W_{ext}$, then:

$$P = \left(2ht_{rhs}f_\varepsilon\right)\frac{\sqrt{2\varepsilon + \varepsilon^2}}{1+\varepsilon} \quad (20)$$

Based on the assumption $f_\varepsilon = 0.5f_u$ and $\varepsilon = 0.5\varepsilon_u$, where f_u and ε_u are respectively the ultimate tensile stress and strain of RHS steel, equation (20) becomes:

$$P = \left(ht_{rhs}f_u\right)\frac{\sqrt{\varepsilon_u + \varepsilon_u^2/4}}{1+\varepsilon_u/2} \quad (21)$$

From equations (16) and (21), the connection stiffness may be obtained as:

$$K = \frac{P}{\Delta} = \frac{2\left(ht_{rhs}f_u\right)}{d}\frac{1}{1+\varepsilon_u/2} \quad (22)$$

Table 2 compares the predicted and test failure modes and loads for the BRE tests (Jarrett and Grantham 1993). For the predictions, the RHS steel fracture strain is taken as 0.20. Also the measured RHS steel ultimate

Table 2. Comparison between prediction and test results for BRE tests.

Test 1 Number	Test results			Predicted results		
	Load (kN)	Mode	Δ_{max} (mm)	Load (kN)	Mode	Δ_{max} (mm)
1	307	a	47	248	a	29
2	146	b	45	227	b	56
3	147	b	33	227	b	33
4	349	a	50	248	a	31
5	219	b	46	227	b	56
6	236	b	44	227	b	33
7	443	c	60	339	c	40
8	425	c	39	339	c	25
9	165	b	78	227	b	56
10	173	b	45	227	b	33
11	295	e	50	339	c	40
12	166	b	37	227	b	33
13	161	b	42	227	b	56
14	336	a	46	248	a	29
15	347	c	43	339	c	42

tensile strength is used. It can be seen that except for test 11, the predicted and test failure modes are the same. For test 11, the test failure mode was fin plate fracture and the predicted failure mode is bolt shear. Furthermore, the predicted failure loads are reasonably close to the test failure loads. The tests that used the thinnest RHS (5 mm, tests 2,3,5,6,9,10,12,13) failed by RHS fracture. For all these tests, equation (21) predicts the same failure load at 227 kN. The test failure loads are lower than this value, except for test 6. The test failure load for test 5 is slightly lower than the predicted using equation (21). Whilst this does not validate the method presented above, it is acceptable as a first order analysis because RHS fracture around the fin plate is affected by residual stress and distortion induced by welding. These factors are not included in the prediction so that equation (21) may be considered to give an upper bound to the test load.

Table 2 also compares the predicted and measured maximum deformation of the connection at fracture. For failure mode (b) (RHS fracture), equation (16) was used with $\varepsilon = 0.5\varepsilon_u = 0.1$. For other failure modes, the calculated deflection using equation (16) was reduced proportionally by the ratio of the predicted failure load of the critical failure mode to the predicted failure load based on RHS fracture. The measured maximum connection deformations tend to be higher than the predicted values.

3.2 Improving connection behaviour for robustness

Although this paper has presented a very simple model of structural behaviour and connection behaviour for assessment of disproportionate collapse, it is possible to identify the main fin plate connection design parameters that would affect structural robustness.

Looking at equation (21), the fracture strength of fin plate connections may be increased by using a long fin plate, thick RHS tube and increased RHS ultimate tensile strength. From equation (22), increasing these three values will also increase the connection stiffness. However, from equation (8), since the required connection tying force resistance will be increased only by the cubic root of the increase in connection stiffness, increasing the connection strength and connection stiffness by increasing these three parameters will result in an overall improvement of structural robustness. Increasing the fracture strain of the RHS steel has the benefit of both increasing the connection strength and reducing the connection axial stiffness, clearly indicating that improved RHS ductility is an important design consideration. It should be pointed out that the RHS steel ductility refers to that at the weld

location. Finally, increasing the RHS width (d) has no effect on the connection strength, but will reduce the connection stiffness, thereby improving structural robustness by reducing the required connection tying resistance.

4 SUMMARY

This paper has presented a method to calculate the required connection tying force resistance to control disproportionate collapse in the event of an internal column removal in steel framed structures. The BRE tests on axial behaviour of fin plate to RHS tubes have been used to demonstrate inherent robustness of RHS framed structures. The following conclusions may be drawn:

1 Most of the connections have sufficient inherent tying resistance to prevent disproportionate collapse in the event of an internal column removal.

2 However, a combination of unfavourable design conditions (high accidental loading, very stiff surrounding structures, high structural efficiency under normal condition) may render some connections incapable of controlling disproportionate collapse. For high risk buildings (Class 3 in the UK), the method in this paper may be used as part of the systematic risk assessment.

3 The simple connection analysis method presented in this paper gives reasonable predictions of the connection strength and maximum deformation for the BRE connection tests. However, more detailed research studies are necessary to refine the prediction model and also to cover other types of connection.

4 For RHS framed structures using fin plate connections, the structural robustness may be improved by using thicker and wider RHS tube, increased RHS ultimate strength, increased RHS ductility in the weld region and longer fin plate. Among them, RHS ductility is the most important.

REFERENCES

BSI (2000), British Standard 5950, Structural use of steelwork in buildings – Part 1: Code of practice for design – rolled and welded sections, British Standards Institution

The Building Regulations (BR 2005), Approved Document A: Structure, Office of The Deputy Prime Minister, UK

Jarrett ND and Grantham RI (1993), Robustness tests on fin plate connections, BRE Technical Consultancy Client Report, Building Research Establishment, UK

Kurobane Y, Packer JA, Wardenier J and Yeomans N (2004), Design guide for structural hollow section column connections, CIDECT Guide No. 9, TUV-Verlag, Germany

Parallel session 4: Static strength of joints

Tubular Structures XI – Packer & Willibald (eds)
© 2006 Taylor & Francis Group, London, ISBN 0-415-40280-8

Through-plate joints to elliptical and circular hollow sections

S. Willibald, J.A. Packer, A.P. Voth & X. Zhao
Department of Civil Engineering, University of Toronto, Canada

ABSTRACT: It is commonly known that branch plate joints to hollow structural section members possess relatively low design resistances as the branch plate connects to the highly flexible hollow section face. To enhance the performance of branch plate joints to hollow section members, the hollow section can be slotted and the plate passed through the hollow section, thereby allowing the plate to connect to two sides of the member. Recently, an experimental study of 12 branch and through plate joints to circular (CHS) and elliptical hollow sections (EHS) has been carried out. The tested joints varied in the orientation of the plate as well as the orientation of the elliptical hollow section. By testing both branch and through plate joints of each variation, the additional gain in strength due to the through plate could be quantified.

1 INTRODUCTION

Branch plate joints are one of the most popular joint types due to their ease of fabrication and handling. Originally, longitudinal branch plate joints were mainly used with I-shaped beams or columns, in which case the branch plate is welded along the middle of the flange, so that the force introduced by the branch plate loading is directly transmitted to the web of the I-section (see Figure 1). Unfortunately, this is not the case for longitudinal branch plate joints to hollow sections. Here, the branch plate is attached to the hollow section face which is very flexible and often deforms excessively, frequently exceeding the deformation limit at relatively low loads. For welded joints of rectangular and circular hollow sections, a serviceability deformation limit of 1% and an ultimate deformation limit of 3% (of the width of the connecting face or the diameter of the chord) have been employed in the past as it was shown that this ultimate deformation limit reasonably corresponds to the yield load of these joints (Lu et al. 1994). Besides

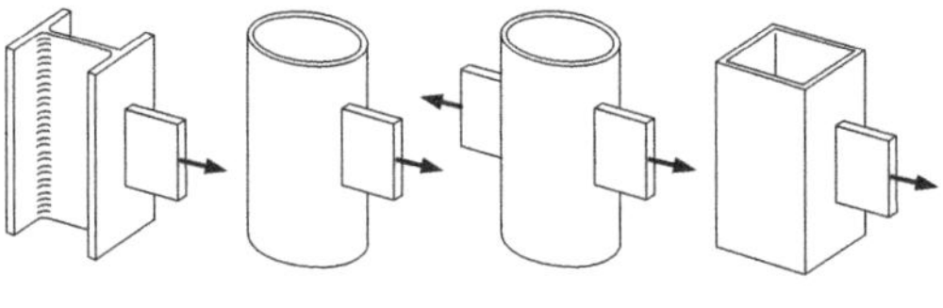

the yield strength or deformation criteria, punching shear of the hollow section connecting face is a further critical limit state which has to be checked, among others. All pertinent limit state checks are summarized in Tables 1 and 2. Generally, the presented formulae have been simplified by considering only loads perpendicular to the hollow section member and disregarding the (generally positive) effect of fillet welds.

1.1 Branch plate to RHS joints

Generally, the orientation (longitudinal or transverse) and width of the branch plate have a major effect on the strength and failure mode of branch plate joints. Joints with transverse plates typically have higher β values than the comparable joints with a longitudinal branch plate. Thus, they are less flexible and can have different failure mechanisms than joints with a longitudinal plate. For branch plate to RHS joints with transverse plates, four basic failure mechanisms have now been identified:

- Chord face plastification;
- Punching shear;
- Chord side wall failure;
- Branch yielding.

The above findings are based principally on research which was carried out by Wardenier et al. (1981) and Davies and Packer (1982). Lu (1997), subsequently found that the yield capacity of the connecting RHS face could be severely lowered in the presence of high normal compressive stresses in

Figure 1.   Longitudinal branch plate joints to I-section, CHS and RHS.

Table 1. Design Resistances of Uniplanar Branch Plate to RHS Joints according to CIDECT (Packer et al. 1992, Kurobane et al. 2004)

Type of Joint	Design Limit State
T- and X-joints – transverse plate	

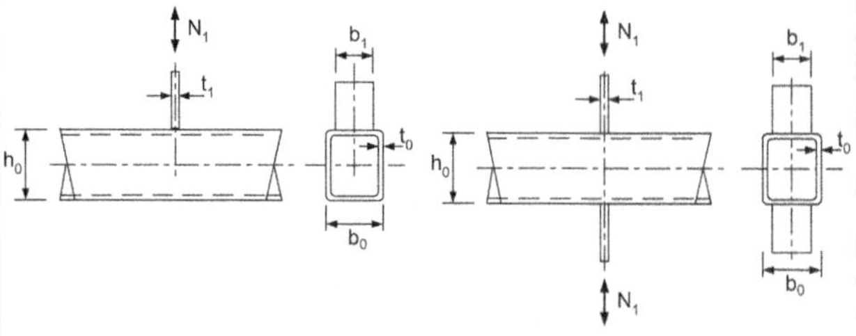

Chord Face Plastification
[when $0.2 \leq \beta \leq 0.8$]

$$N_1^* = (0.5 + 0.7\beta) \frac{4}{\sqrt{1 - 0.9\beta}} f_{y0} t_0^2 f(n)$$

Punching Shear [when $0.8 \leq \beta \leq 1\text{-}1/\gamma$]

$$N_1^* = \frac{f_{y0} t_0}{\sqrt{3}} (2t_1 + 2b_{ep})$$

Chord Side Wall Failure [when $\beta \approx 1.0$]

$$N_1^* = 2 f_{y0} \, t_0 \, (t_1 + 5t_0)$$

Branch Yielding [all β]

$$N_1^* = f_{y1} t_1 b_e$$

T- and X-joints – longitudinal plate

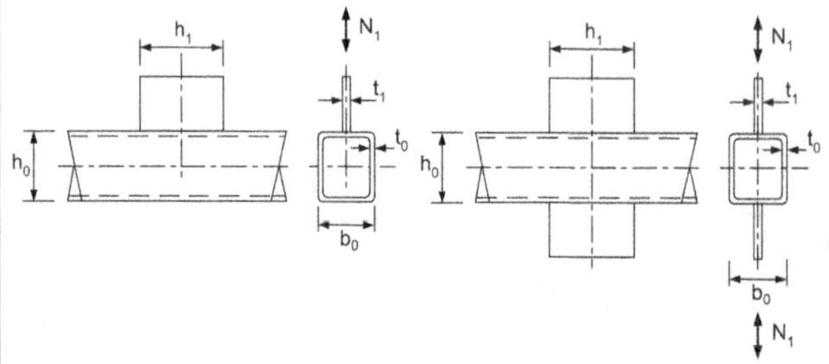

Chord Face Plastification

$$N_1^* = \frac{2 f_{y0} t_0^2}{1 - \beta} \left(\frac{h_1}{b_0} + 2\sqrt{1 - \beta} \sqrt{1 - n^2} \right)$$

Longitudinal through plate joints

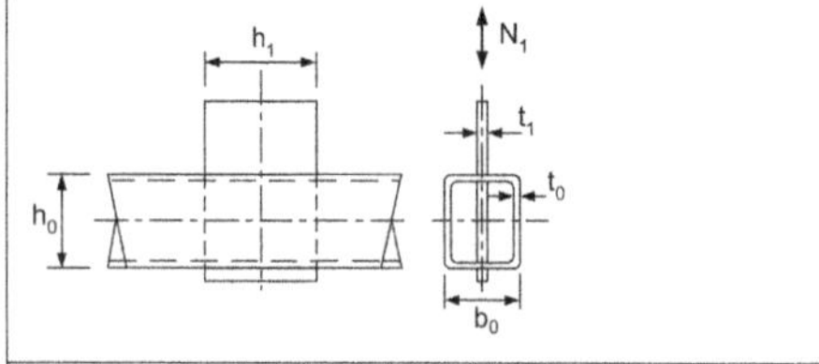

Chord Face Plastification

$$N_1^* = \frac{4 f_{y0} t_0^2}{1 - \beta} \left(\frac{h_1}{b_0} + 2\sqrt{1 - \beta} \sqrt{1 - n^2} \right)$$

Stiffened longitudinal plate (T-stub) joints

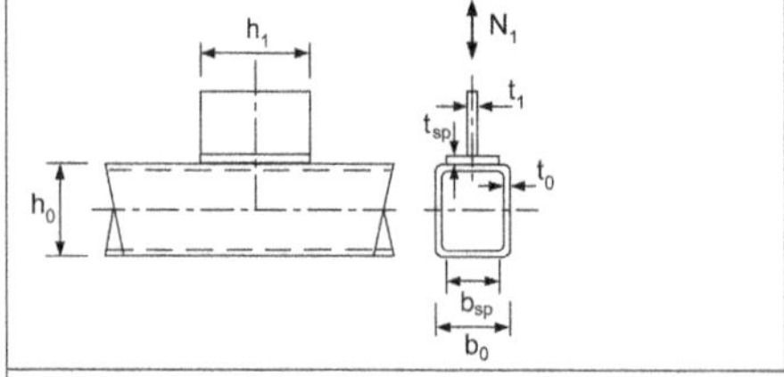

$t_{sp} \geq 0.5 \, t_0 \exp(3\beta^*)$

with: $\beta^* = \dfrac{b_{sp} - t_1}{b_0 - t_0}$

If t_{sp} fulfils above requirement, the joint can be regarded as a RHS-to-RHS T-joint. In the design equations for RHS-to-RHS T-joints, the stiffening plate width b_{sp} is then used for the branch width b_1.

Parameters

$b_e = \dfrac{10}{b_0/t_0} \dfrac{f_{y0} t_0}{f_{y1} t_1} b_1$ but $b_e \leq b_1$ $b_{ep} = \dfrac{10}{b_0/t_0} b_1$ but $b_{ep} \leq b_1$

Transverse plate: $\beta = b_1/b_0$
Longitudinal plate: $\beta = t_1/b_0$

$f(n) = 1 + 1.48n(2\gamma)^{-0.33} - 0.46n^{1.5}(2\gamma)^{(-0.33 - 0.11\beta^2)} \leq 1.0$

$n = \sigma_0/f_{y0}$ and σ_0 is the normal stress in the chord member at the plate joint due to axial load plus bending (if applicable). If plate is inclined to the hollow section, n refers to the "preload" side.

$2\gamma = b_0/t_0$

Range of validity

| Transverse plate: $1.5 \leq 2\gamma \leq 37.5$ | Longitudinal plate: $2\gamma \leq 40$ |

Table 2. Design Resistances of Uniplanar Branch Plate to CHS Joints according to CIDECT (Packer et al. 1991, Kurobane et al. 2004)

Type of Joint	Design Limit State
T- and X-joints – transverse plate	
	Chord Plastification
	$N_1{}^* = \dfrac{5 f_{y0} t_0^2}{1 - 0.81\beta} f(n')$
	Punching Shear [when $b_1 \leq d_0 - 2t_0$]
	$N_1{}^* = 2 t_0 b_1 \dfrac{f_{y0}}{\sqrt{3}}$
T- and X-joints – longitudinal plate	
	Chord Plastification
	$N_1{}^* = 5 f_{y0} t_0^2 (1 + 0.25\eta) f(n')$
	Punching Shear
	$N_1{}^* = 2 t_0 (h_1 + t_1) \cdot \dfrac{f_{y0}}{\sqrt{3}}$
Longitudinal through plate joints	
Note: Conservative postulation only for resistance to chord plastification	Chord Plastification
	$N_1{}^* = 5 f_{y0} t_0^2 (1 + 0.25\eta) f(n')$
	Punching Shear
	$N_1{}^* = 4 t_0 (h_1 + t_1) \cdot \dfrac{f_{y0}}{\sqrt{3}}$

Parameters		
For $n' < 0$ (compression): $\quad f(n') = 1 + 0.3\, n'(1 - n')$ but $f(n') \leq 1.0$ For $n' \geq 0$ (tension): $\quad f(n') = 1.0$ where $n' = \sigma_0/f_{y0}$ and σ_0 is the normal stress in the chord member at the plate joint due to axial load plus bending (if applicable).	$\beta = b_1/d_0$	$\eta = h_1/d_0$

Range of validity
$10 \leq d_0/t_0 \leq 40$ $\eta \leq 4$

the connecting chord, hence this limit state is still included.

For longitudinal plates chord face plastification is critical. The analytical approach to predict this limit state is based on a flexural model using yield line analysis (Cao et al. 1998a). As with the transverse plate joints, the influence of compressive normal stress in the RHS chord member was taken into account (Cao et al. 1998b).

Recently, research on strengthened longitudinal plate-to-RHS joints has been carried out by Kosteski (2001) and Kosteski and Packer (2003a and b).

1.2 *Branch plate to CHS joints*

Work by Wardenier (1982) and Makino (1984) resulted in broadly accepted design equations for longitudinal and transverse branch plate to CHS joints, which can be found in the CIDECT design guides (Wardenier et al. 1991, Kurobane et al. 2004) or Eurocode 3 (CEN 2005).

Unfortunately, all the above research did not cover through plate joints for circular (or elliptical) hollow sections. For through plate joints, it is currently suggested (Kurobane et al. 2004) to use the equations provided for regular branch plate joints, whereby the resistance for the failure mode of punching shear can be doubled but the resistance against plastification of the CHS member is taken to be the same as for the non-through plate joint (see Table 2). The latter approach is very conservative, but is due to the lack of research on this joint type.

Prior research on branch or through plate joints to elliptical hollow sections has not been carried out.

2 EXPERIMENTAL WORK

Experimental research has been performed on a series of branch plate to elliptical hollow section (EHS) welded joints, along with their through plate joint counterparts. A total of six 90° T-joints were tested to failure using a purpose-built testing jig, with the branch plate/through plate loaded in axial tension. The test specimens all used 220 × 110 × 6.3 EHS and included longitudinal plates attached to the wide or narrow side(s) and transverse plates attached to the wide side(s) of EHS members. Another series of six branch plate to CHS welded joints, along with their through plate counterparts, has also been tested to failure in the same manner. These T-joints all had the branch plate/through plate inclined at 90° to the 219 × 4.8 CHS member, but the plate was in turn oriented longitudinal (parallel) to the axis of the tube, transverse (perpendicular) to the axis of the tube, and at a 45° angle (skewed) to the axis of the tube. All 12 joint tests were instrumented and the "capacity", for limit states design, was interpreted in terms of both ultimate strength and an ultimate deformation limit.

2.1 Branch and through plate joints to elliptical hollow sections

A total of six specimens have been tested under quasi-static tension loading. Figure 2 shows the orientation of the hollow sections as well as the gusset or through plates. The notation used in this study is shown in Figure 3.

The EHS members were hot-formed, S355J2H sections (EHS 220 × 110 × 6.3) conforming to EN 10210 (CEN 1994). The steel plate had a nominal yield strength of 300 MPa with a thickness of 19 mm. The welds between the plates and the EHS connecting faces were fillet welds using metal-cored wire and the FCAW-G welding process. The hollow sections and steel plates each came from a single steel stock. The measured dimensions of the test specimens are documented in Table 3. Tensile coupon tests on the EHS

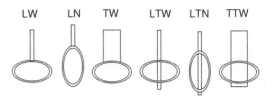

Figure 2. Hollow section and plate orientation of the tested specimens.

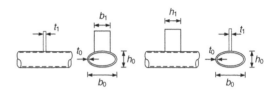

Figure 3. Notation for EHS joints.

Table 3. Measured dimensional properties of tested EHS specimens.

Test	h_0 (mm)	b_0 (mm)	t_0 (mm)	w (mm)	t_1'[a] (mm)	b_1'/h_1'[b] (mm)
LW	111	221	5.93	11	44	119
LTW				11	42	121
LN	221	111		13	44	128
LTN				12	43	122
TW	111	221		12	45	123
TTW				10	41	118

[a] With: $t_1' = t_1 + 2w$
[b] With: $b_1' = b_1 + 2w, h_1' = h_1 + 2w$

Table 4. Measured material properties of tested EHS specimens.

Property	EHS 220 × 110 × 6.3	Plate
f_y (MPa)	421	324
f_u (MPa)	530	493
E (MPa)	215,500	216,500
ε_{ul} (%)	34.7	–

and plate materials were also carried out, and several key material properties are presented in Table 4.

All tests were carried out using a 1,000 kN capacity Universal Testing Machine. Figure 4 shows a typical test set-up of the specimens. In all cases, the branch plate was loaded, in a quasi-static, monotonic manner to failure in axial tension. The load-displacement response of each branch plate-to-EHS joint and the stress distribution in the plate were monitored through the installed LVDTs and strain gauges. All test specimens were whitewashed prior to testing, to visibly show the regions of yielding in the EHS main member.

Figure 4. Test set-up of specimen LTW.

Table 5. Test results and joint strength predictions for joints to elliptical hollow sections.

| | | Chord plastification | | | | |
| | | CHS | | RHS | | |
Test	N_y (kN)	$N_1(b_0)$ (kN)	$N_1(h_0)$ (kN)	$N_1(b_0)$ (kN)	N_u (kN)	N_{ps} (kN)
LW	123	84.0	93.9	86	417	470
LTW	254	84.1	94.2	172	458	940
LN	318	95.4	84.7	133	668	496
LTN	>625[a]	94.4	84.2	258	923	951
TW	128	135	390	105	369	355
TTW	341	130	390	–[b]	467	680
Mean N_1/N_y		0.55	1.12	0.65		
CoV (%)		58.1	101	26.0		

[a] The face-deformation of the specimen could not be measured properly due to yielding in the branch plate
[b] No design method available

Final failure of all specimens was caused by punching shear resulting in a tear-out of the branch or through plate. Most specimens showed excessive deformations prior to final failure and showed distinct yield lines during testing. Table 5 shows the test results (yield strength, ultimate joint strength) as well as the joint resistance predictions according to various calculation methods (for RHS and CHS using formulae in Tables 1 and 2). The yield strength, N_y, is based on the double-tangent method (see Figure 5).

For the calculation of the joint capacity, N_1, the dimensions $b_1'\ (= b_1 + 2w)$, $h_1'\ (= h_1 + 2w)$ and $t_1'\ (= t_1 + 2w)$ have been used as it is indicated in earlier research that the footprint of the branch plate including the welds results in a better representation of the joint behaviour (Davies & Packer 1982; Kosteski and Packer 2003a, 2003b). For design purposes, the weld size is often disregarded resulting in more conservative predictions. Also, in Table 5, the joint capacity, N_1, has been predicted by inserting the EHS dimensions (see Figure 3) into CHS joint formulae in Table 2, hence obtaining two versions depending on whether b_0 or h_0 is used for d_0. Similarly, in Table 5, the joint capacity, N_1, has been predicted by letting b_0 of the EHS (see Figure 3) be b_0 of a RHS and then using the RHS formulae of Table 1. This notion of using EHS dimensions in established formulae for CHS and RHS joints has also been tried by Bortolotti et al. (2003).

Generally, there is little agreement between the predictions of the various design methods and the test results. The predicted punching shear resistance compares well with the ultimate strength for half of the specimens (LW, LTN and TW) but does not agree well for the rest of the specimens (LTW, LN and TTW).

The main finding of these preliminary tests on elliptical branch and through plate joints is that the yield

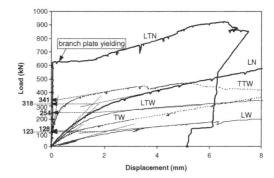

Figure 5. Yield load of joints to elliptical hollow sections (double tangent method).

Table 6. Yield strength comparison of branch and through plate joints to elliptical hollow sections.

Test	LTW/LW	LTN/LN	TTW/TW
N_y (kN) through plate joint	254	>625	341
of: branch plate joint	123	318	128
$\dfrac{N_y \text{ of through plate joint}}{N_y \text{ of branch plate joint}}$	2.07	>1.97	2.66

strength (based on a double-tangent method) can be doubled by using a through plate joint instead of the comparable branch plate joint (see Table 6).

Applying the existing design formulae for branch and through plate joints to RHS and CHS members does not result in satisfying answers. Further research is necessary to find suitable design methods. The design formulae for joints to rectangular hollow sections are based on yield line models that take the stiff corners of the RHS into account. EHS do not posses

Table 7. Measured dimensional properties of tested CHS specimens.

Test	d_0 (mm)	t_0 (mm)	w (mm)	α (°)	b'_1/h'_1 (mm)	t'_1 (mm)
CB0E	220	4.50	11.5	0	123	42.1
CT0E			11.0		122	41.1
CB45E			10.8	45	122	40.6
CT45E			10.1		121	39.2
CB90E			11.0	90	122	41.2
CT90E			9.6		119	38.2

Table 8. Measured material properties of tested CHS specimens.

Property	CHS 219×4.8	Plate
f_y (MPa)	389	326
f_u (MPa)	525	505
E (MPa)	200,000	210,500
ε_{ul} (%)	30.1	37.7

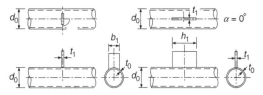

Figure 6. Notation for CHS joints

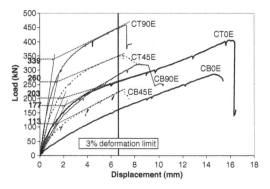

Figure 7. Yield load of joints to round hollow sections (double tangent method).

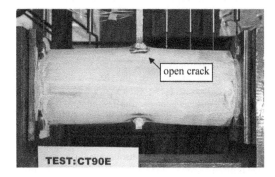

Figure 8. Specimen CT90E at failure.

such very rigid corners. On the other hand the rules for joints to circular hollow sections do not take the changing radii of the EHS into account. Generally, both design methods seem to be too conservative but as the design method for rectangular hollow sections generally results in higher design loads, it seems to give the most adequate answers having a safe (lower than 1) ratio of predicted to measured yield strength and the lowest coefficient of variation (26%) of all possible design methods.

2.2 Branch and through plate joints to round hollow sections

The test series on branch and through plate joints to round hollow sections comprised six specimens. As with the elliptical hollow sections, all specimens were tested under a quasi-static tension loading. Table 7 shows the measured dimensions of the specimens using the notation according to Figure 6. A specimen name starting with CB indicates a branch plate joint and a name starting with CT a through plate joint.

The hollow sections used for the tests were ASTM A500 Grade C cold-formed 219×4.8 mm hollow sections with a minimum specified yield strength of 317 MPa (ASTM 2003). The branch plates were all 100 mm wide with a thickness of 19 mm. The measured material properties of the hollow section and plate are given in Table 8. Fillet welds were used to connect the CHS and the plates (FCAW-G).

All specimens experienced substantial deformations before failure. In all cases, the through plate joint was considerably stiffer than the comparable branch plate joint (see Figure 7). Punching shear of the branch or through plate caused final failure in all

specimens (see Figure 8). In through plate joints, the top of the joint was torn out while the bottom of the joint only buckled into the tube.

Table 9 compares the yield and ultimate strength with the load at the deformation limit as well as the joint strength predictions according to current design recommendations (see Table 2). For the specimens with the plate at a 45° angle to the axis of the tube both the predictions for a longitudinal ($\alpha = 0°$) and transverse ($\alpha = 90°$) plate joint are provided. In accordance with the design recommendations for longitudinal through plate joints, the strength against chord plastification was calculated in the same way for branch and through plate joints, while the resistance against punching shear was doubled for through plate joints compared to their branch plate counterparts.

Table 9. Test results and joint strength predictions for joints to circular hollow sections.

Test	N_y (kN)	$N_{3\%}$ (kN)	N_1 (kN) $\alpha = 0°$	N_1 (kN) $\alpha = 90°$	N_u (kN)	N_{ps}[a] (kN)
			Chord plastification			
CB0E	>50[b]	161	44.9	—[c]	286	249
CT0E	203	259	44.8	—[c]	406	493
CB45E	113	223	44.8	71.5	233	247
CT45E	260	347	44.8	71.0	351	489
CB90E	177	284	—[c]	71.5	320	247
CT90E	339	447	—[c]	70.1	459	481

[a] Punching shear was calculated as $N_{ps} = 2t_0 b_1' f_{y0}/\sqrt{3}$ (branch plates) and $N_{ps} = 4t_0 b_1' f_{y0}/\sqrt{3}$ (through plates). Note that b_1', transverse plate, becomes h_1' for longitudinal plate joints
[b] No clear yield load found
[c] Design method not applicable

Table 10. Yield strength comparison of branch and through plate joints to circular hollow sections.

Test	CT0E/ CB0E	CT45E/ CB45E	CT90E/ CB90E
N_y (kN) through plate joint of:	203	260	339
branch plate joint	>50	113	177
$\dfrac{N_y \text{ of through plate joint}}{N_y \text{ of branch plate joint}}$	<4.06	2.30	1.92

Comparing the yield strength or the load at the 3% deformation limit with the predictions against chord plastification shows that the predictions are all very conservative. On the other hand, the predictions against punching shear are generally not safe for the joints with through plates (where the predicted capacity was doubled relative to the branch plate). For joints with branch plates, the predictions are all reasonably close.

Table 10 compares the yield strengths of branch and through plate joints to round hollow sections. As with the joints to elliptical hollow sections, the yield strength, N_y, of the through plate joint is roughly double the yield strength of the comparable branch plate joint. Branch plate joints are connected to one side of the hollow section and therefore can activate only a small part of the CHS. Through plate joints are connected to twice the area of the CHS compared to the branch plate joints and therefore enhance the force flow from the plate to the CHS.

From literature, it was expected that the yield strength, N_y, and the joint load at the 3% limit, $N_{3\%}$, would roughly compare. In the tested specimens of the presented study the load at the deformation criterion was generally considerably higher than the yield strength. This phenomenon might have been caused by the rather short length of the chord member. The end plates of the specimen are very close to the joint and might have stiffened the joint.

3 CONCLUSIONS

A comprehensive review of current design resistances for RHS and CHS branch plate joints, as well as an experimental study on branch and through plate joints to elliptical and circular hollow sections, has been carried out.

The experimental study on EHS joints shows that the yield strength, N_y, for through plate joints in comparison with branch plate joints is approximately double or greater. Punching shear predictions for through plate joints are generally unsafe whereas the predictions for branch plate joints are close or even conservative when compared to the ultimate joint strength. Existing CHS or RHS design formulae do not adequately describe the joint behaviour but RHS resistance equations could be used as a conservative estimate for design. It is clear from this study that further research is needed to determine a suitable design method that more closely describes the EHS to branch and through plate joint behaviour.

The experimental study on CHS joints shows that predictions given by current design guides for chord plastification are extremely conservative when compared to either the yield strength or the load at the 3% deformation limit of the joint. The high yield or 3% deformation load may be due to the short length of the chord member possibly resulting in a stiffening of the joint. To what degree the chord length affects the joint stiffness is unknown but this will be investigated numerically. As with the EHS joints, the yield strength, N_y, of the joints roughly doubled if a through plate was used instead of a branch plate. Punching shear predictions for the CHS through plate joints are generally unsafe whereas the predictions for CHS branch plate joints are close to the ultimate joint strength.

ACKNOWLEDGEMENTS

Financial support has been provided by CIDECT (Comité International pour le Développement et l'Étude de la Construction Tubulaire) Programme 5BS, the Steel Structures Education Foundation (SSEF), Science and Engineering Research Canada (NSERC) and Deutsche Forschungsgemeinschaft (DFG).

The circular and elliptical hollow sections used in this project have been provided by Copperweld (Canada) and Arcelor Tubes (France) respectively. The plate material has been supplied by IPSCO Inc. (Canada). We are also thankful to Walters Inc. (Hamilton, Ontario, Canada) for fabrication of all plate-to-CHS test specimens presented in this study.

NOMENCLATURE

CHS = circular hollow section
E = Young's modulus
EHS = elliptical hollow section
N_1 = predicted joint capacity
N_1^* = joint limit states design resistance
$N_{3\%}$ = load at 3% displacement
N_{ps} = predicted punching shear strength
N_u = ultimate load of the joint
N_y = yield load of the joint (double-tangent method)
RHS = rectangular hollow section
b_0 = external width of main RHS or EHS member
b_1 = external width of branch member
b_1' = effective external width of branch member $(b_1' = b_1 + 2w)$
b_e = effective width of branch member
b_{ep} = effective punching shear width of branch member
b_{sp} = width of stiffening plate
d_0 = external diameter of main CHS member
f_u = ultimate tensile stress
f_y = yield tensile stress
f_{y0} = yield tensile stress of main hollow section material
f_{y1} = yield tensile stress of branch material
h_0 = external height of main RHS or EHS member
h_1 = external height of branch member
h_1' = effective external height of branch member $(h_1' = h_1 + 2w)$
n = stress ratio in RHS chords
n' = stress ratio in CHS chords
t_0 = thickness of main hollow section member wall
t_1 = thickness of branch member wall
t_1' = effective thickness of branch member $(t_1' = t_1 + 2w)$
t_{sp} = thickness of stiffening plate
w = weld size (leg length)
α = angle between plate and longitudinal axis of hollow section
β = width ratio $(\beta = \frac{b_1}{b_0}, \frac{b_1}{d_0}, \frac{t_1}{b_0})$
β^* = stiffening plate width ratio $(\beta^* = \frac{b_{sp} - t_1}{b_0 - t_0})$
γ = half width to thickness ratio $(\gamma = \frac{b_0}{2t_0}, \frac{d_0}{2t_0})$
ε_{ul} = ultimate strain at rupture
η = ratio of branch member depth to chord diameter $(\eta = \frac{h_1}{d_0})$
σ_0 = normal stress in the chord member at the plate joint due to axial load plus bending (if applicable).

REFERENCES

ASTM. 2003. Standard Specification for Cold-Formed Welded and Seamless Carbon Steel Structural Tubing in Rounds and Shapes. *ASTM-A500-03a*. West Conshohocken: ASTM International.

Bortolotti, E., Jaspart, J.P., Pietrapertosa, C., Nicaud, G., Petitjean, P.D., Grimmault, J.P. & Michard, L. 2003. Testing and Modelling of Welded Joints between Elliptical Hollow Sections. *Proc. 10th Intern. Symp. on Tubular Structures, Madrid*, Spain 2003: 259–266.

Cao, J.J., Packer, J.A. & Kosteski, N. 1998a. Design Guidelines for Longitudinal Plate to HSS Connections. *Journal of Structural Engineering* 124(7): 784–791.

Cao, J.J., Packer, J.A. & Yang, G.J. 1998b. Yield Line Analysis of RHS Connections with Axial Loads. *Journal of Constructional Steel Research* 48(1): 1–25.

CEN. 1994. Hot Finished Structural Hollow Sections of Non-Alloy and Fine Grain Structural Steels – Part 1: Technical Delivery Requirements. *EN 10210-1*. Brussels: European Committee for Standardisation.

CEN. 2005. *Eurocode 3: Design of Steel Structures*. Part 1.8: Design of Joints. *EN 1993-1-8: 2005(E)*. Brussels: European Committee for Standardisation.

Davies, G. & Packer, J.A. 1982. Predicting the Strength of Branch Plate-RHS Connections for Punching Shear. *Canadian Journal of Civil Engineering* 9: 458–467.

Kosteski, N. 2001. *Branch Plate to Rectangular Hollow Section Member Connections*. PhD thesis. Toronto: University of Toronto.

Kosteski, N. & Packer, J.A. 2003a. Welded Tee-to-HSS Connections. *Journal of Structural Engineering* 129(2): 151–159.

Kosteski, N. & Packer, J.A. 2003b. Longitudinal Plate and Through Plate-to-HSS Welded Connections. *Journal of Structural Engineering* 129(4): 478–486.

Kurobane, Y., Packer, J.A., Wardenier, J. & Yeomans, N. 2004. *Design Guide for Structural Hollow Section Column Connections*. Köln: CIDECT and Verlag TÜV Rheinland GmbH. ISBN 3-8249-0802-6.

Lu, L.H. 1997. *The Static Strength of I-beam to Rectangular Hollow Section Column Connections*. PhD thesis. Delft: Delft University of Technology.

Lu, L.H., Winkel, de, G.D., Yu, Y. & Wardenier, J. 1994. Deformation Limit for the Ultimate Strength of Hollow Section Joints. *Proc. 6th Intern. Symposium on Tubular Structures*, Melbourne 1994: 341–347.

Makino, Y. 1984. *Experimental Study on Ultimate Capacity and Deformation for Tubular Joints*. PhD thesis. Osaka: Osaka University (in Japanese).

Packer, J.A., Wardenier, J., Kurobane, Y., Dutta, D. & Yeomans, N. 1992. *Design Guide for Rectangular Hollow Section (RHS) Joints under Predominantly Static Loading*. Köln: CIDECT and Verlag TÜV Rheinland GmbH. ISBN 3-8249-0089-0.

Wardenier, J. 1982. *Hollow Section Joints*. Delft: Delft University Press. ISBN 90-6275-084-2.

Wardenier, J., Davies, G. & Stolle, P. 1981. The Effective Width of Branch Plate to RHS Chord Connections in Cross Joints. *Stevin Report 6-81-6*. Delft: Delft University of Technology.

Wardenier, J., Kurobane, Y., Packer, J.A., Dutta, D. & Yeomans, N. 1991. *Design Guide for Circular Hollow Section (CHS) Joints under Predominantly Static Loading*. Köln: CIDECT and Verlag TÜV Rheinland GmbH. ISBN 3-88585-975-0.

Tubular Structures XI – Packer & Willibald (eds)
© 2006 Taylor & Francis Group, London, ISBN 0-415-40280-8

Evaluation of experimental results on slender RHS K-gap joints

O. Fleischer & R. Puthli

Research Centre for Steel, Timber and Masonry, University of Karlsruhe, Germany

ABSTRACT: Many sections with a width to thickness ratio $2\gamma > 35$ can be found in EN10219-2:2004. However, due to the restrictions in EN1993-1-8, they have normally to be excluded from use. Therefore, for joints with these slender sections, experimental investigations have been carried out at Karlsruhe University to extend the limiting range. The results of these investigations are evaluated according to EN1990:2002 for the maximum loads $N_{i.max}$ from the tests as well as loads $N_{i.u}$ obtained by applying a deformation limit (ultimate loads). Furthermore, gap sizes smaller than those permitted by EN1993-1-8:2005 are also investigated. The results presented in this paper are intended to provide supplementary information to help extend the design rules for slender RHS joints and small gaps $g < g_{min}$. The work is presently being further evaluated using an extended numerical study, which will be presented elsewhere.

1 INTRODUCTION

Experimental investigations were carried out at Karlsruhe University to extend the limitations in EN1993-1-8 of the chord width to thickness ratio $2\gamma = b_0/t_0 \leq 35$ to larger chord slenderness ratios $2\gamma > 35$. The lower bound of the permitted gap size g_{min} of connections using these slender chord sections and a maximum width ratio which excludes braces and welds outside the flat area of the chord section, is always $g_{min} \geq 0.5 \cdot b_0 \cdot (1-\beta)$ and not $g_{min} \geq t_1 + t_2$. However, for the present work, the minimum gap size $g_{w.min}$ is allowed to go down up to $g_{w.min} \geq 4 \cdot t_0$, which is a practicable minimum gap size for weldability of these thin walled sections.

Because the design rules in EN1993-1-8 are based on maximum loads $N_{i.max}$ (overall ultimate loads), these loads are used for the first statistical evaluations of the failure modes observed in the tests. Since chord face failure cannot be observed visually, the deformation criterion from previous work (Lu et al. 1993), which limits the indentation of the chord flange to 3% of the chord width b_0, has been used for further evaluation, on the basis that chord face failure predominates other failure modes at lower loads. With the loads $N_{i.u}$, based on the deformation criterion, a statistical evaluation for chord face failure has been carried out additionally.

These statistical evaluations should provide information on whether if it is possible to use the existing design models of EN1993-1-8 within the extended geometric parameter ranges, or if modifications have to be adopted.

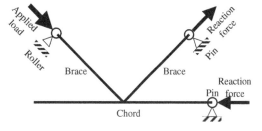

Figure 1. Static system for the tests.

2 EXPERIMENTAL INVESTIGATIONS

In the framework of an ECSC project, 39 tests on axial loaded symmetrical K-joints with gap, made of cold formed RHS (prEN 10219:2004), were carried out at Karlsruhe University.

With this static system, only axial forces are applied, so that a moment loading of the connection is only caused by eccentricity and secondary effects. Prestressing of the chord is not taken into account (Fig. 1). From previous work (Liu et al. 1998) it is known that a system with a chord under axial compression gives the lowest load carrying capacity of the connections, on which basis the static system has been chosen.

2.1 *Geometrical parameters and material properties of the specimens*

In Table 1, the geometrical parameters based on measured dimensions are given. Since the gap sizes were

Table 1. Geometrical parameters, material properties and results from the tests.

KJ	β	2γ	τ	$\dfrac{h_0}{b_0}$	$\dfrac{g'}{g'_{min}}$	Θ_i [°]	$f_{y0.m}/f_y$	$f_{yi.m}/f_y$	$N_{i.max}$ [kN]	FM	$N_{i.u}$ [kN]	$\dfrac{N_{i.max}}{N_{i.Rk}}$	$\dfrac{N_{i.u}}{N_{i.Rk}}$	$\dfrac{N_{i.max}}{N_{i.Rm}}$	$\dfrac{N_{i.u}}{N_{i.Rm}}$
01	0.33	51.7	0.98	0.67	0.59	45	1.27	1.12	452	N.a.	259	1.39	0,80	1,38	0,79
02	0.42	50.0	0.98	0.67	0.67	45	1.27	1.43	529	CW	394	1.24	0,93	1,29	0,96
03	0.50	50.9	0.97	0.67	0.78	45	1.27	1.35	532	CW	532	1.08	1,07	1,08	1,08
04	0.33	51.7	0.98	0.67	0.24	45	1.27	1.12	463	PS	398	1.43	1,23	1,41	1,21
05	0.33	51.7	0.98	0.67	0.48	45	1.27	1.12	421	PS	327	1.30	1,01	1,28	1,00
06	0.33	51.7	0.98	0.67	0.72	45	1.27	1.12	456	N.a.	305	1.40	0,94	1,39	0,93
07	0.42	51.7	1.00	0.67	0.27	45	1.27	1.43	530	PS	$-^{2)}$	1.31	–	1,29	–
08	0.42	50.8	1.00	0.67	0.55	45	1.27	1.43	522	CW	450	1.26	1,08	1,27	1,10
09	0.42	50.0	0.98	0.67	0.82	45	1.27	1.43	513	CW	363	1.20	0,85	1,25	0,88
10	0.50	50.8	0.98	0.67	0.32	45	1.27	1.35	565	N.a.	$-^{2)}$	1.13	–	1,15	–
11	0.50	50.8	0.98	0.67	0.64	45	1.27	1.35	570	N.a.	$-^{2)}$	1.14	–	1,16	–
12	0.50	51.6	0.98	0.67	0.96	45	1.27	1.35	537	N.a.	443	1.10	0,91	1,09	0,90
13	0.50	33.9	0.98	1.50	0.48	45	1.27	1.12	553	PS	395	1.36	0,97	1,38	0,98
14	0.63	33.8	1.02	1.50	0.64	45	1.27	1.43	598	N.a.	$-^{2)}$	1.17	–	1,19	–
17	0.40	41.6	1.04	0.50	0.33	45	1.37	1.34	290	PS	273	1.13	1,06	1,18	1,12
18	0.40	35.1	1.02	0.50	0.40	45	1.35	1.18	338	PS	307	1.02	0,93	1,05	0,95
19	0.40	53.9	0.81	0.51	0.27	45	1.13	1.20	194	EW	174	1.35	1,21	1,11	0,99
21	0.40	35.0	0.53	0.50	0.40	45	1.35	1.20	282	CB	271	0.90	0,82	0,88	0,84
22	0.53	40.6	0.81	1.00	1.05	45	1.29	1.20	222	CW	163	1.17	0,86	1,10	0,81
23	0.53	40.6	0.81	1.00	1.05	45	1.29	1.20	212	CW	151	1.12	0,80	1,05	0,75
25	0.67	42.9	1.00	1.00	0.60	45	1.20	1.20	176	CW	$-^{2)}$	1.36	–	1,20	–
26	0.67	42.9	1.00	1.00	0.60	45	1.20	1.20	201	CW	$-^{2)}$	1.56	–	1,37	–
27	0.67	42.9	1.00	1.00	0.60	45	1.20	1.20	165	CW	$-^{2)}$	1.28	–	1,12	–
29	0.33	49.2	0.92	0.67	0.24	60	1.27	1.12	392	PS	330	1.37	1,16	1,46	1,23
30	0.33	49.1	0.92	0.67	0.48	60	1.27	1.12	343	PS	251	1.20	0,88	1,28	0,94
31	0.33	48.4	0.92	0.67	0.72	60	1.27	1.12	365	PS	274	1.25	0,94	1,36	1,02
32	0.33	47.5	0.94	0.67	0.24	30	1.27	1.12	550	N.a.	460	1.06	0,89	1,19	0,99
33	0.33	49.1	0.95	0.67	0.48	30	1.27	1.12	550	N.a.	384	1.11	0,78	1,19	0,83
34	0.33	48.3	0.94	0.67	0.72	30	1.27	1.12	519	EW	286	1.02	0,56	1,12	0,62
35	0.42	49.3	0.97	0.66	0.27	60	1.27	1.43	390	N.a.	356	1.10	1,00	1,16	1,06
36	0.42	48.5	0.97	0.66	0.55	60	1.27	1.43	414	N.a.	392	1.13	1,07	1,24	1,17
37	0.42	48.3	0.95	0.67	0.82	60	1.27	1.43	332	N.a.	248	0.91	0,68	0,99	0,74
38	0.42	48.3	0.94	0.67	0.27	30	1.27	1.43	581	N.a.	543	0.92	0,86	1,00	0,94
39	0.42	50.0	0.97	0.67	0.55	30	1.27	1.43	558	N.a.	455	0.93	0,76	0,96	0,78
40	0.42	49.3	0.95	0.66	0.82	30	1.27	1.43	533	N.a.	455	0.90	0,74	0,92	0,78
41[1]	0.40	49.9	0.73	0.50	0.27	45	1.10	1.13	265	CB	211	1.44[1]	1,14[1]	1,30	1,03
42[1]	0.40	49.9	0.98	0.50	0.27	45	1.10	1.08	222	EW	$-^{2)}$	1.21[1]	–	1,09	–
43[1]	0.40	42.5	0.83	0.50	0.33	45	1.17	1.08	303	PS	261	1.21[1]	1,05[1]	1,06	0,91
46[1]	0.40	42.5	0.64	0.50	0.33	45	1.17	1.13	265	CB	236	1.06[1]	0,94[1]	0,93	0,83

Annotations: KJ-01 to KJ-40 with steel grade S355 ($f_y = 355\,\text{N/mm}^2$), KJ-41 to KJ-46 with steel grade S460 ($f_y = 460\,\text{N/mm}^2$) N.a. "**N**ot **a**pparent"; CW **C**hord **W**eb Failure; PS **P**unching **S**hear Failure; CB **C**ompression **B**race Failure; EW **E**ffective **W**idth Failure; [1] $0.9 \cdot N_{i.Rk}$ is used (prEN1993-1-12:2004); [2] $\Delta_{CF} = 3\% \cdot b_0$ limit not exceeded.

not measured, nominal values of the dimensionless gap sizes $g' = g/t_0$ and $g'_{min} = g_{min}/t_0$, as defined in EN1993-1-8, are used. Θ_i represents the nominal brace angle and $f_{y0.m}$ & $f_{yi.m}$ refer to the characteristic yield strengths f_y for chord and braces.

2.2 Observed failure modes

Although EN1993-1-8 gives chord face failure as the governing mode for all specimens, other failure modes are also observed in the tests as given below.

2.2.1 Punching shear failure (PS)

For small gap sizes $g < g_{min}$ and a thickness ratio $\tau = t_i/t_0 = 1.0$, punching shear failure beneath the tension brace limits the load carrying capacity of the connections. Only for steel S460 can this failure mode be observed for a thickness ratio $\tau = 0.8$ (Tab. 1).

From Figure 2 it can be seen that the effective length l_{ep} for punching shear given in EN1993-1-8 (Eq. 1) should be reduced as proposed by Wardenier (2002).

$$l_{ep} = \frac{2 \cdot h_i}{\sin \Theta_i} + b_i + b_{ep} \,;\quad b_{ep} = \frac{10}{b_0/t_0} \cdot b_i \leq b_i \qquad (1)$$

Figure 2. Punching shear failure.

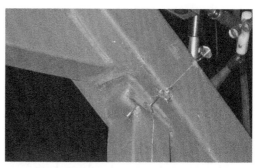

Figure 4. Compression brace failure.

Figure 3. Effective width failure.

where h_i = brace height; b_i = brace width; b_0 = chord width; t_0 = chord thickness; b_{ep} = effective width for punching shear; Θ_i = brace angle.

Due to the high stiffness of the gap region for small gaps, the load transfer occurs predominantly in the gap region. The consequently reduced effective length $l_{ep.red}$ therefore results in smaller punching shear resistances.

2.2.2 Brace Failures – Effective width or compression brace failure (EW or CB)

If the gap size $g < g_{min}$ and the thickness ratio $\tau < 1.0$, tearing in the tension brace or plastification of the compression brace can occur.

As with punching shear failure, the high stiffness of the gap region again causes a reduced effective length $l_{e.red}$ compared to that given in EN1993-1-8 (Eq. 2):

$$l_e = 2 \cdot h_i - 4 \cdot t_i + b_i + b_e \; ; \; b_e = \frac{10}{b_0/t_0} \cdot \frac{f_{y0} \cdot t_0}{f_{yi} \cdot t_i} \cdot b_i \leq b_i \,(2)$$

where t_i = brace thickness; b_e = effective width; f_{y0} and f_{yi} = yield strengths of chord and braces.

The reduced effective length $l_{e.red}$ causes effective width failure on one side of the brace.

2.2.3 Chord Wb Failure (CW)

Especially for high width ratios $\beta \geq 0.42$ and gap sizes other than $g_{w.min}$, distinct deformations of the chord webs have been observed in the tests (Fig. 5).

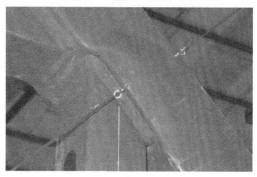

Figure 5. Failure of chord webs.

This failure mode is not covered for K-joints with gaps in EN1993-1-8, so that this may need to be included when extending the rules.

2.3 Maximum $N_{i.max}$ and ultimate loads $N_{i.u}$ from the tests

The load N (Fig. 6) has been applied until either cracking occurs, the applied load decreases (Tab. 1, column FM), or the restrictions of the test setup limits the loading (Tab. 1, failure mode "Not apparent"). These attained loads were taken as maximum loads $N_{i.max}$ of the connections. A load drop with a subsequent increase or a load plateau did not occur in any test.

In addition to the maximum loads $N_{i.max}$, loads based on the deformation criterion from previous work (Lu et al. 1993), which limits the load at indentations $\Delta_{CF} = 3\% \cdot b_0$ on the chord flanges, have been determined from the local displacement measurement (Fig. 6). The loads $N_{i.u}$ so obtained are defined as ultimate loads $N_{i.u}$ of the connections.

2.4 Comparison of test results with mean resistance from previous work for chord face failure

The design resistance $N_{i.Rk}$ calculated with measured dimensions and yield strengths and using EN1993-1-8

231

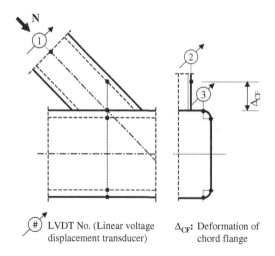

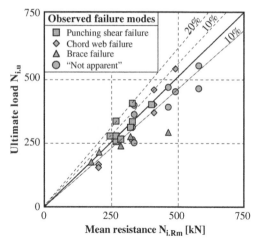

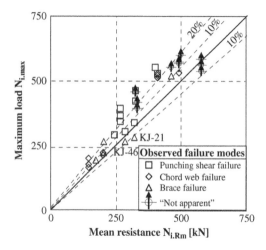

N

① LVDT No. (Linear voltage Δ_{CF}: Deformation of
   displacement transducer)   chord flange

Figure 6. Relative displacement Δ_{CF}.

Figure 8. Mean resistance $N_{i.Rm}$ vs. ultimate load $N_{i.u}$ (deformation limit).

Figure 7. Mean resistance $N_{i.Rm}$ vs. maximum load $N_{i.max}$.

gives chord face failure as the predominant failure mode for all tested specimens. The mean resistance $N_{i.Rm}$ from previous work (Wardenier 1982) is obtained by substituting 10.9 instead of 8.9 in the equation for chord face failure (Eq. 11), where nominal values are used. Therefore, in Figure 7 and Figure 8, this mean resistance $N_{i.Rm}$ is used for comparison.

For small thickness ratios $\tau \leq 0.64$ (Tab 1., KJ-21 and KJ-46), the mean resistance $N_{i.Rm}$ marginally overestimates the maximum load $N_{i.max}$. However, only for KJ-21, which has the smallest investigated thickness ratio of $\tau = 0.53$, this overestimation is 16%. This is because for both tests, effective width failure (KJ-21 – tearing in the tension brace; KJ-46 – plastic failure of the compression brace) has been observed. For brace angles different from $\Theta_i = 45°$ an overestimation can

be observed as well (KJ-37, 39 and 40), but since a failure was not observed for these tests, a subsequent load increase is to be expected. In Figure 7 they are therefore only given for completeness.

The comparison in Figure 7 is only given in order to show that the mean resistance from previous results is on the safe side for all tests except the two cases where brace failure occurs. Therefore, this is not used for further evaluations.

Since failure of hollow section joints is mainly related to the plastification of the chord face around the intersections of the braces and the chord (Lu et al. 1993), but this failure can not be detected visually in the tests, the final failure observed in the tests at a much later stage do not correspond with chord face failure. Therefore, chord face failure has been assumed to be the governing failure mode if the deformation limit is exceeded and no other failure mode is visually observed. However, the statistical evaluation for the individual failure modes is carried out separately in the following sections.

3   STATISTICAL TREATMENT OF THE TEST RESULTS

EN1990 (2002) offers a standardised procedure to investigate experimental test results with reference to existing design models. Since EN1993-1-8 includes design models for every observed failure mode, statistical evaluation of the observed failures with reference to their respective design models have been carried out.

Based on the respective design models, the design resistance $r_t = N_{i.Rk}$ can be calculated using measured dimensions and material properties. By comparison of these resistances with the maximum $r_e = N_{i.max}$ loads

Table 2. Coefficients of the basic variables from literature.

Basic variable X_i	CoV of basic variable V_{Xi}	Reference
b_0, b_i	$V_{b0} = V_{bi} = 0.005$	
t_0, t_i	$V_{t0} = V_{ti} = 0.05$	(Wardenier 1982)
β	V_β can be neglected	
f_{y0}, f_{yi}	$V_{fy0} = V_{fyi} = 0.059$	(Petersen 1993)

(used to investigate the observed failure modes) or ultimate loads $r_e = N_{i.u}$ (used to investigate chord face failure) the mean value of the correction factors b are obtained by linear regression analyses.

If the design models were exact and would consider all influencing parameters, all points would lie on the bisecting lines $r_e = r_t$. Due to variations of the test results $N_{i.max}$ and $N_{i.u}$ as well as of the basic variables X_i, the design resistances r_t and the experimental results r_e will always deviate. These deviations have to be examined, whether it is caused by errors in the test procedure or by the resistance functions.

3.1 Variation of the test results V_δ

With the error terms $\delta_i = r_{ei}/(b \cdot r_{ti})$ for the tests (in the statistical evaluation the index i is used to indicate the tests) and the assumption that the variance s_Δ^2 of $\Delta_i = \ln(\delta_i)$ is an estimation of the basic population σ_Δ^2, the coefficient of variation (CoV) of the error terms V_δ can be calculated:

$$V_\delta = \sqrt{e^{s_\Delta^2} - 1} \qquad (3)$$

3.2 Variations of the basic variables V_{Xi}

Due to the relatively small number of test specimens (number of test $n = 39 < 100$), it is not possible to determine the coefficients of variation of the basic variables V_{Xi} directly from the measurements. Therefore, they are taken from literature or reasonably estimated.

The CoV of the brace angles V_Θ could not be estimated by the measurements, because they were neither measured, nor taken from literature. Based on an assumption for the standard deviation $\sigma_\Theta = 1°$ at a mean $\mu_\Theta = 45°$ of the basic population, the CoV for the brace angles $V_\Theta = \sigma_\Theta/\mu_\Theta = 0.022$ is used.

Since no information is available on the coefficients of variation of the effective widths l_{ep} for punching shear (Eq. 1), l_e (Eq. 2) as well as of the buckling width l_b (Eq. 14) reduced by the buckling factor κ (obtained from the European buckling curve a), it is assumed to be 5% ($V_{lep} = V_{le} = V_{\kappa lb} = 0.05$). From this, the combined influence of the variations of the basic variables V_{rt} can be calculated:

$$V_{rt}^2 = \sum_{i=1}^{j} V_{Xi}^2 \qquad (4)$$

3.3 Determination of characteristic and design loads

The CoV of the design model V_r can be calculated by the combination of the CoV of the error terms V_δ and the basic variables V_{Xi}:

$$V_r^2 = \left(V_\delta^2 + 1\right)\left[\prod_{i=1}^{j}\left(V_{Xi}^2 + 1\right)\right] - 1 \qquad (5)$$

where $j =$ number of basic variables V_{Xi}.

Due to the relatively small number of test results, the statistical distribution of the uncertainties have to be considered for the evaluation as a Student-T-Distribution (EN1990:2002):

$$r_k = b \cdot g_{rt}(\underline{X}) \cdot e^{\left(-k_\infty \cdot \alpha_{rt} \cdot Q_{rt} - k_n \cdot \alpha_\delta \cdot Q_\delta - 0.5 \cdot Q^2\right)} \qquad (6)$$

$$r_d = b \cdot g_{rt}(\underline{X}) \cdot e^{\left(-k_{d.\infty} \cdot \alpha_{rt} \cdot Q_{rt} - k_{d.n} \cdot \alpha_\delta \cdot Q_\delta - 0.5 \cdot Q^2\right)} \qquad (7)$$

where r_k and $r_d =$ characteristic and design resistance; $b =$ mean values of the correction factor; $g_{rt}(\underline{X}) =$ design model with $\underline{X} =$ vector of basic variables; Q_{rt}, Q_δ and $Q =$ standard deviations (EN1990:2002); α_{rt}, α_δ and $\alpha =$ weighting factors (EN1990:2002); k_∞, k_n and $k_{d.\infty}$, $k_{d.n} =$ characteristic and design fractile factors from known values for the basic variables V_{Xi} with $n =$ number of tests (EN1990:2002).

3.4 Recalculation from mean to nominal values

The above evaluations are based on mean (measured) dimensions and yield strengths. A design model usually uses nominal dimensions and characteristic values of the yield strength f_y to determine design loads $N_{i.Rd}$.

In accordance with Wardenier (1982), the assumption that the mean dimensions are equal to the nominal dimensions is made.

Based on a Normal Distribution, the mean value of the yield strength $\bar{f}_y$ could be reduced to the characteristic value f_y by using the CoV of the yield strength V_{fy}:

$$f_y = \bar{f}_y - k \cdot \sigma_{fy} = \bar{f}_y \cdot \left(1 - k \cdot V_{fy}\right) = 0.882 \cdot \bar{f}_y \qquad (8)$$

where $\bar{f}_y =$ mean yield strength of the basic population; $\sigma_{fy} =$ variance of the yield strength of the basic population; $k =$ fractile factor taken as $k = 2$, which is equal to a probability $P(f_{y.m} \leq f_y) = 2.28\%$ (Petersen 1993) with $f_{y.m} =$ measured yield strength.

From the statistical evaluations of the test results and the recalculation using the characteristic value of the yield strength, coefficients can be specified to reduce the design resistance r_t (using measured dimensions and material properties) to the characteristic $\xi_k \cdot r_t$ (Eq. 9) or design $\xi_d \cdot r_t$ (Eq. 10) levels.

$$\xi_k = \frac{b \cdot e^{\left(-k_\infty \cdot \alpha_{rt} \cdot Q_{rt} - k_n \cdot \alpha_\delta \cdot Q_\delta - 0.5 \cdot Q^2\right)}}{0.882} \qquad (9)$$

233

$$\xi_d = \frac{b \cdot e^{\left(-k_{d.\infty} \cdot \alpha_{rt} \cdot Q_{rt} - k_{d.n} \cdot \alpha_{\delta} \cdot Q_{\delta} - 0.5 \cdot Q^2\right)}}{0.882} \qquad (10)$$

The safety factor $\gamma_m = \xi_k / \xi_d$ obtained for the reduced design resistance $\xi_d \cdot r_t$ with reference to the test results r_e can be specified additionally.

4 STATISTICAL ANALYSES

The design resistance $r_t = N_{i.Rk}$ used in the statistical evaluations is determined from the design model for the individual failure modes considered, using measured geometrical dimensions and yield strengths.

To investigate chord face failure, tests which give ultimate loads $N_{i.u}$ obtained from the limitation of the chord flange deformations $\Delta_{CF} \leq 3\% \cdot b_0$ were used for the evaluation. Those tests where the maximum load was reached before the load is limited by the deformation criteria have been excluded from the evaluation. The final failure observed in the tests, which is well beyond maximum load, is therefore not considered in the evaluation for chord face failure.

The statistical evaluation with reference to the observed failure modes (effective width, compression brace, punching shear and chord web failure) were carried out using maximum loads $N_{i.max}$ from the tests with the considered failure mode.

4.1 Evaluation for chord face failure

The design resistance $r_t = N_{i.Rk}$ has been determined for the test specimens based on the design formula for chord face failure (Eq. 11).

$$r_t = \frac{8.9 \cdot f_{y0} \cdot b_0^{0.5} \cdot t_0^{1.5}}{2^{1.5} \cdot \sin \Theta_i} \left(\frac{b_i + h_i}{2 \cdot b_0} \right) \qquad (11)$$

From Figure 9, it can be seen that the chord face failure resistance given in EN1993-1-8 (Eq. 11) has to be reduced to $\xi_d = 0.71$, since a low mean value of the correction factor $b = 0.84$ and a relatively high variation for the error terms $V_\delta = 0.18$ has been determined.

Since the present semi-empirical design formula for chord face failure (Eq. 11) is based on maximum loads $N_{i.max}$ and the valid parameter range had been calibrated for smaller chord slendernesses $2\gamma \leq 35$ and larger gap sizes $g \geq g_{min}$ the reduction factor of $\xi_d = 0.71$ (smaller than 1.0) can be expected.

4.2 Evaluation for punching shear failure

For the tests which failed by punching shear (Tab. 1, failure mode "PS") the design resistance $r_t = N_{i.Rk}$ according to the design formula for punching shear failure in EN1993-1-8 (Eq. 12) has been used.

$$r_t = \frac{f_{y0} \cdot t_0}{\sqrt{3} \cdot \sin \Theta_i} \cdot l_{ep} \qquad (12)$$

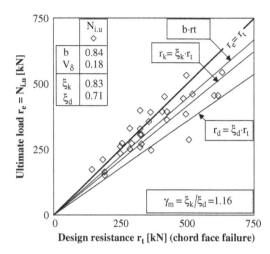

Figure 9. $N_{i.u}$ from tests vs. design resistance $N_{i.Rk}$ for chord face failure.

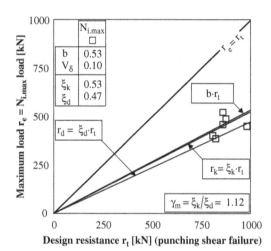

Figure 10. $N_{i.max}$ from tests vs. design resistance $N_{i.Rk}$ for punching shear failure.

where l_{ep} = eff. length for punching shear (Eq. 1).

Since a low mean value of the correction factor $b = 0.53$ is obtained, a high reduction factor of the design resistance $\xi_d = 0.53$ is necessary (Fig. 10). This high reduction can be explained by the reduced effective length $l_{ep.red}$, due to the increased stiffness of the gap region for small gap sizes.

However, the estimation of the coefficient $V_\delta = 0.10$ of the errors is small, so that it can be assumed that the design model of EN1993-1-8 can only be used for joints in the investigated parametric range, if the effective length $l_{ep.red}$ is modified accordingly.

234

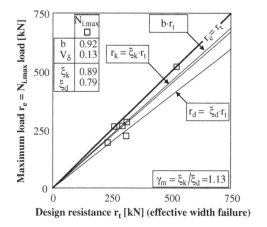

Figure 11. $N_{i.max}$ from tests vs. design resistance $N_{i.Rk}$ for effective width failure.

4.3 Evaluation for brace failure

For the design resistance $r_t = N_{i.Rk}$ from the tests which failed by tearing in the tension brace or plastification of the compression brace (Tab. 1, failure mode "EW" or "CB") the corresponding design model in EN1993-1-8 has been used (Eq. 13).

$$r_t = f_{yi} \cdot t_i \cdot l_e \qquad (13)$$

where $l_e =$ eff. length (Eq. 2).

For effective width failure, the mean value of the correction factor $b = 0.92$ is obtained. Because the coefficient of the errors $V_\delta = 0.13$ is relatively small, the evaluation gives a reduction $\xi_d = 0.79$ of the design model for effective width failure (Eq. 13).

The influence of the gap size on the effective length l_e is not as distinct as observed for punching shear failure. Since the crack propagates parallel to the chord face and not perpendicular to the brace axis as idealised for the formula, the resistance increase with the brace angle Θ_i and the reduction as observed is not as severe as for punching shear.

4.4 Evaluation for chord web failure

Since chord web failure (Tab. 1, failure mode "CW") is not considered for K-joints in EN1993-1-8, the design resistance $r_t = N_{i.Rk}$ is obtained by using the design model for Y-joints in EN1993-1-8 (Eq. 14).

$$r_t = \kappa \cdot \frac{t_0 \cdot f_{y,0}}{\sin \Theta_i} \cdot \left(\frac{2 \cdot h_i}{\sin \Theta_i} + 10 \cdot t_0 \right) = \kappa \cdot \frac{t_0 \cdot f_{y,0}}{\sin \Theta_i} \cdot l_b \qquad (14)$$

where $\kappa =$ reduction factor for buckling obtained from European buckling curve a; $l_b =$ width of the Euler strut.

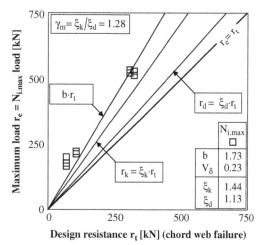

Figure 12. $N_{i.max}$ from tests vs. design resistance $N_{i.Rk}$ for chord web failure.

For chord web failure, a high mean value of the correction factor $b = 1.73$ is obtained, but since the coefficient of the errors $V_\delta = 0.23$ is high, the necessary increase of the design resistance $\xi_d = 1.13$ is moderate.

Especially because of the high coefficient of the errors, it can be assumed that the design model for web failure for Y-joints cannot be used for K-joints without basic modifications.

5 CONCLUSIONS

From the experimental observations and the statistical evaluations of the test results the following conclusions can be drawn for K-joints with slender chord sections $2\gamma \geq 35$ and small gap sizes $g < g_{min}$:

− The observed failure modes and those obtained from EN1993-1-8 do not correspond, which is due to the different γ_m-factors used for the design formulae of the various failure modes (Wardenier 1982).
− Since chord face failure cannot be observed in the tests visually, this failure mode has to be assumed on the basis of limiting the deformation of the chord face to $\Delta_{CF} = 3\% \cdot b_0$.
− The statistical evaluation of the chord face failure based on ultimate loads $N_{i.u}$ shows that a reduction of the present design formula (Eq. 11) of $\xi_d = 0.71$ is required. Since the present design formula for chord face failure is based on maximum loads $N_{i.max}$ and the valid parameter ranges are limited to smaller chord slendernesses and larger gap sizes, the reduction factor obtained is as expected.
− For punching shear failure, a high reduction factor $\xi_d = 0.47$ has to be applied to the design equation in

EN1993-1-8 (Eq. 12). For small gap sizes $g < g_{min}$ the stiffness of the gap region results in reduced effective lengths $l_{ep.red}$, so that punching shear is more likely in such cases. But due to the small coefficient of the errors $V_\delta = 0.10$, the design formula in EN1993-1-8 can only be used for joints in the investigated parametric ranges, if the effective length for punching shear l_{ep} is modified accordingly.

- For effective width failure, the influence of the reduced effective length $l_{e.red}$ can also be observed, but since the crack does not propagate perpendicular to the brace axis (Fig. 3), this influence is only partially compensated. Therefore, a smaller necessary reduction of $\xi_d = 0.79$ to the design formula (Eq. 13) for effective width failure according to EN1993-1-8 is necessary. As for punching shear failure, the design model from EN1993-1-8 can only be used, if the effective length l_e is appropriately modified.

- Since chord web failure is not considered in EN1993-1-8 for K-joints, the design model for Y-joints was used in the statistical evaluation. Because the coefficient of the errors $V_\delta = 0.23$ is high, the use of this design model cannot be proposed for K-joints in the parametric ranges investigated in this paper without basic modifications.

- The use of the reduction factors in its present form would result in a sudden drop in design resistances $N_{i.Rd}$ between joints in the parametric range of EN1993-1-8 and the joint geometries investigated in this paper. The factors will therefore be modified in an ongoing work to avoid this where possible.

- With regard to the FE parameter study already performed but not reported here, it has to be checked from the strains in the gap region, whether the joints have enough deformation capacity, or if cracking is indicated before the load is limited by the deformation criteria (for example see Tab. 1; KJ-7).

6 FUTURE WORK

Based on the results presented in this paper, the results of the FE parameter studies that have already been performed, will be evaluated. With these results, suggestions for a new valid range for the slenderness and gap will be given with appropriate modifications to the design rules given in EN1993-1-8. This will include modifications to existing formulae for chord face, punching shear, chord web and effective width failure. These modifications, supported by the statistical evaluations, are likely to extend the existing rules, for instance by adding a gap function as already used for joints using CHS to avoid the sudden drop in design resistances. This work forms part of the Doctor Thesis of the lead author.

ACKNOWLEDGEMENTS

The authors would like to thank the ECSC (European Community for Steel & Coal) for financial support to carry out the investigations, Rautaruukki Metform, Finland and Voest Alpine, Austria for the supply of materials.

REFERENCES

EN1990:2002-10. 2002. *Basis of structural design*. Brussels: European Committee for Standardisation

EN1993-1-8:2005-7. 2005. *Design of steel structures – Design of joints*. Brussels: European Committee for Standardisation

Fleischer O. & Puthli R. 2003. RHS K joints with b/t ratios and gaps not covered by Eurocode 3. In M.A. Jaurrieta & A. Alonso & J. Chica J.A. (eds), *Proc. 10th intern. symp.*, 207–215, Madrid, 18–20 September 2003. Rotterdam: Balkema

Gulvanessian H. & Calgaro J.-A. & Holický M. 2002. *Designers' Guides to the Eurocodes – Designer Guide to the EN1990* Eurocode: Basis of Structural Design. London: Thomas Telford Publishing

Liu D.K. & Yu Y. & Wardenier J. 1998. Effect of boundary conditiones and chord preload on the strength of RHS uniplanar gap K-joints. In Y.S. Choo & G.J. van der Vegte (eds), *Proc. 8th intern. symp.*, 223–230, Singapore, 26–28. August 1998. Rotterdam: Balkema

Lu L.H. & de Winkel G.D. & Yu Y. & Wardenier J. 1993. Deformation limit for the ultimate strength of hollow section joints. In P. Grundy & A. Holgate & B. Wong (eds), *Proc. 6th intern. symp.* 341–347, Melbourne, 14–16 December 1994. Rotterdam: Balkema

N.N. 2003. *Design rules for cold formed structural hollow sections*. Final Report, Research programme of the Research Fund for Coal and Steel

Packer J.A. & Wardenier J. & Kurobane Y. & Dutta D. & Yeomans N. 1992. *Design guide for hollow sections (RHS) joints under predominantly static loading*. CIDECT (ed.). Köln: Verlag TÜV Rheinland GmbH

Petersen C. 1993. *Stahlbau – Grundlagen der Berechnung und bauliche Ausbildung von Stahlbauten*. Braunschweig: Vieweg Verlag

prEN10219-2:2004-2. *Cold formed welded structural hollow sections of non-alloy and fine grain steels*. Brussels: European Committee for Standardisation

prEN1993-1-12:2004-9. *Design of steel structures – Additional rules for the extension of EN 1993 up to steel grades S 700*. Brussels: European Committee for Standardisation

Puthli R.S. 2002. Hohlprofilkonstruktionen im Geschossbau – Ausblick auf die europäische Normung. In Ulrike Kuhlmann (ed.), *Stahlbau Kalender*, 549–679. Düsseldorf: Ernst&Sohn Verlag

Wardenier J. 1982. *Hollow section joints*. Delft: University Press

Wardenier J. 2001. *Hollow Sections in Structural Applications*. CIDECT (ed.)

Tubular Structures XI – Packer & Willibald (eds)
© 2006 Taylor & Francis Group, London, ISBN 0-415-40280-8

Finite element analysis of slotted end tubular connections

G. Martinez-Saucedo, J.A. Packer & S. Willibald
University of Toronto, Toronto, Canada

X.-L. Zhao
Monash University, Victoria, Australia

ABSTRACT: A total of eight slotted end tubular connection specimens has been previously tested under quasi-static tension and compression loading using three connection types commonly found in practice. A parametric finite element analysis was then undertaken, based on finite element models of these connections, where the responses were verified with the test results. In the finite element modelling, a non-linear time step analysis was performed considering non-linear material properties and with 8-noded solid elements. A maximum equivalent strain was used as the failure criterion with the activation of a "death feature" of the elements. Differences were found between the finite element connection strength and the capacity as predicted by design provisions. Furthermore, these results show that a gradual transition between several failure modes takes place as the weld length increases. Also, the likelihood of developing the full efficiency of the tube net cross-sectional area, if a minimum weld length equal to the distance between welds is used, is illustrated.

1   INTRODUCTION

Gusset plate connections represent one of the easiest methods to connect circular hollow section (CHS) members and examples of this connection type can be found in many steel buildings. For gusset plate to CHS end connections, the most common fabrication details are shown in Figure 1. Fabrication details type A and B require the CHS to be slotted while type C requires further detailing of the gusset plate. Detail A has no return weld at the end of the connection, allowing bigger fabrication tolerances for the slot and thus an easier fit-up.

The research described herein has studied the tensile and compressive behaviour of such CHS connections under quasi-static loading. In one failure mode, circumferential tensile fracture takes place under the influence of shear lag. This phenomenon occurs due to the uneven stress distribution around the circumference of the CHS as a result of the load transfer at the connection. The stresses peak at the points where the CHS is connected to the gusset plate (or actually at the beginning of the weld) and become less as the distance to the weld increases. Therefore, the unconnected circumference of the tube only contributes in part to the capacity of the member. A further tension loading failure mode, tear-out failure along the weld, is also possible.

Research on tubular member to gusset plate connections under static loading started in the early 1990s

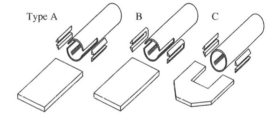

Figure 1.   Fabrication details of experiments and FE models.

(British Steel 1992) with an experimental study on gusset plate connections to CHS, square hollow sections (SHS) and rectangular hollow sections (RHS). An experimental and numerical program on slotted CHS by Cheng et al. (1996), Cheng et al. (1998) and Cheng & Kulak (2000) showed that design codes did not accurately represent the behaviour of slotted CHS connections. Moreover, it was concluded that the shear lag effect was not critical for CHS if the connection length was longer than 1.3 times the tube diameter. Based on the study of slotted SHS and RHS connections, Korol et al. (1994) and Korol (1996) concluded that for connection lengths (L_w) greater than or equal to the distance between the welds (w), shear lag would not affect the connection strength and the maximum load would be the member tensile fracture strength. Therefore, a slightly modified approach for the calculation of the effective net section area was proposed. Recently, studies on slotted CHS connections made of very high strength steel tubes (Ling et al. 2004) and on slotted RHS connections by Wilkinson et al. (2002) have also indicated that the current provisions are too conservative.

For tear-out (block shear) and shear lag, North-American specifications (AISC 2005 & CSA 2001) have gone through numerous modifications (Geschwindner 2004). Meanwhile, Eurocode3 (CEN 2005) does cover block shear, but it only addresses shear lag on bolted connections.

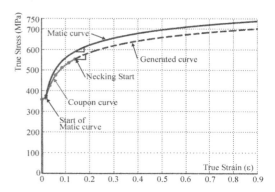

Figure 2. Generation of uniaxial $T\sigma$-$T\varepsilon$ curve.

2 EXPERIMENTAL PROGRAM

The experimentation consisted of six tension tests on gusset plate connections to CHS as well as two compression tests (Martinez-Saucedo et al. 2005). The specimens varied in fabrication detail (Fig. 1) as well as weld length (L_w). All experiments were performed to large-scale and involved duplicate connections at either end of each test specimen. The CHS was a cold-formed 168×4.8 mm with a specified minimum yield stress of 350 MPa. Each specimen was equipped with 10 linear strain gauges to measure the longitudinal strains in the connection region. The overall displacement was monitored by four LVDTs (Linear Variable Differential Transformers) which measured the displacement between the tube centre and the gusset plate.

3 FINITE ELEMENT MODELLING

3.1 Material properties

A non-linear material, multi-linear true stress-true strain ($T\sigma$-$T\varepsilon$) curve, was used to describe the gusset plate, CHS and weld material properties. The generation of the $T\sigma$-$T\varepsilon$ curve based on the engineering stress–strain relations is only suitable prior to the development of necking in the coupon test. Afterwards, the strains are concentrated in the neck region and the stress distribution there changes from a simple uniaxial to a complex triaxial case. In order to generate the $T\sigma$-$T\varepsilon$ curve for a rectangular coupon test in the post-necking region, a method suggested by Matic (1985) has been adapted for use here.

During the laboratory testing of the coupons, the engineering stress–strain relationships were acquired before the coupon tests developed a neck. Afterwards, the clip gauge had to be removed from the coupons, but the load and maximum elongation at rupture were determined for each coupon test.

To complete the material $T\sigma$-$T\varepsilon$ curve, Matic's curve is generated with a starting point corresponding to a zero plastic strain on the coupon test curve data. The generation of the $T\sigma$-$T\varepsilon$ curve in the post-necking region was thus calculated starting from the

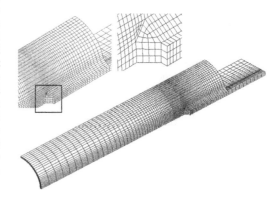

Figure 3. Typical FE model.

necking point (Fig. 2), following the change in the tangent modulus given by Matic's curve.

The best rate of change of the tangent modulus was then determined by an iterative process. For a generated $T\sigma$-$T\varepsilon$ curve, a FE model of the gauge region (emulating the material's coupon test) was analyzed and its load-deformation response was compared with the coupon test response data. This process was repeated until the load and displacement at fracture from the coupon FE model corresponded to the test result.

3.2 Connection modelling

Symmetric boundary conditions in the planes of symmetry allowed the modelling of only one eighth of the test specimens (Fig. 3).

In addition to the symmetric boundary conditions, the nodes at the end of the tube were fixed and the total load acting on the connection was calculated from these nodes. For the meshing process, a refinement of the mesh size was made in areas prone to shear lag. Three elements were used through the tube thickness in all models. In order to ensure load transmission only through the weld, a small gap was left between the plate and the tube elements (this emulates the typical slot oversized for constructional purposes). In all models the tube material properties were used for the welds.

Using the FE program ANSYS, element type SOLID45 was chosen to model the tube, gusset plate and weld materials. This element is defined by eight nodes, each node having three translational degrees of freedom, with large deflection and large strain capabilities.

3.3 Analysis considerations

During the FE analysis of the connections, a non-linear time step analysis was performed by applying incremental displacements to the nodes located at the gusset plate end. This emulates the displacement-control loading throughout the connection tests. The full Newton Raphson method and frontal equation solver were used. Non-linear material properties were considered and geometrical non-linearities were taken into account by allowing large deformations. A uniform reduced integration with hourglass control was applied and shape changes (i.e. area, thickness) were considered as well.

The maximum equivalent strain (ε_{ef}) used to trigger the activation of the elements' "death feature" was obtained by empirical correlations. The use of a ε_{ef} equal to that used in the coupon FE models required considerable deformations to generate large strains in the elements of the FE models. This lack of direct applicability has been related to the difference in the material boundary conditions that exists between the tensile coupons and the connections. Therefore, all connections were analyzed using several ε_{ef} values and the best correlation between experimental and FE analysis results was found using $\varepsilon_{ef} = 0.6$.

3.4 Evaluation of FE models against experiments

The FE models predicted the failure mode and the location of fracture in all connections, and strains recorded in the test specimens also presented good correlation with those in the FE models. The FE analysis results are shown in Table 1.

The slotted gusset plate connections produced the largest distortions in connection shape. Therefore, a limit on the distortion of the tube cross-section is proposed herein to limit the "ultimate capacity" of these connections. A limit of 0.03 D was used, as is popularly recommended for tubular structures (Lu et al. 1994) and now adopted by the International Institute of Welding. The slotted gusset plate connections showed a considerable difference between the distortion limit and the maximum strength. Hence, a distortion limit for the ultimate connection capacity may need to be imposed.

Table 1. Ultimate capacity for tests and FE models.

Spec. Type and No.	Failure Mode	Distortion Limit (kN) N_{uFE-D}	Ultimate Capacity (kN)		
			Test load N_{ux}	FE model N_{uFE}	N_{ux}/N_{uFE}
A1	CF-TO		1031	978	1.06
A2	CF		1151	1130	1.02
A3C	LB		−1145	−1067	1.07
B1	TO		1087	1080	1.01
B2	CF		1211	1216	1.00
C1	CF	858	1107	1081	1.02
C2	CF	902	1196	1235	0.95
C3C	LB		−896	−800	1.09

CF = Circumferential failure; TO = Tear-out failure and LB = Local buckling.

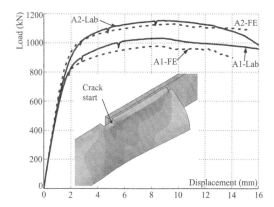

Figure 4. Load-deformation response for connections Type A.

3.4.1 Slotted CHS – slot end not filled (specimens A1 & A2)

Throughout the analyses, the FE models presented a high strain concentration at the weld start which caused yielding and consequent fracture of the tube material. The same failure mode had been previously found for the test specimens. This overall behaviour may be clearly associated with the shear lag phenomenon acting in the connections (Circumferential Tensile Fracture mode). The load-deformation responses for FE models and specimens are shown in Figure 4.

3.4.2 Slotted CHS - slot end filled (weld return) (specimens B1 & B2)

Considering that connection failure started at the weld return toe region and there the toe cracking had its origin in the Heat Affected Zone (HAZ) and continued propagating into the base metal, a FE model was generated considering a HAZ effect and incorporated the change in strength and ductility of the weld as a function of the loading angle (Kulak & Grondin 2002).

In order to model the return weld properties, the engineering stress–strain curve was scaled to describe the properties of a weld loaded at an orientation of 90° to the weld axis. From this new data, a Tσ-Tε curve was generated and then applied to the elements in the weld return region.

A fine mesh was generated in front of the weld return for the HAZ in the tube. During its generation, the tube's material properties were applied but low ductility controlled the material fracture. Thus, the maximum strain in the HAZ was defined to be equivalent to that in the weld oriented at 90°. This low ductility triggered the creation of cracks in the HAZ producing an overall connection behaviour as observed during the test. In general, the FE models described the first stages of the failure modes but the analysis terminated due to excessively high distortion in the dead elements in the weld return region. The von Mises strain distributions in the FE models reproduced those in the specimens during the tests. Moreover, the FE models described a similar load-deformation response curve and they reached a comparable maximum load (Table 1). Finally, good correlation was found between the strain gauge readings of the specimens and strains in the FE models.

3.4.3 Slotted gusset plate to CHS (specimens C1 & C2)

After the elastic response the slotted gusset plate tended to open (under tension loading), inducing a distortion in the tube shape. Figure 5 shows the load-deformation responses.

Although the connections continued to sustain higher loads, the connections were very distorted by the final stages. Thus, the connection capacity was governed by the imposed distortion limit (which was less than the ultimate strength achieved). Fracture in the tubes occurred at the weld end. The failure mode and FE strains at the strain gauge locations were consistent with the results of the test specimens.

3.4.4 Connections under compression load (specimens A3C & C3C)

For these tests, local buckling in the tubes defined the ultimate strength. The load-deformation responses are shown in Figure 6.

Once specimen A3C reached its maximum load the tube developed a local buckle in the slot region, then unloaded in a stable manner. Contact between the plate and the slot end was not achieved as this requires huge deformations. Specimen C3C developed a local buckle across the gross cross-sectional area. The slotted plate bowed inwards exacerbating the tube's local instability and negatively affecting the connection behaviour, but was well captured by the FE model. The FE results found good correlation with the test results.

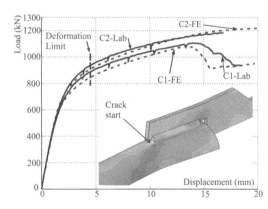

Figure 5. Load-deformation response for connections Type C.

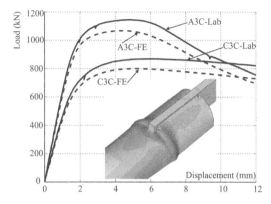

Figure 6. Load-deformation for connections under compression.

4 PARAMETRIC FE ANALYSIS

A parametric analysis was then undertaken using these validated FE models to study the influence of parameters such as: the weld length (L_w), the eccentricity reduced by half the flange-plate thickness ($\bar{x}'$) and the tube diameter-to-thickness ratio (D/t), on the connection strength (Fig. 7).

In total, a further 322 FE connections were modelled. During this parametric analysis, the dimensioning of gusset plates and weld legs was made so as to avoid failure modes other than Tear Out Failure (TO) or Circumferential Tensile Fracture (CF of the tubes. Furthermore, the material properties used throughout these analyses corresponded to the properties used previously during the modelling of the specimens, as these were deemed to be realistic. Several FE models were generated using L_w/w ratios ranging from 0.40 to 1.50. In this range the FE models were able to reproduce pure TO failure, a combination of TO and CF, CF influenced by shear lag and pure CF without shear lag. In addition to this, several D/t ratios based on a CHS with

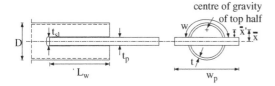

Figure 7. Parameters considered during the parametric analysis.

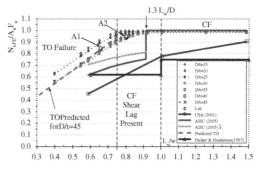

Figure 8. Parametric FE analysis results and experimental results for connection Type A ($N_{uFE}/A_n F_u$).

diameter of 180 mm were used in the generation of the FE models. These D/t ratios corresponded to the range commonly used in practice, such as: 45, 40, 35, 30, 25, 20 and 15.

4.1 Parametric analysis results of slotted CHS – slot end not filled

Failure in this connection type was mainly governed by the growth of a crack in the tube material near the beginning of the weld.

In order to determine the efficiency of the net cross-sectional area of the CHS, the connection strength (N_{uFE}) calculated during the parametric analysis has been normalized with respect to $A_n F_u$. Furthermore, for comparison with current codes or specification recommendations, all resistance factors or partial safety factors were set equal to one.

The results from the parametric analysis show a gradual transition between the TO failure and CF. The transition between these failure modes occurred at a ratio of L_w/w near 0.75 (Fig. 8). However, a lower ratio was found for FE models with a low D/t ratio, which suggests the existence of a correlation between these two parameters.

For FE models with a L_w/w ratio ranging from 0.40 to 0.80, TO failure was found as the governing failure mode. Moreover, these results agreed consistently with the formulae used in AISC (2005) and CSA (2001) to calculate the block shear strength (including weld dimensions). The TO prediction for a tube with D/t = 45 is shown in Figure 8.

The overall results indicate that the full efficiency of the net cross-sectional area can be developed if a ratio of $L_w/w \geq 1.0$ is utilized. The design provisions of AISC recommend the use of a variable efficiency factor for L_w/D ratios <1.3 but for $L_w/D \geq 1.3$ AISC deems that the full section capacity can be achieved. The FE parametric results support this latter rule. However, a considerable difference between the parametric analysis results and AISC took place for L_w/D ratios <1.3. The use of $\bar{x}'$ generally improved the AISC prediction. Nevertheless, this variable efficiency factor is only applicable in the range of L_w/w ratios from 0.75 to 0.91 because TO governs for smaller L_w/w ratios. AISC gives no bounds on the TO failure mode check so this would always be performed in conjunction with

the CF check. The efficiency factor recommended by AISC, for application to the CF limit state check, provides a better solution than CSA (2001) and Packer and Henderson (1997).

Even though the FE models with $L_w/w > 1.0$ developed 100% of $A_n F_u$, the governing failure mode continued to be CF. The strains in the tube material away from the connection remained in the elastic range and the overall deformation was concentrated at the slot region. The reason for this behaviour has been attributed to the high F_y/F_u ratio of the tube material. On FE models with $L_w/w > 1.0$, this high ratio produced an average $A_n F_u/A_g F_y$ ratio near 0.96. Generally, the presence of a low ratio will impede the possibility of developing the gross-section tensile yield strength, thus confining the member deformation in the slot region. While this is still acceptable for statically-loaded connections it has important implications for these connections under cyclic (seismic) loading.

4.2 Parametric analysis results of slotted CHS – slot end filled (weld return)

The behaviour of these connections was generally governed by the formation of a crack in the weld return region which is in turn dependant on the weld length. The fabrication of the weld return ensured that the tensile stress area became equal to the gross cross-sectional area ($A_n = A_g$). The connection strength (N_{uFE}) has been normalized with respect to $A_n F_u$ in Figure 9, where it can be seen that the maximum strength achieved was still only about 0.95 $A_n F_u$, even for very large L_w/w values.

Even though a net section fracture was avoided here, either TO failure or CF through the gross area remained as possible failure modes, with the latter being influenced by shear lag in a small parametric range. The transition between TO and CF occurred for FE models having L_w/w ratios ranging from 0.70 to 0.80. Moreover, this transition and the achievement of the maximum efficiency of the gross cross-sectional

241

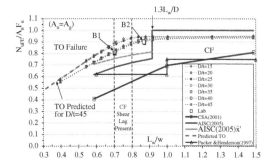

Figure 9. Parametric FE analysis results and experimental results for connection Type B ($N_{uFE}/A_n F_u$).

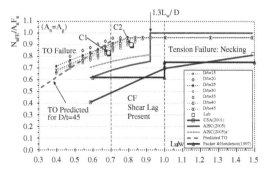

Figure 10. Parametric FE analysis results and experimental results for connection Type C ($N_{uFE}/A_n F_u$).

area were influenced by the D/t ratio. The elimination of shear lag was observed for FE models having a small D/t and L_w/w close to 0.80, but larger L_w/w values applied for thinner tubes. The vertical lines on Figure 9 show only the lower limits for these transitions.

AISC's efficiency factor of 1.0 for connections with L_w/D ratios ≥ 1.3 agrees with the parametric analysis results. However, for L_w/D ratios <1.3 a sudden drop in the connection efficiency is given by AISC whereas there is a gradual change shown by the FE models. The efficiency factors recommended by CSA (2001) and Packer and Henderson (1997) are excessively conservative.

Gross cross-sectional area yielding was achieved for FE models having $L_w/w > 1.0$. However, the amount of deformation sustained by the tubes was limited and their failure was determined by cracking in the weld return region.

4.3 Parametric analysis results of slotted gusset plate to CHS

Although this connection type avoided a reduction in the tube gross cross-sectional area, the strain concentration in the weld region triggered the growth of a crack there, introducing an undesirable failure mechanism. Figure 10 shows the test results and the FE connection tensile strength (N_{uFE}) normalized with respect to $A_n F_u$ (where $A_n = A_g$). Even though this strain concentration is basically determined by the connection weld length (mainly at the elastic response stage), an influence of the gusset plate dimension has been found in the course of this parametric analysis. For FE models with low L_w/w ratios the connection strength was principally controlled by TO failure and the connection strengths were close to the values predicted by design provisions.

The transition between TO and CF occurred in FE models at $L_w/w = 0.70$. For L_w/w ratios between 0.7 and 1.0 the connection strength was limited by shear lag and bowing out of the gusset plate.

For $L_w/w > 1.0$ cracking in the weld region disappeared at ultimate load allowing the formation of a neck at the tube mid-length. This shows that the gradual reduction of strains in the weld region allows large deformations away from the connection region. Despite the generation of a neck for long connections, the ultimate connection strength never exceeded 96% of $A_n F_u$. This has been related to the convergence tolerance during the FE solution process. In most cases, the achievement of the connection ultimate capacity was associated with surpassing the suggested tube distortion limit.

4.4 Connections under compression load

4.4.1 Parametric analysis results of slotted CHS – slot end not filled

The maximum load of these connections was governed by local buckling in the tube slot region. The formation of this local buckle was determined by the tube D/t ratio and the strain concentration at the beginning of the weld, the latter being due to shear lag. The connection strength (N_{uFE}) calculated during the parametric analyses has been normalized with respect to $A_g F_y$ in Figure 11.

For FE models having $L_w/w > 0.92$, the maximum efficiency was determined only by the tube D/t ratio and the slot length since shear lag had no influence.

In most of the FE models, the maximum load was close to the tube distortion limit load and once this limit was exceeded rapid distortion of the tube shape governed the tube behaviour. Thus, no significant difference between the maximum load and the load corresponding to the distortion limit was found.

4.4.2 Parametric analysis results of slotted gusset plate to CHS

For FE models of connections fabricated with a slotted gusset plate, the failure mode of local buckling of the tube gross cross-sectional area was influenced by several factors: a strain concentration at the beginning

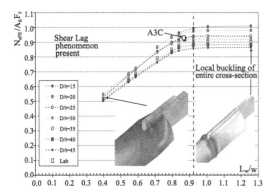

Figure 11. Parametric FE analysis results and experimental results for connection Type A under compression (N_{uFE}/A_gF_y).

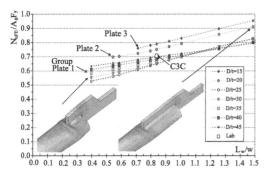

Figure 12. Parametric FE analysis results and experimental results for connection Type C under compression (N_{uFE}/A_gF_y).

of the weld (due to shear lag) which was determined by the L_w/w ratio; the bowing inwards of the gusset plate exacerbating the tube's local stability which is related to the plate's flexural stiffness and load applied; and the tube thickness which defines the D/t ratio and hence the tube local buckling load. The connection strength (N_{uFE}) has been normalized with respect to A_gF_y (Fig. 12).

The D/t ratios for the CHS used herein always corresponded to at least a non-slender section CSA (2001), meaning that the cross-section was sufficiently stocky to avoid elastic local buckling.

In an attempt to reduce the number of parameters having an influence on the FE models' behaviour, a constant plate cross-section at the slot region was tried throughout this parametric analysis. However, this was possible only for FE models with D/t ratios from 25 to 45 (Plate 1). With FE models having a thicker tube, a bigger plate was used (Plates 2 and 3). Due to these differences, the FE analysis results are collated according to their plate properties.

Figure 12 indicates that the factors affecting the connection efficiency continue to be present even for large weld lengths.

Throughout these analyses, the attainment of the maximum load occurred just after surpassing the distortion limit (3% D).

5 CONCLUSIONS

- For slotted CHS connections, without a weld return and loaded in tension, the use of a weld length-to-distance between welds ratio of $L_w/w > 1.0$ allowed the attainment of 100% efficiency of the tube net area (A_nF_u). The inclusion of a *weld return* provides the possibility to eliminate net area fracture and transfer this deformation away from the connection. In general this objective was accomplished for connections with a weld return, as they were capable of attaining their gross-section yield strength (100% A_gF_y).

- In slotted gusset plate connections to CHS loaded in tension, with $L_w/w > 1.0$, the decrease in strain concentration at the weld region allowed the creation of a neck away from the connection. However, the associated deformations in the tube cross-section shape suggest the imposition of a distortion limit to restrict the available connection capacity.

- For the connections loaded under compression, the parametric analysis results have shown the possibility to diminish the influence of shear lag on slotted CHS connections with a ratio of $L_w/w > 0.92$. The gross cross-sectional area efficiency here ranged from 86% to 100% of A_gF_y (this range is due to the net area cross-sectional properties at the slot region).

The behaviour of the slotted gusset plate connection type was less promising and also confirmed the negative effect of gusset plate deformation prevalent with this connection type. However, from these analyses it can be noted that the use of a large gusset plate (with a large moment of inertia) can improve the gross cross-sectional area efficiency, as the FE models with a larger gusset plate reached an efficiency close to 100% of A_gF_y.

- For connections under tension loading, the current design provisions for "block shear" tear out and circumferential tension fracture have been evaluated against the experimental research and parametric analysis results. For the treatment of shear lag, the American Specification (AISC 2005) provides the closest solution to the trend followed by these results. Furthermore, the accuracy of this design method can be improved by reducing the eccentricity of the top half of the connection ($\bar{x}$) by half of the gusset plate thickness (i.e., by using $\bar{x}' = \bar{x} - t_p/2$). Despite this improvement this preferred model is still over-conservative and not representative of the true connection behaviour. For the limit state of block shear failure, the Canadian (CSA 2001) and American AISC specifications use the same design model. However, their application range is not clear and the parametric results have shown that this application range can vary depending on several factors. Based on all these results, a unique and comprehensive design method (covering tear out failure and circumferential tensile fracture) to predict the connection strength, applicable over a full range of weld lengths and tube geometries, is required.

ACKNOWLEDGEMENTS

Financial support for this project has been provided by CIDECT (Comité International pour le Développement et l'Etude de la Construction Tubulaire) Programme 8G, NSERC (Science and Engineering Research Canada) and CONACYT (MEXICO). IPSCO Inc. as well as Atlas Tube generously donated steel material. For fabrication services, the authors gratefully acknowledge Walters Inc. (Ontario).

NOMENCLATURE

A_g = gross cross-sectional area of CHS
A_n = net cross-sectional area of CHS
CHS = Circular Hollow Section
D = outside diameter of CHS
F_y = yield tensile stress
F_u = ultimate tensile stress
LVDT = Linear Variable Differential Transformers
L_w = weld length
N_{uFE} = connection strength from FE analysis
N_{uFE-D} = connection strength at distortion limit
N_{ux} = measured connection strength
RHS = Rectangular Hollow Section
SHS = Square Hollow Section
t = thickness of CHS
t_p = thickness of gusset plate
$T\sigma$-$T\varepsilon$ = uniaxial true stress – true strain curve
w = distance between welds (around tube)
$\bar{x}$ = eccentricity
$\bar{x}'$ = reduced eccentricity ($\bar{x}' = \bar{x} - t_p/2$)
ε_{ef} = equivalent fracture strain

REFERENCES

AISC. 2005. *Specification for structural steel buildings, ANSI/AISC 360-05*. Chicago: American Institute of Steel Construction.

British Steel. 1992. *Slotted end plate connections, Report No. SL/HED/TN/22/-/92/D*. Rotherham: Swinden Laboratories.

CEN. 2005. *Eurocode 3: Design of steel structures – general rules - part 1-8: design of joints, EN1993-1-8: 2005(E)*. Brussels: European Committee for Standardisation.

Cheng J.J.R., Kulak G.L. & Khoo. H. 1996. Shear lag effect in slotted tubular tension members. *Proc. 1st CSCE Structural Specialty Conf.*: 1103–1114.

Cheng J.J.R, Kulak G.L. & Khoo H. 1998. Strength of slotted tubular tension members. *Canadian Journal of Civil Engineering* 25: 982–991.

Cheng J.J.R. & Kulak G.L. 2000. Gusset plate connection to round HSS tension members. *Engineering Journal, AISC* 37(4): 133–139.

CSA. 2001. *Limit states design of steel structures, CAN/CSA-S16-01*. Toronto: Canadian Standards Association.

Geschwindner L.F. 2004. Evolution of shear lag and block shear provisions in the AISC specification. *Proc. ECCS/AISC Workshop, Connections in Steel Structures V*.

Korol R.M., Mirza F.A. & Mirza M.Y. 1994. Investigation of shear lag in slotted HSS tension members. *Proc. 6th Intern. Symp. on Tubular Structures*: 473–482.

Korol R.M. 1996. Shear lag in slotted HSS tension members. *Canadian Journal of Civil Engineering* 23: 1350–1354.

Kulak G.L. & Grondin G.Y. 2002. *Limit states design in structural steel, 7th ed.* Toronto: Canadian Institute of Steel Construction.

Ling T.W., Zhao X.L., Al-Mahaidi R. & Packer J.A. 2004. Tests of slotted gusset-plate connections to very high strength tubes. *Proc. 18th Australasian Conf. on the Mechanics of Structures and Materials*: 1135–1140.

Lu L.H., de Winkel G.D., Yu Y. & Wardenier J. 1994. Deformation limit for the ultimate strength of hollow sections joints. *Proc. 6th Intern. Symp. on Tubular Structures*: 341–347.

Martinez-Saucedo G., Packer J.A. & Willibald S. 2005. *Slotted end connections to hollow sections, CIDECT Report 8G-12/06*. Toronto: University of Toronto.

Matic P. 1985. Numerically predicting ductile material behavior from tensile specimen response. *Theoretical and Applied Fracture Mechanics* 4: 13–28.

Packer J.A. & Henderson J.E. 1997. *Hollow structural section connections and trusses – A design guide, 2nd ed.* Toronto: Canadian Institute of Steel Construction.

Wilkinson T., Petrovski T., Bechara E. & Rubal M. 2002. Experimental investigation of slot lengths in RHS bracing members. *Proc. 3rd Intern. Conf. on Advances in Steel Structures*: 205–212.

Tubular Structures XI – Packer & Willibald (eds)
© 2006 Taylor & Francis Group, London, ISBN 0-415-40280-8

Design of gusset-plate welded connections in structural steel hollow sections

T.W. Ling, X.-L. Zhao & R. Al-Mahaidi
Department of Civil Engineering, Monash University, Clayton, Australia

J.A. Packer
Department of Civil Engineering, University of Toronto, Canada

ABSTRACT: This paper presents an investigation of gusset-plate welded connections in both VHS tubes and SSHS subject to tension. Tensile tests on the connections in VHS tubes were carried out. It was found that shorter weld lengths resulted in block shear tear-out (TO) or shear lag (SL) failure. The existing TO and SL design rules in various standards were examined both for welded VHS tubes and for welded SSHS from other studies. Possible modifications to the existing design rules were postulated and examined. New design rules with the corresponding capacity factors based on a FOSM reliability analysis are proposed both for TO failure and for SL failure. Critical weld lengths are also proposed to predict the failure modes.

1 INTRODUCTION

Gusset-plate welded connections in structural steel hollow sections (SSHS) are connections in which gusset-plates are longitudinally slotted and welded to circular hollow sections (CHS) or rectangular hollow sections (RHS) as shown in Figure 1. The connections can be formed by welding a slotted gusset-plate to an unslotted SSHS or vice versa. This type of connection is often used to attach girders or web members in trusses of buildings or bridge structures.

From the existing studies, three types of failure have been observed. They are block shear tear-out (TO)

failure, shear lag (SL) failure and section tensile failure without any SL effect. It was found that the first two failures occur when a short weld length (L_w) is used. However, there have been few studies of TO and SL failures in gusset-plate welded connections. Moreover, the existing design rules for both TO and SL failures have been developed based on bolted connections and may not be adequate for welded connections. Therefore, these design rules need to be examined further.

This paper presents an investigation of gusset-plate welded connections in very high strength (VHS) tubes, and design comparisons and reliability analyses of gusset-plate welded connections in both VHS tubes and SSHS under tension.

2 EXISTING EXPERIMENTAL WORK

There have been a few studies on gusset-plate welded connections in SSHS under tension. However, not all studies reported the material properties. Thus, only the test results from those studies with reported material properties are used in design comparisons and reliability analyses. These comparisons and analyses are carried out in Section 5.2 for TO failure and in Section 6.2 for SL failure.

The existing studies on gusset-plate welded connections have shown that the use of a short weld length (L_w) may result in TO failure or SL failure. It was found that the failure modes depend on the ratios of L_w/w and L_w/D or L_w/H (Packer & Henderson 1997), where L_w is the weld length, w is the circumferential distance

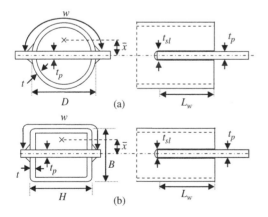

Figure 1. Defined dimensions for a gusset-plate connection in (a) CHS; (b) RHS.

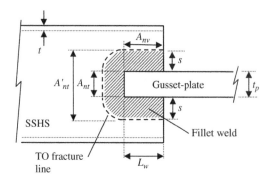

Figure 2. Schematic view of defined tension and shear areas.

between welds, D is the outer diameter of CHS, and H is the distance between welds in RHS as defined in Figure 1.

3 EXISTING DESIGN RULES

As mentioned in Section 1, the existing design rules for TO and SL failures are mainly directed to bolted connections. However, due to the same failure mechanism observed in tests, these design rules may be applicable to gusset-plate welded connections in SSHS. In this paper, only the design rules from the American, Canadian, European and Australian standards are examined. Although the design rules are similar to each other, different capacity (resistance) factors (ϕ) or partial safety factors (γ_M) are adopted in different standards.

For the TO design rules, the gross area in shear (A_{gv}), net area in shear (A_{nv}), net area in tension (A_{nt}) are defined in Figure 2. In welded connections, A_{gv} is always equal to A_{nv}, and A_{nt} is determined excluding a weld leg length (s) as shown in Figure 2. For gusset-plate welded connections with end return welds (ERW) or unslotted SSHS welded to slotted gusset-plate welded connections without ERW, $A_{nt} = 2 \times t_p \times t$. For slotted SSHS welded to unslotted gusset-plate connections without end return welds (ERW), $A_{nt} = 0$.

For the SL design rules, the net cross-sectional area of member (A_n) is equal to the gross cross-sectional area of member (A_g) when end return welds (ERW) or slotted gusset-plates are used. For a slotted SSHS without ERW, $A_n = A_g - 2 \times t_{sl} \times t$, in which t_{sl} is the slot opening width on a SSHS and t is the SSHS wall thickness.

4 TENSILE TESTS ON GUSSET-PLATE WELDED CONNECTIONS IN VERY HIGH STRENGTH TUBES

Tensile tests on gusset-plate welded connections in VHS tubes were conducted at Monash University,

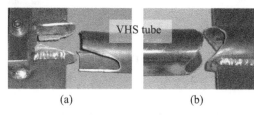

(a)   (b)

Figure 3. Typical failure of (a) TO failure; (b) SL failure.

Australia. All specimens were formed by welding unslotted VHS tubes to slotted gusset-plates longitudinally. Detailed specimen configurations and test set-up were reported in Ling (2005). The VHS tubes used had an average measured yield stress (f_y) and tensile strength (f_u) of 1363 MPa and 1534 MPa, respectively.

The gusset-plates used were 10 mm Grade 350 steel plates with an average measured yield stress and tensile strength of 384 MPa and 537 MPa, respectively. The welding method used was gas tungsten arc welding (GTAW) with a high strength consumable solid copper wire, ER110S-G, which has a nominal 0.2% proof stress and tensile strength of 870 MPa and 940 MPa, respectively.

There were twenty-five specimens that failed by TO failure, and sixteen specimens that failed by SL failure. Figure 3(a) shows a typical TO failure, and Figure 3(b) shows a typical SL failure. The maximum loads will be used to compare with predicted strengths (R_{TO} or R_{SL}) based on the corresponding design rules. Detailed test results were reported in Ling (2005).

5 DESIGN RULES FOR BLOCK SHEAR TEAR-OUT FAILURE

TO failure is a failure along weld contours with an initial crack starting near the slotted end on a parent (base) metal. The existing design rules for TO failure from the American, Canadian, European and Australian standards were examined by comparing the strength predictions with the experimental maximum loads. The comparison and reliability analysis are based on the TO failure specimens of gusset-plate welded connections in VHS tubes in Section 5.1, and in both VHS tubes and SSHS in Section 5.2.

5.1 TO Design for gusset-plate welded connections in very high strength tubes

The strength predictions for TO specimens were determined based on the existing TO design rules from the American (AISC 2005), Canadian (CSA 2001), European (EC3 Part 1.8 2003) and Australian (SAA 1996) standards after removing all capacity (resistance) factors (ϕ) and partial safety factors (γ_M). Since the shear area (A_{nv}) is much larger than the tension area (A_{nt}),

Table 1. Summary of the average mean ratios and COV for the welded VHS tubes that failed by TO failure.

Modification	AISC (2005), CSA (2001), SAA (1996)		EC3 Part 1.8 (2003)	
	Mean	(COV)	Mean	(COV)
Existing design	0.69	(0.09)	0.71	(0.09)
(1): A'_{nt}	0.60	(0.06)	0.61	(0.06)
(2): HAZ	1.39	(0.09)	1.43	(0.09)
(3): Minimum	1.10	(0.09)	1.13	(0.09)
(4): (1) + (2)	1.20	(0.06)	1.22	(0.06)
(5): (1) + (3)	0.95	(0.06)	0.97	(0.06)

Table 2. The capacity (resistance) factors and the load combinations.

Capacity factor ϕ	Load combination		
	Standard	$\gamma_{DL}D_n + \gamma_{LL}L_n$	γ_{LL}/γ_{DL}
0.75	ASCE (2000) Clause 2.3.2	$1.2D_n + 1.6L_n$	1.33
0.90	CSA (2001) Clause 7.2	$1.25D_n + 1.5L_n$	1.20
0.65	SAA (2002) Clause 4.2	$1.2D_n + 1.5L_n$	1.25

the design rules from SAA (1996) are taken as a combination of f_u in tension area with $0.6f_y$ in shear area. This is consistent with the design rules given in other standards.

The actual-to-predicted ratios based on the existing TO design rules for the VHS tube specimens that failed by TO failure are listed in Table 1. The mean ratios showed that the predicted strengths were higher than the actual maximum loads. This suggested that the existing design rules were not adequate, and five possible modifications to the existing design rules were thus postulated.

These modifications are listed in column 1 of Table 1. In Modification 1, a new net tension area, A'_{nt}, was postulated and determined including weld leg lengths (s) as shown in Figure 2. Zhao and Jiao (2004) found a significant reduction in material strengths of VHS tubes due to HAZ softening after welding. Therefore, in Modification 2, a HAZ reduction factor (φ) of 0.50, which is proposed by Jiao and Zhao (2004), was applied to the parent metal strengths (f_u and f_y). For Modification 3, the minimum material yield stress and tensile strength (f_{ym} and f_{um}) in the connection were used, in-stead of the yield stress and tensile strength of parent metal (f_y and f_u). This was because welded connections in VHS tubes were considered as an under-matching condition (i.e. the tensile strength of weld metal is less than that of parent metal). For an under-matching condition, EC3 Part 1.12 (2004) stated that the connection strength should be based on the strength of the weld metal. Modifications 4 and 5 combined Modification 1 with Modifications 2 and with Modification 3 respectively.

The mean ratios of the actual maximum loads to the predicted nominal strengths and their corresponding coefficients of variation (COV) for each modification are summarised in Table 1. The results showed that Modification 5 gave the best results with a mean ratio closest to 1.0 and the lowest COV. Therefore, its result will be used in a subsequent reliability analysis to calibrate the capacity (resistance) factor for the connections.

It is essential for a design capacity (ϕR_n) to be equal to or greater than a design action effect (R_u) based on a certain load combination (e.g. dead load, D_n and live load, L_n). The load combinations and existing capacity (resistance) factors for AISC (2005), CSA (2001) and SAA (1996) listed in Table 2 are used in a reliability analysis. The design rule of EC3 Part 1.8 (2003) is not included in the reliability analysis because it uses two partial safety factors. The reliability analysis is based on the first order second moment (FOSM) method that is described by Ellingwood et al. (1980) and Ravindra and Galambos (1978).

From Modification 5, the design rules can be expressed as:

$$R_{TO} = A'_{nt}f_{um} + 0.6A_{nv}f_{ym} \qquad (1)$$

where f_{ym} and f_{um} are expressed as:

$$f_{ym} = Min\{f_y, f_{yw}\} \qquad (2a)$$

$$f_{um} = Min\{f_u, f_{uw}\} \qquad (2b)$$

The statistical parameters for welded VHS tubes are listed in column 3 of Table 3. P_m and V_P are the mean value and corresponding COV of "professional model" taken as the mean ratio and COV of Modification 5 based on AISC (2005) in Table 1. The mean values of M_m and F_M, which represent the uncertainties in "material strength" and "fabrication", and the corresponding COV (V_M and V_F) are taken from the statistical readings provided by OneSteel Market Mills, Australia. Since all failure occurred in the VHS tubes, this paper assumed that the variables of "material strength" and "fabrication" are either f_y or f_u and the tube wall thickness (t), respectively. To be conservative, the values of M_m and V_M are taken as:

$$M_m = Min\{M_{m,fy}, M_{m,fu}\} \qquad (3a)$$

$$V_M = Max\{V_{M,fy}, V_{M,fu}\} \qquad (3b)$$

In the analysis, the value of a calibrated reliability index ($\beta_{0.25}$) at $D_n/(D_n + L_n)$ of 0.25 as recommended

Table 3. The statistical parameters for the gusset-plate welded connections in VHS tubes and SSHS that failed by TO failure.

Uncertainties	Statistical parameters	AISC (2005), CSA (2001) and SAA (1996)	
		Welded connections in VHS tubes	SSHS connections Inc. VHS tubes
Professional	P_m	0.948	1.087
Provision	V_P	*0.060*	*0.107*
Material	f_y $M_{m,fy}$	**1.003**	**1.003**
Strength*	$V_{M,fy}$	***0.0219***	***0.1000***
	f_u $M_{m,fu}$	1.017	1.017
	$V_{M,fu}$	*0.0087*	*0.0700*
Thickness	F_m	1.002	0.972
(Parent Metal)	V_F	*0.0092*	0.050
Resistance	R_m/R_n	0.953	1.059
	V_R	*0.0645*	*0.1549*

* The values that were used in the reliability analysis are shown in bold.

Table 4. The calibrated results for the TO design based on Modification 5 and welded VHS tubes.

Load combination	Welded VHS tubes			Welded SSHS Inc. VHS tubes		
	$\beta_{0.25}$ based on ϕ	ϕ_p	$\beta_{0.25}$ based on ϕ_p	$\beta_{0.25}$ based on ϕ	ϕ_p	$\beta_{0.25}$ based on ϕ_p
ASCE (2000)	3.20	0.70	3.54	3.04	0.67	3.51
CSA (2001)	2.06	0.74	3.05	2.11	0.72	3.03
SAA (2002)	3.66	0.67	3.51	3.42	0.63	3.55

Table 5. The actual-to-predicted ratios and their corresponding COV based on the existing design rules.

Studies	No. of Tests	Existing design	
		Mean	*(COV)*
VHS Tubes	25	0.69	*(0.09)*
Willibald et al. (2004a)	1	0.97	*(0.00)*
Wilkinson et al. (2002)	5	1.17	*(0.04)*
Zhao et al. (1999)	24	1.32	*(0.06)*
Zhao & Hancock (1995b)	48	1.26	*(0.07)*
SSHS including VHS tubes	103	1.13	*(0.23)*

Table 6. Summary of the average actual-to-predicted ratios and their corresponding COV for 103 tests.

Modification	AISC (2005), CSA (2001) and SAA (1996)	
	Mean	*(COV)*
Existing design	1.13	*(0.23)*
(1): A'_{nt}	0.97	*(0.23)*
(2): HAZ	1.30	*(0.09)*
(3): Minimum	1.28	*(0.13)*
(4): (1) + (2)	1.11	*(0.08)*
(5): (1) + (3)	1.09	*(0.11)*

by Galambos (1995) needs to be equal to or greater than a target reliability index (β_o) based on a given load combination and capacity (resistance) factor (ϕ). To obtain the proposed capacity (resistance) factor (ϕ_P), ϕ was varied until $\beta_{0.25}$ was equal to or slightly greater than β_o. The values of β_o for AISC (2005), CSA (2001) and SAA (1996) are 3.5, 3.0 and 3.5, respectively. The proposed capacity factors, target reliability indices, calibrated reliability indices based on the existing factors and proposed factors are listed in Table 4. The proposed capacity factors were found to be 0.70, 0.74 and 0.67 based on load combinations of ASCE (2000), CSA (2001) and SAA (2002) respectively.

5.2 TO Design for gusset-plate welded connections in structural steel hollow sections

A similar comparison as that in Section 5.1 was carried out for TO specimens from the existing studies that provided material properties. Due to the similarity between design rules of AISC (2005) and EC3 Part 1.8 (2003), only the existing design rules from AISC (2005), CSA (2001) and SAA (1996) after removing of all capacity (resistance) factors were used in the comparison. The studies are listed in column 1 of Table 5, together with the mean ratios of the maximum loads to the predicted nominal strengths for the TO specimens. The mean ratios showed that the predicted strengths were higher than the actual maximum loads for most studies. This suggested that the existing design rules were not adequate, and the same modifications as those described in Section 5.1 were thus used in the comparison. However, the modification based on the HAZ reduction factor was only applied to the connections in VHS tubes. For the studies without reported weld metal properties, a matching condition (i.e. the strength of the weld metal is greater than that of the hollow section) was assumed.

The average mean ratios of the maximum loads to the predicted nominal strengths and their corresponding COV for each modification for the VHS tube study and the existing studies are summarised in Table 6. The results showed that Modification 5 once again gave the best result with a mean ratio closest to 1.0 and the lowest COV. Therefore, its result will be used in the subsequent reliability analysis to calibrate the capacity (resistance) factor for the connections.

The same FOSM method as that described in Section 5.1 was used in the reliability analysis for TO design of gusset-plate welded connections in SSHS based on Equation 1. The statistical parameters for welded SSHS including VHS tubes are summarised in Table 3. The values of P_m and V_P were taken as the average mean ratio and the corresponding COV of Modification 5 in Table 6. Not all studies listed in Table 5 provided the mean values of $M_{m,fy}$, $M_{m,fu}$ and F_M and the corresponding COV. However, there were a few studies that reported these values for cold-formed steels (Rang et al. 1978; Zhao et al. 1999; Zhao & Hancock 1994; Zhao & Hancock 1995a). The minimum mean values and the maximum COV were conservatively taken from both these studies and the VHS tube study (Ling 2005). M_m and V_M were again determined based on Equation 3.

The proposed capacity factors, calibrated reliability indices based on the existing factors and proposed factors are listed in Table 4. The proposed capacity factors were found to be 0.67, 0.72 and 0.63 based on load combinations of ASCE (2000), CSA (2001) and SAA (2002), respectively. A rounded value of 0.70 may be used for TO design in SSHS.

6 DESIGN RULES FOR SHEAR LAG FAILURE

SL failure is a section circumferential failure with the same initial crack as that in TO failure. The existing design rules for SL failure from the American, Canadian and Australian standards were examined by comparing the strength predictions with the experimental maximum loads. The comparison and reliability analysis are based on the SL failure specimens of gusset-plate welded connections in VHS tubes in Section 6.1, and in both VHS tubes and SSHS in Section 6.2.

6.1 Shear lag design for gusset-plate welded connections in very high strength tubes

The strength predictions for SL specimens were determined based on the existing SL design rules in AISC (2005), CSA (2001) and SAA (1998) after removing all capacity (resistance) factors. It should be noted that the additional factor of 0.85 in the design rules of CSA (2001) and SAA (1998) was also removed in the comparison.

The mean ratios based on the existing SL design rules for the VHS tube specimens that failed by SL failure are listed in Table 7. The mean ratios showed that the predicted strengths were higher than the actual maximum loads. This suggested that the existing design rules were not adequate, and five possible modifications to the existing design rules were thus postulated.

These modifications are listed in column 1 of Table 7. In Modification 1, a reduced eccentricity concept ($\bar{x}'$) was used in-stead of $\bar{x}$ as suggested by

Table 7. Summary of the average mean ratios and COV for the welded VHS tubes.

	Average ratios (P_{max}/R_{SL})					
	AISC (2005)		CSA (2001)		SAA (1998)	
Modification	Mean	*(COV)*	Mean	*(COV)*	Mean	*(COV)*
Existing design	0.69	*(0.12)*	0.83	*(0.09)*	0.97	*(0.19)*
(1): $\bar{x}'$	0.66	*(0.06)*	N/A		N/A	
(2): HAZ	1.38	*(0.12)*	1.66	*(0.09)*	1.94	*(0.19)*
(3): Minimum	1.14	*(0.12)*	1.37	*(0.08)*	1.60	*(0.18)*
(4): (1) + (2)	1.31	*(0.06)*	N/A		N/A	
(5): (1) + (3)	1.08	*(0.05)*	N/A		N/A	

Willibald et al. (2004b), and is expressed as:

$$\bar{x}' = \bar{x} - \frac{t_p}{2} \qquad (4)$$

Modifications 2 and 3 were the same as those for TO design to account for the effects of HAZ softening and an under-matching condition, as explained in Section 5.1. Modifications 4 and 5 combined Modification 1 with Modification 2 and with Modification 3 respectively.

The mean ratios and their corresponding COV for each modification for the VHS tube specimens that failed by SL failure are summarised in Table 7. The results showed that Modification 5 once again gave the best result with a mean ratio closest to 1.0 and the lowest COV. Therefore, its result will be used in a subsequent reliability analysis to calibrate the capacity (resistance) factor for the connections.

From Modification 5, the design rules can be expressed as:

$$R_n = A_e f_u = U \cdot A_n f_{um} \qquad (5)$$

where U is determined based on AISC (2005) with $\bar{x}'$ given in Equation 4, and f_{um} given in Equation 2(b).

The same FOSM method as that described in Section 5.1 was used in the reliability analysis for SL design of gusset-plate welded connections in VHS tubes based on Equation 5. The statistical parameters for welded VHS tubes are listed in column 3 of Table 8. P_m and V_P were taken as the mean ratio and COV of Modification 5 based on AISC (2005) in Table 7. The values of M_m and V_M for f_u, and F_t and V_t for tube thickness are given in Table 3 for welded connections in VHS tubes. Since all failures occurred in the VHS tubes, the variables of "fabrication" are considered to be the tube outside diameter (D) and the tube wall thickness (t). The expressions for F_m and V_F were derived in Ling (2005) as follows:

$$F_m \approx F_t \times F_D \approx F_t^2 \qquad (6a)$$

249

Table 8. The statistical parameters for the gusset-plate welded connections in VHS tubes and SSHS.

Uncertainties	Statistical parameters	AISC (2005) Welded connections in VHS tubes	SSHS connections including VHS tubes
Professional model	P_m	1.083	1.091
	V_P	0.053	0.069
Material strength (f_u)	M_m	1.017	1.017
	V_M	0.0087	0.07
Fabrication (thickness)	F_m	1.004	0.945
	V_F	0.013	0.071
Resistance	R_m/R_n	1.106	1.049
	V_R	0.055	0.121

Table 9. The calibrated results for SL design based on Modification 5 of AISC (2005).

Load combination	Welded VHS tubes			Welded SSHS including VHS tubes		
	$\beta_{0.25}$ based on ϕ	ϕ_p	$\beta_{0.25}$ based on ϕ_p	$\beta_{0.25}$ based on ϕ	ϕ_p	$\beta_{0.25}$ based on ϕ_p
ASCE (2000)	4.02	0.82	3.56	3.27	0.71	3.51
CSA (2001)	3.80	0.79	3.53	3.08	0.68	3.52
SAA (2002)	3.76	0.78	3.56	3.04	0.67	3.54

$$V_F \approx \sqrt{V_t^2 + V_D^2} \approx \sqrt{2} \cdot V_t \qquad (6b)$$

The existing capacity (resistance) factor and β_o for SL design in AISC (2005) are taken as 0.75 and 3.5 respectively. The proposed capacity (resistance) factors (ϕ_P), and calibrated reliability indices ($\beta_{0.25}$) based on the existing factors and proposed factors are listed in Table 9. The proposed capacity factors were found to be 0.82, 0.79 and 0.78 based on load combinations of ASCE (2000), CSA (2001) and SAA (2002), respectively.

6.2 *SL Design for gusset-plate welded connections in structural steel hollow sections*

The same comparison as that described in Section 6.1 was carried out to include the SL specimens from the existing studies that provided reported material properties. These studies are listed in column 1 of Table 10, together with the actual-to-predicted mean ratios and COV for the SL specimens based on the existing design rules. The mean ratios showed that the predicted strengths were much smaller than the actual maximum loads for the existing studies except

Table 10. Summary of the actual-to-predicted ratios and their corresponding COV based on the existing design rules.

Studies	No. of Tests	Average ratios (P_{max}/R_{SL})					
		AISC (2005) Mean	*(COV)*	CSA (2001) Mean	*(COV)*	SAA (1998) Mean	*(COV)*
VHS tubes	16	0.69	*(0.12)*	0.83	*(0.09)*	0.97	*(0.19)*
Willibald et al. (2004a)	4	1.26	*(0.04)*	1.52	*(0.07)*	1.80	*(0.05)*
Wilkinson et al. (2002)	3	1.10	*(0.05)*	1.70	*(0.02)*	1.46	*(0.15)*
Cheng and Kulak (2000)	2	1.21	*(0.12)*	1.47	*(0.08)*	1.72	*(0.17)*
SSHS inc. VHS tubes	25	0.87	*(0.3)*	1.10	*(0.33)*	1.22	*(0.32)*

Table 11. Summary of the average actual-to-predicted ratios and COV for welded SSHS including VHS tubes (25 tests).

Modification	Average ratios (P_{max}/R_{SL})					
	AISC (2005) Mean	*(COV)*	CSA (2001) Mean	*(COV)*	SAA (1998) Mean	*(COV)*
Existing design	0.87	*(0.30)*	1.10	*(0.33)*	1.22	*(0.32)*
(1): $\bar{x}'$	0.82	*(0.28)*	N/A		N/A	
(2): HAZ	1.31	*(0.13)*	1.63	*(0.09)*	1.84	*(0.19)*
(3): Minimum	1.16	*(0.11)*	1.44	*(0.10)*	1.62	*(0.17)*
(4): (1) + (2)	1.24	*(0.10)*	N/A		N/A	
(5): (1) + (3)	1.09	*(0.07)*	N/A		N/A	

the welded VHS tube study. This suggested that the existing design rules were too conservative for welded SSHS except VHS tubes, and the same modifications as those described in Section 6.1 were thus used in the comparison. However, the modification based on the HAZ reduction factor was only applied to the connections in VHS tubes. For the studies without reported weld metal properties, a matching condition (i.e. the strength of the weld metal is greater than that of the hollow section) was assumed.

The average actual-to-predicted mean ratios and their corresponding COV for each modification are summarised in Table 11. The results showed that Modification 5 gave the best predictions with a mean ratio closest to 1.0 and the lowest COV. Therefore, its result will be used in the subsequent reliability analysis to calibrate the capacity (resistance) factor for the connections.

The same FOSM method as that described in Section 6.1 was used in the reliability analysis for SL design of gusset-plate welded connections in SSHS including VHS tubes based on Equation 5. The statistical parameters for welded SSHS including VHS tubes are summarised in column 4 of Table 8. The values of P_m and V_P were taken as the average mean ratio and the corresponding COV of Modification 5 based on AISC (2005) in Table 11. Since not all studies listed in Table 10 provided the mean values of M_m and F_t, and the corresponding COV, the same values as in Table 3 for SSHS connections including VHS tubes were used. F_m and V_F were again determined based on Equation 6.

The proposed capacity factors, calibrated reliability indices based on the existing factors and proposed factors are listed in Table 9. The proposed capacity factors were found to be 0.71, 0.68 and 0.67 based on load combinations of ASCE (2000), CSA (2001) and SAA (2002) respectively. A rounded value of 0.70 may be adopted for SL design in both VHS tubes and SSHS.

7 PROPOSED DESIGN RULES

For TO design, the proposed design rule for gusset-plate welded connections in SSHS becomes:

$$\phi_P R_{TO} = \phi \left[A'_{nt} f_{um} + 0.6 A_{nv} f_{ym} \right] \tag{7}$$

where A'_{nt} = new modified net area subject to tension; A_{nv} = net area subject to shear; f_{ym}, f_{um} = minimum strengths as given in Equation 2; ϕ_P = capacity (resistance) factor taken as 0.70.

For SL design, the proposed design rule for gusset-plate welded connections in SSHS becomes:

$$\phi_P R_{SL} = \phi A_e f_{um} = \phi U A_n f_{um} \tag{8}$$

where f_{um} = minimum tensile strength as given in Equation 2b; ϕ_P = capacity (resistance) factor taken as 0.70; U = shear lag factor as determined based on AISC (2005) with the use of $\bar{x}'$ as given in Equation 4.

8 CRITICAL WELD LENGTHS

As mentioned earlier, the use of short weld length will result in TO or SL failure. It was found that the failure modes depend on L_w/D, L_w/H or L_w/w ratios for gusset-plate welded SSHS connections.

The failure modes of specimens both from the VHS tube study and the existing studies are plotted as $P_{max}/A_n f_{um}$ versus L_w/D and L_w/H ratios in Figure 4, and L_w/w ratios in Figure 5. Figure 4(a) shows that TO failure governs when the L_w is less than $0.9D$, which is close to the limit of $L_w = 1.0D$ as given in AISC (2005). The figure also shows that SL failure governs when L_w is greater than $1.1D$. The existing limit of $L_w = 1.3D$ (AISC 2005), beyond which SL failure is not critical, seemed to be not valid as SL failure still occurred up to $L_w = 1.75D$. For the welded connections in RHS, TO failure is dispersed in Figure 4(b) up to a L_w/H-ratio of 1.73. The result suggested that

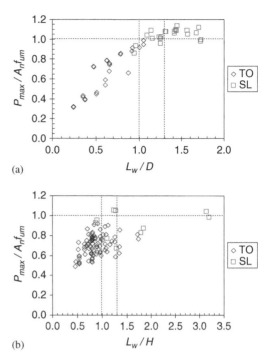

(a)

(b)

Figure 4. $P_{max}/A_n f_{um}$ versus (a) L_w/D ratios for the welded CHS; (b) L_w/H ratios for the welded RHS.

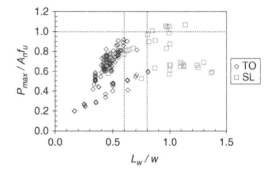

Figure 5. $P_{max}/A_n f_{um}$ versus L_w/w ratios for the slotted gusset-plate connections in SSHS.

L_w/H-ratios cannot be used to predict the failure mode accurately for welded connections in RHS.

The failure modes were more consistent based on L_w/w-ratios as shown in Figure 5. For $L_w \leq 0.6w$, TO failure governs. This is consistent with the existing critical TO limit of $L_w = 0.6w$ as given in Korol (1996) and Packer and Henderson (1997). For $L_w \geq 0.8w$, SL failure governs. In between $0.6w$ and $0.8w$, either TO or SL failure occurs. It is recommended both design rules should be checked within this range. There is insufficient data to verify the recommended limit of

$L_w = 2.0w$, beyond which SL failure is not critical, as given in CSA (2001), Packer and Henderson (1997) and Syam and Chapman (1996). Therefore, further investigation with $L_w/w > 1.5w$ is needed.

9 CONCLUSIONS

The existing design rules for both TO and SL failures were examined and found that they were not adequate for both welded VHS tubes and welded SSHS. Formulae have been proposed for TO and SL failures with a capacity (resistance) factor of 0.70.

A critical TO weld length has been found to be $0.9D$ or $0.6w$ and a critical SL weld length has been found to be $1.1D$ or $0.8w$. The existing critical tensile limit of $L_w = 2.0w$ needs further verification.

ACKNOWLEDGEMENTS

The work reported here was supported by OneSteel Market Mills (Australia), Monash University (MGS & MDS) and an ARC International Linkage Grant.

REFERENCES

AISC 2005. Specification for structural steel buildings. American Institute of Steel Construction, Chicago.

ASCE 2000. Minimum design loads for buildings and other structures. ASCE 7-98, Reston, VA: American Society of Civil Engineers.

Cheng, JJR & Kulak, GL 2000. Gusset plate connection to round HSS tension members. Engineering Journal, AISC 4th Quarter: 133–139.

CSA 2001. Limit states design of steel structures. CAN/CSA-S16-01, Toronto, Canada: Canadian Standards Association.

EC3 Part 1.8 2003. Design of steel structures – Part 1.8: Design of joints. EN 1993-1-8 Draft (Nov 2003) Version, London.

EC3 Part 1.12 2004. Design of steel structures – Part 1.12: Additional rules for the extension of EN 1993 up to steel grades S700. EN 1993-1-12 Stage 34 Draft, London.

Ellingwood, B, Galambos, TV, Mac Gregor, JG & Cornell, CA 1980. Development of a probability-based load criterion for American National Standard A58. In NBS Special Publication 577. Washington, DC.

Galambos, TV 1995. Discussion on the paper entitled "Public safety – is it compromised by new LRFD design standards?" J Struct Engrg, ASCE 121(1): 143–144.

Jiao, H & Zhao, XL 2004. Tension capacity of very high strength (VHS) circular steel tubes after welding.

Advances in Structural Engineering – An Int. Journal 7(4): 85–96.

Korol, RM 1996. Shear lag in slotted HSS tension members. Canadian Journal of Civil Engineering 23: 1350–1354.

Ling, TW 2005. The tensile behaviour of gusset-plate welded connections in very high strength (VHS) tubes. PhD Thesis, Dept of Civil Engrg, Monash University, Melbourne.

Packer, JA & Henderson, JE 1997. Hollow structural section connections and trusses – A design guide. Toronto, Canada: Canadian Institute of Steel Construction.

Rang, TN, Galambos, TV, Yu, WW & Ravindra, MK 1978. Load and resistance factor design of cold-formed steel structural members. In Yu, WW & Senne, JH (eds.), Fourth International Specialty Conference on Cold-Formed Steel Structures, 1–2 June, Missouri.

Ravindra, MK & Galambos, TV 1978. Load and resistance factor design for steel. Journal of the Structural Division, ASCE 104(ST9): 1337–1353.

SAA 1996. Cold-formed steel structures. Australian/New Zealand Standard AS/NZS 4600, Sydney.

SAA 1998. Steel structures. Australian Standard AS 4100, Sydney, Australia: Standards Association of Australia.

SAA 2002. Structural design actions – General principles. Australian/New Zealand Standard AS/NZS 1170.0, Sydney.

Syam, AA & Chapman, BG 1996. Design of structural steel hollow section connections, Volume 1: Design models. Australian Institute of Steel Construction.

Wilkinson, T, Petrovski, T, Bechara, E & Rubal, M 2002. Experimental investigation of slot lengths In RHS bracing members. In Third International Conference on Advances in Steel Structures, Hong Kong.

Willibald, S, Packer, JA, Martinez Saucedo, G & Puthli, RS 2004a. Shear lag in slotted gusset plate connections to tubes. In Amsterdam Workshop: Connections in Steel Structures V, Amsterdam, Netherlands.

Willibald, S, Packer, JA & Martinez Saucedo, G 2004b. Slotted end connections to hollow sections. 2nd Interim Report to CIDECT on Programme 8G, Uni. of Toronto, Canada.

Zhao, XL, Al-Mahaidi, R & Kiew, KP 1999. Longitudinal fillet welds in thin-walled C450 RHS members. Journal of Structural Engineering, ASCE 125(8): 821–828.

Zhao, XL & Hancock, GJ 1994. Tests and design of butt welds and fillet welds in DuraGal RHS members. Research Report No. R702, The University of Sydney, Sydney.

Zhao, XL & Hancock, GJ 1995a. Butt welds and transverse fillet welds in thin cold-formed RHS members. Journal of Structural Engineering, ASCE 121(11): 1674–1682.

Zhao, XL & Hancock, GJ 1995b. Longitudinal fillet welds in thin cold-formed RHS member. Journal of Structural Engineering, ASCE 121(11): 1683–1690.

Zhao, XL & Jiao, H 2004. Recent developments in high strength thin-walled steel tubes. In International Workshop on Thin-Walled Structures: Recent Advances and Future Trends in Thin-Walled Structures Technology, Leicestershire, UK.

Tubular Structures XI – Packer & Willibald (eds)
© *2006 Taylor & Francis Group, London, ISBN 0-415-40280-8*

Strength of slotted rectangular and square HSS tension connections

R.F. Huang
Fluor Canada Ltd., Calgary, Alberta, Canada

R.G. Zhao
Department of Civil Engineering, University of Waterloo, Waterloo, Ontario, Canada

H.A. Khoo
Department of Civil and Environmental Engineering, Carleton University, Ottawa, Ontario, Canada

J.J.R. Cheng
Department of Civil and Environmental Engineering, University of Alberta, Alberta, Canada

ABSTRACT: A research program has been carried out on slotted rectangular (RHS) and square (SHS) structural hollow section connections with and without welding at the end of the gusset plate. The effect of weld length ratio, slot orientation, gusset plate thickness, slot opening length and end welding were investigated. Recommendations are given to improve on provisions to account for shear lag in American and Canadian design standards for slotted RHS and SHS tension connections.

1 INTRODUCTION

Hollow structural sections (HSS) are widely used as welded tension and compression members in bracing and trusses. A commonly employed method to make a slotted connection for a hollow structural section is to slot the tube longitudinally and insert a gusset plate into the slot. The gusset plate is then welded to the tube by longitudinal fillet welds as shown in Figure 1. Welding may or may not be provided around the end of the gusset plate. In both cases, the stress is not distributed uniformly across the section because not all elements of the HSS are directly connected to the gusset plate. Thus, the net section may not be fully effective in carrying the load. This phenomenon is known as shear lag. The effect of shear lag has been found to be related to the ratio of the weld length (L) to the circumferential distance between the longitudinal welds (w) and the ratio of the net section area eccentricity ($\bar{x}$) to the weld length (L). The above parameters are depicted in Figure 1.

In the Canadian Standard CSA-S16-01 (2001), Limit States Design of Steel Structures, the effect of shear lag is accounted for in calculating the effective net area, which is a function of the L/w ratio. In clause 13.2 (a) (iii), the factored tensile resistance, T_r, of a tension member is taken as

$$T_r = 0.85 \phi A_{ne} F_u, \qquad (1)$$

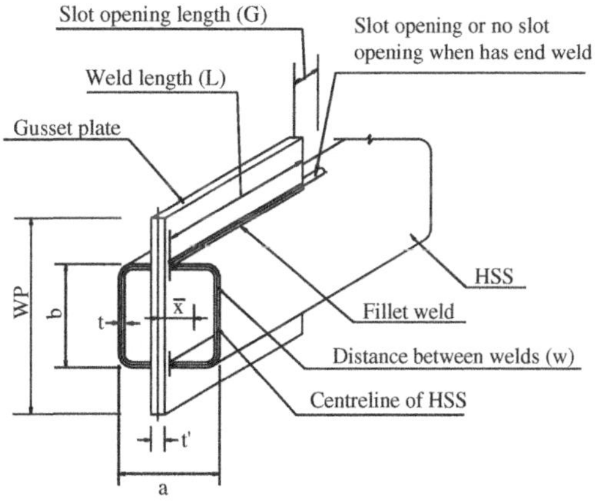

Figure 1. Isometric view of a slotted HSS connection with a gusset plate.

where ϕ is the resistance factor taken as 0.90, A_{ne} is the effective net section area that accounts for shear lag and F_u is the specified minimum tensile strength. When an element of a tension member is connected by longitudinal welds along two parallel edges, the effective net area for that element is taken as

$$A_{n2} = 1.00 \, wt, \qquad \text{for } L \geq 2w, \qquad (2)$$

$$A_{n2} = 0.50 \, wt + 0.25Lt, \qquad \text{for } 2w > L \geq w, \qquad (3)$$

$$A_{n2} = 0.75 \, Lt, \qquad \text{for } w > L, \qquad (4)$$

where t is the thickness of the tension member. For a slotted HSS connection, the effective net section area, A_{ne}, is taken as twice that given by A_{n2}.

The American Specification for Structural Steel Building, ANSI/AISC-360-05 (2005), also has provisions to account for shear lag in calculating the capacity of a tension member. The effective area of a tension member in ANSI/AISC-360-05 is determined from the net area (A_n) by,

$$A_e = A_n U, \qquad (5)$$

where U is the shear lag factor. It is defined for a rectangular HSS with a single concentric gusset plate that has a weld length (L) not less than b as,

$$U = 1 - \frac{\bar{x}}{L}, \qquad (6)$$

with $\bar{x}$ taken as

$$\bar{x} = \frac{a^2 + 2ab}{4(a + b)}, \qquad (7)$$

where a is the overall height of the HSS in the direction perpendicular to the plane of the gusset plate and b is the overall width of the HSS in the direction parallel to the plane of the gusset plate. However, it should be pointed out that the net section eccentricity $(\bar{x})$ calculated using Equation 7 is not representative of the actual net cross-section eccentricity when the HSS aspect ratio (a/b) is less than unity. Thus, a modified net section eccentricity $(\bar{x}^*)$ is adopted in this study. The modified net section eccentricity is taken as the net section eccentricity of an equivalent square HSS with equal circumferential length when the a/b ratio is less than unity. This situation applies to a rectangular HSS connection slotted on its short side. For example, the net section eccentricity for a HSS 127 × 51 slotted on its short side is taken as equal to that for a HSS 89 × 89.

Several studies on slotted HSS connections have previously been carried out. Korol (1996) tested 18 specimens, all of which were slotted square and rectangular HSS connections without end welding. Specimens with the L/w ratio around unity were found to have net section efficiencies close to 1.0. Other specimens with a L/w ratio less than 0.6 have much lower net section efficiencies and failed by block shear tearout. Cheng et al. (1998) tested eight specimens of slotted circular HSS connections with end welding and one specimen without end welding. All specimens achieved net section efficiencies close to unity. Seven specimens that have end welding and with the L/w ratio around 1.0 failed at the mid-length of the HSS. Another study on slotted circular HSS connections with different slotting details was carried out by Willibald et al. (2004). Two of these specimens were fabricated by slotting the gusset plate and four were fabricated by slotting the tube. Among specimens slotted at the tube, two have end welding around the gusset

plate and two without. In all the above three studies, net section efficiencies attained in the tests were much higher than those prescribed by CSA-S16-01 and ANSI/AISC-360-05.

Numerical analyses were also carried out by Cheng et al. (1998) and Willibald et al. (2004). The overall prediction of the numerical simulations in both studies was good with respect to the ultimate strength and the load versus displacement relationship of the connection.

2 EXPERIMENTAL PROGRAM

Two series of slotted RHS and SHS connections were carried at Carleton University. One series by Huang (2005) was on the connection with no end welding and another series by Zhao (2005) was on the connection with end welding.

The objective of the test program was to investigate the strength and behavior of square and rectangular HSS with and without end welding for various slotted connection details. Specimens with no end welding were designed with weld length ratios around 1 to complement test results of Korol (1996). Ultimate load, load versus deformation relationship and failure mode of the slotted HSS connections with and without end welding are examined. Results of these tests are used to develop recommendations to improve on provisions to account for shear lag in the design standards for slotted RHS and SHS tension connections.

2.1 Test specimens

The program consisted of two series of specimens. Details of the test specimens with no end welding are shown in Figure 2. Details for specimens with end welding were similar to those without end welding except that there was no slot opening. In the first specimen series, a total of twenty six specimens consisting of six rectangular HSS and twenty square HSS without end welding were fabricated. These specimens encompassed sixteen different connection configurations. The base configuration of the test specimen consisted of a 5 mm long straight segment of the slot opening and 16 mm thick gusset plates. Three L/w ratios of around 0.79, 1.06 and 1.33 were considered in the test. Three gusset plate nominal thicknesses of 12 mm, 16 mm and 20 mm were also considered. Specimens with a 12 mm or 20 mm gusset plate were tested only for L/w ratios of 0.79 and 1.33. Other than the slot straight segment length of 5 mm, two additional lengths of 25 mm and 50 mm were also tested with the 16 mm thick gusset plate. Duplicate tests were performed for all square HSS specimens.

The second specimen series consisted of four specimens with end welding. One specimen was rectangular

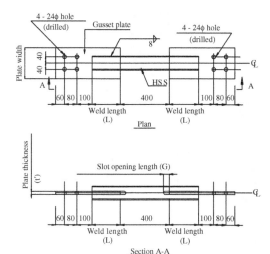

Figure 2. Details of the test specimen with no end welding.

HSS and the other three were square HSS. A continuous weld wrapping around the gusset plate was provided at the end of the slot for these connections. The gusset plate for this series has a nominal thickness of 16 mm. Weld length ratios (L/w) of 1.0 and 0.7 were considered for square HSS specimens while only one ratio of 0.7 was considered for the rectangular HSS specimen.

The hollow structural sections used in this test program consisted of ASTM-A500 Grade C cold-formed non-stress relieved HSS 89 × 89 × 4.8 and HSS 127 × 51 × 4.8. Gusset plates for first specimen series were CAN/CSA-G40.20/G40.21 Grade 350 W plates while gusset plates in the second series were ASTM-A572 Grade 55 plates. For specimens with end welding, the weld length (L) is taken as the length of the straight segment of the longitudinal weld excluding the extra length provided by the end weld. The distance between the welds (w) is taken as the distance along the centreline of the HSS wall from the edge of the slot for a specimen with no end welding, and is taken as half the circumference along the centreline of the HSS wall for a specimen with end welding. Measured specimen dimensions as defined by parameters in Figure 1 are shown in Tables 1 and 2. The thickness t is the average wall thickness of the flat part of HSS.

2.2 Test setup and instrumentation

All specimens were loaded axially through gusset plates that were connected to the top and bottom end fixtures mounted on the testing machine. The top fixture was designed to be able to swivel on a spherical support in order to align with the axis of the specimen during the test. There was no special provision provided to allow for swiveling of the bottom fixture.

Table 1. Dimensions of specimens with no end welding, in mm.

No.	a	b	t	L	G	t
RL5	127.04	51.63	4.52	195.06	14.68	15.72
RS5	51.39	127.30	4.50	194.75	15.83	15.71
S5P16	89.61	89.77	4.42	194.31	14.47	15.70
S5P16R	89.50	89.54	4.41	196.13	14.99	15.82
RL4	127.15	51.59	4.50	155.75	13.76	15.74
RS4	51.45	127.41	4.50	155.88	15.19	15.73
S4P16	89.36	89.56	4.39	156.50	14.16	15.73
S4P16R	89.61	89.51	4.40	156.63	14.24	15.75
RL3	127.08	51.46	4.47	116.13	15.28	15.72
RS3	51.32	127.21	4.50	116.63	15.45	15.81
S3P16	89.93	89.67	4.40	116.50	14.64	15.65
S3P16R	89.42	89.51	4.42	116.56	13.94	15.74
S3P12	89.38	89.56	4.42	118.00	13.51	12.65
S3P12R	89.56	89.50	4.42	118.44	12.70	12.69
S5P12	89.52	89.53	4.41	201.75	9.90	12.69
S5P12R	89.90	89.57	4.41	201.06	12.20	12.69
S3P20	88.82	88.95	4.46	113.50	17.02	19.16
S3P20R	89.39	90.10	4.45	113.94	16.55	19.07
S5P20	89.56	89.59	4.42	190.63	16.23	19.15
S5P20R	89.23	89.55	4.43	190.00	17.74	19.26
S3G25	89.43	89.55	4.40	117.50	33.50	15.76
S3G25R	89.37	89.70	4.42	116.63	34.24	15.70
S3G50	89.18	89.63	4.41	116.69	59.14	15.69
S3G50R	89.37	89.70	4.42	116.00	60.07	15.72
S5G50	89.21	89.64	4.41	196.13	58.83	15.73
S5G50R	89.40	89.69	4.41	195.50	59.32	15.64

Note: G is the overall slot opening length including the semi-circle part.

Table 2. Dimensions of specimens with end welding, in mm.

No.	a	b	t	L	w	t'
S10-a	88.66	89.32	4.42	159.25	165.0	16.55
S10-b	88.78	89.16	4.43	161.88	165.2	16.61
S07	88.81	89.16	4.44	111.88	165.0	16.53
R07	127.27	51.51	4.53	114.63	165.4	16.46

All bolt holes are 24 mm in diameter to accommodate ASTM A325 M22 bolts.

Two LVDTs (linear variable differential transformer) were attached to the gusset plates on both sides of the specimen to measure the axial deformation during the test. Strain gauges were also mounted at the slotted end on the surface of HSS to measure the strain in the longitudinal (loading) direction.

2.3 Test procedure

During the test, readings of strain gauges, LVDT, stroke and load were recorded through a data acquisition system at an interval of 5 seconds. The real-time load versus deformation curve and load versus strain curve displays were monitored during the test.

Table 3. Material properties of test specimens.

Material	Yield strength, Fy (MPa)	Ultimate strength, Fu (MPa)	Elongation (%)
HSS 89 × 89 (1st series)	402.0	485.3	28.0
HSS 127 × 51	380.3	448.0	33.5
HSS 89 × 89 (2nd series)	370.0	439.6	29.2
12 mm plate	319.3	494.8	35.0
16 mm plate (1st series)	337.7	466.0	39.0
20 mm plate	288.7	457.3	39.6
16 mm plate (2nd series)	380.4	561.3	33.6

Stroke control was employed in all the tests. A stroke rate of 1 mm per minute was used in the elastic stage of the test. The first static reading was taken when the load versus deformation curve started to deviate from the straight line. After the first static reading, the loading rate was kept between 1 to 2 mm per minute. Static readings were taken regularly during the inelastic stage of loading. The loading stroke was put on hold for about 30 seconds before a static reading was taken. A test was terminated when fracture occurred in the specimen.

2.4 Ancillary tests

Material properties of HSS and the gusset plates are required in evaluating the net section efficiency of the test specimen and for carrying out the numerical simulation of the test. Tension coupon tests were carried out to obtain material properties of HSS and gusset plates. They were fabricated in accordance to ASTM-E8-04 (2004) and CAN/CSA-G40.20-04 (2004). Coupons for HSS were cut longitudinally from the middle-half of the flat part of HSS that was away from the seam weld and corners.

The tension coupon test was carried out at a stroke rate of 0.5 mm (0.02 inch) per minute. Static readings and transverse deformations were taken at about every 0.5 to 0.8 mm stroke displacement after yielding. Average mechanical properties of both HSS and gusset plates are listed in Table 3. The yield strength of HSS was calculated using the 0.2% offset method.

2.5 Test results and discussion

The effect of shear lag on the test specimen is evaluated in terms of the net section efficiency. In the discussion, the measured test net section efficiency is taken as $P_{uTest}/A_n F_u$, where P_{uTest} is peak static test load, A_n is the net area of the cross-section and F_u is ultimate tensile strength of the flat part of HSS. The ultimate tensile strength of the material was determined from

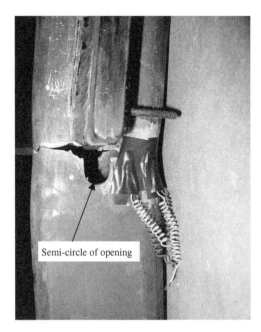

Figure 3. Failure occurred close to the end of the gusset plate.

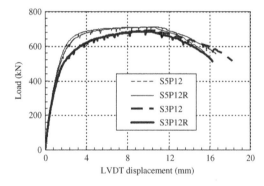

Figure 4. Load versus LVDT displacement curves of selected specimens with no end welding.

tension coupon tests. For specimens with end welding, A_n is taken as the gross cross-section area since there is no opening at the end of the gusset plate.

2.5.1 Specimens with no end welding

All specimens with no end welding fractured at the slot where the cross-section area is the least. The failure occurred either close to the end of the gusset plate or at the end of the semi-circle of the opening. Figure 3 shows an example of the failure that occurred close to the end of the gusset plate. Figure 4 depicts typical load versus LVDT displacement curves of the test.

Test peak loads and net section efficiencies are shown in Table 4. It can be seen in Table 4 that for

Table 4. Test results of specimens with no end welding.

No.	L/w	$\bar{x}^*/L$	a/b	P_{uTest} (kN)	A_nF_u (kN)	P_{uTest}/A_nF_u
RL5	1.33	0.21	2.46	674.6	593.6	1.14
RS5	1.34	0.17	0.40	674.4	586.7	1.15
S5P16	1.33	0.17	1.00	679.9	632.5	1.07
S5P16R	1.34	0.17	1.00	673.3	632.2	1.06
RL4	1.05	0.26	2.46	676.8	595.9	1.14
RS4	1.06	0.21	0.40	652.6	594.4	1.10
S4P16	1.06	0.21	1.00	677.5	630.3	1.07
S4P16R	1.08	0.21	1.00	674.3	625.8	1.08
RL3	0.79	0.35	2.46	615.6	592.7	1.04
RS3	0.80	0.29	0.40	641.5	594.2	1.08
S3P16	0.79	0.29	1.00	650.6	631.5	1.03
S3P16R	0.79	0.29	1.00	652.6	632.9	1.03
S3P12	0.79	0.28	1.00	676.4	647.3	1.04
S3P12R	0.78	0.29	1.00	668.8	649.1	1.03
S5P12	1.34	0.17	1.00	695.6	647.2	1.07
S5P12R	1.34	0.17	1.00	690.9	647.9	1.07
S3P20	0.80	0.29	1.00	621.2	616.8	1.01
S3P20R	0.79	0.30	1.00	634.9	624.8	1.02
S5P20	1.33	0.18	1.00	666.9	617.4	1.08
S5P20R	1.33	0.18	1.00	673.8	618.3	1.09
S3G25	0.80	0.29	1.00	664.0	631.9	1.05
S3G25R	0.79	0.29	1.00	667.7	635.0	1.05
S3G50	0.79	0.29	1.00	665.5	638.0	1.04
S3G50R	0.78	0.29	1.00	650.3	637.8	1.02
S5G50	1.34	0.17	1.00	670.2	631.6	1.06
S5G50R	1.32	0.17	1.00	670.4	632.4	1.06

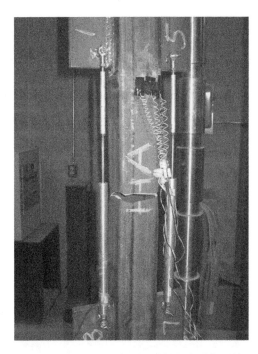

Figure 5. Failure occurred at the mid-length of the HSS.

all cases, the net section efficiency is greater than unity even for the L/w ratio as low as 0.8 and is above 1.05 for the L/w ratio greater than 1.0. There is hardly any improvement in the net section efficiency once the L/w ratio is greater than 1.0. The measured efficiency of the specimens is much higher than the net section efficiency of 0.75 for a L/w ratio of 1.0 and 0.6 for a L/w ratio of 0.8 prescribed by CSA-S16-01. For a square HSS specimen, the net section efficiency was as high as 1.08, and for a rectangular HSS specimen, the net section efficiency was as high as 1.15. A possible reason that some measured net section efficiencies are significantly greater than 1 is because the actual tensile strength of the HSS corner may be much higher than its flat part due to cold-forming in the manufacturing process. However, only the ultimate tensile strength of the flat part of HSS is considered in calculating A_nF_u.

It can also be seen in Table 4 for square specimens, effects of gusset plate thickness (t) and slot opening (G) on the net section efficiency are minimal. Specimens with different gusset plate thicknesses but the same weld length ratio (L/w) have the net section efficiency within 3% of each other. A similar range of net section efficiency difference can be observed for specimens with the same weld length ratio (L/w) but with different slot opening lengths.

Similar to the finding of Korol (1996), the effect of slot orientation on the net section efficiency was found to be minor. Korol (1996) also found that a rectangular HSS specimen slotted on its short side has a better efficiency than on its long side, but the result from this study is inconclusive. This is due to the fact that the effect of slot orientation is more pronounce on specimens with weld length ratios much lower than 1.0, while all specimens in this test program have weld length ratios close to and above 1.0. As can be seen in Table 4, at weld length ratios of around 0.79 and 1.33, specimens slotted on its short side (RS5 and RS3) have a slightly higher net section efficiency than specimens slotted on its long side (RL5 and RL3). But at a weld length ratio of around 1.05, the rectangular specimen slotted on its long side, RL4, has a slightly higher efficiency than RS4.

All square HSS specimens in this test program also have a lower net section efficiency compared to that for rectangular HSS specimens at the same weld length ratio, except for RL3. This can be attributed to the different level of the strength increase at the HSS corner compared to its flat part for square and rectangular HSS.

2.5.2 Specimens with end welding

All square HSS specimens fractured at the mid-length of the specimen after extensive necking. An example of this failure is shown in Figure 5. This means that all square HSS specimens have achieved full efficiency. Unlike square HSS specimens, the rectangular HSS specimen failed at the slotted end. Figure 6 shows the shear failure along the fillet weld of R07. The load versus average LVDT curves for all specimens are also shown in Figure 7. It can be seen that all specimens

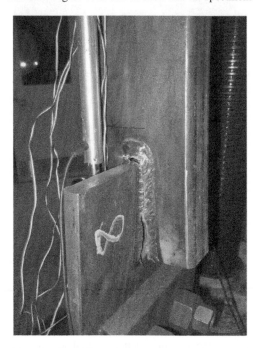

Figure 6. Shear failure along the fillet weld for R07.

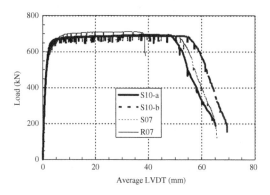

Figure 7. Load versus LVDT displacement curves of specimens with end welding.

Table 5. Test results of specimens with end welding.

No.	L/w	$\bar{x}^*/L$	a/b	P_{uTest} (kN)	$A_n F_u$ (kN)	$P_{uTest}/A_n F_u$
S10-a	0.97	0.21	1.00	668.4	644.9	1.04
S10-b	0.98	0.21	1.00	660.2	646.2	1.02
S07	0.68	0.30	1.00	667.3	647.5	1.03
R07	0.69	0.36	2.46	684.3	674.2	1.01

deformed significantly prior to fracture, even for the rectangular HSS specimen.

Test peak loads and net section efficiencies are shown in Table 5. All specimens have efficiencies slightly greater than unity. Test results also show that there is no significant net section efficiency difference between specimens with the weld length ratios (L/w) of 0.7 and 1.0. But it should be pointed out again that the net section efficiency calculation is based solely on the tensile strength of the flat part of HSS. Comparing specimens with no end welding at the weld length ratio around 1.33, rectangular HSS specimens in general have around a 6% higher efficiency than square HSS specimens. For this reason, the effect of shear lag can be considered to be more severe for R07 than S07 even though both specimens have about the same net section efficiency. The severity of the effect of shear lag can also be deduced from the location of failure, where R07 fractured at the slotted end and S07 failed at the mid-length.

Specimens S07 and R07, with the weld length ratio of 0.7, also have net section efficiencies that are comparable to specimens with no end welding that have a higher weld length ratio of around 0.8. This shows that the effect of shear lag is more severe in a slotted HSS connection without end welding than one with end welding.

3 NUMERICAL SIMULATION

Finite element models were developed according to geometries of test specimens to carry out the numerical simulation of the test. The numerical simulation was carried out with the finite element program ABAQUS (2003). A geometrical and material non-linear analysis was performed in this study. The model for the HSS connection was constructed with both four-node, bi-linear shell element (S4) and eight-node tri-linear solid element (C3D8). Solid element was employed to model HSS only at the region close to the end of the gusset plate in order to reduce the computational time. Figure 8 shows the typical mesh for the square HSS connection without end welding. Only one-eighth of the specimen was modeled due to symmetry. The fillet weld between the hollow structural section wall and the gusset plate is modeled by constraining the displacements and rotations to be equal between the node on the HSS wall to the corresponding node on the gusset plate. Thus, the fillet weld is not explicitly modeled using elements.

The finite element model was loaded by applying uniform displacement at the end of the gusset plate. Failure of the connection was assumed to have occurred and the analysis terminated when the equivalent plastic strain in any part of the model has reached the critical limit of 0.9. This limit was obtained from tension coupon tests of the material. The equivalent plastic strain (ε_{eq}^p) is given by,

$$\varepsilon_{eq}^p = \int d\varepsilon_{eq}^p \text{, with} \tag{8}$$

$$d\varepsilon_{eq}^p = \left(\frac{2}{3}d\varepsilon_{ij}^p d\varepsilon_{ij}^p\right)^{\frac{1}{2}}, \tag{9}$$

where ε_{ij}^p is the plastic strain tensor. All analyses were carried out using a material user-subroutine with ABAQUS so that the analysis can be terminated automatically when the equivalent plastic strain limit has been reached. Material properties used in the analysis were derived from tension coupon tests. A multi-linear true stress versus true plastic strain data were inputted in the analysis for each material. In the analysis, the material was assumed to deform according to the incremental flow theory of plasticity with isotropic hardening. The HSS was also assumed to have a uniform material properties defined by its flat part even

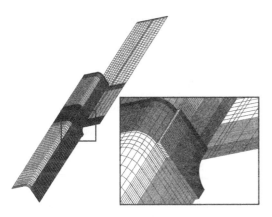

Figure 8. The typical mesh for a square HSS connection without end welding.

Table 6. Simulation results of selected specimens with no end welding.

No.	L/w	$\bar{x}^*/L$	$P_{u\_pred}/A_n F_u$	$P_{uTest}/A_n F_u$
RL5	1.33	0.21	0.98	1.14
RS5	1.34	0.17	1.02	1.15
S5P16	1.33	0.17	1.03	1.07
RL4	1.05	0.26	0.98	1.14
RS4	1.06	0.21	1.02	1.10
S4P16	1.06	0.21	1.03	1.07
RL3	0.79	0.35	0.96	1.04
RS3	0.80	0.29	0.99	1.08
S3P16	0.79	0.29	0.99	1.03
S3P12	0.79	0.28	0.99	1.04
S5P12	1.34	0.17	1.02	1.07
S3P20	0.80	0.29	0.99	1.01
S5P20	1.33	0.18	1.03	1.08

though its corner may have significantly different material properties.

The predicted net section efficiency of a HSS specimen is taken as $P_{u\_pred}/A_n F_u$, where P_{u_pred} is the predicted peak load. Comparisons of test and predicted efficiencies are shown in Table 6 for selected specimens with no end welding. It can be seen that the simulation predicts a lower net section efficiency than that of the test for all specimens. The lower predicted efficiency is the result of neglecting the higher material strength at the HSS corner in the finite element model. The larger discrepancy noticed for the rectangular specimen implies that the amount of strength increase of the HSS corner is much higher for the rectangular HSS than the square HSS.

4 PROPOSED EQUATIONS AND GUILDLINES

4.1 Proposed equations

Figure 9 shows the net section efficiency versus the weld length ratio (L/w) from the current test program, from Korol (1996) and values calculated according to provisions of CSA-S16-01 from Equations 2–4. It can be seen that provisions for shear lag in CSA-S16-01 greatly underestimate the test net section efficiency. An equation to characterize the net section efficiency factor using the weld length ratio (L/w) is thus proposed. The proposed equation is expressed as:

$$U_{pw} = 0.05 + 1.1\left(\frac{L}{w}\right) \leq 1, \tag{10}$$

with an upper limit of 1. The net section efficiency according to Equation 10 is also being compared to the test efficiency in Figure 9. It can be seen that Equation 10 gives a conservative estimate of the test efficiency for all but a small number of specimens from Korol (1996) at the weld length ratio around 1. It should be noted that the net section efficiency reported by Korol (1996) may have been calculated using the static

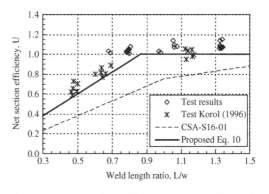

Figure 9. Net section efficiency versus weld length ratio, L/w.

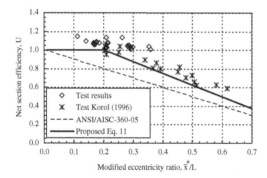

Figure 10. Net section efficiency versus modified eccentricity ratio, $\bar{x}^*/L$.

peak load of the HSS specimen but a non-static ultimate strength of the material. The ultimate strength of the material was determined according to ASTM-E8, which does not has to be the static strength. For this reason, the reported net section efficiencies for some specimens at the weld length ratio around 1 were below unity and were significantly lower than the efficiency observed in this test program at the same weld length ratio. It should also be noted that the predicted efficiency from the simulation in Table 6, P_{u_pred}/A_nF_u, is greater than or close to unity when the weld length ratio of the specimen is greater than 1.0. The efficiency of close to 1 was attained even though the numerical simulation did not consider the possible strength increase at the HSS corner. This gives further support that the reported efficiencies by Korol (1996) may actually be higher.

Figure 10 shows the net section efficiency versus the modified eccentricity ratio. Similar to CSA-S16-01, provisions in ANSI/AISC-360-05 according to Equation 6 have also significantly underestimated the test net section efficiency. Thus, another equation to characterize the net section efficiency factor using the net section eccentricity ratio is proposed. The proposed net section efficiency factor takes the form of:

$$U_{px} = 1.25 - 1.25\left(\frac{\bar{x}}{L}\right) \le 1,\qquad(11)$$

with an upper limit of 1. The net section efficiency according to Equation 11 is also being compared to the test efficiency in Figure 10. Again, the proposed equation gives a conservative estimate of the test efficiency in all but a small number of tests from Korol (1996) at the net section eccentricity ratio of 0.2. These are the same test data from specimens with the weld length ratio around 1 as discussed previously.

5 CONCLUSIONS

1. Provisions for shear lag prescribed in CSA-S16-01 and ANSI/AISC-360-05 are overly conservative

when applied to slotted square and rectangular HSS tension members. Equations 10 and 11 are proposed for calculating the net section efficiency factor that is more representative of the behavior of a slotted HSS tension member.
2. The weld length is the main factor that influences the effect of shear lag. Maximum net section efficiency can be achieved with a weld length ratio (L/w) greater than 1.
3. The thickness of gusset plate and slot opening length only have a minor influence on the strength of slotted HSS connections without end welding.
4. The slotted HSS connection with end welding is always stronger than the one without end welding when all other geometrical parameters are equal. For a slotted square HSS connection with end welding, 100% net section efficiency was achieved with a weld length ratio (L/w) of 0.7.

ACKNOWLEDGEMENT

This research project is funded by the Natural Sciences and Engineering Research Council of Canada and Steel Structures Education Foundation.

REFERENCES

ABAQUS. 2003. ABAQUS version 6.4 Standard User's Manual. Hibbitt, Karlsson & Sorensen Inc., Pawtucket, R.I.
ANSI/AISC. 2005. "Specification for Structural Steel Buildings", ANSI/AISC 360-05, American Institute of Steel Construction, Chicago, Illinois.
ASTM. 2004. Standard Methods for Tension Testing of Metallic Materials: ASTM-E8-04. American Society for Testing and Materials, Philadelphia, Pennsylvania.
Cheng, J.J.R., Kulak, G.L., and Khoo, H. 1998. Strength of slotted tubular tension members. Canadian Journal of Civil Engineering, 25(6): 982–991.
CSA. 2001. Limit States Design of Steel Structures: Standard CAN/CSA-S16-01. Canadian Standards Association, Rexdale, Ontario, Canada.
CSA. 2004. General Requirements for Rolled or Welded Structural Quality Steel: CAN/CSA G40.20-04. Canadian Standards Association, Rexdale, Ontario, Canada.
Huang, R. 2005. An Experimental and Numerical Study of Square and Rectangular HSS Connections. Masters Thesis, Carleton University, Ottawa, Ontario, Canada.
Korol, R.M. 1996. Shear Lag in Slotted HSS Tension Members. Canadian Journal of Civil Engineering, 22(6): 1350–1354.
Willibald, S., Packer, J.A., Saucedo, G. Ma and Puthli, R.S. 2004. Shear Lag in Slotted Gusset Plate Connections to Tubes. ECCS/AISC Workshop, Connections in Steel Structures V: Innovative Steel Connections, Amsterdam, the Netherlands.
Zhao, R. 2005. A Study on Slotted Square and Rectangular Hollow Structural Section Connections. Masters Thesis, Carleton University, Ottawa, Ontario, Canada.

Tubular Structures XI – Packer & Willibald (eds)
© 2006 Taylor & Francis Group, London, ISBN 0-415-40280-8

New ultimate strength formulae for CHS K joints

Y. Kurobane

Kumamoto University, Kurokami, Kumamoto, Japan

ABSTRACT: Many numerical investigations into the ultimate behavior of simple tubular joints have been conducted for the past 10 years. A reanalysis of the new data as well as existing experimental data is attempted to update the past ultimate strength equations. The prediction equations applicable to gap and lap K joints, failing in the chord wall shell bending and brace local buckling modes, are proposed.

1 INTRODUCTION

Many numerical investigations into the ultimate behavior of simple tubular joints have been conducted since the first IIW recommendations were proposed in 1981. A reanalysis of the new data as well as the existing experimental data combined is attempted to update the past ultimate strength equations (Kurobane et al. 1984), on which the IIW Recommendations are based. Above all of interest is if these numerical results can reproduce the past vast amount of experimental database. This report focuses on K joints because the ultimate behavior of K joints is more controversial and complex than that of the other simple joints.

The most common failure mode of K joints is the chord wall shell bending deflection developed under the compression braces. Best predictive equations for the ultimate strengths of the gap and lap K joints failing in this failure mode are calculated using multiple regression analyses. The past equation (Kurobane et al. 1984) used one continuous equation applicable to both gap and lap joints. However, in this proposal, separate formulae are prepared for gap and lap joints. This is because, in recent investigations, the $\tau = t_1/t_0$ ratio was introduced as an important geometrical variable into their numerical studies (See Appendix for symbols). τ gives significant effects on the ultimate behavior of the lap joint while the effect of τ on the gap joint is negligibly small.

The other failure mode common in K joints is the local buckling of the compression braces. This failure mode has not been fully recognized in the existing design recommendations, probably because this was first proposed based only on experimental results of 44 specimens (Kurobane et al. 1986). However, recent numerical results (Dexter et al. 1999, Gazzola et al. 2000) were effective to enhance the reliability and widen the validity range of the existing prediction equation.

2 DEFINITION OF ULTIMATE STRENGTH

The definition of the ultimate strength is identical to that defined in the previous paper (Kurobane et al. 1984). The load-deformation curve of the K-joint frequently shows a multiple curve with two peak loads, the second peak load being usually higher than the first one. The ultimate strength is defined as the first peak load. When the load-deformation curve changes the curvature from a negative to a positive curvature, the load at the point of curvature change is defined as the first peak load.

Recent numerical studies define the ultimate strength as the lowest of the three loads as follows:

1. The first peak load.
2. The load at which the load-deformation curve (or the brace load-chord wall indentation curve) reaches a certain predetermined deformation limit.
3. The load at which the maximum strain in a joint reaches a certain critical value.

Most of the experimental investigations, contrarily, measured only overall deformations of the joints and were found difficult to determine the ultimate strength based on the second criterion. However, the deformation limit governed the ultimate strengths of limited K-joints with certain geometrical and loading conditions. The deformation limit is treated separately from the strength limit and can be used to correct ultimate strength formulae.

The third criterion may be addressing the possibility of ductile tensile cracking starting from weld toes of a joint. However, no reliable numerical approach has been found yet to predict the crack growth from the weld toes of tubular joints. An example of the critical strain observed at the root of surface notches when a very small crack (greater than 0.05 mm in length) appeared is of the order of 100 percent (Toyoda et al. 2001). Recent developments on the ductile crack growth, however, are still at a stage of analyzing simple models like single edge notched bend specimens or axi-symmetric tensile bars (for example Ruggieri 2004).

In this study the ultimate strengths determined by either the 2nd or 3rd criterion are omitted from the database. However, the loads determined by the 2nd or 3rd criterion are frequently very close to the first peak loads. If this latter case applies, the data are included in the database after examining the load deformation curves supplied by the investigators (Dexter 1996, Vegte 2003a).

3 CONSIDERATION OF SIZE EFFECTS

The size of welds has been known to produce a significant effect on the ultimate strength of gap K-joints (Lee et al. 1995, Dexter 1996). The size effect on lap joints is insignificant. When modeling the joints numerically, Lee, Dexter and Gazzola et al. (1996, 1999, 1999a, 2000) assumed the weld profile conforming to the fillet-welded prequalified tubular joints, while Vegte et al. (2003, 2003a, 2004) adopted the weld profile conforming to CJP tubular connections, both being based on AWS D1.1 (2002). The weld leg length (from the brace toe to the weld toe) is equal to the thickness of the brace in the former model, while the leg length in the latter model is nearly nil. The weld leg lengths are somewhat greater than the thickness of the braces for laboratory specimens, although they vary from specimen to specimen. Small-size specimens have relatively larger weld sizes than larger specimens. It is impracticable to fabricate small size tubular joints following the AWS CJP details. (In order to reduce the weld size welders have to reduce the heat-input. This invites a lack of fusion at the roots of the welds.) Possible effects of the weld size on the strength of gap joints are considered by introducing the new procedures as follows:

1. Introducing a new parameter Δd_1 to increase the diameter of the compression brace, in addition to the parameter Δg to decrease the gap size (already used in the existing equation). Both Δd_1 and Δg were determined by the regression analysis and given in mm.
2. The chord outer diameter of the joint models generated by Lee and coworkers is equal to 1000 mm, which is much larger than the chord outer diameters of laboratory specimens. The weld leg length being equal to the thickness of the braces may be a reasonable approximation of laboratory specimens.
3. The chord outer diameter of the joint models generated by Vegte is equal to 406 mm, which is within the size range of laboratory specimens. The joint models in which the weld leg length is nearly equal to 0 should show significantly smaller strength than experimental results. It is required to multiply

dimensions of joints created by Vegte by a certain factor so that his numerical results reproduce the ultimate strength of actual joints.

4. The multiplication factor of 5 is selected here after trial and error approaches so that his numerical results give the same mean strength as that of experimental results as well as the numerical results by the other investigators.

4 CHORD WALL SHELL BENDING IN GAP K JOINTS

The following ultimate strength equation is derived from all the reliable databases for gap K-joints available as of 2005.

$$N_{1,u,pred} = f_0 \cdot f_1 \cdot f_2 \cdot f_3 \cdot f_4 \cdot f_5 \cdot \frac{t_0^2\,\sigma_{y,0}}{\sin\theta_1} \qquad (1)$$

where

$$f_0 = 3.63\left(\frac{d_1 + 10.5}{d_0}\right)^{(0.647+0.00607(2\gamma))}$$

$$f_1 = 1 + \frac{0.0315(2\gamma)^{0.899}}{\exp\left(0.525\dfrac{g-7.14}{t_0}-0.466\right)+1}$$

$$f_2 = (2\gamma)^{0.549}$$

$$f_3 = 1 - 0.286\cos^2\theta_1$$

$$f_4 = 1 + 0.302n_0 - 0.433n_0^2$$

$$f_5 = \left(\frac{\sigma_{y,0}}{\sigma_{u,0}}\right)^{-1.02}$$

The statistics of the observed (experimental and numerical) to predicted ultimate strength ratios are shown below.

	Mean	COV	No. of data set
Test	1.01	0.090	355
FEA	1.00	0.081	173
All data	1.00	0.087	528

The standard deviations were calculated by:

$$\sqrt{\frac{\sum_1^n\left[\left(N_{1,u}/N_{1,u,pred}\right)_i-\left(N_{1,u}/N_{1,u,pred}\right)_m\right]^2}{n-1}}$$

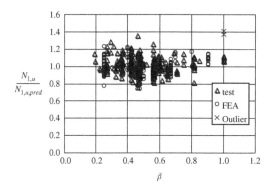

Figure 1. Dimensionless strength plotted against β (gap K joints, Eq. 1).

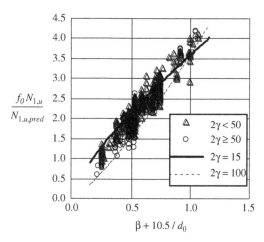

Figure 2. Dimensionless strength showing combined effects of β and 2γ (gap K-joints, Eq. 1).

where the subscript i denotes the i-th data set, while m denotes the mean. The symbol n signifies the number of data sets.

The observed (experimental and numerical) to predicted ultimate strength ratios are plotted against β in Fig. 1. All the test and numerical results distribute randomly about the mean. No systematic tendency is detected in this figure. There are two exceptions to the above conclusion. The two numerical results generated by Dexter (1996), for which $\beta = 1.0$ and 2γ is equal to either 70 or 100, are significantly greater than those predicted by Eq. 1. These two numerical results are shown as outliers in Fig. 1. No other experimental or numerical result exists in the range where $\beta = 1.0$ and $2\gamma \geq 70$. It may be possible that the joint strength increases discontinuously as the geometrical variables approach the above range. However, the numerical models do not take into account the effects of initial imperfections and residual stresses. These effects may be detrimental to the ultimate capacity of thin-walled CHS joints. It is prudent to omit these two joints until more experimental and numerical evidences become available. These two data sets are excluded from the database for regression analyses in this report.

The ultimate strength of the gap K-joint increases with β, which is explained by the exponential function f_0 of Eq. 1. The rate of increase varies with the 2γ ratio. The exponent of f_0 varies from 0.74 to 1.25 as 2γ increases from 15 to 100. The function f_0 is plotted against $(\beta + 10.5/d_0)$ in Fig. 2 for the two cases of $2\gamma = 15$ and $2\gamma = 100$ (heaviest and lightest chords in the database). Observed results are also plotted in the same figure after subdividing them into two groups of $2\gamma < 50$ and $2\gamma \geq 50$. The data points should fall on the space enclosed between the two curves. Although the data scatter, triangular dots are closer to the solid curve, while circular dots are closer to the dashed curve, showing the dependence of the joint strength on β and 2γ both.

The ultimate strength of the gap K-joint increases rather sharply as the two braces approach and intersect with each other. This trend is stronger, the greater the 2γ ratio. The combined dependence of the ultimate strength on g and 2γ is represented by the one smooth function f_1 of Eq. 1. Again, the function f_1 is plotted against $(g - 7.14)/t_0$ in Fig. 3 for the two cases of $2\gamma = 15$ and $2\gamma = 100$. Observed results are also plotted in the same figure after subdividing them into two groups of $2\gamma < 50$ and $2\gamma \geq 50$. Although the data scatter, the function f_1 captures well the behavior unique to the gap K joint.

It is interesting to see that the exponent of the $\sigma_{y,0}/\sigma_{u,0}$ ratio (in the function f_5) becomes nearly equal to -1.0 in Eq. 1. This means that the ultimate strength of the gap K joint is proportional to $\sigma_{u,0}$, rather than to $\sigma_{y,0}$.

It should be noted that Eq. 1 does not include the dimensions related to the tension brace (like d_2 and θ_2). This is because these dimensions were found to be uncorrelated with the strength of the K-joint according to the past extensive analyses. Similarly, the effect of the chord stress on the strength of the K-joint is represented by the function only of the chord stress n_0 on the compression brace side. Note that the ultimate capacity of the K-joint is governed by shell bending deflection of the chord wall under the compression brace.

Vegte et al. (2003) insist that the effect of the chord stress should be represented by the maximum (in absolute value) stress in the chord. In this case the chord stress function is not only a function of n_0 but also of n_0' on the tension brace side. The effect of chord stress can no longer be represented by a simple continuous function of n_0 like f_4. To show this, the chord stress effect (the ratio of the strength of the K-joint to the strength of

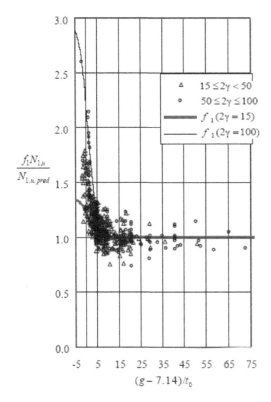

Figure 3. Dimensionless strength showing combined effects of $(g - 7.14)/t_0$ and 2γ (gap K-joints, Eq. 1).

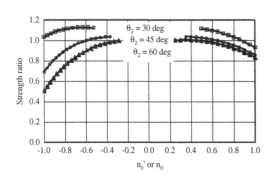

Figure 4. Alternative chord stress functions plotted against maximum stress in chords ($\beta = 0.8$, $2\gamma = 19.2$, $g/d_0 = 0.15$).

the same joint when $n_0' = 0$) is plotted against the maximum chord stress in Fig. 4. The joint models adopted here are identical to one of Vegte's models except that θ_2 is varied as 60°, 45° and 30° while θ_1 is kept at 60°. As this example shows, the chord stress function turns out to be discontinuous and depends not only on the angle between the tension brace and chord but also on either n_0 or n_0'. This would confuse designers.

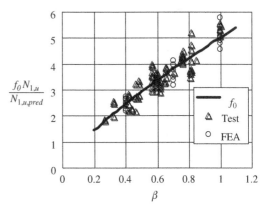

Figure 5. Dimensionless strength showing effects of β (lap K-joints, Eq. 2).

5 CHORD WALL SHELL BENDING IN LAP K JOINTS

The following equation is proposed as the ultimate strength equation for the lap K-joint:

$$N_{1,u,pred} = f_0 \cdot f_1 \cdot f_2 \cdot f_3 \cdot f_4 \cdot f_5 \cdot \frac{t_0^2 \sigma_{y,0}}{\sin\theta_1} \qquad (2)$$

where

$$f_0 = 5.03\,\beta^{0.775}$$

$$f_1 = \left(\frac{\sigma_{y,1}}{\sigma_{y,0}}\tau\right)^{0.352} + 0.152\left(-\frac{g}{d_1}\right)^{1.21}\left(\frac{\sigma_{y,1}}{\sigma_{y,0}}\tau\right)^{3.20}$$

$$f_2 = \left(2\gamma\right)^{0.686}$$

$$f_3 = 1 - 0.620\cos^2\theta_1$$

$$f_4 = 1 + 0.303n - 0.446n^2$$

$$f_5 = \left(\frac{\sigma_{y,0}}{\sigma_{u,0}}\right)^{-0.610}$$

The statistics of the observed (experimental and numerical) to predicted ultimate strength ratios are shown below.

	Mean	COV	No. of data sets
Test	1.01	0.100	84
FEA	1.00	0.076	29
All data	1.00	0.094	113

Figure 5 shows the function f_0 along with the residuals about f_0. No systematic error is found in this figure.

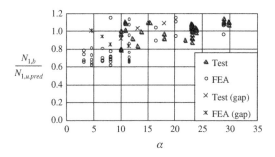

$\dfrac{N_{1,b}}{N_{1,u,pred}}$

▲ Test
○ FEA
× Test (gap)
× FEA (gap)

Figure 6. Local buckling strength to chord wall shell bending strength plotted against α.

6 LOCAL BUCKLING OF COMPRESSION BRACES

The local buckling of the compression braces is the most common failure modes observed in lap K-joints and is observed also in gap K-joints less frequently. The compression braces are squashed showing typical deflection patterns of the brace wall in local areas near the welded joints, frequently accompanying shell bending deflections of the chord and tension brace walls. The ultimate strength of joints failing in the brace local buckling mode is significantly lower than the predicted ultimate strength of the same joints that fail in the chord wall shell-bending mode.

It should be noted that the brace local buckling is not strictly considered as a bifurcation problem. But the failure mode is termed "brace local buckling" because the stability of the brace wall is strongly correlated with the axial compressive stress on the member.

From the 36 data sets generated by Dexter, 2 data sets for gap joints, in which $\beta = 1.0$ and $2\gamma \geq 70$, were deleted for the same reason as that mentioned in Section 4. The ratio of the observed local buckling strength $N_{1,b}$ to the ultimate strength of the same joint predicted by chord wall shell bending strength formula (Eq. 1 or Eq. 2) $N_{1,u,pred}$ is plotted against the local buckling parameter α for all the data sets in Fig. 6. The data for gap joints are distinguished by different marks from the data for lap joints. The local buckling parameter is given as:

$$\alpha = \frac{E}{\sigma_{y1}} \frac{t_1}{d_1} \qquad (3)$$

where $\sigma_{y,1}$ is the yield stress of the compression brace material. The ratio $N_{1,b}/N_{1,u,pred}$ tends to become lower than 1.0 as α gets smaller than about 20 as seen in Fig. 6, although data points scatter widely.

The past investigation (Kurobane, et al. 1986) indicated that the local buckling strength was given as a function of the local buckling parameter α and the chord wall shell bending strength $N_{1,u,pred}$ of each

joint. The past investigation argues that the chord wall shell bending strength influences the brace local buckling strength because a chord wall failure prevents the redistribution of the bending moment at the brace end prior to brace local buckling.

This reanalysis adopts the same mathematical model as the model used in the past investigation. However the new database include many additional data sets especially in the range where α is smaller than 10 (thin-walled braces). A function of β is considered necessary to fit the model with the new database better. The best estimate of the local buckling strength, $N_{1,b,pred}$ is given by the following equation:

$$\frac{N_{1,b,pred}}{A_1 \cdot \sigma_{y,1}} = f_0 \cdot f_1 \cdot f_2 \qquad (4)$$

where

$$f_0 = 0.527\alpha^{0.208}$$

$$f_1 = \beta^{0.141}$$

$$f_2 = \left(\frac{N_{1,u,pred}}{A_1\sigma_{y,1}}\right)^{0.606}$$

Note that both the local buckling strength and chord wall shell bending strength are represented in a non-dimensional form by dividing them by the axial yield load of the compression brace, where A_1 signifies the cross-sectional area of the compression brace.

The statistics of the observed to predicted ultimate strength ratios are shown below:

	Mean	COV	No. of data sets
Test	1.01	0.057	44
FEA	0.996	0.083	47
All data	1.00	0.072	91

It is surprising to see that the COV for the test database is as small as 0.057 and is even smaller than the COV for the FE database. This may be because geometrical variables in the test database distribute over a rather narrow range (See Figs. 7 and 8).

The 2 functions f_0 and f_2 of Eq. 4 are compared with the database in Figs. 7 and 8. Figure 7 shows that the local buckling strength decreases as α decreases. The function f_0 in this figure is equal to non-dimensionalized local buckling strength when $\beta = N_{1,pred}/(A_1\sigma_{y,1}) = 1$. Figure 8 demonstrates how the local buckling strength increases with the chord wall shell bending strength. Figures 7 and 8 suggest that Eq. 4 underestimates the local buckling strength of gap joints by about 10 percent.

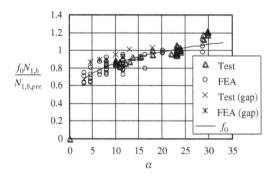

Figure 7. Dimensionless local buckling strength showing effects of α.

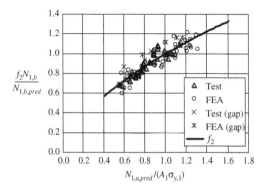

Figure 8. Dimensionless local buckling strength showing effects of chord wall shell bending strength.

7 COMPARISON WITH API FORMULAE FOR CHORD WALL SHELL BENDING STRENGTH

The database for gap K-joints is compared with the new API formulae (API 2003). The factor of safety incorporated in these formulae is taken to be 1.0 here. The observed to predicted ultimate strength ratios are plotted against β in Fig. 9, in which these ratios according to Eq. 1 are also plotted for comparison. The observed to predicted ultimate strength ratios according to the API formulae give the following statistics.

	Mean	COV	No. of data set
Test	1.48	0.209	355
	(1.34)	(0.17)	(161)
FEA	1.33	0.147	173
	(1.14)	(0.11)	(440)
All data	1.43	0.199	528

Note: The figures in parentheses show the values given in the API Recommendations.

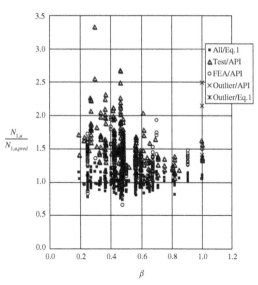

Figure 9. Dimensionless strength plotted against β (gap K-joints, API formulae and Eq. 1).

The following observations can be made about these figures and table.

1. The observed to predicted strength ratios scatter very widely. The predictions according to Eq. 1, in contrast, cluster around the mean value of 1.0.
2. The means of these ratios (the mean biases) according to the API formulae are far from 1.0. The mean for the test database is greater by 11 percent than that for the FE database. This is not surprising because the API formulation does not take into account the weld size effects (See Section 3) and, therefore, because the API formulae underestimate the experimental results.

The API Recommendations make a similar comparison of their formulae with their database. The results are shown in the above table in the parentheses. It is rather confusing to see the differences between the two statistics. Especially the mean bias for the FE database according to the API Recommendations is 16 percent lower than the corresponding value according to the database of this report. It is suspected that the API database is based on numerical models with no welds at the intersections of mid-surface shell models.

When the observed to predicted strength ratios show a bias error and scatter as large as those shown in Fig. 9, it is natural to doubt if the proposed prediction models contain a certain lack of fit. The lack of fit is detected when the proposed gap function Q_g is plotted against g/d_0 and is compared with the variations of $N_{1,u}/N_{1,u,API}$ about the gap function (See Fig. 10). Q_g is the function of g/d_0 only for the range $g/d_0 \geq 0.05$

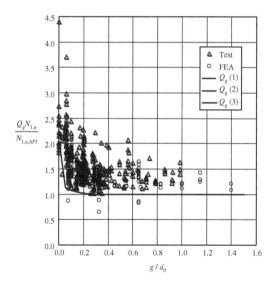

Figure 10. Dimensionless strength showing effects of g/d_0 (gap K-joints API formulae).

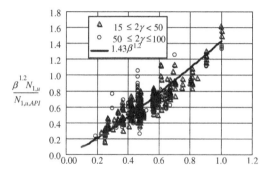

Figure 11. Dimensionless strength showing effects of β (gap K-joints API formulae).

according to the API recommendations. The data scatter most widely for the range $0.05 \le g/d_0 < 0.36$, in which $Q_g = 1 + 0.2(1 - 2.8\,g/d_0)^3$. Many gap K joints exist in this range in actual structures. The deviations of data points from the mean Q_g are caused not only by the pure error but also by the systematic error. Compare Fig. 10 with Fig. 3, which shows that the deviations of data points from the mean are dramatically reduced by introducing the additional variable 2γ into the gap function. When g/d_0 becomes greater than 0.36, the ultimate strength of gap K joints is almost independent of g/d_0. The systematic error mentioned above disappears for this range, apart from the mean bias.

The API formulae use the simple function, $\beta^{1.2}$ to represent the effect of the β ratio on the ultimate strength. This exponent should be a function of 2γ as discussed in Section 4. To see how this simple expression fits with the observed results, a function $1.43\beta^{1.2}$

is compared with the observed results (See Fig. 11). The number 1.43 is the bias factor for all the data (See the previous table). This simple function roughly captures the effect of the β ratio on the ultimate strength. However, a large scatter of data in the middle range of the β ratio ($0.3 < \beta < 0.9$) is one of the shortcomings with the API formulae. Compare Fig. 11 with Fig. 2.

The two numerical results generated by Dexter (1996), for which $\beta = 1.0$ and D/T is equal to either 70 or 100, are also included in Fig. 9 (shown as outliers). The API formulae also show, like Eq. 1, that these two results are significantly greater than the other experimental and numerical results.

8 CONCLUSIONS

The 3 different ultimate strength prediction equations, the one equation applicable to gap K-joints failing in the chord wall shell bending mode, the one equation applicable to lap K joints failing in the chord wall shell bending mode and the one equation applicable to gap and lap K joints failing in the brace local buckling mode, are proposed. These equations show good accuracy with the COV of the observed (experimental and numerical) to predicted strength ratios ($N_{1,u}/N_{1,u,pred}$) varying between 0.072 and 0.094. The means of $N_{1,u}/N_{1,u,pred}$ are equal to 1.0 for both the experimental and numerical results. The numerical database was found to reproduce experimental database very well, provided that the data are processed to take into account the effect of the weld size on the strength of gap K joints. Since all the reliable experimental and numerical results are included in the database, except for a few data that were deleted for physical grounds, the proposed equations are considered reliable and can be used as a basis for developing design equations.

However, the evaluation of the variability of the predicted strength for deriving design equations should be based on the experimental database. This is because numerical models do not consider the following factors that may increase the variability of the ultimate strength:

1. errors in fabrication and unavoidable eccentricities in loading
2. several different boundary conditions in actual structures, which may introduce different primary and secondary bending moments into members
3. initial imperfections
4. residual stresses

The API recommendations do not identify the brace local buckling failure mode but identify only the chord wall shell bending failure mode. The API formulae produce large errors in predicting experimental and numerical results. These errors contain systematic components, especially in their gap function. The API

267

formulae give different mean bias factors varying with the database. Thus, the API formulae are unreliable.

The following subjects still need to be studied:

1. An extension of the formulae for K joints to formulae for multi-planar KK joints. This may be done by following the procedures adopted in the previous investigation (Kurobane et al. 1998).
2. The formulation of design equations, including corrections to allow for the deformation limits.
3. Reexaminations of the database generated by Vegte.

9 APPENDIX NOTATION

0,1,2 = subscripts signifying chord, compression brace and tension brace, respectively
A = cross-sectional area of tube
API = subscript signifying prediction by API formula
b = subscripts signifying local buckling strength
COV = coefficient of variation
d = outside diameter of tubes
g = gap in K-joint between brace toes
N = axial force in member (positive in tension for chord)
n = $N_i/(A_i\sigma_{y,i})$: axial to yield stress ratio in member
$pred$ = subscript signifying prediction
t = wall thickness of tubes
u = subscript signifying ultimate strength
α = local buckling parameter
β = d_1/d_0: diameter ratio
γ = $d_0/(2t_0)$: chord thinness ratio
θ = inplane angle between chord and braces
σ_y = measured yield strength of tube material
σ_u = measured ultimate tensile of tube material
τ = t_1/t_0: ratio of compression brace wall thickness to chord wall thickness

REFERENCES

API, 2003, Proposed updates to API BP2A provisions for tubular joints, American Petroleum Institute SC 2

AWS, 2002, *Structural Welding Code—Steel*, ANSI/AWS D1.1, 18th edition, American Welding Society, Miami, Fla., USA

Dexter, E.M., 1996, Effects of overlap on behavior and strength of steel circular hollow section joints, Ph.D thesis, University of Wales Swansea, Swansea, UK

Dexter, E.M. and Lee, M.M.K., 1999, Static strength of axially loaded tubular K-joints. I: Behavior *J. Structural Engineering*, ASCE, 125 (2), 194–201

Dexter, E.M. and Lee, M.M.K., 1999a, Static strength of axially loaded tubular K-joints. II: Ultimate capacity, *J. Structural Engineering*, ASCE, 125 (2), 202–210

Gazzola, F. and Lee, M.M.K., 2000, Design equation for overlap tubular K-joints under axial loading, *J. Structural Engineering*, ASCE, 126 (7), 798–808

IIW, 1989, Design recommendations for hollow section connections predominantly statically loaded. 2nd edition, International Institute of Welding Commissions XV, IIW Doc. XV-701-89

Kurobane, Y., Makino, Y. and Ochi, K., 1984, Ultimate resistance of unstiffened tubular joints, *J. Structural Engineering*, ASCE, 110(2), 385–400

Kurobane, Y., Ogawa, K., Ochi. K. and Makino, Y., 1986, Local buckling of braces in tubular K-joints, *Thin-Walled Structures* 4, 23–40

Kurobane, Y. and Makino, Y., 1998, Analysis of existing and forthcoming data for multi-planar KK-joints with circular hollow sections, CIDECT Report 5BF-10/98

Lee, M.M.K. and Wilmshurst, S.R., 1995, Numerical modeling of CHS joints with multiplanar double K configuration, *J. Construct Steel Research*, 32, pp. 281–301

Lee, M.M.K. and Dexter, E.M., 1999, A new capacity equation for axially loaded multi-planar tubular joints in offshore structures, Final Report for JIP-Ovalising parameters for multi-planar tubular joints in offshore structures

Makino, Y., Kurobane, Y., Ochi, K., Vegte van der, G.J. and Wilmshurst, S., 1996, Database of Test and Numerical Analysis Results for Unstiffened Tubular Joints, IIIW Doc. XV-E-96-220, Kumamoto Univ., Kumamoto, also accessible at the web site http://www.arch.kumamoto-u.ac.jp/maki_lab/database.html

Ochi, K., Makino, Y. and Kurobane, Y., 1983, Basis for design of unstiffened tubular joints under axial brace loading, Memoirs, Faculty of Engineering, Kumamoto University, Kumamoto, Japan

Ruggieri, C., Numerical investigation of constraint effect on ductile fracture in tensile specimens, *J. Brazilian Soc. of Mechanical Sciences & Engineering*, 26(2),190–199

Toyoda, M., Ohata, M., Yokota, M. M., Yasuda, O. and Hirono, Criterion for ductile cracking for the evaluation of steel structures under large scale cyclic loading, *Proc. 20th Int. Conf. on Offshore Mechanics and Arctic Engineering*, OMAE01/MAT-3103, 2001

Vegte van der, G.J., Liu, D.K., Makino, Y. and Wardenier, J., 2003, New chord load functions for circular hollow section joints, CIDECT Report 5BK-4/03

Vegte van der, G.J., 2003a, private communication

Vegte van der, G.J., 2004, private communication

Wardenier, J. and Koning, de, C.H.M., 1977, Investigation into the static strength of welded Warren type joints made of circular hollow sections, Stevin Report No. 6-77-5, Stevin Laboratory, Delft University of Technology, Delft, the Netherlands

Tubular Structures XI – Packer & Willibald (eds)
© 2006 Taylor & Francis Group, London, ISBN 0-415-40280-8

Tests of stainless steel RHS X-joints

R. Feng & B. Young

Department of Civil Engineering, The University of Hong Kong, Hong Kong

ABSTRACT: This paper describes a test program on cold-formed stainless steel welded tubular X-joints fabricated from square hollow section (SHS) and rectangular hollow section (RHS) brace and chord members. Both high strength stainless steel (duplex and high strength austenitic) and normal strength stainless steel (AISI 304) specimens were tested. The ratio of brace width to chord width of the specimens (β) varied from 0.5 to 1.0 so that failure modes of chord face failure and chord side wall failure were observed. A total of 11 tests was performed. The test results are compared with the design procedures in the Australian/New Zealand Standard for stainless steel structures, CIDECT and Eurocode design rules for carbon steel tubular structures.

1 INTRODUCTION

Cold-formed stainless steel tubular connections are being used increasingly for architectural and structural purposes in recent years. This is due to the aesthetic appearance, high corrosion resistance, ductility property, improved fire resistance and ease of maintenance as well as ease of construction of stainless steel structural members. Typical applications of stainless steel include frameworks in corrosive environments and truss girders in atriums, facade structures in office buildings, offshore platform, canopy structures, wall cladding, and other roof structures. The use of stainless steel as primary structural components is rather limited due to its high material cost and a lack of research in this area. The current design rules for stainless steel tubular connections are mainly based on carbon steel. However, the mechanical properties of stainless steel are clearly different from those of carbon steel. Hence, the stainless steel material is not being fully utilized. To facilitate the use of stainless steel tubular structures, design guidelines should be prepared for stainless steel tubular hollow sections through efficient design.

The experimental investigation of cold-formed stainless steel X- and K-joints fabricated from square hollow sections was carried out by Rasmussen and Young (2001). The X-joints were tested in compression and tension using different ratios of brace width to chord width. The K-joints were tested by varying the ratio of brace width to chord width, the angle between chord and brace members, and the preload applied to the chord. A total of 23 tests was performed and design rules were proposed. Furthermore, Rasmussen and Hasham (2001) conducted an

experimental investigation on cold-formed stainless steel X- and K-joints fabricated from circular hollow sections. The X-joints were tested in compression and tension using three different ratios of brace diameter to chord diameter. The K-joints were tested using three different ratios of brace diameter to chord diameter and three different angles between chord and brace members. A total of 15 tests was performed and design rules were also proposed. The purposes of these tests were to determine the ultimate strength of stainless steel tubular joints and develop design guidelines. In these tests, only the normal strength austenitic stainless steel type 304L was investigated.

Design rules for cold-formed stainless steel tubular joints are available in the Australian/New Zealand Standard (AS/NZS 2001). The Eurocode 3 part 1.8 (EC3 2005) provides design guideline for carbon steel tubular joints. These design guidelines are adopted from the Comite' International pour le Developpement et l'Etude de la Construction Tubulaire (CIDECT) Monograph No. 6 (Packer et al. 1992).

This paper focuses on both high strength and normal strength stainless steel tubular connections, and concerns the strength of welded X-joints of rectangular and square hollow sections (RHS and SHS). The design guidelines in the Eurocode 3 part 1.8 (EC3 2005) for carbon steel are used in this study for stainless steel that incorporate the material properties specific to stainless steel, notably a rounded stress-strain curve with no distinct yield point. As a consequence of the rounded stress-strain curve, deformations of stainless steel joints generally exceed those of carbon steel joints (Rasmussen and Young 2001). Hence, this paper also pays attention to joint deformations. The values of the flange indentation and web deflection of the chord members for the tested specimens are reported. The observed failure modes are also reported in this paper.

2 EXPERIMENTAL INVESTIGATION

2.1 Scope

The strength of X-joints depends mainly on (1) the ratio (β) of brace width to chord width (b_1/b_0); (2) the ratio of brace thickness to chord thickness (t_1/t_0); and (3) the ratio of chord width to chord thickness (b_0/t_0). Tests were performed by applying compression force to the brace members using different values of β ranging from 0.5 to 1.0 (full width joint); t_1/t_0 from 0.5 to 1.5, and b_0/t_0 from 10 to 50, which is beyond the validity range of most current design specifications for tubular connections ($b_0/t_0 \leq 35$).

2.2 Test specimens

The compression tests were performed on cold-formed stainless steel welded tubular X-joints of square and rectangular hollow sections. A total of 11 specimens was tested subjected to axial compression in the brace members. The chord of the specimens was unloaded and free to deform at the ends. All specimens were fabricated with brace members fully welded at right angles to the opposing sides of the continuous chord members. The welded square and rectangular hollow sections consisted of large range of section sizes. For the chord members, the tubular hollow sections have nominal overall flange widths (b_0) ranging from 40 to 200 mm, nominal overall depth of the webs (h_0) from 40 to 200 mm, and nominal thickness (t_0) from 1.5 to 6 mm. For the brace members, the tubular hollow sections have nominal overall flange widths (b_1) ranging from 40 to 150 mm, nominal overall depth of the webs (h_1) from 40 to 200 mm, and nominal thickness (t_1) from 1.5 to 6 mm. The nominal wall thickness of both brace and chord members go beyond the limits of the current design specifications, in which the nominal wall thickness of hollow sections should not be less than 2.5 mm. The length of the chord (L_0) was chosen as $5\,h_0 + h_1$ to ensure that the stresses at the brace and chord intersection are not affected by the ends of the chord. This is because the points of contraflexure on the chord due to the applied load and reactions occur sufficiently far away from the intersection region. The length of the brace (L_1) was chosen as $2.5\,h_1$ to avoid the overall buckling of brace members. The measured cross section dimensions of the specimens are shown in Table 1 using the nomenclature defined in Figs. 1 and 2. The values of width, depth, thickness, and weld leg length are based on the average measurements of all four sides and welds of the intersection for each specimen. The length of brace members was measured from the connecting face of the chord to the supported end. The ends of the brace members of the joints loaded in compression were milled flat to an accuracy of 0.02 mm to ensure full contact between specimen and end plates of the testing machine.

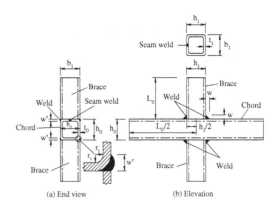

Figure 1. Definition of symbols for welded tubular X-joint.

Table 1. Measured specimen dimensions for tubular X-joints.

Specimen	Chord (mm)					Brace (mm)				Weld (mm)		
	h_0	b_0	t_0	r_i	L_0	h_1	b_1	t_1	L_1	w	w'	β
D-C40 × 2-B40 × 2	40.0	40.2	1.942	2.0	241	39.9	40.2	1.941	99	5.5	8.7	1.0
D-C50 × 1.5-B40 × 2	50.3	50.0	1.543	1.5	290	40.0	40.1	1.969	98	5.0	—	0.80
D-C50 × 1.5-B40 × 2-R	50.2	50.1	1.573	1.5	289	39.9	40.2	1.973	99	5.1	—	0.80
D-C140 × 3-B40 × 2	140.2	80.2	3.332	6.5	737	39.9	40.3	1.957	99	6.6	—	0.50
D-C50 × 1.5-B50 × 1.5	50.2	50.1	1.530	1.5	301	50.1	50.2	1.554	123	4.7	6.8	1.0
D-C140 × 3-B140 × 3	140.0	80.1	3.094	6.5	851	140.1	80.1	3.103	346	6.6	8.5	1.0
H-C150 × 6-B150 × 6	150.3	150.5	5.746	6.0	902	150.3	150.3	5.840	368	9.2	15.5	1.0
H-C110 × 4-B150 × 6	110.3	196.3	3.980	8.5	698	150.3	150.4	5.823	365	9.6	—	0.77
H-C200 × 4-B200 × 4	198.1	108.9	3.983	8.5	1196	197.7	109.6	4.019	494	8.3	12.3	1.0
N-C40 × 2-B40 × 2	40.0	40.1	2.029	2.0	241	40.0	40.1	2.009	99	5.7	9.6	1.0
N-C40 × 4-B40 × 2	40.1	40.0	3.794	4.0	240	40.2	40.1	1.971	98	6.5	11.9	1.0

For all specimens, the seam weld of the chord member was positioned to the connecting face of one of the brace members, except for the specimen with chord member of $110 \times 200 \times 4$, as shown in Figs. 1 and 2.

2.3 Specimen labeling

The specimens are labeled according to their steel types and cross-section dimensions of chord and brace members. For example, the label 'D-C50 × 1.5-B40 × 2-R' defines the following specimen:

- The first letter 'D' indicates that the steel type of the specimen is duplex stainless steel. If the letter is 'H', it refers to high strength austenitic stainless steel. If the letter is 'N', it refers to normal strength austenitic stainless steel type AISI 304.
- The second letter 'C' refers to chord member and the following expression '50 × 1.5' indicates the cross-section dimensions of the chord member, which having nominal overall depth of the webs (h_0) of 50 mm and wall thickness (t_0) of 1.5 mm. The overall flange width is purposely not shown for simplification.
- The third letter 'B' refers to brace member and the following expression '40 × 2' indicates the cross-section dimensions of the brace member, which having nominal overall depth of the webs (h_1) of 40 mm and wall thickness (t_1) of 2 mm. Once again, the overall flange width is purposely not shown.
- If a test is repeated, then the letter 'R' indicates the repeated test.

2.4 Procedure of welding

The welds connecting the chord and brace members were designed according to the AWS D1.1/D1.1M specification (American Welding Society (AWS) 2004) and laid using shielded metal arc welding. The 2.0 and 2.5 mm electrodes of type E308L-17 with nominal 0.2% proof stress, tensile strength, and

elongation of 440 MPa, 570 MPa, and 37%, respectively, were used for welding normal strength stainless steel (AISI 304) specimens. The 2.5, 3.25, and 4.0 mm electrodes of type E2209-17 with nominal 0.2% proof stress, tensile strength, and elongation of 635 MPa, 830 MPa, and 25%, respectively, were used for welding high strength stainless steel (duplex and high strength austenitic) specimens. The electrodes are described in details in the AWS A5.11 specification (AWS 2005). All welds consisted of 2 to 3 runs welding to guarantee that failure of specimens occurred in the chord or brace members rather than the welds. The measured lengths of weld leg w and w' (full width joint) are shown in Table 1.

2.5 Material properties

The specimens were cold-rolled from austenitic stainless steel type AISI 304, high strength austenitic (HSA) and duplex stainless steel sheets. The stainless steel type AISI 304 is considered as normal strength material, whereas the HSA and duplex are considered as high strength material. The brace and chord members with the same dimensions were selected from the same batch of tubes and so could be expected to have similar material properties. In this study, the stainless steel tube specimens were obtained from the same batch of specimens conducted by Zhou and Young (2005) for flexural members. The material properties of the stainless steel tubes were determined by tensile coupon tests and the results are shown in Table 2, which includes the type of stainless steel, the measured initial Young's modulus (E), the static 0.2% tensile proof stress ($\sigma_{0.2}$), the static tensile strength (σ_u), and the elongation after fracture (ε) based on a gauge length of 50 mm. The tensile coupon tests are detailed in Zhou and Young (2005).

2.6 Test rig and procedure

The schematic sketches of the test arrangement are shown in Figs. 3a and b, for the end view and elevation, respectively. Compressive axial force was applied to the specimen by using a servo-controlled hydraulic

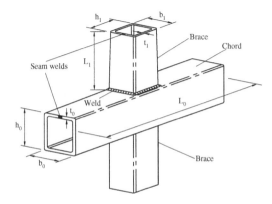

Figure 2. 3D view of welded tubular X-joint.

Table 2. Material properties of stainless steel tubes.

Specimen ($h \times b \times t$)	Type	E (GPa)	$\sigma_{0.2}$ (MPa)	σ_u (MPa)	ε (%)
40 × 40 × 2	Duplex	216	707	827	29
50 × 50 × 1.5	Duplex	200	622	770	37
140 × 80 × 3	Duplex	212	486	736	47
150 × 150 × 6	HSA	194	497	761	52
200 × 110 × 4	HSA	200	503	961	36
110 × 200 × 4	HSA	200	503	961	36
40 × 40 × 2	AISI 304	194	447	704	61
40 × 40 × 4	AISI 304	196	565	725	52

Note: HSA = High strength austenitic.

271

machine. The upper end support was movable to allow tests to be conducted at various specimen dimensions. A top end plate was bolted to the upper end support. This rigid flat bearing plate was restrained from the minor and major axes rotations as well as twist rotations and warping. Hence, this rigid flat bearing plate was considered to be a fixed-ended bearing. The load was applied to the specimen through a special fixed-ended bearing. Initially, the special bearing was free to rotate in any direction and the bottom brace of the specimen was in full contact with the bottom end plate, which was bolted to the special bearing. The ram of the actuator was moved slowly until the top brace of the specimen was in full contact with the top end plate. An initial load of approximately 3 to 5 kN was applied to the specimen. This procedure eliminated any possible gaps between the specimen and the top and bottom end plates. The special bearing was then restrained from rotations and twisting by using the four vertical and two horizontal bolts, respectively. The vertical

and horizontal bolts were used to lock the special bearing into position. Hence, the special bearing became a fixed-ended bearing which was considered to be restrained against both minor and major axes rotations as well as twist rotations and warping. The pure axial compression force without any bending moment was applied to the specimen. Four displacement transducers were positioned on either side of the brace members measuring the vertical deflections at the center of the connecting faces of the chord. The transducers were positioned 20 mm away from the faces of the brace members, as shown in Fig. 4. The flange indentation (u) in the chord was obtained from these transducers. Two displacement transducers were positioned at the center of the chord side wall to record the side wall deflection. The average of these readings was also taken as the chord web deflection (v), as shown in Fig. 4. Two other displacement transducers were positioned diagonally on the bottom end plate to measure the axial shortening of the specimen.

A 1000 kN capacity servo-controlled hydraulic testing machine was used to apply compressive axial force to the specimen. Displacement control was used to drive the hydraulic actuator at a constant speed of 0.2 mm/min for full-width joints and 0.4 mm/min for other tubular joints. The use of displacement control allowed the tests to be continued in the post-ultimate range. The applied loads and readings of displacement transducers were recorded automatically at regular interval by using a data acquisition system. A photograph of the test setup is shown in Fig. 5.

2.7 Test strengths and failure modes

There are three main failure modes observed from the tests of X-joint stainless steel tubes, namely the chord face failure (plastic failure of the chord face), chord side wall failure (crippling or buckling of the chord side wall or chord web failure) and local buckling failure of brace. The failure mode of local buckling failure of brace is similar to that observed in stub column test. According to the design rules in the Eurocode 3 part 1.8 (EC3 2005), the failure modes of welded X-joint of RHS and SHS are defined based on the

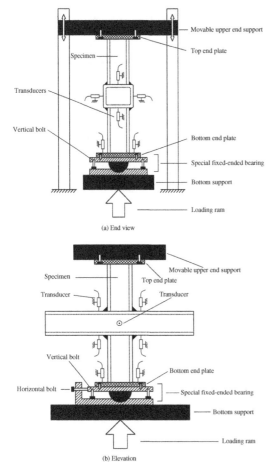

(a) End view

(b) Elevation

Figure 3.   Schematic sketch of welded tubular X-joint test.

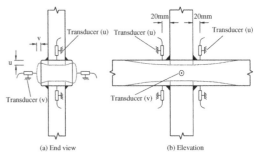

(a) End view   (b) Elevation

Figure 4.   Deformations of welded tubular X-joint.

272

value of brace to chord width ratio (β). The chord face failure occurs when the brace to chord width ratio $\beta \leq 0.85$. This failure mode usually has a post-yield response due to the effect of membrane forces in the chord and strain hardening of the material. The chord

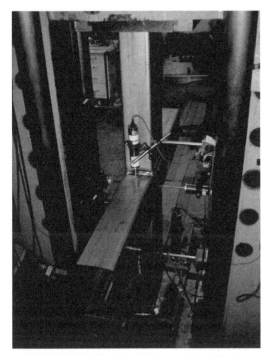

Figure 5. Test setup of welded tubular X-joint of specimen H-C200 × 4-B200 × 4.

side wall failure occurs for the full width tubular X-joint ($\beta = 1$). A clear peak load is normally found for this failure mode. It should be noted that the failure modes observed in the tests are not totally consistent with that defined in the current design specifications based on the value of brace to chord width ratio (β).

The maximum load (N_{1u}) applied to the brace members as well as the observed failure modes are shown in Table 3. All specimens failed by 'plastification of the chord', except for specimen N-C40 × 4-B40 × 2, and the failure generally involved large plastic deformations of the connecting faces and side walls of the chord. The graphs of load (N_1) versus chord flange indentation (u) and load versus chord web deflection (v) are shown in Figs. 6a and b, respectively. However, the curves for specimen N-C40 × 4-B40 × 2 could not be shown on the same graphs, because the deflections (u, v) were negligible when this specimen suddenly failed by local buckling of the brace member. The flange indentation (u) shown in Fig. 6a is the indentation perpendicular to the connecting face of the chord for one flange. The side wall deflection (v) shown in Fig. 6b was obtained using the average reading from the two transducers, measured at the center of the chord. Table 3 shows the values of the flange indentation (u) and the web deflection (v) corresponding at the ultimate load of the test specimens.

3 DESIGN GUIDELINES

3.1 General

Stainless steel does not have a distinct yield stress, therefore the 0.2% proof stress is used as the yield

Table 3. Limit-state loads and failure modes of welded tubular X-joint tests.

Specimen	β	Failure mode	Test strengths N_{1u} (kN)	N_{1s} (kN)	Nominal strengths $N_{1n\sigma0.2}$ (kN)	Ultimate limit state $\dfrac{N_{1u}}{N_{1n\sigma0.2}}$	Serviceability limit state $\dfrac{N_{1s}}{(N_{1n\sigma0.2}/\gamma_{M5})/1.5}$	Flange indentation u (mm)	Web deflection v (mm)
D-C40 × 2-B40 × 2	1.00	B	143.2	141.6	59.7	2.40	3.56	0.11	0.67
D-C50 × 1.5-B40 × 2	0.80	B	47.6	29.5	25.3	1.88	1.75	0.98	4.39
D-C50 × 1.5-B40 × 2-R	0.80	B	47.6	26.9	26.3	1.81	1.53	0.99	4.49
D-C140 × 3-B40 × 2	0.50	A	81.6	17.0	41.4	1.97	0.62	8.80	11.64
D-C50 × 1.5-B50 × 1.5	1.00	B	69.8	65.3	21.7	3.22	4.51	0.20	1.36
D-C140 × 3-B140 × 3	1.00	B	234.1	168.7	65.3	3.58	3.88	0.33	4.18
H-C150 × 6-B150 × 6	1.00	B	898.2	802.6	363.7	2.47	3.31	0.89	4.21
H-C110 × 4-B150 × 6	0.77	A + B	201.1	120.6	118.1	1.70	1.53	8.86	16.85
H-C200 × 4-B200 × 4	1.00	B	383.3	303.8	92.4	4.15	4.93	1.68	13.67
N-C40 × 2-B40 × 2	1.00	B	93.9	92.6	51.6	1.82	2.69	0.14	0.58
N-C40 × 4-B40 × 2	1.00	C	160.1	—	134.5	1.19	—	0.09	0.26
Mean						2.38	2.83		
COV						0.381	0.507		

Note: A = Chord face failure; B = Chord side wall failure; C = Local buckling failure of brace. $\gamma_{M5} = 1.0$.

273

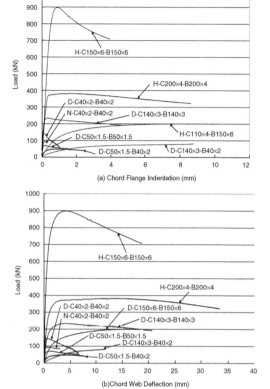

Figure 6. Load versus deformation curves for welded tubular X-joint tests.

stress in this study. The design strengths (N_1) were obtained by replacing the yield stress with the 0.2% proof stress. The design strength (N_1) of square and rectangular hollow section X-joints can be calculated using the following design equations:

For $\beta \leq 0.85$ (Chord face failure):

$$N_1 = \frac{f_{y0}t_0^2}{(1-\beta)\sin\theta}[\frac{2\eta}{\sin\theta} + 4(1-\beta)^{0.5}]f(n)$$

(CIDECT Monograph No. 6: 1992)

$$N_1 = \frac{k_n f_{y0}t_0^2}{(1-\beta)\sin\theta}[\frac{2\eta}{\sin\theta} + 4(1-\beta)^{0.5}]/\gamma_{M5}$$

(EC3 part 1.8: 2005)

$$\phi N_{1n} = \frac{f_{y0}t_0^2}{(1-\beta)\sin\theta}[\frac{2\beta}{\sin\theta} + 4(1-\beta)^{0.5}]k_n(\frac{\phi}{0.9})$$

(AS/NZS 4673: 2001 for square hollow section only)

For $\beta = 1.0$ (Chord side wall failure):

$$N_1 = \frac{f_k t_0}{\sin\theta_1}[\frac{2h_1}{\sin\theta_1} + 10t_0]$$

(CIDECT Monograph No. 6: 1992)

$$N_1 = \frac{f_b t_0}{\sin\theta_1}[\frac{2h_1}{\sin\theta_1} + 10t_0]/\gamma_{M5}$$

(EC3 part 1.8: 2005)

For $\beta \geq 0.85$ (Local buckling failure of brace):

$$N_1 = f_{y1}t_1[2h_1 - 4t_1 + 2b_e]$$

(CIDECT Monograph No. 6: 1992)

$$N_1 = f_{y1}t_1[2h_1 - 4t_1 + 2b_{eff}]/\gamma_{M5}$$

(EC3 part 1.8: 2005)

For $0.85 \leq \beta \leq 1.0$:

Use linear interpolation between the value for chord face failure at $\beta = 0.85$ and the governing value for chord side wall failure at $\beta = 1.0$.

where f_{y0} is the yield stress of the chord, f_{y1} is the yield stress of the brace, θ is the angle between the brace and the chord, η is the ratio of brace depth to chord width (h_1/b_0), k_n and $f(n)$ are the parameters account for the influence of compression chord longitudinal stresses, f_k and f_b are the chord side wall flexural buckling stresses, b_e and b_{eff} are the effective width of the brace.

The design strengths given in CIDECT (Packer et al. 1992) have already incorporated the resistance factor (Φ), which is the products of nominal strengths and resistance factor. The design strengths given in Eurocode 3 part 1.8 (EC3 2005) have also incorporated the partial safety factor (γ_{M5}), where $\gamma_{M5} = 1.0$ is recommended by the Eurocode 3 part 1.8. The design strengths given in Australian/New Zealand Standard for cold-formed stainless steel structures (AS/NZS 4673: 2001) have included the strength reduction factor (ϕ), where $\phi = 0.9$ is adopted. By comparing the design strength equations for X-joint given in these design specifications, it can be generally concluded that the same equations are used, except for the different value of resistance factors. Thus, different nominal strengths would be obtained due to the different value of resistance factors. It should be noted that the Australian/New Zealand Standard does not have design equation for rectangular hollow section X-joint. In this paper, the design equations in the Eurocode 3 part 1.8

(EC3 2005) are used to compute the nominal strengths of the X-joints and compared with the test results.

The design rules of CIDECT and Eurocode 3 part 1.8 (EC3 2005) are limited to the yield stress of 355 MPa and 460 MPa, respectively. These limits are imposed partly because most test data was obtained for joints with yield stresses less than those values and partly because carbon steel joints with yield stresses greater than the limit values may not have adequate ductility. The latter aspect is not of concern in most austenitic stainless steel structures, since austenitic stainless steels generally have high ratios of tensile strength to 0.2% proof stress and high values of elongation after fracture (Rasmussen and Young 2001). The test results showed that these limits of yield stress need not be imposed on welded tubular X-joints fabricated from high strength stainless steel (duplex and high strength austenitic) and normal strength stainless steel (AISI 304).

3.2 Ultimate limit state of X-joints

By using the design rules given in the Eurocode 3 part 1.8 (EC3 2005), the value of the design strength (N_1) is identical to the value of the nominal strength (N_{1n}) since $\gamma_{M5} = 1.0$. For the full width connection in compression $(\beta = 1)$, the design provisions require calculation of the chord side wall flexural buckling stress. In this paper, the buckling curve 'a' having the imperfection factor $\alpha = 0.21$ in the Eurocode 3 part 1.1 (EC3 2005) has been chosen for calculating the buckling stress to obtain the maximum resistance.

The nominal strengths resulting from substituting the measured 0.2% proof stress for the yield stress in the Eurocode 3 part 1.8 (EC3 2005) equations are shown in Table 3 as $N_{1n\sigma0.2}$. Table 3 also shows the ratio $(N_{1u}/N_{1n\sigma0.2})$ of test strengths to nominal strengths based on the 0.2% proof stresses. The values of the ratio are greater than unity indicate conservative design strengths. It follows from the table that the design strengths are conservative for normal strength stainless steel (AISI 304) and very conservative for high strength stainless steel (duplex and high strength austenitic) tubular X-joints when based on the 0.2% proof stress.

3.3 Serviceability limit state of X-joints

The design of stainless steel structures is more likely to be governed by the serviceability limit state than carbon steel structures because the loss of stiffness associated with the low proportionality stress precipitates growth of deformations at loads well below ultimate (Rasmussen and Young 2001). It was proposed in CIDECT recommendations that joint deformations under service loads should be limited to 1% of the chord width (b_0), which has been adopted

in this paper. The loads (N_{1s}) at which the measured deflection $(\max\{u, v\})$ equaled 1% of the chord width are shown in Table 3. These loads are compared with serviceability design loads determined by dividing the design strengths $(N_{1n\sigma0.2}/\gamma_{M5})$ by 1.5, where $\gamma_{M5} = 1.0$ according to the recommendations given in the Eurocode 3 part 1.8 (EC3 2005), and the value of 1.5 is consistent with the recommendations of CIDECT (Packer et al. 1992), as well as the American Institute of Steel Construction (AISC 2005) of using a load factor of 1.5 on the design strength in allowable stress design. The ratio $[N_{1s}/((N_{1n\sigma0.2}/\gamma_{M5})/1.5)]$ of the test serviceability load to design serviceability load based on the 0.2% proof stress are shown in Table 3. The values of the ratio are greater than unity, except for specimen D-C140 × 3-B40 × 2, indicating that the serviceability limit state generally will not be reached if the ultimate strength is calculated as $N_{1n\sigma0.2}$.

The localized deformation taken as $\max\{u, v\}$ in this paper is different from the research (Lu et al. 1994) on deformation limits for carbon steel tubular joints, which simply use u as the measure of localized deformation. Some other research (Cao et al. 1998; Zhao 2000) also suggests that the ultimate load should be limited to the force causing a localized deformation of 3% of the chord width (b_0).

4 CONCLUSIONS

An experimental investigation of cold-formed stainless steel welded tubular X-joints of square and rectangular hollow sections has been presented. The test specimens consisted of different brace to chord width ratio (β), brace to chord thickness ratio (t_1/t_0), and chord width to chord thickness ratio (b_0/t_0). The ultimate strengths, failure modes and load deflection curves for all tests have been reported. The failure mode involved chord face failure, chord side wall failure and local buckling failure of brace.

The test results were compared with the design predictions obtained from the CIDECT and Eurocode for carbon steel tubular structures, and the Australian/New Zealand Standard for stainless steel structures. It is shown that the welded stainless steel tubular X-joints of square and rectangular hollow sections can be conservatively designed using the current design rules for carbon steel and stainless steel tubular joints by replacing the yield stress with the 0.2% proof stress for both high strength stainless steel (duplex and high strength austenitic) and normal strength stainless steel (AISI 304). However, the current design rules are very conservative for the tested specimens. It should be noted that the failure modes observed in the tests are not totally consistent with that defined in the current design specifications based on the value of brace to chord width ratio (β). Further research is still required to define the failure modes based on an appropriate β value and propose accurate design equations for the stainless steel tubular X-joints.

REFERENCES

American Institute of Steel Construction (AISC). (2005). *Specification for Structural Steel Buildings*, ANSI/AISC 360-05, Chicago, Illinois.

American Welding Society (AWS). (2004). *Structural welding code-steel*, AWS D1.1/1.1M: 2004, Miami.

American Welding Society (AWS). (2005). *Specification for Nickel and Nickel-Alloy Welding Electrodes for Shielded Metal Arc Welding*, AWS A5.11/A5.11M: 2005, Miami.

Australian/New Zealand Standard. (2001). *Cold-formed stainless steel structures*, AS/NZS 4673: 2001, Standards Australia, Sydney, Australia.

Cao, J. J., Packer, J. A., and Kosteski, N. (1998). "Parametric finite element study of connections between longitudinal plates and RHS columns." *Proc., 8th Int. Symp. on Tubular Struct., Tubular Structures VIII*, Y. S. Choo and G. J. Van der Vegte, eds., Balkema, Rotterdam, The Netherlands, 645–654.

Eurocode 3 (EC3). (2005). *Design of steel structures – Part 1-1: General rules and rules for buildings*, European Committee for Standardization, EN 1993-1-1: 2005, CEN, Brussels.

Eurocode 3 (EC3). (2005). *Design of steel structures–Part 1-8: Design of joints*, European Committee for Standardization, EN 1993-1-8: 2005, CEN, Brussels.

Lu, L. H., de Winkel, G. D., Yu, Y., and Wardenier, J. (1994). "Deformation limit for the ultimate strength of hollow section joints." *Proc., 6th Int. Symp. on Tubular Struct., Tubular Structures VI*, P. Grundy, A. Holgate, and B. Wong, eds., Balkema, Rotterdam, The Netherlands, 341–347.

Packer, J. A., Wardenier, J., Kurobane, Y., Dutta, D., and Yeomans, N. (1992). *Design guide for rectangular hollow section (RHS) joints under predominantly static loading*, Comite' International pour le Developpement et l'Etude de la Construction Tubulaire (CIDECT), Verlag TUV Rheinland, Cologne, Germany.

Rasmussen, K. J. R., and Young, B. (2001). "Tests of X- and K-joints in SHS Stainless Steel Tubes." *Journal of Structural Engineering*, ASCE, 127(10): 1173–1182.

Rasmussen, K. J. R., and Hasham, A. S. (2001). "Tests of X- and K-joints in CHS Stainless Steel Tubes." *Journal of Structural Engineering*, ASCE, 127(10): 1183–1189.

Zhao, X. L. (2000). "Deformation limit and ultimate strength of welded T-joints in cold-formed RHS sections." *Journal of Constructional Steel Research*, 53(2): 149–165.

Zhou, F., and Young, B. (2005). "Tests of cold-formed stainless steel tubular flexural members." *Thin-Walled Structures*, 43(9): 1325–1337.

Tubular Structures XI – Packer & Willibald (eds)
© *2006 Taylor & Francis Group, London, ISBN 0-415-40280-8*

Design and tests of cold-formed high strength stainless steel tubular sections subjected to web crippling

F. Zhou & B. Young

Department of Civil Engineering, The University of Hong Kong, Hong Kong

ABSTRACT: This paper presents a series of tests on cold-formed *high strength* stainless steel square and rectangular hollow sections subjected to web crippling. The types of stainless steel investigated in this study were high strength austenitic and duplex material. The measured web slenderness value of the hollow sections ranged from 16.5 to 49.6. The tests were carried out under two loading conditions specified in the American Specification and Australian/New Zealand Standard for cold-formed stainless steel structures, namely End-One-Flange and Interior-One-Flange loading conditions.

The web crippling test strengths were compared with the design strengths obtained using the American, Australian/New Zealand and European specifications for stainless steel structures. In addition, the North American Specification for cold-formed carbon steel structural members was also used to predict the web crippling strengths and compared with the test results. Generally, it is shown that the design strengths predicted by the specifications are either unconservative or very conservative. Hence, a unified web crippling equation with new coefficients for cold-formed high strength stainless steel square and rectangular hollow sections is proposed in this paper. It is shown that the proposed web crippling equation is safe and reliable.

1   INTRODUCTION

Web crippling is a form of localised buckling that occurs at points of concentrated loads or supports of structural members. Cold-formed stainless steel square and rectangular hollow sections that are unstiffened against this type of loading could cause structural failure by web crippling. The current web crippling design rules in most of the specifications for cold-formed stainless steel structures are generally empirical in nature. This is because theoretical analysis of web crippling is rather complicated. The mechanical properties of stainless steel are significantly different from those of carbon steel. For carbon and low-alloy steels, the proportional limit is assumed to be at least 70% of the yield point, but for stainless steel the proportional limit ranges from approximately 36 to 60% of the yield strength (Yu 2000).

The web crippling design rules can be found in the American Society of Civil Engineers Specification (ASCE 2002a) for the design of cold-formed stainless steel structural members, Australian/New Zealand Standard (AS/NZS 2001) for cold-formed stainless steel structures and European Code (EC3 1996a) Design of Steel Structures Part 1.4: Supplementary rules for stainless steel. Due to the lack of test data, the web crippling design rules in the aforementioned specifications are based on the test results

of cold-formed carbon steel rather than cold-formed stainless steel members. Therefore, it is important to obtain test data for cold-formed stainless steel sections subjected to web crippling. The web crippling design rules in the ASCE Specification and AS/NZS Standard are based on the experimental investigation of carbon steel with yield stress less than 379 MPa conducted by Hetrakul and Yu (1978). Hence, there is a need to determine the appropriateness of the current design rules for cold-formed high strength stainless steel flexural members against web crippling.

In this paper, the appropriateness of the design rules in the current specifications for cold-formed high strength stainless steel square and rectangular hollow sections subjected to web crippling is investigated. A series of tests subjected to web crippling was conducted under End-One-Flange (EOF) and Interior-One-Flange (IOF) loading conditions. The material of the test specimens consisted of duplex and high strength austenitic stainless steel with the 0.2% compressive proof stress ($\sigma_{0.2}^{TC}$) ranging from 513 to 747 MPa, the 0.2% tensile proof stress ($\sigma_{0.2}^{LT}$) ranging from 448 to 707 MPa and the elongation (ε^{LT}) after fracture based on a gauge length of 50 mm ranging from 29% to 52%. In this study, the stainless steel specimens have higher strength and lower ductility compared with the type 304 stainless steel investigated by Zhou and Young (2006a). The web crippling test strengths were compared with the design strengths obtained using the American (2002a), Australian/New Zealand (2001) and European (1996a) specifications for stainless steel structures. In addition, the unified web crippling equation specified in the North American Specification (NAS 2001a) for cold-formed carbon steel structural members was also used to predict the design strengths. This unified web crippling equation for cold-formed carbon steel was developed by Prabakaran and Schuster (1998) and Beshara and Schuster (2000). A unified web crippling equation with new coefficients for cold-formed high strength stainless steel square and rectangular hollow sections is proposed in this paper. Reliability analysis was performed to assess the current and proposed web crippling design rules.

2 EXPERIMENTAL INVESTIGATION

2.1 Test specimens

A series of tests was conducted on cold-formed *high strength* stainless steel square and rectangular hollow sections subjected to web crippling. The specimens were cold-rolled from duplex and high strength austenitic stainless steel sheets. The specimens consisted of seven different section sizes, having the nominal thicknesses (t) ranged from 1.5 to 6 mm, the nominal overall depth of the webs (d) from 40 to 200 mm, and the nominal flange widths (b_f) from 40 to 150 mm. The measured web slenderness (h/t) value ranged from 16.5 to 49.6. The measured inside corner radius (r_i) ranged from 1.5 to 8.5 mm. The specimen lengths (L) were determined according to the ASCE Specification (2002a) and the AS/NZS Standard (2001). Generally, the clear distance between opposed loads was set to be 1.5 times the overall depth of the web (d) rather than 1.5 times the depth of the flat portion of the web (h), the latter being the minimum specified in the specifications. Tables 1–2 show the measured test specimen dimensions using the nomenclature defined in Fig. 1.

2.2 Specimen labeling

In Tables 1–2, the specimens were labeled such that the loading condition, nominal dimension of the specimen and length of the bearing could be identified from the label. For example, the label "EOF150 × 150 × 6N150" defines the following specimen:

- The first three letters indicate the loading condition of End-One-Flange (EOF) was used in the tests.
- The following symbols are the nominal dimension ($d × b_f × t$) of the specimens in mm. (150 × 150 × 6 means $d = 150$ mm; $b_f = 150$ mm; $t = 6$ mm.)
- The notation "N150" indicates the length of bearing in mm (150 mm).

Table 1. Measured specimen dimensions and experimental ultimate loads for EOF loading condition.

Specimen	d (mm)	b_f (mm)	t (mm)	r_i (mm)	L (mm)	P_{Exp} (kN)
EOF40 × 40 × 2N50	40.1	40.3	1.955	2.0	296	22.3[a]
EOF40 × 40 × 2N25	40.1	40.3	1.934	2.0	247	24.7
EOF50 × 50 × 1.5N50	50.4	50.1	1.545	1.5	335	19.8[a]
EOF50 × 50 × 1.5N25	50.3	50.1	1.539	1.5	254	15.5
EOF150 × 150 × 3N150	150.9	150.4	2.797	4.8	1051	47.2
EOF150 × 150 × 3N75	150.9	150.5	2.792	4.8	750	28.8
EOF150 × 150 × 6N150	150.5	150.4	5.865	6.0	978	184.6[a]
EOF150 × 150 × 6N75	150.5	150.3	5.732	6.0	748	124.3
EOF140 × 80 × 3N75	140.5	80.4	3.080	6.5	719	37.6
EOF140 × 80 × 3N50	140.4	80.4	3.082	6.5	618	33.6
EOF160 × 80 × 3N75	160.9	80.7	2.887	6.0	779	32.4
EOF160 × 80 × 3N50	160.9	80.8	2.887	6.0	679	29.6
EOF200 × 110 × 4N100	204.5	104.8	3.981	8.5	999	60.2
EOF200 × 110 × 4N50	201.0	104.9	3.983	8.5	800	40.1

Note: [a] Shear failure.

Table 2. Measured specimen dimensions and experimental ultimate loads for IOF loading condition.

Specimen	d (mm)	b_f (mm)	t (mm)	r_i (mm)	L (mm)	P_{Exp} (kN)
IOF40 × 40 × 2N50	40.1	40.2	1.940	2.0	370	30.3
IOF40 × 40 × 2N25	40.0	40.2	1.933	2.0	243	27.9
IOF50 × 50 × 1.5N50	50.2	50.1	1.536	1.5	402	21.7
IOF50 × 50 × 1.5N25	50.2	50.1	1.537	1.5	277	19.2
IOF150 × 150 × 3N150	150.7	150.6	2.800	4.8	1205	59.3
IOF150 × 150 × 3N75	150.7	150.5	2.793	4.8	826	51.4
IOF150 × 150 × 6N150	150.3	150.1	5.590	6.0	1199	228.9
IOF150 × 150 × 6N75	150.3	150.1	5.730	6.0	820	207.0
IOF140 × 80 × 3N75	140.3	80.3	3.082	6.5	794	51.4
IOF140 × 80 × 3N50	140.1	80.3	3.081	6.5	668	49.0
IOF160 × 80 × 3N75	160.5	80.9	2.882	6.0	854	52.5
IOF160 × 80 × 3N50	160.5	80.8	2.885	6.0	729	49.1
IOF200 × 110 × 4N100	198.2	109.0	4.006	8.5	1103	98.6
IOF200 × 110 × 4N50	202.6	104.1	3.981	8.5	850	82.3

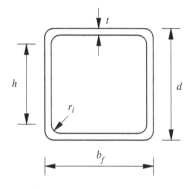

Figure 1. Definition of symbols.

2.3 Material properties

Tensile and compression coupon tests were carried out to determine the material properties of the stainless

Table 3. Material properties obtained from tensile coupon tests.

Section	Type	E^{LT} (GPa)	σ_p^{LT} (MPa)	$\sigma_{0.2}^{LT}$ (MPa)	σ_u^{LT} (MPa)	ε^{LT} (%)
$40 \times 40 \times 2$	Duplex	216	164	707	827	29
$50 \times 50 \times 1.5$	Duplex	200	182	622	770	37
$150 \times 150 \times 3$	HSA[a]	189	155	448	699	52
$150 \times 150 \times 6$	HSA[a]	194	147	497	761	52
$140 \times 80 \times 3$	Duplex	212	199	486	736	47
$160 \times 80 \times 3$	Duplex	208	167	536	766	40
$200 \times 110 \times 4$	HSA[a]	200	150	503	961	36

Note: [a] HSA = High strength austenitic.

Table 4. Material properties obtained from compression coupon tests.

Section	Type	E^{TC} (GPa)	σ_p^{TC} (MPa)	$\sigma_{0.2}^{TC}$ (MPa)
$40 \times 40 \times 2$	Duplex	220	230	747
$50 \times 50 \times 1.5$	Duplex	216	200	652
$150 \times 150 \times 3$	HSA[a]	220	200	547
$150 \times 150 \times 6$	HSA[a]	214	160	678
$140 \times 80 \times 3$	Duplex	220	200	513
$160 \times 80 \times 3$	Duplex	200	210	569
$200 \times 110 \times 4$	HSA[a]	220	200	622

Note: [a] HSA = High strength austenitic.

steel. The tensile coupons were taken from the centre of the face at 90° angle from the weld in the longitudinal direction of the untested specimens. The material properties obtained from the tensile coupon tests are shown in Table 3, which includes the stainless steel type, the measured initial Young's modulus (E^{LT}), the proportional limit (σ_p^{LT}), the static 0.2% tensile proof stress ($\sigma_{0.2}^{LT}$), the static tensile strength (σ_u^{LT}), and the elongation after fracture (ε^{LT}) based on a gauge length of 50 mm. The compression coupons were taken from the centre of the face at 90° angle from the weld in the transverse direction of the untested specimens. The material properties obtained from the compression coupon tests are shown in Table 4, which includes the stainless steel type, the measured initial Young's modulus (E^{TC}), the proportional limit (σ_p^{TC}), and the static 0.2% compression proof stress ($\sigma_{0.2}^{TC}$). The coupon tests are detailed in Zhou and Young (2006b).

2.4 Loading conditions and test rig

The square and rectangular hollow section specimens were tested under the two loading conditions specified in the current American Specification (ASCE 2002a) and Australian/New Zealand Standard (AS/NZS 2001). These loading conditions are EOF and IOF. The test setup and test method are detailed in

Figure 2. Test setup of End-One-Flange (EOF) loading condition.

Zhou and Young (2006b). Photograph of the test setup of EOF loading condition is shown in Fig. 2. All specimens were positioned in the way such that the loaded webs were 90° angle from the weld of the specimens for all loading conditions.

2.5 Test results

The experimental ultimate web crippling loads per web (P_{Exp}) are given in Tables 1–2. For specimens EOF40 × 40 × 2N50, EOF50 × 50 × 1.5N50 and EOF150 × 150 × 6N150 (stockier web having $h/t = 16.5$, 28.7 and 21.6) subjected to EOF loading condition, web crippling was not observed at ultimate load of the tests, but the specimens failed by shear of the sections.

3 RELIABILITY ANALYSIS

The reliability of the web crippling design rules is evaluated using reliability analysis. A target reliability index of 3.0 for stainless steel structural members is recommended as a lower limit in the ASCE Specification (2002a). The design rules are considered to be reliable if the reliability index (β) is greater than 3.0. The resistance (capacity) factor (ϕ_{w1}) for web crippling strength as recommended by the current ASCE Specification, EC3 Code and NAS Specification are shown in Tables 5–6. The load combinations of $1.2DL + 1.6LL$ and $1.35DL + 1.5LL$ as specified in the American Society of Civil Engineers Standard (ASCE 2003) and the European Code, respectively, were used in the reliability analysis, where DL is the

Table 5. Comparison of web crippling test strengths for EOF loading condition.

Section	N (mm)	h/t	r_i/t	N/t	N/h	P_{Exp} (kN)	$\dfrac{P_{Exp}}{P_{ASCE}^{LT}}$	$\dfrac{P_{Exp}}{P_{ASCE}^{TC}}$	$\dfrac{P_{Exp}}{P_{EC3}^{LT}}$	$\dfrac{P_{Exp}}{P_{EC3}^{TC}}$	$\dfrac{P_{Exp}}{P_{NAS}^{LT}}$	$\dfrac{P_{Exp}}{P_{NAS}^{TC}}$	$\dfrac{P_{Exp}}{P_{P}^{LT}}$	$\dfrac{P_{Exp}}{P_{P}^{TC}}$
40 × 40 × 2	50	16.5	1.0	25.6	1.6	22.3[a]	(1.57)	(1.57)	(3.31)	(3.19)	(0.94)	(0.89)	(0.85)	(0.81)
40 × 40 × 2	25	16.7	1.0	12.9	0.8	24.7	1.98	1.98	3.74	3.60	1.31	1.24	1.23	1.16
50 × 50 × 1.5	50	28.7	1.0	32.4	1.1	19.8[a]	(2.17)	(2.17)	(4.95)	(4.65)	(1.45)	(1.38)	(1.29)	(1.22)
50 × 50 × 1.5	25	28.7	1.0	16.2	0.6	15.5	1.95	1.95	3.90	3.67	1.42	1.35	1.29	1.23
150 × 150 × 3	150	48.6	1.7	53.6	1.1	47.2	1.58	1.58	5.07	4.25	1.34	1.10	1.40	1.15
150 × 150 × 3	75	48.7	1.7	26.9	0.6	28.8	1.17	1.17	3.10	2.60	1.04	0.85	1.12	0.91
150 × 150 × 6	150	21.6	1.0	25.6	1.2	184.6[a]	(1.46)	(1.46)	(4.58)	(3.73)	(1.25)	(0.92)	(1.13)	(0.83)
150 × 150 × 6	75	22.2	1.0	13.1	0.6	124.3	1.15	1.15	3.22	2.63	1.08	0.79	1.01	0.74
140 × 80 × 3	75	39.4	2.1	24.4	0.6	37.6	1.35	1.35	3.12	2.98	1.07	1.02	1.28	1.22
140 × 80 × 3	50	39.3	2.1	16.2	0.4	33.6	1.29	1.29	2.79	2.66	1.08	1.03	1.32	1.25
160 × 80 × 3	75	49.6	2.1	26.0	0.5	32.4	1.33	1.33	2.91	2.88	0.95	0.89	1.12	1.06
160 × 80 × 3	50	49.6	2.1	17.3	0.3	29.6	1.30	1.30	2.66	2.63	0.98	0.93	1.18	1.11
200 × 110 × 4	100	45.1	2.1	25.1	0.6	60.2	1.31	1.31	3.16	2.71	1.00	0.81	1.20	0.97
200 × 110 × 4	50	44.2	2.1	12.6	0.3	40.1	0.96	0.96	2.10	1.80	0.81	0.66	1.01	0.81
Mean, P_m							1.40	1.40	3.25	2.95	1.10	0.97	1.20	1.06
COV, V_p							0.228	0.228	0.239	0.226	0.168	0.211	0.105	0.164
Reliability index, β_1							3.38	3.38	4.78	4.66	2.71	2.06	3.93	3.03
Resistance factor, ϕ_{w1}							0.70	0.70	0.91	0.91	0.80	0.80	0.70	0.70
Reliability index, β_2							3.38	3.38	5.60	5.50	3.14	2.45	3.93	3.03
Resistance factor, ϕ_{w2}							0.70	0.70	0.70	0.70	0.70	0.70	0.70	0.70

Note: [a] Shear failure.

Table 6. Comparison of web crippling test strengths for IOF loading condition.

Section	N (mm)	h/t	r_i/t	N/t	N/h	P_{Exp} (kN)	$\dfrac{P_{Exp}}{P_{ASCE}^{LT}}$	$\dfrac{P_{Exp}}{P_{ASCE}^{TC}}$	$\dfrac{P_{Exp}}{P_{EC3}^{LT}}$	$\dfrac{P_{Exp}}{P_{EC3}^{TC}}$	$\dfrac{P_{Exp}}{P_{NAS}^{LT}}$	$\dfrac{P_{Exp}}{P_{NAS}^{TC}}$	$\dfrac{P_{Exp}}{P_{P}^{LT}}$	$\dfrac{P_{Exp}}{P_{P}^{TC}}$
40×40×2	50	16.6	1.0	25.8	1.6	30.3	1.11	1.11	1.52	1.47	0.70	0.66	0.89	0.85
40×40×2	25	16.6	1.0	12.9	0.8	27.9	1.12	1.12	1.71	1.64	0.74	0.70	1.00	0.94
50×50×1.5	50	28.8	1.0	32.6	1.1	21.7	1.24	1.24	1.79	1.68	0.86	0.82	1.08	1.03
50×50×1.5	25	28.7	1.0	16.3	0.6	19.2	1.21	1.21	1.93	1.81	0.88	0.84	1.15	1.10
150×150×3	150	48.4	1.7	53.6	1.1	59.3	1.07	1.00	1.57	1.32	0.98	0.81	1.15	0.94
150×150×3	75	48.6	1.7	26.9	0.6	51.4	1.08	1.00	1.71	1.43	1.01	0.82	1.24	1.02
150×150×6	150	22.7	1.1	26.8	1.2	228.9	1.07	1.02	1.73	1.41	0.91	0.66	1.15	0.84
150×150×6	75	22.1	1.0	13.1	0.6	207.0	1.00	0.95	1.81	1.47	0.89	0.65	1.20	0.88
140×80×3	75	39.3	2.1	24.3	0.6	51.4	0.88	0.86	1.33	1.27	0.81	0.77	1.01	0.95
140×80×3	50	39.3	2.1	16.2	0.4	49.0	0.88	0.87	1.42	1.36	0.84	0.79	1.07	1.02
160×80×3	75	49.5	2.1	26.0	0.5	52.5	1.00	0.99	1.46	1.45	0.85	0.80	1.05	0.98
160×80×3	50	49.5	2.1	17.3	0.3	49.1	0.98	0.97	1.53	1.52	0.86	0.81	1.09	1.03
200×110×4	100	43.2	2.1	25.0	0.6	98.6	0.99	0.95	1.52	1.30	0.89	0.72	1.10	0.89
200×110×4	50	44.6	2.1	12.6	0.3	82.3	0.91	0.87	1.55	1.33	0.86	0.69	1.11	0.90
Mean, P_m							1.04	1.01	1.61	1.46	0.86	0.75	1.09	0.96
COV, V_p							0.107	0.119	0.104	0.108	0.094	0.092	0.081	0.081
Reliability index, β_1							3.41	3.24	3.96	3.56	1.81	1.30	3.77	3.24
Resistance factor, ϕ_{w1}							0.70	0.70	0.91	0.91	0.90	0.90	0.70	0.70
Reliability index, β_2							3.41	3.24	5.09	4.69	2.77	2.27	3.77	3.24
Resistance factor, ϕ_{w2}							0.70	0.70	0.70	0.70	0.70	0.70	0.70	0.70

dead load and LL is the live load. The statistical parameters are obtained from Table F1 of the NAS Specification for web crippling strength, where $M_m = 1.10$, $F_m = 1.00$, $V_M = 0.10$, and $V_F = 0.05$, which are the mean values and coefficients of variation for material properties and fabrication factors. The statistical parameters P_m and V_P are the mean value and coefficient of variation of tested-to-predicted load ratio, respectively, as shown in Tables 5–6. In calculating the reliability index, the correction factor in the NAS Specification (2001a) was used. The respective resistance factor (ϕ_{w1}) and load combinations for the current ASCE Specification, EC3 Code and NAS Specification were used to calculate the corresponding reliability index (β_1). For the purpose of direct comparison, a constant resistant factor (ϕ_{w2}) of 0.7 and a load combination of $1.2DL + 1.6LL$ as specified in the ASCE Specification were used to calculate the reliability index (β_2) for the EC3 Code and NAS Specification, as shown in Tables 5–6. Reliability analysis is detailed in the Commentaries of the ASCE Specification (2002b) and NAS Specification (2001b).

4 COMPARISON OF TEST STRENGTHS WITH CURRENT DESIGN STRENGTHS

The web crippling loads per web obtained from the tests (P_{Exp}) were compared with the nominal web crippling strengths (unfactored design strengths) predicted using the ASCE Specification (2002a), AS/NZS Standard (2001) and EC3 Code (1996a) Part 1.4. The AS/NZS Standard has adopted the web crippling design rules from the ASCE Specification, and no changes are introduced into the web crippling design rules. Hence, the web crippling design strengths predicted by the ASCE Specification and the AS/NZS Standard are identical. The web crippling design rules in the EC3 Code (1996a) Part 1.4: Supplementary rules for stainless steel refers to the web crippling design rules of either the hot-rolled carbon steel EC3 Code (1992) Part 1.1 or the cold-formed carbon steel EC3 Code (1996b) Part 1.3. In this paper, the EC3 Code (1996b) Part 1.3 was used to predict the web crippling design strength. In addition, the test strengths were also compared with the nominal web crippling strengths predicted using the NAS Specification (2001a) for cold-formed carbon steel structural members. Tables 5–6 show the comparison of the test strengths (P_{Exp}) with the unfactored design strengths. The design strengths were calculated using the measured cross-section dimensions and the measured material properties. The longitudinal 0.2% tensile proof stress ($\sigma_{0.2}^{LT}$) and transverse 0.2% compression proof stress ($\sigma_{0.2}^{TC}$) as given in Tables 3 and 4 were used to calculate the design strengths P^{LT} and

P^{TC}, respectively, for the ASCE Specification (2002a), EC3 Code (1996b) Part 1.3 and NAS Specification (2001a).

In Tables 5–6, the web crippling loads per web are generally greater than the predicted strengths using the ASCE Specification and AS/NZS Standard. The values of the tested-to-predicted ratio tend to increase as the yield stress of the material increases. This is because the web crippling strength predicted using the ASCE Specification and AS/NZS Standard for a given section increases as the yield stress increases only up to 459 MPa for EOF loading condition and 631 MPa for IOF loading condition, beyond which the strength decreases as the yield stress increases. In order not to penalise the use of high strength material, the ASCE Specification and AS/NZS Standard conservatively specified that a constant value of $C_3 = 1.34$ is used when the yield stress is greater than or equal to 459 MPa for EOF loading condition, and a constant value of $C_1 = 1.69$ is used when the yield stress is greater than or equal to 631 MPa for IOF loading condition. The coefficients C_1 and C_3 reflect the effect on web crippling strength of the yield stress of the material. Therefore, it is quite conservative to use the current ASCE Specification and AS/NZS Standard to predict the web crippling strength of structural members with high strength material. For example, the value of tested-to-predicted load ratio is 1.98 for the specimen EOF40 × 40 × 2N25 with the 0.2% tensile proof stress of 707 MPa and 0.2% compression proof stress of 747 MPa, as shown in Table 5.

For EC3 Code, the design strengths are generally very conservative for the two loading conditions. The maximum mean value of the tested-to-predicted load ratio is 3.25 with the corresponding coefficient of variation (COV) of 0.239 and the values of $\beta_1 = 4.78$ and $\beta_2 = 5.60$ for the EOF loading condition, as shown in Table 5. For NAS Specification, the design strengths are generally unconservative and unreliable, except for the EOF loading condition calculated using the longitudinal 0.2% tensile proof stress ($\sigma_{0.2}^{LT}$). The minimum mean value of the tested-to-predicted load ratio is 0.75 with the corresponding COV of 0.092 and the values of $\beta_1 = 1.30$ and $\beta_2 = 2.27$ for the IOF loading condition, as shown in Table 6.

5 PROPOSED DESIGN EQUATION

The nominal web crippling strengths calculated using the ASCE Specification (2002a), AS/NZS Standard (2001), EC3 Code (1996b) Part 1.3 and NAS Specification (2001a) are either unconservative or very conservative, except for some cases predicted by the ASCE Specification and AS/NZS Standard as shown in Tables 5–6. Hence, web crippling design equation for cold-formed high strength stainless steel hollow

Table 7. Proposed web crippling design rules for cold-formed stainless steel hollow sections.

Support and flange conditions		Load cases			C	C_R	C_N	C_h	LRFD ϕ_w	Limits Type
Unfastened	Stiffened or partially stiffened flanges	One-Flange loading	End Interior	EOF IOF	5 7	0.40 0.21	0.50 0.26	0.02 0.001	0.70 0.70	HSA[a] and Duplex

Notes: The above coefficients apply when $h/t \leq 50$, $N/t \leq 54$, $N/h \leq 2.0$, $r_i/t \leq 2$ and $\theta = 90°$.
[a]HSA = High strength austenitic.

sections under EOF and IOF loading conditions is proposed in this paper. The unified web crippling equation for carbon steel as specified in the NAS Specification (2001a) is used in this study. The Equation (1) is the proposed unified equation with new coefficients of C, C_R, C_N and C_h as well as using the resistance factor $\phi_w = 0.7$, as shown in Table 7.

$$P_p = Ct^2 f_y \sin\theta \left(1 - C_R\sqrt{\frac{r_i}{t}}\right)\left(1 + C_N\sqrt{\frac{N}{t}}\right)\left(1 - C_h\sqrt{\frac{h}{t}}\right) \quad (1)$$

where C is the coefficient, C_R is the inside corner radius coefficient, C_N is the bearing length coefficient, C_h is the web slenderness coefficient, t is the thickness of the web, f_y is the yield stress ($\sigma_{0.2}$ proof stress), θ is the angle between the plane of the web and the plane of the bearing surface, r_i is the inside corner radius, N is the length of the bearing and h is the depth of the flat portion of the web measured along the plane of the web. The new coefficients were determined based on the test results obtained in this study that included the *high strength* material of duplex and high strength austenitic stainless steel. A similar equation has been proposed by Zhou and Young (2006a) for channel sections, square and rectangular hollow sections of types 304, 3CR12 and 430 stainless steel.

6 COMPARISON OF TEST STRENGTHS WITH PROPOSED DESIGN STRENGTHS

The experimental ultimate web crippling loads per web (P_{Exp}) were compared with the unfactored design strengths calculated using the proposed unified equation. The proposed design strengths were calculated using the measured cross-section dimensions and measured material properties. The resistance factor $\phi_{w1} = 0.7$ for EOF and IOF loading conditions was obtained from reliability analysis. The load combination of 1.2DL + 1.6LL was used to determine the reliability indices (β_1 and β_2) using the corresponding ϕ_{w1} and ϕ_{w2} factors, as shown in Tables 5–6.

The proposed design strengths are generally conservative and reliable for the two loading conditions. The maximum mean value of the tested-to-predicted load ratio is 1.20 with the corresponding COV of 0.105 and the value of $\beta_1 = 3.93$, as shown in Table 5. The minimum mean value of the tested-to-predicted load ratio is 0.96 with the corresponding COV of 0.081 and the value of $\beta_1 = 3.24$, as shown in Table 6. The reliability indices (β_1 and β_2) are greater than the target value for both loading conditions.

7 CONCLUSIONS AND DESIGN RECOMMENDATIONS

The paper presents an experimental investigation of cold-formed *high strength* stainless steel square and rectangular hollow sections subjected to web crippling. The duplex and high strength austenitic stainless steel with the 0.2% tensile proof stress ranged from 448 to 707 MPa and the 0.2% compression proof stress ranged from 513 to 747 MPa were investigated. The test specimens with different plate slenderness of the web were tested, and the web slenderness ranged from 16.5 to 49.6. The specimens were tested under End-One-Flange (EOF) and Interior-One-Flange (IOF) loading conditions in accordance with the ASCE Specification (2002a) for cold-formed stainless steel structures. The flanges of the specimens were not fastened to bearing plates. The test strengths were compared with the design strengths obtained using the current ASCE Specification (2002a), AS/NZS Standard (2001) and EC3 Code (1996a) Part 1.4 for stainless steel structures. The NAS Specification (2001a) for cold-formed carbon steel structural members was also used to predict the design strengths.

It is shown that the design strengths predicted by the stainless steel and carbon steel specifications were either unconservative or very conservative. Therefore, a unified web crippling equation with new coefficients for cold-formed high strength stainless steel hollow sections under EOF and IOF loading conditions has been proposed in this paper. It is shown that the design strengths calculated using the proposed unified equation based on the material properties obtained from either the longitudinal tension or transverse compression coupon tests were generally conservative for the two loading conditions. The proposed design equation is capable of producing reliable limit state design when calibrated with the resistance factor $\phi_w = 0.7$ for EOF and IOF loading conditions. It is also recommended that the web crippling strength of cold-formed high strength stainless steel square and rectangular hollow sections can be calculated using the proposed unified equation, and the material properties of longitudinal tension or transverse compression can be used.

REFERENCES

ASCE. (2003). *Minimum design loads for buildings and other structures*, ASCE Standard 7-02, American Society of Civil Engineers Standard.

ASCE. (2002a). *Specification for the design of cold-formed stainless steel structural members*, American Society of Civil Engineers, SEI/ASCE-8-02, Reston, Virginia.

ASCE. (2002b). *Commentary on Specification for the design of cold-formed stainless steel structural members*, American Society of Civil Engineers, SEI/ASCE-8-02, Reston, Virginia.

AS/NZS. (2001). *Cold-formed stainless steel structures*, Australian/New Zealand Standard, AS/NZS 4673:2001, Standards Australia, Sydney, Australian.

Beshara, B., and Schuster, P.M. (2000). "Web crippling data and calibrations of cold-formed steel members." Final Report, University of Waterloo, Waterloo, Canada.

EC3. (1992). *Eurocode 3: Design of steel structures – Part 1.1: General rules and rules for buildings*, European Committee for Standardization, ENV 1993-1-1, CEN.

EC3. (1996a). *Eurocode 3: Design of steel structures – Part 1.4: General rules – Supplementary rules for stainless steels*, European Committee for Standardization, ENV 1993-1-4, CEN, Brussels.

EC3. (1996b). *Eurocode 3: Design of steel structures – Part 1.3: General rules – Supplementary rules for cold formed thin gauge members and sheeting*, European Committee for Standardization, ENV 1993-1-3, CEN, Brussels.

Hetrakul, N., and Yu, W.W. (1978). "Structural behavior of beam webs subjected to web crippling and a combination of web crippling and bending." Final Report, Civil Engineering. Study 78-4, University of Missouri-Rolla, Rolla, Mo.

NAS. (2001a). *North American Specification for the design of cold-formed steel structural members*, North American Cold-Formed Steel Specification, North American Specification, American Iron and Steel Institute, Washington, D.C.

NAS. (2001b). *Commentary on North American Specification for the design of cold-formed steel structural members*, North American Cold-Formed Steel Specification, North American Specification, American Iron and Steel Institute, Washington, D.C.

Prabakaran, K., and Schuster, P.M. (1998). "Web crippling of cold-formed steel members." *Proceedings of the 14th International Specialty Conference on Cold-formed Steel Structures*, St. Louis, University of Missouri-Rolla, Mo., 151–164.

Yu, W.W. (2000). *Cold-formed steel design*, 3rd Edition, John Wiley and Sons, Inc., New York.

Zhou, F., and Young, B. (2006a). "Cold-formed stainless steel sections subjected to web crippling." *Journal of Structural Engineering*, ASCE. 132(1), 134–144.

Zhou, F., and Young, B. (2006b). "Cold-formed high strength stainless steel tubular sections subjected to web crippling." *Journal of Structural Engineering*, ASCE. (Accepted for publication)

ISTS Kurobane Lecture

Tubular Structures XI – Packer & Willibald (eds)
© *2006 Taylor & Francis Group, London, ISBN 0-415-40280-8*

Punching shear and hot spot stress – Back to the future?

P.W. Marshall
Moonshine Hill Pty., Houston, Texas

ABSTRACT: The early days of offshore construction were by trial and error, i.e. learning from failures. Early research from Texas – cowboy technology – found its way into AWS D1.1 in 1972, in the form of punching shear criteria for ultimate strength (Vp) and fatigue (Curves T&K). Hot spot stress was also introduced (Curve X), as well as fitness-for-purpose ultrasonic testing (class X). Further developments in ultimate strength include the collection of more extensive data bases, design visualization, and design by components. Punching shear also shows up in boundary tractions and mixed mode crack propagation. Further developments in fatigue include hotspot stress *vs.* fracture mechanics, hotspot stress *vs.* notch stress, non-tubular applications, marine hull calibrations, fatigue with shear at welds, and progressive tensile fracture. Tubular structures today can be as reliable as other structural types, and have been mainstreamed into design Codes, e.g. AISC (2005). However, there is no 20,000 cycle guarantee at full static allowable. Bigger and better offshore structures, and bridges at last, enter the scene. Recent standardization work includes: international criteria for materials, welding, inspection, fabrication, and quality[1,2] (superscripts refer to the PPT file, available from mhpsyseng@aol.com).

1 INTRODUCTION

The art and science of welded steel tubular space frame structures has grown up with the offshore platform industry, and is codified in AWS D1.1, as further described by Marshall (1992).

1.1 *Offshore construction*

Offshore construction as we know it today began in 1947 with the installation of the first steel template-type structure in the Gulf of Mexico, for drilling off Louisiana in 20 ft (6 m) of water (Lee 1968). Watch Jimmy Stuart do it in the movie *Thunder Bay*. Today there are 4,000 such structures in the Gulf of Mexico, in water depths up to 1340 ft (400 m), less the 300 or so that failed or were seriously damaged in recent hurricanes Ivan, Katrina, and Rita. Offshore drilling platforms have also been built in US waters off Southern California and Alaska. Internationally, they can be found in every ocean and off shore from every continent except Antarctica.

A welded tubular steel jacket extends from the sea floor to slightly above the waterline. Hollow legs provide a guide for driving piles; these are braced by a space frame that resists lateral loads imposed by nature. The typical jacket is prefabricated onshore in one piece, carried offshore by barge, launched at sea, and set on bottom by ballasting, assisted by a seagoing crane or derrick barge.

The platform foundation is established by driving tubular steel piling through the jacket legs to a penetration of 100 ft–500 ft (30–150 m) into the sea floor. Pilings are attached to the jacket by welding above water or grouting to the annular space between pile and jacket.

A superstructure, or deck section, is set on top to complete the structure. It carries the functional loads for which the structure was built, keeping men and equipment out of reach of wave action.[3]

1.2 *Learning from failures*

In the early days of offshore, the inadequate static strength of welded tubular connections was responsible for a number of premature structural failures (Carter & Marshall 1969). These occurred during severe hurricanes, when the wells were shut in and the personnel evacuated.

When a single-well protector jacket disintegrated and collapsed around its oil well in hurricane Hilda in 1964, all the bracing members pulled out of the jacket leg, leaving a series of holes where their footprints used to be.[4,5] After doing research on tubular connections, the failure was easy to explain, but survival of sister structures was more of a challenge.

Blowouts, fires & collisions do not concern us here today, except for entertainment value.[6,7]

2 COWBOY TECHNOLOGY

Much of the early work on the static and fatigue strength of welded tubular connections was done in the

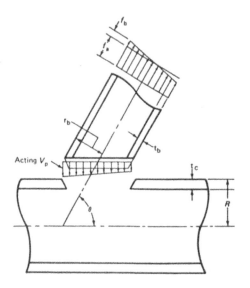

Figure 1.   Concept of punching shear.

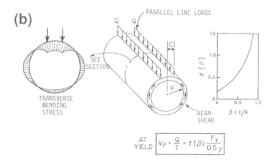

Figure 2.   Closed ring solution.

late 1960s at the University of Texas in Austin, where Yoshi Kurobane was a visiting post-doc, and at Southwest Research Institute in San Antonio. The University of California was a competing center of research, with results outlasting student radicalism of the day.

2.1   Ultimate strength – V_p

Punching shear v_p refers to nominal shear stress on the potential failure surface for the observed pull-out failure mode for tubular joints,[8] as shown in Figure 1. Tau (ratio of branch-to-main member thickness) shows its primarily influence. Tau should always be less than unity for efficient connections. In its simplest form:

acting $v_p = \tau \sin\theta \, (f_a + f_b)$

ultimate $V_p = f(\beta, \text{etc}) \, F_y \, / \, (C \, \gamma^e) \leq F_y / \sqrt{3}$

where F_y is chord yield strength, C is a calibration constant, and other terms are described below.

Actual local stresses are much more complex, as indicated by the equations from shell theory. Punching shear is related to the gradient of shell bending.[11]

Closed form shell stress solutions for simple cases can be reduced to punching shear at first yield. The case of a radial load around the circumference can be reduced to the Kellogg formula ($e = 0.5$), used by designers of pressure piping connections.[9] Line load capacity is dependent on tube thickness and radius-to-thickness ratio (γ). The case of loads along two axial lines has been tabulated by Roark, Figure 2, with the strength also exhibiting a strong dependence on beta (β, analogous to diameter ratio in branch tube connections).[10]

Analytical solutions for intersecting tubes shows a circumferential stress pattern due to ovalizing in the main member, rather similar to the closed ring solution, but spread out along the member axis.[12] For a typical geometry, the shell stress (bending plus membrane) may be eight times the nominal stress in the branch member. The strength trend for punching shear at first yield shows a strong dependence on gamma, but tends to be rather flat as a function of beta in the practical range of 0.4 to 0.8. [13]

Elastic shell theory could not explain the ultimate strengths of tubular joints in tests or observed structural survivals, as yielding occurs at an early stage. Continued loading leads to fully plastic shell bending, load redistribution, strain hardening, triaxial stresses, and ultimately fracture.[14,16]

A plot of the early test data shows that material shear strength only governs for very stocky main members, with the ultimate punching shear V_p declining as the -0.7 power of gamma.[15] More extensive data comparisons show the reliability of the AWS joint design criteria being similar to that of other structural elements.[17] The apparent higher strength for tension loads seen in small tests is neglected in consideration of fracture size effects.

2.2   Fatigue design curves

Repeated overloads during a severe storm, or cumulative damage by wave cycling over the years, can cause fatigue failure in the welded tubular connections of offshore platforms.[18]

Design for fatigue of welded details has been traditionally done in terms of the nominal stress classification method (Gurney 1973, Munse 1978), and curves A, B, C, D, E, and F. The AWS (2004) family of fatigue S-N curves, Figure 3, includes examples of this type, with tubular joint curves DT and ET being relevant to failures in the branch members only.[19]

For pull-out type failure in the main member at a tubular connection, fatigue test data can be plotted in terms of cyclic punching shear to develop a crude design curve.[20] However, the joint strength depends on the loading pattern, with K-joints being stronger than

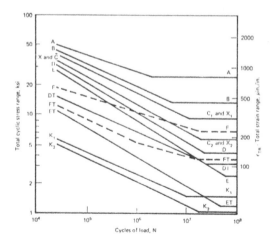

Figure 3.   AWS family of S-N curves.

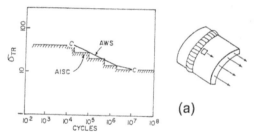

(a)

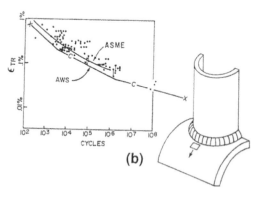

(b)

Figure 4.   (a) Structural stress. (b) Hot spot strain.

T&Y joints, or X-joints. This is because the outward pull of one branch offsets the ovalizing effect of the inward push of the other. Such classification are made heuristically by the designer.[21]

AWS consolidated the S-N curves T and K for the different joint types by introducing an ovalizing parameter alpha, applied as a factor on axial stresses in the branch member: 1.0 for K joints, 1.7 for T&Y, and 2.4 for X joints which get a double dose of ovalizing. For bending in the branch member, in-plane gets a factor of 0.67 and out-of-plane gets 1.5.

2.3   Hot spot stress, curve X

Because branched connections of large-diameter thin-wall tubes can be subject to high stress concentration factors, depending on geometry, their fatigue design is best done in terms of the localized shell bending and membrane stresses, or hot spot strain range, measured in the member adjacent to the joining weld (Toprac 1969). Under repeated plastic loading, these locations actually generate heat.

Large scale model tests of prototype welded components provided a data base of hot spot strain range (peak-to-trough, after shakedown) versus cycles to failure. "Failure" was not just initiation of a tiny crack, but included propagation to a terminal size. Strain gages were applied to hot spot locations identified by "Stresscoat" brittle lacquer. Typical gage size was 3 mm, placed entirely within 6 mm of the weld toe. Microscopic discontinuities and notch effects at the toe of the weld were not measured, but included in the S-N curve.[22,23]

As shown in Figure 4, this protocol placed many different geometries on a common basis, ranging from pressure vessels to welded military components to tubular joints.[24] For structural connections, the measured hot spot stress could be parameterized as the extension of member nominal stress (axial and bending about two axes) times stress concentration factors (SCF).[25]

Finite element shell analysis, with a single-layer mesh, produced comparable SCF results. Thin shells were used by Clough & Greste (1969) at the University of California. Reimer (1976) modeled the gross shape of the weld with isoparametric solid shell elements, but not the weld toe notch effect.[26,27]

These developments enabled systematic fatigue analysis and design of offshore platforms for harsh environments like the North Sea, and for deep water (Kinra 1979).

2.4   Fitness for purpose UT criteria

At tubular T-, Y-, and K-connections, the brace ends are given a saddle-shaped cope, and welds must usually be made from the outside only. Forensic examination of fatigue tests and field failures showed that tubular joints typically fail in the shell of the main member, as long as there is adequate weld throat size at the intersection, despite some incomplete penetration at the root.[28]

Because of the complex T-Y-K geometry, ultrasonic testing (UT) is the only practical means for volumetric weld inspection. Minor root discontinuities and "false alarm" reflectors led to repair attempts, butchering welds which were originally fit for purpose. Finite element results typically show the highest stress to be on the outside surface of both chord and branch, with stresses across the root being less than half as much. Thus, both API RP 2X and AWS D1.1 adopted UT criteria with relaxed criteria at the root of tubular T-Y-K connections (Marshall 1982). This relaxation does not apply to the ID root of butt-welded girth seams, where there is no reduction of stress at the root.[29,30]

3 FURTHER DEVELOPMENTS ON ULTIMATE STRENGTH

For simple tubular joints, punching shear design solved the problem of inadequate static strength (Marshall 1974). Today, design formulas tend to be expressed in limit load format (Wardenier 1982 & 1991, Kurobane et al. 1984), and extended to more complex joints.[31–35] More recent extensions to the data base using inelastic finite element methods, has led to updated API criteria for the static strength of tubular connections (Pecknold 2005).[36]

Revised CIDECT-based formulations also appear in AISC (2005). Criteria comparisons for K-, Y-, and X-joints (Marshall 2004) show all these criteria giving similar results, despite their different formats. Only AWS still retains punching shear.[37,38]

3.1 *Multi-planar joints*

The AWS format (both punching shear and limit load) permits extension of design from the planar "alphabet" joints which make up most of the test database, to the more general case of arbitrary multiplanar joints in tubular space frames. This is done using a computed alpha (ovalizing) parameter which reproduces the planar cases found earlier. Comparisons with multi-planar tests and inelastic finite element analyses are generally favorable, except in the case of symmetrical delta trusses, where transverse gap and combined perimeter effects overwhelm the improvement predicted on the basis of ovalizing.[39,40]

3.2 *Load path visualization*

Failure modes for the chord at tubular connection include ovalizing or general collapse, local chord face plastification, plug pull-out, beam bending, beam shear, sidewall buckling, local buckling due to longitudinal stress, and combinations of these. Punching shear and the equivalent limit load formulations only address the first three. The other modes are usually covered by limitations on D/T and other member proportions.[41]

Failure in the branch member or connecting weld is influenced by uneven distribution of load, especially where ductility is impaired by limited plastic strain capacity, welds which do not match the strength of the connected parts, or the use of non-compact bracing members.

For complex joints, one should investigate cutting planes across which combined loads must be transferred, e.g. between branches of a K-connection, or on the combined footprint of an overlap connection.[43] Acting forces and moments come from satisfying the equations of equilibrium, which must be the complete set. Useful estimates of the limiting capacities of shells can be provided by using membrane shear and punching shear. At equilibrium, some of these resistance elements may not have reached full capacity. As long as ductility is also assured, this "cut and try" approach is justified by the lower bound theorem of plasticity.

Before finite element modeling (FEM) became a feasible practice in design offices, some very large tubular connections were designed by following the loads from branch member, through stiffening gussets and diaphragms, to the chord.[44] Elastic FEM based on first yield is an extremely conservative lower bound. Inelastic analysis with large deflection effects and strain hardening is to be preferred.

"Cut and try" remains a useful tool for reality checks, and was historically practiced by structural engineering students and steel detailers. It is now in danger of becoming a lost art.

3.3 *Design by components*

For a standard range of geometries, the capacities of connection components, e.g. external gussets, can be tabulated, based on the foregoing methods.[45,46] AISC (2005) also provides guidance for the capacity of plates on tubular shells. Other useful and standardizable components are: blind crossings of elements in different planes with an intermediate plate or shell, the strength of cylindrical shells between ring stiffeners or diaphragms, and the strength of the rings themselves (Marshall 1986).[47–49]

4 FURTHER DEVELOPMENTS ON FATIGUE

Expanding databases (UKOSRP 1978, SIMS 1981, Nordhoek & DeBack 1987) provided and confirmed the empirical basis for the fatigue design hot spot S-N curve, with size effect and other modifications (e.g. m = 3) appearing in the UK Department of Energy rules and early drafts of ISO 13819-2 for offshore structures. Somewhat different criteria evolved for fatigue of onshore tubular structures in ISO 14347. The database of Dimitrakis, Lawrence & Mohr (1995) has led to an update of the API fatigue criteria for tubular connections (Marshall et al. 2005). Proposals for AWS D1.1-2008 are considering all these precedents.

Various definitions of hot spot stress, experimental and analytical, have been proposed, with varying degrees of accuracy, consistency with the database, and invariance.[50,51]

Full-scale tubular connection tests at the University of California (from a structure prematurely condemned by UT false alarms, but with workmanlike concave weld profiles) followed the early AWS "close gage" protocol, and provided S-N results consistent with earlier smaller tests, as the un-measured notch effect was the same in both cases.[52]

European researchers sought a more rigorous definition which gave invariant results consistent with SCF from FEM. When worst-case flat weld profiles were used, the un-measured notch effect scaled up with the member thickness and gage placement, resulting in a distinctly adverse size effect. A passionate Transatlantic debate raged for ten years as to whether the low results were due to size or profile.[53] The current understanding is that both are relevant, and that progressive weld improvements can mitigate the low results.[54] Tests of welds with concave profile show that a mild size effect eventually kicks in. Tests with flat profile welds show a severe size effect all the way from 4 mm to 100 mm thickness.[55–57]

4.1 Hot spot stress vs. fracture mechanics

In the seminal JIP on the subject, Hayes (1981) and Grover (1984) describe fracture mechanics fatigue analysis of tubular joints in terms of $K = Y\sigma\sqrt{(\pi a)}$ where σ is structural hot spot stress from uncracked shell FEM and Y is derived as a function of crack depth a from plane strain detailed models of slices through the weld using an embedded "PacMan" crack tip region. AWS D1.1 tubular weld details A, B, C, D, and $\beta = 1$ were modeled for various degrees of weld profile improvement. Weld profile and size effects were evident in elevated Y values over the first 0.1T of crack growth, which accounted for much of the calculated fatigue life. Although beneficial effects of displacement controlled shell behavior (load shedding), finite crack length, and crack curvature for deeper cracks were estimated from previous literature, the basic results were generally conservative, with Y values typically staying above unity. However, fracture mechanics could readily be applied to any tubular joint for which the SCF was known, e.g. from Efthymiou (1988).[58–60]

4.2 Hot spot stress vs. notch stress

Textbook K_f solutions from classical notch stress theory show a clear trend of lower overall fatigue performance, and a steeper size effect, progressing from un-welded components, to flat butt welds, to very convex butt welds, to 45° fillet welds, to 90° corners. The Code rule limiting the height of butt weld profiles mitigates some of their size effect.[61]

The SAE fatigue rules for welded components analyzed by very fine mesh (3 mm ×3 mm elements on a slice straddling the weld) pick up the increasing geometric notch effect as components are scaled up,

and show similar trends of size and profile effects. When applied to a large tubular joint on the Bullwinkle platform, the well-profiled toe of AWS Detail A (150° dihedral) showed better overall performance and a milder size effect than the re-entrant root notch at Detail D (30° dihedral). Detail A results also compared well with the Hayes & Grover fracture mechanics solution, with a terminal flaw depth of 40 mm corresponding to the specified CTOD toughness of the API node steel in the heat affected zone of an unfavorably placed seam weld.[62–64]

5 NON-TUBULAR APPLICATIONS

The hot spot stress method is now being extended to more complex and non-tubular welded connections (Marshall & Wardenier 2005), with publication in AWS D1.1 targeting the 2008 edition.[65]

5.1 IIW structural hot spot stress

By the early 1990s, the technology for detailed fatigue analysis of tubular structures in offshore applications was well established. To extend this for stiffened plate hull construction and onshore structures, Niemi (1993) laid the foundation for four approaches to the fatigue analysis of welded structures in general. Detailed guidance was given on methods of cyclic stress determination for each approach, setting the stage for a massive international data-gathering effort to calibrate the corresponding fatigue resistances.[66–69]

At the 1997 IIW annual assembly in San Francisco, presentations expanded on the four general approaches as follows (FAT-xx gives the fatigue strength at 2 million cycles as xx-MPa stress range, to be used with a S-N curve of standard shape,[70] slope m = 3.0 with flattening beyond 10^7:

1. Nominal beam element stress with multiple S-N curves depending on severity of the detail, ranging from FAT-160 for as-rolled structurals, down to FAT-36 (36 MPa at 2×10^6 cycles) for the worst case considered.
2. Niemi's geometric or structural hot spot stress based on shell bending plus membrane stress (excluding notch effects at the weld toe) and FAT-100 S-N curve, with size and profile adjustments. Hot spot stress can be extracted from a single layer $T \times T$ finite element mesh, or (for standard or parameterized geometries) determined by multiplying the nominal stress by a stress concentration factor (SCF).
3. Notch stress analysis from very detailed solid FEM modeling, with an effective radius of 1 mm at the weld toe or root, and the FAT-225 S-N curve, calibrated on 800 tests (others present suggested 0.25 mm radius as a worst case).

4. Fracture mechanics, with da/dN defined by the Paris law with threshold and fracture limits, and ΔK considering local notch and load shedding effects.

Details of the hot spot method from Hobbacher (2003) are now well-seasoned proposals, expected to be published in final form by now. For the determination of acting structural hot spot stress or stress concentration factor (SCF), two situations are distinguished:

Type (a): stress transverse to weld toe on plate surface (SCF reference points scaled to thickness) involving shell bending and membrane stress, typically FAT-100 (except FAT-90 at load-carrying fillet welds whose shear lag does not show up in the single-layer FEM calculation).

Type (b): stress transverse to weld toe on plate edge (SCF reference points not scaled to thickness) involving geometric singularities in the membrane stress, typically FAT-90 for large attachments.

By way of comparison, the old AWS D1.1 hot spot curves X1 and X2 have fatigue strengths of 100 and 79 MPa, respectively, at 2×10^6 cycles, but with different S-N shapes cris-crossing the IIW curves.

5.2 Boundary tractions

Dong (2004) presents an overview of Battelle's patented approach to fatigue across a broad variety of structural details. The reference stress is the local structural stress, similar to hot spot stress except that it is the sum of linear membrane and shell bending stresses due to equilibrium tractions at the welded boundary, rather than being extrapolated from nearby element surface stresses. Since FEM should always satisfy equilibrium, while element stresses are approximations inherent to their formulation, the method tends to be more accurate and mesh-insensitive, particularly when solid elements are used to model the weld and the mesh size is on the order of $T \times T \times T$. Special protocols allow the method to also be used with thin-shell modeling.[71]

The FEM local structural stress is adjusted for two correction terms, which tend towards unity for a 25 mm (1-inch) thick plain plate with $Y = 1$ in $K = Y\sigma\sqrt{(\pi a)}$. The first term is a size effect correction, with the exponent ranging from -0.17 to -0.22. The second and more important correction term is derived by integrating the fracture mechanics expression for crack growth in a slice through the weld.[72] The latter can account for differences in weld toe notch effect, stress gradients through the thickness, and whether the local situation tends towards load-controlled (onerous where there is no load shedding) or displacement controlled (mitigating for tubular T, Y and K connections or localized gusset singularities in a large box girder or ship hull).

Based on calibration against data from 500 tubular joint tests and 800 non-tubular joint tests, the master S-N curve shows much reduced scatter compared to using raw hot spot stresses. In terms of the corrected stress parameter, the mean fatigue strength at 2×10^6 cycles is 215 MPa, with an S-N slope exponent of $m = 3.3$. However, details of the proprietary Verity® software, catalogue of correction terms, and parameters for the recommended design curve are not being given away.[73]

The method has performed well in round-robin competitions, earning a measure of recognition.

5.3 Marine hull calibrations

ASME (2004), *Specialty Symposium on Integrity of Floating Production, Storage & Offloading (FPSO) Systems*, gives research results and design recommendations for dealing with problems of hull fatigue and extreme loading.[74]

Bergan and Lotsberg (2004) trace the development history of DNV's approach to the hull fatigue problem, from failures in *Alexander Kielland* and *Glas Dowr*, to the present joint industry project (JIP) findings. Their paper is a good starting place for a literature search, with 54 diverse references going back to 1976. They note the paradox of singularity-type details, e.g. gusset tips, where thin-shell mesh refinement can lead to near-singularity results, but are not always on the safe side.[75]

Hot spot stress protocol is linked to the corresponding S-N curve to match fatigue test results. Finite element modeling can be either thin shell or brick elements. The JIP-recommended protocol is to use a $T \times T$ mesh ($T \times 2T$ if the stress gradient is low in one direction), with linear extrapolation to the weld intersection referencing surface stress results at 0.5T and 1.5T away. The resulting hot spot stress is used with the IIW class FAT-90 S-N curve (90 MPa at 2×10^6 cycles) for thicknesses around one inch. Hot spot SCF variations within the recommended protocol ($\sim$10% COV) are small contributors to overall fatigue life uncertainty, compared to S-N scatter, loading uncertainty, and inherent errors in Miner's rule.

5.4 Implementation in AWS D1.1

The AWS Code already has multiple hot spot S-N curves X1 and X2 for tubular applications, so it is not much of a stretch to keep this feature for non-tubular as well. Table 1 tabulates tentative hot spot S-N curves X1, X, Y, and Z, together with suggested size effect rules, which appear to cover the range needed for both tubular and non-tubular applications.[76–78] The solid curves in Figure 5 are for a reference thickness of 1-inch (25 mm), with the dashed lines showing thickness extremes.

The recommended reference points, extrapolation methods and associated S-N curves are summarized in Table 2. Although one method would be preferred,

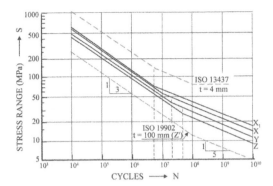

Figure 5. Proposed hot spot S-N curves.

Table 1. Properties of proposed AWS hot spot S-N curves.

	Proposed S-N curves			
	X1	X	Y	Z
Fatigue class (MPa) at $N = 2 \times 10^6$ for t = 25 mm	104	100	90	80
Fatigue class (MPa) at $N = 2 \times 10^6$ for t = 16 mm	114	114	105	96
Knee at N	$N = 5 \times 10^6$	$N = 10^7$	$N = 2 \times 10^7$	$N = 5 \times 10^7$
Slope m	3/5	3/5	3/5	3/5
Size exponent tubular joints t ≥ 16 mm	0.2	0.3	0.3	
t < 16 mm		$0.06 \log N_f$	$0.06 \log N_f$	$0.07 \log N_f$
Size exponent plated details t ≥ 25 mm		0.3	0.35	0.4
t < 25 mm		zero *)	zero *)	zero *)

*) Favorable size effect for the curves for the plated details with t < 25 mm is neglected. This may change after further analysis.

consensus at this stage is difficult because of differences between strongly held preferences. Curve X1 may be used with improved weld profile; curve X is for basic fillet-type weld profiles. Further rules for using the upper *vs.* lower fatigue curve (where two are given) are described in the detailed criteria.[79, 80]

5.5 *Fatigue with shear at welds*

Mainstreaming of FEM stress analysis on a desk-top PC is illustrated by Burley & Roper (2005). Their $T \times T \times T$ mesh is consistent with the foregoing guidelines. They also investigated static ultimate strength, using inelastic analysis. However, including the fatigue effects of shear at the weld turns out to be a surprisingly difficult problem, with a number of proposed solutions

Table 2. Recommended reference points, extrapolation methods and S-N curves.

	Linear extrapolation, fine mesh or gages	Linear extrapolation, $T \times T$ or $T \times T \times T$ mesh	Quadratic extrapolation, fine mesh, gages	Battelle SS method, $T \times T$ or finer
CHS joints	0.4 t, 1.0 t curve **X, X1**	Efthymiou method curve **X, X1**		*) curve **Y**
RHS joints			0.4 t, 0.9 t, 15 t curve **X**	*) curve **Y**
Plated detail (type a)	0.4 t, 1.0 t curve **X, Y**	0.5 t, 1.5 t curve **Y**		*) curve **Y, Z**
Plated detail (type b)	*5 and 15 mm* curve **Y, Z**		*4, 8, and 12 mm* curve **X, Y**	

*) For the indicated details, boundary tractions can be used in general provided the required meshing, equilibrium and distances from weld toe are met. This refers to raw structural stress at the weld toe, membrane plus 80% of shell bending, but without Dong's fracture mechanics correction.

having various degrees of rigor and practicality, but not completely satisfying:[81]

Ignore it. Here we simply use the structural hotspot stress normal to the weld axis, membrane plus shell bending from FEM, or measured surface strains perpendicular to the weld toe. These represent the stresses which are amplified by the notch at the toe of the weld, and which will drive Mode I crack propagation. This approach worked fairly well in early tubular truss connections in offshore platforms, where shell bending in the chord is much larger than the other stresses, and phasing of the load components is defined in terms of applied cyclic wave loads.

In biaxial stress situations, using uni-axial strain can underestimate the true hot spot stress by up to 17%. However, in low cycle fatigue, where stresses exceed yield, the early S-N curve "X" for hot spot stress was based on measured strain range after shakedown. Using elastic SCF can be unconservative here, unless the plastic strains are very localized (e.g. at a notch) and constrained by surrounding elastic material.

Use Von Mises stress. This is today's preferred approach for offshore structures, and was used for Efthymiou's SCF for tubular connections. For a given mode of branch member loading (axial, in-plane bending, out-of-plane bending) the linear trend of the individual stress components at the chord surface is extrapolated to a point at the toe of the weld in a consistent coordinate system; then the stress components are combined according to Von Mises. Although sign is lost in the process, tension *vs.* compression is assigned according to the incoming nominal stress in the brace. Beam shear and torsion are assumed to have negligible effect.

For ship hulls, analyzing shell stiffener to bulkhead connections in terms of just three load components is not as satisfactory. Best practice is to model the connection at the hot spot level of detail, embedded in a global FEM of the entire ship, with stress component phasing and combined Von Mises stress amplitude defined as transfer functions for waves of unit amplitude. Significant stress range is taken as $4\sqrt{m_0}$, where m_0 is the area of the stress spectrum from all the wavelets in a given random sea state.

For the traffic decks of orthotropic box girder bridges, beam shear in the stiffeners is no longer negligible, and its phasing in relation to the other loads is not simply defined. Loss of sign when stress components are squared for Von Mises can be a serious problem. Recourse to using in-service measured stresses and mixed-mode crack propagation calculations have been suggested.

Consider crack propagation. Using fracture mechanics calculations for cracks forming in the notch at the toe of a weld, membrane tension and shell bending are mode I, punching shear is mode II, membrane shear and twisting moment are mode III, and tension parallel to the weld axis has no crack driving effect. Each of these components is affected in a different way by weld toe notch effects, deviations from the linear shell stress assumption, and strain relief from surrounding material as the crack grows larger.

Dong's Verity® software treats these effects with varying degrees of rigor and empiricism, modifying the raw structural hot spot stress with a fracture mechanics I-curve correction as well as a fatigue size effect correction. Different "universal" S-N curves apply to normal stress and shear.

Treat everything as shear in the weld. The above three methods are applicable to fatigue failure in the base metal adjacent to the toe of the weld, which is the usual case for complete joint penetrations welds and for fillet welds which develop the full strength of the connected elements. For undersized fillet welds, which may fail at the weld throat, the use of fatigue curve "F" is suggested in several design Codes. In line with previous discussions of hot spot stress, mesh refinement beyond T × T (or stress averaging over a weld length less than T) should be avoided.[82,83]

Software is available, e.g. Weaver's FEWeld®, to extract load components across a weld line as membrane traction T_j, membrane shear T_w, punching shear T_s. Moment about the weld axis M_w should be avoided on single fillets, and can be resolved into a couple of T_j tractions on double fillets. The combined shear vector T_{jws}, as stress in the weld throat, is used to enter curve "F". If the load components are not in phase, care must be exercised with regard to losing the sign when they are squared in the process of vector addition.

For complex situations, this approach has not been calibrated as extensively as the structural hot spot

method. Because of the flatter S-N curve ($m = 5$, not 3) it becomes less conservative in the low-stress high-cycle range.

Component testing. The nominal stress fatigue categories, the original hot spot curve "X", the new X-Y-Z curves, and Dong's "universal" S-N curve, have all been developed from and/or calibrated on extensive component testing. It is important to maintain consistency between the way the data base was derived and the stress definitions used in analysis. The successful hot spot stress design approaches should work with a wide variety of welded component geometries. In complex or unusual cases, or where a large number of identical components are to be manufactured, component fatigue testing is still justified.

Typical design S-N curves correspond to the 2-sigma lower bound, or 97.7% survival from a large data base, $n > 10$ (n is number of tests). For smaller test programs, IIW defines a protocol to achieve a similar level of reliability, 95% survival at 75% confidence. For example, if testing $n = 4$ samples of a complex structures at $N < 10^6$ (where N is cycles to failure) the standard deviation of log N is taken as 0.25, then the median fatigue life is reduced by a factor of four (2.36σ) from the observed median life of all specimens, or a factor of 2.8 from the life at first failure, after which testing may be terminated.

Other methods. These include mixed mode crack propagation (Qian et al. 2005), combined shear and normal strains in strain-based fatigue, critical plane analysis, and the Wang-Brown criterion (Draper 2001 & 2004).

5.6 *Progressive tensile fracture*

Premature tensile fractures due to shear lag in gusseted connections, and undersized fillet welds in simple tubular joints have several aspects in common: *1)* Progressive tensile failure, ductile tearing or "unzipping" of a weak link as the limit state, arising from uneven distribution of load. *2)* A largely empirical approach to the effect on strength, without a clear understanding of extrapolation limits, e.g. small test vs. big application with different notch toughness. *3)* A wide variety of possible geometries, making analytical approaches, i.e. calibrated FEM protocols, a more attractive and efficient way to cover the possibilities. *4)* Code language that one should investigate "more rational methods." **Fig. 84.**

In the area of progressive fatigue cracking, the hot spot stress method has been calibrated and adapted to cover a wide variety of geometries, ranging from tubular connections to orthotropically stiffened plate structures. Today's powerful computer workstations, FEM software, and capable outsource analysts, are making such analytical modeling increasingly available to designers. Their use for ultimate strength should also be encouraged, but with caveats: *1)* Elastic FEM will produce conservative results, when F_y (or even F_u) for base metal, and shear rupture or transverse breaking strength for welds, are used as the limit states, without accounting for plastic load redistribution. *2)* Elasto-plastic FEM can produce more realistic results, provided that ductility limits are properly recognized. For undersized fillet welds, their load-deformation behavior has already been described in IIW, AWS, and AISC standards. For tensile tearing of base metal adjacent to a weld or discontinuity, stress-strain must be carefully defined. In particular, when analyzed at the weld toe hotspot level, the limiting tensile strain is a function of notch toughness and notch severity, being closer to 2% than 20% (Mohr 2003).

6 TUBULAR STRUCTURES TODAY

Research, testing and practical applications have progressed to the point where simple tubular connections are about as reliable as the other structural elements dealt with in design and fabrication. Once separate tubular provisions were mainstreamed in AWS D1.1 in 1996, and in AISC in 2005. A wealth of additional information is available from CIDECT (www.cidect.org) and the Steel Tube Institute of North America (www.steeltubeinstitute.org).[85]

The Crystal Cathedral in Garden Grove CA is a magnificent tubular space frame structure, designed around 1970. However, upon closer inspection, one discovers that the tubular connections are rather messy, with gusset plates passing through slots which have been extended to relieve tensile strains at the tips.[86–88]

Philip Johnson's more recent Penzoil Building in Houston TX incorporates a similar space frame, but with much cleaner appearance, using simple direct connections designed on the basis of punching shear.[89] Fillet welds at the intersections provide sufficient static strength, and are economical in terms of joint preparation and welder qualification, and their effective throat can be verified without resorting to UT.

6.1 *The 20,000 cycle guarantee*

AISC (2005) Appendix 3 states: "No evaluation of fatigue resistance is required if the number of cycles of application of live load is less than 20,000." In general, this does NOT apply to tubular connections.[90]

Milek's example. A basis for this provision can be found in C2.12.1 of AWS D1.1. For the end of a weld-around cover plate (limited thickness, fatigue category E), applying the static allowable of 210 MPa (for $F_y = 350$ MPa) as a variable live load (R = 0) gives a fatigue life of 36,000 cycles.

However, if the category F fillet weld is stressed to its static allowable of 145 MPa ($0.3 F_{EXX}$ for E70), the fatigue life is reduced to 16,000 cycles. If the thickness exceeds 20 mm, it becomes category E' with a fatigue life of 13,000 cycles. If the variable load is full reversal (R = −1), the fatigue life is only 1,600 cycles for category E' and 250 cycles for category F.

The foregoing results are mitigated for structures with a modest dead load and a non-reversing variable load. Category F satisfies the 20,000 cycle guarantee with 5% dead load; and category E' does so with 12%. An unstated assumption in the use of these nominal stress fatigue categories is that structural connections are stiffened to transfer load primarily by membrane stress, so that there are no substantial uncalculated plate or shell bending stresses.

Tubular joints. Tubular connections transfer load by punching shear and the corresponding shell bending stresses, and utilize local yielding, strain hardening, and load redistribution to reach their design strength. Thus, they exhibit substantially poorer fatigue performance in relation to their static strength than do the cartoon gallery attachment categories. When tubular space frames are designed for use as offshore structures, fatigue assessment is a routine part of their design.

Consider a tee joint of circular hollow sections, 10.75×0.375-inch (270×9.5 mm) branch and 20×1.0-inch (500×25 mm) chord, $F_Y = 50$ ksi (350 MPa). This just satisfies 100% static capacity using the WSD criteria in AWS D1.1, so the 170 tonne allowable static load in the branch produces a nominal stress of 210 MPa. Applying this as a cyclic load (R = 0) against cyclic punching shear fatigue criteria ($\alpha = 1.7$) and extrapolating curve K2 gives a conservative fatigue life estimate of only about 100 cycles, for failure in the chord.

As a second example, consider a tubular joint of rectangular hollow sections, $10 \times 10 \times 3/8$-inch branch ($250 \times 250 \times 9.5$ mm), A500 grade C, fillet welded to a 20×20-inch (500×500 mm) chord, 50 ksi (350 MPa) yield, thickness to be determined. The required fillet weld size is just over ½ inch (13 mm). Using 93% design thickness for the branch, and 210 MPa tensile allowable, the full static capacity load is 180 tonnes, for which AISC K3 (ASD) gives a required chord thickness of 1.3 inches (33 mm). This would likely be a submerged arc welded custom section. Applying the fatigue criteria in ISO 14347, the stress concentration factor (SCF) for branch hot spot location "E" is 11, the hot spot strain range is 1.1%, and the corresponding fatigue life (extrapolated) is 400 cycles. Requiring a complete joint penetration weld at the T-joint only improves the fatigue life to 1100 cycles. Such a connection (CJP weld) would only sustain 20,000 cycles for 38% variable load, 62% dead load. As these load proportions would not be typical for most tubular structure applications, they should be excluded from the 20,000 cycle guarantee.

Using hot spot SCF. While hot spot stress concentration factors have long been a basis for fatigue design in tubular structures, they have only recently been extended to non-tubular connection types. AWS D1.1-2008 will contain such provisions, based on IIW and DNV precedents and the use of T × T finite element meshes. Structural hot spot stress is the local stress (membrane plus shell bending) at the toe of a weld, excluding the notch effect of the weld itself. It enables the designer to assess fatigue damage at more adverse connection configurations not covered by the nominal stress class picture book. It should not be viewed as a refinement of the latter, as mesh refinement only produces singularities in the presence of sharp notches. Indeed, the implied SCF in the class details may be used as a base case for calibrating one's finite element techniques for meshing and stress extraction. The following implied SCF are based on how much a particular detail category degrades the fatigue strength or endurance limit, when nominal stress is used as the basis for calculations.

Category	SCF	Comments on basis ...
C	1.0	Geometric hotspot stress = nominal stress
D	1.4 to 1.7	Geometry or worse notch; backing left on
E	1.7 to 2.2	Still worse: cover plate end, tacked backing
E'	2.5 to 4	Incl. adverse size effect or edge notches
ET	4 to 8	Branch member failure at tubular joint
K1	$7\,\alpha\tau\sin\theta$	Punching shear, chord failure, $\gamma < 25$
K2	ditto	Weld profile $>3/8''$ w/o smooth merge

Using hot spot SCF, the 20,000 cycle guarantee can be satisfied under the following conditions:

420 MPa hot spot stress range
210 MPa nominal stress range and SCF ≤ 2.0
140 MPa nominal stress range and SCF ≤ 3.0

These SCF tables indicate that there is considerable risk of designers inadvertently voiding the guarantee, particularly when there is a low proportion of dead load.

Effect of variable load amplitude. The foregoing is mitigated somewhat if the cyclic load is of variable amplitude, rather than constant cycles, so that there is only one extreme load application at the design allowable. For example, if the stress ranges follow the Rayleigh distribution, or one with similar fatigue damage potential, the 20,000 cycle fatigue life guarantee is satisfied with a nominal stress range at the extreme load of 20 ksi (140 MPa) and an SCF of 6.3 or less.

The Rayleigh distribution is associated with narrow band random dynamic response to spectral energy, such as short term steady state turbulent wind (non-VIV), or wave loading. Long term fatigue damage over the life of the structure may be calculated using a collection of short term loading states, as for marine structures, or field-measured long term stress distributions, as for bridges. Target fatigue lives for such structures are on the order of 100 million cycles, well beyond the scope of the 20,000 cycle guarantee. Thus, explicit fatigue calculations are performed during design for these structures.

6.2 *Bigger and better*

In recent years, fixed platform technology has been refined and extended to include problems of dynamic amplification, fatigue, and sea floor instability, floating ice sheets, as well as earthquake and oceanographic loadings of unprecedented severity. The present state of practice is codified in API RP2A (2000). A final draft international standard for fixed steel structures, ISO 19902 (2006), is also in preparation. Three levels of criticality are recognized, with L-1 being the most critical.

The North Sea Brent platform was designed for 100-ft waves and fabricated in Scotland, using API and AWS criteria, while European standards were still in the drafting stage. The Brent oil field sets the price marker for North Sea crude, and became a target for Greenpeace protests.[91]

Offshore platforms designed for the crushing loads of sea ice sheets locate multiple oil wells inside large diameter tubular legs for protection. Steel platforms were built in Cook Inlet, Alaska in the 1960s for tens of millions of dollars.[92] Concrete platforms, larger but with similar functionality, are being built today off Sakhalin, Russia, costing billions.

The deepest fixed offshore platform is Bullwinkle in 400 m water depth, completed in 1988.[93] Its tubular joint and fracture control practices exemplify standards for L-1 structures (Marshall 1990).

As the water depth increases, the sway mode natural period of a fixed offshore platform increases. For periods greater than 3 to 5 seconds, structures may experience problems with dynamic amplification of wave forces, and fatigue from the resulting cyclic stresses. Compliant towers avoid the wave resonance problem and extend the feasible water depth range of fixed platform architecture by arrangements that give them sway periods ranging from 22 to 30 seconds, with higher whipping mode periods less than about 5 seconds. They have been used for oilfield development in water depths of 300 m to 500 m, with special articulations to reduce their sway mode stiffness, and have been extensively studied for deeper water up to 1000 m.[94]

Compliant (i.e. resilient) restoring forces are provided by buoyancy, guy-lines, or very long flex-piles.

In still deeper water, floating offshore structures are used – tension leg platforms, semi-submersibles and spars.[95] Although tubular in shape, their stiffened-shell construction is more like ship hulls.

6.3 MWIFQ in ISO 19902

Material, Welding, Inspection, Fabrication, and Quality. Use of the proposed new International Standard for fixed steel offshore platforms, ISO 19902, will require support from generic welding standards to address non-offshore-specific topics. The focus of the fabrication and inspection clauses in ISO 19902 is on issues that are unique to the offshore oil and gas industry. The owner and contractor will have new responsibilities required by this arrangement.

ISO 19902 will require that the owner specify the use of a generic fabrication specification for the majority of the provisions applicable to each specific detail of MWIFQ. The standard selected should be one in wide use either domestically or internationally. Significant owner input will be required in order for the fabrication contractor to successfully integrate the generic standard and ISO 19902. The limitations, advantages and disadvantages of all the various options available to the design engineer, owner and contractor are discussed by Campbell (2006).[96]

AWS D1.1, Structural Welding Code-Steel will be the Code of choice for offshore structures in many parts of the world. For European operators, various documents are available, such as EEUMA 158 or NORSOK. In any case, once the selected standard is invoked contractually, the owner must ensure that the relevant clauses in ISO 19902 for material selection, fabrication, and quality management (clauses 18, 19, 20, respectively), along with the associated annexes, are addressed by the contractor. This will require supplementing the contract documents with exceptions, modifications and clarifications needed to smoothly merge 19902 and the selected standard.

A few early platforms with heavily reinforced nodes suffered fractures that were not in the more common punching shear mode. Forensic analysis indicated the possibility of brittle fracture. Charpy statistics showed that ordinary structural steel sometimes fell on the wrong side of service temperatures.

Structural redundancy of typical 8-leg Gulf of Mexico platforms prevented catastrophic collapse, unlike fracture critical structures such as ship hulls and pressure vessels.[97–99]

Toughness classes A, B, & C evolved for the various parts of an offshore jacket, with class A being for the most critical node cans.[100] Its transition temperature is on the good side of the lowest anticipated service temperature (LAST) by a margin of 30°C,

consistent with the Fracture Analysis Diagram for hot spot regions which must strain harden to stresses well above yield.[101] The toughness selection matrix in AWS is the basis for the MC (Material Category) methodology in ISO. Suffix Z-steel is selected where resistance to lamellar tearing is important.[102] X-steel is selected for CTOD prequalification in the heat affected zone.[103,104]

6.4 Bridges at last!

The author's ambition all though school was to become a bridge designer. Now that he has retired from offshore structures, he is retuning to his first love. The Hood Canal link spans in Seattle will link a floating bridge to the shore. They are articulated with spherical and sliding bearings for most degrees of freedom, but are compliant in twist to accommodate roll of the pontoons, and rigid axially to limit expansion joints to about 20-inches movement. A yielding A-frame accommodates larger earthquake motions. Tubular joint design and material selection followed offshore practice. HPS70W steel is used in chords and braces to save weight, but API 2W grade 60 was used for the heavy node cans.[105]

6.5 Node, bent or backing?

In North Sea practice, nodes are pre-fabricated in pressure vessel shops, before assembly of the space frame. This has the advantages of high quality and the ability to perform stress relief heat treatment on the nodes. The disadvantage, aside from scheduling delays and higher cost, is that the critical location moves to single-sided closure welds in the legs and braces, made at the assembly site. The DC (Design Class) methodology in ISO 19902 reflects this pressure-vessel heritage, with more emphasis on leak-before-break toughness criteria based on thickness, rather than on the severity of local stresses and strains.

Bent fabrication refers to the common practice in the Gulf of Mexico and the rest of the offshore world.[106] Entire braces are pre-fabricated with the saddle shaped cope on each end, so that the closure weld occurs at the intersection in T-Y-K connections. All welding must be done from the outside only. Hence all the attention given in AWS to special welder qualification (6GR), specialized UT, and relaxed requirements at the weld root.

European bridge designers using very thick tubes have been reluctant to embrace relaxed requirements at the open root of tubular T-Y-K welds. Their solution is to employ heavy backing, in the form of a second tube with a saddle-shape cope, telescoped into ends the saddle-coped and beveled main brace. Aside from making double work for the fabricator, this approach builds in a huge Northridge-style notch at the root. However, fatigue testing appears to show satisfactory performance.

6.6 Sweating the small stuff

None of the foregoing three approaches is optimal for small onshore tubular structures. Most hiring halls do not have 6GR welders readily available. For branch members thinner than ½-inch (12 mm), effective UT inspection of the T-K-Y intersection is problematic, and complete joint penetration can neither be assured, nor need it be required. The partial joint penetration T-Y-K welds in AWS D1.1 match the static strength of the connected brace, and do not require design checking of the weld itself; only check the chord for punching shear.

For branch members of 3/8-inch (9 mm) or thinner, flat-faced fillet welds usually suffice, and have the advantage of simple inspection to assure adequate throat. The AWS prequalified T-Y-K fillet welds match static strength of mild steel with E-70 welds. They incur a small penalty for fatigue in the brace (curve ET *vs.* DT) and none for the chord. The lazy designer who specifies CJP welds for everything is needlessly costing the project money.[107]

7 CONCLUDING REMARKS

7.1 Design is not just a video game

Designers today seem reluctant to use their intuition and follow load paths through the structural connections. They are more inclined to trust their computer and the "black box" software that comes with it. Design suites are now available that take the designer's geometry definition and produce global frame analysis, Code checks, and detailed drawings without further human intervention.[108] In one sad case, the computer showed a chord thickness change in the middle of a node, with a double-bevel weld symbol that resulted in 50% throat when assembled and welded from the outside. The structure collapsed after just a few cycles of service load.

It is not uncommon for designers to just size the structural members, leaving joint details up to the fabricator. If punching shear has not been checked at the design stage, the selected chord member sizes will not work as simple tubular joints.

Likewise, Ramsgate Ferry, WTC Canopy, and Brokeback Hingebeam are not happy stories either, if they can ever be told.

7.2 Vanity, vanity, vanity ... Wisdom?

Identification of the punching shear failure mode in 1964 field failures was celebrated with a "found art" sculpture from the wreckage, *Hurikan*, exhibited next to one of Calder's.[109] Subsequently it was discovered that Griff Lee had used punching shear (and ½-inch "thick" joint cans) in the 1956 central platform at EI-188, which served as an important pipeline hub for fifty years.[110] An 1857 wrought iron tubular design for offshore lighthouses, used from Carysfort Reef FL to Ship Shoal LA, has survived for well over a century, surpassing the performance record of more modern structures.[111]

More hubris occurred when it was realized that Leonardo DaVinci's theory of bending is essentially the same as recent "original" formulations for grouted tubular T connections, which are very strong in compression, and weaker (but ductile) in tension. Ecole Polytechnique's *Amicus Plato* was awarded in response to the author's Houdremont Lecture at the first ISTS in 1984. He is pleased to be invited back on this occasion, to honor his long-time mentor, Prof. Kurobane.

REFERENCES

AISC. 1997. *Hollow structural sections connections manual.* Chicago: American Institute of Steel Construction/Steel Tube Institute of North America/American Iron and Steel Institute.

AISC. 2005. Specification for structural steel buildings, *ANSI/AISC 360-05.* Chicago: American Institute of Steel Construction.

API. 2004. Ultrasonic and magnetic examination of offshore structural fabrication and guidelines for qualification of technicians. *RP2X.* Washington DC: American Petroleum Institute.

ASME. 2004. *Proc. of Ocean, Offshore and Arctic Engineering (OOAE) Specialty conf. on Integrity of Floating Production, Storage & Offloading Systems (FPSO),* Houston, August 30–September 2, 2004. Houston: International Petroleum Technology Institute.

AWS. 2006. Structural welding code–steel, *D1.1-2006.* Miami: American Welding Society.

Bergan, P.G. & Lotsberg, I. 2004. Advances in fatigue assessment of FPSOs, *OMAE-FPSO'04,* paper 0012.

Burley, T. & Roper, J.R. 2005. Finite element modeling of complex welded structures. *Welding Journal,* December issue: 42.

Campbell, H.C. 2005. The use of ISO 19902 in the fabrication and inspection of fixed offshore structures. *Proc. of Intern. Offshore and Polar Engineering Conf.* (ISOPE 06), San Francisco: May 28–June 2, 2006.

Carter, R., Marshall, P.W. et al. 1969. Materials problems in offshore structures. OTC 1043. *Proc Offshore Technology Conf.,* Houston.

Dimitrakis, S.D., Lawrence, F.V. & Mohr, W.C. 1995. S-N curves for tubular Joints. *Final report to OTJRC of API.*

Dong, P. 2004. The mesh-insensitive structural stress and master S-N method for ship structures, *OMAE-FPSO'04,* paper 0021 (plenary), ibid.

Draper, J. 2001. Introduction to biaxial fatigue. Safe Technology Ltd.

Draper, J. 2004. Fatigue of welded steel joints in *fe-safe,* ibid.

Efthymiou, M. 1988. Development of SCF formulae and generalised influence functions for use in fatigue analysis. *Proc. Offshore Tubular Joint Conference,* Surrey, October 1988.

Greste, O. & Clough, R.W. 1967. Finite element analysis of tubular joints. Berkeley: Univ. of Calif. Structures & Materials Research Rept 67–7.

Grover, G.L. 1984. *Fracture mechanics based fatigue assessment of tubular joints: Development of analysis tools*. Palo Alto: Aptech Engineering Services. (also see OMAE-89)

Gurney, T.M. 1973. On fatigue design rules for welded structures. OTC 1907. *Proc Offshore Technology Conf.*, Houston.

Hayes, D.J. 1981. *Fracture mechanics based fatigue assessment of tubular joints: Review of potential applications*. Palo Alto: Aptech Engineering Services.

Hobbacher, A. 2003. Recommendations for fatigue design of welded joints and components, *XII-1965-03/XV-1127-03*.

ISO. 2002. Fatigue design procedure for welded hollow section joints – Recommendations. *ISO/DIS 14347* (as drafted by IIW-XV-E). Geneva: International Standards Organization.

ISO. 2006. Fixed Steel Offshore Structures. *FDIS 19902*. Former drafts as ISO 13819-2. London: British Standards Institute.

Kinra, R.K. & Marshall, P.W. 1979. Fatigue analysis of the Cognac platform. *Proc Offshore Technology Conf.*, Houston.

Kurobane, Y., Makino, Y. & Ochi, K. 1984. Ultimate resistance of unstiffened tubular joints, *Journ. Structural Engineering* 110(2): 385–400.

Lee, G.C. 1968. Twenty years of platform development. *Offshore*, June issue.

Marshall, P.W. & Toprac, A.A. 1974. Basis for tubular joint design. *Welding Journal*, May issue: 192.

Marshall, P.W. 1974. Basic considerations for tubular joint design in offshore construction. *Bulletin* 193. New York: Welding Research Council.

Marshall, P.W. 1980. Fixed pile-supported Steel offshore platforms. *Journ. of the Structural Division* 107(6): 1083–1094.

Marshall, P.W. 1982. Experience-based fitness-for-purpose ultrasonic reject criteria for tubular structures. *Fitness for Purpose in Welded Construction. Proc.* AWS/WRC/TWI conf.

Marshall, P.W. 1986. Designed of internally stiffened tubular joints. *Safety Criteria in Design of Tubular Structures, Proc IIW-AIJ International Meeting*, Tokyo.

Marshall, P.W. 1990. Fracture control procedures for deep water offshore towers. *Welding Journal* (January).

Marshall, P.W. 1992. *Design of welded tubular connections: Basis and use of AWS code provisions*. Delft: Elsevier. 412. ISBN: 0-444-88201-4.

Marshall, P.W. 1993. Offshore structures. *Constructional Steel Design*. Chapter 6.7. Delft: Elsevier.

Marshall, P.W. 2004. Review of tubular joint criteria. *Proc. Connections in Steel Structures* V. Amsterdam: AISC & ECCS, Bouwen met Stahl.

Marshall, P.W., Bucknell, J. & Mohr, W.C. 2005. Background to new API fatigue provisions, OTC 17295, *Proc. Offshore Technology Conf.*, Houston.

Marshall, P.W. & Wardenier, J. 2005. Tubular versus nontubular hot spot stress methods. *Proc. of Intern. Offshore and Polar Engineering Conf.* (ISOPE 05), Seoul, June 19–24, 2005.

Mohr, W.C. 2003. *Minutes of strain-based design meeting on March 6*. Houston. Edison Welding Institute.

Munse, W.H. 1978. Predicting the fatigue behavior of weldments for random loads. OTC 3300. *Proc. Offshore Technology Conf.*, Houston.

Niemi, E. 1993. Stress determination for fatigue analysis of welded components. *XIII-1221-93*. Villepinte: International Institute of Welding.

Noordhoek, C. & DeBack, J. 1987. *Steel in marine structures*. SIMS-87, Proc. 3rd Intl. Conf. Delft: Elsevier.

Pecknold, D.A., Marshall, P.W. & Bucknell, J. 2005. New API RP2A tubular joint strength design provisions. OTC 17310. *Proc Offshore Technology Conf.*, Houston, Texas.

Qian, X., Dodds, R.H. & Choo, Y.S. 2005. Mode mixity for circular hollow section X-joints with weld toe cracks, Journ. Offshore Mechanical. Arctic Engineering. 127(3): 269–279.

Reimer, R.B. et al. 1976. Improved finite elements for analysis of welded tubular joints. OTC 2642. *Proc. Offshore Tech Conf.*, Houston, Texas.

Roark, R.J. 1954. *Formulas for stress and strain*. New York: McGraw-Hill.

Sherman, D.R. 1976. *Tentative criteria for structural applications of steel tubing and pipe*. AISI Committee of Steel Pipe Producers.

SIMS. 1981. *Steel in Marine Structures*. Luxembourg: ECSC.

Toprac, A.A. 1969. Research in tubular joints: static and fatigue loads. *Proc. Offshore Technology Conf.*, Houston.

UKOSRP. 1978. *European Offshore Steels Research*. Preprints of the Select Seminar, Cambridge.

Wardenier, J. 1979. Design rules for predominantly statically loaded welded joints in circular hollow sections. XIII-918-79/XV-436-79. Villepinte: International Institute of Welding.

Wardenier, J. 1982. *Hollow section joints*. Delft: Delft University Press.

Wardenier, J., Packer, J.A., Dutta, D. & Yeomans, N. 1991. *Design guide for circular hollow section (CHS) joints under predominantly static loading*. Köln: CIDECT and Verlag TÜV Rheinland GmbH.

Yura, J. & Zettlemoyer, N. 1980. Ultimate capacity equations for tubular joints. OTC 3690. *Proc. Offshore Technology Conf.*, Houston.

299

Plenary session B: Code development & applications

Tubular Structures XI – Packer & Willibald (eds)
© 2006 Taylor & Francis Group, London, ISBN 0-415-40280-8

Hollow Structural Sections (HSS) in the 2005 AISC Specification

D.R. Sherman
University of Wisconsin-Milwaukee, USA

ABSTRACT: The unified 2005 Specification of the American Institute of Steel Construction (AISC) is a considerable departure in form from previous editions of the Specification. Its format permits both Load and Resistance Factor Design (LRFD) and Allowable Strength Design (ASD). This is accomplished by presenting nominal resistances for the various limit states and then obtaining the available resistance either by multiplying by a resistance factor for LRFD or dividing by a safety factor for ASD. Provisions for the design of HSS members and connections are included in the 2005 Specification instead of appearing in a separate specification as in the past. Including HSS provisions along with those for other shapes leads to uniformity in design methodology for all structural members, but also produces a complex appearing document. However, the various sections have been subdivided and arranged so that it is possible to extract HSS provisions without being overwhelmed with other requirements. This paper presents an overview of the 2005 Specification format, with emphasis on the new HSS provisions.

1 INTRODUCTION

1.1 *AISC Specifications*

In March 2005, the American Institute of Steel Construction (AISC) issued a new Specification for structural steel buildings (AISC 2005a). The Specification also applies to other structures "designed, fabricated and erected in a manner similar to buildings, with building-like vertical and lateral load resisting elements." The 2005 Specification supersedes previous AISC Specifications including the LRFD Specification (AISC 1999), the ASD Specification (AISC 1989), the Single Angle Specification (AISC 2000a) and the HSS Specification (AISC 2000b.) The Seismic (AISC 2005b) and Nuclear (AISC 2003) Specifications are the only other AISC design standards. The provisions of the previous HSS Specification (AISC 2000b) and were discussed in at a previous symposium (Sherman 2001) and have been included in the 2005 Specification with a few modifications.

1.2 *A unified specification*

The 2005 Specification is referred to as a unified specification since it includes both Load and Resistance Factor Design (LRFD) and Allowable Strength Design (ASD) methodologies. In LRFD the structural analysis is performed at service loads increased by load factors while in ASD the analysis uses service load combinations. The two methodologies are included

by specifying nominal resistances. R_n, for applicable limit states and then obtaining available resistances by multiplying by resistance factors for LRFD or dividing by safety factors for ASD.

$$R_{LRFD} = \phi R_n \tag{1}$$

$$R_{ASD} = R_n/\Omega \tag{2}$$

The original LRFD Specification was calibrated to former ASD design standards by using a live-to-dead load ratio of 3 and load factors of 1.2 for dead load and 1.6 for live load. The design requirements are:

$$LRFD: 1.2D + 1.6L \leq \phi R_n \tag{3}$$

$$ASD: D + L \leq R_n/\Omega \tag{4}$$

where D = dead load; and L = live load.

Dividing Equation 3 by Equation 2 and using L/D equal to 3 yields:

$$\Omega = 1.5/\phi \tag{5}$$

The values of ϕ were determined by reliability analysis in the previous LRFD Specifications and Equation 5 is used to establish the safety factors in the unified specification. Table 1 shows the factors for the key limit states that apply to HSS members and connections. The resistance factors for HSS connections are based on the variability of numerous international tests. They

Table 1. Key resistance and safety factors.

Limit State	ϕ	Ω
Tensile yielding	0.90	1.67
Tensile rupture	0.75	2.00
Column buckling	0.90	1.67
All flexure limit states	0.90	1.67
Beam shear	0.90	1.67
Concrete filled columns	0.75	2.00
Round HSS Connections		
Chord plastification	0.90	1.67
Punching shear	0.95	1.58
Rectangular HSS Connections		
Chord plastification*	1.00	1.50
Punching shear	0.95	1.58
Sidewall yielding	1.00	1.50
Sidewall crippling	0.75	2.00
Sidewall buckling	0.90	1.67
Gap yielding in K	0.90	1.67
Uneven load distribution	0.95	1.58

*0.95 and 1.58 for transverse loads and for overlapping K. 0.90 and 1.67 for gapped K.

vary with the connection configuration, type of loading and the applicable limit states. Table 1 includes the corresponding safety factors for the various resistance factors that appear in the chapter on HSS connection design.

2 SPECIFICATION OVERVIEW

2.1 *Chapters*

The Specification is divided into 13 chapters, A–M. The first two chapters define the scope of the specification and specify the items that must be considered in design, including the options of LRFD or ASD design. The wall slenderness limits for round, square and rectangular HSS are defined in Chapter B. The limits define compact shapes that can achieve the plastic bending moment and uniformly compressed element that achieve the yield stress before local buckling of the element. Sections that exceed these limits are considered to have slender elements that require special design considerations. Based on Australian research (Wilkinson & Hancock 1998), the web slenderness limit for rectangular HSS has been reduced from previous specification, where it was taken to be the same value as for W-shapes.

Another major item in Chapter B is the definition of design wall thickness for HSS. It "…shall be taken equal to 0.93 times the nominal wall thickness for electric-resistance-welded (ERW) HSS…". This is to account for common manufacturing practice in the USA. The A500 material specification for HSS

(ASTM A500-03) permits the wall thickness to be 10% less than the nominal thickness. Tube mills take advantage of this tolerance and consistently produce products with less than the nominal wall thickness unless a contract specifies otherwise.

Chapter C is concerned with stability analysis and design. It specifies the need for second order analysis and prescribes various acceptable methods.

Chapters D–H concern the design of members for tension, compression, flexure, shear, and combined forces including torsion. Chapter I for composite members includes concrete filled HSS. All common types of structural shapes are considered in these chapters, including HSS. Incorporating HSS provisions along with those for other shapes leads to uniformity in design methodology, but also produces a complex appearing document. However, the chapters have been subdivided and arranged so that it is possible to extract HSS provisions without being overwhelmed with other requirements.

Chapter K is devoted to HSS connections. This is new material that did not appear in the previous LRFD and ASD specifications and considerably expands the cases considered in the former HSS specification (AISC 2000b.)

Chapter J deals with criteria for welded and bolted connection that may apply in HSS construction, but does not contain any specific HSS requirements. The remaining chapters pertain to serviceability, fabrication, erection and quality control. The only item in these chapters specifically related to HSS is "When water can collect inside HSS or box members, either during construction or during service, the member shall be sealed, provided with drain holes at the base, or protected by other suitable means."

2.2 *Appendices*

The Specification contains seven appendices. These are alternatives to methods in the body of the Specification or requirements that are infrequently used by building design engineers. Infrequently used appendixes include ponding, fatigue, fire conditions, evaluation of existing structures and stability bracing. Inelastic analysis and the direct analysis method for stability are the two other appendixes. None of these appendices contain information specifically for HSS. The Specification does not consider fatigue in HSS connections.

2.3 *Commentary and user notes*

The Commentary contains background information and references for professionals seeking further understanding of the basis, derivations and limits of the Specification. It is not part of the Specification and the material contained in it is not mandatory in design.

User notes are a new feature in AISC specifications. These are printed in shaded areas in the Specification and are not mandatory requirement. They are intended to give guidance as to when particular provision may not apply. For example, the Section for shear in round HSS contains equations for shear buckling, which seldom controls. A user note following these equations states: "The shear buckling equations, ..., will control for D/t over 100, high strength steels, and long lengths. If the shear strength for standard sections is desired, shear yielding will usually control."

Another helpful feature of the Specification is that when certain words first appear, they are written in italics to indicate that they are defined in a glossary. In the glossary, pipe and tubing say "see HSS." HSS is defined as "Square, rectangular or round structural steel sections produced in accordance with a pipe or tubing product specification."

3   STABILITY ANALYSIS

Although there are no specific requirements for the analysis of HSS structures, these structures are subject to the stability analysis requirements of Chapter C. Several methods are acceptable, but the influence of second-order effects on flexure, shear and axial forces must be considered. Inelastic analysis according to Appendix 1 is acceptable, but the emphasis is that "Any second-order elastic analysis method that considers both P-Δ and P-δ effects may be used." The prescribed methods are:

– amplified first-order elastic analysis using the amplification factors from the previous LRFD Specification (AISC 2000a)
– first-order analysis with added notional lateral loads, provided the required axial load in compression members is less than 0.5 times the yield load
– the direct analysis method in Appendix 7 which includes notional loads and reduced flexural and axial stiffness.

It is inappropriate for this paper to present the details of these methods since this would be a departure from the topic of HSS members. However, it might be noted that in future editions of the specification, it is anticipated that the direct method of analysis will be presented as the preferred method.

4   HSS MEMBER PROVISIONS

4.1   *Tension*

As for other shapes, HSS must be checked for the limit states of yielding and rupture. The nominal resistance for rupture is

$$P_n = F_u A_e \tag{6}$$

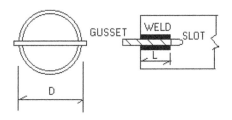

Figure 1.   Round HSS with centered end gusset.

where F_u = the rupture stress; and A_e = effective net area.

$$A_e = A_n U \tag{7}$$

where A_n = net area; and U = shear lag factor.

For slotted HSS welded to a gusset plate, the net area is the gross area minus the product of the thickness and the total width of material that is removed to form the slot. This requirement is the result of practice in the USA not to weld around the end of the gusset plate in order to prevent possible undercutting of the gusset and having to bridge the gap at the end of the slot.

Equations for U based on the outside dimensions of the HSS are included in a table. For rectangular HSS, there is no change from the previous HSS Specification (AISC 2000b). However, based on recent research (Cheng & Kulak 2000) U for round HSS has been changed as follows:

$$L \geq 1.3D, \ U = 1 \tag{8}$$

$$D \leq L \leq 1.3D, \ U = 1 - x/L \tag{9}$$

where x = D/π; D = outside diameter; and L = length of the weld. This implies that the weld cannot be less than D. Details of the end gusset and dimensions are shown in Figure 1.

4.2   *Compression members*

AISC has maintained its policy of having only one column curve for compression members. Therefore, there is no incentive for stress relieving HSS in the USA. The equation for the critical compressive stress is the same as in previous specifications, but the resistance factor has been raised from 0.85 to 0.90 for all compression members including HSS.

The chapter on compression members includes a section for members with slender elements, which was an Appendix in the previous LRFD Specification (AISC 1999.) Since many standard sizes of HSS have slender elements, it is an advantage to have this section in the body of the Specification. The equations for the effective width of slender compression elements in square and rectangular HSS have not changed, nor have the reduction factors used to modify the column equation for rectangular and round HSS.

Table 2. Selection table for the application to HSS of Chapter F sections.

Section in Chap. F	Cross section	Flange slenderness	Web slenderness	Limit states
F7	Square HSS	C, NC, S	C, NC	Y, FLB, WLB
F8	Round HSS	N/A	N/A	Y, LB

Y = yielding; FLB = flange local buckling; WLB = web local buckling; LB = local buckling; C = compact; NC = noncompact; S = slender

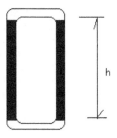

Figure 2. Shear area in a rectangular HSS.

4.3 Flexural members

Chapter F on flexure is the longest and most complex of the chapters on member design. A User Note at the beginning of the chapter serves as a directory showing which sections apply to certain types of members. The directory clearly indicates that provision for rectangular HSS are in Section F7 and for round HSS in F8. There is no need to be familiar with the other sections of the Chapter in order to design HSS flexural members. Table 2 reproduces the HSS portion of the User Note and shows how it identifies the type of members that are covered and which limit states apply.

The provisions for the various limit states are unchanged from the previous specifications. It can be noted that no check is required for lateral-torsional buckling of an HSS beam, even for rectangular HSS bending about the major axis. Serviceability limits for beam deflections will control before lateral-torsional buckling occurs.

4.4 Shear

There are no changes in the HSS provision for the shear yield and shear buckling limit states. The shear provisions for rectangular HSS in Section G5 are the same as for the webs of I-shape beams. The depth of the webs, h, is the defined as the clear distance between the flanges less the inside corner radius as shown in the shaded portion of Figure 2. If the corner radius is not known, h shall be taken as the corresponding outside dimensions minus three times the thickness.

Section G6 defines the nominal shear resistance for round HSS. The limit states of shear yielding and shear buckling have been used in previous AISC specifications.

4.5 Torsion and combined forces

HSS are the only shapes that have specific provisions for resistance to torsion. These appear in Section H3 of Chapter H and have been copied from the previous HSS Specification. For combinations with other loads, the interaction is treated in one of two ways. When the required torsional strength is less than 20% of the available torsional strength, the typical interaction using only axial load and amplified moments is used, the same as for other shapes. For higher required torsional strength, a stress based interaction combining normal stresses with the square of shear stresses is used.

$$(P_r/P_c + M_r/M_c) + (V_r/V_c + T_r/T_c)^2 \leq 1 \qquad (10)$$

where P = axial load; M = moment; V = shear; T = torsion; subscript r = required strength; subscript c = available strength.

4.6 Composite HSS

Chapter I contains new material on concrete filled HSS columns and flexural members. In order to qualify as a composite column

- the steel HSS shall comprise at least 1% of the total composite area
- the maximum flat-width/thickness ratio is 2.26 $(E/F_y)^{0.5}$ compared to $1.40(E/F_y)^{0.5}$ for an unfilled HSS
- the maximum diameter/thickness ratio for a round HSS is $0.15E/F_y$ compared to $0.11E/F_y$ for an unfilled HSS

The nominal compressive strength is the HSS yield strength plus a portion of the concrete strength. An effective column stiffness, which is the HSS column stiffness increased by a portion of the concrete column stiffness, is used to obtain the Euler load in the column strength equation. The tensile strength and shear strength use only the HSS strength.

The flexural strength of a concrete filled HSS is determined from either the superposition of elastic stresses on the composite section or the plastic stress distribution of the steel section alone.

5 HSS CONNECTIONS

The HSS Specification (AISC 2000b) was the only previous AISC specification to contain criteria for

HSS connection. Therefore including Chapter K on HSS connections in the 2005 Specification is a significant addition to the primary AISC Specification. Criteria are included for round and rectangular HSS in the following conditions:

- concentrated forces that are transverse or parallel to the axis of the HSS
- axial forces in truss connections in T-, Y-, Cross- and K-connections, with gaps or overlaps
- round HSS branches with in-plane and out-of-plane moments in branches of T-, Y- and Cross-connections
- rectangular HSS branches with in-plane and out-of-plane moments in T- and Cross-connections
- branches with combined moments and axial forces.

The Commentary references different publications for other configurations, such as multiplanar connections, connections with partially or fully flattened branch member ends, double chord connections, connections with the branch offset from the centerline of the main member or connections with round branches joined to rectangular main members.

The criteria have been taken from the Canadian HSS connection design guide (Packer & Henderson 1997) with a few minor modifications based on more recent testing and research. The LRFD resistance factors are embedded in equations for axial force or moment resistance in the design guide. For the 2005 Specification modifications were made to express the nominal resistance, which is then multiplied by the appropriate resistance factor, ϕ. This was done to permit application of the criteria to ASD design by dividing the nominal resistance by the corresponding safety factor, Ω.

Chapter K has three major changes from the previous HSS specification.

- the scope is expanded to include K-connections with overlaps
- the criteria for truss connections with round HSS are now generally in accord with internationally recognized IIW criteria (1989)
- the criteria for moments in branches also are in accord with IIW criteria.

Chapter K is one of the largest chapters in the 2005 specification.

6 DESIGN MANUAL

In December 2005 AISC issued the 13th edition of a design manual, which is a companion document to the Specification (AISC 2005c). This manual differs from previous manuals in that it must include design aids for both LRFD and ASD. All the classic design tables for members and connections have been retained. Tables such as HSS section properties that are common to

either method are printed in black type. In tables giving load capacities, such as ϕPn and Pn/Ω for HSS columns with various effective lengths, LRFD values appear in blue type and ASD values are printed in a light green shaded areas.

The tables for section properties include ASTM A53 pipe. These are treated in a similar manner to A500 HSS and the design wall thickness of 0.93 times the nominal wall thickness is listed. This accounts for the common practice of HSS manufacturers of downgrading their product and supply it meeting the requirements of ASTM A53 pipe without a pressure test.

In order to control the size of the manual containing both LRFD and ASD aids, design examples are not included in the body of the manual as they were in the past. Instead, a CD is included with the manual that contains a large number of examples illustrating use of the Specification provision and the design aids in the manual.

7 CONCLUSIONS

AISC has reduced the number of specifications related to steel design by issuing a new unified specification which combined both: LRFD and ASD. HSS provisions for member and connection are included in the unified specification. Most of the HSS member design provisions have been taken from the previous HSS Specification. However, there have been some changes in

- the wall slenderness limit for compact rectangular HSS,
- the shear lag factor for round HSS tension members with gusset end connections,
- concrete filled HSS members.

The unified specification is longer and more complex than any of the previous specifications, but most HSS provision are clearly identified in separate sections.

REFERENCES

AISC 1989. Specification for structural steel buildings—allowable stress design and plastic design. Chicago: American Institute of Steel Construction.

AISC 1999. Load and resistance factor design specification for structural steel buildings. Chicago: American Institute of Steel Construction.

AISC 2000a. Load and resistance factor design specification for single-angle members. Chicago: American Institute of Steel Construction.

AISC 2000b. Load and resistance factor design specification for steel hollow structural sections. Chicago: American Institute of Steel Construction.

AISC 2003. Load and resistance factor design specification for steel safety-related structures for nuclear facilities,

ANSI/AISC N690L-03. Chicago: American Institute of Steel Constructions.

AISC 2005a. Specification for structural steel buildings, ANSI/AISC 360-05. Chicago: American Institute of Steel Construction.

AISC 2005b. Seismic provisions for structural steel buildings, ANSI/AISC 341-05. Chicago: American Institute of Steel Construction.

AISC 2005c. Manual of steel construction. Chicago: American Institute of Steel Construction.

ASTM A500-03. Standard specification for cold-formed welded and seamless carbon steel structural tubing in rounds and shapes. Philadelphia: American Society for Testing and Materials.

Cheng, J.J.R. & Kulak, G.L. 2000. Gusset plate connections to round HSS tension members. Engineering Journal, AISC Vol. 37, No.4, 4th Quarter: 133–139. Chicago: American Institute of Steel Construction.

IIW (1989). Design recommendations for hollow section joints – predominantly statically loaded, 2nd edition, IIW document XV-701-89. Helsinki: International Institute of Welding.

Packer, J.A & Henderson, J.E. 1997. Hollow structural section connections and trusses – A design guide, Alliston Ontario: Canadian Institute of Steel Construction.

Sherman, D.R. 2001. HSS design for buildings in the United States. Proceedings of the ninth international symposium on tubular structures: 87–94. Dusseldorf: Balkema.

Wilkinson, T. & Hancock, G.J. 1998. Tests to examine compact web slenderness of cold-formed RHS. Journal of structural engineering, ASCE, Vol. 124, No 10 October: 1166–1174. Reston VA: American Society of Civil Engineers.

Tubular Structures XI – Packer & Willibald (eds)
© 2006 Taylor & Francis Group, London, ISBN 0-415-40280-8

Field joints for tubulars – Some practical considerations

R.H. Keays

Keays Engineering, Melbourne, Australia

ABSTRACT: Trusses fabricated from circular hollow sections are well recognized as being the most efficient solution for long-span structures. Often this efficiency does not translate to economy because the field joints are too expensive or take too long to connect. The author has over 30 years' experience in the design, fabrication, and erection of long-span structures, giving him an insight into the merits or otherwise of different forms of connections. The paper shows examples of different types of welded and bolted field joints, and comments upon the critical aspects of each joint type. A summary of the relative total cost of each different type considered is included.

1 INTRODUCTION

Trusses fabricated from Circular Hollow Section (CHS) chords and webs are well recognized as being the most efficient solution for long-span structures. They have been a feature of many of the sporting grandstands, pedestrian bridges, and aircraft hangars constructed in Australia in recent times. The efficiency comes from the availability of a wide range of sizes, higher strength steel, and the structural efficiency of circular struts in compression. Diseconomies arise from the complexity of shop and field connections.

CIDECT Design Guide 1 (Wardenier et al. 1991) provides detailed guidance on the strength of web/chord joints. Included is a discussion on strategies for minimising connection costs by adjusting the tube diameters and wall thicknesses. One field connection (bolted flange joint) is also documented.

CIDECT Design Guide 7 (Dutta et al. 1998) gives some insight into a number of other field connections. This paper is intended to complement this Design Guide, focussing on the construction issues and on the style of connection preferred in the Australian market.

The Australian market may be different to other markets. In some other markets the design engineers leave the detailed design of connections to the fabricator's design team. In Australia, the design engineer typically dictates the form and detail of each connection, and the fabricator typically prefers to follow these details slavishly. By this, the risk of an inadequate design is entirely shouldered by the design engineer. The advantage of this to the builder is (a) the design risk remains clearly in the domain of the designer, and (b) the steel fabricator can be selected on the basis of price and quality, without need to consider the competence of the fabricator's connection designer. The writer has made a career in connection design, with regular commissions from builders and fabricators to address problems with exhibited designs, devise erection schemes, and forensic studies of structural failures.

The following sections cover the connection details commonly documented for field joints, starting with the simple welded joint and then proceeding to the various forms of bolted joints.

2 FIELD WELDED BUTT JOINTS

Discussion starts with the butt joint where the tube ends are welded one to the other. This is the benchmark for all other joints, as it easiest to design and specify and develops the full strength of the connected members.

In a fabrication shop, the welded butt joint will always be cheaper than the alternatives, because (a) that is what the fabricator does, and (b) no work other than preparing the joint and welding it up is required. Things are different on a construction site:

- Access to the joint is limited.
- The members being connected must be held in place until the weld has been completed and inspected.
- The joint must be welded where it is, and any efficiency that might come from down-hand welding is just not available.
- The workers performing the welds are specialist tradesmen, belonging to a different Trade Union, expect a higher wage, and so have the potential to cause industrial problems.

The ends of the connecting members must be prepared for the butt weld. The typical preparation is as shown on Figure 1. Codes permit different preparations in certain circumstances, but these are not practical in field joints. Preparations for field joints

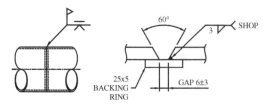

Figure 1. Field welded butt joint.

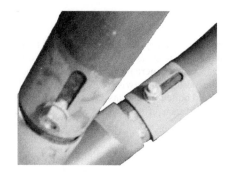

Figure 2. Joint with sliding backing bar.

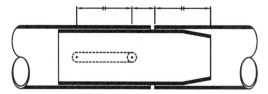

Figure 3. Section on tube and internal sleeve.

must be tolerant of poor fit-up and suitable for welding in any position.

In particular, a backing ring is required. For the chords of a truss, this can be installed in the fabrication shop on one side of the joint. It serves to align the two tubes as well as supporting the liquid weld pool. When erecting a single tube it is easy to slip the end over the backing ring. For truss web members (and chords if not parallel), a sliding backing ring can be used, as illustrated in Figures 2 and 3.

It is important to release each erected member from the crane as soon as practical after it has been positioned. Normally, the erector cannot afford the time required to tack weld the joint, and a temporary bolted connection is needed. Lapped joints with one or two bolts are usually sufficient, as illustrated in Figure 4. If the joint design relies upon the shear capacity of the backing ring for the temporary situation, it is wise to make sure this appears on the shop drawing, so that the welder in the fabrication shop does not skimp on the connection.

In some circumstances alignment of the structure may be critical, and it may be necessary to adjust the joint gaps after erection. Figure 5 shows a photo

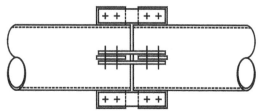

Figure 4. Temporary connection for butt joints.

Figure 5. Adjustable temporary connection.

of the threaded rod connection used on the Cone at Melbourne Central where twelve segments of the circle were required to marry to tight tolerances. The joint gaps were adjusted by tightening/relaxing the nuts on the threaded rods. This joint has some compression capacity if the nuts on the inside of the flange are installed and tightened. This joint is probably excessive for the general run of applications, but finds application where "on the crane hook" time is critical.

3   FIELD WELDED LAP JOINTS

An alternative joint used on some occasions is an external sleeve with fillet welds all round. One variant of this is shown in Figure 6. The primary reason for this style of joint is to replace butt welds by fillet welds. Fillets are generally easier to weld, fit-up is not as critical, and inspection is usually limited to a visual inspection at completion.

Dimensions and details in this joint are appropriate to the full strength of a 355.6×9.5 CHS with $F_y = 350$ MPa; this size is used throughout this paper to provide a basis for comparison of the alternatives.

Temporary fixing of this joint can be achieved by the addition of long pins, as illustrated in Figure 6. After finishing the weld the pins can be driven out and the holes plug welded.

This joint requires more kilograms of weld metal than the butt joint, but half the welds are completed in the shop, so the time on site for fillet welding is

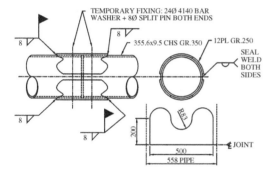

Figure 6. Welded external sleeve joint.

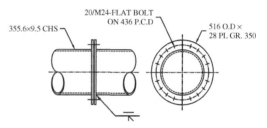

Figure 7. Bolted flange joint without gussets.

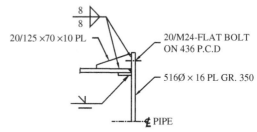

Figure 8. Detail of gusset on alternative flange joint.

less. The surface area damaged by field welding is also greater, which adds to the cost of touch-up painting.

This joint can be adopted where the architecture doesn't require a flush joint, and where the architect doesn't object to the joint's visibility.

4 BOLTED FLANGE JOINTS

Bolted flange joints can be designed to achieve the full strength of the tube in tension, bending, and torsion. They are relatively easy to detail, fabricate and erect. They are, however, very obvious, and there may well be an objection from the architect.

AIJ (2001) provides recommendations for the design of flanged joints with and without gussets. CIDECT Design Guide 9 (Kurobane et al., 2004) has an alternative formula giving similar answers for the bolt sizes and flange thicknesses for the joint without gussets. The AIJ recommendations were used in this example to provide direct comparison between joints with and without gussets. Dimensions shown in Figure 7 are for a full-strength flange without gussets. Figure 8 shows details of the gusset on the alternative.

Adding gussets allows a significant reduction in flange thickness, but has the penalty of the additional welding of the gussets. The addition of the gussets and the effect of weld heat input on the thinner flange make it more likely that the flange will distort beyond acceptable levels after welding. Machining the flange face flat after welding can cure this, but that adds significantly to the cost. The writer's opinion is that gussets are never justified.

The joint strength is limited by plastification of the flange plate; bolt size and number are selected so that bolt strength is not critical. The Japanese F10T bolt used in these examples has an ultimate tensile strength of 981 MPa, marginally less than ISO Property Class 10.9.

The writer has concerns about the use of PC10.9 bolts in tension. Such bolts are more susceptible to stress corrosion cracking and less tolerant to over-tightening than PC8.8 bolts. The recommendation

from the British Standard BS4395 summarized in Owens et al. (1992) is that PC10.9 should not be used in tension joints.

Redundant structures are affected by lack-of-fit forces; tightening the flange bolts to eliminate a gap at a joint is an easy way to induce significant forces. So, even if a structure is designed on an elastic basis with member loads less than the design tension capacity, the flanged joint should always be designed for capacity of the tube.

With curved tubes, the designer needs to be aware of the fabricator's capability. Curving is an art, not an exact science, and frequently results in members not quite the right radius. The tradesman must then cut the tube at the correct angle so that the joint has minimal angular misalignment. Application of brute force in the field (by impact wrench) normally solves this problem. The writer has seen drawings where the designer required measurement of angular misalignment and installation of tapered shims! He has also seen fabricators trial assemble the two halves of such joints before making the tube to flange joint for fear of such misalignment.

In normal circumstances the flange plate completely covers the end of the tube. Sometimes, this is not practical. For large diameter tubes used in light tower masts and windmills, joints such as that shown in Figure 9 can be used. The flange is normally inside the tube, which improves the aerodynamics and aesthetics.

Here, prying action is an integral part of the joint design, and cannot be talked away. Reducing the inside diameter of the flange (by increasing the 225 dimension in Figure 9) reduces the magnitude of prying forces. Countering this is the need to keep the flange

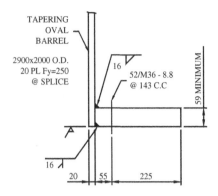

Figure 9.   Internal flange joint for 75 m light tower.

weight to a minimum for general ease of fabrication and erection. There is also a need to leave room for internal access ladders when the tower diameter is less than 1750 mm.

5   BOLTED LAP JOINTS

The joint shown in Figure 10 is a simple bolted connection for a tubular member with pure axial load, but it has almost no strength in bending or torsion.

The thickness of the fin plate is simply a function of the cross-sectional area of the tube, and is typically π times the wall thickness of the tube if it is the same strength material, or 4.4 times if the tube is 350 MPa yield and the plate 250 MPa yield.

The length of the slot for the fin plate is dictated by the shear stress in the wall of the tube, and is a minimum of 1.3 times the tube diameter (for a full-strength joint), or longer if shear lag is taken into account. (In the writer's opinion, shear lag should be considered when the ratio of yield to ultimate exceeds 85%.) The weld size should match the shear strength of the tube.

Where there is an architectural requirement, the width of the fin plate can be made the same as the tube diameter (as shown here), and the fillet weld replaced by a partial penetration butt weld. In such circumstances, the writer suggests using a low-energy weld procedure without a backing bar to achieve a minimum of 85% penetration.

The detail of the end of the fin plate is critical to the ultimate and fatigue capacity of the joint. If there is no weld, the joint will not be sealed, and the tube will fail in the net section at the end of the fin plate. Figure 11 shows one way this can be addressed.

In making the field connection, the double cover plates can be carried loose in the rigger's "man box", or tack welded to the parent plates (one each side). The latter is preferred if the plates weigh more than 10 kg.

Making a simple lap joint can eliminate the double cover plates, providing the eccentricities are taken into account. Figure 12 illustrates the problem. This joint

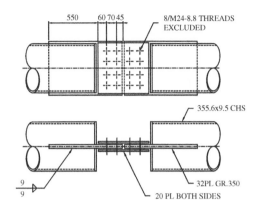

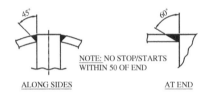

Figure 10.   Fin plate with double cover plates.

Figure 11.   Weld details for fin plates.

![photo](Figure 12)

Figure 12.   Collapsed lap joint.

has less than half the capacity of the double cover plate joint.

An alternative to the slotted fin plate is a Tee Bar connection, as shown in Figure 13. This is simpler to fabricate, but the thickness of the Tee plate required to develop the tube strength and the need for quality welding make this impractical for any joint other than the end of a compression strut with an (l/r) greater than about 120.

6   HIDDEN BOLTED JOINTS

Combining the architect's penchant for hiding construction details, and the builder's phobia for field welding has resulted in the popularity of the "hidden"

312

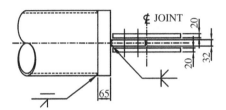

Figure 13.   Full-strength Tee bar joint.

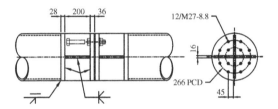

Figure 14.   Cruciform joint with bolts in tension.

joint. The argument for the hidden joint in preference to the external flange is well-expressed in Eekhout (1996), but he cautions, "Hiding requires more skill than revealing."

The typical form of this used in Australia is shown in Figure 14, and illustrated in Figure 15.

There are six shop butt welds in this joint to replace the one butt weld in the field joint. Field welding is expensive, but should not be that much more expensive!

In this extant structure the builder chose to leave the joint visible, which partly defeats the purpose of the hidden joint. It also creates pockets that will collect rainwater, and so reduce the life of the paintwork. The cover plate used in other instances is galvanized sheet steel rolled to a curve, and fixed with self-drilling screws. If the galvanized sheet is kept below 0.5 mm thick, a single sheet can be rolled to a larger radius and then sprung to the final size on installation. Alternatively, two sheets can be used to cover the two halves of the joint. An example appears later in Figure 18.

The thickness of the cap plate on the tube end is determined in the manner used for the Tee Bar Joint, but needs only to be about half as thick because there are two "fin" plates. These two fin plates form a cruciform joint along the axis of the tube.

The cruciform plates give the joint some bending strength and stiffness in both directions. If the plates are sized to achieve the tube strength in tension, the bending capacity is about 40% of the tube's bending capacity. This strength and stiffness should be sufficient in most applications to carry any incidental lateral load and to stabilize the tube against column buckling. Note that the joint has virtually no torsional strength – twisting loads in arches have to be resisted by other means.

The flange plates are designed for bending between the bolts in tension and the supports at the cruciform plates. Like the external flange plate discussed above, the ends must be flat and parallel.

The joint is typically most congested, making sizing and installation of bolts difficult. In the example shown in Figure 14, M24-8.8 bolts were not quite strong enough, and M30-8.8 bolts were too large to fit in the space available. One solution to this problem is shown in Figure 16, which defeats the purpose of using the "hidden" joint.

Figure 15.   Joints on Telstra Dome.

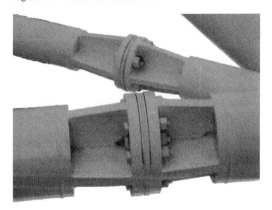

Figure 16.   Cruciform joint with oversize flange.

Other solutions involve the use of larger diameter or higher strength bolts. The writer's concern about the use of PC10.9 bolts was expressed earlier. There have been instances where the very high strength PC12.9 bolts have been used in such joints; these are even more susceptible to stress corrosion cracking (and it did happen!). Larger diameter bolts are more expensive per kN of capacity (AISC, 1996), and introduce a manual-handling hazard because the impact gun weighs over 15 kg.

The writer has heard complaints from riggers that the flush joint is difficult to erect. Normally the rigger will use a "spud wrench" to bring the part being erected into line with the holes in the connecting member. With the bolt holes set deep in the flange, access for the spud wrench is limited, increasing the time the load spends on the crane hook while the flange faces are brought into contact.

313

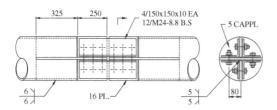

Figure 17.   Cruciform joint with bolts in shear.

Figure 18.   Melbourne sports and aquatic centre roof joint.

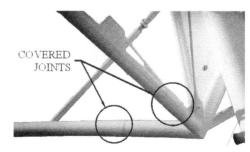

Figure 19.   Joint covered up.

Figure 20.   Southern cross station spine truss joint.

There are variations on the arrangement shown at Figure 14 (above).

- The fabricator's job can be made a little easier by coping the cruciform plates at tube centre, and by reducing the weld between cruciform plates to fillet welds.
- The cruciform plates can be slotted into the pipe wall, eliminating the tee-bar plate. These plates need be only half as long as those in the fin plate connection, as there are two of them.
- The bolts in tension can be replaced by bolts in double shear, with standard angle sections used as connectors.

The combination of these changes is shown on Figure 17.

This arrangement requires care with the detailing of the joint, and transfer of this information from the design drawings to the fabricator and erector.

The joint is efficient in welding, with only (roughly) three times as much weld metal as the field-welded joint. There is, however, the cost of the slots, making the cruciform sections, punching the holes in the angles, and procuring and installing the bolts.

The joint is relatively straightforward for the erector. There are holes available for the spud wrench. The splice angles typically are less than 10 kg. If desired, two angles can be loosely attached to one side of the joint before erection. This gives a "landing deck" for initial securing of the joint. The rigger does need to take care to install all the bolts in a clockwise fashion, so that heads and nuts do not interfere.

This style of joint has been used in recent projects, including the Melbourne Sports and Aquatic Centre (Figures 18 and 19), and Southern Cross Station Roof Spine Trusses (Figure 20).

7 OTHER POSSIBILITIES

The examples above cover most of the practical arrangements for field splices with welded and bolted joints. Pin joints are sometimes used for secondary bracing members or where the architect desires a different expression of the joint. Pin joints are normally made with two plates slotted into the wall of the CHS. As the length of these is dictated by transfer of shear from the wall of the CHS to the fin plates, the slot length will be almost as much as that for the single plate shown earlier at Figure 10.

Figure 21. Folded plate for slotted end connection (from CIDECT 7 Fig 3.14).

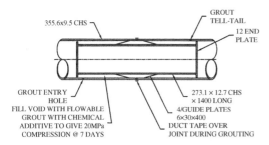

Figure 22. Grouted sleeve joint.

CIDECT 7 includes a folded plate as an alternative to the double plates (Figure 21). The bend radius on the folded plate must be very tight as otherwise the attached plate will be excessively thick. Such tight radii can only be achieved by hot bending.

CIDECT 7 also includes a photograph (Fig. 3.44) of a joint with an internal sleeve and approximately 300 nails each side connecting the tube to the sleeve. This does not seem a practical option for field joints. It is best not to contemplate the OHS ramifications of a man with a high-powered gun shooting nails into a curved surface high above a busy construction site.

This paper has concentrated on joints for CHS, but mention should be made of RHS. Joints shown above can be easily adapted for sections with flat sides. With RHS sections, it is possible to use an internal bolted or pinned sleeve, providing allowance is made for bending effects at the lapped joints, and providing the possibility of internal corrosion of an unsealed section is addressed. Alternatively, blind bolting using Huckbolts (or similar) may be airtight, but these "one-time" bolts present problems for the erector. The erector needs to be able to secure one end of a member in roughly the right place while he checks that the other end is also correctly positioned. Huckbolts cannot be loosely fixed – they rely on the clamping force to deform the blind side of the fastener.

Another style of joint with a radically different means of transfer of load is the grouted sleeve.

Grundy (1991) and others have investigated the use of grouted internal or external sleeves for field connections of tubulars. An expanding grout is used to enhance the nominal shear strength, but the sleeve still needs to be about 5 diameters long. This joint is one being considered for two current constructions in Melbourne, but the chances of success are limited by the skepticism of builders.

8 FATIGUE PERFORMANCE

Fatigue is not normally a problem with structures subject to gravity and wind loads. It can be a problem if there are wind-induced vibrations, and might be a problem if details with poor fatigue performance are used in structures subject to typhoon/cyclone/hurricane loading. Three structures in Melbourne have suffered fatigue failures in joints where wind-induced vibration in tubular members was allowed to continue for months after construction. Fortunately, the cracks were detected before collapse.

The writer has recently studied the fatigue performance of different possible joints as alternatives to the joint shown in Figure 8, using the Classification Method in CIDECT 8 (Zhao et al., 2000), and provisions in Australian Standard AS4100 (1998), and JSSC32 (1993). This particular arrangement is proposed for a large item of mechanical plant subject to a full fatigue cycle with each rotation. The critical detail in this instance is the fillet weld between rib and flange plate, with a fatigue category of 44 MPa (life of 2E6 cycles at this stress range in the parent tube).

Leaving off the ribs and using a thicker flange (Figure 7) improves the fatigue category to 80 MPa, providing bolt fatigue is not allowed to become critical, and the butt weld between tube and flange is full penetration.

Stress concentration at the end of the plate slotted into the tube is the limiting effect for the joints in Figures 10 and 17. With the end detail shown in Figure 11, and limited finishing to eliminate stop/starts and undercut, this has a fatigue category of 65 MPa. This category is likely to be acceptable, as it is typical of tube/tube joints elsewhere in the structure.

Tee-bar joints and eccentric lap joints were not checked in this study, as it was considered they would be no better than the ribbed flange plate. The Tee-bar joint in particular is most intolerant of weld defects at the edge of the Tee-bar and at the adjacent wall of the tube.

It is suggested there would be benefit in some comparative fatigue testing of different joints. Also, diagrams of the typical field joints could be included in the Figures in Table B.1 of CIDECT 8.

9 COMPARISON OF COSTS

Table 1 shows indicative relative costs for some of the joints discussed above. Costs were calculated using Watson (1996) for costs of fabrication and erection, apart from welding costs. Welding costs were based

315

Table 1. Relative costs of different joint styles.

Joint type		Field Butt	Field Fillet	Plain Flange	Gusseted Flange	Fin Plate	Hidden Flange	Hidden Angles
Figure		1&4	6	7	8	10	14	17
Extra mass	kg	12	53	89	70	141	94	109
Weld mass	kg	0.9	1.4	1.5	4.1	1.5	4.6	1.1
Fabrication	hours	3.8	4.6	9.4	13.3	11.1	14.4	9.5
Supply cost	$A	$169	$256	$489	$636	$617	$710	$547
Erection	$A	$83	$83	$83	$83	$83	$104	$83
Field welding	$A	$236	$184	$0	$0	$0	$0	$0
Bolt tightening	$A	$0	$0	$22	$22	$22	$22	$22
Field painting	$A	$70	$108	$15	$15	$15	$15	$15
Total joint cost	$A	$558	$632	$608	$756	$737	$851	$667
Relative cost	%	100%	113%	109%	136%	132%	152%	119%

on a production rate of 1 kg/hour in the workshop, and 0.5 kg/hour in the field. A further allowance of 25% for weld inspection on field butt welds was also included. The numbers shown here are far from precise, but indicative of the relative costs.

This shows that the cheapest joint is the Field Butt Weld. The advantages of the field fillet weld are offset by the extra cost in making the rolled plate sleeve and the larger area of surface requiring painting after welding.

Plain flanges are marginally more expensive than field butt joints, making them the preferred option for those with a phobia for field welding. The Hidden Angle joint is a little more expensive but has the advantage where aesthetics are important.

If this analysis is typical, gusseted flanges, single fin plates, and hidden flanges will not be the preferred connections in any circumstances.

10 CONCLUSIONS

This paper has examined a number of different possible field connections for tubulars, illustrating each with practical details for welds and temporary fixing.

Simple butt-welded joints have the advantage of design simplicity and strength, and will be the cheapest and lightest option providing field-welding costs are contained.

Bolted joints can use bolts in tension or shear. The latter has better ductility, allowing the designer to size the connection for the actual loads when they are significantly less than the member design capacity.

Joints that can be hidden are frequently preferred for their aesthetic merits. Such joints with bolts in tension are relatively expensive and difficult to design. The alternative using four angle sections lapped on cruciform plates is a practical alternative, and has only a limited additional cost.

REFERENCES

Architectural Institute of Japan. 2001. *Recommendation for Design of Connections in Steel Structures* (in Japanese). Tokyo: AIJ.
Australian Institute of Steel Construction. 1996. *Economical Structural Steelwork.* Sydney:AISC.
Australian Standard AS4100. 1998. *Steel Structures.* Sydney: Standard Australia.
Dutta, D., Wardenier, J., Yeomans, N., Sakae, K., Bucak, Ö. & Packer, J.A. 1998. *CIDECT 7: Design Guide for Fabrication, Assembly and Erection of Hollow Section Structures.* Köln: CIDECT.
Eekhout, M. 1996. *Tubular Structures in Architecture.* Köln: CIDECT.
Grundy, P. 1991. *Prestressed Grouted Sleeve Nodes.* Offshore Australia Exhibition and Conference, I-71 to I-76. Melbourne: Monash.
Japanese Society of Steel Construction. 1993. *JSSC Technical Report No.32. Fatigue Design Recommendations for Steel Structures.* Tokyo: JSSC.
Kurobane, Y., Packer, J.A., Wardenier, J. & Yeomans, N. 2004. *CIDECT 9: Design Guide for Structural Hollow Section Column Connections.* Köln: CIDECT.
Owens, G.W., Knowles, P.R. & Dowling, P.J. 1992. *Steel Designers' Manual, 5th Edition (page 665).* London: Blackwell.
Wardenier, J., Kurobane, Y., Packer, J.A., Dutta, D. & Yeomans, N. 1991. *CIDECT 1: Design Guide for Circular Hollow Section (CHS) Joints under Predominantly Static Loading.* Köln: CIDECT.
Watson, K.B., Dallas, S., van der Kreek, N. & Main, T. 1996. Costing of Steelwork from feasibility through to completion. *Steel Construction, Volume 30, No.2.* Sydney: AISC.
Zhao, X.-L., Herion, S., Packer, J.A., Puthli, R.S., Sedlacek, G., Wardenier, J., Weynand, K., van Wingerde, A.M. & Yeomans, N.F. 2000. *CIDECT 8: Design Guide for Circular and Rectangular Hollow Section Welded Joints under Fatigue Loading.* Köln: CIDECT.

Tubular Structures XI – Packer & Willibald (eds)
© 2006 Taylor & Francis Group, London, ISBN 0-415-40280-8

Building bridges with new materials – Footbridge over Bayerstrasse in Munich

C. Ackermann
Ackermann Ingenieure, Munich, Germany

J. Krampen
Vallourec & Mannesmann Tubes, Duesseldorf, Germany

ABSTRACT: For the footbridge spanning Bayerstrasse in Munich, S 690 high-strength steel has been used for the basic construction for the first time (Figure 1). The bridge creates a barrier-free link for pedestrians and cyclists between two parts of town, the Hackerbrücke rapid-transit railway station and the Theresienwiese fairgrounds. A 38-metre hybrid arch construction spans a four-lane main road, two tramlines, and two cycle lanes and pavements.

1 INTRODUCTION

The city of Munich needed a new bridge for visitors coming from the train station on their way to the Theresienwiese. This bridge, situated in downtown Munich, should not only lead the passengers on a save way to and from the Oktoberfest, but it should also be an architectural landmark.

The following paper describes the design and erection of that 38-metre long bridge.

The construction should be as light as possible to adapt to the urban situation.

The use of hollow circular sections of S 690 steel made it possible to design a filigree curved structure and to minimize the dead weight.

The arch is stabilized by a torsionally rigid spatial truss, consisting of a concrete slab as the upper chord, one circular tube as the bottom chord and bracings.

The entire framework of the bridge including the concrete slab was prefabricated, transported by truck to the building site and erected in hours with a mobile crane.

Figure 1. View of bridge, photograph by Jens Weber, Munich.

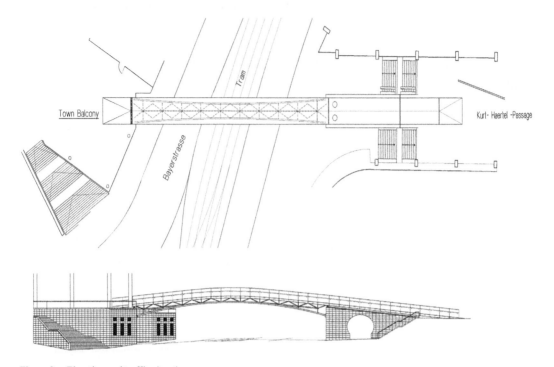

Figure 2. Elevation and traffic situation.

Ground plan

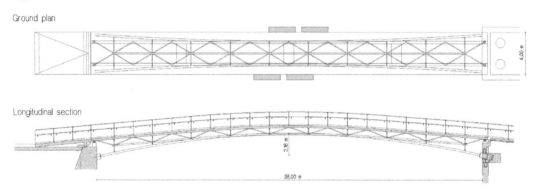

Longitudinal section

38,00 m

Figure 3. Ground plan and longitudinal section of bridge load-bearing structure.

2 ARCHITECTURAL DESIGN

The bridge design creates a barrier-free link for pedestrians between Kurt-Haertel-Passage and Theresienhöhe and establishes a continuous direct footway from the Hackerbrücke rapid-transit railway station right through to Theresienwiese.

The design takes account of the visual relationships along Bayerstrasse to the city centre. This applies to the viewpoints of both pedestrians and motorists. The design of the bridge as a deck bridge deliberately omits structural elements projecting above the walkway slab. The load-bearing members beneath the walkway slab are structurally divided into tension and compression members in order to optimize the input of materials and create a diaphanous load-bearing structure. The design of the walkway slab as a stiffening girder makes it possible to create extremely slim and filigree arches.

Thanks to the transparent, post-free, clamped glass parapet, the horizontal view remains unobstructed.

The two load-bearing principles of the arch and bending girder are each fully butt-welded and form separate units. The coupling of arch and stiffening girder to form a hybrid load-bearing structure is achieved with hinged eyebars and is thus visible.

318

Cross section

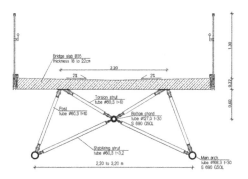

Figure 4. Cross section.

2.1 Load-bearing principle

The coupling of two arches with a bending girder yields a hybrid load-bearing structure (truss arch).

The two-hinged arches (circular steel tube, 168.3 mm diameter, wall thickness 30 mm, of S 690 high-strength steel) are a hyperbolic approximation to the thrust line and discharge the bridge's deadweight by compression to the abutments.

The stiffening girder consists of the 20 cm thick walkway slab (concrete with lightweight aggregates) as the top chord, the bottom chord (circular steel tube, 127.9 mm diameter, wall thickness 30 mm, of S 690 high-strength steel) and the diagonal and upright members (circular steel tube, 60.3 mm diameter, wall thickness 3.2 mm to 17 mm, of S 355 structural steel), which are rigidly welded together.

The stiffening girder absorbs the bending component under asymmetrical loading, distributes the non-uniform load via the coupling to the arch and stabilizes the latter to prevent it from buckling.

The arch and steel girder are coupled with hinged eyebars (circular steel tube, 60.3 mm diameter, wall thickness 3.2 mm to 17 mm, of S 355 structural steel).

2.2 Structural design

Basically, the bridge is structurally designed along the lines of the standard DIN 18800 on steel structures. The bridge is computed with two three-dimensional models: a perfect model and a deflected, imperfect system.

On the perfect model, the characteristic action parameters, bearing pressures and sag are computed according to second order theory.

On the imperfect model, the design action parameters are computed with γ-fold loads according to second order theory. The modulus of elasticity for steel is given with a reduced factor of $\gamma = 1.1$.

With the action parameters calculated from this, the tensions are determined by using elastic-elastic contact analysis and compared to the maximum permitted tension values.

This means that the global stability of the overall system is also demonstrated.

For the individual steel components (posts, columns etc.), stability conforming to DIN 18800, Part 2, is demonstrated with the above-mentioned action parameters.

No national standards were available for the use of high-strength steel in construction.

Not until the draft of the supplementary part 1.12 of Eurocode 3 of 12 September 2004, were high-strength steels included in European standardization.

The structural strength of the nodes of the high-strength tubes was determined by analogy with the CIDECT publication "Design guide for circular hollow section (CHS) joints under predominantly static loading" by Verlag TÜV Rheinland.

Since 12 September 2004, the rules have been extended by the draft of part 1.12 of Eurocode 3 (prEN 1993-1-12: 20xx) to include high-strength steel. A reduction factor of 0.8 is given here. The tests conducted suggest that this factor is on the conservative side. Further investigation to obtain an appropriate factor is desirable.

2.3 Dynamic study

The natural frequency of bridges with spans of 30 to 40 m is close to pedestrian excitation frequency. On this pedestrian bridge, the 1st and 2nd natural values are also within the range of 2 to 3 Hz.

At the design stage, three openings for the attachment of vibration dampers were provided at the points of the greatest vibration amplitudes. The counterbalancing mass was estimated to be 1% of the total dead weight in the static calculation. On completion, vibration tests were carried out with pedestrians with synchronized excitation, and maximum vertical accelerations of up to $1.7 \, \text{m/s}^2$ were measured.

For reasons of comfort and as protection from deliberate rocking, vibration dampers were installed.

2.4 Foundation

The southern impost has a shallow foundation in the new European Patent Office building beneath the "Town Balcony". The arches rest against the surcharges from the columns of the new building.

On the northern side, the arch thrust is transmitted by pressed piles into the load-bearing soil layers. This is because non-load-bearing backfills from earlier construction projects were found here along with a large district heating duct.

Because of the various foundation measures and to allow for subsequent settlement, the impost supports were designed to be adjustable on the northern side. With the aid of hydraulic hand presses, the arch can be pressed back into its original position in the event of yielding foundations. This mechanism was applied and successfully tested during bridge erection.

This adjustment has to be documented in the "client hand book". For upcoming adjustments another 7 cm of displacements are possible, which allow lifting the bridge about 22 cm.

1 st natural frequency = 2.3 Hz

2 nd natural frequency = 2.5 Hz

Figure 5.   Bridge structure's natural frequencies.

2.5   *Bridge parapet*

The bridge is equipped with a clamped glass parapet made of laminated safety glass, consisting of 2×12 mm partially prestressed glass panes with a point-fixed handrail. The top section providing protection from falling in the event of glass breakage has reduced dimensions in order to obstruct the view through the glass as little as possible.

Laminated safety glass panes have been fitted above the tramlines to prevent contact with the overhead lines.

The bridge slab is made of water-impermeable concrete resistant to frost and de-icing salt and does not require any further waterproofing. The client requested a thin-bed coating on the walkway slab for additional protection.

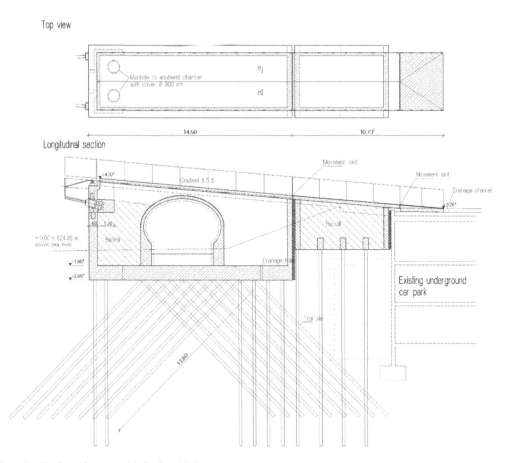

Figure 6.   Northern abutment with displaceable impost support.

320

Water from precipitation is drained openly via a 2% crossfall to the deck centre and via an 8% longitudinal fall to the drainage channels at the ramp transition points.

2.6 Involved in design and construction:

Design of load-bearing structure and project planning:
Dipl.-Ing. Christoph Ackermann, Munich
Beratendes Ingenieurbüro für Bauwesen,
Expert for high-strength steel and glass parapet:
Prof. Dr.-Ing. Ömer Bucak, Munich

Clients
Bayerische Hausbau GmbH, Munich, Dipl.-Ing. Hermann Jung
European Patent Office, Munich, Dipl.-Ing. Thomas Michel

Architecture
Ackermann und Partner, Architects BDA, Munich

Approval authority and granting of permission in individual cases
Building Department of Munich, Land Capital, Hydraulic and Bridge Engineering Unit
Dipl.-Ing. Ulrich Schönemann, Dipl.-Ing. Michael Götschl

Detail of glass parapet

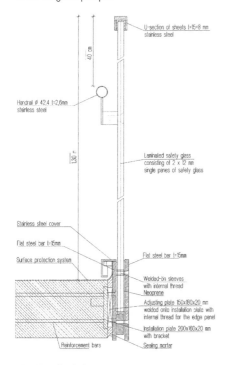

Figure 7.   Detail of glass parapet.

On completion, ownership of the bridge and maintenance pass to the City of Munich.

Test engineer
Prof. Dr.-Ing. Jürgen Feix, Munich

Contractor
Maurer und Söhne GmbH & Co. KG, Munich

Manufacturer of high-strength steel tubes
Vallourec & Mannesmann Deutschland GmbH
Continuous pipe mill, Mülheim

2.7 Time schedule

Urban planning concept: 1980
Start of bridge design: Autumn 1999
Start of construction: August 2004
Completion: June 2005.

Technical data

Span	38 m
Bridge width	4 m
Arch rise	2.16 m
Slenderness (ratio of rise to span)	1:18
Clearance profile	Four road lanes and two tramlines, two cycle paths and pavements Clearance h = 4.5 m
Static system	Hybrid arch system
Weight of structure	96 t
thereof structural steel	18 t (17.5 t reduction in mass due to use of S 690 high-strength steel)
thereof reinforcing steel thereof reinforced concrete slab	7.5 t without abutments 61 t

3   CONCLUSIONS

With the use of circular hollow sections out of high strength steel filigree steel constructions for bridges are possible. A light solution can be designed, which is also economic because of reduced dead weight and low costs for welding.

REFERENCES

DIN 18800. 1990. Steel structures; design and construction
Eurocode 3: Design of steel structures.
    EN 1993-1-1: 2005: General rules and rules for buildings.
    EN 1993-1-8: 2005: Design of joints.
prEN 1993-1-12: Sep 12, 2004: Additional rules for the extension of EN 1993 up to steel grades S 700.
Wardenier, J. et al. 1991. Design guide for circular hollow section (CHS) joints under predominantly static loading. Köln: TÜV-Verlag GmbH.

Tubular Structures XI – Packer & Willibald (eds)
© 2006 Taylor & Francis Group, London, ISBN 0-415-40280-8

Tubular structures in free-form architecture

T.R. Smith
Triodetic Building Products Ltd, Ottawa, Canada

ABSTRACT: Tubular framing was first popularized in space-frame & geodesic dome applications, where long clear spans and light-weight were key requirements. However, latest developments with this technology, using advanced design, fabrication and installation techniques, is now providing novel, complex, 'customized designs', but at 'commodity prices', and more rapidly than traditional structural steel solutions. Tubular framing is proving to be the key to a new era of 'shell-type' or 'free-form' architectural building shapes that are rapidly growing in popularity with developers and designers. This paper illustrates these latest developments in steel and aluminum tube structures, with reference to a unique New York museum roof, where innovative jointing and cladding arrangements achieved unusual geometry, accommodated abnormal loading, and also substantially reduced material, freight & installation costs & time.

1 INTRODUCTION

1.1 Curve appeal

Increasingly we are seeing more creative curved and free-form roof and wall shapes in new building design. The most conspicuous examples include new museums in most cities around the world, designed by Gehry, Calatrava, Cardinal, and other leading architects. These facilities effectively achieve acclaim and notoriety for the designer and builder, but also have proven to be commercially successful as landmark tourist attractions that attract high attendance. Some claim this is more in response to the building design than the exhibits within.

However, curvature is now arising more frequently in roofs & walls in all types of buildings – at airports, medical facilities, community buildings, sporting/entertainment complexes, and commercial centers.

1.2 Rationale

Designers are realizing that the time and cost of replicating (and maintaining) historic building features is prohibitive, and that curvature in some building elements can bring substantial structural efficiencies. The challenge is to capitalize on structural efficiencies, yet not incur fabrication or installation cost premiums that arise with complex concrete formwork, custom trusses, or curved beams (via plate welding, cold or induction bending).

Modular frame nets using metal tube can achieve many different structural configurations & advanced

Figure 1. Strong museum free-form 'Caterpillar' roof architect rendering & Triodetic structure near completion.

design, fabrication and installation technologies, are making these structures affordable.

New York state architects Chantreuil Jenson Stark, envisaged a free-form link roof for the Strong Children's Museum, & assessed the Triodetic system as the key to achieving the desired result. Museum CEO, G. Rollie Adams, describes the caterpillar roof shape as an intriguing complement and ideal companion for the museum's butterfly garden, with whimsical look & feel in keeping with the museum's mission as the National Museum of Play. (Figure 1)

2 STRUCTURAL EFFICIENCIES OF GEOMETRY

To sensibly address free-forms, it is useful to briefly cover the benefits of some basic geometric forms.

2.1 Curvature in one plane

The structural benefits of curvature are well illustrated in history by masonry & timber arches with strong abutments, that span further than heavier horizontal beams. Steel members in a variety of configurations have advanced joint strength & span limits. High tensile steel & aluminum trusses have further expanded span: rise ratios, loading and abutment limitations.

Steel & aluminum tube arranged in 3-Dimensional (triangulated two layer) space frames & barrel vaults offer in-plane and lateral buckling resistance with low structure mass, & continue to provide economical structures spanning to 45 m in a variety of both horizontal & vertical applications. (Figure 2)

2.2 Curvature in two planes

Curvature in two planes is the underpinning of shell structure efficiencies.

A barrel vaults & cylinders with varying cross-section along their longitudinal axes create a family of shapes called 'toroids'. (Figure 3)

These shapes have double curvature affording great strength that allows us to build single layer shells with very large clear spans in two directions.

Provided there is sufficient curvature, loads applied to these thin wall structures will be primarily transmitted through the shell members as a combination of axial compression or axial tension. Curvature can be used to minimize horizontal footing reactions, which would apply for an A-frame roof. Double curvature typically achieves a lighter roof structure and smaller supports, which is very advantageous in building refurbishment projects.

Toroids are in use at Osaka & Dubai airports, London museum, Glasgow science centre & to date Triodetic's largest single layer toroid spans 60 m × 150 m at Escondida Chile. (Figure 3).

The sole purpose of the included steel arches in the Escondida structure is to support the heavy copper ore feeder gantry at the roof apex.

Parabolic arches under symmetric loading are free of bending moment. An arch rotated around its vertical axis produces a dome shape that similarly transmits applied symmetric pressure, axially through the shell wall. Large diameter spherical shells with relatively low moment capacity in the wall behave similarly, provided the wall curvature is sufficient to allow applied forces to resolve into the members at each joint, or if local bending strength is provided to allow stress redistribution throughout the shell.

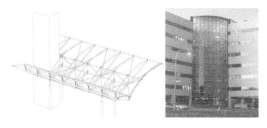

Figure 2. Horizontal & vertical frames with single curvature-Medical Offices Peoria, MD Anderson University Hospital Houston (DBIA & Healthcare Awards of Excellence 2005).

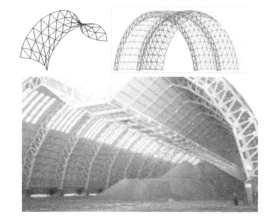

Figure 3. Triodetic Toroids, Escondida, Chile (Worlds largest tube frame toroids, 2006).

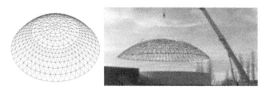

Figure 4. Triodetic Lamella Dome Shell Frames (Worlds first 200ft stainless steel lamella tube dome, PCS Saskatchewan, 2006).

Framing for double-curved surfaces is best achieved with triangular modules. This is a very structurally stable building element that adapts well to complex curves by varying module size & included angles. With modern fabrication efficiencies, geodesic dome geometry has largely been superseded by lamella domes (Figure 4) which allow greater construction efficiencies.

Outward thrust at the dome base is easily restrained by a tension ring, which is very beneficial for low rise structures, and when foundation or support strength is limited.

Dome frames generally require very slender members & often require only 1/3 the mass of

Figure 5. Triodetic Lattice Hypar roof, Marineland Niagara, Ontario.

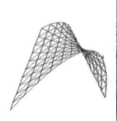

Figure 6. Triodetic Hypars at Tower base, Victorian Arts & Entertainment Centre, Melbourne, Australia.

a conventional A-frame or pyramid roof. Lower material, construction & foundation costs are obvious advantages with dome roofs. Triodetic domes spanning 60 m are in service in architectural & industrial applications, 100 m spans are currently in design.

Hypar surfaces comprise a series of diminishing parabolas, each with apex falling on a curve of increasing or decreasing radius. Double curvature allows these shapes to be designed as shells & Triodetic has designed & built a number of single & multiple Hypar roofs around the world, the largest covering over 8300 sqm. (Figures 5, 6, 7)

2.3 *Amorphous shapes & abnormal loads*

The Toronto Zoo pavilions (Figure 7) achieve nonlinear design by combining irregular hypars in series.

Amorphous geometry can also be approached by first modifying the geometry of regular shapes, such as a symmetrical dome. Domes can be: extended at their base to rest on any shape support (Figure 8); elongated; or non-symmetrical.

The same shape morphing can be applied to cones, toroids, hypars to produce any solution. More complex free-forms combine different sub-shapes, and/or include varying radii. (Figure 9)

All free-forms are well-replicated by triangulated frame nets that divide the overall shape into discrete parts. Stress analysis is then relatively simple & allows us to address abnormal point or line loads, and openings of any shape. A number of available software tools in conjunction with Triodetic's suite of in-house programs allow fast assessment of all load combinations for a variety of geometries & precisely create material manifests to allow accurate costing & automated manufacture.

Conventional I-beams & channels are inefficient for free-forms because loads act eccentrically to the principal axes & connections are cumbersome.

Fabrication of web-flange members for irregular curved shapes from plate or by mechanical or induction curving is inefficient, particularly when compound curves are included in the building shape.

Figure 7. Triodetic hyperbolic parabola (Hypar) shells (T.R. Baines Award for Zoos & Aquariums).

Figure 8. Dome with rectangular base & Triodetic Elongated Dome, Sheraton Hotel/Casino & Hard Rock Café, Niagara, Ontario.

In contrast, circular tubes ideally accommodate loads acting in any direction, and can be more readily fabricated with greater geometrical freedom, particularly when the frame nets have relatively short modules.

3 CONNECTIONS

3.1 *Joint requirements*

In free-forms, the shell members & joints must accommodate an infinite variety of connection geometries, offer sufficient moment capacity to accommodate concentrated loads & openings, & effectively transfer stress from one shell shape to another.

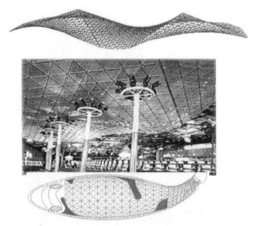

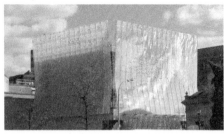

Figure 9. Triodetic free-forms: Pt. Edward Casino, ON; Bass Exhibit, Columbus Center Baltimore, MD, Pittsburgh Museum (2005 American Architecture Award).

Figure 10. Curved I-Beam Framing Marta Herford Museum.

Welded, riveted, bolted or threaded connections (Figure 11) are common in conventional building arrangements, however they each pose some difficulties in free-form applications, including:

(a) Reduced structural efficiency

  – due to deficiencies & tolerances (15–67%)
  – standard sections heavier than necessary
  – gussets required in high stress locations
  – member & connector cross-section & capacity reduced to accept fasteners

Figure 11. Bolted, welded, threaded connections.

Figure 12. Triodetic joint assembly (CMHC 2000 & TCA 2003 Innovation Awards).

(b) Increased design & fabrication time & cost

  – joints individually designed
  – member positioning is not repetitive
  – connectors use more elements & fabrication steps
  – welding, cleaning, inspection

(c) Increased site costs & delays

  – transport, storage, handling of large fabrications
  – simple detailing or fabrication errors can cause costly delays
  – bolting or welding at height is slow, dangerous, & can damage paint finishes
  – site painting raises safety/environmental concerns
  – site inspection required

(d) Appearance & Durability

  – visually un-appealing
  – many corrosion sites
  – not amenable to future modifications or re-use

Triodetic jointing system (Figure 12):

– Proven performance since 1960's
– Offers high axial and bending strength
– Accommodates any structure geometry
– Economical & fast to fabricate and install

Originally developed for space-frames & geodesic domes, the Triodetic joint uses a solid cylindrical aluminum or steel node, with serrated keyway slots. Frame tubes (or other sections) have press-formed ends with a matching serration, formed to the angle required to suit the structure geometry. Each tube is hammered into the (interference-fit) connector slots. The single bolt/nut used to complete each joint is

Figure 13. Joints & multiple tubes in high stress zones.

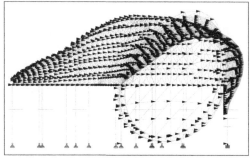

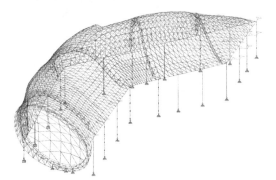

Figure 14. Strong museum frame net.

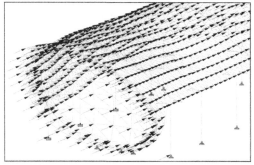

Figure 15. Finite element loading examples.

structurally redundant, but acts to verify that all members are correctly positioned.

3.2 *Joint strength*

The Triodetic joint offers high efficiency (92–100%), design is optimized because stresses are effectively re-distributed throughout the frame net. Componentized structures allow us to use precisely the required material only where it is needed.

Double layer 3-dimensional space-frames and barrel vaults designed with triangular configuration offer good in-plane and out-of-plane rigidity, but in single layer structures, curvature & joint strength becomes more critical.

Tighter radii and curves approaching a hyperbolic shape more effectively minimize bending stresses in the shell, but moment capacity of the members and joints must be sufficient to withstand stress peaks where loads concentrate, where openings occur, and where curvature is shallow. Long spans & abnormal loads are simply accommodated by using larger diameter tube, heavier wall thickness, multiple tubes, or a shorter module. (Figure 13) Connectors are varied in diameter & length to suit, without cost premium.

4 DESIGN

4.1 *Software*

Any surface can be subdivided by a network of lines that each can be described by numerical equations.

Complex shapes just require several equations along each line in each part of the surface until it transitions to another basic geometrical shape. (Figures 14,15).

Geometric and structural design programs have been available for some time to develop & analyze non-linear shapes. More recent advances in computation methods now allow reliable, high speed development of complex, non-symmetrical geometries, & both global & finite element analysis of any load combination.

Common software languages now also allow us to take advantage of previously disparate technologies in geometric & structural analysis. For instance, we now can adapt software developed in structural & fabrication engineering, architecture & graphic design, ship & boat-building, aviation & aerospace.

4.2 *Modeling*

Software accuracy for complex shapes is checked by building physical scale models. These provide a very helpful indication to the customer of the proposed shape, but also allow better consideration of alternate construction plans.

For the Strong museum, Triodetic utilized software used in ship-building, proven for heavy plate and symmetrical hulls. For non symmetrical compound curves, & thin cladding, with panel size limited by coil width, the dimensional differences & higher number of joints had to be assessed for cumulative tolerances.

Figure 16. Museum cladding model.

4.3 *Prototyping*

Wherever possible, Triodetic structures are completely or partially pre-assembled on site at ground level, and crane-lifted into final position.

For the complex geometry in the Strong museum, factory pre-assembly of a full-scale portion of the roof fame & cladding, was undertaken to further validate the design & fabrication software, the fit-up of various frame/deck elements, and to find time saving opportunities, given the imminent opening date for the museum.

Whilst the best free-form framing solution uses triangular elements, cladding geometry must adopt other basic forms for practicality and economy.

Innovative solutions often arise from combining normally unrelated ideas and by looking at problems in a different way. By visualizing the Strong museum roof shape as a jet fuselage, or an inverted ship hull, the best cladding solution arose.

Free-form shapes can be sub-divided into slices, whose double-curved surface can be developed onto a flat two-dimensional plane. Metal cladding is limited by coil width, but the narrower the slices through a free-form, the more readily a 3-D surface can be transposed into a flat plane & the smaller the off-set at abutting panel edges.

By optimizing panel shape we were able to minimize waste & achieve a double-curved surface, with panels curved in only one direction. (Curving panels in two directions for free-forms is impractical because every panel is unique & up-stand edges create practical limits to material flow during curving).

Full-scale prototyping was the only way to assess applicability of existing profiles & cladding fastening & sealing methods. Factory assembly of full-scale roof sections gave birth to a number of important innovations unique to Triodetic, including fabrication equipment that can efficiently create accurate compound curvature in the members and new connection & sealing technologies.

4.4 *Modes of failure*

Free-form structures may look new, but are really combinations of proven structural shapes. However, irrespective of material, section shape or jointing

Figure 17. Frame & cladding prototyping.

system, every structure in human existence has at least one critical mode of failure when over-loaded. If we understand structure behavior when overstressed we can ensure adequate safety margins are used in design for all anticipated service conditions.

Triodetic aluminum extrusion joints have been used with galvanized & painted steel tube members for over 40 years without failure due to material deficiency (corrosion or fatigue). Laboratory testing, industry guidelines & wider experience all confirm the compatibility of aluminum hubs & zinc on galvanized tubes. (Jones 1992 and Doyle & Wright 2000)

Full-scale testing to failure has been performed on members, joints and complete Triodetic structures under combined stresses, which provide a reliable basis for the safety factors that are applied to the axial & buckling limits for all combinations of tube/hub dimension & material.

Finite element analysis allows a limitless variety of structure shapes to be efficiently & reliably designed for any load combination. Testing confirms the Triodetic connector, fabricated in aluminum or steel, is stronger than the members it joins. (Ref.1)

By inspection of a 3-D joint, connecting tubular members that approach from more than one plane, it is clear that over-stress of any single member is resisted by the combined unused axial & bending capacity in the other connecting members. This inherent redundancy is illustrated in space-frame structures that retain structural integrity when isolated members are removed from the frame or when partial frame assemblies are lifted into final position.

Bolted, or welded gusset plate connections may fail by hinging, slip, material tearing or fastener failure. Overload of unrestrained 'I' or 'C' or 'Z' sections results in local buckling or twisting out of plane. Slenderness limits result in stockier members, gussets, heavier, more costly structures.

In Triodetic structures the tube section dimension at each joint is increased in the plane of maximum bending, & joint rotation in that plane cannot occur. Tubes approach each hub at an angle so shear is resolved into axial & bending components. The tight engagement of the tube in the serrated hub keyways amply secures the tube, however joint slip is resisted because of the combined axial & bending forces that apply. Testing has confirmed our design limits ensure joint rotation about the minor stress plane cannot occur, nor have we experienced this behavior in any Triodetic structure in decades of service. In 3-D framing, joint rotation would require simultaneous failure of all members at the joint, which is almost physically impossible, given that they radiate at different angles, carry different loads, in many cases have different section properties, and are restrained by the next connecting node.

The behavior of indeterminate frame structures, with built-in redundancy, under excess load, is to redistribute stress concentrations until equilibrium is restored. Stress will always disperse to the stiffer members. (Every hammock illustrates this principle).

In design, our software shows stress distribution in response to applied loads, or varied geometry. In practice, single layer Triodetic structures spanning 60 m, subject to high dynamic loads, hurricane winds, and seismic conditions effectively transmit these loads as axial tension or compression to stiffer restraining members.

For the Strong museum the free-form roof was required to be completely independent of the existing buildings which it linked. The single layer lattice shell was designed for Rochester's wind & heavy (lake-effect) non-uniform snow loads. A critical design requirement was to provide lateral restraint against outward movement where the free form roof connected to the top of unrestrained columns. Inclusion of three curved lateral trusses illustrates how readily a componentized structure can be modified to accommodate abnormal structural demands & remain cost effective.

5 CONSTRUCTION

5.1 Frame

The adopted assembly method for each free-form or shell structure varies with structure shape, size, and site conditions. Staging space & crane access will dictate how much ground-level pre-assembly is feasible. The construction plan is more often dictated by the project program & assembly can be accelerated by increasing crew size. Typically local unskilled labor is engaged under Triodetic supervision.

The Strong museum free-form link roof (Figures 18, 19, 20) is approximately 20 m span × 70 m (20 m high at its apex). Frame & cladding materials were on site 10 weeks after drawing approvals. A local 5 man crew completed the frame & cladding assembly in 55 days under Triodetic supervision. All work was performed from man-lifts and from on top of the structure & sequenced to allow early access for other trades within the museum.

Figure 18. Strong museum frame assembly.

The lateral trusses were pre-assembled at ground level and required a 25 t crane briefly, for positioning on the support steelwork.

The roof frame comprises over 4500 tube members, all individually made – no two with the same length, end camber, twist or angle.

Purlins allow the transition from the triangulated tube frame to the optimum cladding pattern, dictated to a large extent by coil width. Similarly, no two purlins have the same compound curve.

Structural support to the end glass wall (eye) was achieved using a non-symmetrical dome positioned on edge. This non-traditional dome arrangement had been previously proven in waterpark applications to require less steel than would be required for traditional column-beam construction, less craneage, and avoided site painting & inspection.

Frame assembly follows a simple match-marking system perfected by Triodetic. All members can be lifted by hand & are connections are simply achieved using a mallet & wrench.

A unique feature of the Triodetic system is that the closely spaced frame joints offer high bending stiffness during assembly, allowing installation of new members from on top of the partially built structure, & crane positioning of partial or complete pre-assemblies, even with cladding/glazing attached. Triodetic dome structures are unique in that they can be built from the base to the apex, without jacks or tower cranes (required with other frame systems that are then vulnerable to winds during assembly).

For the Strong museum, time was of the essence, & good access influenced the decision to mostly build in-place.

5.2 Cladding

Free-forms can be clad with galvanized, painted steel, anodized aluminum, copper, stainless steel, conventional shingles, & tensile fabrics. Curved decking has been in use for over 35 years, however achieving leak-proof cladding on free-form shapes requires inclusion

Figure 19. Strong museum cladding assembly.

Figure 21. Strong museum before & after insulation/ fireproofing.

Figure 20. Strong museum 'eye' rendering & construction.

of certain design features to be effective. Experience in metal clad domes has resulted in number of proven alternatives, however the anodized aluminum desired for the Strong museum meant tapered interlocking panels, with purlin supports and batten seal was the best option.

The Strong museum roof uses a proprietary Triodetic cladding system designed with 3 barriers to water ingress. For this project, anodized aluminum cladding is composed of 1,100 individual panels fabricated directly from the design software with curvature, interlocks & up-stands. A gasket/batten system completes the decking assembly. (Figure 19).

For the Strong museum, a challenge was developing a cladding solution for the protruding brow around the 25m high non-circular glass wall (eye). Triodetic developed an innovative hinged panel system that has application in all double curved roofs.

5.3 *Insulation*

Roof arrangements and materials vary for each location and service need & may incorporate plywood panels with insulation bats, pre-cut to the shape of each module, or curved corrugated deck, tapered in-fill panels, or sprayed isoprene and butyl top coat.

For the Strong museum, the Triodetic frame was to be exposed as an internal feature. Once the decking was in place on the frame, the tubes were taped & insulation/fireproofing was spray-applied. Properly sized stand-offs, integral to the frame, achieved the required R-value. Additional fireproofing of the primary structure was not required higher than 26ft above floor level.

6 ENVIRONMENTAL BENEFITS

Triodetic free-form roofs support sustainable building initiatives & are providing community benefit in many diverse applications, such as: support for roof gardens, solar arrays; noise & sight barriers for rooftop mechanicals; dust/odor control from ore stockpiles & tanks; sun & snow protection & rain-screens.

Tangible environmental benefits arise from the materials, processes & technologies employed:

a) Design reduces material use by up to 60%
b) Materials have 60–90% recycled content
c) Manufacture is non-polluting, energy conserving
d) 100% of production excess is recycled
e) Freight is reduced by over 50%
f) Reduced construction time & hazards
g) Ceilings not necessary in some applications
h) Reduced pollution, site damage, maintenance
i) Employment provided for local unskilled labor
j) Easily modified, 100% re-use.

7 CONCLUSIONS

a. Free-forms are a growing trend in architectural and industrial building design because they offer appealing aesthetics & structural economies.
b. Proven design tools are available to allow accurate development of any frame/cladding geometry, and reliable structural analysis of triangulated frame nets. Frame nets with double-curvature offer economy because each member can be sized to suit stresses at any location in the shell.

Structures designed with minimum mass and curvature can afford considerable material cost savings and less costly foundations.

c. Circular metal tube members offer the best solution to the geometric and structural demands of free-form frame nets, because they offer uniform strength irrespective of the direction of applied load.

d. The ready-availability of high quality galvanized steel, stainless steel, & aluminum tube, plus a variety of factory applied finishes, makes these structures suitable for short construction time-frames, & appropriate for a wide spectrum of service conditions.

e. The Triodetic joint accommodates any structure geometry and a variety of tube diameters & wall thicknesses. It also offers high structural efficiency, low fabrication cost, rapid production with consistent quality, and simple, fast assembly for a wide range of building types, structure shapes, loads, locations, and conditions.

f. Triodetic structures are delivered cost-effectively as a compact kit, anywhere around the world, and can be pre-assembled or built in place using unskilled labor. Shipping, site storage & handling, damage, welding, inspection & painting problems common with large conventional structural beam & trusses is avoided.

g. Triodetic decking systems, proven on domes, hypars & toroid shapes, allow free-form structures to be effectively & economically clad with galvanized, painted or stainless steel, aluminum, shingles, tensile fabrics or tiles.

h. Free-form roofs & other tube structures offer the inherent advantages of material conservation through minimum mass design, very high recycled material content, and are 100% re-useable. Many applications provide protection for, and benefit to the environmental.

REFERENCES

Figures 14,15,16. Smith T.R., 2003: Triodetic internal research & development report: Design software & prototyping of single layer shell framing & cladding, ON, Canada.

Jones, D., 1992. Principles & Prevention of Corrosion, McMillan, New York.

Doyle, D.P. & Wright, T. E., 2000. Quantitative Assessment of Atmospheric Galvanic Corrosion, Galvanic Corrosion, ASTM STP 978, H.P. Hack, Ed., ASTM, Philadelphia, 1988, Aluminum Design Manual, The Aluminum Association, Washington D.C.

Tubular Structures XI – Packer & Willibald (eds)
© 2006 Taylor & Francis Group, London, ISBN 0-415-40280-8

Launching a 140 m span tubular truss bridge in Spain

J.M. Simon-Talero & R.M. Merino
Torroja Ingenieria S.L., Madrid, Spain

ABSTRACT: The Guadalfeo bridge is a continuous steel truss bridge. Its spans are $85 + 140 + 140 + 110 + 110$ m long. The deck of the bridge is 10.35 m high and 24 m wide, and the cross section of the girder is made of 5 main chords: three on the upper level and two at the bottom. Nodes of the truss are each 10 m apart and the diagonals connecting each node have circular hollow sections (CHS). On the top of the truss a concrete slab 30 cm thick was placed.

Concerning the construction method, first a span was erected using temporary piers. The next steel truss, 500 m long, was launched from the Southern abutment. The tip deflection of the truss reached about 2000 mm while launching. So, a 40 m high temporary tower balanced by two sets of 22000 kN prestressed cables was implemented. The use of the temporary tower reduced the total amount of steel on the truss to some 380 kg/m^2.

1 INTRODUCTION

1.1 *Location*

Between 1200 and 500 years ago Muslims established a route along the Guadalfeo river and the Izbor river, reaching the southern coast of the Mediterranean Sea. Nowadays this same route is used for escaping from the busy way of life of crowded cities located on the centre and south of Spain to the well known beaches of the southern coast of Spain. From 1991 the Spanish Government is working on different "General Interest Road Improvement Plans". Their main objective is connecting all Spanish principal towns by dual carriageway roads.

Road N-323 crosses the said Guadalfeo river near the toes of the Mount Veleta, one of the highest mountains in Spain, where very special conditions of the surroundings can be found, such as:

– The area around Granada is a geologically active area with definite tectonic activity. In Spain, code NCSE-02 is used for evaluating seismic effects. This code is similar to Eurocode 8, where a statically equivalent action can be used as an approximation of the dynamic effect produced by seismic actions. In doing so, the code proposes for this area a design acceleration equalling 0.22 g.
– A bridge was proposed to be located where the Guadalfeo and Izbor rivers meet. As a consequence, some 15 to 20 m thick alluvial deposits are found. So, a deep foundation using in-situ piers was proposed for the main spans. Attention should be paid to the fact that bored piles 1.8 to 2.0 m in diameter are often used as foundation of similar bridges, but decimetric gravels are expected to be crossed when boring these piles.
– Furthermore, there are some active large landslides on the vicinity of the southern abutment, because of the tectonic activity.
– Central piers are expected to be 90 to 100 m high, because the new road has an alignment to avoid constructing tunnels, increasing the distance from natural terrain to the proposed road profile. The distance between both valley slopes results in a length of 600 m, approximately.
– The construction of a new dam is undergoing 2 km down the river from the site of the proposed bridge. So, in the future the depth of the water will reach 60 to 70 m deep near the bridge.

1.2 *Design criteria*

During the preliminary design phases different solutions were considered, but only medium span bridges were selected because of the foundation conditions, using spans 80 to 170 m long. The number of piers was limited, so that the needed deep foundation did not reach unreasonable proportions. Arch bridges, cable

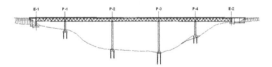

Figure 1. Design elevation of the bridge.

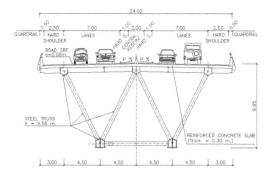

Figure 2. Proposed cross section of the deck.

Figure 3. Cross section of the deck during erection.

stayed bridges and suspension bridges were eliminated because of foundation conditions.

Considering seismic resistance, the material of the deck was analysed. Steel girders were preferred to prestressed concrete girders: concrete decks are approximately 200% heavier than composite girders, and therefore seismic effects are double.

Finally, the existence of a 60 to 70 m deep water body during the construction of the deck was to be taken into account. Different construction methods were studied in connection with selected typologies of the deck. Among them, cantilever construction using temporary wires and launching of the deck were examined.

2 BRIDGE DESCRIPTION

2.1 *Horizontal and vertical alignment*

The bridge over the Guadalfeo river is a five span bridge with a total length of 585 m. Its platform is 24 m wide, holding a dual carriageway with two lanes in each direction.

The horizontal alignment is a very gentle curve, with a radius of 17200 m. Near the northern abutment there is a transition. The cross slope is 8% near the northern abutment; due to the slightly curved horizontal alignment, a cross slope of 2% is used in the rest of the deck.

Concerning the vertical profile, nearly all the deck is placed on a slightly descending alignment sloped 0,20 %. The maximum distance from the upper level of the deck to the ground is about 100 m.

Abutments reach a height of 20 m on the Northern access and some 15 m on the south slope.

2.2 *Deck*

The proposed bridge is a composite truss 585 m long. The deck is a 5 span continuous beam. Spans are $85 + 140 + 140 + 110 + 110$ m.

The cross section of the steel truss is 9.55 m deep. The truss is made of three main beams on the upper level and another two lower main beams. The distance in between main beams is 9 m. The cross section of each bar of the upper level is made of five plates 1000 mm deep and 950 mm wide. Its shape is pentagonal to optimise junction between main beams, transverse beams and diagonals. The cross sections of the two main beams of the lower chord are similar to the above mentioned beams of the upper level, but 1200 mm deep and 1280 mm wide. Additionally, inclined plates were located close to the connection between the webs and the lower horizontal plate. In doing so, a closed partial section was created to distribute the concentrated load from temporary bearings during launching to the webs. By doing so, patch loading effects were diminished. The thickness of webs varies from 15 to 50 mm. Thickness of horizontal plates of main beams varies from 15 to 60 mm.

Groups of four diagonal bars are used for connecting the main beams of the upper level to those of the lower level, with nodes distributed each 10 m. Each diagonal is a circular hollow section (CHS). Their diameter goes from 406 to 609 mm and their thickness is between 6.3 and 35 mm.

Secondary beams are located parallel to main beams of the upper level. An I-shaped section 600 mm deep was proposed. The connection of these longitudinal beams to the main beams is made using transverse I beams located each 10 m.

Therefore, a girder measuring 10×9 m is made by longitudinal secondary beams, main beams of the upper level and transverse I beams. On the top of the girder a reinforced concrete slab 30 cm thick and 24 m wide is placed. Nominal resistance of the on site poured concrete is 30 N/mm^2.

The steel grade for all beams, CHS's and auxiliary elements is S355 ($f_y = 355$ N/mm^2).

2.3 *Substructure*

Piers are made of reinforced concrete with a compression strength of 30 N/mm^2. Their cross section

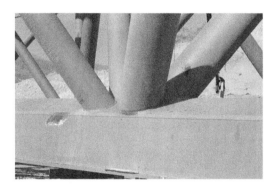

Figure 4. Connection of the main beam of the upper chord with main diagonals.

Figure 5. Pier P3 under construction using sliding formworks.

is a hollow section measuring 4.5×12.0 m at the top. Their transverse dimension increases 2.5% all along the bridge, reaching some 16.5 m on the base of pier P3, which is 90 m approximately high. The wall thickness of the piers is 35 to 40 cm.

The lower part of the pier is made of a stand with pyramidal shape, constituting a transition from the pier itself to the pile cap, which measures $28 \times 12 \times 3$ m. The deep foundations of pier P2 and P3 are made by 17 concrete piles 2 m in diameter, whereas the foundation of piers P1 and P4 are made with 14 piles 2 m in diameter. As it has been mentioned, the piles cross a thick layer of alluvial deposits until the rock layer is reached. The longest piles are 31 m long.

The abutments are made of on site poured 30 N/mm^2 reinforced concrete. The northern abutment is 23 m tall. Foundation of both abutments is directly on superficial layers of adequate resistance.

As mentioned above, seismic effects equivalent to 0.22 g are to be considered in the design and calculations. So, seismic devices were proposed. The two main goals of this design were:

– Eliminating longitudinal loads from the top piers.
– Diminishing global seismic effects to adequate limits.

In order fulfil these two conditions, seismic dampers were proposed. Detailed dynamical calculations were performed for evaluating the forces to be considered for different nominal velocities. Maximum target load was fixed to 15000 kN per abutment and maximum allowed movement was considered to be ± 300 mm.

3 PROPOSED METHOD OF CONSTRUCTION

The highest pier is approximately 90 m. So, sliding formwork was used for erecting all piers (see figure 5).

As mentioned above, it was supposed that water will be flowing underneath the structure before the construction of the deck could be initiated. So, it was proposed to launch the deck. When preliminary calculations were made, a 2000 mm deflection was estimated on the tip of the truss. In order to decrease this high deflection, a temporary tower 40 m high was considered. This tower was balanced by two sets of cables with an adequate strength to resist a load of 22000 kN. In doing so, the maximum vertical deflection was reduced to 850 mm and loads on diagonals and main beams during launching were also reduced. It was not feasible to launch the first span near the north abutment due to its geometry. Additionally, access to the site was possible. So, normal construction was proposed for this first span, erecting steel structures using temporary piers and placing parts of the steel structure with cranes.

4 STRUCTURAL ANALYSIS

Different structural models were used, both for general calculations and for local evaluation of particular effects.

– A finite element model (FEM), including second order analysis, was used for the entire structure. This model was able to consider the possible buckling of steel elements, instability of reinforced concrete sections of piers and contribution of sliding bearings.
– An interactive and time-dependent FEM model was used for studying launching process.
– A general dynamic model was used to estimate the seismic effects on the deck, piers and abutments, considering the contribution of specific implemented seismic devices.
– Different FEM local models were used for studying specific zones of the structure, such as the connections between diagonals and main beams, local bending effects while launching, the connection of the temporary tower to the truss, etc.

Figure 6.   Structure's general model. Launching stages.

While still in the design stage, special care was taken to face three different particular problems:

– Instability of the piers.
– Loads on temporary elements and bearings.
– Evaluation of deflections and loads while launching, specially where these effects conditioned the dimension of the element.

4.1   Instability of piers

Buckling of piers has been studied using a FEM global model including all relevant structural elements: steel truss, reinforced concrete slab of the deck, bearings and piers themselves. Reinforcement, non linear material behaviour and cracking of concrete have been considered on the analysis.

Figure 7 shows the results obtained on the final structure from two different models:

– Case A is the result of a "traditional calculation" used very often for this kind of problems. That is, obtaining loads on the top of each pier using a first order analysis on a global model. Then, these loads are applied to a simplified model and a second order analysis of the single pier is performed.
– Case B is a global calculation using a FEM model able to consider second order effects and also the interaction between the deck, the bearings and the piers.

Results are presented for two relevant design combinations: maximum transverse forces and maximum longitudinal effects. These results are presented in terms of deflections of the top of the pier and of strength (load capacity factor). As it can be deduced from the figure, usage of global second order calculations could be very effective for reducing amount of reinforcement on slender piers, but not to decrease the need for the level of structural capacity of the structure.

4.2   Influence of precamber deflections on flexural effect while launching

Main beams of the truss were fabricated considering expected deflections due to self weight and dead loads.

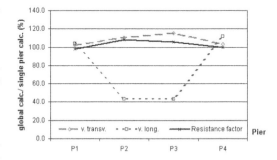

Figure 7.   Instability of piers. Comparison of results of different models.

So, pre-camber on centre of spans 2 and 3, 140 m long, was 260 mm. Using a structural model similar to that on Figure 6, the maximum loads on temporary bearings were estimated for each phase of launching. Figure 8 shows the maximum vertical load calculated for each section, considering three different models:

– Case A: it is supposed to have jacks placed on each temporary bearing. So, level of supports can be adjusted and launching is performed as if the truss would be in a "horizontal" position, with no influences of precamber deflections.
– Case B: it is supposed that no jacks are available. Consequently, precamber deflections of each section produce some flexural effects on the truss.
– Case C: is the proposed method, where some adjustments can be made on particular temporary supports during specific stages of launching.

Figure 7 also includes the "patch loading nominal resistance" of webs of main beams of the lower chords. Results show that some sections are very sensitive to precamber deflection. That is, under certain circumstances, precamber deflections cannot be neglected for estimating loads on temporary launching bearings. Consequently, it is sometimes necessary to use jacks to maintain the loads on bearings under certain limits.

336

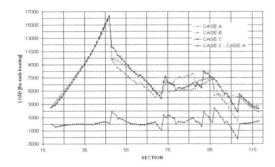

Figure 8. Maximum load per section on temporary launching bearings.

Figure 9. FEM model for evaluating local flexural effects on main beams of the lower chord.

4.3 Local flexural effects on main beams of the lower chords

While launching the deck, main beams are sliding over temporary launching beams. Loads are applied to the truss not only on nodes but also on intermediate points all along main beams of the lower level. Therefore, local shear and bending moments are generated on main beams, working as a continuous beam supported each 10 m. These effects have been found to be the most severe loads on these main beams, so a careful calculation is necessary.

Loads coming up from piers and abutments are applied to the truss using temporary bearings. The dimensions of these bearings are 1500×500 mm. There are two launching bearings per section, able to support a maximum vertical load of 11250 kN. These bearings are neoprene – teflon bearings 30 mm thick. The longitudinal dimensions of the bearings are 1/8 of the total calculation span and bearing thickness is not very high. So, it is not reliable to suppose that the bearing reaction will be constant. It is necessary to take into account the longitudinal local flexural deflection on main beams, as well as the different deformation of the bearing along a distance of 1500 mm. Additionally, the local transverse distribution of the loads between both webs of the cross section is to be considered.

The evaluation of all these structural phenomena is not easy. So, a detailed FEM model was prepared, in order to quantify importance of every effect. Figure 9 shows the local model used for calculations of local effects on the segment of the main beam of the lower chord.

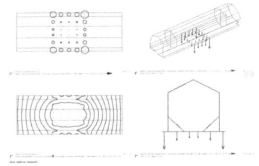

Figure 10. Distribution of reactions on bearing section SI2.

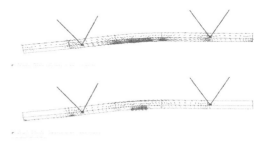

Figure 11. Maximum principal stresses on section SI2 under load equal to 6224 kN.

Figure 10 shows the distribution of reactions on bearings on section SI2. The applied load was 6224 kN, and the maximum vertical compression stress was 200 N/mm², but quickly decreasing to some 96 N/mm².

Figure 11 shows the maximum principal stresses on section SI2. The maximum vertical compression stress is due to patch loading effects. It reaches a value of 103 N/mm². The maximum transverse horizontal stress is produced by the distribution of loads from bearings to the inclined plates and to the webs. Its value is only 122 N/mm². Finally, the maximum longitudinal horizontal stress is produced by local flexural effects when distributing the load from the centre of a 10 m span to both nodes located 5 m far. The stress reaches 318 N/mm².

Table 1 shows a summary of the maximum stresses and the maximum transverse deflection calculated for three different sections of main beams of the lower chords, considering both structural effects: plate bending and membrane stress.

5 MANUFACTURE, ERECTION AND LAUNCHING

5.1 Manufacture and erection

Approximately, 5300 tons of steel have been used to manufacture the steel truss. A total of 3200 tons were

Table 1. Summary of transverse deflections (mm) and principal stresses (N/mm^2) on four different sections while launching.

	SI1	SI2	SI3	SI4
Max stress (N/mm^2)				
Top flange	201	214	217	247
Web (vert. stress)	−218	−103	−72	−112
Web (horiz. stress)	−276	−318	−299	−343
Inclined plate	−215	−300	−292	−334
Bottom flange	−206	−283	−307	−306
Max deflection (mm)				
Web (horiz. deflection)	2.0	1.6	1.2	1.3
Bottom flange (vert. def.)	1.3	1.09	1.1	1.9

Figure 13. Erection of span n. 1.

Figure 12. Upper level connection between chords and diagonals.

used for the main beams of chords, 1400 tons were used for diagonals and 700 tons were used for horizontal bracings. In addition, the temporary tower weights 150 tons and some 80 tons were used for different auxiliary elements. Finally, 460 tons of steel grade S235 were used in a profiled steel sheeting used as formwork for the on site poured reinforced concrete slab.

As it has been mentioned, diagonals have a circular hollow section (CHS). Their diameter goes from 406 to 609 mm and their thickness varies from 6.3 to 35 mm. The steel grade used for all elements is S355, and it has been manufactures and assembled in Spain.

Welded connections were made both in shop and on site. Full penetration butt welds were used for connecting different plates of the five side sections of the main beams of upper and lower chords. Joints between main beams and diagonals were made using fillet welds.

5.2 Launching

Concerning the construction method, the first span was erected using temporary piers (see figure 13) and also using final bearings on abutment 1 and pier 1 for supporting the steel truss. The rest of the steel truss was

Figure 14. Auxiliary tower used in the launch of the steel truss.

launched from the Southern abutment. Because of the length of the main spans, the deflection of the tip of the truss reaches some 2000 mm while launching. To decrease this value, a 40 m tall provisional tower was implemented (see figure 14). This tower was balanced by two sets of prestressed cables. The capacity of these cables was 25000 kN. By doing so, the final deflection of the truss was reduced to 1200 mm, and bending moments and axial forces on the longitudinal beams were also reduced. Cables were tensioned to 45% of its maximum failure load when a 75 m cantilever is reached. Once the pier is reached, tension of cables is decreased to 20% of the initial load.

Profile steel sheeting is located onto main beams of the upper chords all along the width of the deck (24 m) before launching. At the back of the truss, some extra weight was placed during some stages of launching to counterbalance the structure.

Figure 15. General view during launching.

Figure 16. Launching when arriving pier P2.

5.3 Control of launching

The tension on the cables supporting the auxiliary tower was controlled using a load cell near the anchorage of the cables at the top of the tower. In addition to that six strain gauges were located on the base of each leg of the tower to measure the normal force on it.

Two different sections of the truss were also controlled, trying to measure the real stresses on each stage of the launch. In order to do so the main beams and diagonals were studied. On each of the said two truss sections, 19 strain gauges were located: two on each of the three main beams of the upper chord, three on each of the two main beams of the lower chord and seven on the four diagonals of the section.

Also, clinometres were located on each pier, and some temperature probes were located all along the truss.

Finally, the pulling force on launching system was monitored using another load cell connected to the pulling system.

6 CONCLUSIONS

Launching of the deck has been successfully completed on April 2006. Opening to the traffic of the whole section of the Motorway is expected on November 2006.

ACKNOWLEDGEMENTS

Owner:
  Ministry of Public Works, Spain
Design and technical assistance on construction:
  Torroja Ingeniería S.L., Madrid, Spain
    J. M. de Villar
    J. M. Simón-Talero
    R. M. Merino
Steel Manufacturer:
  Acciona – Talleres Centrales, Madrid, Spain
    G. Rodríguez
    M. Sánchez
Direction of construction:
  Ministry of Public Works, Spain
    J. Lorente
Contractor:
  Acciona Infraestructuras, Spain
    P. Torreblanca

TECHNICAL REFERENCES

– Eurocode 3. Design of steel structures (part 1.1, 1.5, 1.8, 2.0)
– RPX-95. Spanish code for the design of composite bridges (Ministry of Public Works, Spain)
– NCSE-02. Spanish code for seismic design
– Guide for the design of joints of CHS and RHS under fatigue actions. (CIDECT, Germany 2003)

Tubular Structures XI – Packer & Willibald (eds)
© 2006 Taylor & Francis Group, London, ISBN 0-415-40280-8

Cost comparison of a tubular truss and a ring-stiffened shell structure for a wind turbine tower

J. Farkas & K. Jármai
University of Miskolc, Miskolc, Hungary

J.H. Negrão
University of Coimbra, Coimbra, Portugal

ABSTRACT: Two structural versions are designed for a 45 m high tower of a 1 MW wind turbine. In the case of the ring-stiffened slightly conical shell constraints on buckling, eigenfrequency and fatigue should be fulfilled and a cost function should be minimized. Loads, diameters and the number of flat stiffeners are given and the shell and stiffener thicknesses are optimized. The cost function relates to the cost of material, cutting of flat stiffeners, forming of plate elements into conical shell, assembly, welding and painting. The four plan tubular truss consists of three parts of different but constant chord distance. It is shown that, because of high bending moments, the bottom part has minimum volume in the case of parallel chords. A suboptimization method is used to obtain the economic circular hollow section sizes for compression struts. Constraint on joint eccentricity, chord plastification, eigenfrequency and fatigue are considered. The cost function parts are as follows: material, cutting and grinding of strut ends, assembly, welding and painting. The comparison shows that the tubular truss has smaller mass and smaller painted surface, thus it is cheaper than the shell structure.

1 INTRODUCTION

Steel towers for wind turbines can be constructed in various structural versions. Ring-stiffened cylindrical shells or tubular trusses are usually applied (Spera 1994). Since the main requirements of engineering structures are the safety, fitness for production and economy, an important problem for designers is the cost comparison of these structural versions.

A cost calculation method has been developed and applied for various welded structures (Farkas & Jármai 2003, Jármai & Farkas 1999). The cost function includes the cost of material, cutting and grinding of tubular member ends, assembly and welding. This cost function has been applied in various problems of minimum cost design, e.g. for a triangular tubular truss (Farkas 2001). Some other problems of economic design are treated in the book of Farkas & Jármai (2003).

This cost calculation method is applied now to two structural versions of a wind turbine tower. The tower is 45 m high, loaded on the top by a factored vertical force of 950 kN (self weight of the nacelle), a bending moment of 997 kNm and a horizontal force of 282 kN from the turbine operation. The tower width is

limited to 2.5 m due to the rotating turbine blades of length 27 m.

Both the shell and the truss structure are constructed from 3 parts each of 15 m in length. The shell parts are designed against shell buckling and panel ring buckling according to the design rules of the Det Norske Veritas (1995). The number of flat ring-stiffeners is determined by the designer to avoid larger ovalization of the cylindrical shell. The three shell parts are joined by bolted connections.

Wind turbines are worldwide used with various capacity and tower height. In the book (Spera 1994) a detailed description is given of various wind turbine towers. Koumousis & Dimou (1995) have treated the minimum volume design of a cylindrical shell tower with varying diameter and thickness considering stress and displacement constraints. Horváth & Tóth (2001) have investigated the most suitable shape of a cylindrical shell tower with variable diameter regarding the natural frequency using the finite element method.

Bazeos et al. (2002) have studied the stability and seismic behaviour of a 38 m high shell tower structure for a 450 kW wind turbine with cylindrical and conical parts of varying diameters and thicknesses. Lavassas et al. (2003) have investigated for gravity, seismic and wind loadings a 1 MW capacity and 44 m high tower of tubular shape with variable diameter and thickness using two different finite element models.

2 RING-STIFFENED SHELL STRUCTURE

For the cost minimization, the procedure already developed by Farkas et al. (2004) to optimize the design of a ring-stiffened cylindrical shell loaded in bending, is used. Design constraints on shell buckling and on local buckling of flat ring-stiffeners are formulated according to DNV (1995) and API (2000) design rules. The wind load acting on the shell tower is calculated according to Eurocode 1 Part 2–4 (EC1) (1999). The wind force and bending moment acting on the top of the 45 m high tower for a 1 MW wind turbine in Greece, is given by Lavassas et al. (2003). The load due to the self-weight of the nacelle is furthermore considered.

To avoid shell ovalization a minimum number of 5 and a maximum number of 15 stiffeners is prescribed. In the shell buckling constraint an imperfection factor as proposed by Farkas (2002) is used, which

expresses the effect of radial shell deformation due to shrinkage of circumferential welds.

Figure 1 shows the diameters, loads, bending moment diagram and the optimum shell thicknesses.

For the three shell segments, the wind loads are as follows: $p_{w1} = 6.334$, $p_{w2} = 6.883$ and $p_{w3} = 6.864$ kN/m. $F_{w1} = 95.01$, $F_{w2} = 103.25$, $F_{w3} = 102.95$ kN.

The factored bending moments due to wind load F_w, are given in Figure 1 (c), with $M_{w0} = 997$ kNm, the safety factor being 1.5. The factored load $F_{w0} = 282$ kN and nacelle (rotor) selfweight $G = 950$ kN.

2.1 Design constraints

2.1.1 Local buckling of the flat ring-stiffeners
The limitation on the height to thickness ratio of a flat ring-stiffener is (API 2000)

$$\frac{h_r}{t_r} \leq 0.375 \sqrt{\frac{E}{f_y}}, \tag{1}$$

where t_r is the ring stiffener thickness to be determined. Using the upper limit to obtain a larger moment of

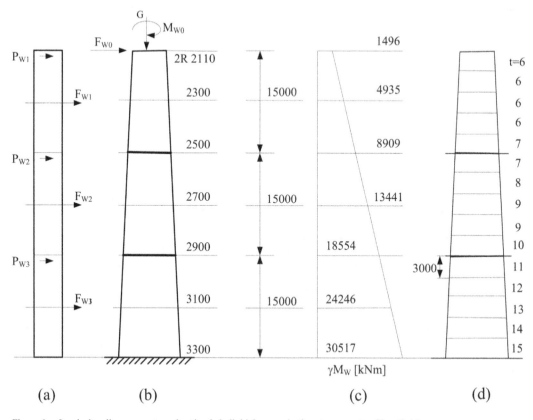

Figure 1. Loads, bending moments and optimal shell thicknesses in three tower parts of length 15 m.

342

inertia, one obtains

$$h_r = 9t_r,$$ (2)

for $E = 2.1 \times 10^5$ MPa and yield stress $f_y = 355$ MPa.

2.1.2 Constraint on local shell buckling (unstiffened)

According to Det Norske Veritas (1995), for the length of one shell part $L = 15$ m, the number of ring stiffeners in one shell part n, R the radius and t the thickness of the shell, γM_W pertaining to the moment on the shell part (see Figure 1(b)):

The sum of the axial and bending stresses should be smaller than the critical buckling stress

$$\sigma_a + \sigma_b = \frac{G}{2R\pi t} + \frac{\gamma M_W}{\pi R^2 t} \leq \sigma_{cr} = \frac{f_y}{\sqrt{1+\lambda^4}},$$ (3)

where

$$\lambda^2 = \frac{f_y}{\sigma_a + \sigma_b}\left(\frac{\sigma_a}{\sigma_{Ea}} + \frac{\sigma_b}{\sigma_{Eb}}\right),$$ (4)

$$\sigma_{Ea} = (1.5 - 50\beta)C_a \frac{\pi^2 E}{12(1-v^2)}\left(\frac{t}{L_r}\right)^2,$$ (5)

$$\sigma_{Eb} = (1.5 - 50\beta)C_b \frac{\pi^2 E}{12(1-v^2)}\left(\frac{t}{L_r}\right)^2,$$ (6)

$$C_a = \sqrt{1+(\rho_a\xi)^2}, C_b = \sqrt{1+(\rho_b\xi)^2},$$ (7, 8)

$$\rho_a = 0.5\left(1+\frac{R}{150t}\right)^{-0.5}, \rho_b = 0.5\left(1+\frac{R}{300t}\right)^{-0.5},$$ (9, 10)

$$\xi = 0.702Z, Z = \frac{L_r^2}{Rt}\sqrt{1-v^2}, L_r = \frac{L}{n+1}$$ (11, 12)

The factor of $(1.5 - 50\beta)$ in Eqs (5,6) express the effect of initial radial shell deformation caused by the shrinkage of circumferential welds and can be calculated as follows (Farkas 2002). The maximum radial deformation of the shell caused by the shrinkage of a circumferential weld is

$$u_{max} = 0.64A_Tt\sqrt{R/t}$$ (13)

where A_Tt is the area of specific strains near the weld. According to previous results of Farkas & Jármai (1997, 1998) for steels

$$A_Tt = 0.844 \times 10^{-3}Q_T \quad (A_Tt \text{ in mm}^2, Q_T \text{ in J/mm}).$$ (14)

For butt welds

$$Q_T = 60.7A_W \quad (A_W \text{ in mm}^2).$$ (15)

When $t \leq 10$ mm, $A_W = 10t.$ (16)

When $t > 10$ mm, $A_W \cong 3.05t^{1.45}.$ (17)

The constants in Eqs. (14–17) are not dimensionless. The shell buckling strength should be multiplied by the imperfection factor, $(1.5 - 50\beta)$ where a reduction factor β is introduced for which

$$0.01 \leq \beta = \frac{u_{max}}{4\sqrt{Rt}} \leq 0.02;$$ (18)

$\beta = 0.01$ for $\frac{u_{max}}{4\sqrt{Rt}} < 0.01$ and

$\beta = 0.02$ for $\frac{u_{max}}{4\sqrt{Rt}} \geq 0.02.$ (19)

Furthermore

$$\xi = 0.6696\frac{L_r^2}{Rt},$$ (20)

From Eq (5) it can be deduced that σ_E does not depend on L_r, since L_r^2 is in the denominator and from (Eq 7), C has L_r^2 in the numerator. The fact that the buckling strength does not depend on the shell length, was first derived by Timoshenko & Gere (1961). Note that this dependence of σ_E on L_r is very small according to the API design rules (2000). It has however been determined that in the case of external pressure, the distance between ring-stiffeners does play an important role (Farkas & Jármai 2003).

2.1.3 Constraint on panel ring buckling

Requirements for a ring stiffener are as follows (DNV 1995):

$$A_r = h_r t_r \geq \left(\frac{2}{Z^2} + 0.06\right)L_r t,$$ (21)

$$I_r = \frac{h_r^3 t_r}{12}\cdot\frac{1+4\omega}{1+\omega} \geq \frac{\sigma_{max}tR_0^4}{500EL_r},$$ (22)

$$R_0 = R - y_G, y_G = \frac{h_r}{2(1+\omega)}$$ (23)

$$\omega = \frac{L_e t}{h_r t_r}$$ (24)

and $L_e = \min\left(L_r, L_{e0} = 1.5\sqrt{Rt}\right).$ (25)

343

2.2 Cost function

A possible manufacturing sequence is as follows:

(1) Manufacture five shell elements with a length of 3 m without rings. Two axial butt welds (Gas Metal Arc Welding with CO_2, GMAW-C) are needed for every shell element. The cost to form a shell element into a slightly conical, near cylindrical shape, is included in the factor K_{F0} described below. From data obtained from the Hungarian production company Jászberényi Aprítógépgyár, Crushing Machine Factory, Jászberény, the time T for bending a plate element of 3 m width can be approximated by the following function:

$$\ln T = 6.85825 - 4.5272t^{-0.5} + 0.0095419D^{0.5}. \quad (26)$$

($4\,mm < t < 40\,mm$ and $1750\,mm < D < 3500\,mm$). In this equation, which also includes the time to form the plate and reduce the initial imperfections due to forming, t is the plate thickness and D is the diameter. The cost for shell formation is thus given by

$$K_{F0} = k_F \Theta_F T \quad (27)$$

where $\Theta_F = 3$ is the difficulty factor indicative of the complexity of fabrication and k_F is the specific manufacturing cost per unit time.

The welding cost of a shell element is (Farkas & Jármai 2003)

$$K_{F1} = k_F \left(\Theta_W \sqrt{\kappa \rho V_1} \right) + k_F [1.3 \times 0.224 \times 10^3 t^2 (2 \times 3000)] \quad (28)$$

where Θ_W is a difficulty factor expressing the complexity of the assembly and κ is the number of elements to be assembled. For the elements of radius R and density ρ to be welded

$\kappa = 2$, $V_1 = 2R\pi t \times 3000$ and $\Theta_W = 2$, where V_1 is the volume of an element.

(2) Weld a complete unstiffened shell part, combining the five elements by using four circumferential butt welds. This implies welding costs of

$$K_{F2} = k_F \left(\Theta_W \sqrt{5 \times 5 \rho V_1} + 1.3 \times 0.2245 \times 10^{-3} t^2 x 4 x 2R\pi \right) \quad (29)$$

for a shell part.

(3) Cut n flat plate rings using acetylene gas. The cutting cost amounts to

$$K_{F3} = k_F \Theta_c C_c t_r^{0.25} L_c, \quad (30)$$

where Θ_c, C_c and L_c are respectively the difficulty factor for cutting, the cutting parameter and the cutting length with values $\Theta_c = 3$, $C_c = 1.1388 \times 10^{-3}$ and

$$L_c \approx 2R\pi n + 2(R - h_r)\pi n$$

for a ring radius R and ring height h_r.

Table 1. Summary of masses and costs.

Shell part	Mass (kg)	Cost without K_p ($)	K_p ($)	Total ($)
Top	5398	12096	6440	18536
Middle	9472	19772	7603	27373
Bottom	15648	30941	8778	39719
Total	30518	62809	22821	85628

(4) Weld n rings into the shell segment with double-sided GMAW-C fillet welds ($2n$ fillet welds):

$$K_{F4} = k_F \left(\Theta_W \sqrt{(n+1)\rho V_2} + 1.3 \times 0.3394 \times 10^{-3} a_W^2 4R\pi n \right) \quad (31)$$

The size of the weld for a ring of thickness t_r is $a_W = 0.5 t_r$, but $a_{Wmin} = 3\,mm$.

The volume of a shell part V_2 is given by

$$V_2 = 5V_1 + 2 \left(R - \frac{h_r}{2} \right) \pi h_r t_r n. \quad (32)$$

The total material cost for a shell part

$$K_M = k_M \rho V_2. \quad (33)$$

The cost of painting

$$K_P = k_P \Theta_P S_P, \quad S_P = 4R\pi 1500 + 5 x 2 x 2 \left(R - \frac{h_r}{2} \right) h_r$$

$$\Theta_P = 2 \quad (34)$$

The total cost for a shell part thus is

$$K = K_M + 5(K_{F0} + K_{F1}) + K_{F2} + K_{F3} + K_{F4} + K_P. \quad (35)$$

The material cost factor is $k_M = 1$ $/kg, the labour cost factor is $k_F = 1$ $/min and the painting cost factor is $kP = 14.4$ $/m^2$, steel density $\rho = 7.85 \times 10^{-6}\,kg/mm^3$.

2.3 Optimization and results

The optimization can be carried out using any appropriate constrained optimization algorithm. Here it was performed using Rosenbrock's search algorithm (Farkas & Jármai 1997). The optimal values of the shell thickness (t) for $n = 5$, which comply with the design constraints and minimize the cost function, are given in Figure 1. The minimal masses and costs are summarized in Table 1.

2.4 Check of eigenfrequency

For the approximate calculation of eigenfrequency a simple model is used: a beam of length L built in

344

at end a free at other to which a concentrated mass m_1 is connected. The circular eigenfrequency for a beam without any concentrated mass at the end can be calculated as

$$\alpha = \frac{1.87^2}{L^2}\sqrt{\frac{EI_x}{m_0}} \qquad (36)$$

For the average shell radius of $R = 1350$ mm, thickness $t = 9$ mm, cross-sectional area $A = 2R\pi t = 7.634 \times 10^{-4}$ mm^2, moment of inertia $I_x = \pi R^3 t = 6.956 \times 10^{10}$ mm^4, specific mass $m_0 = \rho A/g$, $g = 9.81 \times 10^3$ mm/s^2 one obtains $\alpha = 8.45/$s or 1.34 Hz. This value should be modified considering the attached mass according to the value of $m_1/(m_0 L) = 950/305 = 3.1$ using the diagram in Prochnost' (1968) multiplying by $(1.2/1.87)^2 = 0.41$, i.e. $\alpha = 0.55$ Hz, which is larger than the rotor frequency 0.37 Hz, thus the tower satisfies the eigenfrequency requirement.

2.5 Check of fatigue

According to Eurocode 3 Part 1–9 (2002) the fatigue stress range for toe failure for 2×10^6 cycles in the case of longitudinal stiffeners is 71 MPa when the weld length is greater than 100 mm. Lavassas et al. (2003) have given a load spectrum, which gives the reliable wind load for some numbers of cycles. Example for an average number of cycles $N = 10^5$ for a wind speed of 14 m/s $M_W = 1280$ kNm and $F_W = 80$ kN. For the given number of cycles the allowable stress range is 192.7 MPa, dividing by a partial safety factor of 1.35 one obtains 154 MPa. The bending moment is $1280 + 45 \times 80 = 4880$ kNm and the stress $4880 \times 10^6/(\pi R^2 t) = 38$ MPa < 154 MPa, the fatigue constraint is fulfilled.

3 TUBULAR TRUSS STRUCTURE

In the present study a 45 m high welded steel tubular truss tower is designed for a 1 MW turbine. The truss is statically determinate. The distance between parallel chords in the upper part of the tower is limited because of the rotating blades. In the lower part the chord distance can be linearly varied and the inclination angle can be optimized (Fig. 2). Thus, in the optimization procedure the inclination angle and the member dimensions of the lower tower part are sought, which minimize the structural volume and fulfil the design constraints.

The constraints relate to the buckling strength of circular hollow section (CHS) members and to the local strength of welded tubular joints. Seismic behaviour is not treated. In the numerical problem the loads from wind acting on the turbine and from the nacelle mass are selected from the literature. Knowing the member

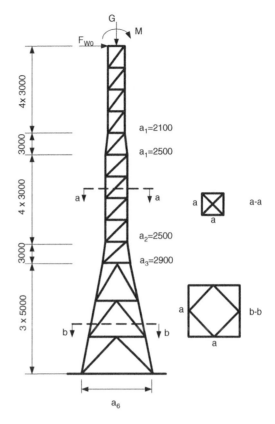

Figure 2. Truss structure of the tower with diaphragms.

forces an iterative suboptimization method is used for the calculation of compression member dimensions.

The cross-section of the truss can be quadratic or triangular. In the case of a triangular cross-section the whole horizontal load should be carried by a truss plane, since the horizontal load direction is variable. Therefore the quadratic cross-section is used. In this case only the half value of the horizontal load is acting on a truss plane.

3.1 Suboptimization problem for the buckling design of a CHS compressed strut

This procedure is described in Farkas & Jármai (1997).

3.2 Design of the upper and middle tower part

The factored loads acting on the tower top according to Lavassas et al. (2003) are as follows: horizontal wind force $F_{w0} = 282$ kN, bending moment $M = 997$ kNm, the nacelle mass according to Spera (1995) $G = 950$ kN. It is sufficient to design one truss plan only, thus, the loads can be halved for it $F_0 = F_{w0}/2 = 141$ kN, $M/2 = 498.5$ kNm and the two forces acting on each chord $G/4 = 237.5$ kN.

Wind forces acting on the middle of the tower parts (Figs. 5, 6, 7): $F_{W1} = 13.9$, $F_{W2} = 23.9$ and $F_{W3} = 19.25$ kN.

3.3 Optimum angle of the lower part

The following analysis can help to understand the existence of an optimal angle in the case of such a combined loading. Three types of loads are acting on the tower as follows: horizontal force and bending moment from wind as well as vertical force from the nacelle mass. The effect of these loads should be analyzed separately.

In the structural volume the most significant part are the chords designed against buckling. Their dimensions are determined by the compression force and the rod length. It is sufficient to analyze the changing of the volume of chords in the function of the inclination angle. It has been shown that, in the case of a truss with parallel chords, an optimum height (distance between the chords) exists, which gives the minimum truss mass (Farkas & Jármai 1997).

Neglecting the branch, a simple analysis shows that this statement is valid also for a cantilever truss with linearly varied height, loaded by a transverse force acting at the truss end. The case of a two-bar truss loaded by a horizontal force is shown in Figure 3. The member force and length is

$$S = \frac{F}{2\sin\alpha}, L = \frac{H}{\cos\alpha} \qquad (37)$$

When α increases, the member force S decreases and its length L increases, thus, an optimum inclination angle exists for the minimum structural volume.

Another case is a simple 3-bar truss loaded by a pair of vertical forces (Fig. 4). In this case

$$S = \frac{Q}{\cos\alpha}, L = \frac{H}{\cos\alpha} \qquad (38)$$

When the angle increases, both the member force S and the length L increases. Thus the optimum angle for the minimum structural volume is 0^0.

From the above mentioned facts it can be concluded that, in the case of horizontal and vertical loads an optimal angle exists and it depends on the ratio of the magnitudes of the two load types. In the case of a high tower with a horizontal force on the top the bending

moment is so large that the optimum inclination angle converges to zero. Detailed calculations show that this optimum angle is about 3^0, thus it is more convenient for fabrication to use parallel chords. We use a bottom part with parallel chords with a distance of 2.9 m (Fig. 7).

3.4 Design of circular hollow sections (CHS) for the three tower parts

The suitable strut profiles are given in Figures 5,6 and 7. The struts of horizontal diaphragms are selected on the basis of the prescription of minimum rod slenderness

$$\lambda = \frac{KL}{r}, r \geq \frac{KL}{\lambda_{min}} \qquad (39)$$

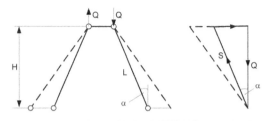

Figure 4. The member force and the bar length when the inclination angle changes in the case of a pair of vertical forces.

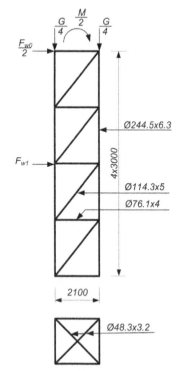

Figure 5. Truss of the top tower part.

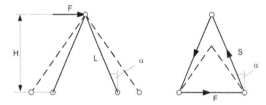

Figure 3. The member force and the bar length when the inclination angle changes in the case of a horizontal force.

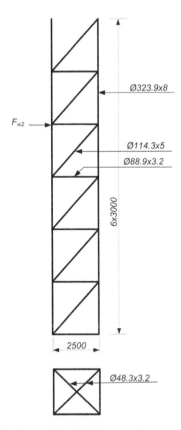

Figure 6.   Truss of the middle tower part.

According to BS5400 $\lambda_{min} = 180$. The strut end restraint factor K is given in Rondal et al.(1992).

The requirements of joint eccentricity and chord plastification (Wardenier et al. 1991) are in all cases fulfilled.

3.5   Check of eigenfrequency

The tower is calculated as a bent beam with a constant average moment of inertia of the middle part $I_x = 4(9910x10^4 + 9860x1250^2) = 6.20x10^{10}$ mm^4. The eigenfrequency calculated with Eq. (36) is

$$\alpha = \frac{1.87^2}{L^2}\sqrt{\frac{EI_x}{m_0}} = 11.2 \ (1.78 \text{ Hz})$$

This value should be modified considering the mass at the beam end $m_1 = 950$ kN. The tower mass is $m_0 L = 175$ kN, thus the modifying coefficient for $m_1/(m_0 L) = 5.4$ is $(0.9/1.87)^2 = 0.23$, i.e. $\alpha = 0.41$ Hz, which is larger than the rotor frequency 0.37 Hz, the tower fulfils the eigenfrequency requirement.

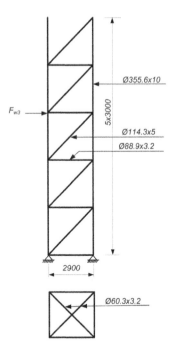

Figure 7.   Truss of the bottom tower part.

3.6   Check of fatigue

The chord members are connected to the base plate by fillet welds and these connections are reinforced by longitudinal attachments for the end of which the fatigue stress range according to Eurocode 3 Part 1–9 (2002) is for 2×10^6 cycles 71 MPa. Using the wind load spectrum given by Lavassas et al. (2003) for an average number of cycles 10^6 the loads are $F_W = 65$ kN and $M_W = 960$ kNm, i.e. the bending moment $45 \times 65 + 960 = 3885$ kNm.

Compression force in a chord $S = 3885/(2 \times 2.9) = 670$ kN and the stress $670 \times 10^3/10.900 = 61$ MPa. Fatigue stress range for 10^6 cycles is 89 MPa, the allowable stress range with a safety factor of 1.25 is $89/1.25 = 71.2$ MPa, this is larger than 61 MPa, thus, the tubular tower is safe against fatigue. Note that the hot-spot stress method cannot be used according Zhao et al. (2001), since the N-type CHS truss is not treated in it. For this joint also stress range of 71 MPa is given.

3.7   Cost calculation

According to Price List (1995) the following profile prices are considered:

The structural mass is

$$G = \rho \sum_i A_i L_i \qquad (40)$$

and the material cost is calculated as

$$K_M = \rho \sum_i k_{Mi} A_i L_i \qquad (41)$$

Cost of cutting and grinding of the tubular member ends can be obtained by

$$K_{CG} = k_F \Theta_{dC} \sum_i \frac{2\pi d_i}{\sin\theta} \left(4.54 + 0.4229 t_i^2\right), \qquad (42)$$

the difficulty (complexity) factor is $\Theta_{dC} = 3$; d_i in m, t_i in mm.
Cost of assembly

$$K_A = k_F \Theta_{dA} \left(\kappa \rho V\right)^{1/2}; \; \Theta_{dA} = 3, \qquad (43)$$

Cost of welding

$$K_W = 1.3 k_F \Theta_{dW} \sum_i C_{Wi} a_{Wi}^n L_{Wi}; \; \Theta_{dW} = 3.5, \qquad (44)$$

for shielded metal arc welding (SMAW) $C_W = 0.7889 \times 10^{-3}$; $a_{Wi} = t_i$ (mm) is the weld size, $n = 2$; $L_{Wi} = \dfrac{2\pi d_i}{\sin\theta_i}$ is the weld length in mm.
Cost of painting

$$K_P = k_P \Theta_P A_P, \Theta_P = 2 \qquad (45)$$

$k_P = 14.4 \times 10^{-6} \; \$/mm^2$

the surface to be painted is

$$A_P = \sum_i d_i \pi L_i \qquad (46)$$

The details of the cost calculation are summarized in Table 3.

Table 2. Material cost factors for available CHS diameters.

d (mm)	k_M (\$/kg)
48.3, 60.3, 76.4	1.0059
88.9, 101.6, 114.3	1.0553
139.7, 168.3, 177.8, 193.7	1.1294
219.1, 244.5, 273.0, 323.9	1.2922
355.6, 406.4	1.3642

4 CONCLUSIONS

Two structural versions of a 45 m high 1 MW wind turbine tower are designed and compared to each other regarding the mass and cost. The stiffened shell structure should be designed with constraints on shell and stiffener buckling, eigenfrequency and fatigue. The shell consists of three slightly conical parts with variable diameters and thicknesses. The three parts are connected with bolted joints the cost of which is neglected.

The tubular truss structure consists also from three parts with different but constant width. The four truss planes are stiffened by horizontal diaphragms constructed from two struts. A suitable suboptimization method is used for the economic design of compressed CHS struts.

It is shown that, in the case of large bending moment from wind loads, the optimum inclination angle of the bottom truss part converges to zero. Therefore the bottom part is constructed also with parallel chords. The tubular joints are checked for chord plastification and eccentricity. The truss is checked for eigenfrequency and fatigue.

The comparison of the two structural versions shows that the tubular truss has smaller mass (17533 compared to 30518 kg, 42 %), smaller surface to be painted and is much cheaper than the shell structure (51161 compared to 85628 \$, 40 %).

ACKNOWLEDGEMENTS

The research work was supported by the Hungarian-Portuguese Intergovernmental S&T Co-operation programme P 6/99. The Hungarian partner is the Ministry of Education, R&D Deputy Undersecretary of State, the Portuguese partner being the Ministry of Universities and Technology, GRICES. The research work was also supported by the Hungarian Scientific Research Foundation grants OTKA T38058, T37941.

REFERENCES

American Petroleum Institute (API).2000. Bulletin 2U. *Bulletin on stability design of cylindrical shells*. 2nd ed. Washington.

Table 3. Costs in \$ of the tubular tower, the surface A_P to be painted in mm^2.

Part	G (kg)	K_M	K_{CG}	K_A	K_W	$A_P \times 10^{-6}$	K_P	K
Top	3437	4139	1936	1180	2514	72.46	2087	11856
Middle	7395	9096	2867	1965	3108	130.11	3747	20783
Bottom	6701	8643	2551	1629	2353	116.17	3346	18522
Total	17533	21878	7354	4774	7975	318.74	9180	51161

Bazeos, N., Hatzigeorgiou, G.D., Hondros, I.D., Karamaneas, H., Karabalis, D.L. & Beskos, D.E. 2002. Static, seismic and stability analyses of a prototype wind turbine steel tower. *Eng. Struct.* 24: 1015–1025.

BS 5400:Part 3: 1982. Steel, concrete and composite bridges. Code of practice for design of steel bridges. BSI, London.

Det Norske Veritas (DNV).1995.*Buckling strength analysis.* Classification Notes No.30.1. Høvik, Norway.

Eurocode 1:1999a. Basis of design and actions on structures. Part 2–1. Actions on structures. Densities, self-weight and imposed loads. ENV 1991-2-1: 1995.

Eurocode 1. 1999b. Part 2–4. Wind loads. ENV 1991-2-4: 1999.

Eurocode 3. 2002. Design of steel structures. Part 1–1. General structural rules. PrEN 1993-1-1:2002.

Eurocode 3. Part 1–9. 2002. Design of steel structures. Fatigue strength of steel structures. CEN, Brussels.

Farkas, J. & Jármai, K. 1997. Analysis and design of metal structures. Rotterdam: Balkema.

Farkas, J. & Jármai, K. 1998. Analysis of some methods for reducing residual beam curvatures due to wekd shrinkage. *Welding in the World* 41 (4): 385–398.

Farkas, J. & Jármai, K. 2001. Height optimization of a triangular CHS truss using an improved cost function. Tubular Structures IX. Proceedings of the 9th International Symposium on Tubular Structures, Düsseldorf. Balkema, Lisse etc. 2001. 429–435.

Farkas, J. 2002. Thickness design of axially compressed unstiffened cylindrical shells with circumferential welds. *Welding in the World* 46 (11/12): 26–29.

Farkas, J. & Jármai, K. 2003. *Economic design of metal structures.* Rotterdam: Millpress.

Jármai, K. & Farkas, J. 1999. Cost calculation and optimization of welded steel structures. *Journal of Constructional Steel Research.* 50: 115–135.

Jármai, K., Farkas, J. & Virág, Z. 2003. Minimum cost design of ring-stiffened cylindrical shells subject to axial compression and external pressure. *5th World Congress of Structural and Multidisciplinary Optimization,* Short papers. Italian Polytechnic Press, Milano, 63–64.

Horváth, G. & Tóth, L. 2001. New methods in wind turbine tower design. *Wind Engineering* 25 (3):171–178.

Koumousis, V.K. & Dimou, C.K. 1995. Optimal design of wind mill towers. Steel Structures – Eurosteel '95. Ed. Kounadis. Rotterdam: Balkema. 443–450.

Lavassas, I., Nikolaidis, G., Zervas, P., Efthimiou, E., Doudoumis, I.N. & Baniotopoulos, C.C. 2003. Analysis and design of the prototype of a steel 1-MW wind turbine tower. *Eng. Struct.* 25: 1097–1106.

Price List 20. 1995. Steel tubes, pipes and hollow sections. Part 1b. *Structural hollow sections.* British Steel Tubes and Pipes.

Prochnost', ustoichivost', kolebaniya. 1968. Moskva, Mashinostroenie. (Strength, stability, vibration. In Russian.)

Rondal, J., Würker, K.-G., Dutta, D., Wardenier, J. & Yeomans, N. 1992. *Structural stability of hollows sections.* Köln: Verlag TÜV Rheinland.

Spera, D.A. ed. 1994. *Wind turbine technology.* New York: ASME Press.

Timoshenko, S.P. & Gere, J.M. 1961. *Theory of elastic stability.* 2nd ed. New York, Toronto, London: McGraw Hill.

Wardenier, J., Kurobane, Y., Packer, J.A., Dutta, D. & Yeomans, N. 1991. *Design guide for circular hollow section (CHS) joints under predominantly static loading.* Köln: Verlag TÜV Rheinland.

Zhao, X-L. et al. (2001) *Design guide No. 8.* Köln: Verlag TÜV Rheinland.

Parallel session 5: Composite construction

Tubular Structures XI – Packer & Willibald (eds)
© *2006 Taylor & Francis Group, London, ISBN 0-415-40280-8*

Nonlinear in-plane behaviour and buckling of concrete-filled steel tubular arches

M.A. Bradford, Y.-L. Pi & R.I. Gilbert
School of Civil & Environmental Engineering, The University of New South Wales, Sydney, Australia

ABSTRACT: This paper concerns the in-plane nonlinear behaviour of shallow, parabolic arches with fixed or pinned bases whose cross-sections comprise of circular steel tubes with an infill of concrete, and that are subjected to uniformly distributed loading. Analytic formulations of the nonlinear equations of in-plane equilibrium are derived, and buckling equations are presented in closed form for antisymmetric and symmetric buckling. The significantly improved structural performance of a tubular arch afforded by the concrete infill is illustrated by an example.

1 INTRODUCTION

Concrete-filled steel tubular arches are an efficient structural form in bridges that are finding extensive contemporary use in China in particular, with efficiencies being attributable to the ability of the lightweight and deployable steel tube to provide formwork for the concrete. Concrete infill is also advantageous in increasing the local buckling coefficient of a circular tube from that of a hollow tube by a factor of $\sqrt{3}$ (Bradford et al. 2002, Bradford 2006).

Parabolic arches are funicular with respect to uniformly distributed loading so that under the assumptions of linearity, a parabolic arch is subjected to compressive actions only. Shallow arches exhibit significant geometric nonlinearity, and the assumption of no bending actions not correct and recourse needs to be made to an appropriate nonlinear theory to predict the behaviour of shallow arches. This paper presents such a theory for shallow parabolic arches that comprise of circular hollow sections filled with concrete, which would be encountered in bridge applications. It builds on the solution for the elastic buckling of a pinned parabolic arch developed by Bradford et al. 2004 by including a composite of a hollow steel tube and a concrete infill, and of fixed supports at the base of the arch. The theory herein considers analytically the nonlinear prebuckling equilibrium, as well as the buckling modes that can occur, and design equations are presented in closed form that treat both antisymmetric and symmetric buckling modes. The results are limited to short-term loading, and shrinkage, creep and thermal effects are not included.

2 NONLINEAR EQUILIBRIUM

2.1 *Differential equilibrium equation*

This paper assumes the validity of the Euler-Bernoulli hypothesis and the arch kinematics is described using two sets of coordinate axis systems: one (*oyz*) being fixed in space and the other ($o^*y^*z^*$) being a moving axis system. The axis *oz* is in the horizontal direction, passing through both ends of the arch and *oy* is vertical, passing through the crown (Figure 1). The centroidal axis is defined by

$$y = \frac{z^2 - (L/2)^2}{2p}, \tag{1}$$

where $p = L^2/8f$ is the parabolic focal parameter, L is the span length and f is the arch rise. The moving axis system has the o^*y^* axis normal to the centroidal axis of the arch before the deformation, and then o^*y^* and o^*z^* rotate to the new position $o_1^*y_1^*$ and $o_1^*z_1^*$, with the rotation between $o^*y^*z^*$ and $o_1^*y_1^*z_1^*$ being described by the rotation matrix (Bradford et al. 2004, 2005)

$$\mathbf{R} = \begin{bmatrix} 1 & \mathrm{d}v^*/\mathrm{d}s \\ -\mathrm{d}v^*/\mathrm{d}s & 1 \end{bmatrix}, \tag{2}$$

where s is the tangential axis along the arch. Since $v = v^* \cos\alpha$ and $\tan\alpha = \mathrm{d}y/\mathrm{d}z$ where α is the angle between the o^*y^* and oy axes and using

$$\mathrm{d}s = \mathrm{d}z\sqrt{1 + (\mathrm{d}y/\mathrm{d}z)^2}, \tag{3}$$

it follows that

$$\frac{dv^*}{ds} = \frac{\sqrt{1+(dy/dz)^2}\cdot dv}{\sqrt{1+(dy/dz)^2}\cdot dz} = \frac{dv}{dz} \qquad (4)$$

and so

$$\mathbf{R} = \begin{bmatrix} 1 & v' \\ -v' & 1 \end{bmatrix}, \qquad (5)$$

where $(\ )' \equiv d(\ )/dz$. The vertical displacements v_P, w_P of a point P on the cross-section can be expressed by those of the centroid (v and w) using

$$\langle v_P, w_P \rangle^T = \langle v, w \rangle^T + \langle y^*, 0 \rangle^T (\mathbf{R}-\mathbf{I}) = \langle v, (w-v'y^*) \rangle^T \quad (6)$$

where y^* is the coordinate of P in the $o^*y^*z^*$ axes.

The longitudinal normal strain at P is obtainable by simplifying the Green strain tensor to

$$\varepsilon = \tfrac{1}{2}\left[(ds^*)^2 - (ds)\right]/(ds)^2 . \qquad (7)$$

Using

$$\begin{aligned}(ds)^2 &= (dz)^2 + (dy)^2; \\ (ds^*)^2 &= (dz + dw_P)^2 + (dy + dv_P)^2 \end{aligned} \qquad (8)$$

then produces (by using Equation 6)

$$\varepsilon = w' + v'y' + \tfrac{1}{2}v'^2 - y^*v'' = w' + \frac{v'z}{p} + \tfrac{1}{2}v'^2 - y^*v'', \quad (9)$$

which assumes that the terms y'^2, $dw \cdot dz$ and higher order terms containing them may be neglected.

The total potential of a concrete filled tubular arch subjected to a vertical uniformly distributed load is

$$V = U + W , \qquad (10)$$

where the total strain energy of the tubular arch is

$$U = \tfrac{1}{2}\left[\int_{V_s} nE_c\varepsilon^2\,dV_s + \int_{V_c} E_c\varepsilon^2\,dV_c \right], \qquad (11)$$

in which E_c is the elastic modulus of the concrete core and $E_s = nE_c$ is the elastic modulus of the steel tube (where n is the modular ratio), V_c is the volume of the concrete core and V_s is the volume of the steel tube. By using $y'^2 \ll 1$ and noting that the axes are centroidal, Equation 11 becomes

$$U = \tfrac{1}{2}\int_{-L/2}^{L/2}\left[AE\varepsilon_m^2 + EIv''^2\right]dz , \qquad (12)$$

where A is the effective or transformed area and I is the transformed second moment of this area, given by

$$A = nA_s + A_c; I = nI_s + I_c, \qquad (13)$$

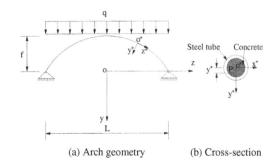

(a) Arch geometry   (b) Cross-section

Figure 1.   Concrete-filled steel tubular parabolic arch rib.

and

$$\varepsilon_m = w' + v'z/p + \tfrac{1}{2}v'^2 \qquad (14)$$

is the membrane strain.

The potential energy of the vertical load q distributed uniformly and horizontal over the span of the arch is

$$W = \int_{-L/2}^{L/2} qv\,dz , \qquad (15)$$

and so the total potential can be written in terms of the functional in the expression

$$V = \int_{-L/2}^{L/2} \mathbf{F}(w',v',v'')dz , \qquad (16)$$

where

$$\mathbf{F} = \tfrac{1}{2}\left(AE\varepsilon_m^2 + EIv''^2\right) - qv . \qquad (17)$$

Equilibrium is defined by $\delta V = 0$. Invoking the Euler-Lagrange equations of variational calculus in the w direction in Equations 16 and 17 produces

$$\varepsilon_m' = 0 , \qquad (18)$$

and so the membrane strain is constant and can be written as

$$\varepsilon_m - \frac{N}{AE}, \qquad (19)$$

where N is the actual axial force in the arch. Using the Euler-Lagrange equations in the v direction in Equations 16 and 17 and using Equation 19 results in the differential equilibrium equation

$$\frac{v^{iv}}{\mu^2} + v'' = \frac{\omega}{p}, \qquad (20)$$

in which

$$\mu^2 = \frac{N}{EI}; \quad \omega = \frac{qp-N}{N}, \qquad (21)$$

with ω representing a dimensionless departure of the actual axial force N from the nominal axial force qp for a shallow parabolic arch for which y'^2 is negligible.

2.2 Nonlinear equilibrium equation

The nonlinear equilibrium equation for a shallow parabolic arch can be established by considering that the constant membrane strain in Equation 19 equals that in Equation 14 when averaged over $z \in [-L/2, L/2]$. Noting that $w(-L/2) = w(L/2) = 0$ and $N/EI = \mu^2 r^2$ (where $r = \sqrt{(I/A)}$ is the transformed radius of gyration), this produces

$$\mu^2 r^2 = \frac{1}{L} \int_{-L/2}^{L/2} \left(v'z/p + \tfrac{1}{2} v'^2 \right) dz . \tag{22}$$

The solution of Equation 20 for a pinned arch (satisfying the kinematic and statical boundary conditions that $v = v'' = 0$ at $z = \pm L/2$) is

$$v = \frac{\omega}{\mu^2 p} \left[\cos\theta \sec\alpha - 1 + \tfrac{1}{2}(\theta^2 - \alpha^2) \right], \tag{23}$$

in which $\alpha = \mu L/2$ and $\theta = \mu z$, while the solution of Equation 20 for a fixed arch (satisfying the kinematic boundary conditions that $v = v' = 0$ at $z = \pm L/2$) is

$$v = \frac{\omega}{\mu^2 p} \left[\alpha(\cos\theta \, \mathrm{cosec}\,\alpha - \cot\alpha) + \tfrac{1}{2}(\theta^2 - \alpha^2) \right]. \tag{24}$$

Using either of Equations 23 or 24 in Equation 22 produces the nonlinear in-plane equation of equilibrium for a shallow concrete-filled parabolic tubular arch in the quadratic transcendental vector format

$$\langle 1, \omega, \omega^2 \rangle \cdot \langle \Phi_1, \Phi_2, \Phi_3 \rangle^{\mathrm{T}} = \langle 1, \omega, \omega^2 \rangle \Phi_{P,F} = 0, \tag{25}$$

in which

$$\Phi_P = \left\{ \begin{array}{c} (\alpha/\lambda)^2 \\[2mm] \dfrac{1}{\alpha^2}\left(1 - \dfrac{\tan\alpha}{\alpha} + \dfrac{\alpha^2}{3} \right) \\[4mm] \dfrac{1}{4\alpha^3}\left(5\alpha - 5\tan\alpha + \alpha\tan^2\alpha + \dfrac{2\alpha^3}{3} \right) \end{array} \right\} \tag{26}$$

if the arch is pinned, and

$$\Phi_F = \left\{ \begin{array}{c} (\alpha/\lambda)^2 \\[2mm] \dfrac{1}{\alpha^2}\left(\alpha\cot\alpha - 1 + \dfrac{\alpha^2}{3} \right) \\[4mm] \dfrac{1}{4\alpha^2}\left(3\alpha\cot\alpha + \alpha^2\cot^2\alpha - 4 + \dfrac{5\alpha^2}{3} \right) \end{array} \right\} \tag{27}$$

if the arch is fixed, in which the modified slenderness is defined by

$$\lambda = \frac{L^2}{4rp} = 2\left(\frac{f}{r} \right). \tag{28}$$

3 IN-PLANE BUCKLING ANALYSIS

3.1 Differential buckling equation

Neutrality of the equilibrium position $\{v,w\}$ requires that $\delta^2 V = 0$ in Equation 10 for any kinematically admissible variations δv and δw, indicating a possible transition from the prebuckled configuration to an adjacent buckled one defined by $\{v + \delta v, w + \delta w\}$. Using the notation $( )_b = \delta( )$, this leads to

$$\delta\left\{ \int_{-L/2}^{L/2} F_b(w_b', v_b, v_b', v_b'') dz \right\} = 0 \tag{29}$$

in which

$$F_b = AE\left(\varepsilon_{mb}^2 + \varepsilon_m v_b'^2 \right) + EIv_b''^2 \tag{30}$$

with

$$\varepsilon_{mb} \equiv \delta\varepsilon_m = w_b' + v_b'z/p + v'v_b' \tag{31}$$

being the membrane strain during buckling. Stationarity of the functional in Equation (29) again leads to

$$\varepsilon_{mb}' = 0, \tag{32}$$

so that the membrane buckling strain is constant, and to

$$\frac{v_b^{iv}}{\mu^2} + v_b'' = \frac{\varepsilon_{mb}}{\mu^2 r^2}\left(v'' + \frac{1}{p} \right). \tag{33}$$

For the arch in question, the symmetric nature of the loading as well as of the boundary conditions dictates that the arch buckles in-plane in either an antisymmetric mode, or a symmetric snap-through mode.

3.2 Antisymmetric buckling

For an antisymmetric buckling mode, v_b is an odd function and v is an even function, so that the three terms in Equation 31 vanish when integrated over the domain $[-L/2, L/2]$ when it is noted $w_b' = 0$ at $z = \pm L/2$. As a result, $\varepsilon_{mb} = 0$, leading to the homogenous differential equation for antisymmetric buckling given by

$$v_b^{iv} + \mu^2 v_b'' = 0, \tag{34}$$

whose solution satisfying $v_b = 0$ at $z = \pm L/2$ is

$$v_b = C(\sin\theta - \theta\sin\alpha/\alpha), \tag{35}$$

where C is a nontrivial amplitude parameter. The eigensolution of Equation 35 for a pinned arch

355

$(v_b = v_b'' = 0)$ is $\sin\alpha = 0$, whilst for a fixed arch $(v_b = v_b' = 0)$ it is $\tan\alpha = \alpha$, which have respective fundamental solutions $\alpha = \pi$ and $\alpha \approx 1.4303\pi$. Substituting $\alpha = \pi$ into the first of Equations 21 produces $N = N_P = 4\pi^2 EI/L^2$ which is the second mode Euler load of a pin-ended column of length L (Trahair & Bradford 1998) and using $\alpha = \pi$ in Equation 25 leads to the buckling load

$$q_{bP}p \approx \left(0.26 \pm 0.74\sqrt{1-0.63\pi^4/\lambda^2}\right)\cdot N_P \qquad (36)$$

for a pinned arch. Similarly for a fixed arch, using $\alpha \approx 1.4303\pi$ produces $N = N_F = 4 \times (1.4303\pi)^2 EI/L^2$ which the second mode "Euler" load for a fixed column of length L) and to the buckling load

$$q_{bF}p \approx \left(0.6 \pm 0.4\sqrt{1-30.686\pi^2/\lambda^2}\right)\cdot N_F \qquad (37)$$

for a fixed arch. In Equations 36 and 37, if $\lambda < 7.83$ then the buckling load for a pinned arch is complex while if $\lambda < 17.40$ that for a fixed arch is complex, so that antisymmetric buckling cannot occur.

3.3 Symmetric buckling

For symmetric buckling, v_b is symmetric and it cannot be argued that $\varepsilon_{mb} \neq 0$. Substituting Equation 23 into Equation 33 for a pinned arch leads to a differential equation for v_b, whose solution satisfying $v_b = v_b'' = 0$ is

$$v_b = \frac{\varepsilon_{mb}}{\mu^4 r^2 p}\left[\left(\frac{1+\omega}{2}\right)\left(\theta^2 - \alpha^2\right) + \left(1+2\omega\right)\left(\cos\theta\sec\alpha - 1\right)\right.$$

$$\left. + \frac{\omega}{2}\left(\theta\sin\theta\sec\alpha - \alpha\cos\theta\tan\alpha\sec\alpha\right)\right]. \qquad (38)$$

Similarly for a fixed arch for which $v_b = v_b' = 0$, the buckling deformation is

$$v_b = \frac{\varepsilon_{mb}}{\mu^4 r^2 p}\left[\left(\frac{1+\omega}{2}\right)\left(\theta^2 - \alpha^2\right) + \left(1+\frac{3\omega}{2}\right)\left(\alpha\cos\theta\csc\alpha\right.\right.$$

$$\left.- \cot\alpha\right) + \frac{\omega}{2}\left(\alpha\theta\sin\theta\csc\alpha + \alpha^2\cos\theta\cot\alpha\csc\alpha\right.$$

$$\left.\left.- \alpha^2\csc^2\alpha\right)\right]. \qquad (39)$$

Since the average buckling membrane strain is constant, using Equation 31 with the appropriate buckling deformation v_b (Equation 38 or 39) leads to the transcendental equation for symmetric buckling as

$$\langle 1, \omega, \omega^2\rangle\Phi_{bP,F} = 0 \qquad (40)$$

in which

$$\Phi_{bP} = \begin{Bmatrix} \Phi_{2P} - \Phi_{1P} \\ 4\Phi_{3P} \\ 2\Phi_{3P} + \Phi_{4P} \end{Bmatrix}; \quad \Phi_{bF} = \begin{Bmatrix} \Phi_{2F} - \Phi_{1F} \\ 4\Phi_{3F} \\ 2\Phi_{3F} + \Phi_{4F} \end{Bmatrix} \qquad (41)$$

for pinned and fixed arches respectively, in which

$$\Phi_{4P} = \frac{7\tan^2\alpha}{8\alpha^2} + \frac{15}{8\alpha^2} - \frac{15\tan\alpha}{8\alpha^3} - \frac{\tan\alpha}{4\alpha} - \frac{\tan^3\alpha}{4\alpha}, \qquad (42)$$

$$\Phi_{4F} = \frac{1}{2} + \frac{\alpha\cot\alpha}{4} + \frac{1}{\alpha^2}\left(\frac{3\alpha^2\cot^2\alpha}{4} + \frac{\alpha^3\cot^3\alpha}{4} - 1\right). \qquad (43)$$

Iterative solutions of Equations 25 and 40 are somewhat complicated, and a proposed approximation is

$$q_{sP}p \approx \left(0.15 + 0.006\lambda^2\right)N_P \qquad (44)$$

for the symmetric buckling of pinned arches, and

$$q_{sP}p \approx \left(0.36 + 0.0011\lambda^2\right)N_F \qquad (45)$$

for the symmetric buckling of fixed arches.

For pinned arches, the slenderness λ that defines a switch between antisymmetric and symmetric buckling modes can be found when $\alpha = \pi$, which leads to $\lambda \approx 9.38$, while using $\alpha = 1.4303\pi$ for a fixed arch leads to the delineating value of λ as $\lambda \approx 18.60$. Since it can be shown for a pinned arch that

$$\lim_{\alpha \to \pi/2} \omega = \frac{q_{sP}p - N}{N} = 0, \qquad (46)$$

the lowest symmetric buckling load $q_{sP}p = N$ is when $\alpha = \pi/2$ and is $q_{sP}p = \pi^2 EI/L^2$, while for a fixed arch

$$\lim_{\alpha \to \pi} \omega = \frac{q_{sF}p - N}{N} = 0, \qquad (47)$$

and the lowest symmetric buckling load $q_{sF}p = N$ is when $\alpha = \pi$ and is $q_{sF}p = 4\pi^2 EI/L^2$. For a pinned arch, the deflection of the crown at the lowest symmetric buckling can be found from Equation 23 when $z = 0$ and $\alpha \to \pi/2$, producing

$$v(0) = \frac{4L^2}{\pi^3 p}\left(1 \pm \sqrt{1 - \frac{\pi^6}{64\lambda^2}}\right), \qquad (48)$$

while the fixed arch counterpart can be found from Equation 24 when $z = 0$ and $\alpha \to \pi$, producing

$$v(0) = \frac{L^2}{\pi p}\left(1 \pm \sqrt{1 - \frac{\pi^4}{\lambda^2}}\right). \qquad (49)$$

It can be seen from Equations 48 and 49 that their solutions are complex when $\lambda < 3.88$ and $\lambda < 9.87$ respectively. Because of this, for a pinned arch it can be concluded that when $\lambda < 3.88$ it does not buckle, when $3.88 \le \lambda < 9.38$ it buckles in a symmetric mode and when $\lambda > 9.38$ it buckles in an antisymmetric mode, while for a fixed arch it can be concluded that when $\lambda < 9.87$ it does not buckle, when $9.87 \le \lambda < 18.60$ it buckles in a symmetric mode and when $\lambda > 18.60$ it buckles in an antisymmetric mode. Advanced finite element studies (Pi et al. 2002) have demonstrated that a pinned arch with $7.83 < \lambda \le 9.38$ buckles symmetrically first and then antisymmetrically on the descending branch of the load-deflection curve, and the same phenomenon occurs for a fixed arch with $17.40 < \lambda \le 18.60$.

The authors have reported experiments on the nonlinear response and buckling of shallow, parabolic reinforced concrete arches under short-term loading (Wang et al. 2005, 2006). Whilst not being of tubular cross-section, the nonlinear response of these members is governed by the same equations derived herein, and the theoretical treatment has been shown (Bradford et al. 2006) to be accurate.

4 ILLUSTRATION

A circular steel tube 457×12.7 CHS (OneSteel 2004) with or without a concrete infill has been chosen to illustrate the results of this paper. The elastic moduli used were $E_s = 200\,\text{kN/mm}^2$ and $E_c = 25\,\text{kN/mm}^2$ and the parabolic arch spanned $L = 40$ m. The buckling loads have been determined for a hollow steel section and for a section with a concrete infill, and the results are shown in Figure 2 for a pin-based arch and in Figure 3 for a fixed-based arch. It can be seen that as the arch becomes shallower and the axial force increases, the buckling loads decrease but the buckling load for the tube with an infill is much greater than that without an infill. The comparison between these buckling loads has been made by plotting the ratio of the buckling load of the filled tube, less its self-weight, to that of the hollow tube, less its self weight, as the filling of the tubes with concrete obviously penalises the structure by increasing the self-weight significantly. Over the range of geometries considered, the provision of a concrete infill increases the buckling load around 40% if the bases are pinned, and around 55% if they are built-in, with the increase in buckling capacity being more marked for shallow arches.

5 CONCLUDING REMARKS

This paper has considered the in-plane nonlinear behaviour, including in-plane buckling, of shallow

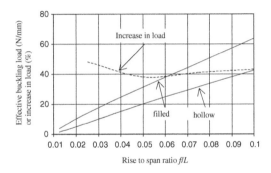

Figure 2. Buckling results for pinned base tubular arch.

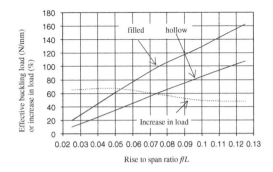

Figure 3. Buckling results for fixed base tubular arch.

parabolic tubular arches comprising of a circular hollow steel section with a concrete infill. Prescriptive equations were derived for the equilibrium between the external loading and the internal actions so that the load-deflection response could be quantified, and buckling loads for both symmetric snap-through and antisymmetric bifurcation modes have been proposed. The use of the equations was illustrated with an example that showed the significant structural improvement in the buckling capacity that can be obtained by use of a concrete infill within a hollow circular steel tube.

ACKNOWLEDGEMENT

The work in this paper was supported by the Australian Research Council through Discovery Projects awarded to the authors.

REFERENCES

Bradford, M.A., Loh, H.Y. & Uy, B. 2002. Slenderness limits for filled circular steel tubes. *Journal of Constructional Steel Research* 58(2): 243–252.

Bradford, M.A., Pi, Y.-L. & Gilbert, R.I. 2004. Nonlinear elastic analysis of shallow, pinned parabolic arches. *17th ASCE*

Engineering Mechanics Conference, Newark, Delaware. Virginia: ASCE: 1–8.

Bradford, M.A., Zhang, Y.X., Gilbert, R.I. & Wang, T. 2005. In-plane nonlinear analysis and buckling of tied circular arches. *Advances in Structural Engineering* (in press).

Bradford, M.A. & Roufegarinejad, A. 2006. Elastic buckling of thin circular tubes with an elastic infill under uniform compression. *11th International Symposium on Tubular Structures*, Quebec City, Canada. Rotterdam: Balkema.

Bradford, M.A., Wang, T., Pi, Y.-L. & Gilbert, R.I. 2006. In-plane stability of parabolic arches with horizontal spring supports. I: Theory. *Journal of Structural Engineering*, ASCE (accepted).

OneSteel Market Mills 2004. *Cold Formed Structural Hollow Sections & Profiles*. Wollongong Australia: OneSteel.

Pi, Y.-L., Bradford, M.A. & Uy, B. 2002. In-plane stability of arches. *International Journal of Solids & Structures* 39: 105–125.

Trahair, N.S. & Bradford, M.A. 1998. *The Behaviour and Design of Steel Structures to AS4100*. 3rd Edn. (Australian), London: E&FN Spon.

Wang, T., Bradford, M.A. & Gilbert, R.I. 2005. Short term experimental study of a shallow reinforced concrete parabolic arch. *Australian Journal of Structural Engineering* 6(1): 53–62.

Wang, T., Bradford, M.A., Gilbert, R.I. & Pi, Y.-L. 2006. In-plane stability of parabolic arches with horizontal spring supports. II: Experiments. *Journal of Structural Engineering*, ASCE (accepted).

Tubular Structures XI – Packer & Willibald (eds)
© 2006 Taylor & Francis Group, London, ISBN 0-415-40280-8

Elastic buckling of thin circular tubes with an elastic infill under uniform compression

M.A. Bradford & A. Roufegarinejad

School of Civil & Environmental Engineering, The University of New South Wales, Sydney, Australia

ABSTRACT: The elastic buckling of a thin-walled cylindrical tube containing an elastic infill is addressed in this paper. It is widely known that the local buckling of a hollow tube can be enhanced by providing a rigid infill (usually concrete), but it is possible to increase the local buckling stress using a less-rigid and less-dense infill that can optimise the strength to weight ratio of the member, which may be advantageous in aerospace applications. The elastic buckling stress is determined using a Ritz-based approach as a function of the stiffness of the infill, and an approximation is proposed for the minimum local buckling stress.

1 INTRODUCTION

Thin-walled circular tubes that are subjected to an axial compressive force find widespread application in engineering structures. It is widely known that if these members are hollow and thin-walled, they can be considered familiarly as Donnell shells whose buckling is influenced by geometric imperfections. Ignoring the effect of these imperfections, the elastic buckling stress derived in Timoshenko & Gere (1961) is $2E(t/d)/\sqrt{[3(1 - v^2)]}$ where E is the Young's modulus, v is the Poisson's ratio, t is the wall thickness and d is the diameter.

Circular hollow section tubes are often filled with concrete in high-rise building construction (Uy 2000). Apart from obvious implications of strength enhancement that accrues to the composite nature of the resulting structural element, the local buckling of the steel tube is enhanced because the rigid concrete infill prevents buckling of the steel tube into the hardened concrete medium. Using an energy method, Loh et al. (2000) and Bradford et al. (2001, 2002) derived the local buckling stress for a thin-walled circular tube filled with a rigid (concrete) infill, and formulated the enhanced slenderness limit for which local buckling and yield occur simultaneously. The local buckling stress was derived in these studies to be $\sqrt{3}$ times that of Timoshenko's solution for a hollow tube. The benign nature of the infill was evident from the increased d/t ratio that delineates a slender section from an effective one (Trahair & Bradford 1998) when the hollow steel tube is filled with concrete.

A circular tube may be filled with a material whose stiffness varies between that of being sensibly rigid (such as with a concrete infill) and hollow (without an infill), so that the enhancement of the elastic buckling stress varies by a factor of 1.0 to approximately $\sqrt{3}$ as the stiffness of the infill increases from zero to infinity. Examples include a tube with a polystyrene fill in aerospace applications, and caisson piles and open caissons in civil engineering applications. This paper derives the local buckling stress of a thin-walled steel cylindrical tube filled in this fashion by formulating the change in total potential during buckling. Recourse is made to a Ritz-based approach and the local buckling stress is expressed as a function of the infill rigidity.

2 ENERGY FORMULATION

Figure 1 shows a cylinder that undergoes "ring buckling" with an infinitesimal deformation w, when the constant axial strain ε_0 that is applied to the section reaches its critical value ε_{cr}. One local buckle "cell" in the x direction of length L is considered, being part of a number of such cells through the length of the tube. The buckling load is found herein by invoking the a Ritz-based procedure that requires knowledge of the strain energy stored due to bending only (U_b), the membrane strain energy due to stretching (U_m) and the work done by the external compressive forces (V) during buckling, as well as the strain energy U_R associated with the buckling of the cylinder wall into elastic restraining medium.

Initially, a shell element has radii of principal curvature in the xz and yz planes of r_{x0} and r_{y0} respectively, and after the buckling deformation these radii of

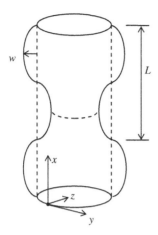

Figure 1.   Ring buckling shape.

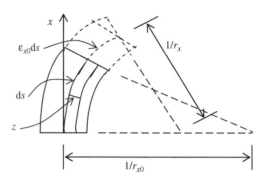

Figure 2.   Element of cylinder in $x-z$ plane.

curvature change to counterpart values of r_x and r_y. In an analogous fashion to flat plates, the strain energy due to bending only is therefore

$$U_b = \tfrac{1}{2} D \int_0^L \int_A \left\{ \left[\left(\frac{1}{r_x} - \frac{1}{r_{x0}} \right) + \left(\frac{1}{r_y} - \frac{1}{r_{y0}} \right) \right]^2 - 2(1-v) \right.$$

$$\left. \left[\left(\frac{1}{r_x} - \frac{1}{r_{x0}} \right) \cdot \left(\frac{1}{r_y} - \frac{1}{r_{y0}} \right) - \left(\frac{1}{r_{xy}} - \frac{1}{r_{xy0}} \right)^2 \right] \right\} \mathrm{d}A\,\mathrm{d}x \quad (1)$$

in which $(1/r_{xy} - 1/r_{xy0})$ is the twist of the element during buckling, A is the area of the cross-section, v is Poisson's ratio and

$$D = \frac{Et^3}{12(1-v^2)} \quad (2)$$

is the rigidity of the shell of thickness t and E is Young's modulus. Because of the axisymmetric nature of the cylinder problem, the strain energy due to bending only reduces to

$$U_b = \tfrac{1}{2} D \cdot 2\pi R t \int_0^L \left(\frac{\partial^2 w}{\partial x^2} \right)^2 \mathrm{d}x, \quad (3)$$

where R is the radius to the mid-plane of the cylinder wall.

By denoting ε_{x0} as the middle surface strain in the x direction and ε_{y0} as the middle surface strain in the y direction, the curved length $\mathrm{d}s$ shown in Figure 2 becomes $\mathrm{d}s(1 + \varepsilon_{x0})$ under a middle surface straining of ε_{x0} and the original length of a laminate distant

z from the mid-plane of $\mathrm{d}s(1 - z/r_{x0})$ then becomes $\mathrm{d}s(1 + \varepsilon_{x0})(1 - z/r_x)$, resulting in a straining of

$$\varepsilon_x = \frac{\mathrm{d}s(1 + \varepsilon_{x0})(1 - z/r_x) - \mathrm{d}s(1 - z/r_{x0})}{\mathrm{d}s(1 - z/r_{x0})}$$

$$= \varepsilon_{x0} - z \left(\frac{1}{r_x} - \frac{1}{r_{x0}} \right), \quad (4)$$

while similarly for the y direction,

$$\varepsilon_y = \frac{\mathrm{d}s(1 + \varepsilon_{y0})(1 - z/r_x) - \mathrm{d}s(1 - z/r_{y0})}{\mathrm{d}s(1 - z/r_{y0})}$$

$$= \varepsilon_{y0} - z \left(\frac{1}{r_y} - \frac{1}{r_{y0}} \right). \quad (5)$$

The plane stress equations of elasticity are

$$\sigma_x = \frac{E}{1-v^2} \left\{ \varepsilon_{x0} - z \left(\frac{1}{r_x} - \frac{1}{r_{x0}} \right) + v \left[\varepsilon_{y0} - z \left(\frac{1}{r_y} - \frac{1}{r_{y0}} \right) \right] \right\} \quad (6)$$

$$\sigma_y = \frac{E}{1-v^2} \left\{ \varepsilon_{y0} - z \left(\frac{1}{r_y} - \frac{1}{r_{y0}} \right) + v \left[\varepsilon_{x0} - z \left(\frac{1}{r_x} - \frac{1}{r_{x0}} \right) \right] \right\}, \quad (7)$$

and these lead to edge forces per unit length of

$$N_x = \int_{-t/2}^{t/2} \sigma_x \,\mathrm{d}z = \frac{Et}{1-v^2}(\varepsilon_{x0} + v\varepsilon_{y0}), \quad (8)$$

$$N_y = \int_{-t/2}^{t/2} \sigma_y \,\mathrm{d}z = \frac{Et}{1-v^2}(\varepsilon_{y0} + v\varepsilon_{x0}), \quad (9)$$

$$N_{xy} = \int_{-t/2}^{t/2} \tau_{xy} \,\mathrm{d}z = \frac{Et}{2(1+v)} \gamma_{xy0}, \quad (10)$$

360

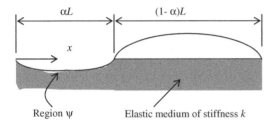

αL (1- α)L

x

Region ψ   Elastic medium of stiffness k

Figure 3. Meridional buckling deformation into and out of elastic infill.

where γ_{xy0} is the shearing strain at the middle surface of the shell. The strain energy due to membrane stretching of the middle surface is then

$$U_m = \tfrac{1}{2}\int_0^L \left(N_x \varepsilon_{x0} + N_y \varepsilon_{y0} + N_{xy}\gamma_{xy} \right)\cdot 2\pi R\, \mathrm{d}x . \qquad (11)$$

For the problem at hand, $\gamma_{xy0} = 0$, and if it is assumed that the force remains constant during local buckling then $N_x = Et\varepsilon_0 = Et(\varepsilon_{x0} + v\varepsilon_{y0})/(1-v^2)$ from Equation 8, producing

$$\varepsilon_0 = \frac{\varepsilon_{x0} + v\varepsilon_{y0}}{1-v^2} . \qquad (12)$$

This result can also be obtained by considering circumferential straining w/R from the relationships

$$\varepsilon_{y0} = w/R - v\varepsilon_0 \quad \text{and} \quad \varepsilon_{x0} = \varepsilon_0 - vw/R . \qquad (13)$$

Using these, the membrane strain energy in Equation 11 reduces to

$$U_m = \tfrac{1}{2}Et\cdot 2\pi R \int_0^L \left(\varepsilon_0^2 + \frac{w^2}{R^2} - 2v\frac{\varepsilon_0 w}{R} \right)\mathrm{d}x . \qquad (14)$$

The work done by the external compressive forces during buckling is equal to the end load $E\varepsilon_0 \cdot 2\pi Rt$ multiplied by the shortening of the axial length due to bending $\tfrac{1}{2}\int (\partial w/\partial x)^2 \mathrm{d}x$ and to the change in axial length caused by the change in strain $\varepsilon_{x0} - \varepsilon_0$ from Equation 13 so that

$$V = 2\pi Etv\varepsilon_0 \int_0^L w\, \mathrm{d}x + \pi ERt\varepsilon_0 \int_0^L \left(\frac{\partial w}{\partial x} \right)^2 \mathrm{d}x . \qquad (15)$$

In Figure 3 which shows a local buckle cell with buckling into the elastic medium, the domain of buckling into the medium is $x \in \Psi = [0, \alpha L]$ and if the

stiffness of the medium is k, then the strain energy associated with this buckle is

$$U_R = \tfrac{1}{2}\int_0^{\alpha L} w^2\, \mathrm{d}x . \qquad (16)$$

Prior to buckling, the strain energy stored is $\tfrac{1}{2}(2\pi Rt)\cdot(E\varepsilon_0)\cdot \varepsilon_0$ and during buckling, this increases by $U_b + U_m + U_R - V$. The total change in potential is therefore

$$\Pi = \pi RD \int_0^L \left(\frac{\partial^2 w}{\partial x^2} \right)^2 \mathrm{d}x + \frac{\pi Et}{R}\int_0^L w^2\, \mathrm{d}x$$
$$+ \tfrac{1}{2}\int_0^{\alpha L} kw^2\, \mathrm{d}x - \pi REt\varepsilon_0 \int_0^L \left(\frac{\partial w}{\partial x} \right)^2 \mathrm{d}x . \qquad (17)$$

In order to determine the buckling strain ε_0, the change in potential Π in Equation 17 must be stationary for all variations of the deformed shape w that satisfies the end conditions for the tube. This becomes a problem in mathematical programming, since the deformation must satisfy the kinematic boundary condition that $w(x = \alpha L) = 0$ whose location is unknown a priori. The buckling formulation therefore reduces to the statement

$$\delta\Pi(w) = 0 \quad \exists\, w(\alpha L) = 0 . \qquad (18)$$

3 RITZ SOLUTION

3.1 Hollow circular tube

The solution for the local buckling of a thin cylindrical tube without considering imperfection sensitivity can be obtained by setting $k = 0$ in Equation 17 and assuming a displacement of the form

$$w = q\sin(n\pi x/L), \qquad (19)$$

which satisfies the kinematic boundary conditions $w(x = 0) = w(x = L) = 0$ exactly. This leads to the buckling solution for nontrivial q that

$$\sigma_{cr} = E\varepsilon_0 = \frac{Et^2}{12(1-v^2)}\left(\frac{n^2\pi^2}{L^2} \right) + \frac{EL^2}{n^2\pi^2 R^2} . \qquad (20)$$

This buckling stress may be minimised according to

$$\frac{\mathrm{d}\sigma_{cr}}{\mathrm{d}(n\pi/L)} = 0 , \qquad (21)$$

producing the solution

$$\left(\sigma_{cr}\right)_{\min} = \frac{E}{\sqrt{3\left(1-v^2\right)}}\cdot\frac{t}{R} \tag{22}$$

when

$$\frac{n\pi}{L} = \left(\frac{Et}{R^2 D}\right)^{1/4}, \tag{23}$$

and which is the same result as in Timoshenko & Gere (1961).

3.2 Concrete-filled circular tube

If the third integral in Equation 17 is ignored, a displacement function that satisfies the kinematic boundary conditions for a rigid infill that $w(x=0)=w(x=L)=0$ and $(dw/dx)(x=0)=(dw/dx)(x=L)=0$ is

$$w = q\sin^2\left(n\pi x/L\right). \tag{24}$$

In accordance with the energy method of solution, this displacement function will lead to an upper bound on the local buckling stress. By using Equation 13 in Equation 17, the local buckling stress becomes

$$\sigma_{cr} = E\varepsilon_0 = \frac{Et^2}{3\left(1-v^2\right)}\left(\frac{n^2\pi^2}{L^2}\right) + \frac{3EL^2}{4n^2\pi^2 R^2}, \tag{25}$$

with the minimisation procedure leading to the result

$$\left(\sigma_{cr}\right)_{\min} = \frac{E}{\sqrt{1-v^2}}\cdot\frac{t}{R} \tag{26}$$

when

$$\frac{n\pi}{L} = \tfrac{1}{2}\left(\frac{3Et}{R^2 D}\right)^{1/4}. \tag{27}$$

This is the same solution as that derived by Loh et al. (2000) and Bradford et al. (2001, 2002).

3.3 Tube with elastic infill

The displacement function chosen to model the buckling mode in Figure 3 is

$$w = \sum_{i=0}^{n} A_i \cos\frac{i\pi x}{L} + B_i \sin\frac{i\pi x}{L}. \tag{28}$$

Because a sine curve was used for a hollow tube and a sine squared curve was used for a tube with a rigid infill, the deformation function in Equation 28 consists of sine plus cosine curves, and the boundary conditions that $w(x=0)=w(x=L)=0$ and $(dw/dx)(x=0)=(dw/dx)(x=L)=0$ were imposed on Equation 28. Hence, in order for this shape to satisfy the boundary conditions at the ends of the local buckle "cell",

$$
\begin{aligned}
w = &\sum_{\substack{i=0\\i\,\text{even}}}^{n-2} q_i\left\{\cos\left(\frac{i\pi x}{L}\right) - \cos\left(\frac{i\pi x}{L}\right)\right\} + \\
&\sum_{\substack{i=0\\i\,\text{odd}}}^{n-3} q_i\left\{\cos\left(\frac{i\pi x}{L}\right) - \cos\left[(n-1)\frac{\pi x}{L}\right]\right\} + \\
&\sum_{\substack{i=2\\i\,\text{even}}}^{n-2} r_i\left\{\sin\left(\frac{i\pi x}{L}\right) - \frac{i\sin\left(\frac{n\pi x}{L}\right)}{n}\right\} + \\
&\sum_{\substack{i=0\\i\,\text{odd}}}^{n-3} r_i\left\{\sin\left(\frac{i\pi x}{L}\right) - \frac{i\sin\left[(n-1)\frac{\pi x}{L}\right]}{n-1}\right\}.
\end{aligned} \tag{29}
$$

Equation 29 can be written conveniently in vector notation as

$$w = \mathbf{f}^{\mathrm{T}}\mathbf{q}, \tag{30}$$

where $\mathbf{q} = \{q_0, q_1, \ldots, r_0, r_1, \ldots\}^{\mathrm{T}}$ are structural freedoms and, for example, when $n=6$ (and there are 9 degrees of freedom),

$$
\begin{aligned}
\mathbf{f} = &\left\langle \left(1-\cos\frac{6\pi x}{L}\right), \left(\cos\frac{\pi x}{L}-\cos\frac{5\pi x}{L}\right), \left(\cos\frac{2\pi x}{L}-\cos\frac{6\pi x}{L}\right), \right. \\
&\left(\cos\frac{3\pi x}{L}-\cos\frac{5\pi x}{L}\right), \left(\cos\frac{4\pi x}{L}-\cos\frac{6\pi x}{L}\right), \\
&\left. \left(\sin\frac{\pi x}{L}-\frac{1}{5}\sin\frac{5\pi x}{L}\right)\right\rangle^{\mathrm{T}}.
\end{aligned} \tag{31}
$$

Equation 31 may be differentiated appropriately with respect to x for use in Equation 17; for instance for the second derivative,

$$\frac{\partial^2 w}{\partial x^2} = \left(\frac{\partial^2}{\partial x^2}\mathbf{f}^{\mathrm{T}}\right)\cdot\mathbf{q} = \mathbf{q}^{\mathrm{T}}\cdot\left(\frac{\partial^2}{\partial x^2}\mathbf{f}\right), \tag{32}$$

$$\left(\frac{\partial^2 w}{\partial x^2}\right)^2 = \mathbf{q}^{\mathrm{T}}\cdot\left[\left(\frac{\partial^2}{\partial x^2}\mathbf{f}\right)\left(\frac{\partial^2}{\partial x^2}\mathbf{f}^{\mathrm{T}}\right)\right]\cdot\mathbf{q}. \tag{33}$$

Doing likewise for the first derivative, as well as for $w^2 = \mathbf{q}^{\mathrm{T}}\cdot[\mathbf{f}\cdot\mathbf{f}^{\mathrm{T}}]\cdot\mathbf{q}$, and substituting into Equation 17 results in

$$\Pi = \tfrac{1}{2}\mathbf{q}^{\mathrm{T}}\mathbf{K}(\alpha)\mathbf{q}, \tag{34}$$

and the condition that $\delta\Pi = 0$ in Equation 34 leads to the familiar eigenvalue equation

$$\mathbf{K}(\alpha)\mathbf{q} = \mathbf{0}. \qquad (35)$$

However, the condition that $w(x = \alpha L) = 0$ needs to be satisfied, which can be stated in vector form using Equation 31 as

$$\mathbf{n}^\mathrm{T}\mathbf{q} = 0, \qquad (36)$$

where

$$\mathbf{n} = \langle (1 - \cos(6\pi\alpha)), (\cos(\pi\alpha) - \cos(5\pi\alpha)),$$
$$(\cos(2\pi\alpha) - \cos(6\pi\alpha)), (\cos(3\pi\alpha) - \cos(5\pi\alpha)), \qquad (37)$$
$$(\cos(4\pi\alpha) - \cos(6\pi\alpha)), (\sin(\pi\alpha) - \tfrac{1}{5}\sin(5\pi\alpha)) \rangle^\mathrm{T}.$$

For a tube with an elastic infill of stiffness k, the matrix $\mathbf{K}(\alpha)$ in Equation 35 can be assembled with an assumed value of $\alpha \in [0^+, 0.5^-]$ and the eigenproblem in Equation 35 is then solved for the local buckling stress σ_{cr} and the eigenvector $\mathbf{q}$. The condition in Equation 36 must then be tested, with iteration on the value of α chosen being undertaken until $\mathbf{n}^\mathrm{T}\mathbf{q} \to 0$ to within a sufficient tolerance on the chosen value of α. The form of the matrix $\mathbf{K}(\alpha)$ can be shown to be

$$\mathbf{K}(\alpha) = \frac{D}{L^2 t}\mathbf{K}_1 + \frac{EL^2}{R^2}\mathbf{K}_2 + \frac{L^2 k}{Rt}\mathbf{K}_3(\alpha) - \sigma_{cr}\mathbf{K}_\sigma \qquad (38)$$

in which $\mathbf{K}_1$, $\mathbf{K}_2$, $\mathbf{K}_\sigma$ contain only numerals, and $\mathbf{K}_3$ contains only numerals and terms in α.

4 SOLUTIONS

Figure 4 shows a typical plot of the local buckling stress σ_{cr} against k, in which the local buckling stress is written as

$$\sigma_{cr} = \omega \frac{E}{\sqrt{1 - \nu^2}} \frac{t}{R}. \qquad (39)$$

It can be seen that for a tube with a rigid infill, the value of $\omega \to 0.9622655$ as $k \to \infty$. The 4% difference between the value of $\omega = 1$ derived by Loh et al. (2000) and Bradford et al. (2001, 2002) in Equation 26 and 0.9622655 obtained herein is attributable to the single harmonic in Equation 24 being an approximation of the true buckled shape and accordingly providing an upper bound solution for the local buckling stress.

Figure 5 shows the solutions for a hollow tube and for a rigidly filled tube, using the formulation of Equation 17. For the hollow tube, k was taken as zero and only the sine terms were used in Equation 28 so that dw/dx was not zero at the ends, while

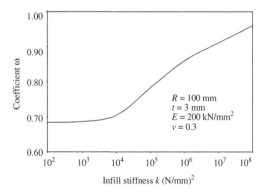

Figure 4. Variation of local buckling coefficient.

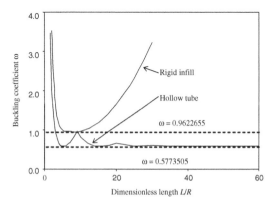

Figure 5. Hollow and rigidly-filled tube local buckling.

for the rigidly-filled tube, linear combinations of sine squared terms were used in Equation 28 using noting that $\sin^2\theta = (1 - \cos 2\theta)/2$ in order to incorporate the cosine terms, and using $k = 10^9$. These solutions used 17 degrees of freedom in the formulation, and the numerical solution for the hollow tube of $\omega = 0.5773505$ is almost identical to Timoshenko's solution of $1/\sqrt{3}$.

As noted earlier, the buckling deformation represented in Equation 29 was chosen to that the deformations and their slopes vanish at the ends of the local buckle cells, with the latter assumption constraining the solution somewhat significantly if this buckling representation is used for a hollow tube. Accordingly, the solution is an upper bound on the local buckling stress, and for tubes with a low rigidity of the infill the solutions overestimate the buckling stress. This can be seen in Figure 4 where for very low values of the infill modulus (less than 10^2), the solution tends to around 0.70 instead of 0.58 and therefore is in error by around 20%.

When Equation 38 is rearranged, it can be shown that if the Poisson's ratio is taken as constant ($\nu = 0.3$

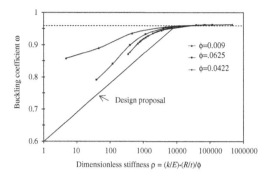

Figure 6. Local buckling coefficient for elastically-filled circular tube.

here), the minimum local buckling stress can be obtained from a relationship of the type

$$\omega = fn\left(\phi = \frac{tR}{L^2}, \quad \rho = \frac{kR}{Et\phi}\right). \quad (40)$$

This generalised relationship is shown in Figure 6, where the buckling coefficient ω is plotted as a function of the dimensionless infill stiffness $kR/(Et\phi)$ for various values of the ratio ϕ, with the abscissa being plotted logarithmically. The curves show a similar trend to those in Figure 4. Based on these curves, a lower bound solution for the local buckling stress can be written as

$$\omega = \begin{cases} 0.6 + 0.0903 \log_{10} \rho & 1 \le \rho < 10^4 \\ 0.962 & \rho \ge 10^4 \end{cases} \quad (41)$$

5 CONCLUSIONS

This paper has considered the local buckling of a thin-walled elastic tube containing an elastic infill. By using a Ritz-based method of solution, the minimum local buckling stress was determined to be expressible as $(\sigma_{cr})_{min} = \omega \cdot Et/[R\sqrt{(1 - v^2)}]$, where the dimensionless buckling coefficient ω is a function of the stiffness of the elastic infill. When the infill has zero stiffness,

$\omega = 1/\sqrt{3}$ which is the solution of Timoshenko & Gere (1961), and when it has an infinite stiffness the numerical procedure here showed that $\omega \approx 0.962$ which is close to the approximate solution of 1.0 derived by Loh et al. (2000) and Bradford et al. (2001, 2002). The displacement functions chosen for the numerical solution allowed for easily-determined buckling solutions, but because they did not satisfy the kinematic boundary conditions at the ends of the tube for small values of the infill stiffness, the results produced in this case overestimated the buckling stress.

Finally, an equation was proposed to approximate the value of the minimum local buckling stress as a function of the dimensionless geometrical parameter $\phi = tR/L^2$ and the dimensionless infill stiffness parameter $\rho = kR/(Et\phi)$. This equation forms a lower bound to the numerical solutions.

ACKNOWLEDGEMENTS

The work in this paper was supported partly by the Australian Research Council through its Federation Fellowship Scheme, and partly by a University of New South Wales Faculty of Engineering Research Grant.

REFERENCES

Bradford, M.A., Loh, H.Y. & Uy, B. 2001. Slenderness limits for CHS sections. *Proceedings of 9th International Symposium on Tubular Structures*, Düsseldorf, Germany, 377–381.

Bradford, M.A., Loh, H.Y. & Uy, B. 2002. Slenderness limits for filled circular steel tubes. *Journal of Constructional Steel Research* 58(2): 243–252.

Loh, H.Y., Bradford, M.A. & Uy, B. 2000. Local buckling of concrete-filled circular steel tubes. *Proceedings of UEF Conference, Composite Construction IV*, Banff, Canada.

Timoshenko, S.P. & Gere, J.M. 1961. *Theory of Elastic Stability*. New York: McGraw Hill.

Trahair, N.S. & Bradford, M.A. 1998. *The Behaviour and Design of Steel Structures to AS4100*. 3rd Edn. (Australian), London: E&FN Spon.

Uy, B. 2000. Strength of concrete filled steel box columns incorporating local buckling. *Journal of Structural Engineering*, ASCE 126(3): 341–352.

Tubular Structures XI – Packer & Willibald (eds)
© 2006 Taylor & Francis Group, London, ISBN 0-415-40280-8

Rehabilitation of tubular members with carbon reinforced polymers

M.V. Seica
Halcrow Yolles and Department of Civil Engineering, University of Toronto, Toronto, Canada

J.A. Packer
Department of Civil Engineering, University of Toronto, Canada

P. Guevara Ramirez
Stantec, Toronto, Canada

S.A.H. Bell
Peter Sheffield, Toronto, Canada

X.-L. Zhao
Department of Civil Engineering, Monash University, Australia

ABSTRACT: Tubular structures have become increasingly popular for steel and aluminium construction, for economic and aesthetic reasons, yet member deterioration and fatigue-cracking of connections is often reported in the offshore industry, in the transportation and agricultural sectors, and with amusement rides. The suitability of such materials to rehabilitate tubular steel members and the structural behaviour of the resulting composite system were investigated experimentally. The programme was tailored primarily to the offshore industry and techniques associated with underwater repair methods. Employing a wet-layup application method, the main focus was on the static flexural performance of steel-FRP composite beams, with principal experimental parameters being the fibre and epoxy material types and the matrix curing conditions. Curing of the specimens was performed both in air and in seawater, the latter simulating the condition for underwater application to offshore structures. The strengthening process was effective as all wrapped composite members exhibited improved structural performance compared to that of a bare tube. The current study has proven the feasibility of rehabilitating tubular steel members with advanced composite materials, underwater.

1 INTRODUCTION

Engineers are faced with a rising level of aging infrastructure which has, in turn, led them to implement new materials and techniques to efficiently combat this problem. The use of fibre reinforced polymers (FRPs) as external reinforcement of concrete structures has shown itself to be a successful alternate method of repair, and has been around for about 20 years. However, research related to FRP applications to steel structures has started fairly recently and there are still few applications in practice.

1.1 Development of composites in construction

The composite industry, as we know it today, was introduced with the production of thermosetting plastics in 1909 and the commercial availability of fibreglass filaments in 1935 (Schwartz 1997, Hollaway 1994). Still, it was not until the 1940s that the composite industry began to bloom. The use of composite technology in aircraft applications flourished and, by the 1950s, FRP boat hulls and FRP car bodies were developed, wherein glass fibres were the major reinforcement used. The higher performance carbon fibre reinforced polymers (CFRPs) were developed in 1963 for specialised applications. Aramid fibres were first developed in 1965 and found acceptance in aerospace and marine applications (Hollaway 1994).

Two of the first known applications of glass fibre reinforced polymers (GFRPs) in civil infrastructure were a dome structure constructed in Benghazi in 1968 and a roof structure at the Dubai Airport in 1972 (Hollaway 1993). In the late 1970s, footbridges using composite materials were constructed in Europe and the USA. Since the early 1990s, there has been an increase in the use of small pultruded structural shapes, wherein structural shapes with constant cross sections are fabricated, for the construction of industrial platforms, pedestrian bridges, latticed transmission towers, and for other applications (Bakis et al. 2002).

Externally-bonded FRP composites applied to reinforced concrete (RC) structures have been used around the world since the mid-1980s. Since then, the application of FRP reinforcement has expanded to include masonry structures, timber and, to a lesser degree, metals (ACI 2002, Hollaway 1993). Nowadays, FRP composites are used in a variety of applications. One particular case is an offshore platform for oil production installed on Shell's Southpass 62 in the Gulf of Mexico in 1986 (Barbero 1998).

In the last 30 years there have been considerable advancements in the use of FRP composites in civil infrastructure which is reflected by the fact that, in the previous decade, the construction industry led the transition from conventional materials to advanced composites by using 30% of all polymers produced (Hollaway and Head 2001). The increasing amount of ongoing research related to this matter indicates that this trend will continue.

2 FRP-TO-STEEL

2.1 Recent research

There have been increasing research efforts concerning the application of FRP to steel structures with very promising results. High stiffness fibres, such as carbon fibres, can effectively enhance the strength and stiffness of steel structures.

The primary application of existing research on FRP-to-steel composites is for the repair of bridges. Many authors (Al-Saidy et al. 2004, Schnerch and Rizkalla 2004, Tavakkolizadeh and Saadatmanesh 2003a,b, Mertz and Gillespie 1996, Sen and Liby 1994) have performed tests on steel-concrete composite girders with emphasis on evaluating the effectiveness of strengthening and repair methods using FRP composites. Other studies have focused on the behaviour of FRP-retrofitted composite girders under fatigue loading (Miller et al. 2001) or have investigated the durability of the rehabilitation methods using composite materials (Mertz and Gillespie 1996). Field experiments on FRP-repaired steel bridges were also performed (Miller et al. 2001).

Although the majority of the published literature has been concerned with the use of composites for bridge repair, other types of structures were also considered. Studies involving the use of FRP applied to tubular steel (Jiao and Zhao 2004, Teng and Hu 2004), to steel polygonal monopoles (Schnerch and Rizkalla 2004) or to cast iron beams and curved steel beams (Garden 2004) have been reported.

One particular un-researched area, however, is the underwater application of composite materials to steel and their resulting behaviour, and this is the focus of this study.

2.2 FRP-to-steel-bonding

Existing literature indicates that debonding and delamination are the main failure modes of the experiments done hitherto with FRP and steel. Some investigators have suggested that, when adhesives are not fully capable of transferring the forces, additional fasteners may be used (Sen and Liby 1994).

Good durability under demanding environmental conditions was demonstrated by the epoxy resin and bonding can be enhanced when the surfaces are treated with silane (Mertz and Gillespie 1996).

In a recent publication, Buyukozturk et al. (2004) concluded that cracks can propagate in the interface of the adhesive with either the steel or CFRP in regions with high stress concentrations.

Bond between steel and CFRP was also studied by Fawzia et al. (2005).

3 EXPERIMENTAL RESEARCH

The objectives of this project were to assess the feasibility of strengthening tubular steel members subjected to bending and to develop an adequate repair method for underwater and in-air applications.

3.1 Carbon fibres

Because of the specific mechanical properties of steel, it was decided to use carbon fibre reinforcement since the modulus of elasticity for carbon is similar to that of steel. The fibres used in the present study were provided by Sika Canada Inc. (Sika) and Fyfe Co. (Fyfe). The Sika products used were SikaWrap® HEX 113C, a bidirectional carbon fibre fabric, and SikaWrap® HEX 230C, a unidirectional carbon fibre fabric. Fyfe supplied Tyfo® SCH-41, a carbon fabric with a fibreglass veil stitched to the one side of the carbon material and Tyfo® SCH-41S, an essentially identical material but without the glass veil.

3.2 Matrix material

The epoxies chosen from both manufacturers were those considered most appropriate for employment under difficult conditions such as overhead and underwater applications. The epoxy resins supplied by Sika were Sikadur® Hex 330 and Sikadur® Hex 306. The Sikadur® Hex 306 was of particular interest to this programme because of its thixotropic nature, since the

Table 1. Measured steel mechanical properties and reported properties.

Coupon location or Origin of data	Static yield strength, $F_{y,st}$ (0.2% Offset method) (MPa)	Ultimate strength, F_u (MPa)	Modulus of elasticity, E (MPa)	Elongation at failure, ε_{ul}(%)
90°	485	546	199,500	27
180°	491	546	192,700	23
270°	471[†]	529	196,600	28
Stub Column	477	n/a	196,600	n/a
Mill Test RT74	478	534	Not Reported	33
Mill Test RT76	487	544	Not Reported	30

[†] Dynamic yield strength.

ability of the epoxy to remain viscous was of importance when considering underwater applications as the resin would tend to stay in the impregnated fibre. A thixotropic material was also selected from Fyfe, the Tyfo® SW-1 Underwater Epoxy Coating. This product was developed specifically to bond fibre products under submerged conditions.

3.3 Tubular steel properties

Round hollow sections were used in this investigation because offshore structures are mainly constructed with circular steel elements. Round HSS168 × 4.8 tubes conforming to CAN/CSA G40.20-04/G40.21-04 (2004) Grade 350W Class C were provided by IPSCO Inc. Tensile mechanical properties measured by the manufacturer are given in Table 1.

In addition to the provided data, tension tests were performed on coupons fabricated from the tubes. The dog-bone shaped coupons, maintaining their natural circular arc profile, were tested according to ASTM E8M (2004) using circular arc insert plates to grip the specimens in the testing machine without flattening their ends. Coupons were cut from three locations around the tube at 90, 180 and 270° measured circumferentially from the longitudinal seam weld. Table 1 also shows the values of the mechanical properties obtained from the tension coupon tests. According to CAN/CSA S16-01 (2001), the tube was a Class 2 section in flexure, implying that it could reach the plastic moment with limited rotation capacity. A stub-column test was also performed on a circular HSS section in accordance with the recommended test procedure (Galambos 1988) and the test results are presented in Table 1, too. It can be seen that the stub-column average material properties are very similar to the values obtained from the tension tests and to the values reported on the mill test certificate.

3.4 CFRP material properties

Coupons of CFRP from both manufacturers were fabricated to perform tension tests according to the ASTM D3039 (2000) standard, to determine the mechanical properties of the supplied composite materials. Large laminates with widths of approximately 300 mm were prepared and then cut to 25.4 mm wide strips, suitable for tensile testing. Specimens of one and two plies were fabricated with thicker, integrated grip tabs.

Tension tests were conducted on six sets of FRP coupons, as listed in Table 2. Each set consisted of five individual coupons loaded in tension until failure, and averages of the five tests are given. Strain gauges were attached to the specimens to determine the modulus of elasticity, and the values of strength and modulus were calculated according to the ASTM D3039 (2000) standard.

Manufacturers often report the properties of cured laminates and these properties, also listed in Table 2, are expected if the laminates are fabricated according to manufacturer-recommended practices under standard conditions. It should be noted that the reported laminate properties were based on laminates using different epoxy adhesives than those used in the present investigation (Bell 2004). It can be seen in Table 2 that, in terms of strength, the results of the SikaWrap® Hex 113C coupons were closer to the expected values, however in the case of the coupons made of SikaWrap® Hex 230C and Tyfo SCH-41 the test results were lower than the manufacturer-reported values. The measured modulus of elasticity was close to reported values for SikaWrap® Hex 113C, whereas the stiffness of the remaining specimens was about 80% of the reported values.

3.5 Test specimen fabrication

Seven 2.4 m long pieces were cut from the steel tubes. The size of the specimens was chosen to suit the available equipment for a four-point bending test employing a 2.2 m span length. As listed in Table 3, one tube was used as a control specimen, whereas the remaining six were wrapped with CFRP materials. Two wrapped specimens were prepared "in-air" (i.e. under standard curing conditions), whereas the other four were wrapped underwater (i.e. under seawater

Table 2. Comparison of measured to reported CFRP coupon properties.

Set	Type of fibre	Measured tensile strength (MPa)	Reported tensile strength (MPa)	Percentage of test/Reported strength values (%)	Measured modulus (MPa)	Reported modulus (MPa)	Percentage of test/Reported modulus values (%)
1	Hex 230C[†]	554	894	62	48,700	61,012	80
2	Hex 230C[†]	520	894	58	52,000	61,012	85
3	Hex 113C[†]	386	456	85	45,300	41,400	109
4	Hex 113C[†]	492	456	108	44,500	41,400	107
5	SCH-41[‡]	543	876	62	62,400	72,400	86
6	SCH-41[‡]	458	876	52	62,600	72,400	86

[†] Sika product.
[‡] Fyfe product.

Table 3. Flexural test programme.

No.	Specimen identification	Test type	Fibre type	Curing conditions	No. of specimens	Comments
First Set of Tests						
1	R	Static	None	None	1	Reference beam
2	Sika-S	Static	Sika	Standard	1	
3	Fyfe-S	Static	Fyfe	Standard	1	
Second Set of Tests						
4	Sika-U	Static	Sika	Seawater	1	
5	Fyfe-U	Static	Fyfe	Seawater	1	
Third Set of Tests						
6	Sika-U-Strapped	Static	Sika	Seawater	1	Nylon strips were used to bind CFRP
7	Sika/Fyfe-U-Strapped	Static	Sika	Seawater	1	Sika fibres used with Fyfe matrix
				Total	7	

Specimen Suffixes: R = Reference; S = Standard Curing ("In-Air" Curing); U = Underwater Curing.

curing conditions). Wrapping CFRP materials underwater involved an additional challenge: adhesion of the wet resin whilst the layers tended to debond from the tube under their own weight. Hence, to ensure adequate bonding of all layers, circumferential nylon ties were used for the last set of two specimens (i.e. 6 and 7) to hold the CFRP sheets in place during curing.

Specimens 2 and 3 (Table 4) were wrapped using Sika and Fyfe fibre systems (fibres and epoxy), respectively, and cured under standard conditions. Two layers of sheets with the fibres oriented longitudinally to the length of the tube were first adhered to each beam and then these were confined with a third layer with the fibres oriented transversely to the tube axis. The circumferential layer was used to confine the longitudinal layers whilst subjected to compressive stresses during bending. It should be noted that, for Specimen 3 which was wrapped with Fyfe fibre systems, the fibre that was bonded directly to the steel was provided with a glass veil, as would be done in typical practical applications to avoid the potential for corrosion. Figure 1 illustrates the typical stacking sequence of

Table 4. Fibre configurations for specimens cured under standard conditions.

	Specimen 2	Specimen 3
Fibre	SikaWrap® HEX 230C	Tyfo® SCH-41, SCH-41S
Epoxy	Sikadur® HEX 306	Tyfo® SW-1
1st Layer	L (0°)	L (0°, SCH-41)
2nd Layer	L (0°)	L (0°, SCH-41S)
3rd Layer	C (90°)	C (90°, SCH-41S)

Legend: L = Longitudinal; C = Circumferential.

the laminates. Specimens 2 and 3 were cured for three weeks in the laboratory, at room temperature.

Specimens 4 and 5 (Table 3) were prepared and repaired in a similar manner to Specimens 2 and 3 described above, with the main difference being that Specimens 4 and 5 were wrapped and cured under artificial seawater. For this purpose, a water tank large enough to hold the two beam specimens was built out of lumber and plywood sheathing, and was insulated

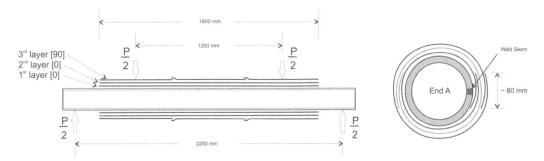

Figure 1. Stacking sequence of the CFRP layers.

with epoxy and polyethylene film to prevent water leakage. Specimens 4 and 5 were left to cure underwater for a period of 10 weeks before being removed from the tank.

To determine the suitability of the adhesives for underwater curing, another set of specimens was prepared. Specimens 6 and 7 (Table 3) were both wrapped using only one type of fibre: SikaWrap® Hex 230C fibres; however, for Specimen 6 the corresponding Sikadur® HEX 306 matrix was employed, whereas the fibres used on Specimen 7 were impregnated using the Fyfe Tyfo® SW-1 matrix, a resin specially-formulated for underwater curing. The wrapping sequence of Specimens 6 and 7 was identical to that described for specimens 4 and 5. In an attempt to improve adhesion of the confining layer, circumferential nylon straps were used around the wrapped tubes to ensure adequate bonding of the composite layers. The second set of underwater-cured specimens was left underwater for a period of eight weeks.

3.6 Experimental setup

The testing arrangement utilised a special-purpose beam jig within a 1,000 kN capacity universal testing machine equipped with a digital data acquisition system. The simply-supported beams were tested in a quasi-static manner under four-point bending until the specimens reached a maximum load and their load-deflection response was on a steady decline. The instrumentation provided during testing included linear variable differential transformers (LVDTs) which were provided to measure the tube displacement. For the first three specimens (i.e. Specimens 1, 2 and 3), three LVDTs were used, with two located at mid-span on each side of the beam to measure the average deflection at the specimens' mid-depth. The third LVDT was used to measure the typical settlement of the supports and hence the net mid-span deflection of the beam was determined. A picture of the Fyfe-wrapped Specimen 3 in the test apparatus is shown in Figure 2.

The second series of tests included Specimens 4 and 5. These specimens were provided with three

Figure 2. CFRP-wrapped beam under test (Specimen 3).

additional LVDTs; one was added to the second support to calculate its movement and the other two transducers were located longitudinally along the steel tube near the ends of the FRP laminate, on the tension side, at each end of the laminate. The latter LVDTs were installed to obtain a measurement of the relative slip between the composite layers and the steel tube for the underwater wrapped tubes. Finally, the third series of tests comprised Specimens 6 and 7 which were tested in a configuration identical to Specimens 4 and 5.

4 EXPERIMENTAL RESULTS

After the specimens were placed into the support apparatus and carefully centred to ensure symmetric loading, the beams were loaded to failure in a quasi-static manner. The load-deflection responses of the specimens tested in the first two stages (i.e. Specimens 1 to 5) are plotted in Figures 3 and 4, wherein the response of the wrapped specimens can be compared to that of the reference beam. The former figure shows the tubes wrapped with the Sika fibre/matrix system, in both standard or "in air" and underwater conditions, whereas the latter presents their counterparts

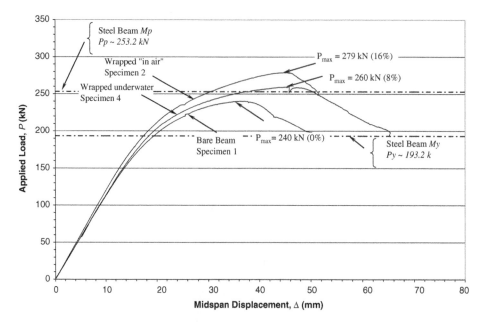

Figure 3.   Load-deflection response of tubular beams wrapped with Sika composite systems.

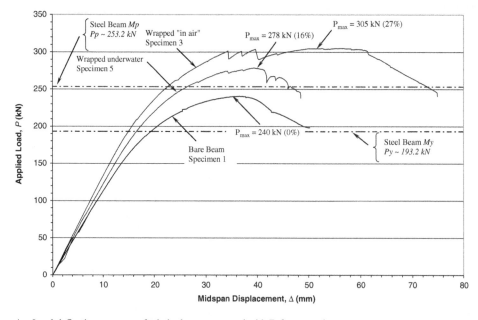

Figure 4.   Load-deflection response of tubular beams wrapped with Fyfe composite systems.

wrapped with the Fyfe CFRP system. It can be seen that, for both curing conditions, the tubes wrapped with Fyfe materials display greater flexural capacity than those wrapped with Sika materials. This was expected since the thickness of the Fyfe laminate, and hence the volume of fibres, was larger than that of the Sika laminate.

All specimens displayed an initially linear-elastic response until an approximate deflection of 15 mm was attained. Then, the beams exhibited an increased inelastic behaviour and, after the maximum load was reached, the specimens continued to deform in a ductile manner whilst sustaining a decreasing load. As expected, the wrapped beams exhibited an increase

Table 5. Experimental results of specimens tested in four-point bending (first and second sets of tests).

Specimen No.	Specimen identification	Elastic flexural stiffness $\times 10^{12}$ Nmm^2 (%)	Ultimate strength kN (%)	Deflection at ultimate load mm (%)
	Reference Beam			
1	R	1.56 (0)	240 (0)	36.6 (0)
	Specimens wrapped with Sika products			
2	Sika-S	1.68 (+7.4)	279 (+16.0)	45.1 (+23.2)
4	Sika-U	1.61 (+3.2)	260 (+8.0)	45.8 (+25.1)
	Specimens wrapped with Fyfe products			
3	Fyfe-S	1.84 (+17.7)	305 (+26.9)	58.1 (+58.7)
5	Fyfe-U	1.73 (+9.7)	278 (+16.0)	39.5 (+8.0)

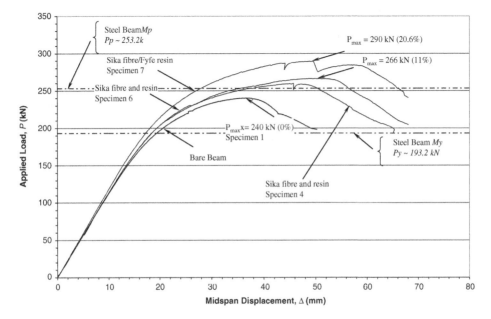

Figure 5. Load-deflection response of tubular beams wrapped with Sika fibres and adhesives from both suppliers.

in stiffness as well as ultimate strength compared to the reference beam. Nevertheless, the improvement was not as considerable for the specimens wrapped underwater, especially for the specimen wrapped with the Sika system. For both systems, the behaviour of the specimens wrapped underwater lay midway between the behaviour of the reference specimen and the specimens cured under standard conditions. The experimental values of the flexural stiffness, ultimate strength and mid-span deflection at ultimate load for all five specimens, as well as the percentage increases for the four wrapped tubular beams relative to the reference beam, are shown in Table 5. The reference beam did not quite reach the theoretical load, P_p, at which the section was expected to achieve the plastic moment (CAN/CSA S16-01 2001).

Figure 5 shows the behaviour of Specimens 6 and 7 which correspond to the third set of tests using Sika materials and adhesives from both suppliers, and it can be seen that the strength and stiffness of Specimen 7 outperformed those of Specimen 6. Of special interest is the fact that Specimen 7 (Sika/Fyfe-U-strapped), which was wrapped underwater, achieved a higher strength than that of standard Specimen 2 (Sika-S), which was repaired "in-air" using all Sika materials. Specimen 7 also outperformed Specimen5 (Fyfe-U), the underwater wrapped specimen using all Fyfe materials, achieving both a higher ultimate load and a more ductile behaviour. This suggests that the combination of Sika fibres with Fyfe underwater matrix may be better suited for seawater application. The "serrated" shape of the load-deflection

Table 6. Experimental results of specimens tested in four-point bending (third set of tests).

Specimen No.	Specimen identification	Elastic flexural stiffness $\times 10^{12}$ Nmm^2 (%)	Ultimate strength kN (%)	Deflection at ultimate load mm (%)
	Reference Beam			
1	R	1.56 (0)	240 (0)	36.6 (0)
	Specimen wrapped with Sika system			
6	Sika-U-strapped	1.6 (+2.7)	266 (+11.0)	50.8 (+39.0)
	Specimen wrapped with Sika fibres and Fyfe matrix			
7	Sika/Fyfe-U-strapped	1.67 (+7.0)	290 (+20.6)	49.4 (+35.0)

curves in the inelastic zone, which was more apparent when the thicker, more rigid Fyfe fibres were used (Figure 4), corresponds to fibre failures, especially on the compression side of the tube (due to buckling and debonding). Table 6 gives a summary of the responses of Specimens 6 and 7.

Figures 3 to 5 indicate that the presence of the CFRP composites extended the inelastic portion of the load-deformation curves, and hence increased the rotation capacity of the beams, relative to the reference beam. This suggests that a Class 2 section can be upgraded to a Class 1 section by CFRP wrapping, which is desirable for member rehabilitation.

The longitudinal slip displacements at both ends of the CFRP laminate varied between 5 and 10 mm. Slippage was minimal initially, then a sudden slip increment of about 0.1 mm occurred, followed by a gradual increase of slippage, as the test progressed. Load-slippage curves indicated that this phenomenon occurred gradually and was not a cause of failure. Inspection of the beams showed that the slippage was due to the peeling of the FRP over a localized area of the tension side (extending over a length of only 100 mm), probably due to a lack of local confinement of the longitudinal layers.

5 CONCLUSIONS

Hitherto, research concerning FRP has been mainly oriented towards strengthening concrete structures. In the present work, the rehabilitation of steel tubular structures with CFRP has been explored, assessing experimentally the possibility of rehabilitating tubular steel flexural members, with emphasis on underwater applications. Seven tubular steel beams were tested in four-point bending. Six tubes were wrapped with CFRP composites from two different manufacturers and the remaining beam was left bare as a reference specimen. The flexural performance of the wrapped tubes was studied for two principal performance parameters: the influence of the fibre/epoxy manufacturer and the curing conditions of the matrix ("in-air" versus underwater).

In general, the ultimate bending strength, flexural stiffness and rotation capacity of the wrapped beams, relative to the reference beam, were increased for both performance parameters under study; however, the composite members wrapped and cured underwater were not able to attain the flexural capacity of those cured in air. For the tubes wrapped and cured in air, ultimate strength increases of 16 and 27% relative to the bare steel beam were achieved, along with increases of 7 and 18% in flexural stiffness. The Fyfe-wrapped specimen displayed higher strength and stiffness improvements than its Sika-wrapped counterpart, mainly because of the increased thickness of the CFRP sheets from the former supplier. When wrapping and curing were performed under seawater, ultimate strength increases for the beams using materials from the two manufacturers were in the range of 8 to 21%. This was accompanied by lower increases in flexural stiffness and rotation capacity, relative to the specimens wrapped and cured in air. No serious debonding problems were found in any of the specimens, suggesting that the fibres bonded adequately regardless of the application and curing conditions involved.

Of special interest was one underwater-wrapped and cured specimen which employed Sika fibres with Fyfe's specially-formulated resin for underwater applications. This was done in an attempt to evaluate the suitability of hybrid materials, especially because the Sika resin appeared not to cure completely in seawater and the Fyfe fibres proved to be too heavy and difficult to manipulate underwater. Indeed, the resulting ultimate strength and flexural stiffness of this specimen were 12 and 4% larger, respectively, than those of the specimen that used all Sika materials. In fact, this hybrid specimen outperformed all other specimens wrapped underwater, and even the specimen wrapped in air with all Sika materials, which confirms that the selection of the matrix is of utmost importance.

Finally, the steel reference beam was a Class 2 section in bending and, although it was expected to reach the plastic moment, it only reached 94% of it. However, all CFRP-wrapped specimens did reach the plastic moment with increased ductility and rotation capacity. It can be concluded that the use of CFRPs to enhance the strength of tubular steel flexural members in air and underwater is feasible.

ACKNOWLEDGEMENTS

Financial support for this project was provided by Science and Engineering Research Canada (NSERC) through an Industrial R&D Fellowship, and by Sika Canada Inc. Materials used in this study were provided by Sika Canada Inc., Fyfe Co. (U.S.A.) and Ipsco Inc. (Canada), and this support is appreciated.

REFERENCES

ACI Committee 440. 2002. *Guide for the design and construction of externally bonded FRP systems for strengthening concrete structures.* ACI 440.2R-02. Farmington Hills, MI: American Concrete Institute.

Al-Saidy, A.H., Klaiber, F.W. & Wipf, T.J. 2004. Repair of steel composite beams with carbon fiber-reinforced polymer plates. *ASCE Journal of Composites for Construction* 8(2): 163–172.

ASTM D3039/D3039M-00. 2000. *Standard test method for tensile properties of polymer matrix composite materials.* West Conshohocken, PA: ASTM.

ASTM E8M-04. 2004. *Standard test methods for tension testing of metallic materials.* West Conshohocken, PA: ASTM.

Bakis, C.E., Bank, L.C., Brown, V.L., Cosenza, E., Davalos, J.F., Lesko, J.J., Machida, A., Rizkalla, S.H. & Triantafillou, T.C. 2002. Fiber-reinforced polymer composites for construction – State-of-the-art review. *ASCE Journal of Composites for Construction* 6(2): 73–87.

Barbero, E.J. Construction. 1998. In Peters ST (Ed.), *Handbook of composites, 2nd Edition.* London, UK: Chapman & Hall.

Bell, S.A.H. 2004. *Feasibility of carbon fibre reinforced polymers for the rehabilitation of steel structures.* Master of Engineering thesis, Toronto, ON: University of Toronto.

Buyukozturk, O., Gunes, O. & Karaca, E. 2004. Progress on understanding debonding problems in reinforced concrete and Steel Members Strengthened using FRP Composites, *Construction and Building Materials (Elsevier)* 18(1): 9–19.

CAN/CSA G40.20-04/G40.21-04. 2004. *General requirements for rolled or welded structural quality steel/structural quality steel.* Toronto, ON: Canadian Standards Association.

CAN/CSA S16-01. 2001. *Limit states design of steel structures.* Toronto, ON: Canadian Standards Association, Canada.

Fawzia, S., Zhao, X.-L., Al-Mahaidi, R. & Rizkalla, S. 2005. Bond characteristics between CFRP and steel plates in double strap joints, *Advances in Steel Construction – An International Journal* 1(2): 17–28.

Galambos, T.V. 1998. *Guide to stability design criteria for metal structures, 4th Edition.* New York, NY: John Wiley & Sons Inc.

Garden, H.N. 2004. Use of advanced composites in civil engineering infrastructure, *Structures and Buildings* 157(6): 357–368.

Hollaway, L.C. 1993. *Polymer composites for civil and structural engineering, 1st Edition.* Glasgow, UK: Blackie Academic and Professional.

Hollaway, L.C. 1994. *Handbook of polymer composites for engineers, 1st Edition.* Cambridge, UK: Woodhead Publishing Limited.

Hollaway, L.C. & Head, P.R. 2001. *Advanced polymer composites and polymers in the civil infrastructure, 1st Edition.* Oxford, UK: Elsevier

Jiao, H. & Zhao X.-L. 2004. CFRP strengthened butt-welded very high strength (VHS) circular steel tubes. *Thin-Walled Structures* 42(7): 963–978.

Mertz, D.R. & Gillespie, J.W. 1996. *Rehabilitation of steel bridge girders through the application of composite materials, NCHRP Rep. No. 93-ID11:* 1–20. Washington, D.C.: Transportation Research Board.

Miller, T.C., Chajes, M.J., Mertz, D.R. & Hastings, J.N. 2001. Strengthening of a steel bridge girder using CFRP plates. *ASCE Journal of Bridge Engineering* 6(6): 514–522.

Schnerch, D. & Rizkalla, S. 2004. *Strengthening of scaled steel-concrete composite girders and steel monopole towers with CFRP, North Carolina State University Report.* Raleigh, NC.

Schwartz, M.M. 1997. *Composite materials, Volume I: Properties, nondestructive testing, and repair, 1st Edition.* Upper Saddle River, NJ: Prentice Hall.

Sen, R. & Liby, L. 1994. *Repair of steel composite bridge sections using CFRP laminates, U.S. Department of Transportation Contract B-7932.* Tampa, FL: University of South Florida,

Tavakkolizadeh, M. & Saadatmanesh, H. 2003a. Strengthening of steel-concrete composite girders using carbon fiber reinforced polymer sheets, *ASCE Journal of Structural Engineering* 129(1): 30–40.

Tavakkolizadeh, M & Saadatmanesh, H. 2003b. Repair of damaged steel-concrete composite girders using carbon fiber-reinforced polymer sheets, *ASCE Journal of Composites for Construction* 7(4): 311–322.

Teng, J.G. & Hu, Y.M. 2004. Suppression of local buckling in steel tubes by FRP jacketing. In: R. Seracino (Ed.), *FRP Composites in Civil Engineering*: 749–753. Tayor & Francis Group, London.

Tubular Structures XI – Packer & Willibald (eds)
© 2006 Taylor & Francis Group, London, ISBN 0-415-40280-8

Numerical modeling of FRP-strengthened long HSS columns

A. Shaat & A. Fam
Queen's University, Kingston, Ontario, Canada

ABSTRACT: This paper presents results of a finite element modeling (FEM) of axially loaded long hollow structural section (HSS) steel columns, strengthened using carbon-fiber reinforced polymer (CFRP) sheet applied in the longitudinal direction. The model provides a lower bound solution as it ignores the contribution of CFRP on the inward buckling face of the column, only at mid-height section, throughout the loading history. The model was verified using experimental results. The non-linear FEM predicted the ultimate loads and failure modes of the columns quite reasonably. The study showed that using 7 layers of 0.54 mm thick CFRP sheets bonded on two opposite sides of an 89 × 89 × 3.2 HSS with 68 slenderness ratio has increased the strength by 22 percent. The model was then used in a parametric study, which showed that the strength gain is significantly affected by the slenderness ratio of the column.

1 INTRODUCTION

Finite element analysis (FEA) of cold-formed structures plays an increasingly important role in modeling, as it is relatively inexpensive and time efficient compared to laboratory tests. Furthermore, it is often difficult to investigate the effects of geometric imperfections and residual stresses of structural members experimentally. In general, FEA is a powerful tool in predicting the ultimate loads and complex failure modes of cold-formed structural members. In addition, geometric imperfections, residual stresses and material non-linearity can easily be included in the FEM.

Steel columns of medium to high slenderness ratios are commonly used in steel structures and are generally governed by overall buckling failure before developing their full plastic capacity. The use of fiber-reinforced polymer (FRP) sheets or strips of adequate stiffness could be quite suitable in such applications, particularly due to their ease of application, excellent corrosion resistance, and fatigue properties (Shaat et al. 2004).

An experimental study was performed by Shaat & Fam (2005) to investigate the effect of adhesively bonded high modulus CFRP longitudinal sheets on the behavior of axially loaded slender HSS columns. It was shown experimentally that CFRP sheets have indeed increased the columns' strengths by up to 23 percent. The study, however, revealed the sensitivity of the axially loaded columns to their inherent geometric out-of-straightness and alignment (imperfections), which affects both the ultimate strength

and lateral displacement of the specimens. Residual stresses are also an important factor, which can affect the behavior of cold-formed columns. While the major parameter intended in the experimental investigation was the effect of the number of CFRP layers, it is believed that the geometric imperfections have also varied inevitably among the specimens. Therefore, no specific correlation could be established experimentally between the amount of strength gain and the amount of CFRP. As such, it was decided to use FEA to model the specimens and isolate the effect of geometric imperfections from the effect of number of CFRP layers, in order to provide an accurate assessment of the strengthening effectiveness of CFRP. The FEA approach was employed by several other researchers to model steel structures retrofitted by FRP materials, mainly for steel girders, and showed excellent results (Sen et al. 2001, Abushaggur and Eldamatty 2003, and Jun Deng et al. 2004). The FEM used in this study was verified against the column test results. As the experimental program was limited to a small number of specimens, the FEM was used to conduct a parametric study to investigate the effects of the number of CFRP layers and their distribution, the initial geometric imperfections and the columns' slenderness ratios.

This paper first summarizes the experimental research program and then a discussion of various aspects of the FEA is presented. This is followed by the results of the FEA in terms of load-displacement curves as well as deformed shapes and failure modes. Finally the parametric study and conclusions are presented.

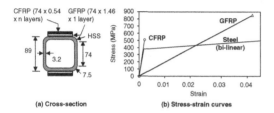

(a) Cross-section (b) Stress-strain curves

Figure 1. Test specimens and material properties.

2 SUMMARY OF EXPERIMENTS

A brief summary of the experimental study is presented in this section, whereas more details can be found elsewhere (Shaat & Fam, in press). The experimental study included 5 slender column tests, conducted on a standard $89 \times 89 \times 3.2$ mm HSS section, as shown in Figure 1(a), with yield strength, F_y, of 380 MPa. The length of the pin-ended columns was 2380 mm, which corresponds to a slenderness ratio (kL/r) of 68. Ultra-high modulus unidirectional carbon fibre sheets were bonded to the specimens in the longitudinal direction. A single composite layer was 0.54 mm thick and had a tensile strength and modulus of 510 MPa and 230 GPa, respectively. A layer of glass-FRP (GFRP) sheet was first installed directly on the steel surface before applying the CFRP layers to prevent direct contact between carbon fibres and steel, which could lead to galvanic corrosion. The GFRP lamina was 1.46 mm thick and had a tensile strength and modulus of 855 MPa and 20 GPa, respectively. The stress-strain curves of steel, CFRP, and GFRP are illustrated in Figure 1(b).

The tested columns included a control (unstrengthened) specimen and three specimens strengthened with one, three and five layers of CFRP, applied on two opposite sides in the plane of overall buckling. The fifth specimen was strengthened with three layers, applied to all four sides of the column. The specimens were given identification codes. For example, 3L-2S indicates three CFRP layers applied on two opposite sides of the column.

The experimental curves of load versus both axial displacement and lateral displacement at mid-length of all specimens are shown in Figures 2 and 3, along with the FEM curves, which will be discussed later.

The gain in axial strength of the CFRP-strengthened specimens ranged from 13 to 23 percent as shown in Table 1. The strength gains, however, did not correlate directly to the number of CFRP layers. As indicated earlier, this was attributed to the variability of geometric imperfections among the specimens, which is usually due to either a slight out of straightness of different values among the specimens (Kulak & Grondin 2002), or minor inevitable misalignments within the test setup, or a combination of both.

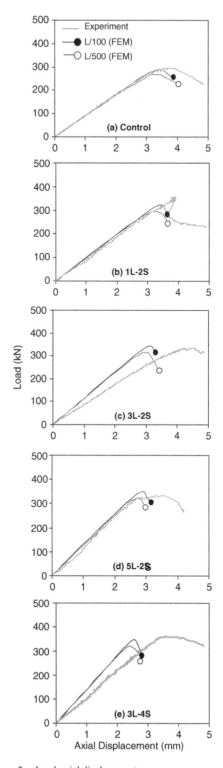

Figure 2. Load-axial displacement responses.

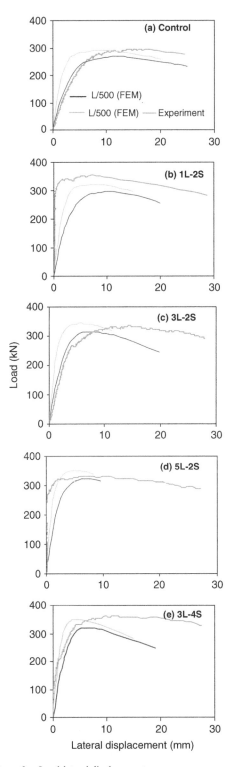

Figure 3. Load-lateral displacement responses.

Table 1. Summary of experimental results.

Specimen I.D.	Maximum load P_{max} (kN)	%age gain in strength	Lateral displacement at P_{max} (mm)
Control	295	—	14.22
1L-2S	355	20	7.17
3L-2S	335	14	14.57
5L-2S	332	13	10.66
3L-4S	362	23	11.26

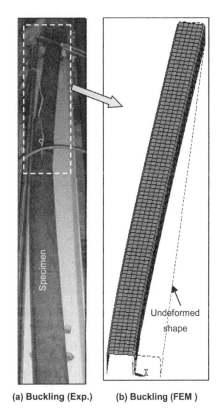

(a) Buckling (Exp.)   (b) Buckling (FEM)

Figure 4. Comparison between deformed shapes.

In all specimens, failure was mainly due to excessive overall buckling of the specimen, as shown in Figure 4(a), followed by a secondary local buckling in the compression side, at or near mid-length of the specimen. For the FRP-strengthened specimens, the secondary local buckling in the compression side was associated with a combined delamination and premature crushing of the FRP sheets.

3 FINITE ELEMENT MODEL

The non-linear finite element analysis program ANSYS was used to model the behavior, ultimate load,

377

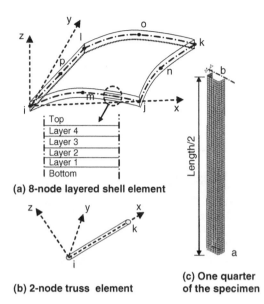

(a) 8-node layered shell element

Top
Layer 4
Layer 3
Layer 2
Layer 1
Bottom

Length/2

(b) 2-node truss element

(c) One quarter of the specimen

Figure 5. Elements types and meshing.

and failure mode of pin-ended HSS long steel columns. The analysis incorporated both geometric and material non-linearities. The centerline dimensions of the cross-sections and the base metal thickness were used in the geometric modeling, based on the measured cross-sectional dimensions of the specimens. The following sections briefly address various aspects of the finite element modeling, such as element type, mesh density, boundary conditions, material properties, geometric imperfections, and residual stresses.

3.1 Elements types and meshing

An eight-node quadrilateral layered shell element (SHELL91) was used for the steel section in this model. The element configurations as well as its coordinate system are shown in Figure 5(a). Each node has six degrees of freedom, namely three translations (U_x, U_y, and U_z) and three rotations (R_x, R_y, and R_z). The multiple layers of the element were utilized to account for the residual stress distribution through the steel wall thickness as will be discussed later in detail. The FRP reinforcing material was modeled using a three-dimensional two-node uniaxial truss element (LINK8), as shown in Figure 5(b). This was deemed appropriate as the thin CFRP layers have insignificant flexural rigidity. Five adjacent longitudinal link elements were modeled parallel to the carbon fiber directions on each strengthened side of the HSS. Each node of LINK 8 element has three degrees of freedom, namely translations in the nodal x, y, and z directions (U_x, U_y, and U_z). A perfect bond between steel and FRP sheets was assumed by defining one node for both SHELL91 and LINK8 elements having

the same coordinates. The stiffness of the CFRP elements on the compression (inward buckling) side, at mid-height section only was set to a very small value (10^{-06}) through out the entire loading history to simulate crushing and debonding of FRP sheets. Therefore, the model represents a lower bound solution.

One quarter of the specimen was modeled, as shown in Figure 5(c), by taking advantage of the double symmetry of the column. Based on the results of a typical procedure of mesh refining, the model shown in Figure 5(c) with 1050 elements is used in all the analyses that followed.

3.2 Loading and boundary conditions

The FEM simulated only one quarter of the specimen by introducing two planes of symmetry, one vertical plane in the longitudinal direction along the full length of the column and another horizontal plane in the transverse direction at column's mid-length (end b, Fig. 5(c)). The two hinged ends of the actual columns were simulated using a thick plate at the model's end (a) with a middle line restrained from translation in the transverse plane. The translational degrees of freedom of this line in the longitudinal direction of the column were released to allow for applying axial loads. The rest of the nodes were free to translate and rotate in all directions.

3.3 Geometric imperfections

As was concluded in the experimental study, the variation of geometric imperfections among different specimens led to inconsistency of results, which resulted in some difficulties in assessing the effect of FRP strengthening technique. Therefore, the effect of two different geometric imperfections, introduced in the FEM at mid-length, was studied. The values of imperfection were Length/500, which is the maximum value permitted by the Canadian code CAN/CSA-S16.1-01, and 50 percent of this value (Length/1000).

3.4 Material properties

The material non-linear properties of steel were incorporated in the FEM by specifying the bi-linear isotropic hardening model, shown in Figure 1(b). Values of Young's modulus and Poisson's ratio of steel were defined as 200 GPa and 0.30, respectively. The tangent modulus for the steel was assumed equal to 0.5 percent of its elastic modulus (Bruneau et al. 1998). For FRP, unidirectional elastic properties were assigned, namely Young's moduli of 20 GPa and 230 GPa for glass- and carbon-FRP, respectively.

3.5 Residual stresses

Residual stresses play an important role in the behavior of steel structures and are normally induced during

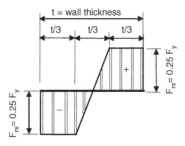

t = wall thickness

Figure 6. Through-thickness residual stress distribution.

the manufacturing process. They typically result in a reduction of the flexural rigidity of slender columns, and consequently, a lower buckling load may result (Weng 1984). Although residual stresses are self-equilibrating, the cross sectional effective moment of inertia will be changed when parts of the section reach their yielding strength. Yielding is initiated first in those portions of the cross section having large compressive residual stresses. An extensive experimental investigation of the residual stresses of hollow structural cold-formed steel sections was performed by Davison and Birkemoe (1983) and Weng and Pekoz (1990). The magnitudes of the measured residual stresses were found to vary, approximately from 25 to 70 percent of the material yield strength, depending on the manufacturing process.

In the proposed FEM, the through-thickness residual stress distribution is idealized as shown in the schematic drawing in Figure 6, as suggested by Davison and Birkemoe (1983) and by Chan et al. (1991), where F_{rs} equals 25 percent of F_y.

4 VERIFICATION OF FEM RESULTS

In this section, comparisons between the experimental results and FEM predictions are presented, as shown in Figures 2 and 3 for the five tested specimens, in terms of both the axial load versus axial and lateral displacements, respectively. The FEM results agree quite well with the experimental results. The relatively higher experimental axial displacements of specimens 3L-2S and 3L-4S are likely due to an unintended slight misalignment of the specimen in the test setup. Also, the very small initial (pre-buckling) lateral displacements of specimens 1L-2S and 5L-2S in the experimental responses are likely due to a nonsymmetrical imperfect shape of the specimens initially, where the measured lateral displacements at mid-length was not necessarily the maximum values of displacement along the length, unlike the FEM results.

The failure mode predicted by the FEM, which is an overall buckling as shown in Figure 4(b), is quite similar to the buckling failure mode observed in the tests (Fig. 4(a)).

Table 2. Summary of experimental results.

Specimen I.D.	ω	Max. load P_{max} (kN)	%gain in strength
68-1000-control	0	291	—
68-1000-1L-2S	0.10	322	11
68-1000-3L-2S	0.26	344	18
68-1000-5L-2S	0.42	350	20
68-1000-7L-2S	0.58	353	21
68-1000-3L-4S	0.52	348	20
68-500-control	0	269	—
68-500-1L-2S	0.10	296	10
68-500-3L-2S	0.26	316	17
68-500-5L-2S	0.42	323	20
68-500-7L-2S	0.58	327	22
68-500-1L-4S	0.20	301	12
68-500-3L-4S	0.52	320	19
68-500-5L-4S	0.84	328	22
68-500-7L-4S	1.16	333	24
90-500-control	0	194	—
90-500-1L-2S	0.10	222	14
90-500-3L-2S	0.26	253	30
90-500-5L-2S	0.42	270	39
90-500-7L-2S	0.58	281	45
160-500-control	0	78	—
160-500-1L-2S	0.10	91	17
160-500-3L-2S	0.26	109	40
160-500-5L-2S	0.42	126	62
160-500-7L-2S	0.58	142	82

5 PARAMETRIC STUDY

The FEM has been used in a parametric study to examine the effects of several parameters on the behavior of HSS steel columns strengthened with CFRP sheets. A total of 25 HSS columns with the same cross sectional dimensions and material properties as those used in the experiments were analyzed. The parameters considered were the number of CFRP layers (1, 3, 5, and 7), distribution of layers (on 2 or 4 sides), geometric imperfection (Length/500 and Length/1000) and columns' slenderness ratios (kL/r = 68, 90, and 160).

The following identification coding system was adopted to distinguish the various cases. The first number represents the slenderness ratio (kL/r), while the second number describes the geometric imperfection. These two numbers are followed by two "number-letter" combinations such as "3L" to identify the number of CFRP layers used and "2S" to indicate the number of strengthened sides. For control columns, the last two "number-letter" combinations are replaced by the word "control". For example, "68-500-5L-2S" describes a strengthened column that has a slenderness ratio of 68, a geometric imperfection of (Length/500), and strengthened with five layers of CFRP applied to two opposite sides of the cross section.

A summary of the FEM results, including the columns strength and their percentage increases compared to the control columns is presented in Table 2.

Figure 7 shows the effect of each parameter on the percentage increases of columns strength. To facilitate the comparisons, a reinforcement index ω is introduced to quantify the amount of FRP reinforcement on the basis of a relative axial stiffness, as given by the following Equation.

$$\omega = \frac{\sum\limits_{i=1}^{n}\left[E_{f_i} A_{f_i} \right]}{E_s A_s} \qquad (1)$$

where E_{fi} = Young's modulus of layer i of FRP ; A_{fi} = area of FRP layer i ; E_s = Young's modulus of steel ; and A_s = cross sectional area of steel section.

5.1 Effect of number of CFRP layers

Table 2 clearly shows that increasing the number of CFRP layers results in increasing the axial strength. As the number of layers increases from one to seven layers, applied on two opposite sides, the column's strength increased from 10 to 22 percent for columns with $kL/r = 68$, from 14 to 45 percent for columns with $kL/r = 90$ and from 17 to 82 percent for columns with $kL/r = 160$. This effect is also demonstrated in Figure 7, where the number of layers is reflected by the reinforcement index (ω).

5.2 Effect of distribution of CFRP layers

The overall buckling behavior of long columns is essentially a flexural problem, in which increasing the moment of inertia of the column will increase its strength. As such, adding FRP layers on two opposite sides of the columns, in the plane of overall buckling is expected to be more efficient than distributing the layers on all four sides. This is clearly reflected in Figure 7(a), where the specimens with FRP layers on two opposite sides showed higher axial strength than those strengthened on all four sides, at any given reinforcement index.

5.3 Effect of initial imperfection (e′)

The initial imperfection of columns greatly affects their axial strength. For example, Table 2 shows strengths of 269 kN for specimen 68-500-control, and 291 kN for specimen 68-1000-control, with imperfections of L/500 and L/1000, respectively. Although both columns have the same length and cross sections they had an 8 percent difference in strength. This emphasizes the crucial role of the initial imperfection in comparisons between columns.

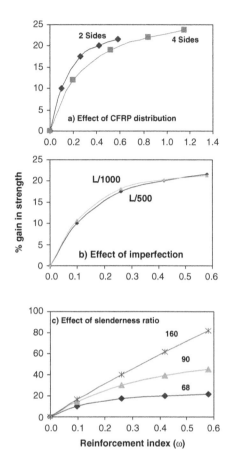

Figure 7. Parametric study.

While the initial imperfection affects the columns' maximum strength, the percentage increase in strength, of the same column, due to FRP reinforcement is quite similar for both investigated levels of imperfection, when each strengthened specimen is compared to its counterpart control specimen, as shown in Figure 7(b).

5.4 Effect of column's slenderness ratio (kL/r)

The effect of the column slenderness ratio was studied by varying the specimens' length (L = 2380 mm, 3150 mm, and 5600 mm), to provide slenderness ratios (kL/r) equal to 68, 90, and 160, respectively. Figure 7(c) shows that the slenderness ratio has a pronounced effect on the effectiveness of the CFRP-strengthening system. For the same CFRP reinforcement index, the strength gain increases substantially as the slenderness ratio is increased. Also, the rate of increase for smaller slenderness ratios (kL/r = 68) is quite non-linear and is also reduced beyond certain number of CFRP layers. On the other hand, for higher slenderness ratios (kL/r = 160), the strength gain becomes almost linearly proportional to the reinforcement index.

6 CONCLUSIONS

In this paper, a numerical investigation has been presented on the behavior of long HSS steel columns strengthened with longitudinal CFRP sheets. A finite element model (FEM) accounting for geometric and material non-linearities has been developed and verified against experimental results. A parametric study has been performed to study the effects of number of CFRP layers, distribution of CFRP, initial imperfection of the steel columns, and the columns' slenderness ratio. The following conclusions are drawn:

- Strengthening of long HSS steel columns using adhesively bonded ultra-high modulus CFRP sheets is a quite effective technique to increase the column's axial strength.
- The strength of control or strengthened long HSS columns is sensitive to the initial inherent imperfections of the column.
- The higher the slenderness ratio of the column, the higher the gain in axial strength due to CFRP strengthening for the same amount of CFRP. For moderate slenderness ratios (68 and 90 for example), there appears to be a certain CFRP reinforcement ratio, beyond which not much gain in strength can be expected. On the other hand, for very high slenderness ratios (160 for example), the gain in strength continuously increases by adding more CFRP layers.
- Bonding CFRP layers to two opposite sides of HSS long column, in the plane of buckling is more efficient than bonding CFRP layers to all four sides.

7 FUTURE WORK

The model will be further enhanced to account for the effect of the progressive failure of the CFRP sheets on the inward buckling side of the column.

ACKNOWLEDGMENT

The authors wish to acknowledge the financial support provided by the Natural Sciences and Engineering Research Council of Canada (NSERC).

REFERENCES

Abushaggur, M., and El Damatty, A. A. 2003. Testing of steel sections retrofitted using FRP sheets. *Annual Conference of the Canadian Society for Civil Engineering, June, 4–7*, Moncton, Nouveau-Brunswick, Canada. (CD-ROM).

ANSYS Inc. ANSYS user's manual, revision 7.1.

Bruneau, M., Uang, C., and Whittaker, A. 1998. *Ductile Design of Steel Structures*. New York. McGraw-Hill.

Canadian Standards Association, CAN/CSA-S16.1-01, *Limit States Design of Steel structures*. Mississauga, Ontario.

Chan, S.L., Kitipornchai, S., and Al-Bermani, F.G.A. 1991. Elasto-plastic analysis of box-columns including local buckling effects. *Journal of Structural Engineering. ASCE.* 117(7): 1946–1962.

Davison, T. A., and Birkemoe, P. C. 1983. Column behaviour of cold-formed hollow structural steel shapes. *Canadian Journal of Civil Engineering.* 10: 125–141.

Jun Deng, Lee M.M.K., and Moy, S.S.J. 2004. Stress analysis of steel beams reinforced with a bonded CFRP plate. *Composite Structures.* 65: 205–215.

Kulak, G.L., and Grondin, G.Y. 2002. *Limit states design in structural steel*. Toronto. Canadian Institute of Steel Construction.

Sen, R., Liby, L., and Mullins, G. 2001. Strengthening steel bridge sections using CFRP laminates. *Composites Part B: engineering.* 32: 309–322.

Shaat, A., Schnerch, D., Fam, A., and Rizkalla, S. 2004. Retrofit of steel structures using fiber-reinforced polymers (FRP): State-of-the-art. *Transportation Research Board (TRB) Annual Meeting. January, 11–15*, Washington. D.C. (CD-ROM-04-4063).

Shaat, A., and Fam, A. 2005. Long HSS columns strengthened using CFRP sheets. *Third International Conference on Construction Materials: Performance, Innovations and Structural Implications (ConMat'05), August 22–24*, Vancouver, Canada.

Weng, C.C. 1984. *Cold-bending of thick steel plates at low R/t ratios*. Master's thesis. Cornell University, at Ithaca, N.Y.

Weng, C.C., and Pekoz, T. 1990. Residual stresses in cold-formed steel members. *Journal of Structural Engineering. ASCE.* 116(6): 1611–1625.

Tubular Structures XI – Packer & Willibald (eds)
© 2006 Taylor & Francis Group, London, ISBN 0-415-40280-8

Mechanical property of concrete filled CFRP-steel tube under uni-axial compression

Y.H. Zhao

Institute of Road and Bridge Engineering, Dalian Maritime University, Dalian, China

Q.Q. Ni

Department of Functional Mechinery & Mechanics, Shishu University, Japan

ABSTRACT: Based on the framework of mechanics of composite materials and the theory of elastic-plasticity, an analytical solution of elastic-plastic responses is developed for a composite system of concrete filled CFRP-steel tube. Under a uni-axial compressive loading the average stresses of 3 constituent materials and the overall effective secant Young's modulus are derived. It is shown that the initial yield stress and ultimate stress of the system is enhanced significantly by bonding with CFRP comparing to the traditional concrete filled steel tube. Also an explicit equation is provided to express the connecting condition of steel tube and concrete core, which shows a remarkable improvement for the confinement of steel tube to concrete.

1 INTRODUCTION

Concrete filled CFRP-steel tube is made by filling concrete into steel tube and wrapping around outside with fiber reinforced polymer (CFRP) sheet. This system is newly proposed to be used as the columns in buildings, bridges and other structures sustaining mainly compressive loadings. The motivation of the suggestion of this new structure is from the traditional concrete filled steel tube system (Han 2000). Concrete filled steel tube offers some distinguished properties which makes it very competitive when used as columns comparing to the ordinary reinforced concrete column (Cai 2003). So it gets broader usage recently in the field of civil engineering. However they are rarely found on underwater structures because of the problem on corrosion resistance of steel tube. For the purpose of gaining even more benefits of this structure while improving its resistance against the environmental corrosion, a new composite column – concrete filled CFRP-steel tube (Wang & Zhao 2003) is suggested.

In the last few years some significant researches (Wang & Zhao 2003, Gu & Zhao, in press, Gu et al. 2004) were preformed for investigating the possibility of this new system. On the earliest work some short cylinders were tested for getting the ultimate loads (Gu et al. 2004). The columns were assumed to be the axial-symmetrical ones under simple compression. The analytical solution based on the method of equilibrium ultimate showed a nice agreement with

the experimental data (Gu & Zhao, in press). Then much more specimens including some longer ones were made for the experimental study on the ability of carrying loads as well as the stability. And manifold loads such as tension and bending and multi-load were applied on tests. Sufficient information from the preliminary work showed the evident improvement of the mechanical properties of the new system comparing with that without CFRP. However the theoretical study is still in need for getting more details and clearer knowledge from the system. And the designers also need the convictive methods to predict the performance of the new system in advance.

This paper is concerned with the determination of the overall elastoplastic behavior for concrete filled CFRP-steel tube under uni-axial compression. Based upon the theoretical framework of mechanics of composite materials and the theory of elastic- plasticity, a series of explicit equations are derived to express the stress-strain response of the constituent materials as well as the overall system. A so called interface connect criterion of steel tube and concrete are also proposed to open up the interface behavior that is usually hardly gained from experiments.

One novel feature of the theory is that despite the nonlinear responses of steel and concrete, the overall stress-strain relation of the system can be readily generated from those of its constituents without any iteration. Thus under a monotonic proportional loading this approach appears to offer simplicity, and is capable of capturing the essential features of the mechanical response for this complicated system.

2 CONSTITUENT PROPERTIES AND ASSUMPTIONS

Let us first specify the material properties of three constituents of the system.

2.1 Concrete core

The property of the concrete under a tri-axial compression is assumed to be nonlinear elastic, following Ottosen's stress-strain response. The secant Young's modulus and the Poison's ratio are given as (Ottosen 1979)

$$E_c = \frac{E_0}{2} - \beta\left(\frac{E_0}{2} - E_{ff}\right) + \sqrt{\left[\frac{E_0}{2} - \beta\left(\frac{E_0}{2} - E_{ff}\right)\right]^2 - \beta E_{ff}^2}$$

$$v_c = 0.36 - (0.36 - v_0)\sqrt{1 - (5\beta - 4)^2} \ (\beta \geq 0.8) \quad (1)$$

In which, E_0, v_0 and E_{ff} are the initial Young's modulus, Poison's ratio and the secant modulus at failure. $\beta = \sigma_3/\sigma_{3f}$ is the nonlinearity index, with σ_3 the actual most compressive principal stress and σ_{3f} the corresponding failure value.

2.2 Steel tube

The steel is taken as elastoplastic, homogeneous, and isotropic material. Under a monotonic, proportional loading, the stress and plastic strain relation is expressed by

$$\sigma_e = \sigma_y + H\left(\varepsilon_e^p\right)^n \quad (2)$$

Where σ_y, H and n are the initial yield stress, the strength coefficient, and the work-hardening exponent, respectively. Following the von Mises' theory, the effective stress and effective plastic strain are given by

$$\sigma_e = \left(\frac{3}{2}\sigma_{ij}'\sigma_{ij}'\right)^{1/2}, \quad \varepsilon_e^p = \left(\frac{2}{3}\varepsilon_{ij}^p\varepsilon_{ij}^{p'}\right)^{1/2}, \quad (3)$$

Taking the total strain to be the sum of the elastic and plastic components, the secant Young's modulus and secant Poison's ratio of the steel are

$$E_s = 1/\left[\frac{1}{E_0} + \frac{\left(\sigma_e - \sigma_y\right)^{1/n}}{H^{1/n}\sigma_e}\right], \quad v_s = \frac{1}{2} - \left(\frac{1}{2} - v_0\right)\frac{E_s}{E_0} \quad (4)$$

the subscript s denotes the steel.

2.3 CFRP tube

CFRP tube is made by immerging carbon fiber sheet into polymer mucus. This, as is usually the case, is taken to be elastic, and sustains only tensile stress in fiber direction because its thickness is ignorable comparing to the diameter of the system. The stress-strain relation in fiber direction follows Hooke's Law as

$$\varepsilon_f = E_f \sigma_f \quad (5)$$

where, takes the subscript f as CFRP.

3 STRESSES OF CONSTITUTENTS

The schematic of Concrete filled CFRP steel system is show in figure 1. When the system is subjected to a uni-axial compression N in the longitudinal direction, a weight means is given by

$$N = \sigma_{zc}A_c + \sigma_{zs}A_s \quad (6)$$

where σ_{zc}, σ_{zs} are the longitudinal stresses of concrete and steel respectively, and A_c, A_s are cross section areas of the concrete core and the steel tube. The stress states of the concrete cylinder, steel tube and CFRP tube are expressed in figure 1a, b, &c respectively. For the possibility of explicit solutions, take the interfaces of both steel-concrete and steel-CFRP as the perfectly bonded ones. Both steel and CFRP tube are considered as thin-walled tubes, with the thickness t_s and t_f. For overcoming the difficulties caused by the nonlinear responses of the hybrid materials, apply the theory of secant moduli for nonlinear analysis. So that the stress-strain relations of respective phases, as well as the whole system can be easily expressed by Hooke's Law through out the loading process.

Under a tri-axial compression the radial and circumferential stresses of concrete cylinder can be written in terms of interfacial stress as

$$\sigma_{rc} = \sigma_{\theta c} = -q_c \quad (7)$$

The corresponding strains are given in terms of secant modulus and Poison's ratio as

$$\varepsilon_{\theta c} = \frac{1}{E_c}\left[(1 - v_c)\sigma_{rc} - v_c\sigma_{zc}\right] \quad (8)$$

$$\varepsilon_{zc} = \frac{1}{E_c}\left[\sigma_{zc} - 2v_c\sigma_{rc}\right] \quad (9)$$

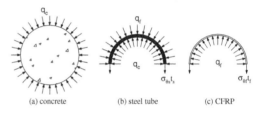

(a) concrete　　(b) steel tube　　(c) CFRP

Figure 1. Schematic diagram of stress state of concrete filled CFRP-steel tube system.

where the secant Young's modulus and Poison's ratio can be found in (1).

The stress-strain relations of a thin-walled steel tube are given by

$$\varepsilon_{\theta s} = \frac{1}{E_s}\left[\sigma_{\theta s} - v_s\sigma_{zs}\right] \tag{10}$$

$$\varepsilon_{zs} = \frac{1}{E_s}\left[\sigma_{zs} - v_s\sigma_{\theta s}\right] \tag{11}$$

with $$\sigma_{\theta s} = \frac{q_c - q_f}{t_s}R \tag{12}$$

where R and t_s denote the radii and thickness of steel tube. The radial stress is neglected for the thin walled tube. Secant Young's modulus E_s and Poison's ratio v_s can be found from (4), in terms of elastic Young's modulus E_0 and Poison's ratio v_0.

CFRP tube is assumed as a membrane structure, in which only tensile stress exists, so we have

$$\sigma_{\theta f} = \frac{q_f}{t_f}(R + t_s) \tag{13}$$

and $$\varepsilon_{\theta f} = \frac{1}{E_f}\sigma_{\theta f} \tag{14}$$

where q_f is the interfacial stress between steel and CFRP, t_f is the thickness of CFRP tube. Then from the compatibility condition of displacements in the interfaces of steel-concrete and steel-CFRP, by applying relations (6)~(14), we get the stresses of the individual parts as

$$\sigma_{zc} = \frac{N}{A_s + A_c}\left(\frac{R + t_s}{R}\right)^2 \{(1 - v_c - 2v_cv_s) + $$
$$+ (1 - \mu_s^2)\left[\theta_{sc} + (1 - _c)\theta_{sf}\right]\}/FM \tag{15}$$

$$q_c = -\frac{N}{A_x + A_c}\left(\frac{R + t_s}{R}\right)^2\frac{(v_c - v_s) + v_c(1 - v_s^2)\theta_{sf}}{FM} \tag{16}$$

$$q_f = \frac{-N}{A_s + A_c}\left(\frac{R + t_s}{R}\right)^2\frac{-\theta_{cf}(1 - v_c - 2v_c^2)v_s - \theta_{sf}v_c(1 - v_s^2)}{FM} \tag{17}$$

$$\sigma_{zs} = \frac{N}{A_s + A_c}\frac{(R + t_s)^2}{Rt_s}\frac{(\theta_{cf} + \theta_{cs})(1 - v_c - 2v_c^2) + (1 - v_cv_s)}{FM} \tag{18}$$

where $\theta_{cf} = \frac{E_ft_f}{E_cR}, \theta_{sf} = \frac{E_ft_f}{E_st_s}, \theta_{cs} = \frac{E_st_s}{E_cR}$ \tag{19}

and $$FM = (1 - v_s^2)\left[\theta_{sf}(1 - v_c) + \theta_{sc}\right] + $$
$$+ 2(\theta_{cf} + \theta_{cs})(1 - \mu_c - 2\mu_c^2) + (3 - \mu_c - 4\mu_c\mu_s)$$

These equations allow one to investigate the mechanical response of the composite system conveniently.

4 GENERAL STRESS-STRAIN RESPONSE OF THE SYSTEM

Consider the composite system now as a whole cylinder body subjected to a simple compression along its longitudinal axis. Then the average stress and strain in the loading direction follow the well known relation, again, in terms of secant modulus, as

$$\sigma = E\varepsilon \tag{20}$$

Note that the longitudinal strains of concrete core and the steel tube and the system as well, caused by the axial compression, should be the same, that is

$$\varepsilon_{zc} = \varepsilon_{zs} = \varepsilon \tag{21}$$

Then from (9) or (11), and note $\sigma = (N)/(A_s + A_c)$, the overall secant Young's modulus in loading direction of the system is then readily derived in terms of material properties of all constituent parts as

$$E = \frac{E_c \cdot FM}{(1 - v_c - 2v_c^2) + (1 - v_s^2)\left[\theta_{sc} + \theta_{sf}(1 - v_c - 2v_c^2)\right]} \tag{22}$$

The above expression allows an overall theoretical analysis during the whole process of loading.

5 CONFINING CONDITION

One distinguish property of concrete is its compressive strength under tri-axial compression is much higher than that under uni-axial loading. So in a concrete filled steel tube, when concrete core is subjected to a longitudinal compression, the steel tube outside would provide a confinement to it. So its stress state becomes tri-axial one, which allows it to stand much higher loading.

However concrete and steel do not connect to each other at the early stage of loading for the difference of their Poison's ratio, $v_c < v_s$. So the interfacial stress $q_c = 0$. With the increase of loading, v_c begins to increase at $\beta = 0.8$ (Eq.1). Here β is the ratio of applied stress and ultimate strength of concrete. When $v_c \geq v_s$, the interface connects, $q_c > 0$, concrete and steel work together and the confinement happens. The analysis suggests that the effective confinement appears quite late (figure 2), $\beta = 0.985$ for present materials.

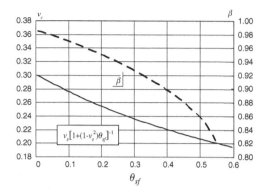

Figure 2. Confining condition of concrete filled CFRP-steel system.

The connect condition of concrete filled CFRP-steel system can be derived from (16) as

$$\nu_c \geq \nu_s \left[1 + \left(1 - \nu_s^2\right)\theta_{sf}\right]^{-1} \tag{23}$$

The above relation shows the necessary ratio for interfacial connection. For a clearer view it is plotted in figure 2. The change of β with the thickness of CFRP is also shown in the same figure, by right hand side coordinate. It is seen that with the increase of thickness of CFRP the confinement appears earlier. The confinement is put ahead remarkably by the bonding of CFRP. The earlier the concrete comes to tri-axial compression state, the more efficient the system is.

6 CONCLUSIONS

Based on some assumptions a set of analytical solutions of stresses of concrete, steel and CFRP sheet are derived in terms of load, geometry of structure and the material properties. The overall secant modulus of the system is also provided, which explain the stress-strain relation of the system for the whole loading process. At last the confinement condition is given as the function of the coefficient $\theta_{cf} = (E_f t_f)/(E_c R)$. In spite of providing higher strength of concrete core, CFRP is proved to advance the confinement remarkably.

REFERENCES

Cai, S.H. 2003. Modern steel tube confined concrete structures, *China Communications Press*. Beijing: China.
Gu, W., Wang, Q.L. & Zhao, Y.H. 2004. Experimental study on concentrically compressed circular concrete filled CFRP-steel composite tubular short columns. *Journal of Shenyang Architectural and Civil Engineering University (Natural Science)*. 2(2): 118~120. (in Chinese).
Gu, W. & Zhao, Y.H. 2006. The analysis on the load carrying capacity of concrete filled CFRP-Steel tubes under axial compression, *Engineering Mechanics(Chinese)*, in press.
Han, L.H. 2000. Concrete filled steel tubular structures. *Science Press*. Beijing: China.
Ottosen, N.S. 1979. Constitutive model for short-time loading of concrete. *J.Engineering Mechanics Division*. 105(EM1): 127~141.
Wang, Q.L and Zhao, Y.H. 2003. A presumption on the concrete filled CFRP-steel composite tube structures. *Journal of Jilin University (Engineering and Technology Edition)*. 33(Sup): 352~355.

Tubular Structures XI – Packer & Willibald (eds)
© *2006 Taylor & Francis Group, London, ISBN 0-415-40280-8*

Numerical analysis of concrete filled tubular beam-columns

J.B.M. Sousa Jr., C.F.D.G. Muniz & A.M.S. Freitas
Department of Civil Engineering, Escola de Minas, UFOP, Ouro Preto-MG, Brazil

ABSTRACT: Steel-Concrete composite beam-columns are formed by the association of a steel profile with concrete, taking best advantage of each material characteristic. In this paper the behavior of composite steel-concrete beam columns, with emphasis on tubular concrete filled sections, is investigated by the finite element method. Different one-dimensional element formulations are considered, e.g. cubic interpolation and co-rotational elements. The implementation allows the consideration of initial lack of straightness of the column, of different stress strain relations (by exact integration of the stiffness at the section level), as well as the physical and geometrical non-linearities involved in the problem. Comparisons with experimental data are presented at the end of the paper. The developed numerical formulation is able to predict accurately the behavior of slender concrete filled tubes under static loading.

1 INTRODUCTION

Composite structures, also called mixed or hybrid structures, combine steel and reinforced concrete to benefit from each material characteristic. Composite construction takes advantage of the speed of construction, light weight and strength of steel, and of the higher mass, stiffness, damping properties and overall economy of reinforced concrete. One of the most suitable structural elements for this combination is the composite column, which in recent years has received much attention by researchers and practicing engineers. Steel-concrete composite columns have been employed in high-rise buildings, bridges, piers and earthquake-resistant structures.

The most common types of composite columns are concrete filled steel tubes and partially or fully encased steel profiles. Encased steel elements have the advantage of higher resistance to fire action, while concrete filled tubes benefit from the increase in strength due to confinement. In general, reinforcement bars may be present, as well as shear connectors to prevent slip.

Most practice codes, such as the widely used AISC (2005), ACI-318(2005) and Eurocode 4 (1994) have incorporated simplified methods for the analysis and design of composite columns. These provisions are generally extrapolated from either reinforced concrete column or steel column design codes (Oehlers and Bradford, 1995). Significant differences between predictions of these codes, however, are an indication that further numerical and experimental research must be carried out to gain understanding about the structural behavior of these elements.

The Finite Element Method (FEM) is the most popular tool for the simulation of structural response. Recent numerical work on composite columns has been focused on either three-dimensional or beam-column elements. Three-dimensional element formulations are able to predict the complex behavior of the composite structures, employing sophisticated material models and interaction models between the materials. Beam-column elements, on the other hand, are simpler to develop and implement and less expensive on computational terms whilst maintaining the precision for engineering design purposes.

The aim of this paper is to develop and implement a numerical procedure for the static analysis of slender concrete-filled composite columns subjected to axial force and bending moment, with emphasis on concrete-filled steel tubes including double skin specimens. The formulation provides a valuable tool for the analysis of composite concrete filled beam-columns, such as circular (CHS) and rectangular (RHS) hollow sections. At the end of the paper, three selected examples from the literature prove the ability of the proposed numerical scheme to provide accurate and reliable results when compared to experimental data. The comparisons are made in terms of the value of the ultimate load as well as in terms of the complete load-displacement behaviour, when the latter is available.

2 NUMERICAL MODEL

Several researchers have carried out numerical investigations on the behavior of concrete-filled tubular columns, and these efforts may be classified in two general types of analysis: (a) full three-dimensional analysis with solid or solid/shell elements (Johansson & Gylltoft 2002; Hu, Huang & Chen 2005) and (b) one-dimensional analysis with beam-column finite elements (Lakshmi & Shanmugam 2002; El-Tawil & Deierlein 2001a,b; Sousa Jr. & Caldas 2005).

Although the full 3D analysis is able to predict several phenomena and provide the most accurate results, it is not practical due to excessive computational cost and access to software. Therefore, beam column elements become an adequate option in terms of precision, ease of implementation and computational effort, provided that some effects such as local buckling and relative slip are not taken into account, and that confinement effects are approximately simulated by modifying on the uniaxial stress-strain relation for concrete.

It should be cited, though, that in a recent work, Liang et al. (2005) have developed a beam-column element able to model local buckling effects by means of a fiber model.

2.1 Cross section formulation

A successful formulation for a numerical analysis of concrete-filled steel tubes with distributed material nonlinearity must take into account the cross sectional response at the integration points. If the section lies on the y–z plane, and the beam axis is along x (see section 2.3), this response is expressed in terms of the section resistant forces N, M_y and M_z, and the generalized section stiffnesses, which are the derivatives of these forces with respect to the deformation parameters ε_0 (axial strain at the centroid) and κ_y, κ_z (curvatures about y and z cross-sectional axes). For plane sections remaining plane after deformation, one has:

$$\varepsilon = \varepsilon_0 + \kappa_y z - \kappa_z y \qquad (1)$$

where

$$\varepsilon_0 = u' + \tfrac{1}{2}v'^2 - \tfrac{1}{2}w'^2 \qquad (2)$$

and

$$\kappa_y = -w'' \qquad \kappa_z = v'' \qquad (3)$$

with the prime denoting differentiation with respect to x. The resistant forces values are obtained by integration of the stresses in the cross section area and depend on the uniaxial material constitutive laws, section geometry and generalized strains:

$$N = \int \sigma(\varepsilon)\, dA$$

$$M_y = \int \sigma(\varepsilon) z\, dA \qquad (4)$$

$$M_z = -\int \sigma(\varepsilon) y\, dA$$

The tangent sectional material moduli (also called section stiffness matrix) are the derivatives of the forces with respect to the generalized deformation variables:

$$k_s = \begin{bmatrix} \partial N/\partial \varepsilon_0 & \partial N/\partial \kappa_y & \partial N/\partial \kappa_z \\ \partial M_y/\partial \varepsilon_0 & \partial M_y/\partial \kappa_y & \partial M_y/\partial \kappa_z \\ \partial M_z/\partial \varepsilon_0 & \partial M_z/\partial \kappa_y & \partial M_z/\partial \kappa_z \end{bmatrix} \qquad (5)$$

The evaluation of Eqs. 4 and 5 is essential in the implementation of beam-column finite elements with distributed inelasticity such as the ones developed in the present work. The most popular method to obtain these responses is the fiber method (El-Tawil & Deierlein 2001a), which simply subdivides the individual material areas into small elements and applies the midpoint integration rule. Other schemes of numerical integration are also possible, such as Gauss or Lobatto quadrature. For irregular domains, however, the geometry usually has to be decomposed in a nontrivial way.

The fiber method is a very powerful tool for the solution of the nonlinear material problem at the cross section. When there is load/strain reversal, for instance, the incremental variables may be stored at each fiber, which is necessary, for example, for seismic analysis and plasticity.

Nevertheless, the fiber method can become quite expensive because the accuracy of the solution depends on the number of fibers in which the cross section is subdivided. Also, in a fiber-based strategy every section should be discretized even if it is still in the elastic range. This makes the analytical integration methods an interesting option for the cases where the load is monotonic (Zupan & Saje 2005).

In this paper, an alternative analytical integration scheme has been employed, using expressions based on transformation of area integrals into line integrals, using Green's theorem. This procedure has been employed, for instance, by Sousa Jr. & Caldas (2005) and performed very well for several different types of composite cross-sections. It is much faster than the fiber method, although it has the disadvantage of not allowing strain reversal, because the evolution variables are not stored along the equilibrium path. Therefore, the present numerical procedure is restricted to monotonic loading.

The cross section is supposed to be composed of different materials, each one with its own uniaxial stress-strain relationship and surrounded by a closed polygon of a generic polygonal shape. To perform the analytical integration, circular or curved section boundaries must be subdivided into straight segments. For circular sections, it has been noticed that 10 segments are enough to provide sufficient accuracy.

The integrals that arise from Eqs. 4 and 5 are of the type

$$I_{ab} = \int y^a z^b\, dA \qquad (6)$$

with a, b integers, and may be transformed in line integrals along the segments that compose the boundaries of each material of the cross section. These line integrals may be expressed further in terms of a single parameter, allowing analytical expressions to be obtained for straight segments (Muniz, 2005). The procedure works with multiple connected domains and also with regions with holes.

2.2 Material model

The cross section analytical integration procedure relies on the availability of uniaxial constitutive laws for concrete and steel. These relationships may be composed of piecewise polynomials of up to the third degree.

For the purpose of this work and in accordance with a lot of published work, e.g. (Lakshmi & Shanmugam 2002), the stress-strain for concrete in compression is composed of a parabolic part from zero to $\varepsilon_1 = -0.002$, as the stress rises from zero to the peak value αf_c, where f_c is the cylinder strength. For strains between ε_1 and ε_{cu}, the maximum allowable value of concrete strain, the stress is kept constant. Tensile stresses in the concrete are neglected, although it would be very simple to define a different constitutive law to consider this possibility. For steel, an elastic perfectly plastic law is adopted.

Concrete confinement was taken into account by taking α equal to 1.00 for CHS, in which the confinement effect is more effective, and 0.85 elsewhere (Spacone & El-Tawil 2004). The strain value ε_{cu}, which is often limited to 0.0035, was left free to increase up to 0.005 in order to take into account the high ductility of confined concrete, which has been verified by experiments.

2.3 Finite Element formulation

The FE formulation developed in this work was based on one-dimensional displacement-based elements, with physical and geometrical nonlinearities taken into account. Two different formulations have been implemented. In both of them the material nonlinearity is dealt with at the cross section level as described in the previous section.

2.4 Cubic Kirchhoff element

The first element is based on the classic cubic transverse displacement formulation, with hermitian interpolation for transverse displacements and quadratic (non-hierarchical) interpolation for the axial displacement. This formulation enables the development of elements capable of modeling biaxial bending and axial force, which is the general case for composite

columns in buildings and bridges. The spatial rotations, however, are considered small so that there is not need to deal with the non-additive rotation tensors.

The element axis is supposed to lie along the x axis. The displacement interpolation is:

$$u = N_u d_u \quad v = N_v d_v \quad w = N_w d_w \tag{7}$$

where N_u is a vector of quadratic shape functions and N_v, N_w are cubic (hermitian) shape functions. The vectors d_u, d_v and d_w collect the nodal values (FE variables). This interpolation strategy means that there is an internal node, with an axial degree of freedom, which prevents membrane locking.

Applying in standard fashion the Principle of Virtual Work and taking into account the displacement interpolation, as well as a set of nonlinear strain-displacement relations, one gets the internal force vector f and the tangent stiffness matrix K.

The internal force is given by

$$f_{int} = \int_0^\ell \begin{bmatrix} N\,N_u' \\ Nv'\,N_v' + M_z\,N_v'' \\ Nw'\,N_w' - M_t\,N_w'' \end{bmatrix} dx \tag{8}$$

where N, M_y and M_z are the resistant forces, and the prime denotes differentiation with respect to x, the element axial coordinate.

Taking the derivative of the internal force vector with respect to the nodal displacements, the tangent stiffness matrix is obtained:

$$K_t = \int_0^\ell \begin{bmatrix} K_u \\ K_v \\ K_w \end{bmatrix} dx \tag{9}$$

where the submatrices are given by

$$K_u = N_u' \left(\frac{\partial N}{\partial d} \right)^T \tag{10}$$

$$K_v = N_v' \left\{ v' \left(\frac{\partial N}{\partial d} \right)^T + N \begin{bmatrix} 0_u & N_v'^T & 0_w \end{bmatrix} \right\} + N_v'' \left(\frac{\partial M_z}{\partial d} \right)^T \tag{11}$$

$$K_w = N_w' \left\{ w' \left(\frac{\partial N}{\partial d} \right)^T + N \begin{bmatrix} 0_u & 0_v & N_v'^T \end{bmatrix} \right\} - N_w'' \left(\frac{\partial M_y}{\partial d} \right)^T \tag{12}$$

In the previous equations N, M_y and M_z are the section forces and the vectors 0_u, 0_v and 0_w are null vectors with the same number of elements as the corresponding shape functions. The vector d collects the nodal displacements d_u, d_v, d_w.

The former expressions assume an initially straight beam. To incorporate the effects of initial out-of-straightness, one must define initial displacement fields v_0 and w_0 which are interpolated in the same way as the previous displacement vectors v and w. A slight modification in the way the bending strains are defined then leads to a tangent stiffness which is very similar to the previous one.

2.5 Corotational elements

The second formulation is based on the kirchhoff corotational plane frame elements presented by Crisfield (1991) and adapted to nonlinear material problems. This second formulation allows only plane problems to be addressed, thus it is not appropriate to the biaxial bending case.

It was verified with numerical tests that the two formulations provide very close results in all the examples, which is to be expected as for composite columns the displacements are not very large. Therefore, for the sake of brevity, the second element formulation will not be described in detail here. It is important to say, however, that the axial displacement is interpolated linearly so that, to avoid membrane locking, an averaged membrane strain is considered (Crisfield, 1991).

2.6 Implementation details

The FE and cross section procedures were implemented by the authors on a general-purpose finite element program, with object-oriented techniques (OO) and the C++ programming language (Martha & Parente Jr. 2005). In the context of OO, a class hierarchy for cross section response was developed and the element class was extended to accommodate the new beam-column elements.

The developed class hierarchy comprises the most usual types of cross sections, i.e. steel profiles, reinforced concrete, concrete encased and concrete-filled hollow sections. User defined section geometries are also allowed.

3 EXAMPLES

In the following, the formulation described in the previous sections is employed to analyze concrete-filled tubular beam-columns. It has been found that both element formulations give almost identical values in every load step. Standard Newton-Raphson or displacement control technique is employed to trace the equilibrium path.

3.1 Example 1

Zeghiche & Chaoui (2005) carried out an experimental program with 27 concrete-filled short and slender circular tubular columns under different load schemes,

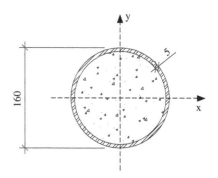

Figure 1. CHS columns tested by Zeghiche & Chaoui (2005).

with axial load combined with single and double curvature bending, and compared their results with the Eurocode 4 predictions. The cross section is indicated at Figure 1. The yield strength of the steel tubes varied from 268 to 283 MPa, and the concrete used had peak stresses from 40 up to 106 MPa. In Table 1 the results of the 27 experiments are displayed, where e_b and e_t indicate load eccentricity at the top and at the bottom of the pinned column; and the load values refer to the experimental and numerical ultimate loads, while t/e is the ratio between the tests and the experimental results.

For the numerical simulation four corotational elements with two integration points were employed. The Kirchhoff element displayed quite similar results.

From Table 1, one can see good agreement between the numerical and experimental values, with a mean t/e ratio equal to 1.07 and a standard deviation equal to 0.10. The largest differences occur for slender columns under single curvature. In that case, inevitable imperfections may have influenced the experimental results obtained with smaller applied load eccentricities (8 mm). As these first order moments are increased, the effect of imperfections becomes less important in the overall column behavior and the results get. It would be possible to take into account some amount of initial out-of-straightness, but as these quantities are not available from the experimental tests, any adopted value would be arbitrary.

3.2 Example 2

Experimental results in double-skin (SHS outer and CHS inner) concrete-filled tubular columns have been published by Han et al. (2004) and have been compared with the results from the present numerical formulation. They tested six different types of specimens, with two tests on each type. The cross section geometry is shown in Figure 2. The average concrete cubic strength was 46.8 MPa and the corresponding cylinder strength was obtained using the Eurocode 2 relationships. The steel yield strengths were 275.9 and 374.5 MPa for the

Table 1. Dimensions, eccentricities at top and bottom and ultimate loads for CHS columns of Example 1.

Height (mm)	e_b, e_t (mm)	Test (kN)	Numeric (kN)	t/n
2000	0, 0	1261	1256	1.00
2500	0, 0	1244	1216	0.98
3000	0, 0	1236	1140	0.92
3500	0, 0	1193	1031	0.86
4000	0, 0	1091	969	0.89
2000	0, 0	1650	1721	1.04
2500	0, 0	1562	1630	1.04
3000	0, 0	1468	1517	1.03
3500	0, 0	1326	1377	1.04
4000	0, 0	1231	1195	0.97
2000	0, 0	2000	2135	1.07
2500	0, 0	1818	2008	1.10
3000	0, 0	1636	1847	1.13
3500	0, 0	1454	1690	1.16
4000	0, 0	1333	1450	1.09
2000	+8, +8	1697	1846	1.09
2000	+16, +16	1394	1572	1.13
2000	+24, +24	1212	1334	1.10
2000	+32, +32	1091	1142	1.05
4000	+8, +8	963	1256	1.30
4000	+16, +16	848	1017	1.20
4000	+24, +24	727	872	1.20
4000	+32, +32	666	756	1.14
2000	+8, −8	1950	2097	1.08
2000	+16, −16	1730	1847	1.07
2000	+24, −24	1480	1628	1.10
2000	+32, −32	1280	1415	1.11

Table 2. Dimensions, eccentricity and ultimate loads for Example 2.

Spec.	Height (mm)	e (mm)	Test (kN)	Numeric (kN)	t/n
1a	1070	4	856	847	0,99
1b	1070	4	872	847	0,97
2a	1070	14	667	692	1,04
2b	1070	14	750	692	0,92
3a	1070	45	480	432	0,9
3b	1070	45	486	432	0,89
4a	2136	0	920	805	0,88
4b	2136	0	868	805	0,93
5a	2136	15,5	596	564	0,95
5b	2136	15,5	570	564	0,99
6a	2136	45	380	359	0,94
6b	2136	45	379	359	0,95

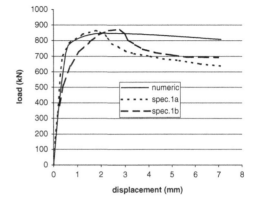

Figure 3. Experimental and numerical load-displacement for specimens 1a, 1b.

It can be seen from the figures that the results display a very good agreement with the experiments not only in terms of the critical load but also for the full load-displacement curve. The average value for the ratio between the numerical and experimental values is 0.95 with a standard deviation 0.05.

The authors believe that the softening branch, especially of specimens 1 and 2, would be better modeled with the adoption of a descending branch in the concrete stress-strain relationship.

3.3 Example 3

Another configuration of a double skin tube (two concentric circular hollow sections with concrete between them) was the subject of an experimental program carried out by Tao et al. (2004). The testing procedure was similar to the one in the previous example. Single curvature loading only was considered.

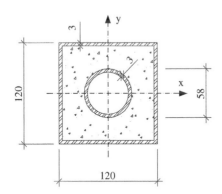

Figure 2. Double skin columns tested by Han (2004).

outer and inner tubes respectively. Both tubes were 3 mm thick. According to the original work, all the columns collapsed due to global buckling.

Four finite elements with two integration points were employed in the analysis and the results for the critical load are shown in Table 2. The complete load-displacement relationship is also available for the 12 experiments and are compared with the numerical results (Figures 3 to 8).

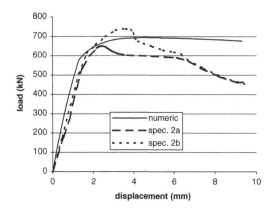

Figure 4. Experimental and numerical load-displacement for specimens 2a, 2b.

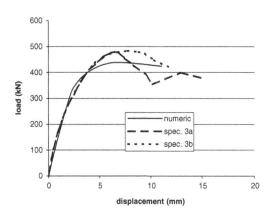

Figure 5. Experimental and numerical load-displacement for specimens 3a, 3b.

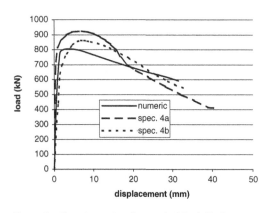

Figure 6. Experimental and numerical load-displacement for specimens 4a, 4b.

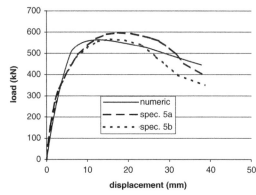

Figure 7. Experimental and numerical load-displacement for specimens 5a, 5b.

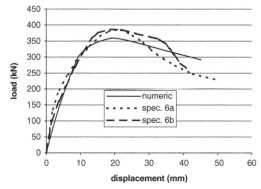

Figure 8. Experimental and numerical load-displacement for specimens 6a, 6b.

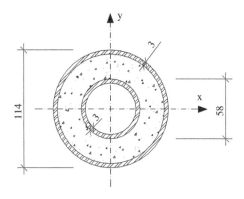

Figure 9. Cross-section for specimens tested by Tao (2004).

The double skin section had external diameters of 114 mm and 58 mm, and both tubes were 3 mm thick (Figure 9). The yield strength was 294.5 MPa for the outer and 374.5 for the inner tube, and the elastic

Table 3. Dimensions, eccentricity and ultimate loads for Example 3.

Spec.	Height (mm)	E (mm)	Test (kN)	Numeric (kN)	t/n
1a	887	4	664	634	0,95
1b	887	4	638	634	0,99
2a	887	14	536	498	0,93
2b	887	14	549	498	0,91
3a	887	45	312	287	0,92
3b	887	45	312	287	0,92
4a	1770	0	620	615	0,99
4b	1770	0	595	615	1,03
5a	1770	15,5	400	400	1,00
5b	1770	15,5	394	400	1,02
6a	1770	45	228	239	1,05
6b	1770	45	227	239	1,05

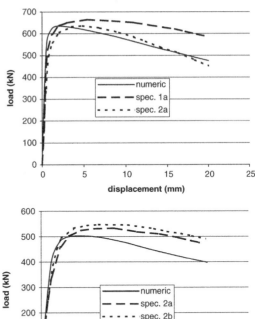

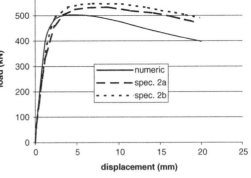

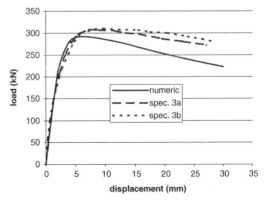

Figure 10. Experimental and numerical load-displacement for specimens 1a, 1b, 2a, 2b, 3a & 3b.

modulus of steel was 200 000 MPa. The peak cubic stress for concrete was 47.4 MPa. The test data and the results for the critical loads are displayed in Table 3. As in the previous example, no local buckling took place, so the present FE formulation is expected to display good results.

Once more there is a very good agreement of the critical load (Figures 10 and 11), and the ratio of numerical and experimental data has a mean value of 0.98 with a standard deviation of 0.05. As in the previous example the softening branch might be better obtained numerically if a softening segment were adopted for the concrete stress-strain curve.

It may be observed that for these columns the numerical results are even better than for the previous example, and that the columns exhibit high ductility after reaching the peak load value.

4 CONCLUSIONS

This paper presents a procedure for the numerical analysis of slender composite steel concrete columns of generic cross section shape. Results were presented with emphasis on concrete filled hollow sections, including double skin columns. The formulation was able to provide very good results for the critical load and for the load-displacement curves when compared to experimental data, with few elements per member and at low computational cost, providing a reliable and robust numerical tool for the analysis of such elements.

As the design codes often give different, and in some cases, conflicting predictions, and some types of cross sections such as the double skin tubes are not covered by the codes, numerical and experimental analyses are of paramount importance to improve the knowledge of these types of structural members.

In the future, the present work may be expanded to take into account several other important aspects such as: (a) analysis of concrete filled columns under fire action (already under development); (b) incorporation of local buckling effects and analyses of complete framed structures with advanced connection modelling; (c) time-dependent (creep and shrinkage) and dynamic effects.

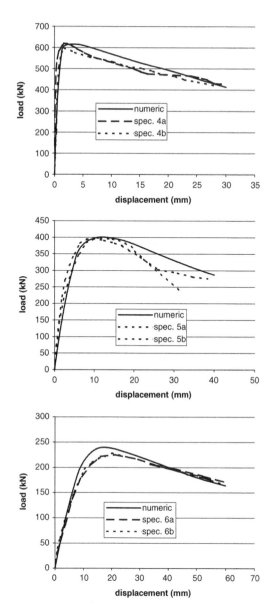

Figure 11. Experimental and numerical load-displacement for specimens 4a, 4b, 5a, 5b, 6a & 6b.

ACKNOWLEDGEMENTS

The authors wish to thank V&M Tubes Brazil, branch of Vallourec-Mannesmann Tubes, for their support in the development of this work.

REFERENCES

ACI-318 (2005). ACI 318-05, Building Code Requirements for Structural Concrete and Commentary. Detroit: American Concrete Institute.

AISC (2005). Specification for Structural Steel Buildings. Chicago: American Institute of Steel Construction.

Crisfield, M.A. (1991). Nonlinear finite element analysis of solids and structures. Vol. 1: essentials. John Wiley & Sons.

El-Tawil, S. & Deierlein, G.G. (2001a). Nonlinear analysis of mixed steel-concrete frames. i: Element formulation. *Journal of Structural Engineering* 127(6), 647–55.

El-Tawil, S. & Deierlein, G.G. (2001b). Nonlinear analysis of mixed steel-concrete frames. ii: Implementation and verification. *Journal of Structural Engineering* 127(6), 656–65.

Han, L.H., Tao, Z., Huang, H. & Zhao, X.L. (2004). Concrete-filled double skin (shs outer and chs inner) steel tubular beam-columns. *Thin-walled Structures* 42, 1329–55.

Hu, H.T., Huang, C.S. & Chen, Z.L. (2005). Finite element analysis of cft columns subjected to an axial compressive force and bending moment in combination. Journal of Constructional Steel Research 61(12), 1692–1712.

Johansson, M. & Gylltoft, K. (2002). Mechanical behavior of circular steel-concrete composite stub columns. *Journal of Structural Engineering* 128(8), 1073–81.

Lakshmi, B. & Shanmugam, N.E. (2002). Nonlinear analysis of in-filled steel-concrete composite columns. *Journal of Structural Engineering* 128(7), 922–33.

Liang, Q.Q., Uy, B. & Richard Liew, J.Y. (2005). Nonlinear analysis of concrete-filled thin-walled steel box columns with local buckling effects. Journal of Constructional Steel Research (in press).

Martha, L.F. & Parente Jr., E. (2005). An object-oriented framework for finite element programming. In WCCM V - Fifth World Congress on Computational Mechanics.

Muniz, C.F.D.G. (2005). Numerical models for the analysis of steel-concrete composite elements (in portuguese). MSc dissertation, Dept. Civil Engnrg., Universidade Federal de Ouro Preto, 35400-000, Ouro Preto-MG, Brazil.

Oehlers, D.J. and Bradford, M.A. (1995). Composite steel and concrete structural members: fundamental behaviour. Oxford: Pergamon Press.

Sousa Jr., J.B.M. & Caldas, R.B. (2005). Numerical analysis of composite steel-concrete columns of arbitrary cross-section. *Journal of Structural Engineering* 131(11), 1721–30.

Spacone, E. & El-Tawil, S. (2004). Nonlinear analysis of steel-concrete composite structures: State of the art. *Journal of Structural Engineering* 130(2), 159–68.

Tao, Z., L.H. Han. & Zhao, X.L. (2004). Behaviour of concrete-filled double skin (chs inner and chs outer) steel tubular columns and beam-columns. *Journal of Constructional Steel Research* 60, 1129–58.

Zeghiche, J. & Chaoui, K. (2005). An experimental behaviour of concrete-filled steel tubular columns. *Journal of Constructional Steel Research* 61, 53–66.

Zupan, D. & Saje, M. (2005). Analytical integration of stress field and tangent material moduli over concrete cross-sections. *Computers & Structures* 42, 1329–55.

Parallel session 6: Fire

Tubular Structures XI – Packer & Willibald (eds)
© 2006 Taylor & Francis Group, London, ISBN 0-415-40280-8

Fire protection of hollow structural sections with intumescent coatings – Case examples

D.F. Falconer & S. Trestain
A/D Fire Protection Systems/ Carboline Fire Protection Division

G.S. Frater
Canadian Steel Construction Council

ABSTRACT: Architects have increasingly taken steel structures and expressed them in their architectural designs leading to the relatively new defined term in steel design codes and handbooks, namely, Architecturally Exposed Structural Steel (AESS). When meeting building code requirements, fire safety for the primary steel structure requires fire protected steel for virtually all non-industrial-type buildings to give the structure fire resistance (known as passive fire protection). When tubular steel, or Hollow Structural Sections (HSS) are selected for an AESS project it can be passively fire protected by external application of intumescent coatings. (HSS can also achieve passive fire protection by protecting the potentially exposed surfaces with conventional sprayed fire resistive materials or gypsum board, or internally by concrete filling or by water filling the void down the centre of a HSS.) Intumescent coatings are paint-like and give the designer an AESS that is decoratively painted in appearance and also fire protected. The intumescent coating's chemical composition reacts due to a fire's heat and expands up to fifty times the applied coating thickness. The expanded ash layer provides an insulating layer between the steel and the heat of the fire. Steel will lose about 50% of its strength at 600°C. To establish the insulating layer's performance under high temperatures, representative steel test specimens are subjected to standard fire exposures in furnaces fired to follow standard time-temperature curves. Fire test standards, such as ASTM E119 (USA), CAN/ULC-S101 (Canada) and ISO 834 (Europe) have test protocols from which hourly fire resistance ratings are determined for the tested assemblies. This paper overviews the use of intumescent coatings on HSS for two projects where fire resistance ratings were required by the applicable building code.

1 INTRODUCTION

Over the last few decades many architects and designers have increasingly designed buildings where structural steelwork is exposed – meant to be seen and not covered up by other elements of the structure. One of the initial examples of an architect using such transparency in their design, namely Helmut Jahn, was the United Airlines Terminal Building at Chicago's O'Hare Airport constructed from 1985 to 1988. In Canada and the United States steel fabricators refer to this type of steel work as Architecturally Exposed Structural Steel (AESS). The American Institute of Steel Construction (AISC, 2005) in Code of Standard Practice for Steel Buildings and Bridges establishes standards for AESS that take into account the need for closer dimensional tolerances and smoother finished surfaces than necessary for typical structural steel framing. Additional guidelines for AESS are published in AISC's Modern Steel Construction (May,

2003) magazine publication and is available via AISC's website, www.aisc.org/aess.

Tubular steel, or Hollow Structural Sections (HSS) are one of the structural members used frequently in AESS. HSS established itself as a popular building member initially due to its attractive appearance, which allowed it to be exposed to view and to be an integral part of a building's design. The aesthetics of HSS have been evident as AESS in building atria, airport terminals, arenas, sport stadiums, convention and shopping centres – to name a few. HSS are also used routinely for other structural applications because of their other attributes such as high strength-to-weight ratios for columns and cost savings for painting or fire protection due to reduced surface areas, i.e., an HSS has about two thirds the surface area of similarly sized I-shaped member.

The continued use of HSS in AESS building projects and its appeal to architects has been aided

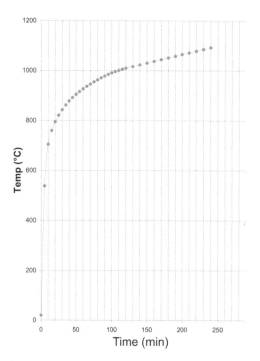

Figure 1. Standard time–temperature curve.

Figure 2. Columns undergoing fire test.

by the ongoing developments in intumescent coatings that provide fire protection to structural steel.

Fire safety requirements in Building Codes stipulate that for occupancies such as assembly, care or detention, business, mercantile, residential, and industrial, a fire resistance rating be determined for material, assembly of materials or a structural member used in building construction. Fire resistance ratings are determined by full-scale fire endurance tests of representative assemblies in accordance with fire-testing standards such as CAN/ULC-S101 in Canada and ASTM E119 in the USA. CAN/ULC-S101 describes a Standard Time-Temperature Curve that provides the average furnace temperature as a function of time for all fire endurance tests. The time–temperature curve rises rapidly to 840°C at 30 minutes and then increases more gradually to 1090°C at four hours (see Figure 1).

This fire exposure is used to evaluate fire protection applied to steel sections used as columns and beams in structural steelwork. The fire test standard specifies the criteria, which the fire protection method must satisfy, in order to qualify for a given time rating: half, one, two, three or four hours fire protection. The purpose of fire protection is to provide thermal insulation to a steel member thereby maintaining its temperature for a required period below a specified temperature criteria in the fire test standard, i.e., 538 and 593°C sectional average temperature for steel columns and beams, respectively. Steel does not burn, but will lose

approximately half its strength around 600°C and at this temperature may fail to support its designed load. In a fire-testing furnace a temperature of 600°C is reached within 7½ minutes of beginning the standard fire exposure. The period of time during which the steel column or beam is able to carry load and stay below the limiting temperature criteria is called the "fire resistance" leading to a hourly fire resistance rating for a particular tested steel assembly comprised of steel structural member(s) and fire protection system (see Figure 2). Tindale (1989) estimates that under the UK building code regulations all the commercial and industrial buildings predominantly require (73%) 1-hour fire resistance, 7% half an hour, and 20% 2-hours and over.

Intumescent coatings are paint-like materials applied to structural steel such as HSS with typical dry film thickness of 1 mm to 3 mm. When fire occurs, an intumescent coating changes dramatically from a relatively thin paint-like coating to what may be described as a meringue-like insulating ash that may be 50 to 100 mm or more in thickness (see Figure 3). It is the intumescent ash that insulates and protects the steel from fire. It's known as a passive fire protection method, i.e., no mechanical, electrical or human action necessary to activate its fire protection. Instead, depending on the formulation, a temperature of 100 to 300°C activates the intumescent chemical reactions.

Generally, the thickness of intumescent ash that develops in response to fire exposure directly relates to the original dry film thickness of the coating. Therefore, for a given steel size, an increase of the applied dry film thickness will increase the fire resistance rating of the column or beam. Similarly, an increase in size of the steel section will increase fire resistance. For that purpose, the size of the steel section is expressed in terms of its A/P, M/D, W/D or Hp/A ratio. These ratios are simply the relation between the cross sectional area (A) or mass (M or W) of the steel section and its heated perimeter (P or D), the latter being that portion of the perimeter of the steel section that would be exposed to heat in the event of a fire. The minimum dry film thickness of an intumescent coating that is required for a given steel size and fire resistance rating, can be found in fire test design information published by testing, certification and other agencies, examples being the Underwriters' Laboratories of Canada (ULC, 2005) List of Equipment and Materials, Fire Resistance, Underwriters Laboratories, Inc. (ULI, 2006) Fire Resistance Directory and Association for Specialist Fire Protection (ASFP, 2004) Fire Protection for Structural Steel in Buildings, commonly referred to as the "Yellow Book." Published design information is proprietary in nature because the fire resistance of a column or beam assembly will depend on the performance, nature and characteristics of the intumescent formulation.

Intumescence is the property of swelling as a reaction to the application of heat. Intumescent coating products are complex formulations that tend to be proprietary, with their thermal properties rarely reported in literature. Lyons (1970) and Vandersall (1971) provide extensive discussion of intumescent coating technology that can be traced from 1938. Fatal fires in the USA space program and on Navy aircraft carriers brought on further impetus for development of intumescent coating technology. These events spurred research and development programs at the National Aeronautical and Space Agency and within the USA military during the 1970's and advanced intumescent coating technology. During this time Underwriters Laboratories Inc. (1969) were listing 2-hour ratings for beams and 1½-hours for columns. Developments in intumescent coating technology since that time have increased ratings to 3-hours and the number of published test designs for thin-film intumescents have also increased. Butler (1997) provides a recent review of chemical and physical phenomena that occur within intumescent coatings upon heating, and various approaches to model intumescent behaviour.

The expansion process found in intumescent coating systems require the interaction of the following four key ingredients (Yandizo et al., 1996):

- Acid source, or material yielding an acid at temperatures of 100 to 300°C, typically ammonium polyphosphate.
- Carbon supplier such as polyhydric compounds (starch) that reacts with the acid to form carbonaceous char.
- Blowing or spumific agent such as melamine that decomposes to liberate large volumes of non-flammable gases (including carbon dioxide, ammonia and water vapour). The gases cause the carbonaceous char to foam, thus producing a meringue-like structure that is highly effective insulator against heat.
- Coating binder or resin that binds together the intumescent ingredients mentioned above and provides

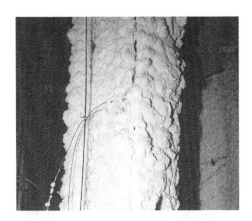

Figure 3. Intumescent ash after fire test exposure.

adhesion to the substrate. The binder aids in the formation of a uniform cellular foam structure by trapping gases given off by the decomposing blowing agents thereby ensuring a controlled expansion of the carbonaceous char. The binder also protects the intumescent ingredients by providing resistance to water, UV light and abrasion.

The advantages of the present day thin-film intumescent fire protection coatings are as follows:

- Exposed, painted steel appearance with topcoats available in a wide range of colours.
- Fire-resistance ratings up to three hours for structural steel.
- Greater compressive strength and impact and abrasion resistance than conventional sprayed fire resistive materials.
- Thin-film allows smaller column footprints and less intrusion into room space.
- Smooth, washable, dust-free finish.
- Accommodates attachments to steel.
- Water-based products are user friendly.

The remainder of this paper overviews the use of intumescent coatings on HSS for two projects where fire resistance ratings were required by the applicable building code.

2   TORONTO EATON CENTRE, MEDIA TOWER, TORONTO, CANADA

Architect: Queen's Quay Architects
Owner: Cadillac Fairview
Structural: Banerjee & Associates Limited
Steel Fabricator: Walters Inc.
Intumescent Applicator: Donalco Inc.
Intumescent Supplier: A/D Fire Protection Systems

The Media Tower addition to the popular Toronto Eaton Centre shopping mall and mixed use complex is a vital component of the revitalized Dundas Square

Figure 4.　Toronto Eaton Centre Media Tower. Photo credit: Kramer Design Associates Limited, the designer and builder of the media signage.

in downtown Toronto. The upper and primarily exterior portion of the new 18-storey structure serves as a multimedia entertainment centre incorporating video displays and lighting effects while the lower portion of the structure serves as a pedestrian entrance from Dundas Square to the Eaton Centre complex and Dundas subway station, part of Toronto's busy transit system (see Figure 4).

Chi-Man Trinh of Queen's Quay Architects, the architect of record, in a conversation on November 29, 2005, said The City of Toronto was heavily involved during the design phase of this project. The City demanded tubular steel for the exterior portion of the media tower so as to avoid an industrial look. Thus tubular steel was used everywhere in the tower (interior and exterior) for aesthetic consistency. The upper section consists of an exterior tubular steel superstructure supporting the video screens and the lower part consists of large interior branching tubular steel "trees" below the roof level. The lower section steel was protected with intumescent fire resistive coating to provide the required 2-hour fire resistance rating. Structural steel included HSS 900 × 25, HSS 400 × 24 and HSS 324 × 13 (see Figure 5).

The tubular steel connections were made with a proprietary connection system from Walters Inc. of Hamilton, Ontario, Canada. This connection system was very successful architecturally as it is a hidden system, complemented by the intumescent coating which could be sprayed directly over the connections.

The very large HSS sections (the "trunks" of the branching interior columns) were not readily available in the domestic market for this project. The steel had

Figure 5.　"Branches" of the tubular steel tree.

Figure 6.　Inside the media tower lower level.

to be imported from an offshore supplier at great difficulty and expense (see Figure 6). Trinh commented, "The tubular look was essential as the City demanded it for the upper exterior superstructure, and it needed to be carried through to the lower section".

Other fire protection methods were considered to achieve the required 2-hour fire resistance rating. The only real aesthetic alternative to the intumescent coating was providing reinforced concrete filled tubular steel. However the intumescent was selected because the concrete-fill concept could not be applied to the smaller branches because of their size and limited fire test data. Thus it was determined that a single system should be used to achieve the ratings throughout and intumescent was the only available system. Trinh also commented that that the engineering support from the intumescent coating manufacturer "turned out to be very important". The strongest feature of the intumescent coating product that was selected was the choice of custom topcoat colours that is not available with other intumescent products (see Figure 7).

Figure 7. The "trunks" of the tubular steel trees.

The Eaton Center complex remained operational throughout the construction period. There were more than ten sequences of construction and rerouting of pedestrian traffic through the mall and Dundas subway station beneath. The intumescent coating application was done at night while the other trades were out of the way.

An aesthetically consistent design was of critical importance for the new Eaton Centre Media Tower. The use of tubular structural steel sections and intumescent fire resistive coating satisfied the design requirements for this popular entertainment and tourist attraction.

3 TORONTO PEARSON INTERNATIONAL AIRPORT, PIER F, TORONTO, CANADA

Architect: Skidmore, Owings & Merrill LLP (SOM)
Owner: Greater Toronto Airports Authority (GTAA)
Structural: Yolles Partnership / Arup & Partners
Steel Fabricator: Walters Inc.
Intumescent Applicator: Donalco Inc.
Intumescent Supplier: A/D Fire Protection Systems

Toronto Pearson International Airport is Canada's busiest airport, with 25.9 million flyers in 2002 and an expected 50 million per year by 2020. To handle this enormous growth, the Greater Toronto Airport Authority (GTAA) initiated a 10-year, $4.4 billion airport expansion plan which includes the development of a new terminal (see Figure 8).

Design of the new passenger terminal building was coordinated between Adamson Associates, Moshe Safdie Associates Ltd. and Skidmore, Owings & Merrill (SOM). The new terminal features extensive

Figure 8. Toronto Pearson International Airport. Pier F in foreground. Photo credit: Greater Toronto Airports Authority.

natural light, large open concourses and architecturally prominent structural tubular steel (see Figures 9, 10 and 11).

SOM's Narin Gobindranauth was involved in the design and construction of the new terminal. In a December 1, 2005 conversation he explained that the decision to extensively use tubular steel was purely aesthetic. "Intumescent fireproofing was considered part of the package," said Gobindranauth, explaining that there was no viable alternate means for achieving the required fire resistance rating while maintaining the exposed look of the steel.

"Tubular steel was used in all public spaces," said. Gobindranauth. Where structural sections were hidden, W-shaped steel with traditional, cementitious fireproofing was employed. Where exposed, all steel was tubular and coated with thin-film intumescent fireproofing.

The quality of intumescent coating finish is greatly affected by the condition of steel substrate, as intumescent coatings do not hide steel imperfections. Walters Inc., the steel fabricator, patched and repaired steel imperfections prior to the application of intumescent fireproofing. As part of a rigorous quality review process, steel was required to be inspected and approved by the architect, project management and intumescent applicator before the spray application could begin.

Gobindranauth was very satisfied with the intumescent finish, commenting that it was "probably the best intumescent finish I've ever seen". Gobindranauth did highlight the need to protect the intumescent coating from damage by heavy construction equipment after application. "It is unusual to be applying an interior finish during construction," he commented.

The application of the intumescent fireproofing product could be done while other trades continued work in the area as the A/D FIREFILM II product used is water-based and has a low Volatile Organic Compound (VOC) content, and did not create a health or safety issue.

In the airport retail areas some portions of the structural tubular steel were hidden within semi-permanent

Figure 9. Inside Pier F. Columns coated with intumescent fireproofing.

Figure 10. Exterior wall of Pier F. Columns coated with intumescent fireproofing.

Figure 11. Closeup of column and bracing details coated with intumescent fireproofing.

fire rated walls. Though fireproofing was not required on these embedded columns, the intumescent fireproofing was still applied to provide flexibility to future renovations by not requiring the application of new fireproofing should the reconfiguration of interior walls expose previously hidden steel.

Tubular (HSS) steel and intumescent coatings are fundamental to the stylish architectural look of Toronto Pearson International Airport's new terminal. The bold architectural statement of exposed tubular steel will shape international travelers' first impression of Toronto for many years.

REFERENCES

AISC. 2003. Architecturally Exposed Structural Steel. Modern Steel Construction, May, AESS Supplement: 1–16. American Institute of Steel Construction, Chicago, Illinois, USA.

AISC. 2005. Code of Standard Practice for Steel Buildings and Bridges, American Institute of Steel Construction, Chicago, Illinois, USA.

ASFP. 2004. Fire Protection for Structural Steel in Buildings, Association for Specialist Fire Protection, Farnham, Surrey, UK.

ASTM. E119. 2000. Standard Test Method for Fire Tests of Building Construction Materials. American Society for Testing Materials, West Conshohocken, Pennsylvania, USA.

Butler, K.M. 1997. Physical Modeling of Intumescent Fire Retardant Polymers. Chapter 15 of Polymeric Foams Science and Technology, ASC Symposium Series 669, American Chemical Society, Washington, District of Columbia, USA.

CAN/ULC-S101. 2004. Standard Methods of Fire Endurance Tests of Building Construction and Materials. Underwriters' Laboratories of Canada, Toronto, Ontario, Canada.

ISO 834. Fire Resistance Tests – Elements of Building Construction. International Standards Organization, Switzerland.

Lyons, J.W. 1970. The Chemistry and Uses of Fire Retardants. John Wiley and Sons Inc. New York, New York, USA.

Tindale, D.J. 1989. Fire Protection for Structural Steelwork. The Structural Engineer, Journal of the Institution of Structural Engineers, 67 (13): A4–A5.

Underwriters Laboratories Inc. 1969. Building Materials List. Chicago, Illinois, USA.

Underwriters Laboratories Inc. 2006. Fire Resistance – Volume 1, Northbrook, Illinois, USA.

Underwriters' Laboratories of Canada. 2006. List of Equipment and Materials, Fire Resistance. Toronto, Ontario, Canada.

Vandersall, H.L. 1971. Intumescent Coating Systems, Their Development and Chemistry. Journal of Fire and Flammability, 2 (April): 97–140.

Yandzio, E., Dowling, J.J. and Newman, G.M. 1996. Structural Fire Design: Off-Site Applied Thin Film Intumescent Coatings. SCI Publication 160, The Steel Construction Institute, Ascot, Berks, UK.

Tubular Structures XI – Packer & Willibald (eds)
© 2006 Taylor & Francis Group, London, ISBN 0-415-40280-8

Fire resistance of SCC filled square hollow structural steel sections

H. Lu

Fuzhou University, Fuzhou, China/Fujian Academy of Building Research, Fuzhou, China

X.-L. Zhao

Department of Civil Engineering, Monash University, Melbourne, Australia

L.H. Han

Department of Civil Engineering, Tsinghua University, Beijing, China

ABSTRACT: Self-consolidating concrete or self-compacting concrete (SCC) can be compacted by its own weight without vibration. There is a potential use of SCC filled hollow structural steel sections (HSS). Fire resistance of such composite sections is an important issue that should be considered in design. In this paper, an investigation on SCC filled HSS stub columns is presented. The cylinder strength of SCC was found to be 100 MPa. Several SCC filled stub columns were tested in a furnace under the condition of standard fire. Numerical simulation of temperature distribution and fire endurance were conducted. The predicted temperature field distribution and fire endurance were compared to those tested.

1 INTRODUCTION

Self-consolidating concrete or self-compacting concrete (SCC) is a relatively new technology in concrete which was developed in Japan in 1980s. The typical characteristic of SCC is that it can be cast and compacted into every part of the mould or frameworks without any segregation by its own weight rather than with the mechanical aid. The use of SCC not only increases the workability of the concrete and reduces the environment impacts (such as noise), but also enhances the quality and durability of the concrete, reduces the construction timetables and costs. In the last decade, SCC has gained more and more widely accepted and applied in Japan, USA, Europe and other countries.

There is a potential use of SCC filled hollow structural steel sections (HSS). The concrete filled hollow structural steel sections are usually acting as columns in structures, which are normally constructed with beams for several stories before the concrete is in-filled in the sections. The columns are relative long in this case. This causes the difficulty of compacting concrete through vibration and SCC can be a good solution. Fire resistance of such composite sections is an important issue that should be considered in the design. Research on SCC filled hollow structural steel sections under ambient temperature was conducted by the authors (Han, et al. 2005). It was found that the

behaviors of SCC filled hollow structural steel sections are similar to those of conventional concrete filled ones. But rare research has reported on SCC filled HSS at elevated temperatures. A few research on SCC columns at elevated temperature show that SCC has greater tendency of spalling than normal strength concrete (NSC) at elevated temperature (Bostrom, 2002 & Persson, 2003). This behavior is very similar to that of high strength concrete (HSC) (Kodur, et al. 2004). Investigation on the behavior of concrete filled HSS in fire has conducted by many researchers. The in-filled concrete severs as heat sink of the composite sections, so the fire endurance of concrete filled HSS is better than that of HSS. In the previous research, it was found that the spalling of core concrete can be prevented by the confinement of the steel sections, but the performance of HSS filled with HSC in fire is worse than that of NSC filled HSS (Kodur, 1998). All of these indicate that there is a necessary of investigation on the fire resistance of SCC filled HSS.

In this paper, experiment investigation and numerical simulation on SCC filled HSS in fire are presented. Six SCC filled square HSS stub columns were tested in standard fire till failure. Fire resistance, failure mode and other experimental results were reported. Numerical simulations were carried out in two steps, i.e. the non-linear analysis of temperature distribution and the analysis of fire endurance. Because there is no data on the thermo and mechanical properties of SCC at elevated temperatures, looking for suitable properties to represent those of SCC is very important. After careful analysis and comparison of the properties of different kind of concrete, the thermo properties of HSC were chosen to represent those of SCC and the mechanical property model developed by Han, et al. (2003b), which is suitable for both NSC and HSC, were used in fire endurance analysis. Finally, the predicted temperature distribution and fire endurance were compared to those tested.

2 EXPERIMENTAL INVESTIGATION

2.1 Specimen

An experimental study was made on 6 SCC filled HSS stub columns in fire. The related parameters of the specimens are shown in Table 1. The length of all specimens was 760 mm. Two kinds of square steel hollow sections, 150 mm × 5 mm and 200 mm × 6 mm, were used in the test. The steel hollow sections are conformed to Australian Standard AS1163(1991).Different levels of load eccentricities and load densities were chosen. The load density (also called the degree of utilization (μ) in Twilt, et al. 1996) is defined as the ratio of the load applied at the fire test to the ultimate capacity at ambient temperature. The load density could go up to 0.6 (Twilt, et al. 1996). Design guidance on minimum cross-sectional dimensions was given in Twilt, et al. 1996 for μ of 0.3 to 0.7. In order to make the specimen have enough fire endurance so as to investigate the behavior of SCC filled HSS in fire, the load density ranging from 0.17 to 0.44 were selected in the current testing program.

2.2 Material properties

Steel coupons were cut longitudinally along the steel hollow sections. The elastic modulus, yield stress and ultimate stress of the steel were obtained using the standards tensile coupon tests. The mechanical properties of the steel are listed in Table 2.

The most prominent difference between SCC and normal concrete is the properties of fresh concrete. In order to achieve the self-consolidating effect, the mix proportion of SCC is usually different from normal concrete, and some techniques are used to test the flowability of the concrete. The mix proportion of SCC in test is listed in Table 3. It can be seen that higher amount of superplaticizer was used so as to improve the flowability of the concrete and eliminate the possibility of segregation. The coarse aggregate used was basalt. It is a kind of silica stone.

Two typical techniques were used to test the workability of the SCC, slump cone test and L-Box test. In the slum cone test, the slump, the slump flow and the time the concrete getting to the circle with diameter of 500 mm (T_{50}) were recorded. An L-Box was used to simulate the ability of a SCC sample to flow through reinforcement net by its own weight. The schematic diagram of L-Box is shown in Figure 1. The flow time from the sliding door to 40 cm (T_{40}), the flow speed, and other data were recorded.

The fresh properties of the SCC mixture are shown in Table 4.

Concrete cylinders were also cast and cured in conditions similar to the related specimens. The cylinder compressive strength (f_{cu}) of the concrete at 28 days and at the day of test was 90 and 99 MPa, respectively.

Table 2. Mechanical properties of steel.

HSS	Elastic modulus (GPa)	Yield stress (MPa)	Ultimate stress (MPa)
150 × 5	197	486	558
200 × 6	189	467	544

Table 3. Mix proportion of self-consolidating concrete (kg/m^3).

Water	Cement	Fly ash	Sand	Coarse aggregate	Superplasticizer
178	380	170	776	831	11

Table 1. Information of the specimens.

Specimen	Width (mm)	Thickness (mm)	Load eccentricity (mm)	Load (kN)	Ultimate capacity (kN)	Load density	Fire endurance (min)	Predicted fire endurance (min)
S1R2E0	150	5.0	0	486	2787	0.17	>90	61
S1R4E0	150	5.0	0	1216	2787	0.44	26	31
S1R4E2	150	5.0	25	808	2049	0.39	55	35
S2R3E0	200	6.0	0	1226	4702	0.26	55	53
S2R4E0	200	6.0	0	1800	4702	0.38	43	36
S2R4E1	200	6.0	25	1350	3702	0.36	43	40

2.3 Experimental process

Firstly, the hollow structural steel sections were cut to the designated length. A hole with a diameter of 20 mm for venting vapor in concrete was drilled at the middle height of every specimen. In order to monitor the temperature in the specimen, three thermocouples were embedded in every specimen. The location of the thermocouples in specimen is shown in Figure 2. One was at the interface of steel and concrete, the other two were at the 1/4 width and at the center of the section. The specimen was placed vertically on to a plywood plate. Silica glue was used to attach the tube to the plywood plate. Finally, the self-consolidating concrete was poured into the specimen from the open side of the specimen without any vibration.

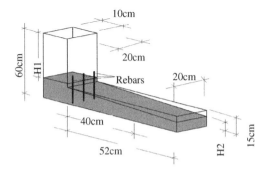

Figure 1. Schematic diagram of L-Box.

Table 4. Fresh properties of the self-consolidating concrete.

Slump (mm)	Slump flow (mm)	T50 (s)	T40 (s)	H1 (mm)	H2 (mm)	Flow speed (mm/s)
273	695~740	3.1	3.6	530	60	110.5

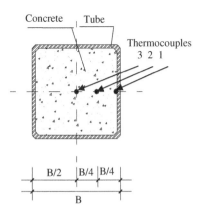

Figure 2. Location of thermocouples in specimen.

Specimens were tested in the gas furnace and the 500 ton reaction frame in Civil Engineering Laboratory at Monash University. Photograph of the fire test rig is shown in Figure 3. The fire test was conformed to the standard AS 1530.4 (1997). Specimen was loaded to the designated level, before it was heated up. The fire temperature prescribed by AS1530.4 (1997) is expressed as following formulation:

$$T_t - T_0 = 345\log(8t+1) \tag{1}$$

Where, T_t and T_0 are furnace temperature at time t and initial furnace temperature, in degree Celsius and t is time into the test, measured from the ignition of furnace, in minutes.

All the specimens were tested till failure. The failure criteria were the axial deformation or the deformation rate of the specimen that met the value prescribed in AS1530.4 (1997).

2.4 Experimental results

The temperatures in the specimens and the axial deformations of specimens versus fire exposed time are shown in Figure 4 and Figure 5, respectively. Specimen S1R2E0 was tested to 90 minutes in fire, but it had not reached its fire endurance. Because the total time for temperature elevating was set to 90 minutes, the fire was extinguished by the control unit and the test stopped.

It can be seen that the temperature elevation in the concrete of larger size specimen is slower than that of smaller size specimen. At the same time, the temperatures in the concrete are much lower than that in interface especially for ones with larger sectional size. From Figure 5, it can be clearly seen that the axial deformation of the specimen in fire usually can be divided into three stages. In the first stage, the expansion deformation of the specimen is greater than the compressive deformation. In the second stage, the compressive deformation accelerates and overcomes the expansion. This is due to the deterioration of concrete and steel at elevated temperature. When

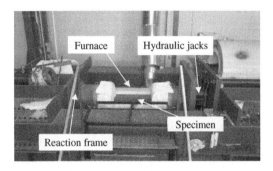

Figure 3. Test rig.

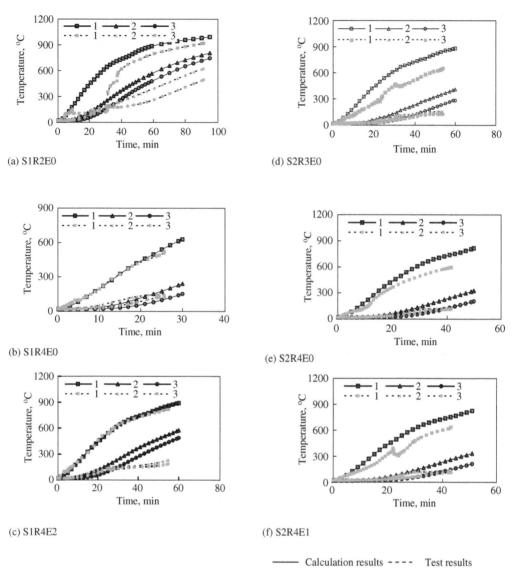

(a) S1R2E0

(b) S1R4E0

(c) S1R4E2

(d) S2R3E0

(e) S2R4E0

(f) S2R4E1

——— Calculation results - - - - Test results

Figure 4.   Temperature in the specimen.

the specimen nears its fire endurance, the compressive deformation and the deformation rate sharply increase. At this moment, the specimen reaches its fire endurance because it can no longer withstand the load applied on it. Fire endurance of the specimens is listed in Table 1.

A typical failure mode is shown in Figure 6. It can be seen that there is an outward buckling of hollow steel section near the mid height of the specimen. Figure 7 shows the core concrete in a typical specimen after test. It was found that the core concrete near the mid height of the specimen, which was corresponding

to the location of hollow steel section buckling, was damaged. Because the temperature in the hollow steel section is higher than that in the core concrete and the mechanical property deterioration of steel is faster than that of concrete at elevated temperature, the buckling of the steel hollow section could have happened before the concrete crashed. The buckling of the steel tube makes the hollow steel section lose its bearing capacity, load that previously borne by the hollow steel section transfers to the core concrete. At the same time, the buckling of the steel tube leads the loss of confinement on the core concrete. The load bearing capacity of

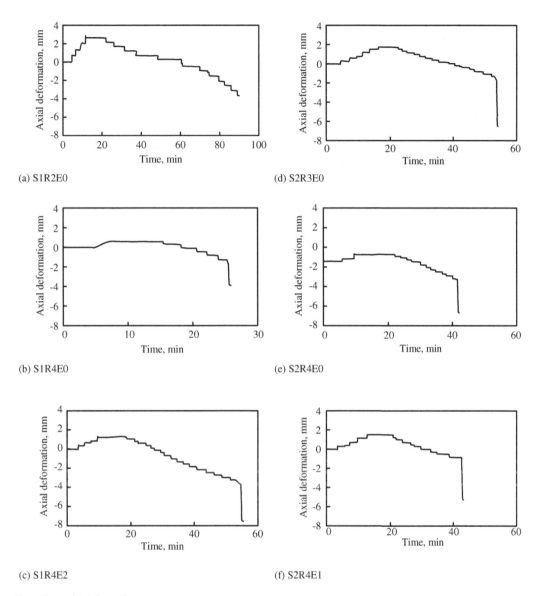

(a) S1R2E0

(d) S2R3E0

(b) S1R4E0

(e) S2R4E0

(c) S1R4E2

(f) S2R4E1

Figure 5. Axial deformation.

the concrete is further weakened. Finally, the specimen collapses quickly.

2.5 Discussion

It can be found that fire endurances of the specimens range from 26 to 90 minutes. Generally speaking, such endurances can not meet the requirements of related design codes. Means have to be used to enhance the fire endurances, such as the commonly used fire protection coatings method. From the test results, it can be seen that the fire endurance increases with the decrease of load density ratio in fire. Using of HSC in hollow steel section may be a way to reduce load density ratio so that the fire endurance can be increased. Taking the specimen S1R2E0 as an example, if the specimen is designed under ambient condition with concrete strength, f_c of 30 MPa, its axially ultimate bearing capacity is 1944 kN (CEN, part 1-1 1994). If the load under fire condition is 486 kN, the load density ratio in fire would be 0.25. In order to reduce the load density ratio in fire, substituting the concrete for high strength SCC with the strength of 99 MPa. In this case, the load density ratio in fire reduces to 0.17. Its fire

407

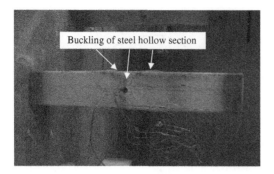

Figure 6. Failure model of typical specimen.

Figure 7. Core concrete after test.

endurance can increase to greater 90 minutes, according to the test. The load density for S1R4E0 would become 0.64 if f_c of 30 MPa was used. Similarly for S2R3E0 which has a fire resistance about one hour, if f_c of 30 MPa was used, a load of 1226 kN would give a load density of 0.4.

3 NUMERICAL SIMULATION

The thermo and mechanical properties of concrete and steel vary with the temperatures. The thermo and the mechanical properties are assumed to be uncoupled. So the numerical analysis of concrete filled HSS in fire is usually divided into two consecutive steps, temperature filed analysis and fire endurance analysis. In the first step, the temperature distribution in the sections in fire derives from solving heat conductive differential equation. The fire endurance analysis is a mechanical analysis process. It simulates the component under structural loads and fire loads from the beginning of the fire till the failure of the component during exposed to fire. The temperature distribution in the section derived from temperature filed analysis serves as one of the input conditions in the second step analysis. The thermo and mechanical properties of concrete and steel at elevated temperature are needed in the analysis of concrete filled HSS in fire. There are

many material property models for steel and various types of concrete (such as NSC and HSC) at elevated temperature proposed by researchers, such as models proposed by Lie, (1994) & Kodur, et al. (2004). But there is no data on the thermo and mechanical properties of SCC at elevated temperature at the present time. So in this numerical simulation, the thermo and mechanical property models were chosen to represent those of SCC through careful analysis and comparison of the properties of different kind of concrete. The predicted temperature distribution and fire endurance of the specimen were compared to the experimental results.

3.1 Temperature field simulation

In the previous investigation on the temperature distribution in SCC columns, it was found that the thermo coefficients for HSC are suitable for predicting the temperatures in SCC columns (Lu, et al. 2005). The HSC thermo properties, proposed by Kodur, et al. (2004), are used to analysis SCC filled HSS. A nonlinear temperature field analysis model (Han, et al. 2003a) for concrete filled HSS at elevated temperature was adopted in the temperature field analysis.

The predicted temperatures at different depth in the specimens corresponding to the location of thermocouples in tested specimens are shown in Figure 4. Comparing the predicted and measured temperatures, it can be seen that there is a general agreement between the calculated and measured temperatures. The temperature difference may be attributed to the migration of moisture when the columns were heated (Lu, et al. 2005). The vaporizing of the moisture has been considered in the simulation, but the moisture migration was not included in the model.

From the comparison of the predicted and test temperatures in the specimen, it can be found that the thermo property coefficients of HSC are suitable for representing those of SCC in the temperature distribution analysis of SCC filled HSS.

3.2 Fire endurance prediction

The mechanical model developed by Han, et al. (2003b) was used to calculation the fire endurance. This is a simplified model for concrete filled HSS fire endurance calculation. In the calculation of fire endurance of concrete filled HSS, several assumptions were adopted in the simplified model, such as plane section remains plane, the deflection curve is in the shape of semi-sine curve, no slip happens between concrete and steel. Detailed description of the model can refer to Han, et al. (2003b) and Han, (2004).

The mechanical properties, or stress and strain relationship and thermo expansion coefficient, of steel and SCC at elevated temperature are needed in the calculation of fire endurance. The steel mechanical properties proposed by Lie, (1994) were adopted in the analysis. Due to the lack of SCC mechanical properties at elevated temperature, appropriate property model should be chosen to represent that of SCC. After careful comparison of different type of concrete mechanical property model, the property model for core concrete in HSS at elevated temperature was chosen (Han, et al. 2003b). This concrete property model was used to prediction both of the NSC and HSC filled HSS and was verified by the test results. In this model, the advantage of the confinement of hollow steel section on the core concrete is considered.

The predicted fire endurance of the specimens is listed in Table 1. 4 out of 6 of the specimens' predicted to experimental fire endurance ratios are within 20%. The predicted fire endurance generally agrees the test result. This indicates that the adopted mechanical property model for concrete at elevated temperature is appropriate for representation that of SCC.

4 SUMMARY

Six SCC filled square hollow structural steel sections were tested under standard fire. The fire resistances range from about 30 minutes to more than 90 minutes depending on the load density and load eccentricity. Numerical simulations were conducted to predict the temperature field and fire resistance. Based on this limited experimental data, reasonable agreement between experimental and calculated results was obtained. More experiment is needed to further verify the numerical model and thoroughly understand the behavior of SCC filled HSS in fire.

ACKNOWLEDGEMENT

The investigation presented was financially supported by Australian Research Council and the National Natural Science Foundation of China (No.50425823). The authors would like thank Smorgon Steel Group Limited and Degussa Construction Chemicals Australia Pty Ltd for providing hollow structural steel sections and superplasticizers.

REFERENCES

Standards Australia 1991. *Structural steel hollow sections, AS 1163*. Sydney: Standards Australia.

Standards Australia 2002. *Structural design actions – General principles, AS 1170*. Sydney: Standards Australia.

Standards Australia 1997. *Methods for fire tests on building materials, components and structures – Combustibility test for materials, AS 1530.4*. Sydney: Standards Australia.

Bostrom, L. 2002. The performance of some self compacting concretes when exposed to fire. *SP Fire Technology, SP Report 2002:23*, Sweden: Swedish National Testing and Research Institute.

CEN 1994. Eurocode 4: *Design of composite steel and concrete structures, Part 1-1: General rules and rules for buildings (together with United Kingdom National Application Document). DD ENV 1994-1-1:1994*, Brussels: European Committee for Standardization.

CEN 1994. Eurocode 4: *Design of composite steel and concrete structures, Part 1-2: General rules-structural fire design. DD ENV 1994-1-2:1994*, Brussels: European Committee for Standardization.

China Ministry of Construction 2002. *Load code for the design of building structures GB 50009*, Beijing: China Architecture & Building Press.

Han, L.H., Xu, L. & Zhao X.L. 2003a. Temperature field analysis of concrete-filled steel tubes, *Advances in Structural Engineering-An International Journal*, 6(2), 121–133.

Han, L.H., Zhao, X.L., Yang, Y.F. & Feng, J.B. 2003b. Experimental study and calculation of fire resistance of concrete-filled hollow steel columns, *Journal of Structural Engineering, ASCE*, 1239(3), 346–356.

Han, L.H. 2004. *Concrete filled tubular structures-from theory to practice*, Beijing: Science Press.

Han, L.H., Yao, G.H. & Zhao, X.L. 2005, Tests and calculations for hollow structural steel (HSS) stub columns filled with self-consolidating concrete (SCC), *Journal of Constructional Steel Research, 61(9), 1241–1269.*

Kodur, V.K.R. 1998. Performance of high strength concrete-filled steel columns exposed to fire, *Canadian Journal of Civil Engineering*, 25(6), pp. 975–981.

Kodur, V.K.R., Wang, T.C. & Cheng, F.P. 2004. Predicting the fire resistance behaviour of high strength concrete columns, *Cement & Concrete Composites*, 26(3), 141–153.

Lie, T.T. 1994. Fire resistance of circular steel columns filled with bar-reinforced concrete, *Journal of Structural Engineering, ASCE*, 120(5), 1489–1509.

Lu, H., Zhao, X.L. & Han, L.H. 2005. Temperatures distribution in self-compacting concrete columns, *4th Australasian Congress on Applied Mechanics*, Melbourne: Institute of Materials Engineering Australasia Ltd.

Persson, B. 2003. Self-compacting concrete at fire temperatures, *Research Report, Division of Building Materials*, Sweden: Lund Institute of Technology.

Twilt, L., Hass, R., Klingsch, W., Edwards, E. & Dutta, D. 1996. Design guide for structural hollow section columns. exposed to fire, Germany: Verlag TUV.

Tubular Structures XI – Packer & Willibald (eds)
© 2006 Taylor & Francis Group, London, ISBN 0-415-40280-8

Numerical studies on HSC-filled steel columns exposed to fire

P. Schaumann & O. Bahr
Institute for Steel Construction, University of Hannover, Hannover, Germany

V.K.R. Kodur
Department of Civil and Environmental Engineering, Michigan State University, East Lansing, MI, USA

ABSTRACT: In recent years, the use of Hollow Structural Section (HSS) steel columns filled with high strength concrete (HSC) is becoming more popular due to numerous advantages they offer over traditional columns. Apart from the aesthetic point of view, this composite system offers structural advantages. The concrete filling improves the load-bearing capacity at ambient temperature and can significantly enhance the fire endurance of the column. However, while the design rules for steel columns filled with normal strength concrete (NSC) are well established, there are many uncertainties for the filling with HSC. Thus, a numerical study is performed using the computer program 'BoFIRE' to investigate the behavior of HSS steel columns filled with HSC at both room and elevated temperatures. The test variables included column slenderness, load eccentricities, concrete compressive strength and cross-sectional shape.

1 INTRODUCTION

The concrete filling of Hollow Structural Section (HSS) steel columns has two main beneficial effects. At ambient temperature, the load-bearing capacity is significantly increased. Thus, it is possible to reduce the dimensions of the cross-section, which results in enhanced usable space in the building. Under fire load, the advantageous thermal properties of concrete lead to increased fire endurance. Therefore additional external fire protection for the steel might be superfluous.

Though the fire performance of HSS columns filled with NSC is well established (Kodur & Lie 1995; Kodur & MacKinnon 2000), there are many uncertainties for HSC-filled steel columns (Kodur 1998; Kodur & McGrath 2003). This is mainly attributed to the material properties of HSC at elevated temperatures that are not that well established. Therefore two different approaches dealing with HSC were considered for the numerical analysis of HSS columns. These are the Canadian provisions according to the standard CAN/CSA-S16-01 and to the work of the researchers Kodur & Sultan (2003) as well as Cheng, Kodur & Wang (2004). In addition, the European material properties according to the codes Eurocode 2, part 1-2 (prEN 1992-1-2) for concrete structures and Eurocode 4, part 1-2 (prEN 1994-1-2) for composite structures were considered.

The provided material models were implemented in the FEM program 'BoFIRE'. A series of parametric studies were carried out on typical HSC-filled HSS steel columns to investigate their behavior at both ambient and elevated temperatures. The studies included the column slenderness, load eccentricities, concrete compressive strength and cross-sectional shape.

2 COMPUTER PROGRAM 'BOFIRE'

All parametric studies were carried out using the transient, non-linear, incremental computer code 'BoFIRE'. This computer program is based on the finite element formulation and written by Schaumann (1984) and further developed by Upmeyer (2001) and Kettner (2005). It is capable of predicting thermal and structural behavior of both steel and composite structures exposed to fire. The program is based on the following principle:

$$R(t) \geq S(t) \tag{1}$$

where $R(t) = $ resistance at time t; $S(t) = $ effect of mechanical action at time of fire exposure t.

The load-bearing capacity of structures $R(t)$, which are charged by a mechanical load $S(t)$ while exposed to fire, depends on the modification of the material properties, such as decreasing of strength and elastic modulus affected by heat. Thus, the procedure for determining the remaining bearing capacity of structures is based on a numerical calculation model coupling the thermal and mechanical response at various time steps.

At first the thermal response takes place. In this stage, the fire temperature and the temperature distribution of the cross-section are computed. According to the temperature distribution, the modification of the material properties caused by temperature can be computed. Subsequently, the mechanical response is calculated where deformation and remaining strength of the members are determined. These results are compared to the applied load on the column and it is verified whether the structure still has sufficient load-bearing capacity. This procedure is repeated for various time steps until the resistance of the member is less than the applied load, which represents failure of the column. The duration to failure is taken as the fire resistance period of the column.

2.1 Thermal response

In 'BoFIRE', the temperature field is calculated using the Fourier differential equation for heat conduction:

$$- \text{div} \left(\lambda \times \text{grad} \, \theta \right) + c \times \rho \times \dot{\theta} - f = 0 \qquad (2)$$

where $\lambda =$ thermal conductivity; $\theta =$ temperature; $c =$ heat capacity; $\rho =$ density; $\dot{\theta} =$ derivation of temperature with respect to time; $f =$ heat source.

Caused by the modifications of material properties due to heat exposure, the differential equation becomes transient since the temperature field gets inhomogeneous. Thus, that equation has to be solved numerically. In the following, the basis of that method will be described according to Kettner (2005).

A mathematical transformation of Equation 2 results in the weak formulation of the differential equation:

$$\int_{\Omega} \lambda \times \text{grad} \, \theta : \text{grad} \, \delta\theta \, dA + \int_{\Gamma} q \times \delta\theta \times n \, dS + ...$$
$$... + \int_{\Omega} \rho \times c \times \dot{\theta} \times \delta\theta \, dA = 0 \qquad (3)$$

where $\Omega =$ area; $\Gamma =$ boundary of considered area; $q =$ heat flux; $n =$ normal vector on the boundary.

For the solution of the weak form, bi-linear shape functions on a four node isoparametric element according to Equation 4 are used.

$$N_i = \frac{1}{4} \times \left(1 \pm \eta\right) \times \left(1 \pm \xi\right) \qquad (4)$$

The approach is presented on the left side of Figure 1. An example for mesh generation with BoFIRE is shown at the right side of Figure 1. Moreover, the computer program 'BoFIRE' also recognizes different material properties as a function of temperature including that of fire protection materials.

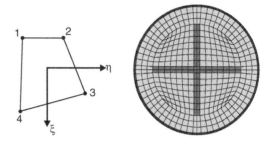

Figure 1. Four node isoparametric element (left) and mesh for a concrete-filled HSS column with embedded X-shape profile (right).

2.2 Mechanical response

It is possible to calculate all types of cross-sections and linear structures as beams, columns or plane frames taking second order theory into account. The calculation is based on the Bernoulli hypothesis for plain state of strains. Shear deformations are not considered.

Due to the nonlinear material properties, cross-sectional values and internal forces depend on the temperature field and strains into the cross-section. The strains are calculated by the balance of internal and external forces. The solution of the incremental system equation is given by Schaumann (1984):

$$\Delta \underline{S}_L - \Delta \underline{S}_{th} = \left(\underline{K}_t^I - \underline{K}_{t_0}^I \right) \times \underline{v}_{t_0} + \Delta \underline{K}^{II} \times \underline{v}_{t_0} + ...$$
$$... \left(\underline{K}_t^I + \underline{K}_t^{II} \right) \times \Delta \underline{v} \qquad (5)$$

where $\Delta \underline{S}_L =$ difference between external forces per time increment; $\Delta \underline{S}_{th} =$ difference between thermal strains per time increment; $(\underline{K}_t^I - \underline{K}_{t_0}^I) \times \underline{v}_{t0} =$ difference of system matrix stiffness (elastic portion); $\Delta \underline{K}^{II} \times \underline{v}_{t0} =$ difference of system matrix stiffness (geometric portion according to second order theory); $(\underline{K}_t^I + \underline{K}_t^{II}) \times \Delta \underline{v} =$ difference of deformations per time increment.

At first, the internal force variables and deformations caused by the external forces $\Delta \underline{S}_L$ are computed in one or more increments. In a parallel calculation the temperature field is established as previously described. Because of the incremental procedure it is possible to linearize the influence of non-linear material behavior and temperature distribution.

3 HSC MATERIAL PROPERTIES

The carried out investigations are based on both Canadian (CAN3-A23.3-M94) and European HSC material properties. The considered codes show differences regarding mechanical as well as thermal properties. Because of its importance for the computation of heated cross-sections, the stress-strain relationship at elevated temperatures is compared. The investigation

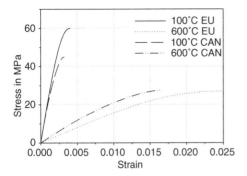

Figure 2. Comparison between Canadian (CAN) and European (EU) stress-strain relationship at elevated temperatures.

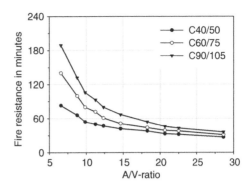

Figure 3. Effect of A/V-ratio on fire resistance period of HSC-filled HSS columns.

is carried out for the case of HSC-filling with siliceous aggregate concrete of 60 MPa strength. To show the characteristics, the ascending branch of the stress-strain relationships is presented only for 100°C and 600°C in Figure 2 according to the code regulations. It is obvious that the Canadian regulations are more conservative since the peak stress is significantly reduced for temperature of 100°C. In contrast to this, the European code prEN 1992-1-2 does not diminish the peak stress for the same temperature. This is also true for temperature of 200°C, which is not presented in Figure 2. For greater temperature of 600°C, the peak stress according to both codes is almost equal. However, the Canadian code assumes a far more brittle HSC behavior since the strain at peak stress is less than the corresponding European value. The differences for other temperatures are less pronounced and thus not presented.

4 EFFECT OF SECTION SIZE

A massive column will show an improved fire resistance over a leaner column (Lie & Kodur 1996), which is expressed in the A/V-ratio, where 'A' is the area of the surface and 'V' is the volume of a member both per unit length.

To illustrate the effect of section size on fire resistance of concrete-filled HSS columns, a parametric study was carried out on ten different circular HSS columns filled with HSC. For excluding effects of slenderness, the selected columns have a length of only 0.50 m (short columns) with fixed-fixed end conditions. The yield strength of the steel tube is assumed to be 235 MPa. As filling, concretes with cylindrical compressive strength of 40, 60 or 90 MPa with calcareous aggregates are chosen. These strengths are equivalent to 50, 75 and 105 MPa based on cube strength of concrete as measured in Europe. The steel sections were selected such that they contribute about one third to the total load-bearing capacity at room temperature. The calculation is carried out with

material properties according to Eurocodes and a load level of 50% at room temperature. The interrelationship between the fire resistance period of a column and its A/V-ratio is apparent in Figure 3.

It is obvious that a column with low A/V ratio ex-poses a relatively small surface to the fire delaying its heating. Thus, the use of HSC as HSS-filling is in particular interesting for massive columns since its thermal properties do not play crucial role as in the case of leaner columns. This results in remarkably enhanced fire resistance. Yet this advantage diminishes with rising A/V-ratios. For the given load level, the application of HSC might not be reasonable for ratios exceeding a value of 15. The parametric study is once more repeated using the Canadian material properties.

For a direct comparison between both approaches, the average of the computed fire resistance period for the tens columns according to the Canadian approach is divided by the analogous value calculated according to the European standard. The ratio for each concrete class is presented in Figure 4. It is apparent that the Canadian results become more conservative with rising concrete strength.

5 EFFECT OF CROSS-SECTION SHAPE

The fire resistance of a column is also influenced by the shape of the cross-section. To illustrate this effect, a numerical study is carried out. Consequently, three of the previous investigated circular HSS cross-sections are taken as the base of comparison. Since it was shown that the A/V-ratio has crucial influence on the fire endurance, the square HSS columns have the same A/V-ratio as the circular HSS columns. In addition, the wall thickness of the square HSS columns is calculated such that the resulting steel area for the square section is the same as for circular cross-section. This is necessary to account for the decreasing material properties of steel at elevated temperatures. The examined columns can be found in Table 1.

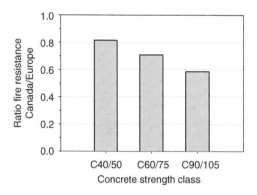

Figure 4. Comparison between European and Canadian material properties for varying A/V-ratios.

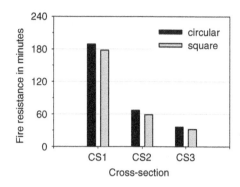

Figure 5. Comparison between circular and square cross-sections with same A/V-ratio.

Table 1. Parameters for numerical study.

Cross-section			Load for C90/105	
		A/V-ratio	Circular	Square
Circular	Square	m^{-1}	kN	kN
CS1 C610 × 16	SQ610 × 12.5	7	15,316	20,628
CS2 C273 × 8	SQ273 × 6.2	15	3117	4214
CS3 C139.7 × 4	SQ139.7 × 3.1	29	813	1103

The HSS columns, which have a load level of 50% at room temperature, are filled with calcareous C90/105 concrete. The first value denotes the compressive strength of the concrete in MPa based on cylinder strength, whereas the second value is based on cube strength. Second-order effects are excluded with a chosen column length of only 0.50 m (short columns) with fixed-fixed end conditions.

The fire endurance for the different HSS steel columns with HSC-filling is presented in Figure 5. It is apparent that the circular cross-sections show a slightly better fire resistance period than the square cross-sections with the same A/V-ratio. Whereas the difference between the massive C610 × 16 mm and SQ610 × 12.5 mm cross-sections is only 6% (CS1), it amounts to about 11% for two other examples (CS2 and CS3).

The lower fire resistance in square columns can be explained based on the fact that in square cross-section corners are heated from two sides which leads to an increased supply of heat and thus earlier failure. This is illustrated in Figure 6. In the case of square columns, the temperature at a given distance from the edge is higher at the corner than that along the sides caused by the heat transfer from two sides.

In contrast to this, the circular cross-section shows a uniform temperature distribution along the radial lines. However, the investigation illustrates that the difference between the two cross-sections is moderate.

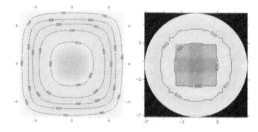

Figure 6. Temperature field for square SQ139.7 × 3.1 mm and circular cross-section C139.7 × 4 mm (CS3) after 32 minutes.

Therefore only circular cross-sections are considered in the remaining case-studies. Considering that square cross-sections show lower fire endurance of about 10%, the numerical results for the circular HSS columns in general allow fire resistance assessment of square cross-sections with comparable A/V-ratio.

6 EFFECT OF SLENDERNESS

Columns are primarily compression members, however they are susceptible to bending with respect to column slenderness and load eccentricity. Thus, a numerical study was carried out to set a reasonable limitation for the use of HSC as HSS-filling.

The parametric study was carried out with a circular cross-section C219.1 × 6 mm and with hinged end conditions. A HSS column was filled with calcareous aggregate concrete with a compressive strength of 40, 60, 80 or 100 MPa based on cylinder strength. This corresponds to compressive strength of 50, 75, 95 and 115 MPa based on cube strength.

With regard to the tube, conventional yield strength of 235 MPa and improved yield strength of 355 MPa are chosen for the first two cross-sections. The third cross-section consists of a tube with yield strength of 235 MPa and additional 4 × 25 mm reinforcement bars.

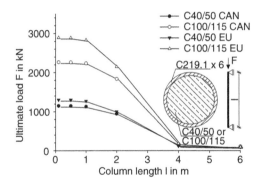

Figure 7. Ultimate load for HSC-filled HSS column with tube yield strength of 235 MPa under centric load using Canadian (CAN) and European (EU) material properties at 30 minutes fire exposure.

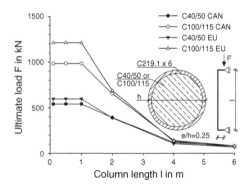

Figure 8. Ultimate load for HSC-filled HSS column with tube yield strength of 235 MPa and eccentric load of e/h = 0.25 using Canadian (CAN) and European (EU) material properties at 30 minutes fire exposure.

The influence of slenderness is examined by varying the column length from 0.1 to 6 m with an assumed imperfection of L/1000. Both the Canadian and European material properties for HSC are used. The cross-sections are heated 30 minutes according to the ISO time-temperature curve. This standard curve according to the European code Eurocode 1, part 1-2 (prEN 1991-1-2) is very similar to the Canadian standard time-temperature curve, which is defined by the code ULC-S101. The short heating period of 30 minutes already shows the unfavorable fire endurance of slender HSC-filled HSS columns. For longer fire exposure times, the decline of ultimate load would be even more rapid.

After the heating process, the load is gradually increased until failure occurs.

6.1 *Calculation with tube yield strength of 235 MPa*

Results from the analysis are plotted in Figure 7, which shows the ultimate load-bearing capacity at 30 minutes as a function of column length. As expected, load carrying capacity decreases with increased length of the column. The Canadian code provisions are more conservative for both NSC and HSC under centric load, which is also illustrated in Figure 7. This is in particular true for the HSC class C100/115 as the European results for non-slender columns exceed the Canadian predictions about 25%.

Nevertheless, the differences between the approaches become less important for column lengths exceeding 2 m where buckling failure is crucial. In view of the fact that the stress-strain relationships compute very comparable results, the elastic modulus is very similar, too. This means in conclusion that no great differences with respect to buckling can be expected.

The results for moderate eccentric loads with e/h = 0.25 are given in Figure 8. The Canadian code provisions still calculate the more conservative results.

Nevertheless, the difference between the approaches is insignificant for column lengths exceeding 2 m.

Due to its low tensile strength, the beneficial effect of concrete with high compressive strength is sharply reduced for slender columns and load eccentricities causing bending moments. Thus, the use of HSC as filling for HSS columns should be limited to non-slender columns with only moderate load eccentricity.

6.2 *Calculation with tube yield strength of 355 MPa*

The average steel temperature of the tube is 860°C for a fire exposure of 30 minutes. Thus, the load share of the steel tube is insignificant since its strength and elastic modulus are considerably reduced. The outcome is that the HSS column with the improved tube yield strength of 355 MPa carries only slightly greater loads than the cross-section with a steel tube having yield strength of only 235 MPa. Therefore the results are not presented in detail.

6.3 *Calculation with tube yield strength of 235 MPa and reinforcement bars*

The addition of reinforcement bars in concrete filling is helpful in improving fire resistance. This can be seen in Figure 9, where the HSS column with the tube of improved yield strength of 355 MPa is compared to the tube with yield strength of 235 MPa and additional 4 × 25 mm reinforcement bars.

With a concrete cover of 50 mm, the reinforcement bars are to some degree protected against heating. The outcome is an average temperature of 500°C in reinforcement bars remarkably reducing the elastic modulus to about 60% of its initial value according to the European code EN 1994-1-2. Nevertheless, in comparison to the directly fire-exposed steel tube with an average temperature of about 860°C and corresponding reduction to only 8% of the initial elastic modulus, the beneficial effect of the column concrete cover is

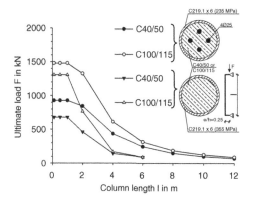

Figure 9. Comparison of ultimate loads between HSS column with tube yield strength of 235 MPa and additional reinforcement bars and tube of improved yield strength of 355 MPa both under eccentric load of e/h = 0.25.

evident. Thus, the cross-section with additional bars shows superior behavior for slender columns since the elastic modulus is crucial for the buckling failure.

While the reasonable use of cross-sections with the improved tube is restricted to about 3 m for the examined column, cross-sections with reinforcement bars are able to bear a comparable amount of load up to a column length of approximately 6 m. In addition, the reinforcement bars allow the HSC to develop its beneficial effect on the load-bearing behavior since the difference between the concrete strength classes are more pronounced than for the cross-section with improved tube yield strength of 355 MPa.

7   CONCLUSIONS

Based on the information presented, the following conclusions can be drawn:

- HSC-filling is particularly beneficial for massive columns with low A/V-ratio as the material properties of HSC at elevated temperatures do not play crucial role as in the case of leaner columns.
- Circular concrete-filled HSS steel columns provide higher fire resistance than square concrete-filled HSS columns of similar area of cross-section.
- The use of HSC as HSS-filling is reasonable for non-slender columns with only moderate load eccentricity. This recommendation is result of investigation on fire exposure of 30 minutes. Since the disadvantageous load-bearing behavior of HSC-filling increases with rising temperatures, the conclusion is also valid for higher fire resistance classes.
- The fire resistance can be significantly enhanced by arranging reinforcement bars into the cross-section. The use of tubes with higher yield strength has very limited effect on the fire performance.

REFERENCES

Canadian Standards Association 1994. Code for the Design of Concrete Structures for Buildings (CAN3-A23.3-M94). Mississauga, Canada.

Canadian Standards Association 2001. Limit States Design of Steel Structures (CAN/CSA-S16-01). Mississauga, Canada.

Canadian Standards Association 1989, Standard Methods of Fire Endurance Tests of Building Construction and Materials (ULC-S101-M89). Toronto, Canada.

European Committee for Standardization (CEN). prEN 1991-1-2 (Eurocode 1) 2002. Actions on structures, Part 1-2: Actions on structures exposed to fire. Brussels, Belgium.

European Committee for Standardization (CEN). prEN 1992-1-2 (Eurocode 2) 2002. Design of concrete structures, Part 1-2: General rules – Structural fire design. Brussels, Belgium.

European Committee for Standardization (CEN). prEN 1994-1-2 (Eurocode 4) 2004. Design of composite steel and concrete structures, Part 1-2: General rules – Structural fire design. Brussels, Belgium.

Cheng, F.-P., Kodur, V.K.R. & Wang, T.-C. 2004. Stress-strain curves for high strength concrete at elevated temperatures. Journal of Materials in Civil Engineering 16(1): 84–94.

Kettner, F. 2005. Investigations on the load-bearing behavior of composite columns under fire conditions. Institute for Steel Construction, University of Hannover, Germany.

Kodur, V.K.R. & Lie, T.T. 1995. Fire performance of concrete-filled hollow steel columns. Journal of Fire Protection Engineering 7(3): 89–98.

Kodur, V.K.R. 1998. Performance of high strength concrete-filled steel columns exposed to fire. Canadian Journal of Civil Engineering 25: 975–981.

Kodur, V.K.R. & MacKinnon, D.H. 2000. Fire endurance of concrete-filled hollow structural steel columns. AISC Steel Construction Journal 37(1): 13–24.

Kodur, V.K.R. & McGrath, R. 2003. Fire endurance of high strength concrete columns. Fire Technology 39(1): 73–87.

Kodur, V.K.R. & Sultan, M.A. 2003. Effect of temperature on thermal properties of high-strength concrete. Journal of Materials in Civil Engineering 15(2): 101–107.

Kodur, V.K.R. 2003. Fire resistance design guidelines for high strength concrete columns. ASCE/SFPE Specialty Conference of Designing Structures for Fire and JFPE, Baltimore, MD, USA.

Lie, T.T. & Kodur, V.K.R. 1996. Fire resistance of steel columns filled with bar-reinforced concrete. Journal of Structural Engineering 122(1): 30–36.

Schaumann, P. 1984. Computation of steel members and frames exposed to fire (in German: 'Zur Berechnung stählerner Bauteile und Rahmentragwerke unter Brandbeanspruchung'). Technisch-wissenschaftliche Mitteilungen Nr. 84-4. Institut für konstruktiven Ingenieurbau, Ruhr-Universität Bochum, Germany.

Upmeyer, J. 2001. Fire design of partially encased composite columns by ultimate fire loads (in German: 'Nachweis der Brandsicherheit von kammerbetonierten Verbundbauteilen über Grenzbrandlasten'). Schriftenreihe des Instituts für Stahlbau der Universität Hannover, Heft 19, Germany.

Tubular Structures XI – Packer & Willibald (eds)
© 2006 Taylor & Francis Group, London, ISBN 0-415-40280-8

The behaviour of rotationally restrained concrete filled tubular columns in fire

J. Ding & Y.C. Wang

School of Mechanical, Aerospace and Civil Engineering (MACE), The University of Manchester, Manchester, UK

ABSTRACT: Concrete filled tubular (CFT) columns are widely used all over the world, due to their significant advantages in construction speed and high fire resistance. Columns are critical members of a structure. Because of various interactions between a column and its adjacent members in a complete structure, the loads and boundary conditions of the column change under fire conditions. Such variations should be evaluated to enable proper design of the column as part of a frame structure. This paper presents the results of a series of numerical studies on the behaviour of rotationally restrained CFT columns exposed to the standard fire condition. The numerical simulations are conducted using a general finite element analysis package ANSYS. The finite element model is first validated by comparing with the results of a series of full-scale fire tests on rotationally restrained CFT column assemblies conducted at the University of Manchester. The test assembly consisted of a CFT column jointed by a pair of short beams from two sides using extended end plate connections. The column was loaded axially and the applied loads on the beams varied to give different amounts of initial bending moment in the column. By comparing the failure times of the column assemblies to the isolated columns under different bending moments, it was conclude that the design column bending moment may be taken as that arising from the unbalanced beam load acting on the surfaces of the steel tube. Another objective of the numerical simulations is to investigate the effect of rotational restraint on column effective length, which confirmed the recommendation in EN 1994-1-2. In addition, the numerical simulations confirm that the slip between the steel tube and the concrete core has no significant influence on the failure time of CFT columns.

1 INTRODUCTION

The advantages of concrete filled tubular (CFT) columns are numerous, including attractive appearance, structural efficiency, reduced column footing, fast construction and high fire resistance without external fire protection. Due to their wide use in buildings, the fire performance of CFT columns has been extensively studied by a number of researchers (Wang & Davies 2003a, Kimura et al. 1990, Lie 1992).

Columns are critical members of a structure. Because of various interactions between a column and its adjacent members in a complete structure, the loads and boundary conditions of the column change under fire conditions. It has long been recognized that the fire performance of isolated structural members is different from that of whole structures. However, most of the current studies have concentrated on the fire performance of isolated CFT columns. Therefore, it is necessary to evaluate the interactions between a column and the adjacent structure, so that the columns can be properly designed as part of a frame structure under fire conditions.

This paper first describes the finite element model generated using a general finite element analysis package ANSYS. The finite element model is validated by comparing with the results of a series of full-scale fire tests on rotationally restrained CFT column assemblies recently conducted at the University of Manchester (Wang & Davies 2003a).

Then the effect of bending moment on the column fire resistance has been investigated using the validated finite element model. The results of the column assemblies subjected to different combined loads show that the bending moments transferred from the beams to the column affects the failure time of the CFT columns and should be included in the calculations. By comparing these results to the isolated columns under different bending moment, it is suggested that the design column bending moment may be taken as that arising from the unbalanced beam loads acting on the surfaces of the steel tube.

A series of numerical studies on the behaviour of rotationally restrained CFT columns exposed to the standard fire condition is carried out using the same finite element method. As a result, the failure time of each column is determined, as well as the effective length of the column. These results confirm the recommendation in EN 1994-1-2 that for columns continuous at one end, the effective length should be taken as 0.7 times the column height.

2 VALIDATION OF THE FINITE ELEMENT MODEL

2.1 Description of the fire tests

A series of full-scale fire tests on rotationally restrained CFT column assemblies has been conducted at the University of Manchester. Two series of column assemblies have been fire tested. Figure 1 shows a schematic arrangement of the test set up. In each test, a hot-finished rectangular steel tube of grade S275 and dimension of either RHS 200 × 100 × 5 mm or RHS 200 × 100 × 12.5 mm, filled with nominally C30 concrete, was connected to a pair of steel beams at one end using extended end plate connections. Details of the test setup can be found in the related papers (Wang & Davies 2003a, b).

As shown in Figure 1, the column was connected to the reaction frame at the foot by a roller bar, which acts as a pin joint about the minor axis of the column. Connections between the restraining beams and the reaction frame were sliding joints. And the column top was connected to the loading system outside the furnace which acts as a pin joint about the minor axis, but can move along the column height direction. Figure 2 shows the dimensions of the test assembly and Figure 3 shows the boundary conditions of the test assembly.

Each series of tests consist of four combinations of two levels of total axial load in the column, with either equal or unequal loads on the connected beams. Two levels of the total column axial load, representing about 30% and 70% of the column compressive strength at ambient temperature, were applied. Due to different thickness of the steel tubes, the two series of fire tests show different behaviour. There was no local buckling

in columns using the thicker tube. For columns using the thinner tube, local buckling was observed. The tests using RHS 200 × 100 × 5 mm are simulated by finite element method in this paper.

2.2 Finite element analyses

As mentioned before, finite element analysis package ANSYS has been used to conduct this simulation (Ding & Wang 2005). Since temperature distributions in a CFT column is highly non-uniform and it is difficult to record the concrete temperature during a fire test, the finite element analyses involves both thermal and structural parts.

It should be noted that the material properties are non-linear and temperature dependent. The material properties at elevated temperatures recommended in EN 1994-1-2 (CEN 2004) are adopted in this paper. The following material properties have been used to simulate the fire tests:

- Concrete cylinder strength at ambient temperature is 60 N/mm^2.
- Yield strength of steel at ambient temperature is 300 N/mm^2.

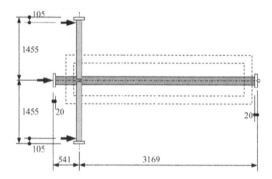

Figure 2. Dimensions of test assembly.

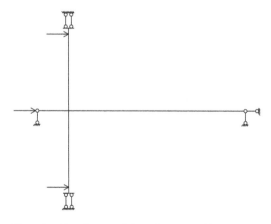

Figure 3. Boundary conditions of test assembly.

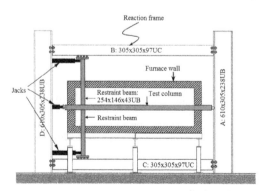

Figure 1. Schematic arrangement of test.

418

2.2.1 Thermal model

In general transient distribution within a structural member is described by a differential equation with 3 variables for space coordinates and 1 variable for the time. For practical analysis, the column is treated as uniformly heated lengthwise in the analyses. Thus, the temperature distribution within a composite column cross section can be analyzed by using a 2-D model. Because of symmetry of the geometry and the boundary conditions, only one quadrant is analyzed, as shown in Figure 4.

Non-linearity due to the temperature dependent material properties and boundary conditions (convection and radiation) are taken into account. This study adopted the material thermal properties for steel and concrete recommended in Eurocode 4 Part 1.2 and the recommended heat transfer coefficients given in Eurocode 1 Part 1.2 (CEN 2000). The moisture content in the concrete is chosen as 4%.

Due to high thermal mass of the concrete filled column assemblies relative to the small size of the furnace, the recorded average furnace temperature was lower than the standard fire curve. The fire temperature-time relationships for ANSYS simulations are the average recorded furnace temperatures.

2.2.2 Structural model

By taking advantage of symmetry, only half of the assembly is analyzed. As shown in Figure 5, the model is symmetrical according to the x-z plane. 3-D solid elements are used to model steel tube and concrete core, while beam elements are used for the restraint beams and shell elements for the endplates. The boundary conditions shown in Figure 3 are used in the model. Contact elements have been used to model the surface between the steel tube and the concrete core, which can simulate the composite action between steel and concrete, so that slip between steel and concrete is allowed.

Two properties are required for the contact elements which are Coulomb coefficient and maximum bond stress between steel and concrete. There is no information of these two values at elevated temperatures at the moment. Therefore, the values at ambient temperature are taken in the model, where Coulomb coefficient equals to 0.8 and maximum bond stress is chosen to be 0.5 MPa (Wium & Lebet 1994).

Figure 6 shows details of the extended end plate connection. Considering the difficulty of modelling the bolts, the end plates are simply treated as welded to the steel tube.

2.3 Comparing with test results

Finite element simulations have been performed for four tested assemblies with cross-section RHS $200 \times 100 \times 5$ mm. Table 1 gives details of the applied loads of each test, as well as the test and finite element analyses results. The FEA results are acceptable considering the various uncertainties inherent in the test data, such as the heating condition along the height of the column, the degree of rotational restraint at the column ends, the unintentional eccentricity of axial load, the initial imperfection of the column, etc.

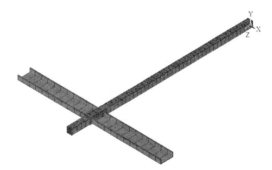

Figure 5. 3-D Structural model.

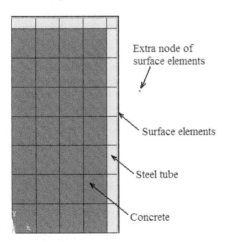

Figure 4. 2-D thermal model.

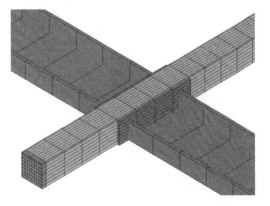

Figure 6. Details of beam to column connections.

Table 1. A summary of test and FEA results.

| Test no. | Applied loads (kN) | | | Test | FEA |
	Upper beam	Column	Lower beam	Failure time (s)	Failure time (s)
SCR41	37.5	272.5	0	2620	2079
SCR42	37.5	235	37.5	2700	2038
SCR43	88	632	0	1860	1277
SCR44	88	544	88	920	1034

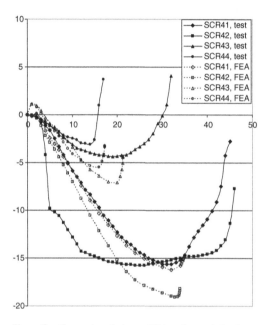

Figure 7. Comparison between FEA and recorded column axial deformations.

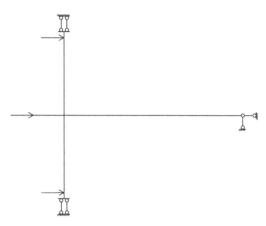

Figure 8. Boundary conditions used in parametrical analyses.

Table 2. List of combined loads and failure time of calculated column assemblies.

| Column no. | Applied loads (kN) | | | FEA |
	Upper beam	Column	Lower beam	Failure time (min)
SCR41a	37.5	235	37.5	34.58
SCR41b	37.5	272.5	0	33.73
SCR41c	75	235	0	33.21
SCR43a	88	544	88	21.64
SCR43b	88	632	0	20.13
SCR43c	176	544	0	19.09

Figure 7 compares the FEA predicted and the measured column axial deformations. Although there are differences between the curves, they appear to have the same trend. Specially, the curves of test assembly SCR41 agree with each other very well. The differences might be caused by the imperfection of the assemblies and the supports, non-uniform heating inside the furnace and unknown of the real material strength. From the above results, this finite element model is considered to be validated to be able to predict the fire resistance of the CFT columns.

3 EFFECT OF BENDING MOMENT

Because extended end plate connections were used in the tests, bending moments were transferred from the loaded beams to the test column. It has been noticed from the tests that the column bending moments decrease during fire exposure. To investigate the effect of bending moment on the column fire resistance, a set of parametrical analyses has been carried out.

Test asemblies SCR41 and SCR43 are investigated in this parametrical study. The same finite element model described in section 2 has been used, but with slightly different boundary conditions which are shown in Figure 8. Also the recorded furnace temperatures are used in the calculations. The column is loaded axially and the applied loads on the beams are varying to give different amounts of initial bending moment in the column. Table 2 lists the applied loads and the FEA results of each assembly.

It can be seen from the results in Table 2 that the bending moments transferred from the beams to the column have some effect on the failure time of the CFT columns and should be included in the calculations. The test results showed that the column bending moments decrease during fire exposure. It is important to know how much bending moments have been transferred from the beams to the column at the fire limit state for design purpose. Therefore, the isolated column model is created to evaluate the bending moment.

Table 3. Failure time of isolated columns with different bending moments.

Column no.	Bending moment 1		Bending moment 2	
	$M = N \cdot e^*$ (kNm)	Failure time (min)	M_i (kNm)	Failure time (min)
SCR41a	0	33.63		
SCR41b	1.875	31.44	11.91	29.43
SCR41c	3.75	30.95	23.80	27.60
SCR43a	0	21.76		
SCR43b	4.4	20.07	27.95	15.22
SCR43c	8.8	19.13	55.84	10.56

*Bending moment 1 = unbalanced beam load x connection eccentricity from column centre line (50 mm).

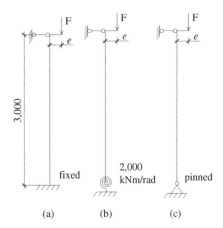

Figure 9. Three different rotational restraints.

Table 4. Rotational stiffness of the columns.

Section no.	Section size (mm)	Column height (mm)	Rotational stiffness (kNm/rad)
A	RHS $200 \times 100 \times 5$	3000	1,425
B	SHS $200 \times 200 \times 6.3$	3000	9,583
C	SHS $200 \times 200 \times 12.5$	3000	13,867

The isolated column model has the height of 3169 mm with pin joint at one end and fixed joint at another end. The column is subjected to combined axial load and bending moment. Contact elements are used in the model to simulate the composite action between steel and concrete.

The following two values of column bending moment are considered: the unbalanced beam loads acting on the faces of the steel tube and the initial bending moment in the subframe column at ambient temperature. Adopting the first value allows all connections to be treated as pinned. The second value assumes that the bending moment remains as the initial value under fire condition. The calculated results are listed in Table 3.

By comparing the results of the column assemblies in Table 2 with the results of the isolated columns in Table 3, it is found that the isolated column subjected to the first bending moment value gives similar failure time as the corresponding column assembly. When the initial bending moments in the subframe columns at ambient temperature were applied to the isolated columns, much lower column failure times are obtained. Therefore, the first bending moment value which allows for all connections to be treated as pinned is suggested in column design calculations.

4 NUMERICAL STUDIES

To investigate the effect of rotational restraint on column effective length, a series of numerical studies on rotationally restrained CFT columns exposed to standard fire condition has been carried out by using the same finite element method. The finite element model has been further simplified. Only the column itself is modelled, with pin joint at one end and various rotational restraints at another end. The column is free to expand. Figure 9 shows details of the load and supports. The columns are heated according to ISO-834 standard fire curve.

Three cross-sections are chosen for the numerical study, which are RHS $200 \times 100 \times 5$ mm, SHS $200 \times 200 \times 6.3$ mm and SHS $200 \times 200 \times 12.5$ mm. All columns have the same height, which is 3 m. The column model with cross-section size RHS $200 \times 100 \times 5$ mm is designed to fail about the minor axis. Therefore, the three different columns reflect different slenderness. They also have different rotational stiffness, which is calculated as 3EI/L, shown in Table 4.

The load ratio R is the ratio of the applied axial load to the column's strength at ambient temperature. Column strengths at ambient temperature were calculated using EN 1994-1-1 (CEN 1992), assuming an effective length ratio of 1.0. Two values of load ratio were used in this study, 0.3 and 0.7, representing the lower and upper bounds on the load ratio. The load was applied to the column head either centrally or with an eccentricity $e = 0.1B$, where B is the width of the cross-section about the bending axis. Two different finite element models are used. One includes contact elements to model the bond between steel and concrete, so that slip between steel and concrete is allowed. Another one assumes fully bond between steel and concrete, so that no slip is allowed between steel and concrete.

Table 5. FEA results of RHS 200 × 100 × 5 mm.

Column no.	R	e (mm)	F (kN)	Effective length L_e	Rotational stiffness (kNm/rad)	Failure time (min) With slip	Failure time (min) Without slip
A1a	0.3	0	268	0.7 L	fixed	21.18	21.62
A1b	0.3	0	268	0.7 L	2,000	21.11	21.47
A1c	0.3	0	268	L	pinned	18.72	15.84
A2a	0.3	10	199	0.7 L	fixed	23.27	23.67
A2b	0.3	10	199	0.7 L	2,000	23.21	23.55
A2c	0.3	10	199	L	pinned	20.56	18.12
A3a	0.7	0	626	0.7 L	fixed	7.08	8.04
A3b	0.7	0	626	0.7 L	2,000	6.98	7.74
A3c	0.7	0	626	L	pinned	5.91	5.43
A4a	0.7	10	465	0.7 L	fixed	11.87	11.84
A4b	0.7	10	465	0.7 L	2,000	11.50	11.40
A4c	0.7	10	465	L	pinned	7.96	6.40

* L is the height of the column.

Table 6. FEA results of SHS 200 × 200 × 6.3 mm.

Column no.	R	e (mm)	F (kN)	Effective length L_e	Rotational stiffness (kNm/rad)	Failure time (min) With slip	Failure time (min) Without slip
B1a	0.3	0	660	0.7 L	fixed	38.87	39.24
B1b	0.3	0	660	0.7 L	2,000	37.70	37.99
B1c	0.3	0	660	L	pinned	25.40	26.86
B2a	0.3	20	508	0.7 L	fixed	38.15	39.03
B2b	0.3	20	508	0.7 L	2,000	36.93	37.89
B2c	0.3	20	508	L	pinned	26.36	27.19
B3a	0.7	0	1540	0.7 L	fixed	18.58	17.73
B3b	0.7	0	1540	0.7 L	2,000	17.77	16.62
B3c	0.7	0	1540	L	pinned	7.65	7.48
B4a	0.7	20	1185	0.7 L	fixed	17.68	18.27
B4b	0.7	20	1185	0.7 L	2,000	17.56	17.39
B4c	0.7	20	1185	L	pinned	10.77	10.56

Table 7. FEA results of SHS 200 × 200 × 12.5 mm.

Column no.	R	e (mm)	F (kN)	Effective length L_e	Rotational stiffness (kNm/rad)	Failure time (min) With slip	Failure time (min) Without slip
C1a	0.3	0	969	0.7 L	fixed	33.98	34.00
C1b	0.3	0	969	0.7 L	2,000	33.26	33.37
C1c	0.3	0	969	L	pinned	27.72	28.16
C2a	0.3	20	740	0.7 L	fixed	33.92	35.06
C2b	0.3	20	740	0.7 L	2,000	33.37	34.60
C2c	0.3	20	740	L	pinned	29.52	29.82
C3a	0.7	0	2261	0.7 L	fixed	19.72	18.44
C3b	0.7	0	2261	0.7 L	2,000	18.70	16.54
C3c	0.7	0	2261	L	pinned	8.97	9.45
C4a	0.7	20	1728	0.7 L	fixed	21.42	21.42
C4b	0.7	20	1728	0.7 L	2,000	20.17	21.14
C4c	0.7	20	1728	L	pinned	13.68	13.55

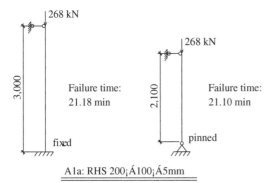

A1a: RHS 200¡Á100¡Á5mm

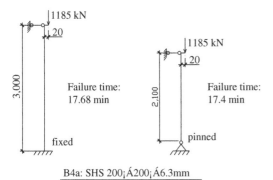

B4a: SHS 200¡Á200¡Á6.3mm

C3a: SHS 200¡Á200¡Á12.5mm

Figure 10. Structural analyses conducted to check the column effective length at the fire limit state (check drawing).

This numerical study considers three different rotational restraints, which are fixed, 2,000 kNm/rad and pinned. The materials used are grade S275 steel and C30 concrete. The results of finite element analyses for different cross-sections are listed in Tables 5–7 respectively.

Column No. in Tables 5–7 consists of three letters, the first capital letter refers to the cross-section type, the number in the middle refers to different load cases which include different load and eccentricity combinations, the last letter refers to three different rotational

restraints. For example, Column No. 'A1a' means the column with cross-section RHS $200 \times 100 \times 5$ mm fixed at one end subjected to load case 1 which has load ratio 0.3 and 0 eccentricity.

The column effective length listed in Table 5–7 are determined by the deformed column shape, which is taken as the distance from the largest rotation point to the pin end. To check the validity of the effective length measured from the finite element model, the failure times of column A1a, B4a and C3a have been compared with the failure times of the similar pin-ended columns with height of $0.7 L = 2.1$ m, but subjected to the same load combinations. The results are displayed graphically in Figure 10. It can be seen clearly that the failure times of the simply supported columns with a height of 0.7 L under the same load and eccentricity are almost same as the ones from the rotationally restraint columns.

From the results in Table 5–7, it has been found that different rotational restraints have effect on the column effective length. As a result, stiffness of the rotational restraint affects the failure time of the CFT column. The results show that the columns with one fixed end or a rotationally restraint end with the stiffness of 2,000 KNm/rad and pinned at another end have an effective length of 0.7 L, where L is the height of the column. The rotational stiffness of a CFT column is usually higher than 2,000 KNm/rad, see Table 4. Therefore, the effective length ratio of columns continuous at one end should be taken as 0.7, which confirms the recommendation in EN 1994-1-2.

In addition, by comparing the results calculated using two different finite element models, with or without slip between the steel tube and the concrete core, it is confirmed that slip has no significant influence on the failure time of CFT columns.

5 CONCLUSIONS

This paper presents a 3-D finite element modelling of CFT column. A validation of this FEM has been conducted by comparing the calculated results with the results of a series of full-scale fire tests on rotationally restrained CFT column assemblies conducted at the University of Manchester.

It has been noticed from the tests that the column bending moments decrease during fire exposure. Some investigations have been carried out on the column assemblies subjected to different bending moments. It can be seen from the results that the bending moments in the column affect the failure time of the CFT columns and should be included in the calculations. By comparing the failure time of the column assemblies to the isolated columns with different bending moments, it is suggested that the design column bending moment may be taken as that from the unbalanced beam loads acting on the surfaces of the CFT column, which will allow all connections to be treated as pinned under fire conditions as far as inducing bending moment is concerned.

This paper also presents the results of a series of numerical studies on the behaviour of rotationally restrained CFT columns exposed to the standard fire condition. It has been found that with different rotational restraints, the column will have different effective lengths. Thus, the stiffness of the rotational restraint affects the failure time of the CFT column in fire. It has also been noticed that for columns continuous at one end, the effective length should be taken as 0.7 times the column height.

In addition, the numerical simulations confirm that the slip between the steel tube and the concrete core has no significant influence on the failure time of CFT columns.

REFERENCES

Ding, J. & Wang, Y.C. 2005. Finite element analysis of concrete filled steel columns in fire, *Fourth international conference on advances in steel structures, June, 2005.* Shanghai, China.

European Committee for Standardisation (CEN). 1992. Eurocode 4: Design of composite steel and concrete structures, *Part 1.1: General rules and rules for buildings.* London: British Standards Institution.

European Committee for Standardisation (CEN). 2000. Eurocode 1: Basis of Design and Actions on Structures, *Part 1.2: Actions on Structures – Actions on Structures Exposed to Fire.* London: British Standards Institution.

European Committee for Standardisation (CEN). 2004. Eurocode 4: Design of Composite Steel and Concrete Structures, *Part 1.2: Structural Fire Design.* London: British Standards Institution.

Kimura, M., Ohta, H., Kaneko, H. & Kodaira, A. 1990. Fujinaka H. *Fire resistance of concrete-filled square steel tubular columns subjected to combined loads.* Takenaka Technical Research Report 43: 47–54.

Lie, T.T. & Chabot, M. 1992. *Experimental studies on the fire resistance of hollow steel columns filled with plain concrete.* Internal Report No. 611. National Research Council of Canada.

Wang, Y.C. 1999. The effects of structural continuity on the fire resistance of concrete filled columns in non-sway columns. *Journal of Constructional Steel Research* 50: 177–97.

Wang, Y.C. & Davies, J.M. 2003a. An experimental study of the fire performance of non-sway loaded concrete-filled steel tubular column assemblies with extended end plate connections. *Journal of Constructional Steel Research* 59: 819–838.

Wang, Y.C. & Davies, J.M. 2003b. Fire tests of non-sway loaded and rotationally restrained steel column assemblies. *Journal of Constructional Steel Research* 59: 359–383.

Wium, J.A. & Lebet, J.P. 1994. Simplified calculation method for force transfer in composite columns. *Journal of Structural Engineering* 120: 728–746.

Tubular Structures XI – Packer & Willibald (eds)
© 2006 Taylor & Francis Group, London, ISBN 0-415-40280-8

Shear behavior of fin plates to tubular columns at ambient and elevated temperatures

M.H. Jones & Y.C. Wang
University of Manchester, Manchester, UK

ABSTRACT: Following recent events such as the World Trade Center building collapse and the Cardington large scale structural fire research program, the fire behavior of connections has now become an important research subject. This paper reports the preliminary results of an experimental and numerical investigation to understand the behavior of various components of simple shear connections to steel tube and concrete-filled tubular (CFT) columns at elevated temperatures, which compliments the CIDECT sponsored research project on fire performance and robustness of steel beam – CFT column structural assemblies, which has recently started at the University of Manchester. Experimental studies are now being conducted in an electrically heated kiln to collect data on the behavior of fin plate, end plate and reverse channel connectors connected to short steel and CFT columns under tension or shear force in the connectors at various elevated temperatures. This paper will mainly present the results of a parametric numerical investigation of the shear behavior of fin plates at elevated temperatures.

1 INTRODUCTION

Following events such as the World Trade Center building collapse and the Cardington structural fire research program, the fire behavior of connections has become an important research subject. Traditionally, steel structures have been designed for their performance in fire with reference to the standard fire resistance test results of idealized structural elements. This 'prescriptive' approach assumes the need for fire protection of steel structures and the thicknesses of such fire protections are duly prescribed in various industrial standards.

Recent research into the fire protection requirements of both steel and composite structures has focused upon the development of a 'performance-based' approach to such designs which endeavors to take account of more specific circumstances which may affect a steel structure in fire. With regard to the specific effects of the interactions between different structural elements the philosophy of the performance-based approach is now being applied in order to develop simplified computation methods to describe the behavior of such connections. The European standard, Eurocode 3 for steel structures (CEN 2003), enshrines some of these computation methods for a variety of welded and bolted connections between steel I-beam and H-columns.

The design method presented in Eurocode 3 Part 1.8 (CEN 2003), hereinafter referred to as the 'component method', is founded upon the principle that a beam-column connection may be reduced to a series of discrete elements whose behavior is dependent upon their location within the connection, for example the connection behavior in shear or tension. The various components are isolated and treated as individual 'spring' systems with unique stiffness and strength. The individual components may then be added together to describe the global connection behavior.

Extensive research has been conducted into the behavior of the connection types mentioned above and standards such as Eurocode are readily available for such designs. However, there is a comparative lack of research into the component behavior of simple welded connections to hollow or concrete-filled tubular (CFT) columns. CFTs are used increasingly in tall, multi-storey buildings as an alternative to bare steel or reinforced columns. They allow for a comparatively reduced column cross-sectional area and also possess inherent fire-resistance properties due to the insulative qualities of the concrete in-fill. The aim of this research program is to extend the component method of Eurocode to encompass certain connections to CFTs. This paper presents the results of preliminary experimental and numerical investigations into the shear behavior of welded fin plate – tubular connections.

The component method described in Eurocode 3 does not encompass welded fin plate connections. However, guidance is given as to the design plastic shear resistance, $V_{pl,Rd}$, of a plate of cross-section, A. This resistance is defined by Equation 1.

$$V_{pl,Rd} = A(f_y/\sqrt{3})$$ (1)

where f_y is the yield stress of the steel.

Eurocode also allows for the influence of combined bending and shear force by noting the reduction in plastic moment capacity due to the presence of shear. This is encapsulated in Equations 2 and 3.

$$M_{V.Rd} = M_{pl.Rd}\left(1 - \left[\frac{2V_{Sd,fin}}{V_{pl,Rd}} - 1\right]^2\right)$$ (2)

$$M_{V.Rd} = V_{Sd,fin}.l$$ (3)

where $V_{Sd,fin}$ = resistance in fin plate due to bending and shear; l = lever arm length; $M_{V.Rd}$ = reduced design plastic resistance moment allowing for shear force; and $M_{pl.Rd}$ = design plastic moment resistance.

The Eurocode guidance does not, however, take account of the effect of tubular column properties upon the behavior of such a fin plate under shear load when connected to a tubular column. The research presented in this paper is intended to fill this gap in the research field.

Although no design guide has yet been found by the authors concerning the interaction between fin plate and steel hollow section (SHS) column under shear load, there have been several experimental investigations into such connections at ambient temperatures. White & Fang (1966) conducted research into the behavior, including shear performance, of five types of welded connections to square hollow section columns. The closed nature of the column cross-section led White and Fang to conclude that welding was the only practical method for fastening the connection to the tube. The research highlighted the significance of the ratio of width of tube wall to tube thickness. It was noted that as this ratio increases, connections fastened directly to the tube wall tended to become more flexible. Simple plate connections, fillet welded to the tube wall, were observed to produce distortion of the tube wall as the plate rotated under load. When such connections were loaded directly in shear White and Fang encountered several failure modes in different specimens including local buckling of the tube, weld tearing and web crippling of the connected beam. Sherman (1995) conducted a large series of tests upon realistically loaded framing connections between SHS

columns and wide flange beams under predominant shear load. These tests expanded upon those conducted by White & Fang. One limit state was identified for the column face under such load. This limit state was punching shear failure due to the attached beam end rotation when the shear tab connection was thick relative to the SHS wall. The criterion to avoid this failure mode, as described by Packer & Henderson (1997), is to ensure that the tension resistance of the shear tab under axial load (per unit length) is less than the shear resistance of the SHS wall along two planes (per unit length). Thus the thickness of the plate, t_p, is limited with respect to the tube wall thickness, t_O, by the following relationship.

$$t_p < \left(\frac{F_{uO}}{F_{yp}}\right)t_O$$ (4)

This paper presents the preliminary results of an investigation intended to address the current lack of understanding of the influence of SHS column properties upon the shear behavior of simple connections at both ambient and elevated temperatures.

2 SHEAR TEST PROGRAMME

A series of shear tests has been started in order to investigate the interaction between welded connections and both filled and unfilled CFT columns. The tests and results presented in this paper investigate the interaction between welded fin plate connections and unfilled square hollow section columns.

2.1 Test set-up

The test specimens consist of T-stubs cut from a UB $203 \times 102 \times 23$ section fillet welded at the web to a SHS column. The column length is 450 mm and the fin plate is welded halfway along the column length such that there is an equal amount of bare column face above and below the fin plate. Thus the machined web of the T-stub acts as a fin plate connection. This arrangement is depicted in Figures 1 and 2.

The thickness of the T-stub web, effectively a fin plate, is 6 mm. The depth of the web is 100 mm. The SHS column is 5 mm thick with a width of 200 mm. The test specimen is anchored to the reaction frame at the top via a short I-section and at the bottom via a 30 mm thick steel strap as shown in Figure 2 and close-up in Figure 3.

Four M20 grade 8.8 bolts connect the T-stub flange to the steel strap. The applied shear force is transmitted to the connection via a hydraulic load jack which pushes one end of the top cross-beam which is pinned at the other end.

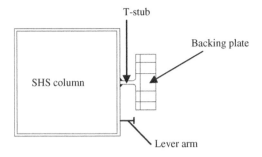

Figure 1.   Cross-sectional overview of shear test specimen.

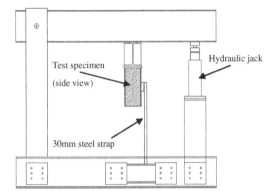

Figure 2.   Test rig with specimen mounted for shear test.

Figure 3.   Detail of fin plate connection.

2.2   *Test results*

Two ambient temperature tests have so far been conducted. In the first test, the lever arm was 60 mm long, this being the distance from the column face to the plane of applied load at the interface between the steel

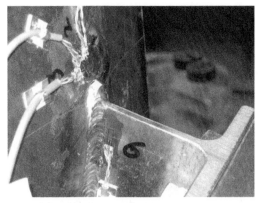

Figure 4.   Weld pull-out at top of fin plate connection.

Figure 5. Experimental shear failure of the fin plate connection.

strap and the T-stub flange. During testing, it was found that the column face was under combined bending and shear, instead of being subject to predominantly shear force.

Consequently the connection was rotating excessively and the test was terminated due to the weld pulling out from the column face at the top of the fin plate, which is pictured in Figure 4. The failure load was 84.1 kN, below the theoretical value, $V_{Sd,fin}$ (cf. Equation 2), which was calculated at 104.1 kN. It was considered that a lever arm of 60 mm was unrealistically long, therefore, to reduce excessive rotation of the fin plate connection during future tests and to ensure that the shear is the primary action in the connection, another test was done with a reduced lever arm of 30 mm.

As expected, the modified connection specimen was observed to fracture in shear through the web of the T-stub as shown in Figure 5. The fracture line occurred from the web root at the top of the connection diagonally to the opposite corner at the bottom of the fin plate. Compared to the T-stub web the tube wall did not deform significantly indicating that the failure of the specimen was due almost entirely to shear in the T-stub web. The recorded failure load at the fin plate was 189.4 kN. The calculated shear resistance using Equation 2 is 131.8 kN.

3 FINITE ELEMENT (FE) MODEL

ABAQUS (2004) is used to model the aforementioned tests and to perform parametric numerical studies. The parametric studies are performed to investigate the following aspects of connection behavior:

- The influence of ratio of fin plate thickness, t_f, to column thickness, t_c, for a fixed value of t_f of 6 mm and a fixed connection lever arm of 30 mm.
- The influence of lever arm length for different restrained connections.
- The effect of elevated temperatures with varying t_f/t_c ratio for a fixed lever arm of 30 mm.

These parameters were identified for investigation for the following reasons. Firstly, to confirm the observation from the initial experimental test with 60 mm lever arm where the failure mode was weld pull-out due to excessive rotation of the connection rather than shear in the plate. From this observation the authors anticipated that lever arm was an influential factor in determining the failure mode of the connection.

Secondly, noting the observation by White & Fang that connection flexibility increased with increased ratio of tube wall width to tube thickness, the t_f/t_c ratio was varied in order to examine the effect of this ratio upon connection failure mode and to estimate the influence of this ratio on connection shear failure load.

Thirdly, the effect of elevated temperatures upon such connections was investigated, to provide guidance on future experimental studies to be carried out by the authors.

3.1 Material properties

Tensile tests were performed upon three coupons cut from the UB section web and three coupons from the SHS tube. The results of these tensile tests are summarized in Tables 1 and 2. The values detailed in Tables 1 and 2 were averaged and used in conjunction with the engineering stress-strain curves obtained from the Instron testing machine to define the material models in ABAQUS.

3.2 Comparison of FE results with tests and model refinement

ABAQUS has the facility to analyze non-linear 3D models created using solid or shell elements. Initially

Table 1. SHS tensile coupon test results.

Test no.	Young's modulus, E (GPa)	Ultimate tensile stress, f_u (MPa)	Yield stress at proportional limit, f_P (MPa)
1	237.742	460.696	363.722
2	233.546	453.307	338.965
3	234.311	453.587	353.467

Table 2. Beam web tensile coupon test results.

Test no.	Young's modulus, E (GPa)	Ultimate tensile stress, f_u (MPa)	Yield stress at proportional limit, f_P (MPa)
1	193.994	464.818	357.880
2	199.985	454.816	338.535
3	214.298	451.790	338.748

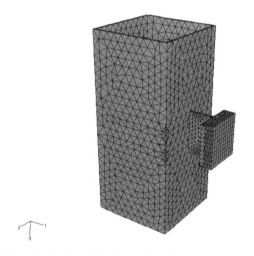

Figure 6. Typical solid element FE model.

a model was created using non-linear 20-node solid elements (designation C3D20), which is shown in Figure 6. To consider the restraining effect of the steel strap, the movement of the T-stub flange perpendicular to the column face is prevented.

For the test specimen with lever arm = 60 mm the solid element model reached the failure point at a load of 98.1 kN, which is in excellent agreement with the calculated theoretical value $V_{Sd,fin}$ of 104.1 kN. At this point on the fin plate shear force – bending moment interaction curve the bending moment is substantial with $M_{V.Rd}/M_{pl.Rd}$ ratio = 0.904. For the test specimen with a lever arm of 30 mm, the solid element model reached the failure point at an equivalent

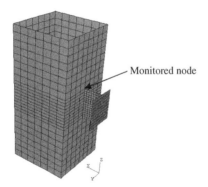

Figure 7. Typical shell element model.

Table 3. Results of parametric study.

t_f/t_c	$L_N/V_{Sd,fin}$	Monitored nodal displacement (mm)	Primary yield zone
0.3	1.186	0.19	fin plate
0.6	1.139	0.70	fin plate
1.2	1.063	1.44	fin plate
1.5	1.025	1.65	fin plate
2	0.977	1.94	fin plate
2.4	0.949	2.13	fin plate
3	0.929	2.59	fin plate
4	0.811	1.68	column face
6	0.494	0.12	column face

load of 145 kN in the connection, which is only 76% of the experimental load (189.4 kN). This FE model value is consistent with the theoretical failure load of 131.8 kN, calculated using Equation 2. At this load the corresponding $M_{V.Rd}/M_{pl.Rd}$ ratio is 0.565. The yield contour in the FE model was observed to be greatest at the top of the fin plate near to the T-stub web root radius, which agrees with the observed yield pattern from the experimental shear test.

The reduction in numerical failure load compared to the experimental failure load may be attributed to using the engineering stress-strain relationship rather than the true stress-strain relationship of the fin plate steel, in the FE model.

Although the solid element model described above yielded results reasonably consistent with the experimental results the time taken to complete the solution was around 5 hours. Reducing the mesh density reduced this time to around 2 hours without significantly affecting the simulation results. However, in order to drastically cut computational time so as to perform the parametric studies a simpler model constructed from non-linear 8-noded shell elements (designation S8R5) was developed, which is shown in Figure 7. For the test with 30 mm lever arm, the failure load observed with a relatively fine mesh was 140.1 kN, almost the same as that obtained when using solid elements and in good agreement with theory.

3.3 Results of parametric FE analyses

The same material properties as in the previous section were used in the parametric studies.

3.3.1 Effects of varying t_f/t_c

In this study the backing plate is restrained in the plane of the fin plate perpendicular to the loading direction. The fin plate thickness, t_f, is fixed at 6 mm and the ratio of t_f to column wall thickness, t_c, varied from 0.3 ($t_c = 20$ mm) to 6 ($t_c = 1$ mm). The lever arm is fixed as 30 mm. Table 3 presents the variation of numerical failure load, L_N, to the theoretical failure load due

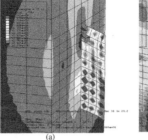

Figure 8. Contour plots depicting different failure yield modes for t_f/t_c ratios of (a) 0.6 and (b) 6.

to bending and shear of the plate, $V_{Sd,fin}$, of a plate of similar dimensions was recorded in order to determine the effect of changing the thickness ratio on the connection failure mode and failure load.

$V_{Sd,fin}$ was calculated using Equation 2. Table 3 also gives the displacement perpendicular to the initial column face position of a single node at the top of the fin plate-tube connection (see Figure 7) which was monitored in order to measure the degree to which the column wall was pulling out. For each model the location of the primary yield zone is given in Table 3 to indicate the connection failure mode.

As the t_f/t_c ratio was increased it was observed that yield was transferred from the fin plate to the column face. As this occurs the $L_N/V_{Sd,fin}$ ratio also decreases since the failure mode is no longer purely shear failure of the fin plate but is greatly influenced by the column face flexibility. This is shown in Figure 8 for t_f/t_c ratios of 0.6 and 6. from Figure 8(a), when the t_f/t_c ratio is smaller, the failure mode is observed to be primarily shear failure of the fin plate with some pull out of the column face. When the t_f/t_c ratio is high, the failure mode is completely shear deformation in the column face as shown in Figure 8(b).

The change of failure mode from the fin plate to the column face occurs at $t_f/t_c = 3$. at lower t_f/t_c values, Equation 2 appears to predict results that are reasonably close to the ABAQUS simulation results. However, at $t_f/t_c > 3$ the correlation between L_N and $V_{Sd,fin}$ is poor.

3.3.2 Varying the lever arm

In the analysis described in the previous section the T-stub flange was restrained from movement perpendicular to the column face in order to represent the experimental test set-up. As a result the nodal pull-out at the top of the connection was restrained by a compressive force and was consequently very small. In realistic construction, the fin plate could be unrestrained.

Figure 9 depicts the variation of $L_N/V_{Sd,fin}$ with increasing t_f/t_c ratio for unconstrained fin plate models with lever arm $l = 30$ and 60 mm. The curve for the constrained fin plate model, $l = 30$ mm, is included for reference. For realistic t_f/t_c ratios of less than 2 the constrained model gives results which are in good agreement with the theory. However, it may be observed from Figure 9 that the effect of removing the fin plate restraint is to significantly reduce the $L_N/V_{Sd,fin}$ ratio with increasing t_f/t_c. The failure mode is no longer shear in the fin plate but has progressed to rotational failure of the column face due to fin plate rotation. This rotational failure is more pronounced with increased lever arm and leads to greatly increased pull-out of the column face at the top of the connection as Figure 10 shows.

3.3.3 Conclusions from FE analysis at ambient temperature

From the above analyses the following observations can be made

- As t_f/t_c is increased in the constrained fin plate model the failure mode changes from fin plate shear for $t_f/t_c < 3$ to yield in the column face due to connection rotation.
- Removing the fin plate constraint for a similar lever arm length reduces the t_f/t_c ratio at which yield of the column face due to connection rotation becomes the dominant failure mode.
- Increasing the lever arm reduces the failure load of the connection and increases column face deformation.
- Calculating the fin plate shear capacity should include the effect of column tube thickness, which tends to reduce the fin plate connection shear strength as t_f/t_c increases.

3.3.4 Proposal for a new design method

A new design method is proposed which will take into account the effect of shear failure of the column face in addition to combined shear and bending failure of

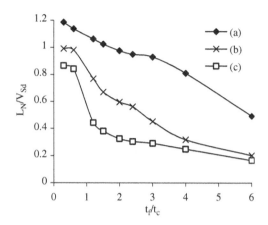

Figure 9. $L_N/V_{Sd,fin}$ vs. t_f/t_c for (a) $l = 30$ mm: constrained fin plate and (b) $l = 30$ mm and (c) $l = 60$ mm: unconstrained fin plate.

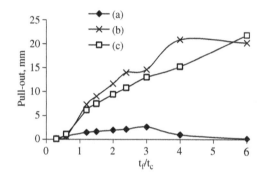

Figure 10. Nodal pull-out vs. t_f/t_c for (a) $l = 30$ mm: constrained fin plate and (b) $l = 30$ mm and (c) $l = 60$ mm: unconstrained fin plate.

the fin plate. The shear capacity of the column face is calculated using the following equation.

$$V_{Sd,SHS} = 2Dt_{SHS}\left(f_y/\sqrt{3}\right) \qquad (5)$$

Where $D =$ depth of the fin plate connection; and $t_{SHS} =$ steel tube thickness.

This shear failure of the column, $V_{Sd,SHS}$, is then combined with the failure strength of the fin plate, $V_{Sd,fin}$ (calculated using equation 2) by using the following equation:

$$\frac{1}{V_{Sd,com}} = \frac{1}{V_{Sd,fin}} + \frac{1}{V_{Sd,SHS}} \qquad (6)$$

The curves depicted in Figure 9 were modified to incorporate $V_{Sd,com}$ and the modified curves are shown in Figure 11.

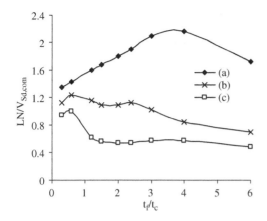

Figure 11. $L_N/V_{Sd,com}$ vs. t_f/t_c for (a) $l = 30$ mm: constrained fin plate and (b) $l = 30$ mm and (c) $l = 60$ mm: unconstrained fin plate.

Figure 11 shows that for realistic t_f/t_c ratios (<3) the calculation of $V_{Sd,com}$ is conservative for the constrained model case (lever arm $= 30$ mm) and in good agreement with the numerical results for the unrestrained model with lever arm of 30 mm. For the unconstrained model with lever arm $= 60$ mm, the ABAQUS simulation results are much lower than those calculated using equation (6). This indicates that the effect of bending should be considered when calculating the shear resistance of the column. Equations (5) and (6) will be refined to improve correlation with the numerical results.

3.4 Elevated temperature FE analysis

A model representing the experimental shear test was used to investigate the effects of elevated temperatures upon the failure mode and failure load of the model for various t_f/t_c ratios. The elevated temperature numerical analysis presented in this paper has thus far concentrated upon one lever arm case, $l = 30$ mm, of a 6 mm thick fin plate welded to a 200×200 mm SHS column.

3.4.1 Material properties
Eurocode 3 Part 1.2 (CEN 2001) allows for the influence of elevated temperature upon steelwork by employing reduction factors applicable to the effective yield strength, f_y; the proportional limit of the stress-strain curve, f_p, and the Young's modulus, E, of the steel at temperature, θ, relative to the same properties at ambient temperature. The elevated temperatures are applied to the whole model as a field in an ABAQUS analysis step. The temperatures analyzed in the current analysis and the applicable reduction factors as detailed in Eurocode 3 Part 1.2 (CEN 2001) are tabulated in Table 4.

Table 4. Eurocode material property reduction factors for steel at elevated temperature.

Temperature (°C)	$f_{y,\theta}/f_y$	$f_{p,\theta}/f_p$	E_θ/E
20	1.0	1.0	1.0
200	1.0	0.807	0.9
300	1.0	0.613	0.8
400	1.0	0.420	0.7
600	0.47	0.180	0.310
800	0.11	0.050	0.090

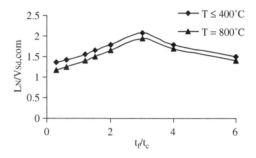

Figure 12. $L_N/V_{Sd,com}$ vs. t_f/t_c for ambient and elevated temperatures.

3.4.2 Results of elevated temperature analysis
The same range of t_f/t_c ratios was investigated as in the ambient analysis described earlier with lever arm $= 30$ mm. Figure 12 depicts the variation of numerical shear load normalized with respect to the combined theoretical shear load, $L_N/V_{Sd,com}$, with increasing t_f/t_c ratio. The elevated temperature curves differ very little from both the ambient case and each other. The behavior of the connection under a uniformly applied temperature field is very similar to the ambient analysis. For the case considered in this paper, the proposed model gives close and safe predictions to the ABAQUS simulation results.

4 FUTURE EXPERIMENTAL WORK

The experimental and numerical work presented herein are part of a research project to enhance understanding of component behavior of connections to CFT columns both at ambient and elevated temperature which will include the following experimental investigations.

- The experimental and numerical analysis thus far carried out into the interaction between fin plate and SHS column under shear load is to be extended to CFT columns both at ambient and elevated temperatures. The geometry of such connections will be developed with regard to the observations presented in this paper in order to confirm the results of the numerical analysis. As well as welded fin plate connections, tests will also be performed on welded reverse channel connections as well as a new endplate blind bolting technique known as Molabolt (www.molabolt.co.uk).

• Tensile behavior of the connections.

5 SUMMARY

This paper has presented the results of an experimental and numerical study of shear behavior of fin plates welded to steel tubes at ambient and elevated temperatures. Based on the results of this study, the following main conclusions may be drawn:

• The behavior of the fin plate connection in shear is affected by the restraint condition to the fin plate. Without restraining the fin plate perpendicular to the tubular face, the fin plate shear strength is much lower. For restrained fin plate connection, if the plate thickness to tube thickness ratio is within the realistic range (<3) so that failure is in the fin plate, Equation 6 gives conservative results for the fin plate strength.
• The fin plate shear strength is greatly affected by the steel tube. The higher the fin plate thickness to tube thickness ratio, the greater the reduction in fin plate strength compared to the theoretical fin plate shear strength without considering the steel tube.
• Consideration of the failure due to bending of the column face should be made when considering the overall failure load of the connection.
• A simple method is proposed in this paper to combine the failure load of the fin plate with that of the steel tube in pure shear. The modified calculation results are much better correlated to the ABAQUS simulation results. However, the model will be

further refined to take into account bending in the steel tube.

• On the assumption that the steel tube and fin plate increase in temperature uniformly, the elevated temperature results follow the same pattern as at ambient temperature.

ACKNOWLEDGEMENTS

This work was made possible by the sponsorship of an EPSRC industrial CASE award to the first author in which the industrial partner is Corus Plc. The authors would like to thank Mr. Andrew Orton of Corus for his interest in the research. Thanks are also due to Mr. Jim Gee of the Structures Division, University of Manchester, for fabricating the test specimens and assisting with the tests.

REFERENCES

CEN. 2001. Eurocode 3: Design of steel structures, Part 1.2: General rules – structural fire design. Draft for development, European committee for standardization. Document DD ENV 1993-1-2: 2001.
CEN. 2003. Eurocode 3: Design of steel structures, Part 1.8: Design of joints. European committee for standardization. Document prEN 1993-1-8: 2003.
Hibbit, Karlsson & Sorensen, inc. 2004. ABAQUS manuals version 6.3. Providence RI, USA.
Packer, J.J. & Henderson, J.E. 1997. *Hollow structural section connections and trusses: a design guide, 2nd Ed.*, Canadian Institute of Steel Construction.
Sherman, D.R. 1995. Simple framing connections to HSS columns. *Proc. American Institute of Steel Construction, National steel construction conference, San Antonio, Texas, USA*. 30.1–30.16.
White, R.N. & Fang, P.J. 1966. Framing connections for square structural tubing. *Journal of the structural division*, American Society of Civil Engineers 92(ST2): 175–194.

Tubular Structures XI – Packer & Willibald (eds)
© *2006 Taylor & Francis Group, London, ISBN 0-415-40280-8*

Buckling of circular steel arches subjected to fire loading

M.A. Bradford

School of Civil & Environmental Engineering, The University of New South Wales, Sydney, Australia

ABSTRACT: This paper derives a geometrically nonlinear formulation for pin-ended circular steel tubular arches, typically in a cladded roofing structure, at elevated temperature. An energy method is used to derive the in-plane equations of equilibrium in differential form, which are solved to produce an analytic solution in closed form, as well as the elastic buckling load under thermal loading, where the thermal strain is treated as a non-mechanical strain in the derivation. It is shown that the arch may buckle in an antisymmetric or in a symmetric snap-through mode, depending primarily on its slenderness and the external load on the arch, and the paper presents prescriptive equations for the buckling loads of the arch as a function of the temperature of the fire. The application of the equations to design is discussed.

1 INTRODUCTION

Tubular steel arches are aesthetic elements used in many engineering structures, such as in the support of roof cladding for open-plan sporting facilities, vehicle shelters and the like. This is because they are lightweight members subjected to predominant axial compression only, so that they offer efficient design solutions. One application to a large-span railway station roofing system is shown in Figure 1 (Bradford & Pi 2003). Steel arches in these structures may be subjected to thermal loading induced by fire, and their behaviour is influenced greatly by this fire loading. The effect can be very significant if the arch is shallow, as the response of the arch is highly nonlinear geometrically, and material nonlinearities are important owing to the degradation of the yield strength of the steel with an increase in temperature.

When an arch is in the configuration depicted in Figure 2 with the fire loading as shown, a pressure differential can be developed owing to the confinement of the heated portion of the structure, which may produce a net suction loading on the arch. This hydrostatic loading is normal to the arch, and results in a compression in the arch rib and so the influence of this compressive force on possible arch buckling must be borne in mind. Of course, arches of the type shown in Figure 1 are necessarily designed to resist wind-induced inward radial loading that is produced by internal suction and/or external pressure loading (Bradford & Pi 2003). This paper considers the scenario of the arch subjected to elevated temperature in combination with a pressure-induced suction.

Despite the ubiquity of arches as structural elements, the in-plane analysis of arches including geometric nonlinearity is complex (Pi et al. 2002), and it was recently shown that four different and quite independent solutions exist for the problem (Bradford et al. 2005, Tong et al. 2005). This paper makes general recourse to the formulation for the nonlinearity

Figure 1. Spencer street station redevelopment, Melbourne, Australia (spencer street redevelopment authority).

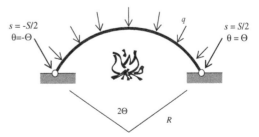

$s = -S/2$
$\theta = -\Theta$

q

$s = S/2$
$\theta = \Theta$

2Θ

R

Figure 2. Steel arch under thermal action.

presented by Pi et al. (2002) (as is done in Bradford et al. 2006) to formulate the problem and to derive its solution, as an extension of the treatment of Bradford (2006). The results are discussed subsequently, with the practical ramifications being discussed.

2 IN-PLANE EQUILIBRIUM

2.1 Strain-displacement relationship

This paper assumes the validity of the Euler-Bernoulli hypothesis, from which Pi & Trahair (1998) determined the strain-displacement relationship for a point $P \in \Omega$ ($\in \Re^2$) on the cross-section of the arch to be

$$\varepsilon_p = \widetilde{w}' + \tfrac{1}{2}\widetilde{v}'^2 + \tfrac{1}{2}\widetilde{w}'^2 - y\left\{\frac{\widetilde{v}''(1+\widetilde{w}') - \widetilde{v}'\widetilde{w}''}{\sqrt{(1+\widetilde{w}')^2 + \widetilde{v}'^2}} + \kappa_0\left[\sqrt{(1+\widetilde{w}')^2 + \widetilde{v}'^2} - 1\right]\right\} \quad (1)$$

in which v and w are the radial and tangential displacements respectively, $\kappa_0 = R^{-1}$ is the initial curvature of the arch, y is the radial coordinate of P relative to the centroid of Ω, $(\cdot)' \equiv d(\cdot)/ds$, $s \in [-S/2, S/2]$ (Figure 2) is the tangential coordinate of the arch profile of total length S, and in which

$$\widetilde{v}' = v' + w\kappa_0 \quad \text{and} \quad \widetilde{w}' = w' - v\kappa_0. \quad (2)$$

Equation 1 is clearly very complex, and the formulation used here (and that of Pi et al. 2002) can be derived more easily by simplifying it to be

$$\varepsilon_p \approx \widetilde{w}' + \tfrac{1}{2}v'^2 - yv''. \quad (3)$$

Assuming the application of a uniform temperature strain ε_T with respect to s ($d\varepsilon_T/ds = 0$), the total strain can be decomposed into mechanical (ε_e) and non-mechanical (thermal) (ε_T) strains, so that

$$\varepsilon_p = \varepsilon_T + \varepsilon_e = w' - \frac{v}{R} + \tfrac{1}{2}v'^2 - yv''. \quad (4)$$

2.2 Nonlinear equilibrium equation

The mechanical strain ε_e produces a mechanical stress (in the elastic range) of $\sigma = E \cdot \varepsilon_e$, in which E is the temperature-dependent elastic modulus. This stress is then

$$\sigma = E\left(w' - v/R + \tfrac{1}{2}v'^2 - \varepsilon_T - yv''\right). \quad (5)$$

Invoking the statement of the principle of virtual work for the arch produces

$$\delta\Pi = \int_{Vol} \delta\varepsilon_p \cdot \sigma \, d(Vol) = \int_{-S/2}^{S/2} q\,\delta v\,ds$$

$$\forall \, \delta w, \delta v, \delta w', \delta v', \delta v'', \quad (6)$$

where $\delta w, \delta v, \delta w', \delta v'$ and $\delta v''$ are kinematically admissible variations of the arch deformations, and Vol is the volume of the arch. It is assumed here that the arch

axes are centroidal so that $\int_\Omega y\,dA = 0$ and that the arch temperature varies through the depth of Ω according to

$$\varepsilon_T = \varepsilon_{T0} + y\nabla_T, \quad (7)$$

where ∇_T is the (assumed linear) thermal strain gradient, given as $\nabla_T = (T_{\text{internal}} - T_{\text{external}}) \cdot \alpha_T/h$ in which T is the temperature relative to ambient, h is the depth of the cross-section and α_T is the (assumed constant) coefficient of thermal expansion.

Equation 7 may be integrated twice by parts to produce

$$AE\varepsilon_m \delta w\Big|_{-\frac{S}{2}}^{\frac{S}{2}} + \left(AE\varepsilon_m v' - EIv'''\right)\delta v\Big|_{-\frac{S}{2}}^{\frac{S}{2}} + EI\left(\nabla_T + v''\right)\delta v'\Big|_{-\frac{S}{2}}^{\frac{S}{2}}$$

$$+ \int_{-S/2}^{S/2}\left[EIv^{iv} - AE\left(v'\varepsilon_m\right)' - \left(\frac{AE}{R}\right)\varepsilon_m - q\right]\delta v\,ds$$

$$- \int_{-S/2}^{S/2} AE\varepsilon_m'\delta w\,ds = 0, \quad (8)$$

where A is the area of the cross-section Ω and I is the second moment of this area, so that AE is the axial stiffness and EI is the flexural stiffness, and where

$$\varepsilon_m = w' - \frac{v}{R} + \tfrac{1}{2}v'^2 - \varepsilon_{T0} \quad (9)$$

is the membrane strain defined at the centroid of the locus of the cross-sections. Since the variations δw are arbitrary,

$$\varepsilon_m' = 0 \quad \Rightarrow \quad \varepsilon_m = -\frac{N}{AE} = \text{const} \quad (10)$$

in which N is the axial force in the arch (which is constant with respect to s) and because the variations δv are also arbitrary in Equation 8,

$$EIv^{iv} + N\left(v'' + 1/R\right) - q = 0. \quad (11)$$

The kinematic boundary conditions for the arch are $w = 0$ and $\delta w = 0$ at $s = \pm S/2$, and $v = 0$ and $\delta v = 0$ at $s = \pm S/2$. Using these in Equation 8 produces the additional static boundary conditions that

$$-EI\left(v'' + \nabla_T\right)\Big|^{S/2} = 0 \quad \text{and} \quad -EI\left(v'' + \nabla_T\right)\Big|_{-S/2} = 0. \quad (12)$$

The solution of the fourth order differential Equation (11) subject to the boundary conditions $v(\theta = -\Theta) = v(\theta = \Theta) = 0$ and $v''(\theta = \pm\Theta) = -\nabla_T$, in which

$$\mu^2 = \frac{N}{EI}, \quad \theta = \mu s \quad \text{and} \quad \alpha = \frac{\mu S}{2}, \quad (13)$$

is

$$v = \frac{\omega}{\mu^2 R}\left[\left(\frac{\cos\theta}{\cos\alpha} - 1\right)\left(1 + R\nabla_T\right) + \tfrac{1}{2}\left(\theta^2 - \alpha^2\right)\right], \quad (14)$$

434

where

$$\omega = (qR - N)/N \qquad (15)$$

is the dimensionless force in the arch.

The nonlinear equation of equilibrium can be obtained by reasoning that the membrane strain in Equations 9 and 10 when averaged over the domain $s \in [-S/2, S/2]$ must equal $-N/(AE) = -\mu^2 r^2$, in which $r = \sqrt{(I/A)}$ is the radius of gyration of the arch cross-section. For a pinned arch, $\int w\prime \mathrm{d}s = w(-\Theta) - w(\Theta) = 0$, so the equation of equilibrium can be found from

$$\int_{-\Theta}^{\Theta} \left(\tfrac{1}{2} v\prime^2 - v/R\right) \mathrm{d}\theta = 2\Theta\left(\varepsilon_{T0} - \mu^2 r^2\right). \qquad (16)$$

Using Equation 14 and its derivative in Equation 16 leads to the equation of equilibrium in the quadratic equation, after some mathematical manipulation, which is given by

$$\gamma_0 + \gamma_1\omega + \gamma_2\omega^2 = 0. \qquad (17)$$

Equation 17 is transcendental, because the coefficients γ_0, γ_1 and γ_2 are functions of the load parameter $\alpha = \mu S/2 \propto \sqrt{N}$. These coefficients are given by

$$\gamma_0 = \frac{R\nabla_T}{2}(R\nabla_T + 4)(\alpha - \tan\alpha) + \frac{R\nabla_T}{2}\alpha\tan^2\alpha$$
$$+ \left(\frac{\alpha}{\lambda_s}\right)^2 - 4\varepsilon_{T0}/\Theta^2, \qquad (18)$$

$$\gamma_1 = \frac{1}{3}\left(1 + \frac{3R\nabla_T}{2}\cdot\frac{\tan^2\alpha}{\alpha^2}\right) + \frac{1}{\alpha^2}\left(1 + \frac{3R\nabla_T}{2}\right).$$
$$\left(1 - \frac{\tan\alpha}{\alpha}\right), \quad \text{and} \qquad (19)$$

$$\gamma_2 = \frac{1}{6} + \frac{1}{4\alpha^2}\left(5 + \tan^2\alpha - \frac{5\tan\alpha}{\alpha}\right), \qquad (20)$$

in which

$$\lambda_s = \frac{\Theta S}{4r} = \frac{S^2}{4rR} \qquad (21)$$

is a modified slenderness parameter. For the non-thermal case that $\varepsilon_{T0} = 0$ and $\nabla_T = 0$, then

$$\gamma_0 = \left(\frac{\alpha}{\lambda_s}\right)^2 \quad \text{and} \quad \gamma_1 = \frac{1}{3} + \frac{1}{\alpha^2}\left(1 - \frac{\tan\alpha}{\alpha}\right), \qquad (22)$$

with γ_2 being given by Equation 20, and these are the same as those derived by Pi et al. (2002) and Bradford et al. (2006) for an arch without thermal loading.

3 IN-PLANE BUCKLING

3.1 Differential equation of buckling

Neutrality of the equilibrium position defined by $\{v, w\}$ requires that the second variation of the total potential vanishes for any kinematically admissible variations δv and δw, indicating a possible transition from the prebuckled configuration to an adjacent buckled one defined by $\{v + \delta v,\ w + \delta w\}$. In the prebuckled configuration, the total potential may be written as

$$\Pi = \tfrac{1}{2}\int_{Vol} E\varepsilon_P^2 \,\mathrm{d}(Vol) - \int_{-S/2}^{S/2} q v \,\mathrm{d}s, \qquad (23)$$

so that equilibrium requires that the variation

$$\delta\left\{\int_{-S/2}^{S/2} \mathsf{F}(w_b', v_b, v_b', v_b'')\mathrm{d}s\right\} = 0. \qquad (24)$$

Using the Euler-Lagrange equations of variational calculus on the functional F leads to the differential equation of equilibrium (11). In the buckled configuration, however, and using the notation $(\ )_b = \delta(\ )$ leads to

$$\delta\left\{\int_{-S/2}^{S/2} \mathsf{F}_b(w_b', v_b, v_b', v_b'')\mathrm{d}z\right\} = 0, \qquad (25)$$

in which

$$\mathsf{F}_b = AE\left(\varepsilon_{mb}^2 + \varepsilon_m v_b'^2\right) + EIv_b''^2, \qquad (26)$$

with

$$\varepsilon_{mb} \equiv \delta\varepsilon_m = w_b' - v_b/R + v'v_b' \qquad (27)$$

being the membrane strain during buckling. For $\{\cdot\}$ in Equation 25 to be stationary,

$$\varepsilon_{mb}' = 0 \qquad (28)$$

so that the membrane buckling strain is constant, as well as the Euler-Lagrange equations requiring that

$$\frac{v_b^{iv}}{\mu^2} + v_b'' = \frac{\varepsilon_{mb}}{\mu^2 r^2}\left(v'' + \frac{1}{R}\right). \qquad (29)$$

3.2 Antisymmetric buckling

Antisymmetric buckling represents a bifurcation from the prebuckling symmetric prebuckled path to an adjacent antisymmetric mode. From Equation 28, the constant buckling membrane strain is

$$\varepsilon_{mb} = \tfrac{1}{S}\int_{-S/2}^{S/2} \varepsilon_{mb}\,\mathrm{d}s. \qquad (30)$$

Since v_b and $v\prime v\prime_b$ in Equation 27 are odd functions within the interval $[-S/2,\ S/2]$, $\varepsilon_{mb} = 0$ for antisymmetric buckling, and so

$$v_b^{iv}/\mu^2 + v_b'' = 0 \qquad (31)$$

whose solution that satisfies $v_b = v_b''$ at $s = \pm S/2$ produces the eigenvalue condition that $\sin \alpha = 0$, for which

$$\alpha = \pi. \tag{32}$$

3.3 Symmetric buckling

Using Equation 14 in Equation 29 produces

$$\frac{v_b^{iv}}{\mu^2} + v_b'' = \frac{\varepsilon_{mb}}{\mu^2 r^2 R}\left[1 + \omega\left(1 - \frac{\cos\theta}{\cos\alpha} + \frac{\nabla_T R\cos\theta}{\cos\alpha}\right)\right]. \tag{33}$$

This equation may be solved by using Equation 30 and cancelling the nonzero ε_{mb} from both sides of the resulting equation. This process leads to a formulation for symmetric buckling similar to that in Equation 17, but the analytic derivation of the coefficients can be overly complicated.

3.4 Buckling check

The load-deflection response under a thermal regime that defines the thermal strain ε_{T0}, as well as the dimensionless thermal gradient parameter $R\nabla_T$ and external load q, can be determined using Equation 17, with the path parameter α on the load-displacement locus defining the in-plane configuration of the arch. This equation can be solved iteratively (Waterloo Maple 2005). When the condition that $\alpha = \pi$ is attained, antisymmetric buckling may occur. The current equilibrium state at α can be checked for symmetric buckling by using the argument of Section 3.3. However, since the nonlinear symmetric response is traced with the parameter α, a peak in the load-deflection response signifying symmetric or snap-through buckling is evident on the load-deflection curve, using Equation 14 (with $\theta = 0$ if the crown deflection is sought).

It should be noted further that the condition of temperature-dependent first yield must also be checked. The axial force in a cross-section is

$$N = 4\alpha^2 EI/S^2, \tag{34}$$

and first yield occurs when the axial force reaches

$$N = A\sigma_y, \tag{35}$$

where E is temperature-dependent and σ_y is the temperature-dependent yield stress. The condition that the equilibrium state is elastic is then defined by

$$\alpha \le \frac{2\lambda_s \varepsilon_y^{0.5}}{\Theta}, \tag{36}$$

where $\varepsilon_y = \sigma_y/E$ is taken as the temperature-dependent first yield strain.

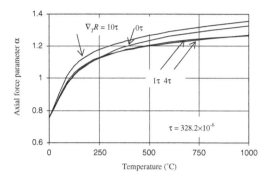

Figure 3. Axial force for $q = 1.2$ N/mm.

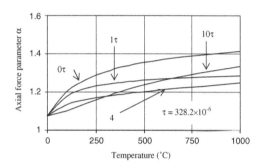

Figure 4. Axial force for $q = 4.8$ N/mm.

4 ILLUSTRATION

The nonlinear response has been determined for a 457×9.5 CHS tubular arch section with a radius $R = 150$ m and an arc length $S = 30$ m. The centroidal thermal strain was taken as $\varepsilon_{T0} = 10^{-5}T$ and the thermal gradient as being based on $(T_{\text{internal}} - T_{\text{external}}) = 0.1T$, so that $\nabla_T = 2.188 \times 10^{-9}T\,\text{mm}^{-1}$ producing a dimensionless thermal gradient parameter $R\nabla_T = \tau = 328.2 \times 10^{-6}T$. Two values of q were chosen: viz. 1.2 N/mm and 4.8 N/mm. These suctions are typical of those caused by air speeds at lower temperatures and of a suction with twice this air speed at a higher temperature.

Figure 3 shows the variation of the axial force parameter α with the temperature T for $q = 1.2$ N/mm and Figure 4 plots the same relationship for $q = 4.8$ N/mm.

The relationship between the temperature and the dimensionless crown deflection v_c/R is shown in Figure 5 for $q = 1.2$ N/mm and in Figure 6 for $q = 4.8$ N/mm, where v_c is the value of v in Equation 14 when $\theta = 0$. These figures shown that the structural response does not display any nearing of a limit point in the range of temperatures considered, indicating that symmetric snap-through is not imminent and moreover the axial force parameter α shown in Figures 3 and 4

436

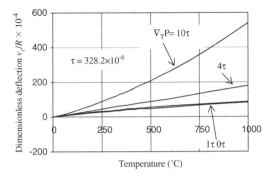

Figure 5. Crown deflection for $q = 1.2$ N/mm.

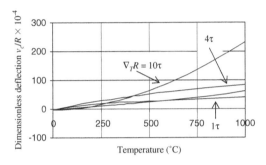

Figure 6. Crown deflection for $q = 4.8$ N/mm.

is always less than π so that antisymmetric buckling is not reached. The deflection of the arch without thermal loading is small, as predicted by linear theory for which it is zero, and when the temperature increases the deflections caused by the nonlinear geometric effects with the load q, and with the thermal gradient ∇_T interact significantly and nonlinearly. Using the Australian values of E and σ_y as functions of T (Standards Australia 1998) for this particular arch with the yield criterion of Equation 36 shows that the structural behaviour is elastic throughout.

5 CONCLUDING REMARKS

This paper has presented a method for the nonlinear elastic analysis of pinned tubular arches in typical contemporary cladding structures at elevated temperatures. The technique allowed for the highly nonlinear prebuckling behaviour of shallow arches. It was shown for the flat arches examined that buckling at elevated temperatures does not occur when uniform thermal straining and a strain gradient through the cross-section interact, and that first yield at elevated temperature is unlikely to occur. The analytic representation of the solution identifies the parameters that govern the highly nonlinear behaviour immediately, such as the dimensionless thermal gradient $R\nabla_T$.

ACKNOWLEDGEMENT

The work in this paper was supported by the Australian Research Council through a Federation Fellowship awarded to the author.

REFERENCES

Bradford, M.A. 2006. In-plane nonlinear behaviour of circular pinned arches with elastic restraints under thermal loading. *International Journal of Structural Stability and Dynamics* 6 (in press).

Bradford, M.A. & Pi, Y.-L. 2003. Strength design procedure for steel arches for the roof of Southern Cross Station Melbourne. Unisearch Report J057618 (Commercial in Confidence), University of New South Wales, Australia, February.

Bradford, M.A., Tong, G., Pi, Y.-L. & Tin-Loi, F. 2005. Nonlinear in-plane buckling analysis of deep circular arches incorporating transverse stresses. I. Theory (submitted for publication).

Pi, Y.-L. & Trahair, N.S. 1998. Non-linear buckling and post-buckling of elastic arches. *Engineering Structures* 20(7): 571–579.

Bradford, M.A., Pi, Y.-L. & Gilbert, R.I. 2006. Nonlinear in-plane behaviour of concrete-filled steel tubular arches. *11th International Symposium on Tubular Structures*, Quebec City, Canada. Rotterdam: Balkema.

Pi, Y.-L., Bradford, M.A. & Uy, B. 2002. In-plane stability of arches. *International Journal of Solids & Structures* 39: 105–125.

Tong, G., Pi, Y.-L., Bradford, M.A. & Tin-Loi, F. 2005. Non-linear in-plane buckling analysis of deep circular arches incorporating transverse stresses. II. Applications (submitted for publication).

Standards Australia 1998. *AS4100 Steel Structures*. Sydney: SA.

Waterloo Maple Inc 2005. *Maple 10 User Manual*. Waterloo, Canada: Maplesoft.

APPENDIX: PRINCIPAL NOTATION

A	Area of cross-section.
E	Temperature-dependent elastic modulus.
F	Functional in variational calculus.
I	Second moment of area of cross-section.
N	Axial force in cross-section.
P	Point on cross-section of coordinates (y, s).
q	Sustained radial uniformly distributed load.
R	Radius of arch.
r	Radius of gyration $= \sqrt{(I/A)}$
S	Total arc length of arch.
T	Temperature above ambient (°C).
v, w	Deformations of arch in y and s directions, respectively.
$\tilde{v}', \tilde{w}'$	Deformations defined in Equation 2.
Vol	Volume of arch member.
y, s	Transverse and tangential axes of arch, respectively.

α	Axial force parameter $= \mu S/2$.	σ_y	Temperature-dependent yield stress.
ε_e	Mechanical strain.	τ	Dimensionless thermal gradient $R\nabla_T$.
ε_P	Longitudinal strain at point P.	Θ	Half included angle subtended by arch.
ε_m	Membrane strain.	θ	Dimensionless length $= \mu s$.
ε_T	Thermal strain.	μ	$\sqrt{(N/EI)}$.
ε_{T0}	Membrane thermal strain.	Ω	Cross-section domain.
ε_y	Temperature-dependent yield strain.	ω	Dimensionless load $= (qR - N)/N$.
$\gamma_0, \gamma_1, \gamma_2$	Coefficients defined in Equations 18 to 20.	$(\ )$	$\mathrm{d}(\ )/\mathrm{d}s$.
κ_0	Initial curvature $= 1/R$.	$(\ )_b$	Value of $(\ )$ at buckling.
λ_s	Modified slenderness in Equation 21.	∇_T	Thermal gradient across cross-section.
Π	Total potential.		

Special session: CIDECT President's Student Awards

Research Award Competition

Tubular Structures XI – Packer & Willibald (eds)
© 2006 Taylor & Francis Group, London, ISBN 0-415-40280-8

Buckling of members with rectangular hollow sections

P. Kaim

ILF Consulting Engineers, Innsbruck, Austria

ABSTRACT: For the member buckling check EN 1993-1-1 Method 2 distinguishes between members which are not susceptible to torsional deformations and members which are susceptible to torsional deformations. During the development of the new buckling formula, a study was necessary to clarify to which group pertain members with rectangular hollow sections (RHS). This paper deals of members with a buckling behaviour, which is at the limit between a failure with or without torsional deformations. This is the case for members with RHS with a high height-to-width ratio (h/b). Thereby, the influence of this ratio on the buckling behaviour of members with RHS is studied. The aim of the study is, to provide limiting values for the dimensions of the cross-section shape, for which lateral-torsional buckling is not relevant. Existing limiting values in codes or proposals are compared. For the investigation, numerical simulations are carried out on the basis of a geometrically and materially nonlinear analysis of the imperfect structure (GMNIA). Finally new limiting values are recommended and a proposal is made, how to proceed with hollow sections, if these values are exceeded.

1 INTRODUCTION

EN1993-1-1 Method 2 provides formulae for lateral-torsional buckling and formulae for flexural-buckling. Flexural buckling means in-plane and out-of-plane buckling without torsional deformations. Which set of formulae should be applied depends on the distinction, whether the members are susceptible to torsional deformations or not. It is not defined where the border between these two classifications is.

At first view, it is obvious, that hollow sections pertain to the second group. But what happens, if they get higher and narrower? At a certain geometry the member will also have the tendency to fail in LT-buckling (see Fig. 1).

If the section is stocky, for pure bending no buckling occurs and the cross-section check is relevant, while for bending with axial compression flexural buckling is the governing design check. For a straight forward use of the Eurocode it is necessary to provide safe-sided limiting values for the cross-section dimensions, for which LT-buckling is not relevant. Further knowledge is necessary how to proceed in a buckling check if the above limits are exceeded.

2 CONCEPTS IN EXISTING CODES

For pure bending some recommendations exist, dealing with LT-buckling of beams. These are limiting values for the length of the member, which depend on the section geometry h/b and determine whether a LT-buckling check is needed or not. The length is expressed by the slenderness λ_z or by the length-height ratio L/h. The values further depend on the steel grade. If these values are not exceeded, no member stability check is necessary and the cross-section check is relevant.

The British Standard BS5950-1:2001 gives a table for the limiting slenderness $\lambda_{z,lim}$ for a range of the h/b ratio from 1.25 to 4, depending on the yield stress f_y. The Australian Standard AS4100 provides a formula for $\lambda_{z,lim}$, depending on the b/h ratio, the moment diagram and the yield stress f_y.

Another proposal was provided by CIDECT. In the CIDECT publication, Rondal (1992), a formula for limiting values L/h is given, depending on the b/h ratio and the steel grade. The limiting values are given as

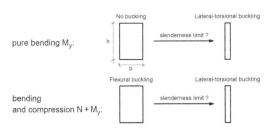

Figure 1. Failure modes for RHS members.

follows for the British Standard

$$\lambda_{z,lim} = f(\frac{h}{b}) \frac{275}{f_y} \quad (1)$$

the Australian Standard,

$$\lambda_{z,lim} = (1800 + 1500\psi) \frac{b}{h} \frac{250}{f_y} \quad (2)$$

and the CIDECT publication

$$\frac{L}{h} = \frac{113400}{f_y} \frac{\gamma^2}{1+3\gamma} \sqrt{\frac{3+\gamma}{1+\gamma}} \quad \text{using} \quad \gamma = \frac{b}{h} \quad (3)$$

The limiting values in these three proposals are quite different. They are all recalculated of a code specific limiting slenderness $\bar{\lambda}_{LT,lim}$, for which the reduction factor for LT-buckling χ_{LT} is 1.0. The expressions in BS5950 and that by CIDECT are based on $\bar{\lambda}_{LT}$ equal to 0.4, while the formula in AS4100 is based on λ_{LT} equal to 0.4.

3 MEMBER SUBJECT TO PURE BENDING

For pure bending the relevant design check could either be the cross-section check or the LT-buckling check (see Fig. 1). This depends on the length of the beam and the dimensions of the cross-section. The buckling behaviour of RHS members is studied by ultimate strength calculations. This study is the basis to recommend improved limiting values.

3.1 Study of the buckling behaviour

The limiting values, given in Chapter 2, are only based on the plateau values of LT-buckling curves given in code specifications. Since these code specific curves are not necessarily related to RHS members, the real buckling behaviour of RHS will be investigated. In the following, a geometrically and materially nonlinear analysis with imperfections (GMNIA) is performed. A single span beam with end-fork conditions was assumed. Since the ABAQUS beam elements are not capable of analysing the failure of members with these sections in LT-buckling, the NLBEAM3D program by Salzgeber (2000) has been used.

RHS with a width of 10 cm and a flange and web thickness of 1 cm are studied in the range of an h/b ratio between 1 and 10 for a slenderness $\bar{\lambda}_z$ between 0.1 and 10. This corresponds to a length of up to 38 m for the h/b ratio of 1 and a length of up to 46 m for the h/b ratio of 10. It was assumed, that no local buckling occurs. The influence of the shear stress on the plastification of the material was neglected. The studied sections are given in Fig. 2.

h/b-ratio RHS-Section	1	3	5	10
h/b/t [cm]	10/10/1	30/10/1	50/10/1	100/10/1

Figure 2. Studied RHS dimensions.

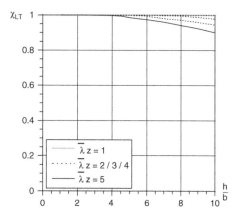

Figure 3. LT-buckling of RHS (GMNIA) – χ_{LT} depending on $\bar{\lambda}_z$ and h/b.

It should be noted, that the scope of the parameters h/b and $\bar{\lambda}_z$ is far wider than needed for practical relevance. However, this wide range has been chosen in order to visualize a significant amount of reduction due to LT-buckling and to find buckling curves extending in $\bar{\lambda}_z$ to the range of the Euler-curve. It will turn out, that the results for the high parameters rather clarify the theoretical behaviour of RHS than can be used for practical design. If just the practically relevant range of values h/b and $\bar{\lambda}_z$ were considered, most of the stability effects looked for in the investigation could not clearly be illustrated.

The resulting ultimate moment M_{ult} is shown in Fig. 3 in terms of the buckling reduction factor

$$\chi_{LT} = \frac{M_{ult}}{M_{y,pl}} \quad (4)$$

The results of the GMNIA analysis are plotted depending on the h/b-ratio with the curve parameter $\bar{\lambda}_z$ varying from 1 to 5. Using the above results, recommendations for limiting values can be made, based on the real buckling behaviour according to the GMNIA analysis.

The limiting value means, that the beam has the fully plastic bending moment resistance. However, in a nonlinear analysis without assuming strain hardening, the fully plastic moment cannot exactly be reached. Therefore, the following study is based on beams, where the buckling load is 99% or 95% of the fully plastic

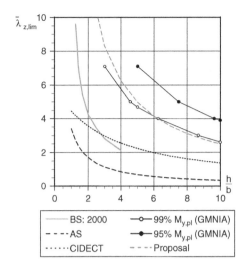

Figure 4. Comparison GMNIA – codes – proposal.

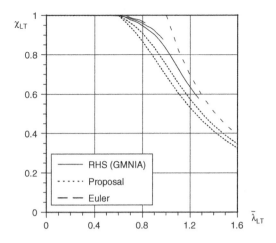

Figure 5. Comparison GMNIA – proposal.

moment capacity. For the studied parameter range the pairs of data $\bar{\lambda}_z$ and h/b are collected from Fig. 3, where χ_{LT} is 0.99 or 0.95. The result is given in Fig. 4 and is compared with the existing code formulae, given in the equations (1) to (3).

The British Standard provides values only up to an h/b ratio of 4. The diagram shows, that the differences between the existing code formulae and the GMNIA results are significant. Even if 99% of the fully plastic moment capacity is presumed, the limiting values of the codes are only about half the values, given by the GMNIA analysis.

3.2 Recommendation of new limiting values

The results in Section 3.1 show, that a GMNIA based proposal can give much higher values than the existing ones. For the limiting values a hyperbola (like in AS4100) is proposed, which fits the 99% curve. It gives a limiting slenderness $\bar{\lambda}_{z,lim}$, depending on the h/b ratio. The parameter of the curve has to be calibrated and statistically evaluated. In equation (5) the factor 25 has been chosen. The proposed curve is illustrated in Fig. 4.

$$\bar{\lambda}_{z,lim} = \frac{25}{(h/b)} \qquad (5)$$

For other steel grades than S235, the values $\bar{\lambda}_{z,lim}$ need to be adapted by the factor $\sqrt{235/f_y}$.

3.3 LT-buckling check

If the above proposed limiting values are exceeded, the member will fail in LT-buckling and cannot reach the fully plastic moment capacity. Then a LT-buckling check, like for other sections, is necessary. In the Eurocode, no specification is made, which curve should be used for RHS.

The results of the GMNIA study are now related to the slenderness $\bar{\lambda}_{LT}$ (see Fig. 5). The curves in the diagram for the different sections end at different slendernesses $\bar{\lambda}_{LT}$, due to the limitation of the range of $\bar{\lambda}_z$ up to 10. It can be stated, that the curves for all sections are very close together. Also the Euler hyperbola for the perfect elastic beam is shown in the diagram. The buckling curves of the RHS tend to the Euler curve for high slendernesses. In contrary to the LT-buckling curves for I-sections, now for RHS the plateau length increases up to a slenderness $\bar{\lambda}_{LT}$ of about 0.6. This indicates that a potential buckling curve for RHS would be much higher than the design curves for I-sections in the present codes.

The results of the GMNIA analysis have been compared with the LT-buckling curves of the Eurocode for I-sections with the plateau value of 0.4 (not shown here). The comparison has shown that even the highest Eurocode curve gives a very conservative prediction of the buckling load.

For the design, either the above mentioned LT-buckling curve can be applied for all kinds of RHS (hot finished or cold formed), or if desired, a specific LT-buckling curve for members with RHS can be defined with a plateau value $\bar{\lambda}_{LT,0} = 0.6$. The coefficient β in the equation for χ_{LT} in Section 6.3.2.3 of EN1993-1-1 is chosen as unity, because it fits best to the GMNIA curves. The imperfection factor α_{LT} can be chosen as 0.21 (cp. EC curve a) or as 0.34 (cp. EC curve b), if a distinction is necessary for hot finished or cold formed RHS. Both are illustrated in Fig. 5 and compared with the GMNIA results.

Using the proposed buckling curves, it has to be noted, that in case of very high and narrow sections it has to be ensured, that no local buckling occurs and the beam theory is still valid.

4 MEMBER SUBJECT TO BENDING AND COMPRESSION

For members with RHS subject to bending and axial compression, the relevant design check could be either the flexural buckling check or the LT-buckling check (see Fig. 1). Which failure mode occurs depends on the length of the member, the dimensions of the cross-section and the magnitude of the axial force.

The limiting values in the existing codes are only valid for pure bending. The question arises whether these values are also valid for combined bending and compression? If not, how should these values be modified? In the following studies, an answer will be given on the basis of GMNIA calculations. Finally, a recommendation is given how to proceed in the LT-buckling check, if the limiting values are exceeded.

4.1 Study of the buckling behaviour

Also for the combined action of axial compression and bending, the influence of the torsional deformations has been studied for different h/b ratios and slendernesses. Here, only the case of h/b = 10 is presented for three different slendernesses $\bar{\lambda}_z$ (1, 2 and 3). The members are subject to a uniformly distributed load and a compression force. For more results see also Kaim (2004).

The ultimate strength is determined in two ways: on the one hand the RHS-member is calculated with continuous stiff torsional restraints (to simulate flexural buckling) and on the other hand it is unrestrained (to simulate LT-buckling). A comparison of these two analyses shows the influence of the torsional deformations. GMNIA analyses of I-sections with continuous stiff torsional restraints have also been used to calibrate the formulae for flexural buckling in EN1993-1-1.

The results of the numerical simulations of the unrestrained RHS-members are given in Fig. 6 by the bold grey lines. The results of the members with the continuous stiff torsional restraints are given by the dotted lines. If the curves for the restrained and the unrestrained members are almost coincident, the unrestrained beam-column fails with nearly no torsional deformation, i.e. it fails in flexural buckling about both axes (FByz). In this case the set of formulae for members not susceptible to torsional deformations may be applied.

4.2 Recommendation of limiting values

For members subject to pure bending, the criterion for the GMNIA based limiting values is the χ_{LT} value,

i.e. the relation between the ultimate strength and the cross-section capacity. The criterion for limiting values for the combined action of bending and compression could be the relation between the ultimate strength for a failure in LTB and in FByz. Since this depends on the effect of the axial compression, the relation is read off at the position of the interaction curve, where the difference reaches a maximum. This is sufficiently accur-ate at the point where the reference loads $M_{y,pl}$ and $\chi_z N_{pl}$ are multiplied with the same load amplification factor Λ. This leads to the following ultimate loads:

$$M_{ult} = \Lambda \times M_{y,pl} \text{ and } N_{ult} = \Lambda \times \chi_z \times N_{y,pl} \qquad (6)$$

This is illustrated in Fig. 6 for a uniform bending moment in case of the slenderness $\bar{\lambda}_z = 3$ and the h/b ratio of 10. The black arrows on the ordinate and the abscissa indicate the reference loads. The grey arrow and the shorter black arrow show the load amplification factors for FByz (Λ_{FB}) and for LTB (Λ_{LTB}), respectively. For the other slendernesses the relevant points are illustrated in the diagram by black and not-filled points, respectively.

For a proposal of a limiting value for beam-columns, an extensive parametric study has been performed by Kaim (2004) on members with a uniform moment diagram, h/b ratios of 1, 3, 5 and 10 and slendernesses in the range of $\bar{\lambda}_z$ between 0.2 and 5. The parametric study has shown that also for a combined loading with bending and compression the susceptibility to torsional deformations is very small. A reduction of the ultimate load due to torsional deformations could only be observed for very high $\bar{\lambda}_z$ or h/b ratios.

Using these results, limiting values can be proposed on a GMNIA basis. The procedure is similar to that in Section 3.2. The pairs of data $\bar{\lambda}_z$ and h/b are collected, where the load amplification factor Λ_{LTB} is 99% or 95% of Λ_{FB}. If the difference between Λ_{LTB} and Λ_{FB} is sufficiently small, the design formulae for flexural buckling can be used for the buckling check. The results of the parametric study are given in Fig. 7.

As before for pure bending, also for bending and compression a hyperbola is proposed for the limiting values, which has to be calibrated on GMNIA results. Here, a curve is chosen, that agrees well with the numerical simulations at a level of 99% for the ratio between Λ_{LTB} and Λ_{FB}. It is given in equation (7) and illustrated in Fig. 7.

$$\bar{\lambda}_{z,lim} = \frac{10}{(h/b)} \qquad (7)$$

The comparison of the numerical results and the proposal, shown in Fig. 7, is based on the most unfavourable load combination of bending moments and axial compression. As the real behaviour depend

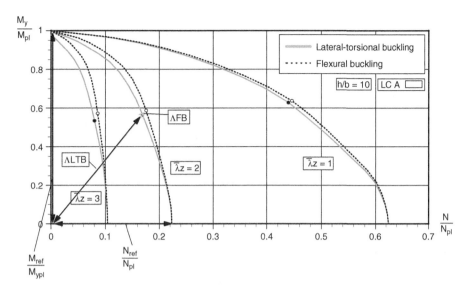

Figure 6. N-My-interaction for h/b ratio of 10 (uniform moment).

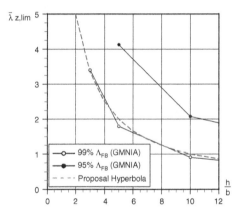

Figure 7. Comparison GMNIA – proposals.

members with I-sections, it has to be studied how they work for the design of members with RHS. Therefore, the results of the numerical simulations have been compared with the Eurocode formulae.

The existing buckling curves χ_{LT} and interaction factor k_{LT} as derived for I-sections have been used. It turned out that the existing formulae give very conservative results, due to the different interaction behaviour of I-sections and RHS. The N-M_y-interaction curve for RHS shows a distinct curved form (not shown here), similar to the flexural buckling behaviour, while the design curves are nearly linear. This is the characteristic behaviour of an I-section, on which they have been calibrated. For a more economic design, the buckling curve presented in Section 3.3 with a plateau value of 0.6 can be used together with the existing interaction factors k_{LT}.

on the portion of the axial compression, for a lower axial force the limiting values could be raised. As the proposed limiting values already cover sufficiently the frequently used dimensions, no compression-dependent values are suggested. However, as a consequence, there is no direct transition to the case of pure bending.

5 VERIFICATIONS OF THE SIMULATIONS

In this section the above numerical simulations with beam elements are verified with numerical simulations using shell elements. Furthermore, numerical results are compared with test results.

4.3 *LT-buckling check*

If the above proposed limiting values are exceeded, the design check for members susceptible to torsional deformations (LTB) is relevant. As the LT-buckling formulae in EN1993-1-1 have been calibrated on

5.1 *Verifications with shell elements*

All the above presented numerical simulations have been carried out by the NLBEAM3D program, using beam elements. The influence of the shear on the plastification cannot be modelled with these beam

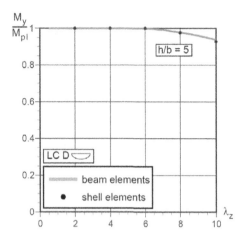

Figure 8. Comparison shell – beam elements (LT-buckling curve for pure bending).

5.1.2 Bending and compression

The influence of the methods of analysis is also studied for the combined loading of axial compression and bending moments. The slenderness $\bar{\lambda}_z$ is equal to 2. The left diagram of Fig. 9 shows the N-My-interaction for h/b = 2. The results of the two methods are almost identical. For a h/b ratio of 5 in the range of high axial forces, there is a small difference between the ultimate loads, calculated with beam and shell elements. Since the h/t ratio is 50, local buckling obviously has an influence under pure compression, but not under pure bending.

5.2 Comparison with tests

For a verification, the LT-buckling tests and FE-calculations, given by Zhao (1995) are simulated for comparison with the here used FE-model. The comparison is given in Fig. 10. The left diagram shows the results of Zhao. Tests on three beams with RHS have been made and verified with a FE-analysis. The height is 75 mm, the width 25 mm and the thickness 2.5 mm. The slendernesses $\bar{\lambda}_z$ are approximately 2.0, 4.0 and 7.0. The members with $\bar{\lambda}_z$ equal to 4.0 and 7.0 are simulated, using the NLBEAM3D program.

According to Zhao, the geometrical imperfections are an initial lateral imperfection of L/1000 and a twist of 0.04 radians. The load-deflection paths of the NLBEAM3D simulations are given in the right diagram by the grey lines. They well coincide with the tests and the simulations of Zhao; however, they give a slightly higher ultimate load. This is in accordance with the tests of Zhao, because the tests have been terminated due to large deflections, a bit before reaching the ultimate load of the beam.

elements. The parts of the stiffness matrix, related to transverse shear and St.Venant torsion, are always calculated on an elastic basis. Furthermore, the shear stresses of the transverse forces and the St.Venant torsional moment do not influence the plastification of the cross-sectional fibres, because they do not contribute to the von-Mises stress. However, their significance should be low for the studied load cases of uniform moment or uniform load, because the shear stresses reach their maximum at the supports of the member, while the longitudinal stresses reach their maximum at midspan.

Another effect, which cannot be modelled by beam elements, is local buckling and cross-sectional deformations. Especially cross-sections with high h/b ratios and accordingly thin-walled cross-sections could be affected by local buckling. For these two reasons, i.e. shear stress and local buckling, a numerical simulation has been made with members modelled with shell elements. In both cases residual stresses and an initial parabolic bow imperfection of L/1000 is applied. With regard to the focus of the study on beam-column behaviour, an extension to global-local behaviour with specific local imperfections has not been considered here.

5.1.1 Pure bending

The studied beams are subject to pure bending and have a length according to the range of $\bar{\lambda}_z$ from 2 to 10. The results are presented in Fig. 8 for a h/b ratio of 5 and are compared with the buckling curves determined with beam elements. The results of both calculations are nearly coincident and confirm the inferior significance of the shear stresses.

6 SUMMARY

From the different results of the study it may be concluded that unrestrained RHS members under pure bending or bending plus axial compression theoretically show a tendency to LT-buckling, in principle analogously to open sections. A new LT-buckling curve was derived.

However, in the range of h/b ratios and slendernesses of practical relevance torsional deformations usually do not affect the load-carrying capacity of RHS members. The study showed that for pure bending and h/b ratios up to 5 and slendernesses $\bar{\lambda}_z$ up to 5 no buckling effects occur at all. Two simple formulae for limiting values have been proposed, one for pure bending and one for bending and compression. The proposed limiting values, therefore, considerably simplify the design of RHS members, which under pure bending need only a cross-section check and for bending and axial compression can be carried out by the flexural buckling formulae alone.

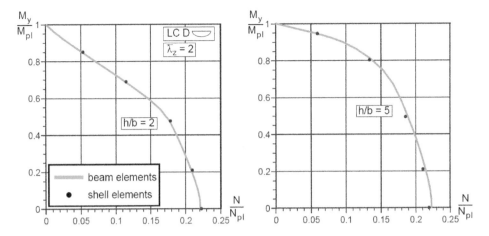

Figure 9. Comparison shell – beam elements (N-My-interaction).

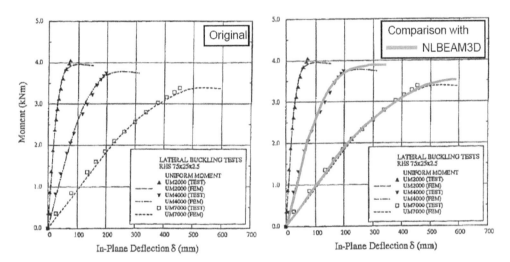

Figure 10. Comparison with the tests of Zhao (1995).

REFERENCES

Kaim, P. 2004. Spatial bucking behaviour of steel members under bending and compression, Graz University of Technology, Dissertation thesis

Rondal, J. 1992. Knick-und Beulverhalten von Hohlprofilen (rund und rechteckig), Verlag TÜV Rheinland Köln, Comité International pour le Développement et l'Etude de la Construction Tubulaire CIDECT

Salzgeber, G. 2000. Nichtlineare Berechnung von räumlichen Stabtragwerken aus Stahl, Graz University of Technology, Dissertation thesis

Zhao, X.L. 1995. Lateral buckling of cold-formed RHS beams, Proceding Structural Stability and Design, Kitipornchai, Hancock & Bradford (eds), Balkema Rotterdam.

449

Tubular Structures XI – Packer & Willibald (eds)
© 2006 Taylor & Francis Group, London, ISBN 0-415-40280-8

On the definition of a material model, stiffness and ultimate strength for cold-formed structural hollow section X-joints

V.K. Pellikka

Lappeenranta University of Technology, Finland

ABSTRACT: The ultimate strength and stiffness of X-joints fabricated from hot- and cold-formed structural hollow sections was studied using non-linear finite element methods. Numerical results were compared to test results obtained from four full-scale joints tested at −40°C. Plate bending and section distortion tests at room temperature were also performed. The main objective was to understand the material model requirements needed for finite element analysis. Isotropic and anisotropic material models for the cold formed corner regions were investigated. It was concluded that it was not necessary to use an anisotropic material model. The higher yield strength in the corner region of a cold-formed section and the larger corner radius influenced the joint stiffness (Pellikka 2004).

1 INTRODUCTION

Due to their good structural properties and aesthetic appearance, structural hollow sections (SHS) are widely used in mechanical and civil engineering applications. Many experimental tests and numerical analyses on the capacity and fracture mechanics of structural hollow sections joints have been performed. However, most of those investigations involved joints fabricated using hot-formed structural hollow section (HFSHS) members and testing has been at room or elevated temperatures. Only limited test data is available for cold-formed structural hollow section (CFSHS) joints at lower temperatures (Soininen 1996).

In X-joints, brace load will be transmitted from the flange to the web of the chord member causing high stresses in corner area. Material and geometric features of the corner area are, therefore, very important for X-joints. The SHS corner highlights the main differences between HFSHS and CFSHS. In CFSHS, material properties of the strain hardened corners differ from the properties of the flat faces. Material in the corner typically has higher yield strength but also lower plastic strain at fracture. Due to the fabrication process, the corner radius of CFSHS is typically larger than in an HFSHS with the same nominal size. Also, corner region residual stresses due to cold forming may be large.

In X-joints the corner areas of the chord are loaded mainly by transverse stresses independent of β. Hence, the transverse material properties in the chord corner are critical. However, in design guidance documents the load and deformation capacities of X-joints are usually defined based on material properties measured in the longitudinal direction of chord member. Longitudinal tensile test specimens can easily be taken from the area of flat face and also from the corner area. In CFSHS, most of the strain hardening is transverse and the implications of this possible anisotropy are difficult to determine.

Strain hardening increases the material strength but decreases the plastic deformation capacity. In addition, the heat input due to welding can alter the material properties and the residual stresses. These effects may increase the probability of brittle fracture as the failure mode for an X-joint fabricated with CFSHS and hence restrictions for welding on corner areas are given (EN 2003).

Theoretically, the larger corner radius of a CFSHS chord member makes the transverse face of the brace member more effective than in the case of HFSHS when the width ratio, β, between the brace and chord members is near unity. This is due to the increased stiffness of a curved shell as opposed to a flat shell. If β is small, e.g., $\beta \leq 0.85$, the failure mechanism changes, but again the larger radius of chord member increases the stiffness and strength of the chord flange. Consequently, independent β, the larger corner radius of the chord member for CFSHS should increase the ultimate load carrying capacity of the X-joint if other dimensions remain unchanged.

For $\beta \leq 0.85$, the elastic-plastic capacity of the X-joint depends more on the local bending load-carrying capacity than on the membrane load-carrying capacity of the chord flange. The residual stresses through the wall thickness of the cold-formed flange influence the force-displacement curve of the X-joint. For this reason bending strength, rather than tensile strength, was judged to better describe the material properties for analysis purposes.

Considering all the previously described phenomena, the maximum load-carrying capacity and deformation behaviour of an X-joint is so complex that numerical simulation must be calibrated based on experiments. An extensive test program of X-joints of different geometries and at different temperatures was carried out at Lappeenranta University of Technology. In total 20 joints were tested at temperatures ranging from $+20°C \ldots -60°C$. In these tests the load and deformation capacities for X-joints fabricated from HFSHS and CFSHS were measured. The current paper focuses on four of the joints tested at $-40°C$. Two of the joints were fabricated from HFSHS and two nominally identical joints were fabricated from CFSHS. FE-analyses were used to define the theoretical ultimate load carrying capacity and expected failure mechanism. Experimental results have also been compared with analytical joint strength capacities calculated using current design guidance documents (EC3) (SFS-ENV 1992). Failure analyzes of the observed fractures have been performed.

2 EXPERIMENTAL STUDY

The effect of corner radius on the ultimate capacity and fracture behaviour of X-joints has been studied experimentally. The loading fixture is illustrated in Figure 1. Measured material properties of the test specimens are given in Table 1 and nominal dimensions of the test specimens in Tables 2–4. Whenever possible the chord and brace members of the X-joints were joined with fillet welds. However, for $\beta = 1.0$ joints, side butt welds were used. In all cases a manual MAG welding process was used. The joints were prepared according to the instructions given in design guidance documents (N37B 2002).

Experimental investigations did not reveal any significant differences in the ultimate strength capacity of joints fabricated from HFSHS and those fabricated from CFSHS. Even though testing was performed at $-40°C$, all joints fulfilled the requirements in EC3. The ultimate deformation capacities are more difficult to compare, because the failure of the CFSHS joints was remote from the welded region. In all joints evaluated here, even for those exhibiting brittle fracture, the plastic deformation capacity was always larger than the 1% value that is normally considered as good design limits for $\beta = 1$ and $\beta < 1$ joints. Some significant differences in the failure mechanisms, however, were observed. The ultimate failure mode for CFSHS was in the brace member due to large plastic deformation. In HFSHS the failure was located in the region of the welded joint.

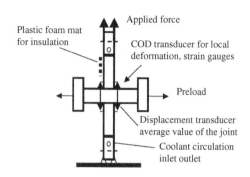

Figure 1. Schematic test set up.

Table 1. Measured material properties.

SHS	Cold or hot formed	$f_{0.2}$ [MPa]	f_u [MPa]	A_5 [%]	$T_{50\%}$ T-L [C]	$T_{50\%}$ L-T [C]
$100 \times 100 \times 10$	CF	468	551	30.0	-10	-15
$100 \times 100 \times 10$	HF	424	561	35.1	-40	-70
$50 \times 50 \times 5$	CF	480	541	30.0	$\ldots$	-465
$50 \times 50 \times 5$	HF	561	641	28.3	$\ldots$	-100

Table 2. Chords of the analyzed X-joints.

ID.	Chord	r_0 [mm]	t_0 [mm]
X18	CF $100 \times 100 \times 10$	26.2	9.93
X20	CF $100 \times 100 \times 10$	26.2	9.97
X22	HF $100 \times 100 \times 10$	10.5	9.93
X23	HF $100 \times 100 \times 10$	10.5	9.97

Table 3. Braces of the analyzed X-joints.

ID.	Brace	r_1 [mm]	t_1 [mm]
X18	CF $100 \times 100 \times 10$	26.2	9.93
X20	CF $50 \times 50 \times 5$	9.5	4.90
X22	HF $100 \times 100 \times 10$	10.5	9.93
X23	HF $50 \times 50 \times 5$	5.7	4.90

In order to define the bending load carrying capacity of the chord flange, a bending test was performed on a strip cut from the flange of the CFSHS. This test also served as a simple comparison study for the finite element analyses (FEA) and was used to assess the potential effect of through thickness residual stresses on the bending capacity. A photograph of the test set up is shown in Figure 2.

ID.	β	Weld	Throat thickness [mm]
X18	1.0	Single bevel	10
X20	0.5	Fillet	5.5
X22	1.0	Single bevel	10
X23	0.5	Fillet	5.5

Table 4. Properties of the analyzed joints.

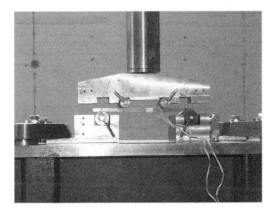

Figure 2. Bending test.

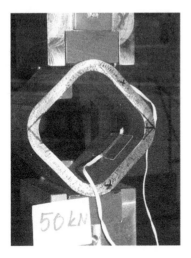

Figure 3. Distortion test.

One SHS distortion test was carried out in order to study material properties in the transverse direction of the cross section. In this test the transverse elastic-plastic behaviour of the material in the corner region was derived. A photograph of the distortion test set up is shown in Figure 3. In both of these tests, strain values were measured by strain gauges. Bend test and distortion test materials were taken from the same tube from

which the X-joint test specimen X18 was fabricated. Tests were executed at room temperature. In the distortion test, deformations occurred not only in the corner areas, but also in the area of flat faces. This indicates the effect of strain hardened material in corners of the CFSHS.

3 FINITE ELEMENT ANALYSES

Finite element analyses were performed in order to assess the assumptions regarding the effects of fabrication process on the strength capacity of the X-joints. The four joints from the experimental tests reported here were analyzed using nonlinear FEA. Geometrically, the major differences between the CFSHS and the HFSHS X-joints were the SHS corner geometries. Dimensions for each model are seen in the Tables 2–4.

Due to the three-fold symmetry, it was necessary to model only 1/8 of the X-joint. Eight node solid elements were used for most of the model, but some tetra elements were placed in the transition zone. Two material models were included in the analyses of the CFSHS joint in order to take into account the different material properties in the corner areas and the flat sections of the SHS. The FEA meshes are shown in Figure 4. For the X18 and X20 meshes, the regions of different material properties are indicated with different shading. The effect of the corner radius alone was assessed by using the precise geometries from HFSHS and CFSHS joints but with an identical material model for both, i.e. properties from the cold formed section.

The boundary conditions used reflected those in the X-joint laboratory experiments. The preload of the chord member was modelled as a constant load rather than as a displacement. Large displacements were allowed in the FE analyses. A von Mises yield criterion was used and a Newton-Raphson iteration method was applied. The MCS-Marc FEA package was used for analyzing (MSC 1997).

Material σ-ε models for FEA were defined based on tensile tests with test coupons taken from the flat faces of the sections and, in the case of CFSHS, also from longitudinal coupons taken from the corner areas of the section. For HFSHS the materials was assumed to be homogeneous. The stress-strain curve obtained from coupon tests for X18 is shown in Figure 5. For FEA the true stress strain curve is used in place of the engineering stress strain curve measured during tensile testing. The true stress strain curve is inputted into the FEA program as a piecewise linear curve. The open symbols in Figure 5 indicate the stress-strain combinations used to define the piecewise linear curves.

In Figure 6 the FEA models of the bending and distortion tests are shown. The bending deformation of the strip that was observed when cutting a strip from the CFSHS due to residual stress relief was included

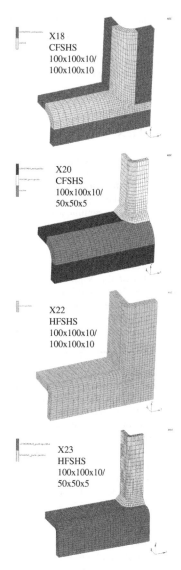

X18
CFSHS
100x100x10/
100x100x10

X20
CFSHS
100x100x10/
50x50x5

X22
HFSHS
100x100x10/
100x100x10

X23
HFSHS
100x100x10/
50x50x5

Figure 4. The FEA meshes of the joints.

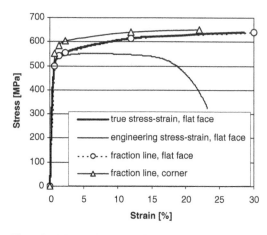

Figure 5. Measured σ-ε curve for the SHS flat face and true stress-strain curves for flat and corner sections of specimen X18. The point to point linear curves used in FEA are also indicated.

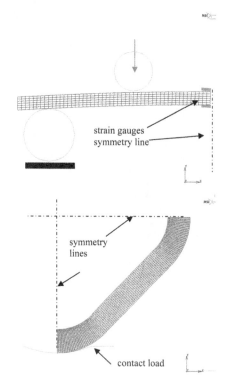

Figure 6. FEA models of the bending and SHS distortion tests for assessing the material σ-ε models.

in the FEA model. Due the extensive cold working of the CFSHS corners, the material was expected to be anisotropic. However, FEA of the SHS distortion tests showed that anisotropy had little influence on the force-deformation curves and on the ultimate deformation capacity. Therefore, all further FEA was performed using an isotropic material model, but with different σ-ε behaviour for material on the flat edges and in the corner region of a section.

Results from the bending and SHS distortion evaluations are illustrated in Figure 7. The experimental test curves in those figures are based on measured forces and strains. From the bending test curve, Figure 7a, it

can be seen that, in the region near first yielding, the FE results using the measured σ-ε model slightly underestimates the load when compared to the experimental test. In the region of high plastic strain the FEA results

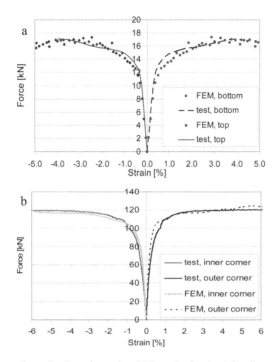

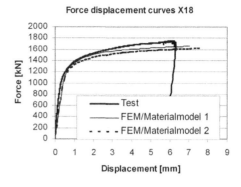

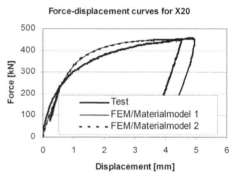

Figure 7. Experimental and FE results for the a) bending and b) SHS distortion tests.

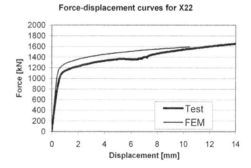

were not fully stable due to convergence problems, but the lower bound solutions were in good agreement with the experimental results. Similar correlation would be found, if the force-deformation curve would be plotted.

Based on a comparison of the bending and SHS distortion test results and the simulation of these tests using FEA, it was judged that σ-ε model defined by tensile testing of a coupon taken from the longitudinal SHS flange was sufficient. The materials, therefore, was assumed to be isotropic.

The FE based force-displacement results for the four X-joint tests are seen in Figure 8. For comparison, the experimental test results are plotted in the same figures. The X-axes in these figures represent the joint displacement as illustrated in Figure 1. For the CFSHS, material model 1 means that the corner and flat region of the SHS were modelled as having different material properties. Material model 2 indicates that the whole section was modelled as being homogenous.

In joint X18 the FEA results and test results differ from each other only slightly. Analysis with strain hardened corner material gives slightly higher ultimate strength.

In joint X20 the shapes of the experimental and FEA based load-displacement curves are different. This might be the result of residual stresses in the

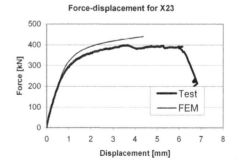

Figure 8. Force-joint displacement curves for X-joints in comparison.

Force-displacement curves

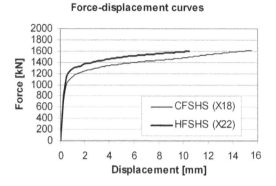

Force-displacement curves

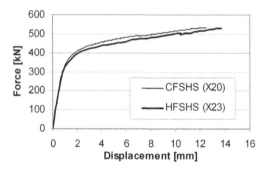

Figure 9.   Effects of corner radius.

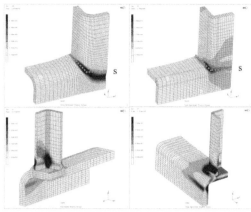

Figure 10.   Distribution of strain in X-joints.

chord member. It was found that the exact material model used in the corner of the chord member had little effect on the ultimate strength of the joint because final failure was due to brace member yielding.

The effects of corner radius were studied numerically and the results are seen in the Figure 9. In these analyses the same material σ-ε curve was used for all models. Only the SHS geometries were modified. All the dimensions are held constant except that the measured corner radii from the respective CFSHS and HFSHS were used. Figure 8a shows the result for X-joint with $\beta = 1.0$. This shows that the predicted strength of the HFSHS was slightly greater for all displacements. Further evaluation revealed that the sharper corner of the brace member meant that the flat faces of the section are larger and, thus there is a better load transfer between chord and brace flanges. In Figure 9b the same type of analysis was done for joints with $\beta = 0.5$. Here it can be noticed that the CFSHS had slightly greater predicted strength. This was due to the increased stiffness of the chord flanges caused by the greater radius.

Total equivalent strain distributions of FEM models are shown in Figure 10. During laboratory testing, local strains near the weld were monitored with strain gauges as shown in Figure 11. Strain results from FEA

and measured results from the strain gauges are also shown in Figure 11.

The initiation point of the failure in joints was studied by examining strains in the joint area as illustrated in Figure 10. In the HFSFS joint, the largest strains were located near the corner of the brace member. In the CFSHS joint, the area of largest strains was distributed more widely starting from the web of the chord and continuing over the corner of the brace. In $\beta = 0.5$ joints the largest strains were located across the cross section of the brace member. For HFSHS joints, the web of the chord member was subject to higher bending strains than was the web of the CFSHS joint.

The sharper corners of the HFSHS resulted in a significantly larger strain concentration in the corner of the X-joint. For the CFSHS the estimated maximum strain at 1580 kN was less than 14% while for the HFSHS at the same load the strain was more than 25%. Strains in HFSHS brace and in CFSHS brace as a function of weld coordinate are presented in Figure 12. The numbering of the curves indicates loading applied.

Experimental investigations did not reveal any significant differences in the ultimate strength capacity of joints fabricated from HFSHS and those fabricated from CFSHS and all joints fulfill the requirements obtained in EC3. The ultimate deformation capacities are more difficult to compare, because the failure of the CFSHS joints was remote from the welded region. In all joints evaluated here, even for those exhibiting brittle fracture, the plastic deformation capacity was always larger than the 1% value that is normally considered as good design limits for $\beta = 1$ and $\beta < 1$ joints. Some significant differences in the failure mechanisms, however, were observed. The ultimate failure mode for CFSHS was in the brace member due to large plastic deformation. In HFSHS the failure was located in the region of the welded joint.

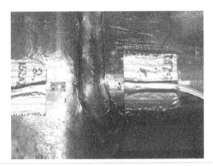

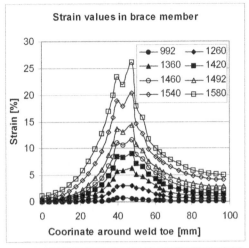

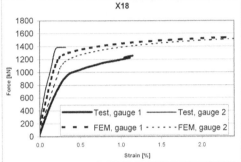

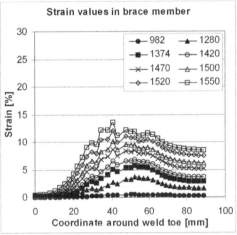

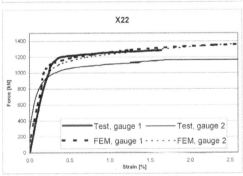

Figure 11.   Local strains of joints X18 and X22.

Figure 12.   Strain values in HFSHS brace (a) and in CFSHS brace (b). Numbering of the curves indicates the loading [kN] applied.

4   CONCLUSIONS

The effect of the geometry and material properties on the ultimate strength of welded X-joints fabricated from CFSHS and HFSHS profiles has been studied experimentally and using non-linear finite element analysis. Numerical simulation and experimental tests focused on the importance of this corner effect and the separate role geometry, yield strength and fracture toughness on the total fracture strength of the joint. The following conclusions can be made:

(1) Equations found in design guidance document, EC3, could be used to estimate the capacity of CHSHS and HFSHS X-joints at temperatures as low as −40°C.

(2) Material properties in corner area of CFSHS seemed to be only a slightly anisotropic. Material properties in corner differ from properties of flat face more significantly, but the force-displacement behaviour of the X-joint could be adequately modelled using the assumption of isotropic material behaviour.

(3) Smoother geometry of CFSHS X-joint produced smaller local strains near the weld.

(4) The corner radius of the SHS has a small but clear effect on the behaviour of the X-joint.

(5) As β decreases, a chord member with large radius has greater flange stiffness and strength. When $\beta = 0.5$, the corner radius of the brace member is less critical than for $\beta = 1.0$ and the larger corner

radius of the cold-formed chord member slightly increases the ultimate capacity of the joint.

REFERENCES

Pellikka V., On the definition of material model, stiffness and ultimate strength for cold-formed structural hollow section X-joint. Master's thesis 2004, Lappeenranta University of Technology. 2004. (in Finnish)

Soininen R., Fracture behaviour and assessment of design requirement against fracture in welded steel structures made of cold-formed rectangular hollow sections. PhD thesis, Lappeenranta University of Technology. 1996

prEN1993-1-8. Design of steel structures, Part 1-8: Design of joints. European Committee for Standardization, 2003

SFS-ENV 1993-1-1:1992. Eurocode 3. Design of steel structures. Part 1-1. General rules and rules for buildings. European prestandard. CEN, European Committee for Standardization

N37B. Execution of steel structures and aluminium structures. prEN1090, Chapter 7. Welding. European Committee for Standardization, 2002

MSC.Marc 1997. Volume A. Theory and User Information. Version 1997. Santa Ana, MSC.Software Corporation

NOMENCLATURE

A_5 = percentage elongation at fracture of the gauge length

f_u = ultimate tensile strength [N/mm^2]

f_y = yield strength (of chord member) [N/mm^2]

$f_{y,1}$ = yield strength of brace member [N/mm^2]

$f_{y,0.2}$ = yield strength relating 0.2% plastic deformation [N/mm^2]

r_0 = outer radius of corner in chord member [mm]

r_1 = outer radius of corner in brace member [mm]

s = distance along perimeter [mm]

$T_{50\%}$ = temperature, which obtain 50% ductile area of fractured surface [°C]

t_0 = wall thickness of chord member [mm]

t_1 = wall thickness of brace member [mm]

β = width ratio = b_1/b_0

ε = strain

σ = stress [N/mm^2]

Tubular Structures XI – Packer & Willibald (eds)
© 2006 Taylor & Francis Group, London, ISBN 0-415-40280-8

Numerical fatigue crack growth analyses of thick-walled CHS T-joints

M. Oomens

Faculty of Civil Engineering and Geosciences, Delft University of Technology, Delft, The Netherlands

ABSTRACT: The present paper concerns fatigue crack growth analyses on steel thick-walled CHS T-joints, which are characterized low values for the joint parameter 2γ and contain a significant surface crack. A linear elastic fracture mechanics (LEFM) approach is adopted and a three-dimensional finite element model is used to determine stress intensity factors. The Paris' law is applied to relate stress intensity factors to crack growth. The fatigue crack shape evolution is simulated and evaluated for different loading conditions, joint parameters 2γ and initial crack shapes as part of a pilot study. Subsequently, the effect of these research variables on the remaining fatigue strength is also reported.

1 INTRODUCTION

The fatigue strength of a welded tubular joint containing already a fatigue crack can be estimated by means of the theory of fracture mechanics. Fracture mechanics analyses have been conducted primarily on thin-walled circular hollow section (CHS) joints. These joints own relatively high values for joint parameter 2γ, which is defined as the chord diameter to wall thickness ratio (D/T). Nowadays, thick-walled tubular joints can be found in many structures subjected to fatigue loading such as railway bridges. These joints are often applied to increase fatigue life by decreasing nominal stresses, but possible fatigue damage during service life may remain an issue.

The fatigue crack growth in thick-walled joints may develop differently from thin-walled joints due to differences in stress fields inside the joints. Numerical research can give more insight into the influence of certain joint and crack parameters on crack evolution and remaining fatigue strength.

This paper presents the results of conducted numerical fatigue crack growth analyses on thick-walled CHS T-joints. The crack shape evolution is simulated for different loading conditions, values for joint parameter 2γ and initial crack shapes. The remaining fatigue strength is also evaluated.

2 SCOPE OF NUMERICAL STUDY

Fatigue crack growth analyses are carried out on three different thick-walled CHS T-joints. The geometrical definitions for a CHS T-joint are given in Figure 1 and the geometrical properties of the analyzed T-joints

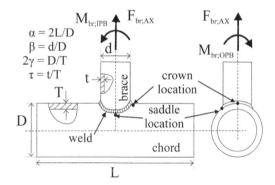

Figure 1. Definitions for CHS T-joint.

Table 1. Geometric properties of analyzed CHS T-joints.

T-joint Nr.	D (mm)	d (mm)	T (mm)	t (mm)	β	2γ	τ	α
1	500	250	50.0	25.0	0.5	10	0.5	15
2	500	250	35.7	17.9	0.5	14	0.5	15
3	500	250	27.8	13.9	0.5	18	0.5	15

are tabulated in Table 1. The outer diameters D, inner diameter d and the joint parameters β, τ and α are held constant during the analyses. Joint parameter 2γ is chosen as a research variable and equals 10, 14 or 18. Both ends of the chord member are fully clamped. The end of the brace member is free supported allowing a lateral movement of the brace end.

Three basic load cases are considered; axial loading ($F_{br;AX}$), in-plane bending ($M_{br;IPB}$) and out-of-plane bending ($M_{br;OPB}$). They all act on the brace member of the joint. Axial loading ($F_{br;AX}$) results in a nominal cross-sectional stress of 20 N/mm². Both bending loads result in a nominal stress of 20 N/mm² in the outer fibers of the cross-section.

Only surface cracks are considered in this study. The crack location depends on the applied load case. In case of axial loading ($F_{br;AX}$) and out-of-plane bending ($M_{br;OPB}$), the surface crack is located at the saddle location of the chord member. When in-plane bending ($M_{br;IPB}$) is concerned, the surface crack is situated on the crown location. These locations are also indicated in Figure 1 and correspondent with the hot spot stress locations for non-cracked joints under the same loading and boundary conditions.

3 THE NUMERICAL MODEL

The numerical model consists of a simplified model of the semi-elliptical surface crack, a finite element model to obtain the required stress intensity factors (SIF's) and a crack growth model to relate the calculated stress intensity factors to crack growth. A linear elastic fracture mechanics (LEFM) approach is adopted.

3.1 Semi-elliptical surface crack

As shown in Figure 2 for a flat plate, the surface crack is simplified to a semi-elliptical surface crack with crack depth a and semi crack length c. The curve of the crack front resembles that of a semi-ellipse. The stress intensity factors (SIF's) for the crack tip at the deepest point (denoted with K_a) and at both crack ends (denoted with K_c) are considered. It is assumed that the surface crack maintains a semi-elliptical shape during the entire fatigue crack growth simulation and that the SIF's at the two concerned crack tip locations determine only crack growth. However, in reality the highest SIF's may be found between these two typical crack tip locations.

The required SIF's for a specific crack configuration are obtained with the help of finite element program MSC.Marc2003 by calculating the strain energy release rates according to the G-integral method.

Equation 1 is used to convert the strain energy release rate G to the equivalent SIF, whereby assuming a condition of plane strain everywhere along the crack front.

$$K = \sqrt{\frac{G \times E}{1 - n^2}} \tag{1}$$

The Young's Modulus E is set to 210,000 N/mm² and the Poisson's ratio v equals 0.3. A linear elastic material model is applied. The effects of a mixed mode loading (mode I, II and III) situation, which is present around the surface crack, are taken automatically into account

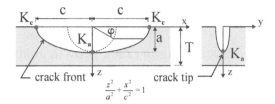

Figure 2. Definition of semi-elliptical surface crack.

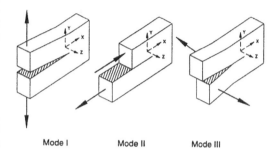

Figure 3. Basic crack modes for fracture mechanics.

by MSC.Marc2003. The individual crack modes are visualized in Figure 3.

3.2 Finite element model

A three-dimensional finite element method (FEM) model is used to obtain the strain energy release rates belonging to a specific cracked T-joint. The FEM models are made with the help of mesh generation program MSC.Mentat2003.

The FEM model concerns a CHS T-joint with an actual semi-elliptical surface crack at the saddle or the crown location of the chord member (Oomens 2005). The deepest point of the surface crack is located precisely at the crown or saddle location. The geometry of the full penetration weld is simulated in compliance with the American Welding Society (AWS 2000) provisions. An example of a FEM model is given in Figures 4 and 5.

The FEM model is constructed completely of solid elements with quadratic shape functions. A reduced integration scheme with Gaussian integration points is applied. An appropriate mesh refinement is applied for the region containing the crack. Special elements within software package MSC.Marc2003 are used to transfer the nodal loads and the nodal reaction forces to the finite elements next to the section ends. These special elements own a mathematical infinite stiffness, but they do not cause irregularities in the stiffness matrix.

3.3 Fatigue crack growth simulation method

The Paris' law is used to calculate the increase in number of load cycles ΔN following from a certain

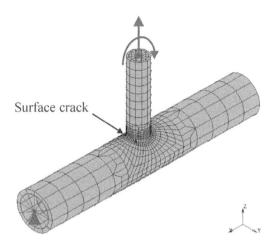

Surface crack

Figure 4. Example of finite element model of cracked T-joint.

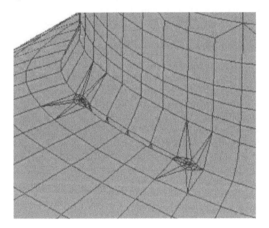

Figure 5. Detailed view of surface crack in FEM model.

increase of crack size. The constant C and m in the Paris' law is $1.42 \cdot 10^{-12}$ (mm and N) and 2.71 respectively. The procedure for executing fatigue crack growth simulations (Dijkstra 1997) is summarized below.

1. A crack with an initial crack depth a_i of 3.57 mm (1/10 of T for $2\gamma = 14$) is assumed as a starting point for all analyses. Crack initiation is not concerned in this study. Most analyses are carried out assuming an initial crack length of 17.9 mm and an initial a/c-ratio of 0.2. For other analyses an initial a/c-ratio of 0.1 and 0.4 is assumed resulting in an initial half crack length c of 35.7 and 8.93 mm respectively. The values for K_a and K_c for the initial crack are calculated by means of the finite element method and Equation 1.

2. An increase of crack depth Δa as step size is assumed and the corresponding number of cycles

ΔN (increase of fatigue strength) can be calculated by using the Paris' law.

$$\Delta N = \frac{\Delta a}{C \left(\Delta K_a \right)^m} \tag{2}$$

Step size Δa equals 1/10 part of the chord wall thickness with exception of the first step. Therefore, the absolute step size Δa is dependent on joint parameter 2γ.

3. The increase of crack length Δc is calculated by using again the Paris' law.

$$\Delta c = C \left(\Delta K_c \right)^m \cdot \Delta N \tag{3}$$

4. The number of cycles (fatigue strength) is increased with ΔN.

$$N_{i+1} = N_i + \Delta N \tag{4}$$

5. The new crack dimensions are obtained by increasing crack depth a as well as half crack length c.

$$a_{i+1} = a_i + \Delta a \quad \text{and} \quad c_{i+1} = c_i + \Delta c \tag{5a,b}$$

6. When the new crack dimensions are known, a new FEM model will be generated to calculate stress intensity factors K_a and K_c for the new crack configuration.

7. Steps 2 to 6, forming together a simulation cycle, will be repeated until a predefined final crack depth a_f is reached. The final crack depth is assumed 90% of the chord wall thickness.

4 VALIDATION

The numerical model has been validated for thin-walled CHS T-joints (Oomens 2005). Unfortunately, the validation could not been carried out for thick-walled T-joints. To the author's knowledge, suitable experimental crack growth data were not available at that moment. The observed fatigue crack growth of a thin walled T-joint, referred as Specimen B3 (Noordhoek 1984), is simulated in order to compare the numerically obtained fatigue strength with experimental results. The simulations are executed for three different combinations of the constants C and m to be used in the Paris' law, because it is not known which specific values are applicable for Specimen B3. The applied values for the constants can be found in Table 2.

The FEM results for crack growth in depth direction a are presented in Figure 6 together with the experimental results. A similar comparison for crack growth in length direction c is made in Figure 7. Both figures show that a reliable fatigue crack growth simulation is possible, but the results are dependent upon the chosen values for the constants C and m. For both crack growth directions, the best agreement with experimental results is achieved by applying set 1. The values of set 1 are used for the fatigue crack growth analyses

Table 2. Used constants in Paris' law for crack growth.

Set Nr.	C (mm and N)	m
1	$1.42 \cdot 10^{-12}$	2.71
2	$3.34 \cdot 10^{-13}$	2.92
3	$1.832 \cdot 10^{-13}$	3.00

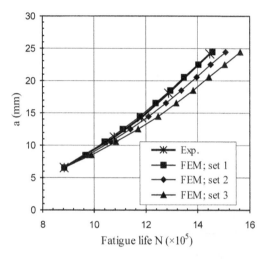

Figure 6. Comparison for crack growth in depth direction a.

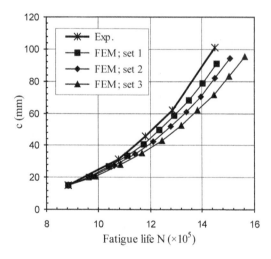

Figure 7. Comparison for crack growth in length direction c.

in this study. Note that applying a smaller step size will probably move the curves somewhat to the left. Then, another set can give a better agreement with the experimental results.

The deviations for the crack growth in length direction are larger then for in depth direction. This difference in agreement is reasonable, because the agreement between the FEM model and the tested joint is far better for the deepest point of the crack than for both crack ends next to the weld. A regularly shaped weld toe is indirectly assumed for the FEM model, while Specimen B3 owns a very irregularly shaped weld toe.

The numerical model under predicts crack growth c. If another set of constants is used, the underrating is also applicable for crack growth in depth direction. The predicted remaining fatigue strength will be overrated, resulting in non-conservative results. Adjustments, such as applying corrections factors, to the fracture mechanics model are a possibility to improve the accuracy of the numerical model, but then more validations are needed, while the required crack growth data are not available.

Because of the achieved good results and the lack of suitable experimental crack growth data for thick-walled CHS T-joints, it is assumed for the time being that the finite element model is useable and reliably for analyses on thick-walled T-joints.

5 CRACK GROWTH SIMULATIONS

Crack growth analyses are conducted to investigate the crack shape evolution during fatigue crack growth. The crack shape aspect ratio, defined as the crack depth a to half crack length c ratio (a/c-ratio), is used in this study as an indicator for the crack shape evolution. Firstly, an initial a/c-ratio of 0.2 is assumed for the simulations, but several simulations are subsequently repeated for a different initial crack shape aspect ratio.

5.1 Crack shape evolutions

Figure 8 presents the simulated crack shape evolutions for different 2γ-values and basic load cases. The graphs in Figure 8 show clearly that the a/c-ratio evolves during crack growth. Assuming a constant a/c-ratio for a fatigue strength estimation analysis seems therefore not to be realistic.

For all basic load cases and 2γ-values, the aspect ratio increases rapidly in the beginning of the crack growth simulation. After reaching the top value, the a/c-ratio decreases almost linearly. However, in case of in-plane bending the a/c-ratio is practically constant after passing a crack depth of about 30 mm. The crack shape aspect ratio at the final crack depth a_f is almost the same for all three 2γ-values in case of axial loading ($F_{br;AX}$) and amounts then about 0.28, while the final a/c-ratio varies between 0.32 and 0.24 for the other two basic load cases.

Furthermore, it appears from Figure 8 that a higher 2γ-value results into a smaller a/c-ratio for the final crack. However, the differences are too small to speak about a significant influence.

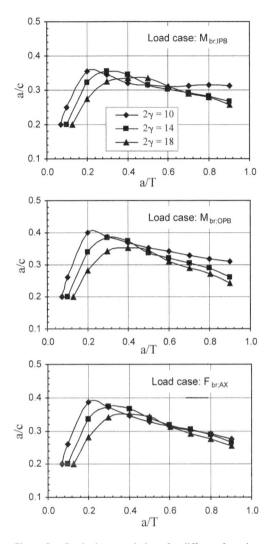

Figure 8. Crack shape evolutions for different 2γ-values and load cases.

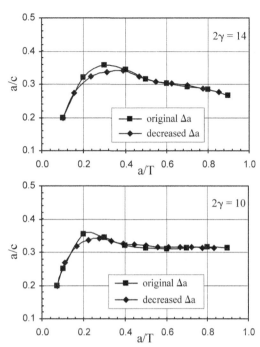

Figure 9. Influence of decreased step size Δa on evolution of a/c-ratio.

5.2 Influence of crack depth increase Δa

It is suspected that the observed evolutions of the a/c-ratio (Fig. 8) is influenced by the applied step size Δa. Additional simulations are carried out for $2\gamma = 14$ and $2\gamma = 10$, whereby the step size Δa is reduced. The analyses are only carried out for load case $M_{br;IPB}$. The results are compared with those for the original step size in Figure 9.

The decrease of step size Δa has lowered the tops of the graphs for both 2γ-values, but it has no influence on the final crack shape aspect ratio. In general, the reduction of the step size has no significant effect on the crack shape evolution. Future research should make clear if further reduction of step size Δa will still result into a significant influence on the evolution of the a/c-ratio for small cracks.

5.3 Influence of initial a/c-ratio

So far, an initial a/c-ratio of 0.2 is assumed for the crack growth analyses, but the question arises whether a different initial a/c-ratio will give similar crack shape evolutions. Observations of various thin-walled CHS joints (Kam 1991) have shown significant smaller a/c-ratios in the beginning of the crack growth phase. The same observations may be expected for thick-walled joints.

Low values for the initial crack shape aspect ratio are preferred when conducting numerical analyses with the used FEM model. Obtaining a reliable mesh proves to be difficult in case of narrow surface cracks. Badly shaped finite elements may come in existence, because of the strong and local curvature of the semi-elliptical crack front close to the crack ends. Therefore it is interesting to know the consequences of assuming a too low initial crack shape aspect for the sake of modelling conveniences.

Additional analyses are carried out for $2\gamma = 14$, whereby the T-joint is subjected to in-plane bending ($M_{br;IPB}$). This time, initial a/c-ratios of 0.1 and 0.4

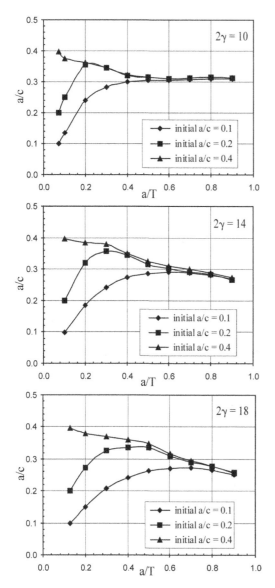

Figure 10. Crack shape evolutions for load case $M_{br:IPB}$, three 2γ-values and initial a/c-ratios.

to grow in such way the least energy is needed. The convergence emphasizes therefore the reliability of the numerical model.

Furthermore, it seems at first sight that the moment of achieving convergence is dependent on the 2γ-value owned by the T-joint. The convergence appears to be achieved faster for a low 2γ-value. However, when the presentation of the x-axis is changed from non-dimensional scale a/T to the absolute crack depth a, the rate of convergence appears to be almost independent of joint parameter 2γ. The convergence is then obtained at a crack depth close to 25 mm. For a joint with $2\gamma = 10$ the convergence is achieved at a crack depth of around 60% of the chord wall thickness T. This is equal to a crack depth of 25 mm (50×0.6). When joint parameter 2γ equals 14, the convergence is reached at 70% of the chord wall thickness meaning a crack depth of 25 mm (35.7×0.7). Finally, the same crack depth can be calculated for $2\gamma = 18$, based on a full convergence at 90% of the chord wall thickness T. From these observations it is expected that for similar cracked T-joints with $2\gamma > 18$ full convergence is not obtained before reaching final crack depth, when the difference between initial crack depth a_i and the final crack depth is less than 0.75T.

6 FATIGUE STRENGTH CALCULATIONS

The fatigue strength of the considered joints can also be derived from the numerical output. In this study, the fatigue strength N_{ref} is defined as the required number of cycles according to the numerical model to accomplish a crack growth starting from the initial crack until the moment the crack has grown through 90% of the chord wall thickness. As mentioned before, crack initiation is not taken into account.

6.1 Influence of load case and γ-value

The calculated fatigue strengths are presented in Figure 11 for three load cases as function of the joint parameter 2γ. They are also tabulated in Table 3.

Figure 11 and Table 3 shows both a strong dependence of the calculated fatigue strength on the applied load case and joint parameter 2γ. A low value for the joint parameter 2γ results in a significant larger fatigue strength for all three load cases. A decreased 2γ-value means an increase of the chord and brace wall thicknesses, when assuming a constant outer chord diameter. More material is then available for the crack to grow through. Besides that, the accompanying increase of the chord's moment of inertia will have a reducing effect on the stress level. On the other hand, the increased total force on the brace member due to constant nominal stress of 20 N/mm² results in higher stresses next to the cracked area. All together,

are assumed. The results are compared with those for an initial crack shape aspect ratio of 0.2 in Figure 10. The evolution of the a/c-ratio is displayed as function of the non dimensional a/T-ratio.

It appears that the assumed initial a/c-ratio does not determine the final crack shape, because in each subfigure the graphs converge nicely.

The observed convergences in Figure 10 seem to be reasonably. It indicates that some state of equilibrium is reached during the simulation. The crack tends

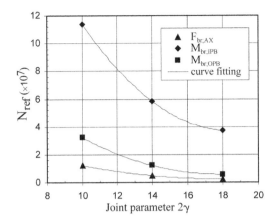

Figure 11. Calculated fatigue strength as function of joint parameter 2γ for different load cases.

Table 3. Calculated fatigue strengths for basic load cases.

Load case	$2\gamma = 10$		$2\gamma = 14$		$2\gamma = 18$	
	N_{ref} ($\times 10^7$)	%*	N_{ref} ($\times 10^7$)	%*	N_{ref} ($\times 10^7$)	%*
$F_{br;AX}$	1.23	11	0.487	8.4	0.237	6.4
$M_{br;IPB}$	11.4	100	5.83	100	3.72	100
$M_{br;OPB}$	3.22	28	1.22	21	0.588	15

* Percentage of fatigue strength for load case $M_{br;IPB}$.

Table 4. Calculated fatigue strengths for three initial a/c-ratios.

Initial a/c-ratio	$2\gamma = 10$		$2\gamma = 14$		$2\gamma = 18$	
	N_{ref} ($\times 10^7$)	%*	N_{ref} ($\times 10^7$)	%*	N_{ref} ($\times 10^7$)	%*
0.1	10.1	89	4.79	82	2.85	77
0.2	11.4	100	5.83	100	3.72	100
0.4	12.5	110	6.53	112	4.22	113

* Percentage of fatigue strength for initial a/c-ratio of 0.2.

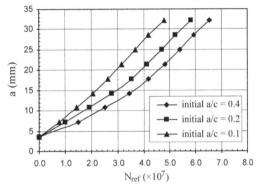

Figure 12. Fatigue strength for T-joint with $2\gamma = 14$, load case $M_{br;IPB}$ and different initial crack shape aspect ratios.

it will take more load cycles to obtain the final crack shape in case of small 2γ-values.

Furthermore, it appears that, independent of the 2γ-value, axial loading ($F_{br;AX}$) involves the smallest fatigue strength, while in-plane bending ($M_{br;IPB}$) results into the largest fatigue strength. The fatigue strengths for the load case of in-plane bending are significant larger than for the other two load cases. In-plane bending is therefore favorable upon axial loading and out-of-plane bending when concerning fatigue strength.

In contrast to the observed crack shape evolutions, the calculated fatigue strength is dependent on the applied level of nominal stresses. The stress intensity factor is related linearly to the nominal stresses in the brace member. The fatigue strengths, which are given in Table 3 for a nominal stress of 20 N/mm², can be converted to other stress levels by applying a multiplication factor. Equation 6 can be used to calculate the fatigue strength N_{ref*} for a deviating nominal stress level.

$$N_{ref*} = N_{ref} \left(\frac{20}{\sigma_{nom}} \right)^{2.71} \quad (6)$$

6.2 Influence of crack depth increase Δa

A reduction of the step size Δa results into a lower fatigue strength of about 7 % for $2\gamma = 14$ and $2\gamma = 10$. Further reduction of step size will improve the accuracy of the model, but it also means more work due to the complex FEM model.

6.3 Influence of initial a/c-ratio

From a practical point of view it is interesting to investigate the effect of assuming a different initial a/c-ratio on the calculated fatigue strength. The obtained fatigue strengths are tabulated in Table 4. Figure 12 shows the calculated fatigue strength N_{ref} for a T-joint with $2\gamma = 14$. Similar graphs can be drawn for $2\gamma = 10$ and $2\gamma = 18$, but they are not given.

Table 4 and Figure 12 show both clearly that the fatigue strength is dependent of the assumed initial crack shape aspect ratio. For all 2γ-values, a lower initial a/c-ratio result into more fatigue strength for the T-joint. This effect is reasonable as will be explained later. When taking the fatigue strength belonging to an initial a/c-ratio of 0.2 as reference (100 %), the deviations in fatigue strength vary between 10 and 23 %. The initial crack shape aspect ratio should therefore be

465

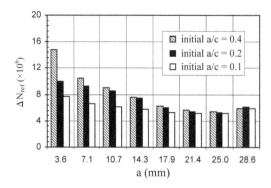

Figure 13. Increase of fatigue strength per simulation cycle as function of crack depth for load case $M_{br;IPB}$, $2\gamma = 14$ and different initial crack shape aspect ratios.

chosen with caution. Otherwise, the remaining fatigue strength may be largely underestimated due to a too low initial a/c-ratio.

The observed differences in fatigue strength are caused primarily in the beginning of the crack growth evolution as shown in Figure 13. A smaller a/c-ratio means a larger initial crack size for a constant initial crack depth. It also means higher stress intensity factors and thus a lower increase of fatigue strength ΔN_{ref} according to the Paris' law. As the crack growth continues the differences in increase of fatigue strength disappear gradually, because the crack shape evolutions show convergence in Figure 10.

7 CONCLUSIONS

Numerical fatigue crack growth analyses are performed on thick-walled CHS T-joints containing a surface crack. Different values for joint parameter 2γ, basic load cases and initial crack shapes are considered and their influence of the crack shape evolution and remaining fatigue life are investigated.

The analyses have shown that the crack shape aspect ratio (crack depth a to half crack length c ratio) evolves during fatigue crack growth. However, the final crack shape for a crack depth of 90% chord wall thickness is nearly independent on the assumed initial crack shape.

The assumed initial crack shape influences the remaining fatigue strength. A larger value for the crack shape aspect ratio results into more fatigue strength for the T-joint in case of a constant initial crack depth. In order to conduct a reliably fatigue strength analysis, the assumed initial crack should be chosen carefully to prevent calculating too overrated remaining fatigue strength.

Furthermore, the calculated remaining fatigue strength is dependent significantly on the joint parameter γ and the applied load case. Decreasing joint parameter 2γ gives the CHS T-joint more remaining fatigue strength. In-plane bending involves the most fatigue strength for the joint and is therefore favourable upon out-of-plane bending and axial loading of the brace member.

REFERENCES

A.W.S. 2000. Structural Welding Code-Steel, AWS D1.1:2000. American Welding Society Inc., Miami, USA.

Dijkstra, O.D. & Straalen van, I.J. 1997. Fracture mechanics and fatigue of welded structures, *Proceeding International Institute of Welding Conference*: 225–239.

Kam, J.C.P. & Ma, C.N. 1991. Crack shape evolution in tubular welded joints, *NDT&E International*, Vol 24 (16): 291–302.

Noordhoek, C. & Verheul, A. 1984. Comparison of the ACPD method of in-depth fatigue crack growth monitoring with the crack front marking technique, *Stevinreport 6-84-13, TU Delft*.

Oomens, M., Romeijn, A., Wardenier, J. & Dijkstra, O.D. 2005. Development and Validation of a 3-D Fracture Mechanics Model for Thick-walled CHS T-joints, *Proceedings of the Fifteenth (2005) International Offshore & Polar Engineering Conference, Seoul, Korea*, Vol 4: 330–340.

Design Award Competition

Tubular Structures XI – Packer & Willibald (eds)
© 2006 Taylor & Francis Group, London, ISBN 0-415-40280-8

Arsenal London Stadium

C. Drebes & M. Rüster
TU Darmstadt, Darmstadt, Germany

1 TASK

Since 2001, the FC Arsenal London has been planning a new Stadium, which is to replace the since 1913 used Highbury Stadium as playground for the "Gunners". The new building is to be planned and built in direct environment of the old Highbury stadium and will be finished in 2006.

The task consisted of designing this new stadium for the club as a pure football stadium. The draft has to fulfill all requirements of a modern football stadium. Special attention within this draft had to be placed on atmospheric qualities inside and outside of the stadium. Furthermore attention had to be paid to a good organization of the different functions and especially to the integration of the construction into the architectural draft. The stadium should have a capacity of 60,000 seats and has to fulfill the requirements for international competitions based on FIFA regulations.

2 DESIGN

The planning area is in the center of London, in the district of Islington. Islington consists of small parceled workers' housing, built in the typically English architectural style with heights of up to three floors. This district of London has approximately 181,000 inhabitants. The population structure in Islington changed in the course of time. Today, a share of 55% of council housing is facing private real estates in the value of up to 3 million Pound.

The football club FC Arsenal London, which has developed in the course of its history from a workers'

Figure 1. Perspective.

club to a world class football team, is identity-giving for the population. The most distinct points in the closer surrounding of the planning area are the still used Highbury stadium and three high-rise buildings with a height of approximately 60 m. Due to the fact that Islington is in the center of London and it has no direct access to any motorway, a good and fast access is only provided by suburban public transport. The public transport stops are in proximity of the planning area. The current building structure on the planning area is exclusively industrial and is surrounded by railway tracks of the London underground and the national rail. Due to the inner-city location and the crowded roads of London, parking space is at a premium in the whole city. In Islington especially, there were never any parking options since the beginning of the football club. The arrival of the fans via car would change London's traffic situation to the worse. So the journey of the fans is made nearly always by public transport. This arrival situation is seen as basic principal for the designing of the new stadium, therefore, new parking sites have been ruled out completely. Only for VIP guests, a car-park is integrated in the basements. This situation formed the basis for the orientation of the stadium and also the idea to separate the different fan groups from each other before they enter the stadium area. This will be realized by two different underground lines being separated from each other. The first one stops at the Arsenal underground station in the North and the second one stops at the Highbury station in the South. While the "Gunners", the fans of the FC Arsenal London, arrive via the Arsenal underground station and are led by structural measures to their traditional "north tribune", the guest fans reach the stadium area via the Highbury station in the South. The desired fan splitting is emphasized by the entrance situation, which makes entering of the inner area of the stadium only possible through the entrances either in the north or in the south.

Special attention has been put on the atmospheric impression that the supporter will get when entering the stadium. The arriving fan can already have

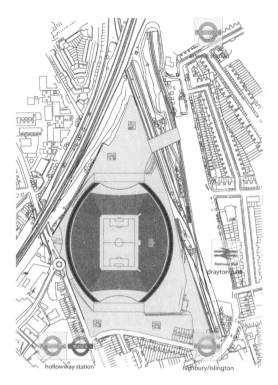

Figure 2.   Ground plan.

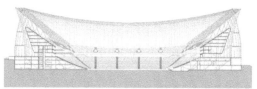

Figure 3.   Cross-section.

a first glance of the inner stadium when entering the grounds. The closer he gets to the stadium entrance the more views are granted, whereby the anticipation for the game is immensely increased. This atmospheric effect causes the supporter to be quasi-sucked into the stadium. The filigree columns, which separate the upper ranks from the lower ranks, make it possible to look through the whole stadium. The view giving range to the stadium inside is framed by the arc-shaped entrance situation. After entering the stadium the fan is situated on the central distribution level which leads around the stadium and follows the alignment of the roof on the outside edge. The inner border of this level forms the upper edge of the lower ranks, whereby the complete access of the lower ranks is ensured. The same path is used in a possible evacuation of the stadium as escape route. The upper ranks are accessed over ramps, which can be reached from the central distribution level and which follow the curved lines of the arches. Over these ramps, the visitors get to the different distribution levels from which the front and side tribunes can be reached. From the distribution levels the fans get to the stadium interior through mouth openings in the tribunes. In the case of a necessary evacuation, the supporters can escape through the mouth openings and the ramps to the outside of the stadium. The VIP range forms an exception in the

accessibility. The visitor of the VIP range arrives, if he does not reach the stadium via the public transport system, by car via the VIP parking area in the basement of the stadium. Independent stairways, which serve as escape routes in an emergency, and lifts lead from there into the exclusive VIP level, which is between the upper tier and the lower tier. The VIP tiers are only on the side tribunes. The VIP zone is distributed over two floors, like a wedge between the balconies. In these two "VIP areas" the demanded number of VIP places has been accomplished.

If the tribunes are regarded in the plan view, the development from a rectangular shape at the playground to a swung form at the roof can be seen. The characteristics of traditional English tribunes are the proximity to the playing field and the players as well as a barrier-free view to the field and the game. This produces an atmosphere which can carry away the fan in a special way and brings the support of the fans particularly closer to the player. This tribune type was a distinguishing characteristic of the Highbury stadium and therefore it has been integrated in the planned draft in order to offer the fans and the players a similar atmosphere. From the rectangular form the tribunes develop to a curved form so that the optimal range of vision which has an ellipsoidal layout can be achieved. The optimal range of vision has been determined by the FIFA and describes the maximum useful distance between the spectator and the field. Besides the specification of the optimal range of vision, the quality of the views from the different ranks to the playing field was also of decisive importance. So the lower ranks describe a parabolic curve and the upper ranks a straight line with an angle of 34°. This makes it possible to have a good view from every single place in the stadium and to remain in the optimal range of vision. So that the desired number of 60,000 seats can be realized.

Based on these two rules of stadium design, two arches facing themselves evolved as the optimal roof structure. This is caused by the fact that the arches can, due to their curvature and their necessary height, follow the course of the edge of upper ranks perfectly. Thereby, the stadium reaches a max. height of 60 m in the middle of the arches and orientates itself on the height of the high-rise buildings existing in the direct vicinity.

Figure 4. Nightshot.

In their wish to create an overwhelming atmosphere most stadiums built today give the impression of a multipurpose gym rather than that of a football stadium due to their relatively narrow building design. Therefore, it gives the spectators the impression of football as a kind of indoor sport and the basic idea of football as an open air sport is getting lost.

From the very first, the tribune's body of the designed stadium was thus decoupled from the roof construction, to completely avoid this closed building appearance but without loosing the "boiling pot" atmosphere.

The roof was intended to obtain an "airy" impression, and therefore special attention was paid to make the roof design as weightless as possible. The overall concept has been derived from a hanging cloth which can possibly hang over a long distance without any heavy or huge constructive elements.

3 SUPPORTING STRUCTURE

The concept of the hanging cloth was realised in the roof structure by using a cable construction with suspension and guy cables covered with a light roof membrane using Plexiglas panels. The cable construction is tightened and held by two arches which are led lengthwise across the stadium. The roof structure orientates itself on the form of the double curved arches and thereby forms an inherently stable, hyper parabolic surface. The loads from the roof are transferred by the arches in horizontal direction and then transmitted in vertical direction by columns below the arches. The horizontal loads from the roof are always absorbed in the axis of the arches, at the connection points between the carrying cables and the arches itself. This is possible because of the doubly bended form of the arches. The arches transfer the horizontal loads into the reinforced concrete foundations as normal forces. The foundations also form the external walls of the basements. The two arches are connected with each other by these foundation walls, whereby the horizontal load portion in transverse direction is balanced. The horizontal load portions in longitudinal direction is neutralized by a tension cable under the playing field, which is designed as a reinforced concrete base plate

Figure 5. Isometry.

and couples the foundation of one end of the arches, with the foundation of the other end of the arches. The weight of the head tribunes works supportively against the horizontal loads in longitudinal direction.

In order to minimize the weight of the arches, the arches have been designed as a latticed truss built out of structural steel hollow sections. The arches are built as triple-chords, have a static height of 3 m and use profiles with the dimensions of 323.9×8 mm. Due to the enormous span of the arches of over 280 m the remaining weight of the arches is transferred by the earlier-mentioned columns (Fig. 5). The arches are simply placed upon the columns. They are connected with the pillars via steel clips and therefore possess the capability of moving freely in horizontal direction without causing any reaction forces within the construction (Fig. 6). The chosen supporting structure prevents a possible uplift of the roof which can be caused by a suction effect due to wind loads. To allow the columns to handle compression as well as tension forces, they are made from hollow structural sections with the dimension of 600×300 mm. The columns reach a height of approximately 60 m in the middle of

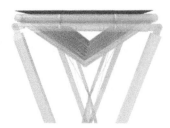

Figure 6.   Detail.

Figure 7.   Detail.

the arches. The ramps, which are following the course of the arches of the roof construction and which enable the access to the upper ranks, partition and thereby prevent the columns from buckling (Fig. 7).

The tension forces existing in the columns are evened out by the weight of the upper ranks of the longitudinal tribunes connected by a latticed framework placed in the basements. Thereby, an uplift of the columns due to unfavourable loading is prevented. The tribunes consist of precast concrete parts and are carried by the columns which are placed with a centre-to-centre distance of 11 m and are designed as a steel framework. The tribunes are connected with the columns below the arches via the basements. The upper ranks are stiffened in transverse direction by the three-dimensional truss columns. In the longitudinal direction the reinforcement takes place by connecting the ramps with the distribution levels.

Tubular Structures XI – Packer & Willibald (eds)
© *2006 Taylor & Francis Group, London, ISBN 0-415-40280-8*

Sportwelt Hamburg – City-Nord

T. Burgmer
Technical University of Darmstadt, Germany

ABSTRACT: Despite of the decreasing German population, Hamburg is a growing city. Urban redevelopment of certain quarters, such as the "Hafen-City" ("Harbour-City") with its future iconic building "Elbe Philharmonic", contributes to the city's fresh and modern image. The task of the diploma project "Sportwelt Hamburg" aims to create such an icon for the business district "City-Nord". Therefore the new building, home to several sports and recreation facilities, not only has to provide perfect circumstances for all sorts of sports pitches, gyms and gastronomy. The quality of space, the internal impressions, the construction and the range of materials have to be appropriate to make the building a memorable landmark in an upcoming city.

1 INTRODUCTION

The office district "City-Nord" is located in the north of Hamburg. In the south it is connected to one of Hamburg's biggest parks, the "Stadtpark Winterhude". It has been designed based upon the ideas of the "Athens Charta" from 1933.

Its status shall be enhanced by developing a complex for sports activities and recreation. This would give the opportunity of health and fitness training, recreation and social events not only to the employees of the local companies but as well to the habitants of the nearby residential areas. As a result the social life would be diversified and the location factors would be improved.

Figure 1. Aerial photograph of Hamburg's City-Nord. The site is highlighted (cf. arrow).

The design has to consider that a new, iconic building may have considerable influence on the area's future development by creating a positive, modern image.

2 CONCEPT

2.1 Crucial requirements

In addition to the creation of an iconic architecture, one of the most important tasks is to assemble rooms of extremely different sizes, such as indoor beach volleyball, basketball or tennis courts on the one hand and on the other hand changing rooms and sanitary facilities, in one homogeneous building. Further more this building should not look like a regular gymnasium.

2.2 Solution

The solution proposed by this project is a structure which suits both spatial and constructive needs. It creates smaller rooms with a high complexity. In terms of construction it is able to surround the bigger halls without any pillars, achieving a very efficient use of space.

The basic element is an octahedron, cut at its corners and reduced to steel tube edges. By connecting many of these elements to a three-dimensional structure various units of space can be created. Where needed wall- or floor elements are added to the structure. To create the larger rooms (such as indoor-courts) some of the octahedrons are left out (cf. larger and smaller bubbles within a sponge).

Through cutting off the octahedron's corners every single element looses its rigidness. But the addition

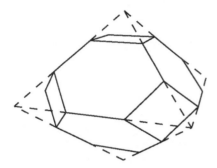

Figure 2.   The basic element.

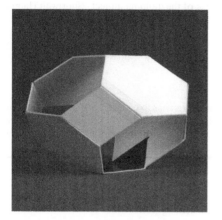

Figure 3.   Working process model of the basic element.

Figure 4.   Working process model of the basic element, showing the steel tube edges and the added wall and floor elements.

of multiple elements leads to neutralized horizontal forces within the structure. Only at its edges, i.e. the external and internal façades, the structure needs additional reinforcement. That is provided by rigid

Figure 5.   Working process model of various elements in addition, giving an idea of the complexity of internal spaces.

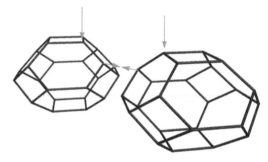

Figure 6.   Sketch schematically showing vertical and horizontal forces.

wall elements and a surrounding mesh within the façade.

3   CONSTRUCTION

The whole constructive framework consists of tubular profiles, due to their advantages in terms of dimensions and connections. As usual the nodes and joints are crucial. The horizontal beams are box sections. They allow easy connections for purlins and finishes (ceiling and floor).

Because of the unusual angles, connecting different wall elements is fairly complicated. That is why the diagonal beams are pipes. Again the reason is efficiency of the connections.

4   APPROACH

The design process was basically determined by experimental approaches to numerous three-dimensional structures. By switching between digital and physical models, the architectonical and the constructive capacity of these structures were determined.

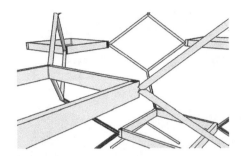

Figure 7. Sketch showing the use of box section profiles and circular profiles.

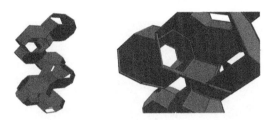

Figure 8. Digital 3D-study – vertical addition.

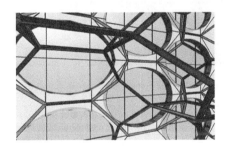

Figure 9. Digital 3D-study – internal spaces.

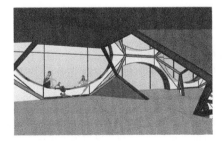

Figure 10. Digital 3D-study – internal spaces.

5 DIE SPORTWELT – THE SPORTSWORLD

5.1 *Form and façade*

Developing the façade of the building there was one crucial question: How would such a structure, endless

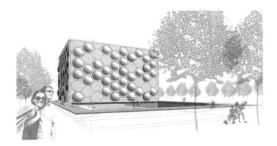

Figure 11. Perspective view.

Figure 12. West elevation.

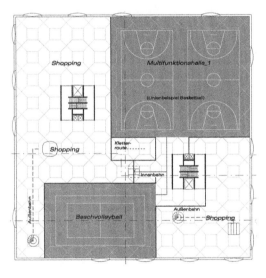

Figure 13. Schematic plan – Level 1.

in theory, come to an end? After another series of studies and taking the urban context into account, "rough cuts" seem to be the appropriate solution. That means the three-dimensional mesh ends with plane surfaces and in a simple geometrical shape. At the same time this has to be regarded as a quotation from the surrounding buildings, reducing their large number of geometries to one of the basic forms in plan, the square.

5.2 *Internal organisation*

Due to the high complexity of the structure, the experience of using the building's facilities is something completely new. Most of its visitors will never have seen anything comparable before. Therefore "Die

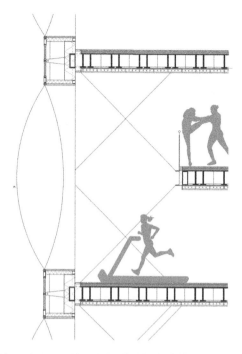

Figure 16. Model photograph – external view at night.

Figure 14. Façade section – The openings in the façade are made of ETFE-Pillows. They allow wide spans and small loads.

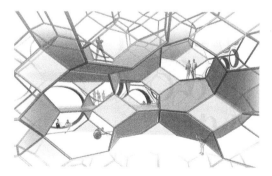

Figure 17. Model photograph – external view at night.

Figure 15. Perspective internal view.

the mesh, which is a complete unusual kind of space. Therefore the layout of the functions and the openings in the façade are organized in a special manner, providing certain views to different points of orientation.

Sportwelt" will not only attract people from the surrounding quarters but as well those who search for an inspiringly different environment for their recreational activities.

On the other hand the complexity might bring difficulties of internal orientation with it. To put it short, people might find it hard to find their way through

6 CONCLUSION

The connection between architectural and structural benefits is the system's basic achievement. However, the present result must not be regarded as a final solution. It is rather a frozen state of a process, which might lead to a completely new kind of architectural space. That is the most interesting and exciting conclusion. Constructive structures do not necessarily restrict architectural ideas but can lead to new spaces, just by combining both approaches. That is not news but has often been forgotten.

Tubular Structures XI – Packer & Willibald (eds)
© 2006 Taylor & Francis Group, London, ISBN 0-415-40280-8

City stadium 2006 – Cinadium

M.K.A. Staubach & H. Mott
University of Applied Sciences, Darmstadt, Germany

ABSTRACT: The City stadium is a mobile unit for public spaces. Our studies have been initiated by the outstanding event of the Soccer World Cup which takes place in Germany in 2006. The construction's initial task is to show live broadcasts of the soccer games on wide screen. Referring to its most important quality – to host open air events involving cinematic projections in a stadium like atmosphere – it is called "Cinadium".

1 INTRODUCTION

1.1 *General information*

The described project is the result of a teamwork assignment in structural construction taught and supervised by Professor Uwe Laske and Professor Marcin Orawiec at the University of Applied Sciences in Darmstadt.

1.2 *Task and objective*

The project has been inspired by the Soccer World Cup representing one of the most important sports and media events in the World. The idea is based on the societal significance of huge sport events like World Championships and Olympic Games as well as common ticketing problems. "Cinadium" is supposed to offer all fans a chance to witness such special events adequately, even if they were not lucky to obtain a ticket. Regardless of the individual financial situation, people are invited to collectively join in and feel the unifying spirit.

1.3 *General conditions*

Certain conditions for the construction of the City stadium had to be considered and fulfilled. Some of the most important were efficient and fast erection, minimum weight and easy transport, compact storage, modules to be added flexibly, reversible sun and rain shelter, 400 seats, standing room for 1600 people, wide screen with stage as well as counters to provide refreshments.

2 USAGE

While the World Cup represents a unique event, the stadium is designed for further usage. The nature of the mobile construction enables easy set-up at diverse locations and varying causes such as music-, theatre-, cinema- or information events as well as festivals and so on. In the long run, the concept of "Cinadium" is to contribute to the revitalization of urban city culture.

3 ECONOMIC ASPECTS

The requirements of frequent transport and set-up were not the only reasons to find a construction as light as possible while having a maximum of stability. The predominant elements used for the structure are circular hollow sections made of stainless steel grade V2A. They are welded to form triangular girders to achieve the best possible structural efficiency.

The modularity of the girders and stands elements allow a fast set-up but also flexible extension and therefore better options for a specific usage (see Figure 1). "Cinadium's" multipurpose design also improves its marketing value. Repeated production of identical (basic) module elements finally causes a significant reduction of production costs.

4 DESIGN CONCEPT

The basic idea was to transfer the mood of the stadium atmosphere onto a public place. One important point is the right form of the curved seating areas to achieve mutual perception which stimulates the communication within (Figure 2).

The distances from the stands to the wide screen as well as the view angles onto it have been carefully examined. We wanted to make sure that a good view is offered from all seats. Therefore, the first row starts at a seat level of 1.65 m taking into account the standing space in front of it. The following rows

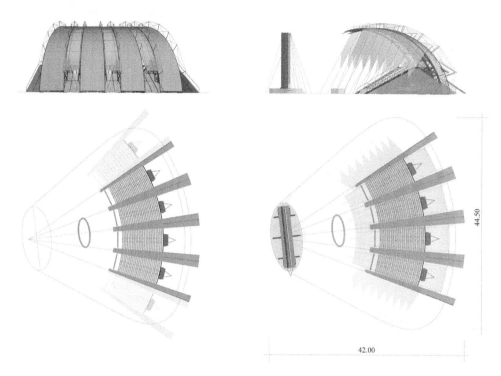

Figure 1. Views showing back, side, top and modularity.

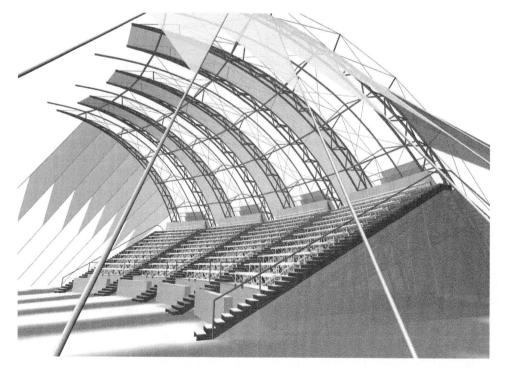

Figure 2. Perspective view showing curved seating areas.

478

Figure 3. Spectators view.

are raised 40 cm each. The vast cantilever roof construction provides a free view at the wide screen and creates an atmosphere which keeps the spirit of the crowd inside (Figure 3). The overall construction with external dimensions of 42.00×44.50 m concentrates around a central bar offering refreshments to the surrounding audience. The different translucent green membranes and lamellas forming the outer skin resemble light-flooded and protecting leaves (Figure 4). To realize the intended shape for the girders looking like branches circular hollow sections were used.

Figure 4. Branches and leaves.

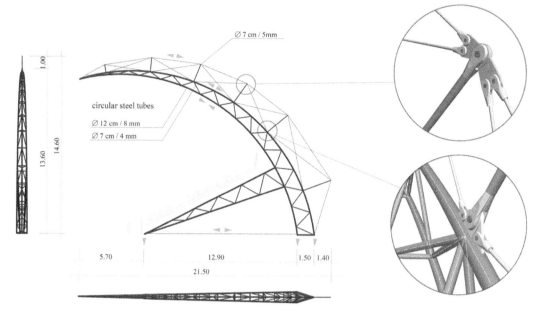

Figure 5. Construction of curved cantilever.

5 CONSTRUCTION

5.1 Girder

The developed girder is the result of intense studies to efficiently transfer loads into the ground while using a minimum of material.

During the design process we took great care of the fact that there was no possibility to permanently anchor the structure using massive foundations. For the functioning of the mobile structure it is very vital to ensure its mobility and ease of erection in public places by not having a drastic impact along with radical modifications of the construction site's surface.

Another essential factor was to guarantee the afore-mentioned clear view onto the wide screen which finally gave the impulse to use a cantilever roof construction.

The girder is designed as curved triangular (truss) cantilever and gets thinner to the free end in the horizontal and vertical axis (Figure 5). Two of the chord hollow sections are on the bottom to withstand pressure forces. The third circular hollow section is located on top to transfer the tensile forces. The brace members are also made of circular steel tubes and have welded joints. This kind of connection gives the truss its unique stability.

The most severe moment loading develops at the connection point of the curved and the straight part of the construction. The straight part prevents the curved one from tilting and therefore allows the curved girder

Figure 6. Model.

to project over the audience. The additional traction rope on the bottom further reduces the moments at the connection point.

Although the triangular truss cantilever is able to take all the existing forces a traction rope made of steel wire is added above it to guarantee the constructions stability even during a larger wind load attack. The final design of the truss cantilever enables it to trans-fer all forces vertically and securely into the ground

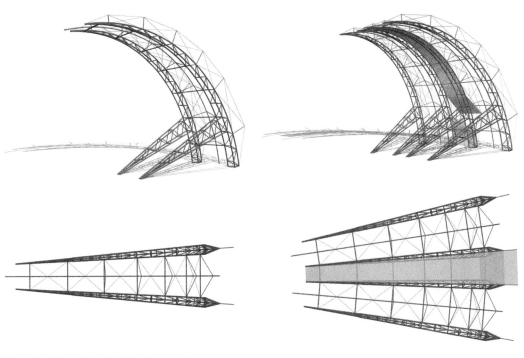

Figure 7. Connection of girders to segments.

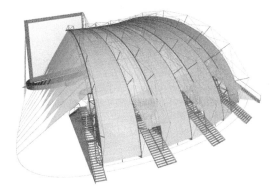

Figure 8. Perspective view.

Figure 9. Tribune segments.

at the base points. The results of our research have been constantly examined and analysed with the help of models (Figure 6).

Two identical girders are connected to one another with hollow sections at the lower truss joints to form a segment (Figure 7). The segments are also connected to each other, the outer ones being wired on the construction site like a circus tent (Figure 8).

5.2 Tribune segments

To achieve fast set-up and transport, the tribune segments are mounted on flat bed trailers and primarily consist of welded hollow sections (Figure 9). Each segment offers seating for about 109 people and is divided into three sub segments. The middle is permanently connected to the flat bed trailer by a hinge at the front, whereas the outer parts are hinged to the middle one. They fold upwards to achieve the maximum transport width of 2.50 m. In the unfolded position the rear width measures about 7.50 m. (Figure 10 shows an abstract scheme of the setup process). The tribune automatically rises to its final position being elevated by two hydraulic jacks (Figures 11 and 12). Supporting

481

Figure 10.   Scheme of tribune setup.

Figure 11.   Tribune mounted on flat bed trailer.

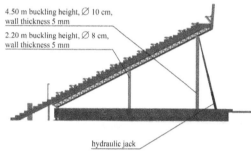

Figure 12.   Side view of tribune segment.

columns in form of steel tubes (hinged under the scaffold of the stands and designed to resist buckling) unfold into a vertical position and click in place when the stands are hydraulically lowered a few centimetres.

Because the stands' segments load transfer is independent from the roof structure (and the other way around) it is possible to use the different elements individually according to the required capacities and requirements.

5.3 *Widescreen*

The screen is also designed for rapid set-up. Just like the tribune segments it is entirely mounted onto a flat bed truck and can be erected without the help of a crane (Figures 13 and 14). Six hydraulic feet reach out to compensate any uneven spots in the ground. The two outer girders are permanently hinged to the vehicle and also rise hydraulically from their horizontal storage position. Once they have reached the vertical position they will guide the middle girder to the top. It is pulled upwards by two thin steel ropes running over the top of the outer vertical girders by a jack that is also located and concealed in the base. While the middle

girder is moving upwards the actual screen fabric is unrolled from its storing place also positioned in the base. Finally the screen is wired to the ground with steel ropes that were attached to the top of the girders before their erection.

5.4 *Lamellas*

The lamellas are made of light, hardwearing, translucent, UV- and weatherproof, Fiberglas-reinforced plastic.

They were developed to solve the most complex problems of the construction. First, they are needed to create a reversible cover as their design allows them to be stored compactly (Figures 15 till 18).

Secondly, they show intelligent behaviour for both main load cases of potential wind loads.

When wind attacks the structures outer surface the lamellas remain closed and prevent the penetration of wind and rain. The arising weight of the wind is lead securely into the ground through the cantilevers and their additional spans (shown in Figure 19 on the left).

When wind attacks from the inner side the arising pressure causes the lamellas to open (like a wind sock). This prevents a sail effect and reduces the wind load to a minimum (see Figure 19 on the right).

Figure 13.   Erection of screen and detail view.

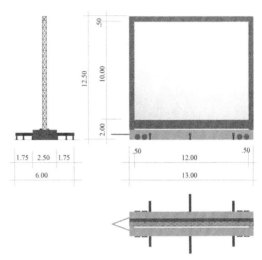

Figure 14.   Views of screen scaffold.

Figure 15.   Section of lamellas.

Figure 16.   Lamellas in stored, closed and open position.

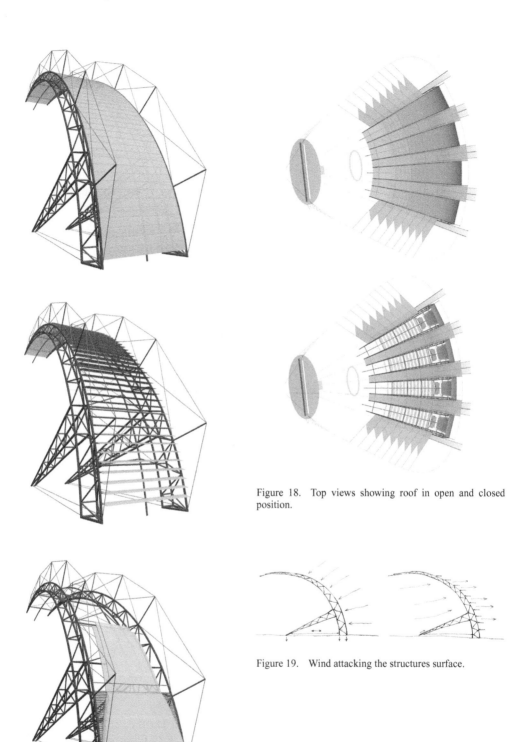

Figure 18. Top views showing roof in open and closed position.

Figure 19. Wind attacking the structures surface.

Figure 17. Mounted lamellas.

Parallel session 7: Composite construction

Tubular Structures XI – Packer & Willibald (eds)
© 2006 Taylor & Francis Group, London, ISBN 0-415-40280-8

Experimental and analytical investigation of vanadium micro-alloyed concrete-filled tube-concrete footing connections

A.M. Kingsley, T.S. Williams, D.L. Lehman & C.W. Roeder
University of Washington, U.S.A.

ABSTRACT: Use of high-strength steel structural members can reduce the material required for a given quantity of resistance. However, reduced sections can be susceptible to local instability and result in increased deflections. In concrete-filled tube (CFT) columns, the full tensile strength of the steel can utilized because the concrete fill restrains buckling of the tube and reduces member deflections. An ongoing research study was initiated to develop reliable connections for CFT columns constructed using for high-strength, vanadium micro-alloyed steel (CFVST) columns. The current research investigates the performance and constructability of embedded-type CFVST column-concrete footing connections. The experimental results indicate that the connection contributes significantly to the global response. To verify the results, nonlinear finite element models were developed, which included material nonlinearities and damage to the concrete in the footing and joint. The analytical and experimental results were used to develop preliminary recommendations for analysis and design of CFT column-to-footing connections.

1 INTRODUCTION

Concrete filled steel tubes (CFT) are useful and efficient structural members. Steel tubes have large moments of inertia and radii of gyration in all directions, providing large buckling and bending capacities while minimizing cost and material. The tube serves as formwork and eliminates the need for reinforcement, which reduces the labor and material required for construction. In addition, the steel tube enhances the shear resistance and confines the concrete, therefore increasing the compressive strain capacity of the concrete and the displacement ductility of the element. In addition to providing increased flexural stiffness and compressive resistance, the concrete fill restrains local buckling of the tube. In comparison with traditional reinforced concrete construction comparable strengths may be achieved with lighter and more slender CFT columns. Multi-story buildings develop large compressive loads due to the accumulation of gravity loads over the height of the building, and CFT columns are particularly attractive for the lower story columns.

In a moment-resisting frame or a bridge, large plastic rotation demands are expected in the column-to-footing connection. Therefore a robust base connection capable of transferring the full moment demand and sustaining the cyclic plastic rotation demand is required for reliable CFT construction in moderate to high seismic zones. Previous research of CFT column base connections indicates that the connection

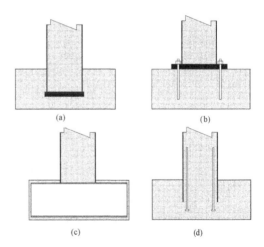

Figure 1. Base Connections (a) Embedded (b) Exposed (c) Embedded Steel (d) Reinforcing Steel.

can be categorized into one of the following types, as indicated in Figure 1: (a) embedded (b) exposed baseplate (c) embedded structural-steel and (d) reinforcing transfer bar.

Examples of base connections structural steel connections are shown in Figure 1. Embedded steel connections (Figure 1c) include research by Marson and Bruneau (2004) in which a pair of steel channels were used to connect the steel tube to the footing. Japanese researchers have investigated exposed (Figure 1b) and reinforcing steel (Figure 1d) column base connections; the use of reinforcing transfer steel provides enhanced strength and drift capacity relative to a standard base plate connection however it requires the use of column reinforcing steel, as depicted in Figure 1d. (e.g., Hitaka, Suita, and Kato 2004, Kadoya, Kawaguchi, and Morino 2005, and Morino et al. 2003). The embedded connection, shown in Figure 1a, presents a practical and efficient connection detail. Research shows that rectangular CFT columns with embedment depths (l_e) of 1.0D can achieve the theoretical flexural strength and large drifts with minimal damage to the foundation (Hsu and Lin 2003). The Architectural Institute of Japan (AIJ) design considers the connection to be fixed if the l_e is at least 2D (Morino et al., 2003). The embedded connection was considered in detail in this study.

2 AISC DESIGN PROVISIONS FOR CFT MEMBERS

Current provisions by the American Institute of Steel Construction (AISC, 2005) provide design recommendations for CFT members, including limitations on the material strengths and member geometry, and guidelines for the prediction of CFT strength and stiffness.

For the purposes of strength determination, concrete compressive strength, f'_c, must be greater than 3 ksi, and less than 10 ksi. Higher values of f'_c are acceptable for determination of stiffness. The yield strength of structural steel components, F_y must be no greater than 75 ksi. To be considered a composite section, at least 1% of the total composite cross-sectional area must be structural steel. In addition, the following limit is placed on the diameter-to-thickness ratio, D/t, for circular tubes:

$$D/t \le 0.15 \frac{E_s}{F_y} \tag{1}$$

The effective stiffness, EI_{eff}, of the composite cross-section is defined as

$$EI_{eff} = E_s I_s + E_s I_{sr} + C_3 E_c I_c \tag{2}$$

$$C_3 = 0.6 + 2\left(\frac{A_s}{A_c + A_s}\right) \le 0.9 \tag{3}$$

where E_s and E_c are the elastic moduli of the steel and concrete, respectively; I_s, I_{sr} and I_c are the moments of inertia of the steel tube, steel reinforcement (if present), and concrete, respectively; and A_s and A_c are the cross-sectional areas of steel and concrete, respectively.

Two methods are provided to determine the flexural strength of CFT members subjected to combined loading: the plastic stress distribution method and the strain compatibility method. The plastic-stress distribution method assumes that the steel has reached its yield stress in tension and compression, and that a compressive stress of $0.95f'_c$ is uniformly distributed in the compression region of the concrete for circular sections. The strain compatibility method assumes a linear distribution of strain across the section and the strength is estimated using appropriate constitutive relations for the concrete and steel and a limiting concrete compression strain of 0.003 in/in. Both methods assume the tensile capacity of the concrete is zero.

3 ANALYTICAL INVESTIGATION

A simple connection scheme and construction method for an embedded column-footing connection was studied in detail for the analytical and experimental portions of this research. Rather than a base plate, a steel annular ring is welded to the base of the steel tube to provide anchorage of the column into the footing and is intended to transfer the column forces into the connection. The footing is reinforced with flexural and shear reinforcement and may be placed in one or two lifts. Anchor bolts through the annular ring provide stability during construction and are not intended for load transfer.

A numerical model of the embedded CFT-footing connection described above was developed to investigate the effects of varying connection parameters on the response of the connection to aid in the development of an experimental test matrix. A series of finite element analyses were performed to determine the effects of footing depth, embedment length, annular ring dimensions, steel-concrete friction, and footing reinforcement on the behavior of the CFT-footing connection.

The MSC.Marc finite element analysis program was used to perform each analysis. Eight-node brick elements were used to model the steel tube, concrete fill, and concrete in the footing. Rebar elements, which define a plane of distributed reinforcement with uniaxial stiffness, were used to represent flexural reinforcement planes within the footing. The vertical reinforcement was modeled discretely as a series of two-node, uniaxial truss elements. Deflections in the shear reinforcement elements were controlled by the displacement of the surrounding footing elements. Linear elastic material models defined the behavior of the materials in the CFT column-footing connection model. Contact between the steel tube, concrete fill, and concrete footing was modeled using the contact option in MSC.Marc. The contact bodies were defined as initially touching; separation between bodies was allowed and penetration, or overlap of bodies, was prevented. Bond and sliding behavior were modeled as Coulomb friction behavior.

The results of the elastic finite element analyses lead to the following conclusions on CFT-footing connection response:

- The annular ring is necessary to develop anchorage of the column in the footing; however, the width and thickness of the ring do not appear to have significant effect on connection response.
- In the presence of the annular ring, and with linear elastic material behavior, changes in friction do not significantly influence stresses calculated in the steel tube or footing.
- Vertical reinforcement reduces the stress demands in the embedded portion of the steel tube and distributes stresses caused by column anchorage through the footing.

The most significant parameter was the embedment depth. Four models with $1.8D$ deep footings and $0.6D$, $0.9D$, $1.2D$ and $1.5D$ embedment, respectively, were analyzed and compared. Embedment length showed a significant effect on the stresses (normalized to the concrete strength and the square root of the concrete strength, in psi) in the footing, as shown in Table 1. Models with embedment depths of $0.6D$ and $0.9D$ embedment yielded at approximately 1% drift before yielding; the models with larger embedment depths yielding occurred at approximately 1.5% drift, indicating better development of column forces. The maximum and minimum principle stress at the bearing surface increased slightly with embedment. The stress at the 45° plane decreased with increased embedment. This stress was approximately four times greater in the $0.6D$ embedded connection compared to the $1.5D$ connection, suggesting a reduced tendency for developing cone pull-out with increased embedment.

To gain a better understanding of the connection response, the finite element model described above was modified to included nonlinear, inelastic material properties in the concrete and the steel tube. Effects of confinement and a smeared cracking model were applied to concrete elements, and steel plasticity was considered.

The inelastic analyses indicated that that the connection response is highly nonlinear, even at low drifts. Damage in the connection was concentrated around the embedded portion of the tube; although, maximum principle stresses exceeding the critical cracking stress are widely distributed through the footing. Yielding and permanent deformation were predicted in the tube base and annular ring, as well as high compression stresses at the column-footing interface at low and moderate drifts. Concentrated cracking strains at the surface of the footing parallel to the direction of loading imply that cracking may occur in this direction.

4 EXPERIMENTAL TEST PROGRAM

To evaluate the feasibility and effectiveness of embedded connections for high-strength, vanadium microalloyed steel tubes, a series of large-scale specimens were tested. The experimental study described herein was developed to evaluate the influence of the embedded length footing vertical (shear) reinforcement, and the connection construction procedure on the inelastic cyclic response of the connection. In contrast to previous research, this study focused on using a welded flange, or annular ring, rather than an end plate.

Table 2 lists the study parameters and measured material properties for each specimen. The shallowest embedment was chosen to permit a shallow footing depth, and to evaluate the adequacy of smaller embedment depths in low to moderate seismic zones. The larger embedment depth was expected to achieve the load capacity and displacement ductility needed in regions of high seismicity.

Figure 2 shows the geometry and reinforcement common to all test specimens. The specimens represent approximately a full-scale building column or a half-scale bridge pier. The steel tubes were 508 mm in diameter and 6.4-mm thick, resulting in a diameter-to-thickness (D/t) ratio of 80. This value exceeds the D/t ratio limit in the AISC 2005 provisions (D/t = 62 for F_y = 70 ksi). The welded flange was 160-mm wide

Table 1. Stresses at 1.5% drift for various embedment depths.

| | Principle stress in footing | | | | | |
| | 45° plane | | Below flange | | Bearing surface | |
$\frac{l_e}{D}$	$\frac{\sigma_1}{\sqrt{f'_c}}$	$\frac{\sigma_3}{\sqrt{f'_c}}$	$\frac{\sigma_1}{\sqrt{f'_c}}$	$\frac{\sigma_3}{\sqrt{f'_c}}$	$\frac{\sigma_1}{\sqrt{f'_c}}$	$\frac{\sigma_3}{\sqrt{f'_c}}$
0.6	12.55	−0.07	40.0	−2.11	112.5	−3.92
0.9	7.42	−0.09	32.9	−1.39	125.4	−4.2
1.2	5.04	−0.12	26.5	−0.95	125.4	−4.32
1.5	2.84	−0.10	49.2	−0.28	128.7	−4.35

Table 2. Experimental parameters and material properties.

Specimen	l_e/D	ρ_v	$\rho_{l,p}$ (%)	f'_c (Mpa)	f_y (Mpa)	f_u (Mpa)
I	0.6	0	0.1	75.8		
II	0.6	0.3%	0.1	75.8	526	603
III	0.9	0.3%	0.1	69.2		
IV*	0.5	0.3%	0.1	69.2		

*Specimen IV was constructed using a grouted connection procedure.

489

and 6.4-mm thick, and it projected 102 mm outside and 51 mm inside of the tube.

The dimensions of the footing were 1.93 m × 1.73 m in plan and 0.61-mm deep. Footing dimensions and reinforcement were selected to permit transfer of the column forces, reflect current footing design in regions of high seismicity in the US, and to minimize the effect of the footing size on the behavior or mode of failure of the connection. For all specimens, the horizontal reinforcement at the top and the bottom of the footing consisted of No. 6 (19-mm diameter) bars spaced at 102 mm, both at the top and bottom of the footing in the direction of loading ($\rho_{l,l}$). In specimens I and II, the horizontal reinforcement in the direction perpendicular to the direction of loading ($\rho_{l,p}$) consisted of No. 4 (13-mm diameter) bars spaced at 225 mm at the top and bottom of the footing. In Specimen III, flexural reinforcement perpendicular to the plane of loading was increased to match the reinforcement in the direction of loading, or No. 6 bars spaced at 102 mm.

The vertical reinforcement in the footing was No. 3 (9.5-mm diameter) bars detailed with standard seismic hooks. Variations in the vertical reinforcement ratio (ρ_v) represented the range of reinforcement found in construction in all seismic zones. In Specimen I, vertical footing reinforcement was not included. The vertical reinforcement in Specimens II and III represented the maximum level of vertical reinforcement required for design of bridges in regions of high seismicity.

A fourth specimen was tested to evaluate the potential structural benefits and efficiency of an alternative construction procedure. This specimen utilized a grouted connection, in which the footing was cast with a void left for the CFT column, the column was placed, and the connection region surrounding the tube as filled with high-strength fiber-reinforced grout.

A self-reacting test frame was constructed to test the cantilever column-footing specimens. The test frame was placed below a Universal Testing Machine, which was used to apply a constant axial load of 1.824 MN to the specimens (approximately 10% of the gross cross-sectional capacity). A 979-kN horizontal actuator was used to apply cyclic lateral loading. The displacement history was based on ATC-24 protocol (ATC 1992). Column displacement and tube strains at various locations along the column height were measured throughout the test.

5 EXPERIMENTAL RESULTS

The observed performance (or damage) states and the corresponding drift levels are tabulated for each specimen in Table 3 and are illustrated in Figure 3.

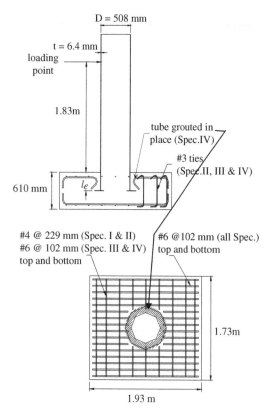

Figure 2. Geometry and reinforcement of CFT-footing specimens.

Table 3. Specimen response (% drift and horizontal resistance) at identified performance states.

Performance state	Specimen			
	I	II	III	IV
Interface cracking	0.4% 326 kN	0.5% 293 kN	0.5% 290 kN	0.5% 263 kN
Bisecting cracking	1% 540 kN	0.9% 473 kN	1.3% 597 kN	1% 401 kN
Diagonal cracking	1.7% 572 kN	1.5% 520 kN	x	1.8% 545 kN
Pull-out	4% 550 kN	7% 369 kN	x	6.6% 537 kN
Tube yielding	x	1.5% 520 kN	1.3% 597 kN	2.8% 617 kN
Tube buckling	x	x	4% 684 kN	x
Tube tearing	x	x	6% 535 kN	x
Grout uplift	x	x	x	4% 617 kN

x: Performance state was not observed.

490

All of the specimens initially developed small cracks at the column-footing interface which resulted in a gap there, as illustrated in Figure 3a. With larger drift demands, the cracks spread from the column interface into the footing. Four primary cracks bisected the footing as illustrated in Figure 4b. In Specimens I, II, and IV the cracks propagated through the depth of the footing, and diagonal cracks formed as well

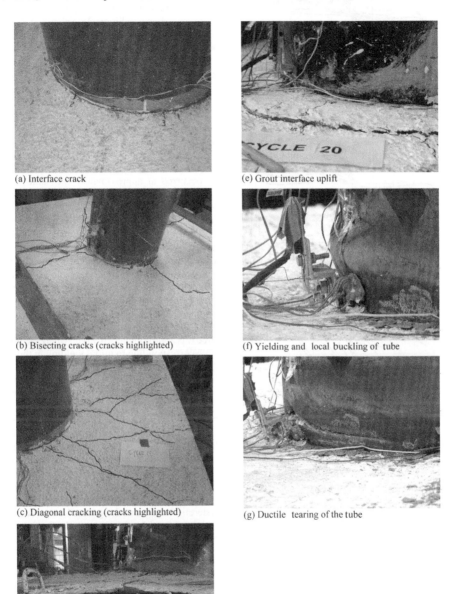

(a) Interface crack

(b) Bisecting cracks (cracks highlighted)

(c) Diagonal cracking (cracks highlighted)

(d) Pull-out of footing

(e) Grout interface uplift

(f) Yielding and local buckling of tube

(g) Ductile tearing of the tube

Figure 3.   Column-footing connection performance states.

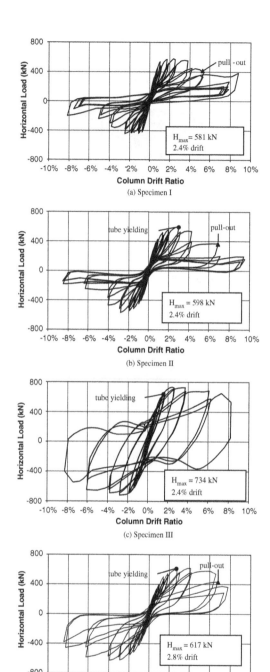

Figure 4. Horizontal load–drift response.

(Figure 4c). Specimen II exhibited a crack pattern of many narrow cracks, while Specimen I had fewer, wider cracks in the foundation. These specimens also sustained pull-out failure of the concrete as illustrated

in Figure 3d, where pull-out is defined as 12.7 mm (0.5 in.) or more of upward movement of the concrete in the footing. Prior to pull-out failure, Specimen IV exhibited damage and uplift at the interface between the grouted region of the connection and the concrete footing, as shown in Figure 4e. Specimen III sustained significant yielding of the tube, and as result, local buckling (Figure 3f) and subsequent ductile tearing of the steel tube (Figure 3g) occurred. Specimen III sustained limited damage to the footing.

The cyclic load-drift response for each specimen is shown in Figures 4a-d. The response of each specimen was similar to a drift ratio of approximately 1%, which indicates that embedment depth, shear reinforcement, and construction procedure had little effect on the initial behavior and stiffness of the connection.

Initial yielding of the tube in Specimens II, III and IV occurred at approximately the same load level, as indicated in Figure 4b-d. (Tube yielding was not detected in Specimen I.) The large drift level at tube yielding in Specimens II and IV compared to Specimen III can be attributed to a large component of deflection caused by cracks opening in the reinforced concrete footing. The cyclic response indicates that the presence of vertical reinforcement in Specimen II improved the symmetry of the hysteretic response relative to Specimen I, however it did not change the failure mode (comparison of Figures 6a and b). Although Specimen I did not achieve the yield force, the maximum force was approximately $0.97H_y$, where H_y is the shear force corresponding to initial yielding in the tube. The strength degradation in Specimen II was more severe than that recorded for Specimen I; this unexpected trend may be due to the demand that resulted from the symmetric response of Specimen II.

Specimen III achieved a significant increase in load capacity and ductility, and exhibited reduced strength degradation with cyclic loading. Specimen III also had significantly larger energy dissipation capacity. Specimen IV did not achieve increased load capacity relative to Specimens I and II. However, the specimen exhibited increased deformation capacity and fuller hysteretic loops, and the grouted connection delayed the initiation of damage states in the footing compared to the first two specimens.

6 COMPARISON TO AISC DESIGN EXPRESSIONS

The experimental results were compared with the current AISC design recommendations for flexural stiffness and strength of CFT members described above. The maximum experimental moment, M_{meas}, is defined as

$$M_{meas} = H \times L + P \times \Delta \qquad (4)$$

492

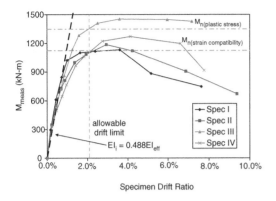

Figure 5. Comparison of experimental specimen response to column design parameters.

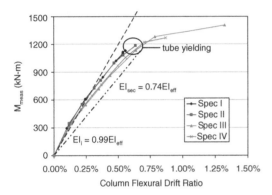

Figure 6. Experimental column flexural response.

where L is the column length, P is the applied axial load, and Δ is the horizontal displacement of the column. The calculated moment strengths are compared to the moment-drift ratio envelopes of each specimen in Figure 5. All of the specimens were able to resist loads greater than the nominal CFT capacity calculated using the strain compatibility method.

The initial stiffness of the measured response is compared with the design stiffness, EI_{eff}, defined in Eqs. 2 and 3. Each of the column-footing specimens exhibited an initial stiffness that was approximately 49% of the design stiffness. The vertical dotted line in Figure 6 indicates the maximum permissible story drift of 2% for a 5-story or taller building as recommended by NEHRP 2003 seismic provisions (FEMA 450). Each specimen was able to reach this story-drift limit without a decrease in load carrying capacity.

The contribution of column bending to the total specimen drift was calculated by integrating the tube strains over the column height. The moment-column flexural drift envelopes for each specimen are compared to the design stiffness in Figure 6. The specimens exhibited initial column stiffness approximately equal

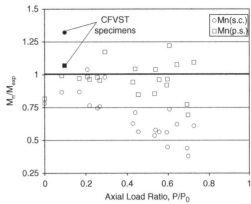

Figure 7. Ratio of predicted to experimental flexural strengths.

to the design stiffness. The secant column stiffness, EI_{sec}, is the slope of the line extending from the origin. The EI_{sec} value is a reasonable approximation of column response prior to yielding. The EI_{sec} value provides an estimate of the yield point but may underestimate the experimental moment by as much as 25% for a give column drift ratio.

Further evaluation of the AISC flexural strength expressions was achieved by considering additional published experimental data. Gaines (2000) and Marson and Bruneau (2000) present databases of experimental tests designed to study the flexural response of CFT beam-columns. From these studies, a series of CFT specimens with a wide range of D/t ratios, material strengths, and axial load ratios (P/P_0) were selected. The nominal flexural strength of the experimental CFT was calculated using both of the recommended methods ($M_{n(s.c.)}$ and $M_{n(p.s.)}$)and compared to the experimental flexural capacity for each specimen, M_{exp}. Figure 7 shows the results of this comparison. For both methods, the ratio of predicted to experimental strengths varied over a wide range. However, this range is centered at zero for the plastic strain distribution method, while the strain compatibility method consistently predicted a strength that was less than the measured flexural strength. In addition, the strain compatibility method was less accurate for specimens with larger axial load ratios. No other trends in accuracy were observed over the experimental parameters considered.

7 CONCLUSIONS AND FUTURE WORK

The test results indicate that the proposed annular-ring embedded connection is effective and practical. Although the significant damage noted at large story drift with the shorter embedment depth suggests that use of a shorter embedment depth (0.6D) may be more appropriate for low and moderate seismic zones. The specimen with the longer embedment depth is capable of achieving drift capacities far in excess of the maximum seismic design drifts without degradation of the system, and this connection is an appropriate detail for high seismic zones and other extreme loading conditions. Although termed the "longer" embedment depth, this recommended length is shorter than the current recommendation of 1.5D to 2D (Morino et al., 2003, Hitaka, Suita, and Kato, 2004). Analysis of the experimental results indicates that the CFVST construction methods and tested connections are a promising improvement for future structural engineering applications, including resistance to extreme loads with significant ductility demands.

An analytical model which can accurately predict the nonlinear response of the experimental connections is a valuable tool in assessing and developing design recommendations for an adequate column-footing connection. The inelastic finite element studies performed provided adequate qualitative predictions of the experimental connection behavior and progression of damage. However, the accuracy of the numerical results in predicting the full experimental response is still under investigation.

The current test series represent a limited study regarding the relationship between the D/t ratio, the steel strength (F_y) and the required embedment depth ratio, l_e/D. Because these parameters determine the tension force demand, which in part determines the embedment length requirements, a thorough study of these parameters is required. In addition, it is expected that the axial load ratio and the deformation of the footing can influence the force transfer mechanism and resulting damage pattern. Further experimental and analytical research is needed to address these issues and to develop design expressions for CFVST columns and their connections.

ACKNOWLEDGMENTS

Research was sponsored by the Army Research Laboratory and was accomplished under Cooperative Agreement Number DAAD19-03-2-0036. The views and conclusions contained in this document are those of the authors and should not be interpreted as representing the official policies, either expressed or implied, of the Army Research Laboratory or the U.S. Government. The U.S. Government is authorized to reproduce and distribute reprints for Government purposes notwithstanding any copyright notation heron. The authors gratefully acknowledge the financial support of the Army Research Laboratory, and the advice and assistance provided by J. Tirpak and C. Kramer of the Advanced Technology Institute, and the members of the Vanadium Technology Partnership. The spiral welded tubes were manufactured by Northwest Pipe in Portland Oregon.

REFERENCES

American Institute of Steel Construction (AISC) (2005). *Specification for Structural Steel Buildings*, AISC, Chicago, Illinois.

Federal Emergency Management Agency (FEMA) (2003). *NEHRP Recommended Provisions for Seismic Regulations for New Buildings and Other Structures (FEMA 450)*, FEMA, Washington D.C.

Hitaka, T., Suita, K. and Kato, M. (2003) "CFT Column Base Design and Practice in Japan," *Proceedings of the International Workshop on Steel and Concrete Composite Construction (IWSCCC-2003)*, Report No. NCREE-0.-0.26, Taipei, Taiwan, October 8-8, 2003, pp. 291–290.

Hsu, H. and Lin, H. (2003) "Performance of Concrete-Filled Tube Base Connections Under Repeated Loading," *Proceedings of the International Workshop on Steel and Concrete Composite Construction (IWSCCC-2003)*, Report No. NCREE-0.-0.26, October 8-8, 2003, Taipei, Taiwan, pp. 291–299.

Kadoya, H., Kawaguchi, J. and Morino, S. (2005) "Experimental Study on Strength and Stiffness of Bare Type CFT Column Base with Central Reinforcing Bars", *Composite Construction in Steel and Concrete V* Ed. Roberto Leon and Jörg Lange, United Engineering Foundation, Inc., July 2004.

Marson, J. and Bruneau, M. (2004) "Cyclic Testing of Concrete-Filled Circular Steel Bridge Piers Having Encased Fixed-Base Detail", *ASCE Journal of Bridge Engineering*, Vol.9, No.1, pp. 14–23.

Morino, S., Kawaguchi, J., Tsuji, A. and Kadoya, H. (2003) "Strength and Stiffness of CFT Semi-Embedded Type Column Base," *Proceedings of ASSCCA 2003*, Sydney, Australia, A.A. Balkema, Sydney, Australia.

Tubular Structures XI – Packer & Willibald (eds)
© 2006 Taylor & Francis Group, London, ISBN 0-415-40280-8

Axial capacity of concrete filled stainless steel circular columns

D. Lam & C. Roach
School of Civil Engineering, University of Leeds, Leeds, UK

ABSTRACT: Concrete filled tubes (CFT) have been widely used in structures throughout the world. This increase in its use is due to the significant advantages that concrete filled steel columns offer in comparison to more traditional steel or concrete constructions. Steel – concrete composite columns combined and make use of their constituent material properties resulted in reduction of the column sizes, which ultimately lead to significant economic savings. This reduction in column size can provide substantial benefits where floor space is at a premium. The use of stainless steel column filled with concrete is new and innovative, not only provides the advantage mentioned above, but also durability associated with the stainless steel material. This paper concentrates on the axial capacity of the concrete filled stainless steel circular columns. A series of tests was performed to consider the behaviour of short composite stainless steel columns under axial compressive loading, covering austenitic stainless steel CFT with normal and high strength concrete. Comparisons with Eurocode 4, ACI-318 and the findings of this research were made and comment.

1 INTRODUCTION

Modern engineering is for ever pushing the limits of materials and construction. Led by modern economies and drives in engineering investments, expectations for both economical and functional designs are ever increasing. The past decade has seen a boom in the use of composite constructions particularly in the Fast East and Australia. While extensive research has been conducted on steel and concrete in the last century, it is only in the past decades that research has focused on composite constructions.

The use of composite materials in engineering offers many significant advantages over traditional singular material construction. Steel members have the advantage of high tensile strength and ductility while the benefit of concrete is high strength and stiffness, thus the combination of the two is far superior to one alone. The use of composite materials can also offer huge cost savings in terms of formwork and therefore lead-in times as well as allowing designers to maximise the workable floor area.

The use of stainless steels and high strength steels in composite construction is relatively new, and research is ongoing, the outcome of these researches would offer superior advantages in a highly competitive market. While both the designers and industry are eager to use such materials, little is known about their true behaviour, particularly as regards composite stainless steel construction. Stainless steel offers both improved durability and aesthetic qualities that appeal to both architects and designers. While building codes exist for carbon steel composite constructions, little is known about the behaviour and capacity of stainless steels used in composite column constructions.

2 PAST RESEARCH

Composite materials have been widely used in construction for many years due to the advantages of combining steel and concrete. A state of the art review was reported by Shanmugam & Lakshmi (2001) on steel-concrete composite columns highlighting that significant research in this area over the past 15 years. The majority of this research has focused on the use of carbon steel encased concrete columns, fire and seismic resistance performance and on various models that can predict the capacity and buckling behaviour of such columns. However, little research has been conducted on the behaviour of composite circular CFT columns using stainless steel sections.

Other research involving the use of carbon steel was carried out by Schneider (1998) who carried out an experimental and analytical study on the behaviour of short CFT concentrically loaded in compression. The tests investigated the effect of steel tube shape and wall thickness with D/t ratios between 17 and 50 on the ultimate strength. The results suggested that circular tubes offer substantial post yield strength, stiffness and ductility which are not available in other more conventional square or rectangular cross sections.

The results also confirmed that current design specifications were adequate to predict the yield load under most conditions. He noted the fact that under moderate loads, the steel tube expands more than the concrete core since Poisson's ratio is higher for the steel section, suggesting that the steel may offer little confinement under certain conditions. This was quantified by tests to show that significant confinement is not present for most specimens until the axial load reaches 92% of the yield strength of the column.

Lam & Giakoumelis (2004) conducted a series of test on the axial capacity of concrete filled circular hollow sections (CHS) to investigate the effect of concrete strength, tube thickness and bond strength between the interfaces. The tubes D/t ratios ranged from 22.9 to 30.5, all containing concrete strengths of 30, 60 or 100 MPa. The results showed that concrete shrinkage is critical for high strength concrete while the effect is negligible for normal strength concrete. The results were then compared with existing Eurocode 4, Australian and American standards, with all codes predicting lower than measured axial capacity, with Eurocode 4 giving the best results. The results showed that the circular tubular columns have an advantage over the other cross sections as they have a uniform flexural stiffness in all directions, the infill concrete prevented local buckling of the steel shell and itself in a confined state is able to sustain higher stresses.

Knowles & Park (1969) undertook a study on both square and circular columns with D/t ratios of 15, 22 and 59. It was found that the larger than expected column capacity with L/D < 11 was due to increase in concrete strength due to the triaxial confinement stress created by the steel tubes. It was observed that the concrete began to increase in volume due to micro cracking, thus inducing concrete confinement by the steel tube wall. This increase in capacity was noted for CFT short columns only. For larger columns with higher L/D ratios, buckling occurred before the necessary stain required to cause micro cracking was able to achieve.

Sakino et al (1998) tested 18 CFT specimens with D/t ratios between 18 and 192, with different loading conditions; on the concrete and carbon steel simultaneously, the concrete alone and the concrete and steel with a greased column. Results indicated that when the concrete and steel were loaded simultaneously the tube provided no confinement until post yield behaviour. It was also found that although the axial stiffness of the concrete only loaded columns were half that of the other CFT's the columns actually obtained a greater yield and ultimate axial load capacity.

Concrete compaction also plays a vital role in the performance of CFTs. This was suggested by Han & Yang (2001). Tests of concrete filled carbon steel tubes were carried out to determine the influence of compaction methods of column capacity and investigate the influence on section capacities and concrete modulus. The test confirmed that the better compaction resulted in higher sections capacities and highlighted the importance of concrete compaction on composite concrete filled CHS. Han & Yao (2004) offered an alternative approach by investigating the experimental behaviour of thin walled hollow structural steel (HSS) columns filled with self consolidating concrete (SCC). It was recognised that SCC is being increasingly used in CFT columns as it can avoid vibro-compaction while still achieving good consolidation. It was suggested that SCC's will be increasingly used in the industry and noted that future research is needed in this area.

O'Shea & bridge (1997) conducted research into carbon steel thin-walled tubes with high strength concrete in fill. The strength enhancement provided by the confined composite and the resulting improved ductility were examined. A high strength infill (114 MPa) was used, with d/t ratios ranging between 55 and 200. Tests were carried out on thin walled tubes, under differing loading conditions, while results were compared to strength models in design standards. It was found that the concrete infill has little effect on the local buckling strength of the steel tube, with Eurocode 4 giving a good strength estimate when confinement is ignored. It was also found that increased strength due to confinement can be achieved if the concrete is loaded and the steel is not bonded to the concrete.

Gardner & Nethercot (2004) investigated the material, cross section and member behaviour of stainless steel hollow sections. Tests included looking at the basic material properties and stress-strain curves of the square, rectangular and circular hollow sections. The results were used to develop a relationship between slenderness, deformation capacity and hence propose a new design approach for stainless steel structures. The paper pointed out that stainless steels differ from carbon steel in the basic material stress-strain curve with the stainless steel curve possessing a rounded stress-strain plot and no definitive sharp yield point. The results showed that the Eurocode 3 prediction was conservative on average by 20% and does not adequately reflect the physical properties of stainless steels. While this approach was enthusiastic with the use of hollow sections, no design parameters or empirical relationships have been examined for composite concrete filled stainless steel hollow sections.

Lakshmi & Shanmugam (2002) proposed a method to predict the behaviour of in-filled columns using moment-curvature-thrust relationships. Square, rectangular and circular cross sections of compact steel tubes filled with concrete were examined. The model developed accurately predicted the observed experimental results.

Al-Rodan (2004) investigated a wide range of experimental data associated with carbon steels to investigate the applicability of two widely used design codes. The specimens included short and slender concrete filled square tubes made with both normal and high strength concrete. The study highlighted the large differences between BS5400 and EC4. It was found that the results from the use of EC4 were generally closer than that of the BS5400. The paper noted that finite element methods tend to produce significantly improved results. Wang (1999) carried out tests on slender composite columns and compared the results with the EC4 and BS5400 codes, he found that EC4 provided the best estimate, with both the EC4 and BS5400 being conservative. This was however using carbon steel and not stainless steels.

Brauns (1999) carried out an analysis of the stress state existing in concrete filled columns using carbon steel sections. Similar stress states would also exist when using stainless steels. The paper pointed out that the stress state inside the column is determined by the dependence of the modulus of elasticity and Poisson's ration on the stress level in the concrete. The effect of confinement occurs at high stress levels when the concrete is acting in compression and the steel in tension, while the ultimate state on the material strength is not attained for all parts simultaneously. He also pointed out that the governing working condition of a composite column and the subsequent failure may be due to small steel thickness, the loading eccentricities and fire resistance.

Hossian (2003) carried out an experimental and analytical investigation into the behaviour of thin walled composite tube columns under axial compression using carbon steel. The investigation looked at failure modes, strain characteristics, load deformation responses, using lightweight volcanic pumice, as well as various geometry parameters on the columns. He noted that using thin walled composites has significant advantages including ease of construction and providing formwork as well as acting as reinforcement in the service stage. He found that the short columns failed due to the formation of successive local buckles associated with plastic yielding occurring between adjacent buckles. It was found that the axial capacity was 30% higher than the sum of the individual components.

While all this research reflects the need for research into composite concrete-steel circular hollow sections (CHS), little is known about the behaviour of composite columns when replacing the carbon steel with stainless steel.

3 EXPERIMENT

All the specimens were tested under axial compression to determine the failure modes. 8 specimens were

Table 1. Properties of composite columns.

Ref	D × t (mm)	d/t	f_y (MPa)	Concrete
C1	104.0 × 2.0	52	440	C30
C2	114.3 × 6.02	19	270	C30
C3	104.0 × 2.0	52	440	C60
C4	114.3 × 6.02	19	270	C60
C5	104.0 × 2.0	52	440	C100
C6	114.3 × 6.02	19	270	C100
C7	104.0 × 2.0	52	440	–
C8	114.3 × 6.02	19	270	–

tested, all 300 mm long to reduce the effect of slenderness. The thick samples were 114 mm in diameter with 6 mm wall thickness, while the thin samples were 104 mm in diameter and 2 mm wall thickness.

Three strain gauges were fitted to each sample at mid height. Two linear variable transducers were fitted at diametrically opposite positions at the side of the column to measure and record the deflection. All the data was then recorded and logged. LVDT readings were initially taken both without loading as well as under a residual loading. This was done to ensure that no significant eccentric loading occurred. While minute eccentricities may have occurred, every effort was made to reduce the effects of an eccentric loading and ensure the columns were concentrically loaded. The properties of the specimens are shown in Table 1.

Clamps were placed at each ends of all the thin walled specimens to reduce any end effects as well as to ensure that the failure would occur closer to the location of the strain gauges. The clamps themselves were placed fractionally lower than the top of the specimen to ensure that no load was transferred to the cylinder shell by the clamps.

Plaster was used to ensure a uniform loading. Minimum amounts of plaster were used to reduce the influence of the plaster in the initial stages of the loading. The plaster was placed on the top and bottom of the cylinder and allowed to harden under a residual stress of 20 kN. After the plaster had hardened, the LVDT's were placed at geometrically opposite sides of the cylinder. Stain gauges were initialised and loading commenced. Figure 1 shows the experimental apparatus and a specimen during test. Each cylinder was loaded manually at a rate of 3 kN/s with data taken at zero load, and then at the residual load of 20 kN and then approximately 50 kN intervals thereafter until the concrete capacity was reached. Data was taken at intervals of 10 kN until failure where successive data points were logged closely to determine the failure curve.

After the testing was complete, the samples were taken from the machine for inspection. This along with the recorded data was used for analysis. The specimens

Figure 1. Test set up and instrumentations.

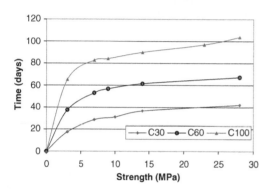

Figure 2. Concrete strength development.

were then photographed to record the failure mode shape.

4 CONCRETE PROPERTIES

The concrete strengths used were 30, 60 and 100 MPa. The concrete was tested during the development stages. Figure 2 shows the strength development for the 30, 60 and 100 MPa concrete samples over a period of 28 days. Before any test began, 2 standard concrete cubes and 2 standard cylinders were tested to determine the concrete strength. Table 2 shows the concrete strength of the specimens at the test day.

5 STAINLESS STEEL PROPERTIES

Coupons were cut from the stainless steel tubes and tested to determine the tensile strength of the stainless steel. The coupon was cut form the side of the cylinder, and milled to specification. Some flattening of the ends occurred while gripping the specimen but this was well away from the 'neck' of the sample. The results were

Table 2. Concrete strength of the test specimens.

Concrete	Test days	Type of test	Compressive strength (MPa)
C30	28	Cube	42
	28	Cylinder	31
C60	28	Cube	67
	28	Cylinder	49
C100	23	Cube	97
	23	Cylinder	65

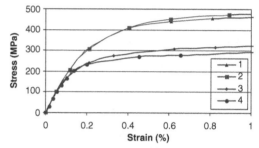

Figure 3. Stress vs. strain curves of the stainless steel tubes.

recorded and the tensile strength of the stainless steel was found. Figure 3 shows both the stress vs. strain curves of the thick and thin tubes. 1 and 2 denote the samples from the 104 mmϕ × 2 mm tubes, while 3 and 4 denote the samples from the 114 mmϕ × 6 mm tubes. The 0.2% proof stress was found to be 440 MPa for the thin tube and 270 MPa for the thick tube respectively.

6 TEST RESULTS & DISCUSSION

The typical mode of failure was local buckling (outward folding) as shown in Figure 4. The buckling of the composite stainless steel concrete-filled CHS stub column is mainly caused by the expansion of the concrete under axial load. The steel sections provided confinement for the concrete as it expands and helped maintaining its strength and ductility.

Figures 5 and 6 show the axial load vs. displacement curves for the 2 mm and 6 mm CHS columns with various infill concrete strength. The results showed the clear advantage of composite stainless steel columns over the bare steel counterpart. It can be observed from the stub column tests that the specimens with 6 mm thick stainless steel circular hollow sections behaved in a more ductile manner which would suggested better confinement is provided by the steel sections. In addition, composite columns with C30 & C60 concrete infill behaved much more ductile at failure as compare

Figure 4. Typical failure shape of 2 mm thick specimens.

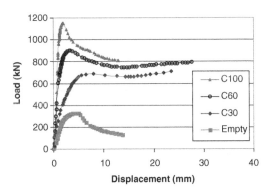

Figure 5. Load vs. displacement curves of 2 mm CHS columns.

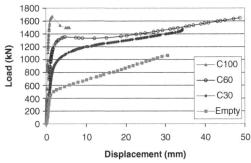

Figure 6. Load vs. displacement curves of 6 mm CHS columns.

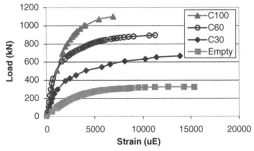

Figure 7. Load vs. strain curves of 2 mm CHS columns.

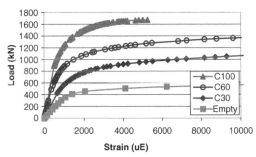

Figure 8. Load vs. strain curves of 6 mm CHS columns.

with the composite columns filled with C100 concrete. Figures 7 and 8 show the load vs. strain curves of the composite columns.

7 COMPARISON WITH THE DESIGN CODES

No code currently exist for the use of composite stainless steel columns with concrete infill hence the codes for carbon steel concrete filled tubes were used. The Eurocode 4 (2005) and ACI standards (1995) were used to compare with the experimental results. The Eurocode 4 incorporates the use of partially and fully encased steel and concrete sections both with and without reinforcement and takes into accounts the increases in concrete capacity due to confinement by the steel sections.

In accordance to Eurocode 4, the composite column capacity is governed by:

$$N_{u0} = A_a \eta_2 f_y + A_c f_c \left(1 + \eta_1 \left(t/D\right)\left(f_y/f_{cyl}\right)\right)$$

499

Table 3. Comparison of test results.

Ref	Test N_{test}	EC4 N_{u0}	ACI N_u
C1	699	712.8	525.7
C2	1593	1151.8	807.8
C3	901	847.9	667.1
C4	1648	1293.2	955.7
C5	1133	968.7	792.8
C6	1674	1419.3	1087.2
C7	328	282.1	282.1
C8	1062	553.1	553.1

where:

A_a Cross sectional area of the steel tube
D External diameter of the steel tube
f_c Compressive strength of the concrete
f_{cyl} Compressive strength of concrete cylinder
f_y Yield strength of the steel tube
N_{u0} Ultimate axial strength of column
t Thickness of steel tube
η_1 Coefficient of concrete confinement
η_2 Coefficient of steel confinement

While the American ACI codes does not take into account the increase in capacity cause by the concrete confinement.

The ACI code determines the column strength by:

$$N_u = A_c f_c + A_s f_y$$

where:

A_c Cross sectional area of the concrete
A_s Cross sectional area of the steel tube
f_c Compressive strength of the concrete
f_y Yield strength of the steel tube
N_u Ultimate squash load

Table 3 shows the comparison of the test results with the Eurocode 4 and the ACI standards. The results showed the Eurocode 4 gives a better estimation of the axial capacity of the stainless steel concrete filled composite than the ACI standards as not confinement strength of the concrete is being considered by the ACI standards. The comparison of the Eurocode 4 and ACI standards with the experimental results of the 2 mm and 6 mm concrete filled CHS columns are shown in Figures 9 and 10.

8 CONCLUSIONS

This paper presents the findings of a series of short circular concrete filled stainless steel columns tested under axial load. The results showed that composite columns filled with high strength concrete, the peak load was achieved after small shortening, whereas for

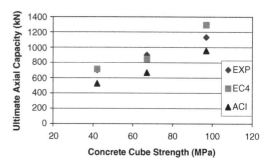

Figure 9. Comparison of EC4 & ACI with test results of 2 mm CHS composite columns.

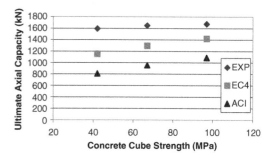

Figure 10. Comparison of EC4 & ACI with test results of 6 mm CHS composite columns.

normal strength concrete, the ultimate capacity was gained with large displacement. This inherent capacity can has significant implications in building design when disproportional and accidental collapse is being considered.

In general, the 6 mm thick specimen provided higher confinement pressure than the 2 mm thick tubes and behaved more ductile at failure. The Eurocode 4 gives the better estimate for the column capacity than the ACI standards and provides a good approximation to the actual column strength of the stainless steel composite columns.

ACKNOWLEDGEMENTS

The authors would like to acknowledge the support provided by Outokumpu (formerly Avesta Polarit, UK) for supplying the steel specimens. The skilled assistance provided by the technical staff in the School of Civil Engineering at Leeds University is also appreciated.

REFERENCES

ACI 318 (1995), Building code requirements for structural concrete, Detroit; American Concrete Institute.

Al-Rodan A. (2004), Comparison between BS5400 and EC4 for concrete filled steel tubular columns. Advances in structural engineering, Vol 7, No. 2, pp. 113–127

Brauns J. (1999), Analysis of stress state in concrete filled steel column. Journal of constructional steel research, Vol 49; pp. 189–196

Eurocode 4, EN1994-1-1 (2005), Design of composite steel and concrete structures, British Standards Institution; London

Gardner L. & Nethercot D.A. (2004), Experiments on stainless steel hollow Sections. Journal of Constructional Steel Research, Vol. 60, pp. 1291–1332

Han L.H. & Yang Y. (2001), Influence of concrete compaction on the behaviour of concrete filled steel tubes with rectangular sections. Advances in structural engineering, Vol 4, No. 2, pp. 93–108

Han L.H. & Yao G.H. (2004), Experimental behaviour of thin walled hollow structural steel (HSS) columns filled with self consolidating concrete (SCC). Journal of Thin Walled Structures, Vol. 42, pp. 1357–1377

Hossain K. (2003), Axial load behaviour of thin walled composite columns. Composites: Part B, Vol. 34; pp. 715–725

Knowles R.B. & Park R. (1969), Strength of concrete-filled steel tubular columns. Journal of Structural Engineering, ASCE, Vol. 95, No.2, pp. 2565–2587

Lakshmi B. & Shanmugam N.E. (2002), Non-linear analysis of in-filled steel-concrete composite columns. Journal of structural engineering; ASCE, Vol. 128, No. 7, pp. 922–933

Lam D. & Giakoumelis G. (2004), Axial Capacity of circular concrete filled tube columns. Journal of constructional steel research, Vol. 60, pp. 1049–1068

O'Shea M.D. & Bridge R.Q. (1997), Circular thin walled tubes with high strength concrete infill. Composite construction in steel and concrete III American Society of Civil Engineers; pp. 780–793

Sakino K., Tomii M. & Watanabe K.(1998), Sustaining load capacity of plain concrete stub columns by circular steel tubes. Conference on concrete filled steel tubular construction; pp. 112–118

Schneider S.P. (1998), Axially loaded concrete-filled steel tubes. Journal of Structural Engineering, ASCE, Vol. 10, pp. 1125–1138

Shanmugam N.E. & Lakshmi B. (2001), State of the art report on steel-concrete composite columns. Journal of constructional steel research, Vol 57, pp. 1041–1080.

Wang Y.C. (1999), Tests on slender composite columns. Journal of constructional steel research, Vol 49; pp. 25–41

Tubular Structures XI – Packer & Willibald (eds)
© 2006 Taylor & Francis Group, London, ISBN 0-415-40280-8

A numerical and experimental study of Concrete Filled Tubular (CFT) columns with high strength concrete

M.L. Romero & J.L. Bonet
Universidad Politécnica de Valencia, Valencia, Spain

J.M. Portoles
Universitat Jaume I, Castellon, Spain

S. Ivorra
Universidad de Alicante, Alicante, Spain

ABSTRACT: In recent years an increment in the utilization of concrete tubular columns was produced due to its high stiffness, ductility and fire resistance. On the other hand the use of high strength concrete (HSC) is more common due to the advances in the technology. The use of this material presents different advantages, mainly in elements subjected to high compressions as building supports or bridge columns.

However, there is a notably lack of knowledge in the behavior of high strength concrete filled tubular columns which produces that the existing simplified design models for normal strength concretes are not valid for them. This paper presents the initial results of a research project, where a numerical and an experimental study of high strength CFT's is performed.

1 INTRODUCTION

The utilization of hollow steel sections is well-known in multi-story buildings because a substantial reduction of the cross section is obtained. Moreover, the high strength concrete is spent more and more mainly for precast concrete structures, but the influence of the high strength on this type of columns (CFT's) is not well studied.

The actual simplified design methods are similar to the methods for reinforced concrete, assuming the steel section as an additional layer of reinforcement. Each country has its different code of design for composite sections (Japan, Australia, Canada, United States, Europe, etc.). The design code of the Euro-Code 4 (1992) (allows only the utilization of concrete with strength lower than 50 MPa (cylinder strength); therefore for high strength concretes the method and the interaction diagrams are not valid. Furthermore, as the section is reduced, for an equal length of the element, the slenderness is increased and the buckling is more relevant.

1.1 *Normal strength concrete*

The utilization in Europe of normal strength concrete filled tubular columns is well-known some decades ago when appeared the first monograph from the CIDECT (1970) simplifying its applicability for practical engineers. Later research works gave rise to the monograph n°5, CIDECT(1979). All this documents were the base to make the Euro-Code 4 (1992), with a special section for CFT's. In Spain, the ICT (Instituto de la Construcción Tubular) published a practical monograph, CIDECT (1998), with the idea to make easier the design of this type of sections.

1.2 *High strength concrete*

However, there are not a lot of investigations regarding high performance materials for CFT's, focused in the buckling. The research on high strength concrete (HSC) has demonstrated that the tensile capacity does not increase in the same proportion as the compression capacity. For hollow sections filled with concrete the tension problem is not as much important because the concrete cannot split of. Therefore in this type of section is where more advantage is taken.

1.2.1 *Experimental tests*
The more important contributions are concentrated in the last 5–10 years:

Grauers [1993] performed experimental tests over 23 short columns and 23 slender columns, stating that the methods of the different codes were valid but should be extended in order to analyze the effect of other parameters. Bergman (1994) studied the confinement mainly for normal strength concrete and partially for high strength concrete, but applying only axial load.

Aboutaba et al. (1999) compared classical columns of HSC with concrete filled tubular columns with HSC. Those ones presented more lateral stiffness and ductility.

Certainly, the research of Rangan & Joyce (1992) and Kilpatrick & Rangan (1999) has advanced a lot in the field. They presented experimental results from 9 columns for uniaxial bending and 24 columns for double curvature. However they used small sections and proposed a simplified design method regarding the Australian code, which does not follow the same hypothesis than the Eurocode-4, as i.e. the buckling diagrams.

Liu et al. (2003) compared experimentally the capacity of 22 rectangular sections with the different codes (AISC, ACI, EC4) and they concluded that the Euro-Code 4 was on the unsafe side while other codes over-designed the sections.

Varma et al. (2002, 2003, 2004) studied initially the behavior of the square and rectangular tubular columns and recently they have investigated cyclic loads and the existence of plastic hinges.

Gourley et al. (2001) have published a research report about the state of the art of concrete filled tubular columns.

1.2.2 Numerical models

Concerning the numerical models, it can be stated that there are not a lot of specific studies that applied the finite element method or sectional analysis to this type of structure. Most of them as Hu et al.(2003), Lu et al. (2000), and Shams & Saadeghvaziri (1999) study normal strength concretes. Only, recently Varma et al. (2005) have implemented a fibber model applied to square tubular sections but for short columns, without taking into account the buckling.

There is also an important research work from Johansson and Gylltoft (2001) where the the effects of three ways to apply a load to a columns is investigated numerically and experimentally.

If a good sectional characterization (moment-curvature) was performed, it can be inferred that the actual simplified methods are valid as a first approach to study the strength of these supports.

Few months ago, Zeghiche and Chaoui (2005) have published a small study for circular sections following this procedure. They affirmed that more numerical and experimental tests should be performed to check the validity of the buckling design methods of the EC4 for high strength concrete and double curvature.

Due to that the authors are performing a research project to study the effect of high strength concrete

in the buckling. It has three parts: experimental study, one-dimensional numerical model and three-dimensional model. In this paper the initial results of the experimental and one-dimensional (1-D) part is presented.

For the 1-D part a nonlinear finite element numerical model for circular concrete filled tubular sections will be presented. The method has to be computationally efficient and must represent the behavior of such columns, taking into account the effect of high strength concrete and second order effects. The model is validated with experiments from different authors from the bibliography.

Also an experimental study of circular concrete filled tubular sections is presented. The experimental tests selected corresponds to circular tubular columns filled with concrete (CFT) with pinned supports at both ends subjected to axial load and uniaxial bending. In these tests the eccentricity of the load at the ends is fixed and the maximum axial load of the column is evaluated. In the date of the congress only the half of the tests will be carried out, but all the numerical tests are now accomplished.

2 NUMERICAL MODEL

In this section the numerical model based on the finite element method is illustrated.

2.1 Formulation

The finite element selected is a classical one-dimensional 13 degrees of freedom (d.o.f.'s) element. It has 6 d.o.f.'s at each node (three displacement and three rotations), and a longitudinal degree of freedom in the mid-span to represent a non-constant strain distribution to represent the cracking. The Navier-Bernoulli hypothesis is accepted for the formulation. Also perfect bond between the concrete and structural steel is assumed. In the model the local buckling of the hollow steel section is neglected by stating at least the minimum thickness pointed out in the EC-4 (1992) (Art 4.8.2.4.):

$$D/t: 90 \cdot \varepsilon^2 \qquad (1)$$

where: D external diameter of the circular section
      t thickness of the circular hollow steel section
      ε $= \sqrt{235/f_y}$
      f_y yielding stress of the structural steel (MPa)

The model includes the second order effects by the formulation of large strains of the element (using the nonlinear deformation matrix $-B_L-$ and the geometric stiffness matrix –K_g–) and large displacements (by stating the force equilibrium in the deformed shape). The model automatically obtains the maximum load by using the total potential energy analysis "V" of the structure, Gutiérrez et al. (1983); detecting if the structure reaches an stable equilibrium position ($\delta^2 V > 0$), unstable equilibrium ($\delta^2 V < 0$) or instability ($\delta^2 V = 0$). Moreover, the Newton-Raphson method was selected to solve the nonlinear system of equations for a known load level. An arc-length method, Crisfield (1981), was used for displacement control. A more extensive definition of the finite element model can be found at Romero et al. (2005).

2.2 Constitutive equations of the materials

A bilinear (elastic-plastic) stress-strain diagram for the structural steel was assumed, EC-4 (1992) (Art 3.3.4)). The equation proposed by the CEB (1990) for the Model Code was used for the columns filled with normal strength concretes ($f_c \leq 50$ MPa); for the cases of high strength concrete ($f_c > 50$ MPa) was selected the equation from the CEB-FIP (1995). The tension-stiffening effect was considered by a gradual unload method, Bonet (2001). The permanent deformations due to cyclic loads were not included in the model. Therefore it is assumed that the maximum load is not dependent on the adopted path.

2.3 Cross section integration

The classical section integration of a circular (or annular) section is performed by a decomposition of the cross section into layers. However, this integration procedure has a high computational cost because it needs a large amount of information to characterize the section and a lot of calculations to obtain an admissible error. Also, due to the lack of an exact adjustment of the layers to the geometry, some convergence problems in the nonlinear iteration solution could appear due to the sharp variation of the neutral axis.

Because of that, this paper proposes a numerical algorithm to integrate the stress field of tubular circular sections filled with normal and high strength concrete (CFT) using the Gauss-Legendre quadrature.

The internal forces from the section integration are obtained as the addition of the concrete and structural steel internal forces:

$$N = N_c + N_a$$
$$M_y = M_{cy} + M_{ay} \quad\quad (2)$$
$$M_z = M_{cz} + M_{az}$$

where: (N_c, M_{cy}, M_{cz}) concrete internal foces
       (N_a, M_{ay}, M_{az}) structural steel internal forces

On the other hand, the tangent stiffness matrix of the section (D_t) is obtained adding the constitutive

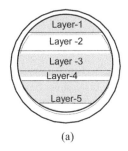

(a)

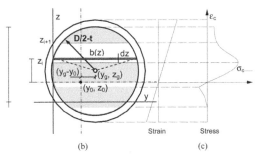

(b)   (c)

Figure 1. Numerical integration of concrete. Circular section (a) Decomposition in wide layers (b) Transformation into a path integral (c) Concrete stress field.

matrix of the concrete (D_{ct}) and the steel constitutive matrix (D_{at}).

The steps followed in the numerical algorithm are: decompose the section in "wide" layers, Figure 1.a, transform them in a path integral and later on to evaluate the integral by a Gauss-Legendre quadrature. To evaluate the internal forces of the cross section or the constitutive matrix, the terms of the concrete and steel are performed separately.

2.3.1 Concrete stress integration

The internal forces from the stresses of a circular concrete section can be obtained by the next equations (Figure 1.b):

$$N_c = \iint_{A_c} \sigma_c(y,z) dA_c = \int_{z_g+D/2-t-x}^{z_g+D/2-t} b(z) \cdot \sigma_c(z) dz$$

$$M_{cy} = \iint_{A_c} \sigma_c(y,z) \cdot (z-z_0) dA_c$$

$$= \int_{z_g+D/2-t-x}^{z_g+D/2-t} b(z) \cdot (z-z_0) \cdot \sigma_c(z) dz \quad\quad (3)$$

$$M_{cz} = \iint_{A_c} \sigma_c(y,z) \cdot (y_g-y_0) dA_c$$

$$= \int_{z_g+D/2-t-x}^{z_g+D/2-t} b(z) \cdot (y_g-y_0) \cdot \sigma_c(z) dz$$

where: A_c is the area of the concrete integration region; (y_0, z_0) are the coordinates of the stress reference centre; (y_g, z_g) are the circular centre of gravity coordinates; $\sigma_c(z)$ is the concrete stress in a fibber with a coordinate "z"; $b(z)$ is the width of the section in terms of "z":

$$b(z) = 2 \cdot \sqrt{(D/2 - t - z + z_g) \cdot (D/2 - t + z - z_g)} \quad (4)$$

The y-axis has been chosen parallel to neutral axis in the equations. As the stress field has a predominant direction (perpendicular to the neutral axis); the integral of the stresses in the concrete compression zone A_c can be transformed into a path integral in terms of z.

The components of the concrete tangent stiffness matrix (D_{ct}) can be computed in the same way, where E_{ct} is the tangent elastic modulus of the concrete. For example:

$$D_{ct}(1,1) = \iint_{A_c} E_{ct}(y,z) \cdot dA_c = \int_{z_g + D/2 - t - x}^{z_g + D/2 - t} b(z) \cdot E_{ct}(z) \cdot dz \quad (5)$$

2.3.2 Stress integration using the gauss-lengedre quadrature

Consequently, the integral of stresses over the integration are of concrete can be reduced to a path integral using the Green's theorem:

$$\iint_{A_c} f(y,z) \cdot dy \cdot dz = \int_{a_1}^{a_2} h(z)dz \quad (6)$$

To perform each integral, the Gauss-Legendre method is used. The next coordinate transformation is required (Figure 1.b):

$$z = \frac{x}{2} \cdot (\xi + 1) + \left[z_g + \frac{D}{2} - t - x \right] \quad \Rightarrow \quad dz = \frac{x}{2} d\xi \quad (7)$$

Hence, the relating integrals (equation 3 and 5) are computed using the next expressions:

$$\iint_{A_c} f(y,z) \cdot dy \cdot dz = \int_{a_1}^{a_2} h(z)dz = \frac{x}{2} \sum_{k=1}^{npg} \omega_k \cdot h(\xi_k) \quad (8)$$

where: ξ_k is the value of the curvilinear coordinate of the Gauss point "k", ω_k is the weight associated to this Gauss point and "npg" is in the number of Gauss points used in the integration.

2.3.3 Stress integration using "wide layers"

The accuracy of the integral depends on: the number of Gauss points used, the shape of the concrete constitutive equation and on the shape of the section (circular, in this case).

When the mathematical function to integrate ($h(z)$) is not adjusted to a small order polynomial or it is defined by branches, it is necessary to use a large amount of Gauss point to achieve an acceptable accuracy (Bonet et al. 2005). For these cases is better to subdivide the integration area into wide layers parallel to the neutral axis. Thus, for instance, the implementation of this method for the typical stress-strain concrete

relationship could be performed using five wide layers (Figure 1a).

In this case, for each wide layer the coordinate transformation is performed using next equation (Figure 1.b):

$$z = \frac{(z_{i+1} - z_i)}{2} \cdot (\xi + 1) + z_i \quad \Rightarrow \quad dz = \frac{(z_{i+1} - z_i)}{2} d\xi \quad (9)$$

Therefore, the internal forces and the constitutive matrix of the concrete section are obtained as the addition of the integrals of each layer.

$$\iint_{A_c} f(y,z) \cdot dy \cdot dz = \int_{a_1}^{a_2} h(z)dz$$

$$= \sum_{i=1}^{n^\circ layers} \left[\frac{(z_{i+1} - z_i)}{2} \cdot \sum_{k=1}^{npg} \omega_k \cdot h(\xi_k) \right]_i \quad (10)$$

2.3.4 Stress integration of the structural steel

The internal forces (N_a, M_{ay}, M_{az}) and the constitutive matrix of the tubular section (Dat) are computed by superposition of two circular sections with a radius (D/2) and (D/2 − t) respectively. For this case next

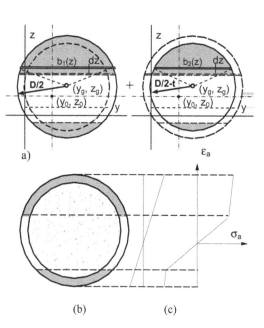

a)

(b)   (c)

Figure 2. Numerical integration method of the annular steel section: superposition of two sub-sections with radius (D/2) and (D/2 − t) Decomposition into two sub-sections (b) Definition of the wide layers: criteria defined by the stress field (c) stress field in the steel.

506

equations are used:

$$N_a = N_a(D/2) - N_a(D/2 - t)$$
$$M_{ay} = M_{ay}(D/2) - M_{ay}(D/2 - t) \quad (11)$$
$$M_{az} = M_{az}(D/2) - M_{az}(D/2 - t)$$
$$D_{at} = D_{at}(D/2) - D_{at}(D/2 - t) \quad (12)$$

The next procedure is followed for each of the subsections: to decompose the section into wide layers, to transform it in a path integral and to evaluate the integral using the Gauss-Legendre quadrature. Figure 2 shows the implementation of this method for the classical constitutive steel equation using three wide layers for each sub-section.

2.4 Verification of the method

In order to verify the accuracy degree of the implement numerical model, a comparison with 78 experimental tests from the literature is performed in this paper.

Table 1. Variation of the parameters for the selected experimental tests.

	Parameter	Range
$\alpha = 1$	Slenderness (L/D)	7.95–31.61
	Relative eccentricity (e/D)	0.05–0.49
	f_c (MPa)	23–102
	f_y (MPa)	193–435
$\alpha \neq 1$	Slenderness (L/D)	12.48–21.43
	Eccentricities ratio (α)	0.75–(−1)
	e_{01}/D	0–0.49
	e_{02}/D	0.10–0.49
	f_c (MPa)	54–102
	f_y (MPa)	268–410

The accuracy degree is obtained from the next equation:

$$\xi = \frac{P_{u,test}}{P_{u,SN}} \quad (13)$$

where $P_{u,test}$ is the ultimate axial load in the experimental test (TEST) and $P_{u,NS}$ is the ultimate axial load in the numerical simulation (NS).

The experimental tests selected corresponds to circular tubular columns filled with concrete (CFT) with pinned supports at both ends subjected to axial load and uniaxial bending.

In these tests the eccentricity of the load at the ends is fixed and the maximum axial load of the column is evaluated. In 52 of the selected experiments of the bibliography, the eccentricity is equal at both ends ($\alpha = 1$) while in 26 tests the applied eccentricities are different ($\alpha \neq 1$), where α is the ratio between both eccentricities:

$$\alpha = \frac{e_{01}}{e_{02}} \quad (14)$$

where: e_{01} and e_{02} are the first order eccentricities of both ends of the support (e_{02} is the higher of them in absolute value). Table 1 presents the variation of all the parameters for the experimental tests.

Table 2 shows the authors that have developed those tests, and the accuracy degree obtained with the implemented numerical model both for $\alpha = 1$ and $\alpha \neq 1$. Also the accuracy that each author obtains with his own numerical simulation is presented.

From this table can be inferred that the proposed model obtains a better accuracy degree in comparison

Table 2. Results of the verification of the model.

	Authors	N°	Numerical model				Author			
			ξ_m	V.C.	ξ_{max}	ξ_{min}	ξ_m	V.C.	ξ_{max}	ξ_{min}
$\alpha = 1$	Zeghiche (2005)	8	1	0.03	1.04	0.96	1.01	0.02	1.03	0.97
	Rangan 1992	9	0.99	0.15	1.34	0.83	1.17	0.17	1.59	1.04
	Neogi 1969	18	1.10	0.15	1.49	0.89	1.17	0.17	1.68	0.99
	Kilpatrick 1999	17	0.97	0.07	1.13	0.88	–	–	–	–
	All	52	1.02	0.13	1.49	0.83	1.13	0.16	1.68	0.97
$\alpha \neq 1$	Zeghiche (2005)	4	1.03	0.02	1.04	1.00	0.95	0.02	0.97	0.92
	Kilpatrick 1999	22	0.96	0.08	1.18	0.87	–	–	–	–
	All	26	0.97	0.08	1.18	0.87	0.95	0.02	0.97	0.92
		78	1.01	0.12	1.49	0.83	1.11	0.16	1.68	0.97

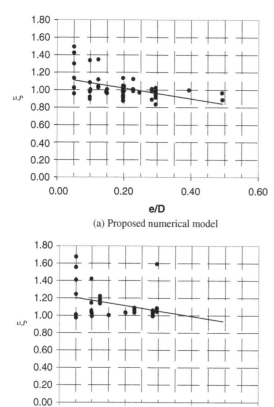

(a) Proposed numerical model

(b)Other Authors models

Figure 3. Accuracy degree in terms of the relative eccentricity (e/D) for α = 1 (a) Proposed numerical model (b) Authors.

with the different authors. A mean value close to one is reached and in the safe side (1.01), and with an acceptable variation coefficient (0.12). The accuracy is similar for eccentricities equal at both ends ($\alpha = 1$) and different ($\alpha \neq 1$).

Figure 3.a presents the accuracy degree (for the case of equal eccentricities at the ends) in terms of the relative eccentricity (e/D); also a trend line is presented. As can be shown the numerical model proposed tends toward the safe side when the relative eccentricity (e/D) diminishes. This behavior is also observed for the other authors, but with a higher error (figure 3.b). In conclusion the 1-D should be improved for e/D > 0.3.

3 EXPERIMENTAL PROGRAM

Forty-four test specimens of NSC and HSC columns will be studied in this experimental program; see

Table 3. They are designed to investigate the effects of four main parameters on their behavior: slenderness (Length/Diameter), ratio D/t (Diameter/thickness), strength of concrete (Fck) and eccentricity. The column lengths, L, are 2135 and 3135 mm and the cross-sections are circular with a 100, 125, 159 and 200 mm of outer diameter. The thickness of the steel tubes are 3, 4 and 6 mm. The strength of concrete varies from 30, 60 to 90 MPa and the axial load is applied with two eccentricities: 20 and 50 mm. All of the tests are going to be performed in the laboratory of the Department of Mechanical Engineering and Construction of the Universitat Jaume I in Castellon, Spain.

3.1 Materials

Concrete
All columns were cast using concrete batched in the laboratory. The concrete compressive strength fck was determined both from the cylinder and cubic compressive tests at 28 days. The specimens were cast in a inclined position.

Steel
The steel used was S275JR. The yield strength fy, the ultimate strength fu, the strain at hardening, the ultimate strain and modulus of elasticity E of the steel were obtained from the Eurocode 2.

3.2 Test setup

The specimens are tested in a special 2000 kN capacity testing machine. The eccentricity of the applied compressive load was equal at both ends, so the columns are subjected to single curvature bending. It was necessary to built up special assemblages at the pinned ends to apply the load eccentrically, Figure 4. Figure 5 presents a general view of the test. Five LVDTs were used to measure symmetrically the deflection of the column at midlength(L/2), and also at four additional levels, Figure 6. The strains were measured at the mid-span section using strain gages.

4 RESULTS

All the 1-D numerical tests are accomplished, Table 3, but on the date of the congress only the half of the tests (one per week) will be carried out.

Also the first two experimental tests can be presented right now (December 2005), Table 4.

Table 4 presents the maximum load and the corresponding displacements of both the numerical and experimental cases. Also the error (defined in the previous section) is computed. From this table (but with the carefulness of not having yet all the cases carried out) it can be inferred that the tendency shown in the Figure 3.a is also valid. For a ratio e/D = 0.2 the 1-D

Table 3. Tests for CFT columns, and numerical results.

Test	D (mm)	t (mm)	D/t	L (Mm)	L/D	fck (MPa)	e (mm)	e/D	Pmax (kN)	dy (m)
									Numerical	
1	100	3	33.33	2135	21.35	**17**	20	0,20	173,6	0,015
2	100	3	33.33	2135	21.35	**32**	50	0,50	127,7	0,022
3	100	3	33.33	2135	21.35	60	20	0,20	250,7	0,019
4	100	3	33.33	2135	21.35	60	50	0,50	146,7	0,023
5	100	3	33.33	2135	21.35	90	20	0,20	287,1	0,022
6	100	3	33.33	2135	21.35	90	50	0,50	158,3	0,023
7	100	3	33.33	3135	31.35	30	20	0,20	139,4	0,021
8	100	3	33.33	3135	31.35	30	50	0,50	94,3	0,032
9	100	3	33.33	3135	31.35	60	20	0,20	160,7	0,021
10	100	3	33.33	3135	31.35	60	50	0,50	110,8	0,039
11	100	3	33.33	3135	31.35	90	20	0,20	173,9	0,022
12	100	3	33.33	3135	31.35	90	50	0,50	115,4	0,034
13	100	6	16.67	2135	21.35	30	20	0,20	273,0	0,016
14	100	6	16.67	2135	21.35	30	50	0,50	175,4	0,023
15	100	6	16.67	2135	21.35	60	20	0,20	318,7	0,018
16	100	6	16.67	2135	21.35	60	50	0,50	198,3	0,023
17	100	6	16.67	2135	21.35	90	20	0,20	353,8	0,021
18	100	6	16.67	2135	21.35	90	50	0,50	212,6	0,025
19	100	6	16.67	3135	31.35	30	20	0,20	195,7	0,021
20	100	6	16.67	3135	31.35	30	50	0,50	133,9	0,032
21	100	6	16.67	3135	31.35	60	20	0,20	230,0	0,034
22	100	6	16.67	3135	31.35	60	50	0,50	146,1	0,033
23	100	6	16.67	3135	31.35	90	20	0,20	245,4	0,038
24	100	6	16.67	3135	31.35	90	50	0,50	153,5	0,033
25	200	3	66.67	2135	10.68	30	20	0,10	965,0	0,008
26	200	3	66.67	2135	10.68	30	50	0,5	680,2	0,012
27	200	3	66.67	2135	10.68	60	20	0,10	1412,7	0,007
28	200	3	66.67	2135	10.68	60	50	0,25	926,9	0,010
29	200	3	66.67	2135	10.68	90	20	0,10	1841,2	0,008
30	200	3	66.67	2135	10.68	90	50	0,25	1174,3	0,011
31	200	3	66.67	3135	15.68	30	20	0,10	739,0	0,007
32	200	3	66.67	3135	15.68	30	50	0,25	488,3	0,011
33	200	3	66.67	3135	15.68	60	20	0,10	1249,0	0,016
34	200	3	66.67	3135	15.68	60	50	0,25	814,3	0,022
35	200	3	66.67	3135	15.68	90	20	0,10	1575,2	0,019
36	200	3	66.67	3135	15.68	90	50	0,25	983,8	0,025
37	125	4	31.25	3135	25.08	60	20	0,16	386,2	0,030
38	125	4	31.25	3135	25.08	60	50	0,40	248,3	0,035
39	125	4	31.25	3135	25.08	90	20	0,16	431,7	0,034
40	125	4	31.25	3135	25.08	90	50	0,40	268,0	0,036
41	159	6	26.5	3135	19.72	60	20	0,13	882,6	0,022
42	159	6	26.5	3135	19.72	60	50	0,31	593,4	0,028
43	159	6	26.5	3135	19.72	90	20	0,13	1008,8	0,026
44	159	6	26.5	3135	19.72	90	50	0,31	665,7	0,032

finite element model is on the safe side, but for a ratio e/D = 0.5 the model does not predict well the column behavior.

Figure 7 presents the horizontal displacement at mid-span versus the vertical load applied for both cases. From them and Table 4 it can be concluded that the numerical model is more rigid than the experimental one. It needs to be improved.

5 CONCLUSIONS

In this paper a nonlinear numerical model for the analysis of concrete filled tubular columns using the finite element method is presented. The novelty of the model is focussed in the numerical integration of the cross section using the Gauss-legendre quadrature. The numerical model implemented has presented a suitable accuracy degree in comparison with 78 experimental tests from the literature. Also an experimental program and the design of the setup are presented. The numerical model is more rigid and needs to be improved for the cases of higher eccentricities, where the displacement corresponding to the maximum load is not well predicted.

This issue is very important to create new buckling curves for high strength concrete.

Figure 4. Special assemblage to apply the load eccentrically.

Figure 5. Test setup.

ACKNOWLEDGEMENTS

The authors wish to express their sincere gratitude to the Spanish "Generalitat Valenciana" for help provided

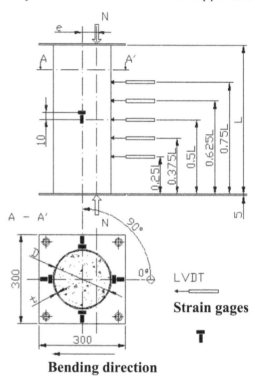

Bending direction

Figure 6. Instrumentation.

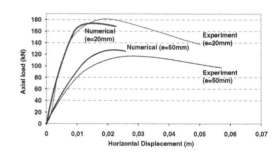

Figure 7. Load-displacement curve of Test 1 and 2.

Table 4. Comparison of numerical and experimental tests.

N°	D (mm)	t (mm)	L (Mm)	(fck)	e (mm)	e/D	Pmax (kN) Numerical	dy (mm)	Pmax (kN) Experim.	dy (mm)	Pm (kN) Error ξ	dy (mm)
1	100	3	2135	**17**	20	0,20	173	15	181	18	1.04	1.19
2	100	3	2135	**32**	50	0,50	127	22	117	26	0.92	1.18

510

through project GV04/11/2004, and to the "ICT, Instituto de Construccion Tubular", partner of CIDECT, for their advice.

REFERENCES

Aboutaba, R.S. & Machado, R.I. 1999, Seismic resistance of steel-tubed high-strength reinforced-concrete columns, J STRUCT ENG-ASCE 125(5): 485–494.

Bergmann, R. 1994. Load introduction in composite columns filled with High strength concrete, Proceedings of the 6th Int Symposium on Tubular Structures, Monash University, Melbourne, Australia.

Bonet, J.L. 2001. Método simplificado de cálculo de soportes esbeltos de hormigón armado de sección rectangular sometidos a compresión y flexión biaxial, PhD Thesis, Civil Engineering Dept., Technical University of Valencia.

CIDECT. 1979. Monograph n° 1: Concrete filled hollow section steel columns design manual. British edition.

CIDECT. 1979. Monograph n° 5: Calcules Poteaux en Proliles Creux remplis de Beton.

CIDECT. 1998. Guía de Diseño para columnas de perfiles tubulares rellenos de hormigón bajo cargas cíclicas estáticas y dinámicas. TUV-Verlag.

Comité Euro-internacional du beton. 1991. CEB-FIB Model Code 1990. C.E.B. Bulletin N° 203–204 and 205.

Comité Euro-internacional du beton. 1995. High Performance Concrete. Recommended extensions to the Model Code 90 research needs. C.E.B.. Bulletin N° 228.

Crisfield, M.A. 1981. A fast incremental/iterative solution procedure that handles "snap-through". Computers & Structures 13(1–3): 55–62.

Eurocode 4. 1992. Proyecto de Estructuras Mixtas de Hormigón y Acero, Parte 1-1: reglas generales y reglas para edificación.(in Spanish).

Gourley, B.C., Tort, C., Hajjar, J.F. & Schiller, P.H. 2001. A Synopsis of Studies of the Monotonic and Cyclic Behavior of Concrete-Filled Steel Tube Beam-Columns. Structural Engineering Report No. ST-01-4. Department of Civil Engineering. University of Minnesota, Minneapolis, Minnesota.

Grauers, M. 1993. Composite columns of hollow sections filled with high strength concrete. Research report. Chalmers University of Technology, Goteborg.

Gutiérrez, G. & Sanmartin, A. 1983. Influencia de las imperfecciones en la carga crítica de estructuras de entramados planos. Hormigón y Acero 147: 85–100.

Hu, H.T., Huang, C.S., Wu, M.H., et al. 2003. Nonlinear analysis of axially loaded concrete-filled tube columns with confinement effect, Journal of Structural Engineering-ASCE 129(9): 1322–1329.

Johansson, M. & Gylltoft, K. 2001. Structural behavior of slender circular steel – concrete composite columns under various means of load application. Steel and Composite Structures 1(4):393–410.

Kilpatrick, A.E. & Rangan, B.V. 1999. Tests on High-Strength Concrete-Filled Steel Tubular Columns. ACI Structural Journal 96(2): 268–275.

Liu, D.L., Gho, W.M. & Yuan, H. 2003. Ultimate capacity of high-strength rectangular concrete-filled steel hollow section stub columns. Journal of Constructional Steel Research 59(11): 1499–1515.

L.u. X.L., Yu, Y., Kiyoshi, T., et al. 2000. Nonlinear analysis on concrete-filled rectangular tubular composite columns. Structural Engineering Mech. 10(6): 577–587.

Neogi, P.K., Sen, H.K. & Chapman, J.C. 1969. Concrete-Filled Tubular Steel Columns under Eccentric Loading. Structural Engineer 47(5): 187–195.

Rangan, B. & Joyce, M. 1992. Strength Of Eccentrically Loaded Slender Steel Tubular Columns Filled With High-Strength Concrete, ACI Structural Journal 89(6): 676–681.

Romero, M.L., Bonet, J.L., Ivorra, S. & Hospitaler, A. 2005. A Numerical Study Of Concrete Filled Tubular Columns With High Strength Concrete, Proceedings of the Tenth International Conference on Civil, Structural and Environmental Engineering Computing, B.H.V. Topping, (Editor), Civil-Comp Press, Stirling, United Kingdom, paper 45.

Shams, M. & Saadeghvaziri, M.A. 1999. Nonlinear response of concrete-filled steel tubular columns under axial loading. ACI Structural Journal 96(6): 1009–1017.

Srinivasan, C.N. 2003. Discussion of "Experimental behavior of high strength square concrete-filled steel tube beam-columns" by Amit H. Varma, James M. Ricles, Richard Sause, and Le-Wu Lu, Journal of Structural Engineering-ASCE 129(8): 1285–1286.

Varma, A.H., Ricles, J.M., Sause, R., et al. 2002. Experimental behavior of high strength square concrete-filled steel tube beam-columns, Journal of Structural Engineering-ASCE 128(3): 309–318.

Varma, A.H., Ricles J.M., Sause, R., et al. 2004. Seismic behavior and design of high-strength square concrete-filled steel tube beam columns, Journal of Structural Engineering-ASCE 130(2): 169–179.

Varma, A.H., Sause, R., Ricles, J.M., et al. 2005. Development and validation of fiber model for high-strength square concrete-filled steel tube beam-columns. ACI Structural Journal 102(1): 73–84.

Zeghiche, J. & Chaoui, K. 2005. An experimental behaviour of concrete-filled steel tubular columns. Journal of Constructional Steel Research 61(1): 53–66.

Parallel session 8: Cast steel

Tubular Structures XI – Packer & Willibald (eds)
© 2006 Taylor & Francis Group, London, ISBN 0-415-40280-8

Connecting tubular structures with steel castings

D. Poweleit

Steel Founders' Society of America, Crystal Lake, IL, U.S.A

ABSTRACT: Steel castings have the potential to provide new opportunities for reducing cost, improving performance, and facilitating unique structures. Steel castings are currently used for high performance and critical applications in the railroad, construction, mining, and pump/valve industries. Recent research has shown steel castings can be used in seismic applications for building construction and perform well beyond requirements. Steel castings provide open geometry, manufacturing flexibility, equivalent mechanical properties to wrought material, and good weldability. Steel Founders' Society of America (SFSA) and American Institute of Steel Construction (AISC) have partnered with the University of Arizona to develop casting applications, such as modular and seismic connectors, for building construction through National Science Foundation sponsorship.

1 INTRODUCTION

1.1 Steel Founders' Society of America (SFSA)

SFSA has been serving steel foundries for over 100 years. SFSA is a not-for-profit trade association that promotes and develops steel castings. SFSA develops new technology, coordinates the exchange of technical and operating experience, expands steel casting applications, and is known world-wide for its reference literature.

1.2 Background

Steel castings are not commonly used in building construction. In the past, castings have been used in the building industry and for bridges. There are occasional uses but the ordinary application of steel castings to create steel structures that are pervasive in industrial equipment are absent from building construction. Steel castings are used in safety critical applications and in harsh demanding environment to carry significant loads. They are commonly welded into a fabricated structure. Castings give users unlimited potential for steel geometry. In industrial equipment, steel castings are used as connectors to perform demanding tasks while holding the other structural elements together. The use of steel castings as connectors in building construction can be an attractive option to improve the performance of the connection while lowering the total cost of the structure.

Connections are a critical feature of building construction. While the connectors are rarely more than 5% of the total weight, they are typically 60% of the cost. Moment and special bracing connections are

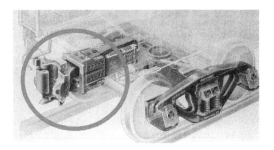

Figure 1. Standard railcar coupler.

especially difficult and costly. The cost for steel construction in 2001 was 25% material, 33% shop, 28% erector, and 14% other. Labor related costs exceed 60% of the cost of construction (Geschwindner, 2002). Modular cast connectors could reduce labor and erection costs, improve performance, decrease erection time, easily transfer loads from one shape beam or column to another, enhance reliability of the connection, and reduce engineering and detailing costs. In seismic applications, they could improve the safety and reliability of the structure cost-effectively. For special architectural features, they could provide innovative and attractive transitions between shapes or unique designs.

One common application of steel castings is couplers for the railroad industry, Figure 1. As an example of the severe requirement, the electric utilities use large dedicated trains to move coal from the Powder River Basin to the Midwest. These trains are commonly made up of 120 freight cars that each weigh 130,000 kg, for a total train weight of over 15,600 Mg. This train is powered by two sets of 8,950 kW locomotives for a total of 17,900 kW. These trains cross the Rocky Mountains in the dead of winter at temperatures well below freezing on tracks that are remote and may have rocks or other debris on the track. This whole system of 15,600 Mg and 17,900 kW is connected in the center by one set of steel castings in the form of a coupler. The example of a railroad coupler illustrates that steel castings can clearly meet the requirements listed for a good connection. They are capable of safe, economical, reliable, simple, repetitive performance.

If connections are a problem, and steel castings offer an answer, why are castings not being used? The American Institute of Steel Construction (AISC) and the Steel Founders' Society of America (SFSA) have formed a joint task committee to explore the possibilities of using steel castings in steel building construction. Most of the problems in using steel castings in building applications seem to be related to a lack of understanding. Steel foundries do not understand the requirements or needs of the building construction industry. Designers, fabricators and erectors do not understand the use of castings. A brief overview of steel casting use, method of manufacturing, properties, and purchase requirements may be useful.

2 STEEL CASTING

2.1 *Applications*

Steel castings are used broadly in industrial equipment as connectors. The example of a railroad coupler already discussed is a good example. The railroad industry uses steel castings extensively for the trucks, wheels, and corners. Fifth wheels for large over the road trucks are steel cast connectors. Caterpillar aggressively used steel castings in its 797 mine haul truck. The truck can carry 325 Mg and has a diesel engine that provides 2500 kW (Caterpillar, 2003). Caterpillar used steel castings to make the entire load-bearing frame for improved durability and resistance to impact loads. Whether as connectors for structural components or power transmission parts, steel castings are commonly used in industrial equipment (SFSA, 1995).

One example that may be easier to relate to steel structures is the use of steel castings for valves and fittings. Steel castings can have complex internal passages, which is a unique capability compared to other manufacturing processes. This is why castings are used for blocks or heads in automotive engine applications. In fittings for piping, castings not only provide a flow path, they are a structural connector of pipe, a hollow structural shape. The structural design for fittings is seen as straight forward since the fitting is a successful structural component as long as it outlives the pipe.

The cast steel fitting cross section is maintained larger than the pipe cross-section and this assures that the pipe will fail first.

The reason steel castings are attractive as connectors is the freedom of design. Any shape that can be imagined can be cast. Frequently, a casting cannot be made effectively not because the design was too aggressive but because it was too timid. Often a fabrication design that is inadequate is sent to the foundry to see if it can be made as a casting. This normally causes manufacturing problems. And changing the manufacturing process will not overcome inherent performance characteristics of a design. Cast parts are best manufactured when designed for the casting process. One reason that castings are seen as problems is because the foundry is often asked to make poor designs successful and so the lead-time is long and the cost is high. Good casting design allows weight to be reduced, cost to be lowered and performance to be improved.

Steel castings are expensive sources of steel but cheap suppliers of geometry. Good applications of steel castings are details that have many parts with high fabrication costs, poor material utilization and performance limits based on section size or geometry limits. The flexibility of casting allows material to be placed where needed and material to be removed where it is not needed. Castings like big sweeping curves, non-uniform sections, and complex geometry.

One structural example of the use of steel castings to give shape and performance in a steel structure is their use as nodes in offshore oil platforms, Figure 2. The casting is designed to perform in a demanding environment of high stress and corrosive atmosphere, it weighs 20% less than a fabricated connection, and moves the welds to the tubular structures outside the high load regions of the structure to prevent failure of a welded joint in a high stress region. The steel cast nodes are designed so the pipes drive the loads into the casting. The casting geometry and section size are tailored to survive the requirements of the application (Marston, 1991). Similar to welding procedures that rely on the same principles of metal solidification as castings, casting procedures call out testing requirements to ensure mechanical performance. First article tests typically call out x-ray and magnetic particle inspection to ensure the quality level of the casting. Therefore, ensuring quality and performance of a casting is very similar to that of a weldment.

2.2 *Process*

All steel is cast. Traditional integrated mills melted ore in blast furnaces, converting blast furnace iron to steel and cast steel ingots. The ingots are rolled into plate or bar and then finally rolled into the desired structural shape. Mini-mills melt steel scrap in electric furnaces and then continuously cast the steel into bars that are directly rolled into the structural shape. Foundries melt

Figure 2. Cast node and offshore oil platform.

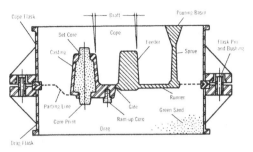

Figure 3. Sand mold terminology.

steel scrap in electric furnaces like mini-mills but cast the steel directly to shape in molds.

Molds are made of sand held together by a binder. Molds were traditionally made of sand with clay and water as a binder but now the use of organic polymers to hold the sand together. The desired shape is first made as a pattern of wood, metal or plastic. The pattern is made oversize to compensate for the change in size of the metal cooling to room temperature. The pattern forms the mold cavity that holds the molten steel during solidification and cooling. Most molds are made in two halves, the top half is called a cope and the bottom is called the drag. The joint between the cope and drag is called the parting line. The pattern must be removed from the mold at the parting line without damage to the pattern or mold. This requires draft of about 1° on the pattern tooling. Draft is required in all split manufacturing methods like casting or forging. Sometimes reorientation of the shape in the mold can avoid the need for draft on some surfaces, for example an offset parting on an "L" bracket.

Making loose pieces of sand called cores can create features that cannot be made in the mold by the pattern. The use of cores in molds allows castings to be made hollow or with features that cannot be formed in the mold. Cores add cost in manufacturing, reduce the ability of the casting to hold tolerances, but may be necessary to form key features that provide the geometry designed. A design may be modified to reduce the number of cores by consolidating features, using offset parting lines or through changes in component design.

The top half of the mold must be held in position or it will float when the liquid steel is poured into the mold and the steel will run out at the parting line. The cores used to form features must also be held in place to locate the feature on the casting and to prevent the core from floating before the liquid steel solidifies. Often the core is made long to extend into the mold and the pattern is modified to create a pocket to hold the core. This pocket is called a core print and allows the foundry to remove the spent sand of the core from the casting, locates and holds the core, and allows the core feature to be inspected.

In addition to using sand to make a mold, a ceramic shell can also be utilized for the production of casting steel. This process is referred to as investment (or lost-wax) casting. A positive, or replica of the part, is made of wax. This is then dipped in a ceramic slurry several times to build up a shell. The shell is then dried and the wax is melted out in an autoclave. The cavity is now ready to be filled with molten steel. After the steel solidifies, the ceramic shell is broken away from the casting. This process is typically used for smaller parts with finer detail, ones that require no draft, and those that have complex geometries that would otherwise require extensive coring.

The mold has a flow path for the liquid steel to allow it to fill the mold cavity without damage to the mold shape or metal quality. This flow path is called a gating system. The sprue allows the liquid steel to drop into the lower parts of the mold and then gates and runners transfer the steel to the mold cavity. A typical sand mold with associated terminology is shown in Figure 3.

When steel solidifies it shrinks. The mold must include risers to make the casting sound. A casting geometry with an isolated heavy section such as a flange or boss would have a shrinkage cavity in the center. Risers are placed on heavy sections of the casting to overcome the shrinkage in the part and hold it in the

517

riser. It may be necessary or desirable to add taper to a section or reorient a heavy section to make sure the casting is sound. Castings are evaluated for soundness during design using computer simulation much like how finite element analysis is used to evaluate structures for service performance. Optimizing the design and rigging for casting the part through directional solidification will decrease cost.

After solidification, the casting is removed from the mold and shot blasted to remove the sand from the surface and internal cavities. The gating system and risers are cut off. The casting is heat treated to the properties desired. The casting is inspected to ensure it meets the requirements of the purchaser.

2.3 *Material*

Many believe that cast steel is brittle because the cast iron that is commonly used in automotive and household goods, like cookware, easily cracks. However, the properties of steel are very different from iron. Steel castings can meet or exceed the ductility, toughness, or weldability of rolled steels. Technically, all steel is cast. Designers generally think of design requirements in terms of strength, but the design is commonly constrained by modulus, fatigue, toughness or ductility. Increasing the strength of steel normally reduces the ductility, toughness, and weldability. It is often more desirable in steel casting design to use a lower strength grade and increase the section size or modify the shape. The design freedom makes castings an attractive way to obtain the best fabrication and material performance and the needed component stiffness and strength.

Rolled sections of steel have their structure elongated in the direction of rolling. The strength and ductility is improved in that direction but they are reduced across the rolling direction, Figure 4. The lack of a rolling direction in steel castings gives them uniform properties in all directions. Rolling steel cold can also strengthen the steel but reduces ductility and toughness. Cast steel grades achieve the same trade off by alloying and heat treatment.

Steel castings are used in demanding applications that are safety critical, highly specified, and performance demanding. A railroad coupler is a good example of a common application that is critical. Castings are used in high-pressure applications in nuclear power plants. The use of steel castings in pressure containing systems is common and specified in the ASME Boiler and Pressure Vessel Code.

One aspect of the ASME code is the requirement that suppliers develop and demonstrate a weld procedure including welded properties for the components and materials they supply. The cast carbon steels that would be used in building construction are already well known and established in the Code, including their design requirements and welded properties.

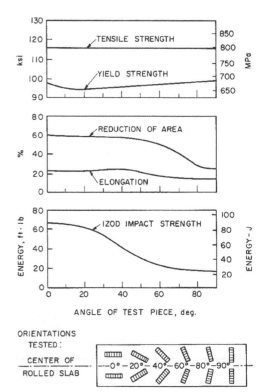

Figure 4.   Orientation properties.

2.4 *Requirements*

The biggest advantage in quality that forged or rolled shapes have over steel castings is their ability to begin with a simple optimal casting. The ingot or bar can be easily inspected prior to rolling or forging. The use of casting processes to make uniquely designed shapes requires inspection that is correlated to the casting process, part design, and performance requirements. Often the purchaser of steel castings uses nondestructive examination, mechanical testing, and engineering analysis to ensure the desired reliability.

Steel casting producers routinely test each heat of steel to make sure it meets the mechanical properties required in the material specification. The heat is also analyzed chemically to certify that it meets the standard. Other specialized tests can be required like low temperature impact testing when service performance requirements dictate. The dominant material used in building construction is carbon steel because of its reliable properties, low cost and ease of fabrication. One common grade used for building construction in rolled sections is ASTM Specifications A36. The use of steel castings is permitted in building construction using material from either ASTM A27 grade 65-35 or ASTM A148 grade 80-50 (AISC, 1998). The properties of carbon steel depend on the composition and heat treatment. Because designers use yield strength as a basic property in design, often material is ordered to higher strength without considering the advantage in castings of using a lower strength material with optimum ductility and weldability. Since the load-carrying cross-section can be increased to accommodate lower strengths, the casting can be supplied in the highest ductility with strength levels that are compatible with the rolled structural shapes. This use of cast carbon steel in its optimal condition makes sure that the casting will perform safely and reliably and that excessive loads will cause failure to occur first in the rolled section familiar to the designer. The use of ASTM A27 grade 65-35 in the normalized and tempered condition will give a strong ductile weldable steel.

Cast steel alloys are available in a wide range of options. One can increase different performance characteristics such as corrosion resistance and wear resistance through alloying and heat treatment. Mechanical properties such as strength and elongation can likewise be adjusted. ASTM A216 grade WCB offers a good option to balance strength, weldability, ductility, and cost (yield strength of 1.7 MPa and elongation of 22%). If a 50 ksi steel is the only factor in material selection, ASTM A954 grade SC8620 class 80/50 can be utilized (yield strength 2.4 MPa and elongation of 22%). For corrosion resistant applications, ASTM A743 grade CF-8M (similar to wrought 316) is a good option (yield strength 1.4 MPa and elongation of 30%). Since mechanical properties of a part are based on material and geometry, it is possible to design a cast steel connector with lower material yield strength than the wrought sections it connects by utilizing the geometry available through casting. This provides the added benefit of making the casting more ductile and more weldable than the wrought material. This practice is commonly used to design the casting to out live the wrought material it joins.

Traditionally, nondestructive testing has been used to certify casting quality. Soundness is verified through the use of radiographic inspection. Surface quality is evaluated using magnetic particle inspection. More recently, the use of computer simulation of solidification of the casting integrated with finite element analysis of its performance has been used to design optimal casting configurations. The development of these tools allows the designer to ensure that critical areas of the part meet requirements while ensuring the most economical means of manufacturing the whole part. Casting design tutorials are available on the web (SFSA, 2006).

2.5 Design

There are six key factors in casting design (Gwyn, 2003). These factors are based on physics and govern the castings service performance, manufacturability, and cost. Understanding and utilizing these six casting factors in addition to the freedom of geometry results in cast components that excel in their applications. The first four factors are related to casting properties and include fluid life, solidification shrinkage type and volume, inclusion formation tendency, and pouring temperature. The other two factors are related to structural characteristics and include section modulus and modulus of elasticity.

Fluid life refers to the dynamic change in property of the molten metal as it enters the mold. As the metal enters the mold it transfers heat, forms a skin of solidified metal, and reduces in fluidity. The characteristic is not only governed by pouring temperature but also chemistry of the alloy. Fluid life impacts the minimum section thickness, details such as lettering on surface, and transition geometry. Section thickness is governed both by the material cast and the length of section. Carbon and low alloy steels tend to have lower fluid life. Thus, details of the part are placed in the drag half of the mold where the liquid metal enters and is at its hottest. Cast letters on steel castings need to be larger. Softer transitions, large radii and good taper, are also utilized to maximize fluid life. This has the added benefit of having geometry that can more readily handle loads and reduces stress concentration points.

The two major forms of shrinkage in making a casting occur when the liquid changes to a solid and when the solid cools to room temperature. Both of these forms of shrinkage must be accounted for in order to achieve dimensionally accurate parts. The shrinkage in volume as the solid cools is referred to as "patternmakers contraction". The way the mold and cores are designed largely influence this. Steel shrinks at approximately 6 mm per 0.3 m. Risers, or reservoirs of liquid metal, are required in the casting process to feed the shrinkage that would otherwise lead to shrinkage cavities in the casting. Steel has a larger amount of shrinkage and directional solidification. Directional solidification refers to the fact that steel will solidify from the further point first and work its way back to the point of origin. Therefore, if good risers and taper are incorporated into the mold design, steel castings can have good internal soundness. Geometry plays a role in solidification due to its effect on heat transfer. Thus, geometry can be used to improve solidification in a part.

Nonmetallic inclusions form from reoxidation during the melting and pouring process, and from components of refractory. Nonmetallic inclusions form when air is entrained in the molten metal as it enters the mold or when the steel is moved from the furnace to pouring ladle. The oxygen in the air chemically reacts and precipitates out of the metal as an inclusion. Inclusions from refractory come from the material that is used to line the furnace or transfer vessels for pouring the steel. Steel has a moderate tendency to form inclusions. It is also important to make certain the metal enters the mold in with minimum turbulence. Turbulence in the liquid metal causes more air to be entrained and results in additional inclusions. Good gating design along with smooth casting geometry reduce turbulence. Inclusions can be minimized through good foundry practices.

Pouring temperature is simply the temperature of the liquid metal when it is poured into the mold. Higher melting point metals like steel are more aggressive on refractories and start to approach the limit of the sand mold materials. Steel is typically poured at temperatures a little under 1650°C. Heat transfer impacts how the part will solidify.

Design engineers typically design parts based on allowable stress and deflection. Both the stress and deflection capabilities of a part are governed by the geometry. Bending stress, torsional shear, and deflection calculations are all based on the area moment of inertia of the cross section. Therefore, utilizing clever geometry can yield a part that is capable of meeting specified mechanical requirements. Designing cross sections as I-beams or C-channels utilize geometry to increase strength and reduce deflection. Figure 5 shows how different cross sections provide different maximum stresses for a sample part. Note, that the "omega" cross section design handles stresses second best to the solid shape. It is easy to cast omega geometry versus trying to create this as a fabrication, utilizing this geometry reduces both weight and cost. Steel is a very robust material and doesn't necessarily require a lot of geometry to help make the part perform at the desired level. However, one can make use of steel's excellent mechanical performance by making innovative cast parts that can absorb shocks or severe loading.

The modulus of elasticity is based solely on the material itself. It is based on the type of alloy and its metallurgy. Heat treatment does not affect modulus; it does affect the yield strength. Steels have a modulus around 1,436 MPa, which puts them near the top compared to other metals. The larger the modulus the smaller the deflection of a part.

From the six casting design factors, some general rules of thumb can be deduced. First, if isolated heavy sections are required to meet form/fit/function, then these sections will need to be fed. The formation of isotherms also demonstrates how it is desirable to have these areas in the risers as these are not part of the final part and are removed. Junctions within a casting should be designed not to add mass. If a junction is required, tapering the section, adding feeding to the section, or dimpling the area can all be leveraged to ensure a sound casting. Changing section thickness in a casting is best handled with smooth, easy transitions. Adding taper and having a large radius help to accomplish this. Reducing the number of undercuts and other

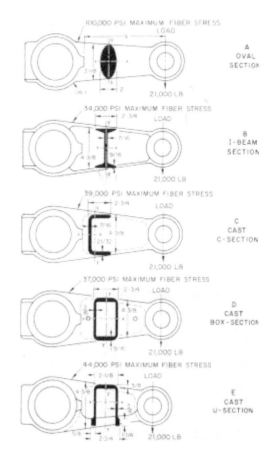

Figure 5. Impact of cross section shape on stress.

sections that would require cores helps to minimize cost. Specifying as-cast tolerances when allowable is also important for minimizing cost. Related to this, considering what other post-processing will be required and how it will be fixtured also influences final cost of the part. Working directly and concurrently with a foundry will ensure a part is optimized for both casting and the end use.

3 BUILDING CONSTRUCTION

3.1 *Applications*

Steel castings have been shown to be capable of demanding service in building construction. One example is the development of a modular connection by the ATLSS Center at Lehigh University, Figure 6. A self-aligning beam to column connection was designed to improve safety and productivity in erection. This self-aligning connection used a wedge

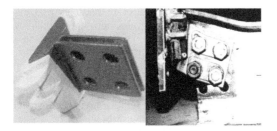

Figure 6. ATLSS connector and test.

Figure 7. ICF Kaiser bolted connection and test.

shaped extension on the beam that slid into a wedge shaped slot on the column. The manufacture of the complex wedge and slot was accomplished with steel castings. The ATLSS connection was subjected to full-scale mechanical performance tests. When loaded beyond the design requirement, it finally deformed plastically and did not fail catastrophically. Additional information on the test and the results can be found in the reference (Fleischman, 1993).

A gusseted reinforced "L" bracket was designed as a carbon steel casting and was tested for earthquake retrofitting of damaged and undamaged structures in California, Figure 7. The connection was designed by ICF Kaiser to be installed where welds had failed by bolting it to the bottom of the beam column connection. A prototype was cast, tested, and approved by the State of California. The test demonstrated that the cast connector would survive the maximum load required. Additional information on the test and the results can be found in the reference (Bleiman, 2003).

3.2 Research and development

An example of steel casting advantages for high performance complex connections is shown in some recent work by Robert Fleischman at the University of Arizona for designing seismic connectors (Fleischman, 2002). Since castings can have non-uniform walls and contain complex features, they can be designed to locate the strain deformation of a loaded structure. A cast modular node was produced that looks nominally like a reinforced welded connection. In reality, the casting process allows the intersection of the beam and column to be increased and the column and

Figure 8. Panel Zone seismic connector and test.

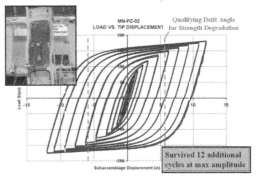

Figure 9. Panel Zone FEMA test results.

panel section tailored to absorb the deformation with little transferred load to the beam or column. The welds can be made outside of the node and high stress region. This Panel Zone part, as-cast and after test, is shown in Figure 8. The Panel Zone part is designed with a weaker panel section that facilitates energy dissipation. The challenge is to limit the plastic deformation of the panel section otherwise curvature or kinking with occur in the columns. "Dog bone" features were incorporated into the part to allow for deflection within the node and at controlled locations. Cast geometry efficiently provides the functionality of the part. A graph of the FEMA cyclic test is shown in Figure 9.

A cast Modular Connector (MC) was also developed and evaluated using finite element modeling and casting solidification simulation to provide an effective design (Fleischman, 2002). The casting prototype for the cast modular node is shown in Figure 10. A full size test subassembly was fabricated and subjected to the Federal Emergency Management Agency (FEMA) – 350 cyclic test protocol. The cast modular node exceeded the requirement greatly as can be seen in Figure 11. Note, after being taken to failure, which was well beyond test requirements, the casting failed

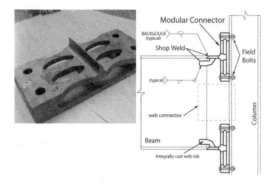

Figure 10. Cast modular connector.

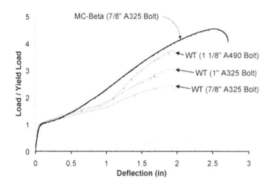

Figure 11. Test results for modular connector.

in a ductile manner. The MC design allows for the cast connector to absorb energy and has a higher capable load strength compared to a standard connection with a bigger bolt because it mitigates prying forces. As with the PZ part, "dog bone" geometry is used to control the location and amount of yielding.

4 CONCLUSIONS

Steel casting may provide new opportunities for lower cost, improved performance, and unique designs.

An improved understanding of steel castings will facilitate their use by designers and fabricators in building construction. The ability to make complex shapes repetitively can allow the design of modular connectors that are reasonable in cost, and reduce shop and erection costs. The ability to tailor geometry, customize steel properties, and integrate castings by welding allows improved performance. The freedom of geometry, size and complexity allows the designers flexibility that is unprecedented. Steel castings offer architectural and structural flexibility that will challenge building designers' imagination.

REFERENCES

AISC 1998. *Load & Resistance Factor Design Volume 1, Structural Members, Specifications & Codes, Manual of Steel Construction.* 6–26 to 6–28.
American Iron and Steel Institute website 2003. *Industry statistics.*
Bleiman, David. *The ICF Kaiser Bolted Connection* CD 2003.
Caterpillar website 2003. *Products and Services.*
Fleischman, R. B., Viscomi, B. V., & Lu L. W 1993. ATLSS Connection and Automated Construction. *ATLSS Report No. 93-02.*
Fleischman, R. B. 2002. Development of Cast Modular Connections for Seismic-resistant Steel Frames. *SFSA 56th Technical and Operating Conference.* Session 1.6.
Geschwindner, Louis F. 2002. Practical Steel Design: 2 to 20 Stories, A Tutorial for Practicing Engineers. *North American Steel Construction Conference.* Tutorial 1.
Gwyn, Michael A. 2003. Cost-Effective Casting Design. *American Foundry Society.*
Johnson, Andy 2002. Castings in Building Construction: Breakthrough Opportunity? *SFSA 56th Technical and Operating Conference.* Session 1.5.
Marston, G. J. 1991. Better Cast then Fabricated. *SFSA 45th Technical and Operating Conference.* Keynote.
SFSA 1995. *Steel Casting Handbook.*
SFSA website 2006. *Casting Design Tutorials.*

Tubular Structures XI – Packer & Willibald (eds)
© *2006 Taylor & Francis Group, London, ISBN 0-415-40280-8*

Cast steel nodes in tubular construction – Canadian experience

J.C. de Oliveira, S. Willibald, J.A. Packer & C. Christopoulos
Department of Civil Engineering, University of Toronto, Canada

T. Verhey
Walters Inc., Hamilton, Ontario, Canada

ABSTRACT: Cast steel nodes have been used for some time in offshore steel structures and recently there has been a resurgence of interest in steel castings for onshore tubular structures. This has taken place principally in Europe, where cast nodes have now gained acceptance in road and railway bridges that are susceptible to fatigue. In North America, the potential for revival of steel castings in onshore construction has just been realized. Tubular structures, in particular, are an ideal application and castings can provide viable solutions to tubular connection design problems, particularly under seismic and fatigue loadings. A summary of Canadian experience with cast steel nodes is given, including descriptions of both an in-progress project involving cast steel nodes used for a tree-like roof supporting structure at the University of Guelph, Ontario as well as current research being carried out at the University of Toronto on cast steel connectors for tubular hollow sections which are subjected to cyclic inelastic loading.

1   INTRODUCTION

Steel castings are rarely used in conventional building construction in North America. Even in Europe, significant use of castings is typically reserved for high profile buildings where the desire for architectural flair is the driving force behind the building's design. Castings have had greater success in the European bridge industry, where castings can be found in many pedestrian, highway, and railroad bridges, particularly where hollow structural sections (HSS) have been used (Veselcic et al. 2003). In the offshore construction industry, castings have been used extensively since the early 1980s, where they represent an enormous improvement to welded connections in tubular steel offshore platforms.

There is justification for the successful use of castings in these situations – the casting process allows for almost any shape to be manufactured and, unlike wrought steel, cast steel is isotropic in strength and toughness. This makes the use of castings ideal in fatigue critical applications, particularly when regions of geometric complexity coincide with the heat-affected zone of a weld. In these cases, the use of a cast steel node relocates the weld to a region of lower stress and improves transitional geometry, both of which significantly reduce stress concentrations and greatly improve the fatigue life of the connection (Lomax 1982).

The manufacture of unique cast steel connections is typically cost prohibitive; cost savings realized from the use of castings are primarily derived from savings in fabrication cost for highly complex geometry and in high repetition. Thus, it has been shown that the use of castings in typical onshore tubular structures should be restricted to those situations where aesthetics are a main concern (Glijnis et al. 2003). However, it is still believed that the use of castings can provide cost savings in structures where a complex connection is repeated a number of times. Further, when a connection is subjected to cyclic inelastic loading, the use of a cast steel seismic fuse can provide improved energy dissipation (Fleischman et al. 2004).

2   CANADIAN EXPERIENCE WITH STEEL CASTINGS

2.1   *CDP Capital Centre, Montréal, Québec*

Canadian experience with steel castings in building construction is limited. It is believed that only one Canadian building structure, the CDP Capital Centre located in Montréal, Québec, features significant use of structural castings. Completed in 2003, this building's eight-storey open atrium features a glass façade that is supported by tubular steel trusses. The trusses incorporate cast steel nodes (Figure 1) that were cast at

Figure 1. Steel castings featured in the eight-storey trusses of the CDP Capital Centre, Montréal, Québec.

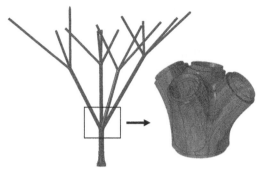

Figure 2. Tree structure and enlarged view of the lowermost branching point.

the foundry Métallurgie Castech Inc., Thetford Mines, Québec.

With this limited experience, Canadian structural engineers are hesitant to specify steel castings in building construction. Many engineers wrongly assume that cast steel exhibits the same brittle response as cast iron. There is also the misconception that all castings exhibit critical defects and are unsafe for structural use. While defects can occur during casting solidification, modern software packages allow the foundry to simulate fluid flow and solidification of the molten steel within the casting mould. This enables the foundry to greatly reduce the likelihood of developing a defect within the casting. Many Canadian structural engineers are unaware of the modern foundry's capability of producing sound and reliable castings that can be produced to conform to a wide range of standards for mechanical properties.

2.2 New Science Complex at the University of Guelph, Ontario

Construction of the New Science Complex (Phase 2) at the University of Guelph began in March 2005. Designed by Robbie/Young & Wright Architects, the Science Complex will provide new laboratory and research space for chemical and biological sciences. Completion of the project is set for spring 2007.

The complex includes an open atrium featuring a glass roof supported by a single tree-like structure, a rendering of which is shown in Figure 2. The tree is approximately 25 m in height and will be fabricated using HSS welded to cast steel nodes at the tree's branching points. Each cast steel node is approximately 1 metre in height and depth and weighs up to 1.1 tons.

The steel fabricator, Walters Inc. (Hamilton, Ontario, Canada), was responsible for the design of the connections. Although aesthetics were the driving

force behind the decision to use cast steel nodes rather than weld-fabricated connections, machining the nodes from stock steel as well as using laminated profiled plates were options that were also considered but were ruled out due to excessive cost.

Design of the nodes began with the creation of suitable and aesthetically pleasing exterior geometry. Connector wall thickness was then optimized to ensure a safe but also economical design using finite element stress analysis. Stress analysis was carried out using the finite element software package CATIA ME2. Due to the complexity of the shape, free meshing using tetrahedral elements was employed. In addition to the branching point, 300 mm long flanges, representing short lengths of the attached hollow structural sections, were included in the model to ensure realistic load transfer into the casting (Figure 3).

Boundary conditions for the finite element model consisted of fixing the end of the lower trunk-flange and equally distributing the applied loads from each upper branch to the ends of the respective branch flanges. Loading of the cast node was based on both design loads as provided by the structural engineer as well as the condition that the node be stronger than the connected HSS to ensure member rather than "connection" failure. As all stresses had to remain below or just reach yield stress, linear elastic stress analysis was sufficient.

Initial wall thickness for each node was chosen based on the results of the finite element stress analysis. As is typical in the design of castings, solidification analysis carried out by the foundry revealed a need to increase the initial wall thickness toward the base of each of the nodes to facilitate directional solidification. Final node wall thickness varied from 25.4 mm (top branching point) to 63.5 mm (bottom branching point).

German foundry Friedrich Wilhelms-Hütte GmbH, Mülheim an der Ruhr was selected to produce the cast steel nodes. Friedrich Wilhelms-Hütte has great experience with structural castings, providing high

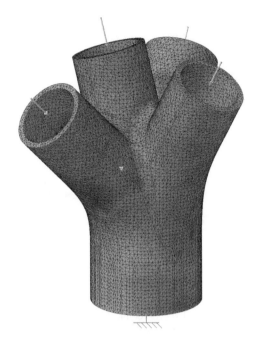

Figure 3. Finite element model of the bottom branch point (includes 300 mm long flanges).

Figure 4. Node immediately after casting (unfinished surface).

fabrication standards and offering a wide range of material and structural testing of their products. Test procedures included non-destructive testing (visual, ultrasonic, liquid penetrant, and magnetic particle examination) as well as tensile, bending, and notch toughness (Charpy) testing of the cast material.

Walters Inc. specified a cast material based on ASTM A27 (2005a) Grade 70/40, slightly modified to

Figure 5. Final, smooth-finished branch connection.

improve the material's weldability (lower carbon and silicon content were specified to eliminate the need for weld preheating). Ultimately, the material utilized for the castings was G20Mn 5 N according to EN 10213-3 (CEN 1996). For a thickness up to 50 mm, G20Mn 5 N has a specified minimum yield strength of 300 MPa and an ultimate strength of 500 MPa, which is similar to the prescribed strength of typical HSS in Canada. The actual yield and ultimate strengths of the casting are 401 MPa and 580 MPa respectively, as indicated by the foundry's tensile testing of coupons cast from the same heat as that of the nodes. The measured notch toughness of the cast material was 117 J at room temperature. Figure 4 shows the casting after its removal from the mould. Figure 5 shows the final surface finish of a completed node. Site erection of the tree was in progress at the time of submission of this paper (Figure 6).

3 NEW RESEARCH: CAST STEEL SEISMIC CONNECTORS FOR TUBULAR BRACES

3.1 Background

Concentrically braced frames (CBF) are an efficient means of providing lateral support in low- and medium-rise steel structures. By specifying simple shear connections throughout the structure, fabrication cost and erection time are both greatly reduced in comparison to other structural systems. The nature of the bracing system, typically consisting of several diagonal braces located intermittently throughout the

Figure 6. Site erection of the tree structure by Walters Inc.

(a) (b)

Figure 7. (a) Modern seismic RHS brace connection (Photo courtesy of Professor R. Tremblay, École Polytechnique de Montréal, Canada). (b) Proposed cast steel connector for CHS braces.

structure, allows for great design versatility and erection simplicity. Design flexibility is further increased by the variety of concentric brace configurations that are at the designer's disposal.

The diagonal brace members of a CBF are subjected to either compression or tensile forces with the application of lateral load on the structure, which can be induced by wind pressure or ground motion. Typical bracing members include but are not limited to angles, double angles, rectangular (RHS) and circular (CHS) hollow sections. HSS are a particularly popular selection for lateral bracing members because of their efficiency in carrying compressive loads, their improved aesthetic appearance, and because of the wide range of section sizes that are available.

In the event of a design level earthquake, brace members experience cyclic inelastic loading, repeatedly yielding in tension and buckling in compression, thereby dissipating seismic energy. Recent seismic events have shown the susceptibility of slotted tube to gusset connections to brittle connection failure (Tremblay et al. 1996). Initially, CHS were considered best suited for seismic brace applications due to their improved inelastic response over RHS tubes owing to the reduced degree of cold working involved in their

manufacture. However, connection issues relating to net section fracture at gusset connections led to the increased use of RHS braces for reasons associated with ease of connection reinforcement. Thus, modern CBFs feature RHS braces with connections that are significantly reinforced; as a result, these tube connections are expensive and unsightly, with performance being sacrificed to accommodate connection reinforcement (Figure 7).

New research at the University of Toronto is focused on the use of cast steel connectors for tubular brace elements subjected to cyclic inelastic loading. For this application, a casting provides an efficient connection between a single gusset plate that is connected to the building structure and a CHS brace member as shown in Figure 7.

3.2 Benefits of using a cast steel seismic connector for tubular braces

The geometric freedom that casting manufacturing offers allows the design of a connector that is able to develop loads in excess of the yield capacity of the brace. The casting can then be designed to use a full-penetration weld to connect to a range of CHS members of variable wall thickness but constant outer diameter. Thus, the cast connector is an improvement to the weld-fabricated slotted tube to gusset connection as unreinforced CHS brace members can be used without risk of net section fracture during a seismic event. Further, as a single connector type can be utilized throughout an entire structure (as brace wall thickness can be varied to achieve the required lateral storey strength), and for both static and dynamically loaded braces, the connectors can be mass-produced, greatly

526

reducing the manufacturing cost per cast connector. Another advantage to using a pre-engineered cast steel connector is that the fabricator need not individually design every tube-to-gusset connection as is required when specifying typical weld-fabricated connections.

3.3 Design of a cast steel seismic connector

Design of the cast connector had to satisfy all of the requirements for dynamically loaded connections set out in CSA (2001a). Further, in typical concentrically braced frames designed for seismic applications, the brace members themselves are the energy absorbing elements. Therefore, according to the principles of capacity design, the cast connector had to be designed to remain elastic during the formation of the following energy absorbing mechanisms:

1. tensile yielding of the brace member,
2. elastic buckling of the brace, or
3. plastic hinging of the brace at midspan due to overall or local inelastic buckling.

Plastic hinges at the brace ends were designed to form in the gusset plates welded to the building connection region during brace buckling. This limits the bending demand put on the connector during brace buckling and allows for the control of the buckling direction of the brace.

3.4 Selection of a cast steel alloy for the connector

Design of the connector first required the selection of a suitable cast steel alloy. Although the current Canadian structural steel design specification (CSA 2001a) references only two standards for cast steel (ASTM A27 2005a, and ASTM A148 2005b), there are many other ASTM standards for cast steel which are suitable for structural use. ASTM A958 (2000) is a specification for carbon and low-alloy steel casting grades having a chemical composition similar to that of a standard wrought steel grade, with each casting grade being heat treated such that the cast steel's tensile properties conform to those of a specific wrought steel grade. Since the cast connector is to be welded to HSS brace members having mechanical properties similar to ASTM A572 (2004) Grade 50, the connector was made according to ASTM A958 Grade SC8620 Class 80/50. According to the ASTM specification, this casting material provides a minimum yield stress of 345 MPa, a minimum ultimate tensile strength of 550 MPa, and a minimum elongation in 50 mm of 22% with a reduction of area of 35%. ASTM A958 was selected over ASTM A27 or ASTM A148 because A27 and A148 are general specifications that allow the foundry to make a variety of cast materials that comply with the specification but may not be most suitable for structural use due to other properties

(i.e. weldability or ductility), whereas A958 is a more stringent standard, therefore ensuring a higher quality casting.

It was requested of the foundry that tensile tests be carried out for each heat and that the cast steel have a minimum Charpy V-Notch (CVN) impact test value of 27 J at −20°C. To ensure the casting was sound, it was subject to the following acceptance criteria: visual examination per ASTM A802 (2001) Level I, non-destructive examination per ASTM A903 (2003) Level III, and ultrasonic examination per ASTM A609 (2002) Level 3. Welding between the CHS brace and the cast connector did not require any special provisions for preheating or weld preparation. Welding was carried out in accordance with CSA W59 (2003). CSA (2001a) stipulates that for dynamically loaded connections weld filler material must have a CVN test value of 27 J at −30°C as certified in accordance with CSA W48 (2001b). The quality of the weld was also confirmed using visual examination and ultrasonic testing.

3.5 3-dimensional modelling and finite element stress analysis of the cast steel connector

Once a cast material grade had been selected, the connector was designed first in 2-dimensions and finally in 3-dimensions using SolidWorks 2005. As previously mentioned, both connecting ends of the casting were designed according to CSA requirements for dynamically loaded connections and to ensure elastic behaviour during yielding or buckling of the brace. Transitional geometry between the boundaries was simply created by inserting smooth curves (*splines*) over a transitional length of approximately one tube diameter. As a result, finite element analysis was required to verify that the connector would not yield in the transitional region of the casting during inelastic deformation of the connected brace.

Finite element (FE) analysis was carried out using ANSYS Workbench v. 9.0. Only one-eighth of a full brace assembly was modelled, with symmetry boundary conditions used to emulate the remaining assembly. Solid bodies were meshed with higher order 3-D 10-node tetrahedral solid elements (SOLID187) using mapped face meshing. These elements have quadratic displacement behaviour and are best suited for modelling solid bodies that are curved or have irregular boundaries (Moaveni 1999). Non-linear analysis was carried out by applying incremental displacements to the bearing faces of the bolt holes. In reality, the bolts are pretensioned resulting in load transfer through distributed frictional stresses between the cast tabs and the gusset plate, however, application of displacement in this manner adequately emulated static displacement-control loading of the connection assembly and also produced conservatively large stress concentrations at

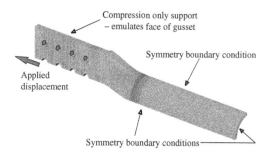

Figure 8. Finite element model used for stress analysis of the cast steel seismic connector.

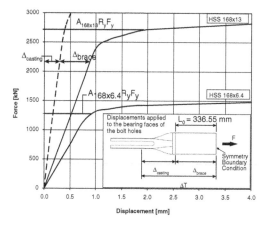

Figure 9. Tensile load-displacement results from FE analysis.

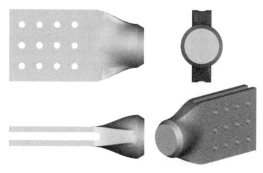

Figure 10. Final connector geometry (bolt holes, connecting nose, and gusset fitting gap to be machined after casting).

Figure 11. One-half of the positive pattern used to form the negative sand mould.

the bolt holes. Non-linear material properties were considered and geometrical non-linearities and shape change during loading were taken into account by allowing large deformations. Reduced integration was used for the formulation of the local stiffness matrix of each element, thereby preventing volumetric mesh locking of the elements within the solid bodies of the model. Figure 8 shows the finite element model of the connector, full penetration groove weld, and brace segment.

Finite element analysis confirmed that when the brace assembly was loaded, inelastic deformations were localized in the brace member up to the probable yield capacity of the brace as shown in Figure 9.

The initial design was forwarded to Canada Alloy Casting Company (CAC), the foundry partner retained for this project, in the form of drawings and a 3-dimensional solid model in stereolithography (*.STL) file format. The drawings indicated surfaces to be machined and stipulated that critical cast dimensions have dimensional tolerance in accordance with ISO 8062 (1994) Grade 8, with all other cast dimensions to have a dimensional tolerance corresponding to ISO 8062 Grade 10. CAC suggested several design

changes to improve the castability of the connector. The changes were adopted, with an increase in connector weight of approximately 18-percent. The final connector geometry is shown in Figure 10. The connectors were cast using the sand casting process. Figure 11 shows one-half of the positive pattern CAC developed to form the sand mould later used for casting.

3.6 Conclusions

Pseudo-dynamic and static testing of tubular brace assemblies, including gusset plates and cast connectors, as well as an investigation on the difference between the stipulated material properties and the casting's actual mechanical properties, are ongoing. This research will confirm that the use of cast steel connectors is a viable, elegant, and cost effective solution for connecting to dynamically loaded HSS braces.

ACKNOWLEDGEMENTS

For the research project on seismic connectors for tubular braces, financial support has been provided by MMO (Materials and Manufacturing Ontario) and NSERC (Natural Sciences and Engineering Research Council of Canada). Canada Alloy Casting Company (CAC) provided steel casting services and Atlas Tube donated HSS material. The authors gratefully acknowledge Walters Inc. (Ontario) for fabrication of test specimens for the tubular brace connector project. The collaboration of Professors C. Ravindran (Ryerson University) and A. McLean (University of Toronto) is much appreciated.

REFERENCES

ASTM. 2000. Standard Specification for Steel Castings, Carbon, and Alloy, with Tensile Requirements, Chemical Requirements Similar to Standard Wrought Grades. *ASTM-A958-00*. West Conshohocken: ASTM International.

ASTM. 2001.Standard Practice for Steel Castings, Surface Acceptance Standards, Visual Examination. *ASTM-A802/ A802M-95*. West Conshohocken: ASTM International.

ASTM. 2002.Standard Practice for Castings, Carbon, Low-Alloy, and Martensitic Stainless Steel, Ultrasonic Examination Thereof. *ASTM-A609/A609M-91*. West Conshohocken: ASTM International.

ASTM. 2003. Standard Specification for Steel Castings, Surface Acceptance Standards, Magnetic Particle and Liquid Penetrant Inspection. *ASTM-A903/A903M-99*. West Conshohocken: ASTM International.

ASTM. 2004. Standard Specification for High-Strength Low-Alloy Columbium-Vanadium Structural Steel. *ASTM-A572/A572M-04*. West Conshohocken: ASTM International.

ASTM. 2005a. Standard Specification for Steel Castings, Carbon, for General Application. *ASTM-A27/A27M-05*. West Conshohocken: ASTM International.

ASTM. 2005b. Standard Specification for Steel Castings, High Strength, for Structural Purposes. *ASTM-A148/ A148M-05*. West Conshohocken: ASTM International.

CEN. 1996. Technical delivery conditions for steel castings for pressure purposes – Steels for use at low temperatures. *EN 10213-3*. Brussels: European Committee for Standardisation.

CSA. 2001a. Limit states design of steel structures, *CAN/ CSA-S16-01*. Toronto: Canadian Standards Association.

CSA. 2001b. Filler Metals and Allied Materials for Metal Arc Welding. *CAN/CSA-W48-01*. Toronto: Canadian Standards Association.

CSA. 2003. Welded Steel Construction (Metal Arc Welding). *CAN/CSA-W59-03*. Toronto: Canadian Standards Association.

Fleischman R.B., Sumer A. & Xuejun L. 2004. Development of modular connections for steel special moment frames. *Proc. 2004 Structures Congress – Building on the Past Securing the Future*: 821–829.

Glijnis P.C. & Crommentuyn J. 2003. To cast or not to cast. *Proc. 10th Intern. Symp. on Tubular Structures*: 129–134.

ISO. 1994. Castings – System of dimensional tolerances and machining allowances. *ISO 8062:1994*. Geneva: International Organization for Standardization.

Lomax K.B. 1982. Forgings and castings. *Materials for the process industries: papers originally prepared for a Conf. held 1982*: 23–34. London: Mechanical Engineering Publications.

Moaveni S. 1999. *Finite Element Analysis: Theory and application with ANSYS*. New Jersey: Prentice Hall.

Tremblay R., Bruneau M., Nakashima M., Prion H.G.L., Filiatrault A. & Devall R. 1996. Seismic design of steel buildings: Lessons from the 1995 Hyogoken-Nanbu earthquake, *Canadian Journal of Civil Engineering* 23: 727–756.

Veselcic M., Herion S. & Puthli R. 2003. Cast steel in tubular bridges – new applications and technologies. *Proc. 10th Intern. Symp. on Tubular Structures*: 135–142.

Tubular Structures XI – Packer & Willibald (eds)
© 2006 Taylor & Francis Group, London, ISBN 0-415-40280-8

Comparison between a cast steel joint and a welded circular hollow section joint

L.W. Tong, H.Z. Zheng & Y.Y. Chen

Department of Building Engineering, Tongji University, P.R. China

ABSTRACT: Experimental comparison was performed between a cast steel joint and a welded circular hollow section joint in tubular truss structures to investigate their behaviour and difference. The cast steel joint was 70 mm thick, whereas the welded circular hollow section joint was fabricated using 30 mm thickness steel plate and stiffeners was set inside the chord member at the joint. The two joints were experimented under the same testing condition. Strain gauges were arranged to measure stress distributions in the joints. Finite element analysis was carried out to get more information on the behaviour of the joints. The experimental and numerical results showed both joints had enough strength to bear design loads. This means welded circular hollow section joints are also alternative approach for large and complicated joints of space trusses, besides cast steel joints.

1 INTRODUCTION

Circular hollow section structures have been widely used in large-span roof trusses in China due to their excellence performance. Design and fabrication of the connections between hollow section members are important tasks for tubular structures. Welded tubular joints are the most commonly used connection type for hollow section structures, on which tubular members are welded together directly by sawing and cutting the end of the tubes. Cast steel joints are another practical alternative, although which is less favorable. The advantage of cast steel joints is more freedom in shaping, which leads to an optimum shape from aesthetic and structural viewpoint. Furthermore, cast steel joints need less welding than welded tubular joints. However, the high cost of cast steel joints is an important disadvantage to hindrance their applications.

The cost of cast steel per weight unit is considerably higher than that of welded joint due to the expense of pattern and core box used in casting process (Lin & Liu, 2005). The design strength of the useable structural cast steel in China, which refers to GS-20Mn5 of German Standard DIN 17182, is 230 MPa. It is lower than design strength of Q345 (from 250 MPa to 315 MPa, depending on the thickness of the plate), which is one of the most commonly used structural alloy steel in China. Lower strength of cast steel leads to heavier self weight of cast steel joint, which would increase the cost of the joint and the whole structure. However, due to the different fabrication process of two kinds of connections, the maximum feasible thickness of a cast steel joint is much larger than that of a welded tubular joint. For many structural engineers, increasing tubular section thickness is the most straightforward and favourite method to ensure the safety of a tubular connection, although which may not be the most economical and effective way. It leads to a common belief in China that cast steel joint is the only choice in some case, especially for complicated connections of space truss members with large diameter. In fact, cast steel joints are adopted in some projects in China just because the feasibility of welded tubular joint are doubted. Investigation on the applicability of welded tubular joint for large-sized and complicated connections and the comparison between the two kinds of joints are desirable.

This paper deals with the behavior and the difference of a cast steel joint and a welded circular hollow section joint. Both joints were multiplanar joints with full size based on a real engineering project Haerbin International Conference and Exhibition Center of China. The two joints were designed so that they could act completely same in the structure. They were tested under the same loading and constrain conditions. Finite element analysis was carried out to get more information on behavior of the both joints.

2 TEST SPECIMENS

Two full scale test specimens were designed and fabricated. One was made of cast steel, whereas the other of welded circular hollow section. They were the same

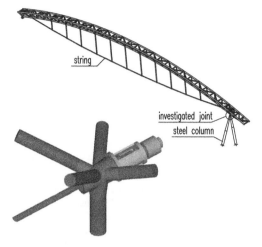

Figure 1. Location of investigated joints.

Table 1. Mechanic property of cast steel.

Yield strength	275 MPa
Tensile strength	450 MPa
Design strength	230 MPa
Modulus of elasticity	206000 MPa
Elongation rate	22%

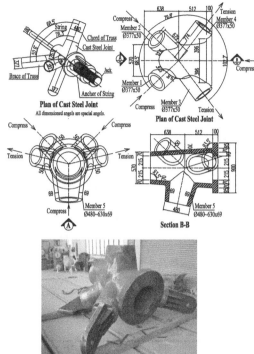

Figure 2. Cast steel joint.

Table 2. Forces on specimens.

Members	1	2	3	4	5	String
Axial force (kN)	−2348	−2348	2305	2305	−1810	6000

joints of a circular hollow section roof truss. Six tubular members intersect at the joint and a string was fastened at the joint, as shown in Figure 1. The diameters of the braces, the axial length of members and the angles between the members were kept the same for both joints, so that they function totally the same in the roof truss.

The cast steel joint was fabricated using cast steel refering to GS-20Mn5 according to German Standard DIN 17182, whose design value of strength conforms closely to Q235, Chinese Code GB50017. The nominal properties of the cast steel are listed in Table 1.

Figure 2 illustrates the details of the cast steel specimen. The maximum nominal thickness of the members was 70 mm. Fillets with radius of 50 mm were introduced between the intersections of the members. The ends of the cast steel joint members 1~5 were supposed to be welded with roof truss members by butt welding on site. The axial length of the joint members was determined by requirement of welding process. The cast steel joint was inspected using non-destructive testing methods to ensure no unacceptable defect. The surface of the cast joint was fairly rough and sand holes could be observed clearly. Thickness and diameters of the sections were a little uneven. The self weight of the cast steel joint was 2140 kg.

Assuming continuous chords and pin-ended braces, structural analysis for whole roof truss was carried out to determine the forces in the members. The structural analysis results showed that members 1, 2, 5 were axial compressed and members 3, 4 were axial tensioned under the most unfavorable design loads combination. The forces applied to the specimen in the experiment were amplified to 1.2 times of the design value to ensure that the joint had enough safety redundancy. The forces were listed in Table 2.

The welded circular hollow section joint was made of rolled tubes, whose material was structural alloy steel Q345B. Its design strength is 295 MPa according to Chinese Code GB50017. The nominal thickness of the chord was 30 mm, which was the reasonable maximum thickness for a rolled circular tube. Two internal ring stiffeners and eight plate stiffeners were introduced inside the chord to reinforce the joint. The welded circular joint was fabricated in accordance with Chinese Code JGJ 81-2002. Figure 3 gives the detail of the welded circular hollow section joint. The self weight of welded joint was 1290 kg. Loads applied on

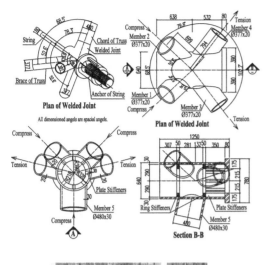

Plan of Welded Joint

All dimensioned angels are spacial angels.

Plan of Welded Joint

Section B-B

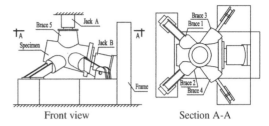

Front view   Section A-A

Figure 4. Test set-up.

Figure 5. Test scene.

Figure 3. Welded circular hollow section joint.

the welded joint were the same as those of cast steel joint, as listed in Table 2.

The total prices of welded joint and cast steel joint were 9030 RMB and 25680 RMB respectively. The expense of the welded joint was only 0.35 times of the cast steel joint. Two factors make the welded joint cost considerably lower than the cast steel joint. (1) Higher design strength of structural steel Q345B decreased self weight of the welded joint. Even the advantage of shaping of the cast steel could not compensate this factor by far. The self weight of the welded steel joint was 0.60 times of the cast steel joint. (2) The cost for per unit weight of material of welded joint is lower than that of the cast steel joint. The unit price for the cast steel was about 12000 RMB per ton in China. The unit price for the structural steel Q345B was about 7000 RMB per ton, including material, fabrication and transportation, which was only 0.58 times of the cast steel. Although welded joint needs more welding cost than cast steel joint, it is a more economical choice in most case (Pieter & Jooster, 2003).

3 TEST SET-UP

Figure 4 shows the test set-up for the both specimens. As a self-balanced system, a special frame was designed and fabricated in order to apply the loads to the joints. Member 5 was set upward and the axial load was applied at the end of the member by Jack A to simulate the reaction force sustained by the member 5 in the roof truss. The tension force caused by the string was simulated by Jack B. The contacting area between Jack B and the end-plate of the chord was equal to that between the anchor of the string used in the project and the end-plate of the chord. Members 1~4 were prolonged to the surface of the frame by plate bars. The ends of member 1 and member 2 were blocked by plates welded on the frame after the specimen was installed on the frame to bear the horizontal shearing force. The ends of member 3 and member 4 were connected with the frame by pins respectively. Jack A and B were loaded synchronized in 10 steps and unloaded synchronized in 3 steps.

In order to get the strain distribution in the cast steel joint, sixty strain rosettes (designated as T) were installed on the surface of the specimen. Most of them were located near the intersections between members, where the maximum stress was expected to occur. Fifteen one-way strain gauges (designated as S), 3 for each member, were used to measure the axial forces in members 1~5.

As for the test of welded circular hollow section joint, the arrangement of strain gauges was nearly similar to the cast steel joint, but slightly modified.

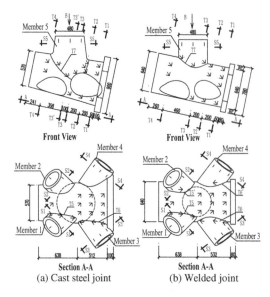

Front View   **Front View**

Section A-A   **Section A-A**

(a) Cast steel joint   (b) Welded joint

Figure 6.   Arrangement of strain gauges.

Sixty-four strain gauge rosettes (designated as T) and fifteen one-way strain gauges were placed on the surface of the welded joint. The layout of strain gauges for both joints is shown in Figure 6.

4   TEST RESULTS AND DISCUSS

No failure or crack was observed in both joints. The strength theory based on Von-Mises stress (VMS) was used as a criterion for judging whether the cast steel and the structural alloy steel would become yielding under the loading condition. In order to get VMS in the specimens, the stress components perpendicular to the surface of the tubular wall were not taken into account due to the difficulty of measurement, although these stress components might not be ignored because of the large thickness of the sections.

For both specimens, the real axial forces in members 1~5 measured through one-way strain gauges were in good agreement with those listed in Table 2. It means that loads were applied as expected. The measured strain along the cross sections T1~T4 in both specimens were fairly symmetrical about the axis of the chords, which conformed to the two specimens having geometrical symmetry with the chord axis.

Figure 7 illustrates the VMS distribution in the cast joint. Except one point in cross section T5, VMS on all the measured points didn't exceed the design strength of the cast steel (230 MPa). Although VMS (343 MPa) for one point in cross section T5 was higher than nominal yield strength (275 MPa), the load-strain curve based on the gauge rosettes at the point was in the linear condition and no residential deformation existed

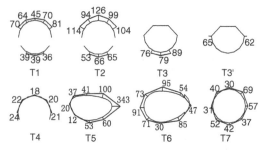

Figure 7.   VMS distributions in the cast steel joint.

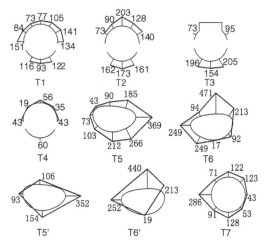

Figure 8.   VMS distributions in the welded joint.

after the test completed. It can be concluded that the point didn't yield.

Figure 8 illustrates the VMS distribution in the welded joint. VMS didn't exceed the design strength of Q345 (290 MPa) except 4 points, whose VMS were 369 MPa (T5), 471 MPa (T6), 352 MPa (T5′) and 440 MPa (T6′) respectively. The four points were located on the brace members 1~4 near the intersection between these members and the chord member.

The above test results show that both the cast steel joint and welded hollow section joint have enough load carrying capacity. The VMS in the welded circular hollow section joint was obviously higher than that in the cast steel joint. This does not mean that the cast steel joint is more safe and reliable than the welded joint, since the cast steel joint was thicker and consumed more steel than the welded circular hollow section joint. If the section thickness of members 1~4 of the welded joint were increased, for example, from 20 mm to 30 mm, its safety and reliability could be improved greatly while the self-weight of the joint increased only 150 kg.

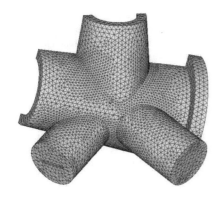

Figure 9. FE model for cast steel joint.

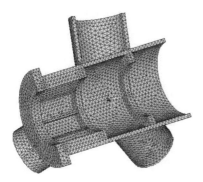

Figure 10. FE model for welded joint.

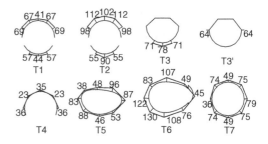

Figure 11. Calculated VMS distributions in cast steel joint.

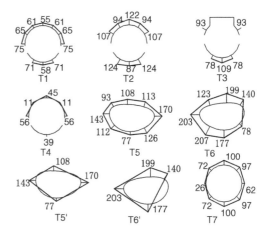

Figure 12. Calculated VMS distributions in welded joint.

5 FINITE ELEMENT ANALYSIS

Finite element (FE) analysis was carried out to know more detail about the stress distributions in both specimens. The stress component along the thickness of the tubular wall can not be measured and then considered in measured VMS. FE analysis is able to check how the stress component is in quantity.

Ten-node triangular pyramid elements of the software package, ANSYS, were adopted to model half of the specimens. The loads and constrains were applied according to the actual condition. Convergence analysis was conducted to find out suitable element size. Figures 9 and 10 present the FE models for the both joints.

Figures 11 and 12 illustrate the calculated VMS distributions in which the stress components perpendicular to the surface of the tubular wall were considered. Comparisons between calculated and experimental measured VMS for the both joints was shown in Figure 13. Both the FE analysis and the comparison with experimental data indicate that:

(1) The difference between calculated Von Mises stresses considering the stress component along

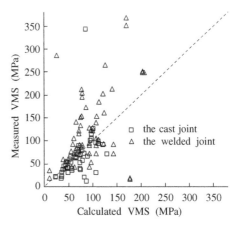

Figure 13. Comparison between calculated VMS and experimental VMS.

thickness of the tubular wall and not considering that component was less than 5% for both kinds of joints. It means that the stress component along thickness of the tubular wall can be neglected.

535

(2) The calculated VMS was in generally acceptable agreement with the measured VMS. Behaviour of complicated tubular truss joints can be understood well by means of FE analysis.

6 CONCLUSIONS

Both the experimental investigation and the FE analysis of two kinds of joints, the cast joint and the welded circular hollow section joint, were presented. Full sized specimens were used and tested under the same loading and constrain conditions. The following conclusions are drawn:

(1) Both the cast steel joint and the welded circular hollow section joint have sufficient load-carrying ability.
(2) It is reliable to design complicated multiplanar tubular joints by means FE analysis.
(3) It is really a good way to use cast steel joints in the complicated connections between chord and brace members with large diameters in space tubular trusses, where the joints are expected to have high loading capacity. However, the kind of joints is much more expensive as it consumes more steel material with lower strength and is fabricated using special casting process.

(4) The welded hollow section joint is also suitable for the complicated connections of the space tubular truss members with large diameters if it is designed and welded reasonably, and it has lower cost. However, some additional stiffeners must be arranged inside the kind of joint to reinforce the joint as the plate thickness of the joint is much less than that of the cast steel joint.
(5) The materials of the two joints were different. The welded joint is overloaded, whereas the cast joint is designed conservatively. So the conclusions are only valid for the Chinese market using Chinese standards.

REFERENCES

Pieter C. Glijnis, Jooster Cromentuyn, 2003, To Cast or not to Cast, Proceeding of the 10th International Symposium on Tubular Structures, Madrid, Spain, pp. 129–134.

Lin Yan, Liu Xiliang, 2005, Design of Cast Steel Joint and Its Project Use, Industrial Construction (in Chinese), Vol. 38: pp. 20–30.

Technical Specification for Welding of steel structure of building, Chinese Code JGJ 81-2002.

Code for Design of Steel Structures, Chinese Code GB50017-2003.

Parallel session 9: Static strength of joints

Tubular Structures XI – Packer & Willibald (eds)
© 2006 Taylor & Francis Group, London, ISBN 0-415-40280-8

Effect of chord loads on the strength of RHS uniplanar gap K-joints

D.K. Liu & J. Wardenier

Delft University of Technology, Delft, The Netherlands

ABSTRACT: In current design codes, the equations accounting for the effects of chord stress in circular (CHS) and rectangular (RHS) hollow section joints are inconsistent. For CHS joints, the chord stress function is based on the chord pre-stress while for RHS joints the maximum chord stress is used. In the framework of a previous CIDECT programme 5BK, the chord load effects for CHS joints were analyzed in order to establish a chord stress function based on the maximum chord stress, being consistent for both CHS and RHS joints. The second objective of the CIDECT programme, was to re-analyze existing chord stress functions for CHS and RHS joints in order to establish a general form of chord stress function which can be applied for different CHS and RHS joints. In a new CIDECT programme the existing chord stress functions for RHS gap K-joints are analyzed. In this paper, the numerical results of this study for the chord load effects for different RHS gap K-joints are presented and compared with chord stress functions of the CIDECT Design recommendations.

1 INTRODUCTION

In current design rules, insufficient emphasis is put on the consistency of various design equations. For circular hollow section (CHS) joints, the external chord "pre-load" (i.e. the additional load in the chord which is not necessary to resist the horizontal components of the brace forces) is used to account for the effects of chord loading. However, for rectangular hollow section (RHS) joints, the chord stress formulation is based on the maximum chord stress i.e. the stresses as a result of axial forces and (where applicable) bending moments.

To a designer, it is confusing that different approaches should be used for different categories of joints, which may lead to misinterpretations and errors. Hence, in the framework of a CIDECT (Comité International pour le Développement et l'Étude de la Construction Tubulaire) programme, it was proposed to re-analyze the effects of chord stress on the ultimate strength of tubular joints in order to establish a chord stress formulation as a function of the maximum chord stress, consistent for CHS and RHS joints. The various available chord stress functions for different CHS and RHS joints are mostly based on individual experimental results for individual joint configurations. A second objective of the CIDECT programs is to establish a general form of the chord stress function which can be applied for different CHS and RHS joints.

In the first phase of the first CIDECT programme 5BK (Van der Vegte et al., 2001), the influence of chord stress on the axial strength of CHS X-joints was re-analyzed. The ultimate loads and existing experimental data for X-joints were then compared with the existing chord stress functions (Togo, 1967, Wardenier et al., 1991, IIW, 1989, Kurobane et al., 1984, API RP2A, 1993, Boone et al., 1982 and Weinstein & Yura, 1985).

In the second phase of the programme (Van der Vegte et al., 2002), additional data on axially loaded uniplanar gap K-joints subjected to chord pre-load were generated, using numerical methods. Similar to the evaluation of the results for X-joints, the numerical results of uniplanar K-joints were compared with the chord pre-stress functions adopted by (i) CIDECT Design Guide No. 1 (Wardenier et al., 1991), (ii) Pecknold et al. (2001) and (iii) ISO Draft (Dier & Lalanic, 1998).

A general form of chord stress function, indicated in Eq. 1, was established for CHS joints using the maximum chord stress ratio n as primary variable in line with the current chord stress functions adopted for RHS joints. The accuracy of the proposed chord stress functions is assessed by comparing the predictions with the available experimental results and FE data. (Van der Vegte et al., 2003, 2005a, 2005b).

$$f(n) = (1.0 - |n|^A)^{(B + C\beta + F\beta^2)} \qquad (1)$$

Liu et al. (2002) presented an overview of all available chord stress functions for CHS and RHS joints. A general form of chord stress function was generated based on available chord stress functions for RHS joints (Liu et al., 2004).

It is found that additional data of chord stress effects for RHS gap K-joints was required to make a general form of chord stress function. In this new CIDECT program a finite element (FE) parametric study has been initiated in order to generate the data. Nine combinations of brace width to chord width ratios β (varying from 0.4 to 0.8) and chord width to thickness ratio 2γ ($15 \leq 2\gamma \leq 35$) were analyzed numerically for different values of the chord pre-stress ratio.

In this paper, the numerical results of chord stress effects for RHS gap K-joints are presented and compared with the existing function in the CIDECT Design Guide No. 3 (Packer et al., 1992).

2 JOINT AND CHORD LOADS

2.1 Configurations of the joint

The joint configuration is shown in Figure 1. To investigate the possible interaction between chord loads and the chord width to chord thickness ratio 2γ as well as the width ratio β different joints listed in Table 1 are numerically studied for various chord loads.

The configuration of the joint with $\theta = 45°$ is shown in Figure 1. The nominal dimensions are as follows: $b_0 = 150$ mm, $t_1 = t_2 = 4.3$ mm, t_0 and b_i are determined for different values of β and $\tilde{\gamma}$.

2.2 Boundary conditions

The load and boundary conditions applied at the brace and chord ends are intended to reflect the boundary conditions a joint encounters within a girder. Various load and boundary conditions have been applied in tests and/or numerical calculations of hollow section K-joints. Based on the results of Liu & Wardenier (1998a, b) the boundary conditions indicated in Figure 2 are used in the analyses.

2.3 Finite element modelling

Due to symmetry of the joint only half of the joint was modelled. Based on the experience from massive

Table 1. Joint dimensions (chord dimension: $150 \times 150 \times t_0$).

No.	θ_i	Geometrical parameters β	2γ	Gap & eccentricity (mm) g	e
1	45°	0.4	15	100	17.40
2	45°	0.4	25	100	17.40
3	45°	0.4	30	100	17.40
4	45°	0.6	15	57.4	17.40
5	45°	0.6	25	57.4	17.40
6	45°	0.6	35	57.4	17.40
7	45°	0.8	15	15	17.40
8	45°	0.8	25	15	17.40
9	45°	0.8	35	15	17.40

previous numerical analyses eight noded isoparametric doubly curved thick shell elements (MARC element type 22) were used to model the chord, the braces as well as the welds.

A typical finite element mesh (for $\theta = 45°$) used in the analyses is shown in Figure 3. Both geometrical and material nonlinearity, simulated by true stress-strain relationship, was used throughout the study. It is noted that the steel grade S355 with a design value $f_y = 355$ N/mm^2 is used for the chord, the braces and the welds. However, for all cases of $2\gamma = 15$ and for $\beta = 0.8$ with $2\gamma = 25$ a steel grade with a higher yield stress is used for the brace in the analyses to prevent brace buckling or full yielding.

Thick end plates were included in the model at the chord and brace ends to apply the loads. Corner radii were modelled for both the chord and the braces. The mesh has been calibrated by comparing numerical results with the experimental results for uniplanar K-joints (Wardenier & Stark 1978).

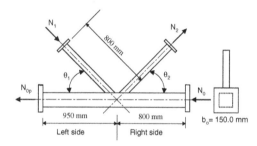

Figure 1. Joint configurations.

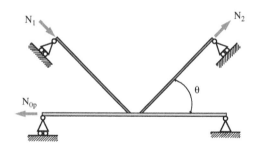

Figure 2. Boundary conditions.

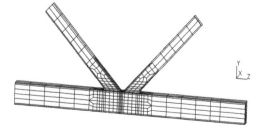

Figure 3. Finite element modelling.

3 RESULTS AND DISCUSSIONS

3.1 Numerical results

Nine combinations of β (varying from 0.4 to 0.8) and 2γ ($15 \leq 2\gamma \leq 35$) were analyzed numerically for different values of the chord stress ratio. The numerical results are given in Tables 2 to 4 for different joint configurations. The numerical results are not analyzed yet due to the time limitation for presenting this paper. Only the results related to the chord face yielding failure mode are indicated. Sometimes the joint fails in other modes.

In the tables, N_0/A_0f_{y0}(C) represents the non-dimensional chord stress ratio at the right side of the chord, while N_0/A_0f_{y0}(T) represents the non-dimensional chord stress ratio at the left side of the chord, see Figure 1. $N_1/N_{1,0}$ represents the ratio where N_1 is the joint capacity while $N_{1,0}$ is the joint capacity in case of a chord stress ratio N_0/A_0f_{y0} which is close to zero.

Totally about four different chord failure modes were found in the FE analysis:

- Chord face yielding (CFY),
- Chord face yielding plus chord side wall yielding (CFY + side wall),
- Chord section failure (CSF),
- Chord failure (CF).

Chord face yielding is defined as yielding of the chord face with yield lines.

Chord face yielding plus chord side wall yielding indicates a failure mode where yielding is not only observed at the chord face but also at the chord side wall.

Chord section failure means that the chord section is yielding due to the interaction of the brace axial loading, the chord stress and the secondary bending moments.

Chord failure means chord member failure for complete yielding (left side or right side of the chord) due to a high chord load.

The numerical results related to the chord failure modes CSF and CF are member failures and will be analysed separately and be initially excluded from the analysis.

Based on the numerical results the relationship between the chord stress ratio and the joint capacity is plotted in Figure 4 for joints with $\beta = 0.4$, in Figure 5 for joints with $\beta = 0.6$ and in Figure 6 for joints with $\beta = 0.8$. It is noted that for joint with $\beta = 0.8$ the relationship between the chord stress ratio and the joint capacity is plotted only for the case of $2\gamma = 35$. The chord stress functions recommended in the CIDECT Design Guide No. 3 (Packer et al., 1992) are also plotted in the figures.

For information, the joint capacities are also plotted against to the indentations of the chord face for different chord loads for joints with $\beta = 0.4$ and $2\gamma = 15$ in Figure 7 and for joints with $\beta = 0.4$ and $2\gamma = 30$ in Figure 8, respectively.

Table 2a. Numerical results for joint with $\beta = 0.4$ and $2\gamma = 15$.

N_0/A_0f_{y0} (C)	N_0/A_0f_{y0} (T)	$N_1/N_{1,0}$	Max (3%)	Failure mode
−0.89		0.51	max	**CFY + CSF**
−0.80		0.80	max	CFY
−0.59		0.94	3%	CFY
−0.49		0.97	3%	CFY
−0.26		0.99	3%	CFY
−0.08	0.18	1.00	3%	CFY
0.00	0.26	1.00	3%	CFY
0.11	0.37	1.00	3%	CFY
0.46	0.70	0.96	3%	CFY + side wall
0.69	**0.90**	0.83	3%	**CFY + CSF**
0.75	**0.93**	0.74	3%	**CFY + CSF**

Table 2b. Numerical results for joint with $\beta = 0.4$ and $2\gamma = 25$.

N_0/A_0f_{y0} (C)	N_0/A_0f_{y0} (T)	$N_1/N_{1,0}$	Max (3%)	Failure mode
−0.93		0.42	max	**CSF**
−0.88		0.50	max	CFY + side wall
−0.74		0.69	max	CFY + side wall
−0.66		0.78	3%	CFY
−0.53		0.86	3%	CFY
−0.36		0.93	3%	CFY
−0.26		0.95	3%	CFY
−0.16		0.97	3%	CFY
0.00	0.16	1.00	3%	CFY
0.30	0.46	1.02	3%	CFY
0.71	0.85	0.95	3%	CFY + side wall
0.85	**0.96**	0.72	3%	**CSF**

Table 2c. Numerical results for joint with $\beta = 0.4$ and $2\gamma = 30$.

N_0/A_0f_{y0} (C)	N_0/A_0f_{y0} (T)	$N_1/N_{1,0}$	Max (3%)	Failure mode
−0.90		0.37	max	**CSF**
−0.75		0.61	max	CFY + side wall
−0.64		0.73	3%/max	CFY + side wall
−0.44		0.87	3%	CFY
−0.30		0.94	3%	CFY
−0.13		0.99	3%	CFY
−0.04	0.09	1.00	3%	CFY
0.15	0.28	1.03	3%	CFY
0.25	0.38	1.04	3%	CFY
0.46	0.58	1.06	3%	CFY
0.63	0.76	1.04	3%	CFY + side wall
0.88	**0.96**	0.66	max	**CSF**

Table 3a. Numerical results for joint with $\beta = 0.6$ and $2\gamma = 15$.

N_0/A_0f_{y0} (C)	N_0/A_0f_{y0} (T)	$N_1/N_{1,0}$	Max (3%)	Failure mode
−0.91	−0.61	0.61	max	**CSF**
−0.86	−0.33	0.75	max	**CSF**
−0.76	−0.31	0.93	3%	CFY + side wall
−0.63	−0.16	0.97	3%	CFY + side wall
−0.48	0.00	0.99	3%	CFY + side wall
−0.31	0.17	1.00	3%	CFY + side wall
−0.15	0.34	1.01	3%	CFY + side wall
0.02	0.50	1.00	3%	CFY + side wall
0.19	0.65	0.98	3%	CFY + side wall
0.46	**0.87**	0.88	3%	**CSF**
0.62	**0.96**	0.72	3%	**CF**

Table 3b. Numerical results for joint with $\beta = 0.6$ and $2\gamma = 25$.

N_0/A_0f_{y0} (C)	N_0/A_0f_{y0} (T)	$N_1/N_{1,0}$	Max (3%)	Failure mode
−0.93	−0.78	0.45	max	**CSF**
−0.90	−0.72	0.55	max	**CSF**
−0.84	−0.62	0.71	max	**CSF**
−0.65	−0.38	0.86	3%	CFY + side wall
−0.51	−0.20	0.92	3%	CFY + side wall
−0.30	0.00	0.97	3%	CFY + side wall
−0.20	0.11	0.98	3%	CFY
−0.09	0.22	0.99	3%	CFY
0.13	0.44	1.01	3%	CFY
0.34	0.65	1.00	3%	CFY + side wall
0.54	0.83	0.95	3%	CFY + side wall
0.74	**0.96**	0.74	max	**CF**

Table 3c. Numerical results for joint with $\beta = 0.6$ and $2\gamma = 35$.

N_0/A_0f_{y0} (C)	N_0/A_0f_{y0} (T)	$N_1/N_{1,0}$	Max (3%)	Failure mode
−0.84	−0.71	0.54	max	**CSF**
−0.76	−0.47	0.61	max	CFY + side wall
−0.64	−0.47	0.70	max	CFY + side wall
−0.48	−0.27	0.83	3%	CFY
−0.36	−0.15	0.89	3%	CFY
−0.23	0.00	0.94	3%	CFY
−0.07	0.16	0.99	3%	CFY
−0.02	0.22	1.00	3%	CFY
0.10	0.34	1.02	3%	CFY
0.27	0.52	1.05	3%	CFY
0.46	0.71	1.06	3%	CFY + side wall
0.73	**0.96**	0.96	3%	**CSF**

Table 4a. Numerical results for joint with $\beta = 0.8$ and $2\gamma = 15$.

N_0/A_0f_{y0} (C)	N_0/A_0f_{y0} (T)	$N_1/N_{1,0}$	Max (3%)	Failure mode
−0.97		0.65	max	**CSF + CF**
−0.87	0.07	1.07	3%/max	**CSF**
−0.82	0.13	1.09	3%/max	**CSF**
−0.75	0.20	1.09	max	CFY + side wall
−0.61	0.33	1.08	max	CFY + side wall
−0.43	0.65	1.07	max	CFY + side wall
0.02	0.88	1.00	max	**CSF**
0.06	0.91	0.99	max	**CSF**
0.28	0.96	0.79	max	**CSF + CF**

Table 4b. Numerical results for joint with $\beta = 0.8$ and $2\gamma = 25$.

N_0/A_0f_{y0} (C)	N_0/A_0f_{y0} (T)	$N_1/N_{1,0}$	Max (3%)	Failure mode
−0.96		0.67	max	**CSF + CF**
−0.94		0.82	max	**CSF**
−0.84		0.99	3%	**CSF**
−0.72	0.15	1.03	3%	CFY + side wall
−0.58	0.31	1.05	3%	CFY + side wall
−0.26	0.61	1.04	3%	CFY + side wall
0.01	0.82	1.00	3%	**CSF**
0.28	0.97	0.82	special	**CF**

Table 4c. Numerical results for joint with $\beta = 0.8$ and $2\gamma = 35$.

N_0/A_0f_{y0} (C)	N_0/A_0f_{y0} (T)	$N_1/N_{1,0}$	Max (3%)	Failure mode
−0.94		0.70	max	CFY + side wall
−0.89		0.81	max	CFY + side wall
−0.80		0.90	max	CFY + side wall
−0.63		0.97	max	CFY + side wall
−0.44	0.23	1.00	3%	CFY + side wall
−0.20	0.46	1.01	3%	CFY + side wall
0.04	0.68	1.00	3%	CFY + side wall
0.25	0.87	0.96	3%	CFY + side wall
0.42	0.96	0.85	max	**CF**

close to a parabolic curve suggested by Van der Vegte et al. (Van der Vegte et al., 2003, 2005).

Figures 7 and 8 show that due to the eccentricity and deformations secondary bending moments have been introduced. The numerical results are definitely affected by these secondary bending moments. That is also the reason for the failure mode CSF.

For joints with $\beta = 0.4$ the chord stress function recommended in the CIDECT Design Guide No. 3 (Packer et al., 1992) and based on (Wardenier, 1982) gives a good lower bound except for high chord stress ratios $|N_0/A_0f_{y0}| > 0.9$), where the CIDECT recommendations are slightly optimistic. The numerical

3.2 Observations and discussions

Figures 4,5 and 6 show that the relationships of chord stress effects against the chord stress ratio are very

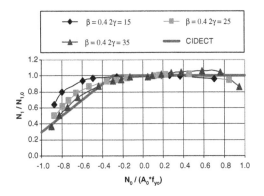

Figure 4. Chord stress effects for joint with $\beta = 0.4$.

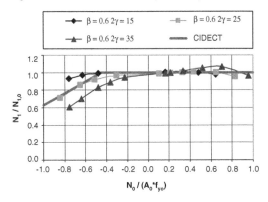

Figure 5. Chord stress effects for joint with $\beta = 0.6$.

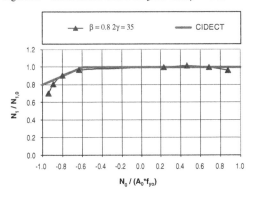

Figure 6. Chord stress effects for joint with $\beta = 0.8$.

results for these high chord stress ratios will be further analyzed in detail.

For joint with $\beta = 0.6$ the chord stress function recommended by the CIDECT Design Guide agrees very well with the numerical results for $2\gamma = 25$ but for $2\gamma = 35$ it is too optimistic.

For joint with $\beta = 0.8$ and $2\gamma = 35$ the chord stress function recommended by the CIDECT Design Guide agrees well with the numerical results except for high

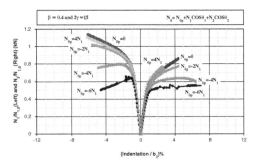

Figure 7. Load deformation plots for joints with $\beta = 0.4$ and $2\gamma = 15$.

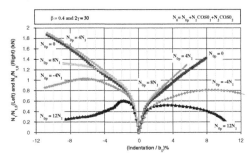

Figure 8. Load deformation plots for joints with $\beta = 0.4$ and $2\gamma = 30$.

chord stress ratios $(-0.9 > N_0/A_0f_{y0} > 0.9)$, where the CIDECT chord stress function is too optimistic.

For high tensile chord loads a reduction in the joint capacity exists but this is smaller than that observed in T, Y and X joints (Liu et al., 2002 and 2004).

4 CONCLUSIONS

From this numerical research the following conclusions can be drawn:

- For $\beta = 0.4$ the recommendation in the CIDECT Design Guide No. 3 (Packer et al., 1992) and based on (Wardenier, 1982) gives a lower bound.
- For $\beta = 0.6$ the numerical results for $2\gamma \leq 25$ are in line or higher than those based on the recommendation in the CIDECT Design Guide No 3. For $2\gamma > 25$ they are considerable lower.
- For $\beta = 0.8$ and $2\gamma = 35$ the numerical results are in lined with those based on the recommendation in the CIDECT Design Guide No. 3 except for high chord stress ratios. The results which showed failure of the chord due to the interaction of the brace axial load, the chord load and the secondary bending moments (CSF) have to be analyzed more in detail.

The results of this study will be used for the determination of a chord stress function for K gap-joints of rectangular hollow sections (RHS).

The results show that this function can have a similar form as that proposed by van der Vegte et al. for CHS joints (Van der Vegte et al., 2003, 2005).

REFERENCES

American Petroleum Institute (1993). Recommended Practice for Planning Designing and Constructing Fixed Offshore Platforms – Working Stress Design, API RP 2A-WSD, 20th Edition.

Boone T.J., Yura J.A. and Hoadley P.W. (1982). Chord stress effects on the ultimate strength of tubular connections, PMFSEL Report 82-1, University of Texas, Austin.

Dier A.F. and Lalani, M. (1998). New code formulation for tubular joint static strength, *Proc. 8th International Symposium on Tubular Structures*, Singapore.

IIW (International Institute of Welding Subcommission XV-E) (1989). Design Recommendations for Hollow Section Joints – Predominantly Statically Loaded, 2nd Edition. IIW Doc. XV-701-89, International Institute of Welding, Annual Assembly, Helsinki, Finland.

Koning C.H.M. de and Wardenier J. (1979). Tests on welded joints in complete girders made of square hollow sections, *Stevin Report* 6-79-4.

Kurobane Y., Makino Y. and Ocji K. (1984). Ultimate resistance of unstiffened tubular joints, *Journal of Structural Engineering*, Vol 110, No. 2.

Liu D.K. and Wardenier J. (1998a). Effect of axial chord force in chord members on the strength of RHS uniplanar gap K-joints, *Proc. 5th Pacific Structural Steel Conference*, Seoul, Korea.

Liu D.K. and Wardenier J. (1998b). Effect of boundary conditions and chord load on the strength of RHS multiplanar gap KK- joints, *Proc. 8th International Symposium on Tubular Structures*, Singapore.

Liu D.K., Wardenier J. and Vegte G.J. van der (2002). Survey of chord load functions for hollow section joints, CIDECT Report 5BK-4/02.

Liu D.K., Wardenier J. and Vegte G.J. van der (2004). New chord stress functions for rectangular hollow section joints, *14th International Offshore and Polar Engineering Conference*, Toulon, France.

Lu L.H., de Winkel G.D., Yu Y. and Wardenier, J. (1994). Ultimate deformation limit for tubular joints, Tubular Structures VI, *Proc. of the Sixth International Symposium on Tubular Structures*, Melbourne, Australia.

Packer J.A., Wardenier J., Kurobane Y., Dutta D. and Yeomans N. (1992). *Design Guide for Rectangular Hollow Section (RHS) Joints under Predominantly Static Loading*, Published by Verlag TUV Rheinland GmbH, Köln, Germany, ISBN 3-8249-0089-0.

Pecknold D.A., Park J.B. and Koppenhoefer K.C. (2001). Ultimate strength of gap K tubular joints with chord preloads, *Proc. 20th International Conference on Offshore Mechanics and Arctic Engineering*, Rio de Janeiro, Brazil.

Togo T. (1967). Experimental study on mechanical behavior of tubular joints, Osaka University, (in Japanese).

Vegte G.J. van der, Makino Y., Choo Y.S. and Wardenier J. (2001). The influence of chord stress on the ultimate strength of axially loaded uniplanar X-joints, *Proc. 9th International Symposium on Tubular Structures*, Düsseldorf, Germany, pp.165–174.

Vegte G.J. van der, Makino Y. and Wardenier J. (2002). The effect of chord pre-load on the static strength of uniplanar tubular K-joints, *Proc. 12th International Offshore and Polar Engineering Conference*, Kitakyushu, Japan, Vol. IV, pp. 1–10.

Vegte G.J. van der, Liu D.K., Makino Y. and Wardenier J. (2003). New chord load functions for circular hollow section joints, CIDECT Report 5BK-4/03, Stevin Report 6.03.1, Faculty of Civil Engineering, Delft University of Technology.

Vegte G.J. van der and Makino Y. (2005a). New strength formulation for tubular K-joint with gap, [Oral presentation].*15th International Offshore and Polar Engineering Conference*, Seoul, Korea.

Vegte G.J. van der and Makino Y. (2005b). Ultimate strength formulation for axially loaded CHS uniplanar T-joints, *Proc. 15th International Offshore and Polar Engineering Conference*, Seoul, Korea.

Wardenier J. and Stark J.W.B. (1978). The static strength of welded lattice girder joints in structural hollow sections, Stevin Report 6-78-4. Faculty of Civil Engineering, Delft University of Technology.

Wardenier J. (1982). *Hollow Section Joints*. Delft University Press, The Netherlands.

Wardenier J., Kurobane Y., Packer J.A., Dutta D. and Yeomans N. (1991). *Design Guide for Circular Hollow Section (CHS) Joints under Predominantly Static Loading*, Published by Verlag TUV Rheinland GmbH, Köln, Germany, ISBN 3-88585-975-0.

Weinstein R. and Yura J.A. (1985). The effect of chord stress on the static strength of DT tubular connections, PMFSEL Report 85-1, University of Texas, Austin.

NOMENCLATURE

CHS	circular hollow sections
RHS	rectangular hollow sections
A_0	cross sectional area of the chord
N_i	axial brace load (i = 1 or 2)
$N_{1,0}$	axial load capacity of the compression brace (1) corresponding to nearly zero compressive stress in the chord
N_0	axial chord load
N_{0p}	external axial chord load
b_0	external chord width
n	$N_0/(A_0f_{y0})$
e	eccentricity
f_{y0}	yield stress of the chord
g	gap size
n	non-dimensional maximum chord pre-stress ratio
t_i	wall thickness of the brace i(i = 1 or 2)
t_0	wall thickness of the chord
β	width ratio between braces and chord
γ	half width to thickness ratio of the chord, $\gamma = b_0/2t_0$
θ_i	angle between a brace member i (i = 1, 2) and the chord

Tubular Structures XI – Packer & Willibald (eds)
© 2006 Taylor & Francis Group, London, ISBN 0-415-40280-8

A steel tubular column-shearhead system for concrete flat plates

A.H. Orton
Corus Tubes, UK

ABSTRACT: The use of steel tubular columns with concrete flat plates appears to offer significant practical benefits. At present the use of thinner, more economic slabs and the provision of service holes adjacent to the column are hampered by the problem of shear transfer at column heads and, in seismic areas, the use of flat plates is not favoured at all, because of the limited strength and ductility of the slab to column joint. This paper proposes a new structural arrangement to address this issue, using steel tubes. The advantages of the proposed system, are discussed. Details of the proposed column shearhead device are presented, as well as a discussion of the factors involved in its design. A programme of testing of such a device for use in seismic areas is outlined.

1 INTRODUCTION

1.1 *Economic background*

Concrete construction is established worldwide and in many countries, for reasons of economy, it is the principal material of construction. However composite construction, here meaning the use of steel sections with concrete, still appears relatively undeveloped, even in those countries that already make substantial use of steel and concrete separately. In spite of the high unit cost of steelwork, as it may appear in some countries, the use of steel in concrete elements will often result in major improvements in overall costs and performance.

Steel hollow structural sections (HSS) acting compositely with concrete are frequently used as columns but have substantial benefits, too, as beams. In the application proposed in this paper, concrete filled tubes (CFT) are used as columns but also as shearheads, which are embedded within the depth of a concrete flat plate floor. The advantages of this combination of materials apply in both seismic and non-seismic areas but particularly so in seismic areas, because the CFT column has excellent ductility and energy absorption. There is a pressing need to develop structural solutions for seismic areas that are effective in resisting the forces but economic in use. In those countries that already make extensive use of concrete, this means that in order to be of practical importance any new solutions should only involve modest changes to existing construction practices.

1.2 *Concrete columns and flat plate floors*

Concrete flat plate floors are popular on account of the simple formwork required and the minimum overall

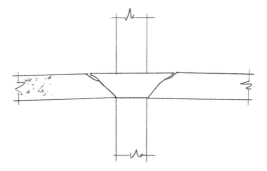

Figure 1. Schematic section through reinforced concrete flat plate indicating deformation and crack pattern at point of failure due to shear under vertical load.

floor depths that can be achieved. However, arguably, the full benefits of this form of construction, that is flat soffits, thin slabs, somewhat safer failure modes and fast construction procedures have not yet been fully achieved. This can be attributed to the critical influence of shear and bending at the column heads, which has necessitated the use of downstand beams, drop panels, shear reinforcement or thicker slabs than would otherwise have been necessary (Figure 1). In seismic areas, the use of flat plates is virtually excluded at present, because of the problem of shear transfer at the slab-column joints (Figure 2).

Concrete wide beam-column frames have been proposed for use in seismic areas, with certain restrictions (Gentry and Wight 1994) but require high levels of reinforcement and good site control to ensure the necessary ductility at joints. Also, columns in seismic areas will require ductility, to cope with high local

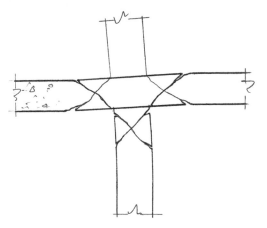

Figure 2. Schematic section through reinforced concrete flat plate structure showing possible deformation and crack patterns at failure due to shear under seismic loading.

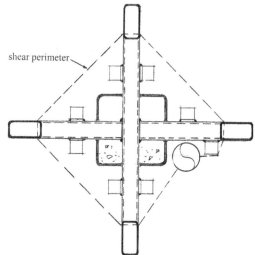

Figure 3. Plan on a typical square HSS column and HSS shearhead showing shelf angles on shearhead arms and the assumed design concrete shear perimeter when there is an opening for vertical services.

demands, even when column strengths have been factored, for example to meet code provisions aimed at inducing plastic yielding only in the beam elements. These demands arise because of overstrength, higher frequency modes and other effects not allowed for in conventional analysis.

2 NON-SEISMIC APPLICATIONS

2.1 Failure mechanisms in existing types of flat plate in non-seismic areas

Flat plate floors have had wide application in non-seismic areas for offices and residential buildings. The determinant factor in sizing of flat plate concrete floors under gravity and wind loads is the highly stressed area of the slab in the immediate vicinity of the supporting columns (Figure 1); the joint volume that is common to the slab and column is highly constrained and the stresses here are not usually critical. The maximum punching shear stresses, that arise in the slab near the column, increase if unbalanced moment is transferred into the column. All known reported failures have involved sudden collapse due to shear failure in this area. The failures have been attributed to understrength concrete, poor maintenance of the concrete, design oversights or defective construction control, including the incorrect positioning of top reinforcement over the column, which seriously weakens the shear resistance.

2.2 Column-shearhead system in non-seismic areas

The proposed column-shearhead system uses CFT shearheads over the columns in conjunction with CFT columns. This is a modified version of some existing shearhead arrangements. At interior columns, these shearheads have a cruciform shape in plan (Figure 3). The critical perimeter for shear on the concrete is taken just inside the perimeter that links the ends of the CFT shearhead (Corley and Hawkins 1968); this assumes that failure first takes place by yield, in bending, of the steel shearhead and bar reinforcement, rather than by shear in the concrete. ACI 352.1R-89 (ACI 1997) places a limit on the basic shear strength of concrete when the ratio of this perimeter length to the effective depth of the slab exceeds 20.

In the chosen arrangement, all shear transfer into the columns is taken by the shearhead arms, which have shelf angles welded onto them near the column faces to collect the increased shear at these points. Other shear transfer into the shearheads is effected by strut and tie action in the concrete or by bottom reinforcement bars that pass through the CFT shearhead and are suitably sized for stiffness and strength. (Figure 4); this arrangement maintains the integrity of the slab across the shearhead arms and enables it to develop a suitable period of fire resistance.

2.3 Application of new HSS shearhead system in non-seismic areas

For the designer, use of a shearhead allows thinner slabs, a way of meeting structural integrity requirements and the option of running vertical services against the column, within the shearhead perimeter. The openings for services reduce the critical shear

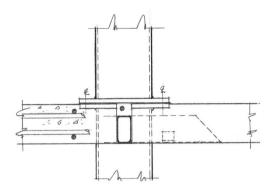

Figure 4. Section through slab at column-slab junction showing column splice with one shearhead arm shown dotted on right-hand side.

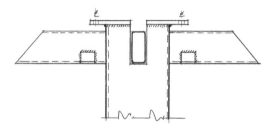

Figure 5. Elevation on steel HSS column and HSS shearhead, which is of cruciform shape and fits into slots in the column.

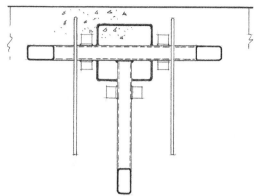

Figure 6. Plan on column-shearhead detail with a square HSS column at edge of concrete slab, indicating typical positions of bar reinforcement for torsion.

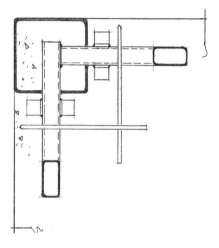

Figure 7. Plan on column-shearhead detail with a square HSS column at corner of concrete slab, indicating typical positions of bar reinforcement for torsion.

perimeter depending on their size and position (Figure 3). In thin flat plates, use of vertical stirrups and shear studs is not possible, because of difficulties in anchoring them, and a shearhead is the only reliable option. For the contractor, the shearhead provides reliable support to bars and prestressing strand in the top layers, faster forming of the columns and improved safety.

For a flat plate concrete slab with 250 mm × 6.3 mm square HSS columns, filled with C40 concrete, on a 6.25 m × 6.25 m grid, a 175 mm thick slab with a 120 × 60 HSS shearhead at the columns is sufficient in shear using, for example, the provisions of *ACI 352.1R*; this assumes a uniformly distributed live load of 3.5 kN/m^2 and a dead load of 5 kN/m^2. For a two storey building, with storey heights of 4.2 metres, the unprotected, unreinforced column could have a one hour fire rating, under a standard fire curve. The size of any column, which does not have external fire protection, can be reduced, whilst maintaining a one hour or other fire rating, by adding reinforcement.

The use of a shearhead facilitates the transfer of moment into the columns from the slab, so that flat plates can be made to work as frames carrying lateral load. At the edge columns, moment may be transferred into the column by flexure and torsion over the whole width of the slab column strip or the shearhead width, whichever is less; the shearhead arm at right angles to the slab edge may also be used to carry sagging moment (Figure 6). At corner positions the shearhead is used in a similar way (Figure 7). At interior columns with shearheads, the moment transfer is also improved and can be assumed to occur over the same increased widths.

In normal reinforced concrete construction, by comparison, the transfer of moment is limited by the high strains needed, and the cracking that thus develops, before the reinforcement becomes effective. This cracking also reduces the stiffness of joints, to well below calculated elastic values, and aggravates P-delta effects (Darvall and Allen 1984).

3 APPLICATION IN SEISMIC AREAS

3.1 *Introduction*

Historically, in seismic areas, reinforced concrete buildings have been prone to failure because of unreliable behavior at the beam-column joints under earthquake loads. A common cause is shear failure in the columns (Figure 2). For example in a reinforced concrete frame office building that was monitored during the 1994 Northridge earthquake, the primary cause of failure was found to be insufficient shear capacity of the concrete columns, which were then unable to develop their flexural strength (Li & Jirsa 1998). It is found that joints in concrete, whose strength depends on shear, torsion or bond, have only limited resistance when subject to repeated deformation within the inelastic range (ATC 1997). However it is precisely these factors that are critical at the flat plate-column joint.

Reinforced concrete, even new and correctly detailed, will only have reliable ductility in certain deformation mechanisms, which essentially are those controlled by flexure. As columns under earthquake loadings can be subject to significant and often unknown inelastic demand, the proposed use of CFT columns and shearheads should provide superior performance compared to reinforced concrete, particularly in preventing soft storey collapse.

The general arrangement of the shearhead in seismic areas is similar to that used in non-seismic areas. However additional stiffness, strength and ductility is now needed to cope with the larger and dynamic nature of the seismic forces that may occur (Figure 8). As in the case of non-seismic loading, the shelf angles attached to the shearhead arms are available to take downward acting shear; the flange plate at the column splice is available to carry reversals of this shear force.

The column-shearhead joint has to provide a ductile connection that either:

(a) has the necessary strength and ductility to resist the effects of vertical and horizontal gravity and earthquake loads when it is used as part of a primary frame taking bending moment at the joints (referred to here as the 'ductile moment' joint) or

(b) has the strength and ductility to resist the effects only of vertical gravity and earthquake loads (the 'secondary' joint), where it is part of a secondary system and lateral loads are taken separately, for example by a shearwall.

In the first case, concentrated plastic hinges would need to develop under earthquake loads to limit the effective loads on the frame. In the second case, a lower joint ductility would suffice, sufficient to sustain the cyclic deflections imposed on it without significant loss of vertical load carrying capacity.

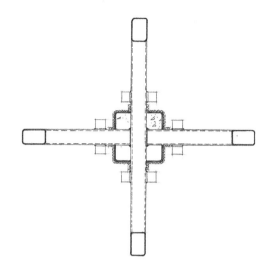

Figure 8. Plan on column-shearhead detail for seismic areas indicating partial isolation of slab from column when ductile rotation occurs in the shearhead arms.

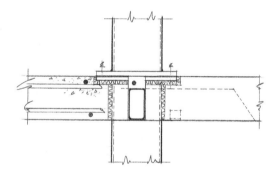

Figure 9. Vertical section through slab with partial depth shearhead arms, indicating isolation of slab from column when ductile rotation occurs in the shearhead arms.

3.2 *Ductile moment joints in seismic areas*

Where frame action is required, a ductile moment joint is designed at the slab-column junction and plastic hinges would form in the slab and shearhead arms adjacent to the column face; this maintains a strong column – weak beam strength distribution. The slab is partially insulated from the column by low modulus material to limit the transfer of forces into the column from the slab and to allow a better estimation of the magnitude of these forces, allowing for overstrength of the shearhead arms (Figures 8 & 9). When necessary, the shearhead arms may occupy the full depth of the slab with the reinforcement made to pass through holes near the top of the arm.

Under seismic loads, the CFT shearhead-column arrangement has advantages over normal reinforced concrete flat plates and wide beam-slab floors: firstly by substantially reducing the loss in the stiffness of the slab-column joint; this may be reduced, by up to two-thirds of the uncracked value, by cracking in the slab near the column (Hwang and Moehle 1993); secondly by helping to prevent the degradation in strength and deformability that occurs in RC joints under repeated cyclic loading and large deformations; and thirdly by ensuring that there is a ductile failure mode, with hinges in the shearhead adjacent to the column face, sagging moments in the span being kept elastic by bar reinforcement or the use of the shearhead in composite action. By anchoring column bars around the shearhead arms, column bar slip in the joint, which reduces stiffness and energy dissipation, can be prevented.

The HSS column:shearhead arrangement is designed to give increased ductility and energy absorption in the column and the shearhead; in both cases there is concrete inside the tube that is fully constrained. As in non-seismic applications, a better transfer of moment into the columns is achieved.

According to the classification given in *ACI 318-02*, a CFT shearhead joint in a seismic area would be a Type 2 connection and would require sustained strength under reversals of load in the inelastic range. Following the rules of *ACI 352.1R-89* (ACI 1997) and assuming that flexural yielding was present, the allowable shear at the critical perimeter for the concrete, just inside the shearhead arms, would be reduced by a factor of 0.75 compared to the basic shear stress; in addition, for a shear force acting in conjunction with an inelastic transfer of moment into the column, the allowable shear would be reduced, by a factor of 0.5.

3.3 *Analysis of frames with ductile moment joints*

In the type of application envisaged for a ductile moment joint, which would essentially be limited to low-rise buildings, simple methods of analysis are to be preferred. Use of a linear-elastic method of analysis with pseudo-lateral forces would be chosen on these grounds. However where behaviour needs more accurate prediction, for example to meet a specified performance level, a displacement method may be preferred, given that displacements are the most convenient indicators of damage in the non-linear range. In the reinforced concrete frame office building that was monitored during the 1994 Northridge earthquake, by careful choice of stiffnesses, a pushover analysis with a triangular loading pattern, was found to correlate well with the recorded pattern of damage (Li & Jirsa 1998). Discussion of the limitations of the various methods of analysis is given elsewhere (ASCE 2000).

A particularly convenient method of displacement analysis is the Yield Displacement method, in which the capacity of the structure is based on a non-linear static pushover analysis with the demand assumed to be given by a Yield Point spectra for the specified earthquake hazard; the Yield Point spectra is a constant ductility spectra, re-plotted from elastic and ductility modified response spectra, in which yield strength is plotted against yield displacement so that the required strength to limit ductility response can be read off directly (Aschheim 2002). The effect of the use of this method is to shortcut many cycles of iteration in the design process.

For a frame with concrete flat plate floors and HSS shearhead columns, the depth of the slab is almost invariable, for a given span and loading arrangement, so that the yield displacement at the top of the frame would also be nearly constant, as a proportion of height, and is largely independent of the strength of the slab and number of floors, assuming that beam and column proportions, relative to span or storey height, are maintained throughout the height of the building. Columns are assumed to be stronger than the combined strength of the slab and shearhead arms at the column face, where the plastic hinges form.

3.4 *Secondary joints in seismic areas*

In principle, the ductile element of the shearhead-column joint may be designed into the column or into the shearhead. However where a secondary joint is used in medium or high rise buildings, so that columns are large, plastic hinges would need to form in the shearhead arm, as in the case of the ductile moment joint. For joints in low-rise frames, where the columns are smaller and the overall stiffness of the slab-beam would exceed that of the column, a possible alternative is to allow the plastic hinge to form in the column itself. In this case, the flange plates at the column splice, at top of slab level, (and it is assumed that, for flat plate construction, there will need to be column splices at each floor level) already provide a partial strength plastic connection.

Joints must have sufficient ductility to meet the specified performance level, that is, in most cases, a set maximum inter-storey drift, correlated with the tolerable damage, for a set probability of its occurrence.

3.5 *Analysis of systems with secondary joints*

The column-shearhead secondary joint is used within a concrete flat plate slab when the lateral loads on it are taken by a separate system, for example a braced frame. In these conditions, the loading on the joint is much reduced compared to the ductile moment joint and the requirement reduces to that of carrying gravity and vertical seismic loads at the specified displacement ductility. The joint can be designed so as to minimize the forces induced in it by lateral deflections at the specified limit, for example by limiting the depth of the shearhead arms.

4 DEVELOPMENT IN SEISMIC AREAS

4.1 Test programme

Development work is to be undertaken with the object of facilitating the reliable design of concrete flat plates in seismic areas. A test program is at present being planned to establish the characteristics of the two types of joint arrangement described, that is the 'ductile moment' joint and the 'secondary' joint. The tests will be on a large scale sub-assembly under monotonic and cyclic load with deformation control; the tests will monitor the complete load-deformation cycles up to failure, according to ECCS recommended testing procedures, for example, so that the yield point, stiffnesses, resistance ratios and absorbed energy can be calculated for the load cycles undertaken. The cyclic tests will enable the construction of a backbone curve, from which input design values, to be used in analysis, for the yield load, initial stiffness, post-yield stiffness, ductility, and lateral force at the target displacement, for the vertical load applied in the span, may be established. The damage, or a damage function, can be correlated with the monotonic and cyclic deformations and energy inputs, as the tests proceed.

4.2 Analysis of test results

Tests under reversed cyclic loads can establish the shape of the hysteresis loops and the degradation in load capacity, stiffness and ductility that result from these loads and these can be used in a pushover analysis. However this set of information does not adequately describe or predict the point of failure, if this is due to low-cycle fatigue. Previous work on HSS braces under reversed axial load indicated that brittle fracture, due to low-cycle fatigue, was the critical factor (Elghazouli et al 2004) and this may be the case too for the column-shearhead, even though the loading regime is different. If this is the case, data may be required on the allowable stress or strain ranges against the number of load cycles experienced. In order to delay the onset of brittle fracture as much as possible, it is proposed that hot finished HSS sections are used in all tests on the proposed column:shearhead system.

5 CONCLUSION

A proposal has been made to exploit some of the unique properties of the steel tube, in this case concrete filled, in order to provide a thinner and more ductile form of the flat plate floor, in all non-seismic areas. In all those seismic areas where concrete is the preferred material of construction, it is hoped to provide an effective and economic solution both for buildings having a separate system for lateral load and for low-rise buildings in which the flat plate forms part of a moment frame.

ACKNOWLEDGEMENTS

The comments of Edmund Booth, Alastair Hughes as well as those of Dr Ahmed Elghazouli and Dr Robert Vollum of Imperial College London, where the testing work is to be undertaken, are gratefully acknowledged.

REFERENCES

ACI 1997. Recommendations for design of slab-column connections in monolithic reinforced concrete structures ACI 352.1R-89. Report of ASCI-ASCE Committee 352. Detroit, Michigan: American Concrete Institute.

ASCE 2000. Prestandard and commentary for the seismic rehabilitation of buildings (FEMA publication 356). Reston, Virginia: American Society of Civil Engineers for the Federal Emergency Management Agency, Washington DC.

Ascheim, M. 2002. Seismic design based on the yield displacement. Earthquake Spectra 18(4): 581–600. Oakland, California: Earthquake Engineering Research Institute.

ATC 1997. NEHRP commentary on the guidelines for the seismic rehabilitation of buildings (FEMA publication 274). Redwood City, California: Applied Technology Council for the Building Seismic Safety Council, Washington DC.

Corley, W.G. & Hawkins, N.M. 1968. Shearhead reinforcement for slabs. ACI Journal 65: 811–824.

Darvall, P. & Allen, F. 1984. Lateral load effective width of flat plates with drop panels. ACI Journal 81(6):613–617.

Elghazouli, A.G., Broderick, B.M., Goggins, J., Mouzakis, H., Carydis, P., Boukamp, J. & Plumier, A. 2004. Shake table testing and seismic performance evaluation of bracing members, Paper 2651. In Proc.13th World Conference on Earthquake Engineering, Vancouver BC, 1–6 August 2004. CD publication of the International Association of Earthquake Engineering.

Gentry, T.R. & Wight, J.K. 1994. Wide beam-column connections under earthquake-type loading. Earthquake Spectra 10(4): 675–703. Oakland, California: Earthquake Engineering Research Institute.

Hwang, S.J. & Moehle, J.P. 1993. An experimental study of flat plate structures under vertical and lateral loads, Report no. UCB/EERC 93/03. University of California at Berkeley: Earthquake Engineering Research Center.

Li, Y.R. & Jirsa, J.O. 1998. Nonlinear analyses of an instrumented structure damaged in the 1994 Nothridge earthquake. Earthquake Spectra 14(2): 265–283. Oakland, California: Earthquake Engineering Research Institute.

Tubular Structures XI – Packer & Willibald (eds)
© 2006 Taylor & Francis Group, London, ISBN 0-415-40280-8

A joints solution by laser cutting in the chords of CHS structures

Ph. Boeraeve, B. Ernotte & J. Dehard
Hemes, Gramme Institute of Engineering, Liege, Belgium

E. Bortolotti
Arcelor, Arcelor Research Centre, Liege, Belgium

Ph. Zieleman
TSC CPI, Yutz, France

J.M. Babin
Chaillous, Nantes, France

ABSTRACT: This paper presents an alternative to the traditional joints in tubular structures with complex end shapes. The technique consists in first laser cutting holes in the chords and then pushing the plane cut diagonals, into those holes before welding. Some tests have been achieved in order to study the behavior of such joints and to establish a comparison with traditional joints. After a FEM model calibration, a parametrical study was carried out to evaluate the performance of the precut joints compared to the traditional ones. A technological study has, however, shown all the possibilities offered by this new technique. A lot of applications from architectural buildings to mechanical applications could benefit from the present technique.

1 INTRODUCTION

Traditionally, joints in tubular structures are welded, and the ends of diagonals have to be cut precisely to the geometry of the joint. For CHS this may lead to complex, out-of-plane cuts (Figure 1).

The positioning of such tubes before welding is a delicate job and requires the conception and construction of a jig. This explains why tubular structures are rather expensive.

A new method, called precut method, simplifies the operations mentioned above (Figure 2).

The precut method consists in first laser cutting holes in the chords and then pushing the plane cut diagonals into those holes before welding.

2 NONLINEAR FEM MODEL

2.1 Materials

The stress-strain curve of the steel tubes is bi-linear. Failure of the diagonals by yielding in compression or tension was avoided artificially by adopting a perfectly elastic constitutive law for those elements.

The weld is considered as perfectly elastic, because it is usually stronger than the tubes and a failure of the

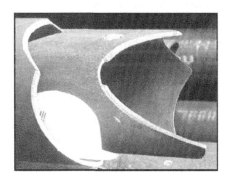

Figure 1. Complex end cut in CHS.

Figure 2. Traditional and Precut method for CHS joints.

weld should normally not be influenced by precutting the chord. This assumption was also taken by Van Der Vegte (1995).

2.2 FEM Program

ABAQUS was used for the FEM study, with 4 nodes shell elements (S4R, reduced integration). The "Static General" study was chosen with geometric nonlinearities. Due to the type of chosen elements, the punching shear failure may not be predicted by the model, but that failure won't be affected by precutting the chords.

3 CALIBRATION STUDY

3.1 Aim

A calibration study has been achieved in order to check the model.

3.2 Comparison with tests

Let's take an example of an X joint where both verticals are submitted to a tensile load P. By symmetry, only 1/8th of the joint is modeled. During the incremental loading, a P-Deformation curve is recorded, where P is the load applied to the verticals, and the deformation is taken as the elongation of a gauge length of 100 mm (see Figure 3) divided by the diameter of the chord.

Figure 4 shows the results of the FEM model and tests on traditional and precut joints. This example corresponds to a width ratio between the diagonal and the chord of 0.6.

Other comparisons have been made on width ratios of 0.3 and 0.9, and also in compression. All the comparisons have led to the same conclusions : the model is able to predict accurately the behaviour of the tests, provided that no punching shear failure is expected. As discussed later, the useful part of the diagram corresponds to a deformation limited to 3%. In that region, the FEM model results are very close to the test results.

4 PARAMETRIC STUDY

4.1 Aim

The aim of this study is to compare the resistance of precut and traditional joints. The final objective is to define formulas of correction factors to apply for the design of precut joints on the basis of design methods for traditional joints. As design methods may change with time, no reference will be made to existing design formulas. As our correction factors apply to the actual resistance of the joint, the application of those factors to any existing design method will be safe. Plane joints of types T, X, K and N have been studied, as well as 3D joints of types XX, TT and KK.

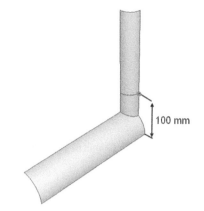

Figure 3. X joint model and gauge length.

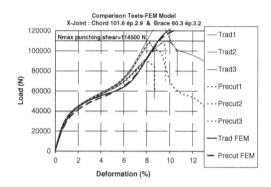

Figure 4. Example of comparison between tests (Trad1 to Trad3 and Precut1 to Precut3) and FEM model results.

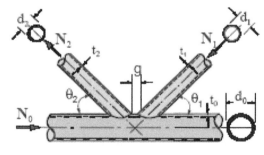

Figure 5. Parameters for type K joints.

4.2 Parameters

The main parameters of the study were:

- the width ratio between the diagonal and the chord $\beta = d_1/d_0$ (Figure 5),
- the width to thickness ratio of the chord $\gamma = d_0/t_0$,
- for N joints, two types of loading have been considered : diagonal in tension or in compression,

552

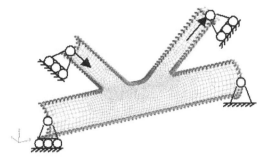

Figure 6. Support conditions for K joints.

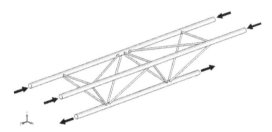

Figure 7. Compression and tension in the chord due to bending.

- XX joints have been studied with symmetric and non symmetric loading (load ratios J of −1, −0.6 and +1),
- in K joints, two values of the angle between the diagonal and the chord have been taken: 45° and 60°,
- in TT joints, two values of the angle between the diagonals have been studied: 90° and 60°,
- in KK joints, diagonals making an angle of 45° with the chord, and 60° with the diagonal of the same side of a cross-section in the chord have been studied,
- To isolate the influence of the geometric parameters, chords were considered unloaded except in N, K, KK joints, where the supports conditions induced a reaction force in one part of the chord (Figure 6).

4.3 Load in the chord

The normal situation for the chord is to be loaded on both sides of the joint (Figure 7). An exploratory study of the influence of an additional tensile($n' < 0$) or compression($n' > 0$) load in the chord has been achieved in the K and KK joints. The variable n', equal to the maximum load in the chord divided by the plastic load of the chord, has been taken as a parameter. Values of n' considered were −1, −0.6, −0.3, 0, +0.3, +0.6 and +1.

4.4 Scope of the study

The configurations of Table 1 have been analyzed and the combinations of chord-diagonal have been

Table 1. Configurations of this study (dimensions in mm).

Joint	Chord	Diagonal
T90°	169.3 × 3 168.3 × 4 114.3 × 3 114.3 × 5	33.7 × 2 70 × 2 101.6 × 3.6 139.7 × 3.6 168.3 × 4
X90° & X45°	160 × 12 160 × 6 160 × 4 101.6 × 2.9	33.7 × 3.2 40 × (4 & 5) 60.3 × 3.2 80 × (4, 6 & 8) 88.9 × 3.2 120 × (4, 6 & 10) 140 × (4 & 6) 160 × 12
K45° & K60°	219.1 × 3 219.1 × 5 219.1 × 10 88.9 × 7.1 88.9 × 3 88.9 × 2	48.3 × 3 48.3 × 5 76.1 × 3 76.1 × 5 114.3 × 3 114.3 × 5 139.7 × 3 139.7 × 5 168,7 × 3 168,7 × 5 193,7 × 3 193,7 × 5
N45°	219.1 × 3 219.1 × 5	48.3 × (3 & 5) 76.1 × (3 & 5) 114.3 × (3 & 5)
XX	160 × 4 160 × 6 160 × 12	40 × (4 & 5) 80 × (4, 6 & 8) 95 × (4, 6 & 10)
TT90° & TT60°	168.3 × 4 168.3 × 8	33.7 × 2 55 × 2 70 × 2 101.6 × 3.6
KK45°	168.3 × 4 193.7 × 5 219.1 × 3 219.1 × 5 219.1 × 10	48.3 × (3 & 5) 55 × 4 76.1 × (3, 4 & 5) 82.5 × 4 88.9 × (3 & 5)

553

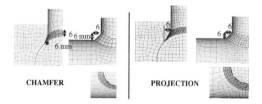

CHAMFER   PROJECTION

Figure 8.   Two ways of modelling the weld.

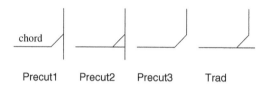

chord

Precut1   Precut2   Precut3   Trad

Figure 9.   Ways of connecting the weld to the chord and diagonal.

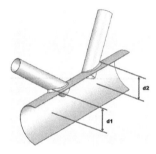

Figure 10.   Gauge lengths for type K joints.

chosen in order to cover the whole range of realistic β and γ.

4.5   Modelling the weld

A preliminary study has been achieved in order to consider two ways of modelling the weld.

In the first one, called CHAMFER, the weld extends tangentially to the chord, to the same distance as it has at the top.

In the second one, called PROJECTION, the weld extends horizontally everywhere to the same distance. As a consequence, for high values of β, the weld becomes very large and unrealistic.

In the parametric study, it has thus been decided to model the weld by means of the "chamfer" method.

Moreover, the members and the weld may be connected by different ways in the FEM model.

After a short comparison study, we finally decided to choose the Precut3 and Trad models for the parametric study, because it is safer not to consider the extension of the diagonal into the chord, because it has a stiffening effect.

4.6   Failure criteria

A Load-Deformation curve is plotted, where P is the load applied to the diagonals, and the deformation is taken as the largest radial displacement d on the chord, taken from the center of its section divided by the diameter of the chord.

The failure load is taken arbitrary as the load achieved when the deformation reaches 3%, except when a maximum in the curve is observed before. That rule is a classical one already taken by Wardenier (1982) and Van Der Vegte (1995).

5   RESULTS

5.1   Reduction factors formulas

Table 2 summarizes the formulas that have been found a safe prediction of the reduction factor of the failure load of a precut joint compared to a traditional joint.

$$N_{Rd\ PRECUT} = f(p) * N_{Rd\ TRAD} \qquad (1)$$

where $N_{RdPRECUT}$ = failure load of the precut joint; $f(p)$ = reduction factor formula; and $N_{Rd\ TRAD}$ = failure load of the traditional joint.

6   TECHNOLOGICAL STUDY

Several beams made of traditional and precut joints have been built in order to compare the preparation time and to be able to give good practice rule.

6.1   Plane and 3D truss beams

Those beams have been galvanized and the precut method has shown some advantages:

- less preparation time needed,
- no degassing holes are needed for Zn coating,
- less Zn retention in the diagonals due to the better circulation of the molten Zn in the tubes,
- for fire resistance, it should be possible, with precut joints to irrigate the structure.

6.2   Cambered beams

The problem for cambered beams, is that laser cutting machines only accept straight members. As the chords must be bend to realize an arch, a prospective study has been realized.

The idea is to leave small links to the parts to remove. Those links help to keep the good geometry of the holes when cambering the chords. After bending those links are cut manually.

The joints in those beams were made from chords 168.3 mm × (5 or 3 mm) with $\beta = 0.52$ and 0.59.

Table 2. Proposed formulas for the reduction factor of the failure load of precut joints.

Joint	Reduction factor
T	$f(p) = 1 - 0.16\beta^2$
X	$f(p) = 1 - 0.03\,e^{1.5\beta}$
K, N	$f(p)_{n'=0} = -0.25\beta^2 + 0.13\beta + 0.89$ $f(p)_{n'<0} = -0.09\beta + 0.90$ $f(p)_{n'>0} = -0.42\beta + 0.97$
XX	$f(p) = -1.65\beta^2 + 0.9\beta + 0.82$
TT	$f(p) = 1 - 0.16\,\beta^2$ (as for T-joints)
KK	$f(p) = 1.01 - 0.315\beta$

Notes:
- For K joints, the lowest reduction factors have been found for the 45° configuration.
- For N joints, the same formula as for K joints has been adopted because it was always safe.
- The diagrams that have led to those formulas are grouped on the next page (Figure 11 to 18).
- A compression load in the chord has been found to have an important influence on the failure load for K joints, but not for KK joints. This is probably due to the smaller range of β (0.22 τ o 0.45) in the KK joints. In KK joints, β must be limited, for geometrical reasons, to allow the precut method to be feasible. Further work should be necessary to study that influence more widely for all the other types of joints and a wider range of dimensions of tubes.

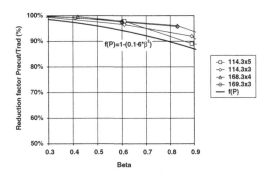

Figure 11. Results of T joints.

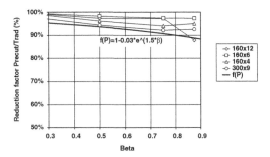

Figure 12. Results of X joints.

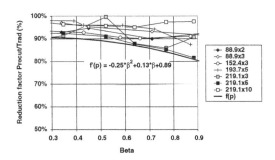

Figure 13. Results of K joints, no load in the chord.

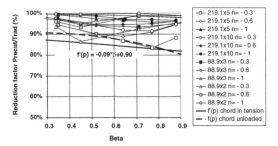

Figure 14. Results of K joints, chord in tension.

555

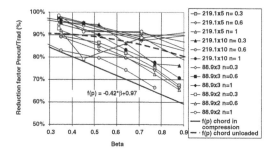

Figure 15. Results of K joints, chord in compression.

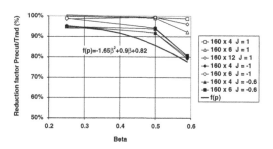

Figure 16. Results of XX joints.

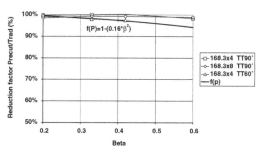

Figure 17. Results of TT joints.

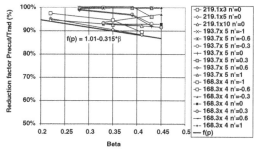

Figure 18. Results of KK joints.

Links description

4 links whose length are: 8 mm for laterals 12 mm for longitudinals	Sufficient for holes on the shortening side of the cambered chord (could be limited to 4 links of 8 mm).
8 links whose length are: 8 mm for laterals 12 mm for longitudinals	Links are well positioned for holes on the stretching side of the cambered chord. Symmetry of the links is important. For a chord thickness of 3 mm, link length should be increased (e.g. 10/14 mm)

The conclusions were:

• No problem encountered while $\beta \leq 0.6$.
• Camber radius: no problem until a value of 7 m with a chord diameter of 168.3 mm.
• Holes deformation during cambering: if on the shortening side, a shortening of the hole may occur, a tolerance gap must be foreseen in order to facilitate the penetration of the diagonals into the holes.

Figure 19. 3D cambered beam.

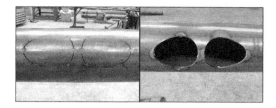

Figure 20. Process for cambered chords.

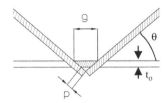

Figure 21.   Gap between holes.

Figure 22.   Types of laser cuts.

7   CONCEPTION HINTS FOR PRECUT JOINTS

7.1   *Width ratio*

For practical reasons, the width ratio should be limited to 0.85.

7.2   *Gap*

A minimum gap should exist in order to allow the diagonals to penetrate the chords of at least p.

$$g \geq 2 \frac{p.\sin\theta + t_0}{tg\theta} \quad \text{with } (p \geq t_0) \qquad (2)$$

7.3   *Weld*

The laser cutting of the chords may be realized in different ways : the cut made in the chord may be perpendicular to its axis, or parallel to the diagonal or, eventually, a combination of both.

Solution 1 requires more weld (Figure 22). Solution 2 is faster to weld but a minimum gap must be present to facilitate the positioning (a 2 mm gap was found to be sufficient).

In precut joints, the thickness of the diagonals should be limited to 8 mm, because for greater thicknesses, the weld would not be penetrating. The angles of the lasercut should lie between 30° and 90°. Beyond 60°, the cut can be perpendicular to the axis of the chord, because the gorge to fill with the weld becomes small enough.

7.4   *Workshop*

The precut method needs less preparation than the traditional method, because most of them are realized automatically by the laser cut machine. The ends of the diagonals are cut by saw, perpendicularly or skewed to the axis of the diagonal.

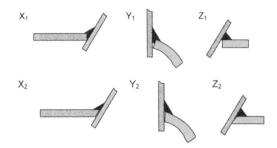

Figure 23.   Types of welds in laser cut chords.

Figure 24.   Laser cut perpendicular to the chord axis.

The positioning of the diagonals before welding is greatly simplified in the case of precut joints, it simplifies the jig and that makes the interest of the method.

The holes should be foreseen for a diameter 2 mm greater than the diameter of the diagonal, in order to allow some adjustment when positioning.

Laser cutting machines can also be used to mark the tubes, in order to simplify the positioning of the tubes, especially in 3D structures.

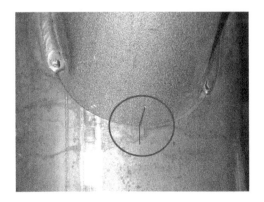

557

8 CONCLUSIONS

Precutting the chords presents some advantages:

- simplification of the operations in the workshop,
- better protection in case of Zn coating.

Those advantages have a cost : the study has shown that precutting the chords lead, generally, to a decrease of the failure load.

That loss of strength increases width ratio β, but no correlation was found with the width to thickness ratio γ.

Formulas has been developed to estimate the loss of strength, compared to traditional joints. Those formulas can thus be applied to any existing code design formula for traditional joints.

ACKNOWLEDGMENTS

This research was sponsored by the Research Ministry of the Region Wallonne of Belgium and by ARCELOR. Laser cutting was made thanks to TSC CPI, France.

REFERENCES

Grundy P., Holgate A. & Wong B., 1994, *Tubular Structures VI*, A.A. Balkema,

Jaurrieta M.A., Alonso A., Chica J.A., 2001, *Tubular Structures X* , A.A. Balkema,

Packer J.A., Wardenier J., Kurobane Y., Dutta D., Yeomans N., 1991, *Guide de dimensionnement – Assemblages de sections creuses circulaires (CHS) sous chargement statique prédominant*, Comité International pour le Développement et l'Etude de la Construction Tubulaire,

Van Der Vegte G.J., 1995,*The Static Strengh of Uniplanar Tubular T- and X-Joints*, Delft University Press,

Wardenier J., 1982, *Hollow Section Joints*, Delft University Press,

Wardenier J., 2000, *Hollow Sections in Structural Applications*, Comité International pour le Développement et l'Etude de la Construction Tubulaire (CIDECT).

Tubular Structures XI – Packer & Willibald (eds)
© 2006 Taylor & Francis Group, London, ISBN 0-415-40280-8

Hollow flange channel bolted web side plate connection tests

Y. Cao & T. Wilkinson
The University of Sydney, Sydney, NSW, Australia

ABSTRACT: This paper describes tests on bolted Web Side Plate (WSP) connections of LiteSteel Beams. The LiteSteel Beam (LSB) is an innovative structural section manufactured by Smorgon Steel Tube Mills (SSTM) using the dual Electrical Resistance Welding process to create a channel section with hollow flanges. The hybrid section utilises beneficial section properties of hollow sections and the traditional open channel section profile for improved connectivity. The aim of these tests is to investigate the WSP connection behaviour on LSB and verify the suitability of designing WSP connection using the current design models.

This paper describes a total number of 6 full scaled tests performed on 2 different sizes of LSB and 2 configurations of bolt sizes and numbers. All tests performed were on single row of bolts and the number of bolts ranged from 2 to 3. A new failure mode associated with bearing and shear in the LSB web was observed. The result also showed that the joint rotation impacts the connection shear capacity – significantly in cases of deeper WSP connections.

1 INTRODUCTION

The LiteSteel Beam (LSB) is a channel section with hollow flanges as shown in Figure 1. It has been developed by Smorgon Steel Tube Mills (SSTM) using the dual Electrical Resistance Welding process. The section is cold-formed using a single steel strip with a nominal yield stress of 380 MPa. The finished product, due to the cold-forming process, has flanges with nominal yield stress of 450 MPa and web with nominal yield stress of 380 MPa. A range of LSB sections from $125 \times 45 \times 1.6$ LSB to $300 \times 75 \times 3.0$ LSB are being manufactured.

The LSB is being used primarily as members for flooring systems (bearer or joist), rafters, roof beams and the like. Due to its unique shape, some new failure modes that are not covered in the current design codes could occur and therefore research into the behaviour of the LSB has been conducted and at the time of this paper is still being conducted. This paper describes some tests of bolted web side plate (WSP) connections of LSB.

The web side plate (WSP) connection is also referred to as "fin plate connection", "single plate framing connection" (Kulak et al. (1987), Richard et al. (1980)), "single plate shear connection" and "shear tab" (Astaneh et al. (2002)). A bolted WSP connection consists of a number of elements including the supporting member, side plate and the supported member. It is constructed by welding the side plate onto the supporting member and bolting the web of the supported member onto the side plate through prefabricated holes. A typical WSP connection is shown in Figure 2.

A review of research conducted in WSP connections before 1994 may be found in Syam (1994). More recent papers on WSP connections research include

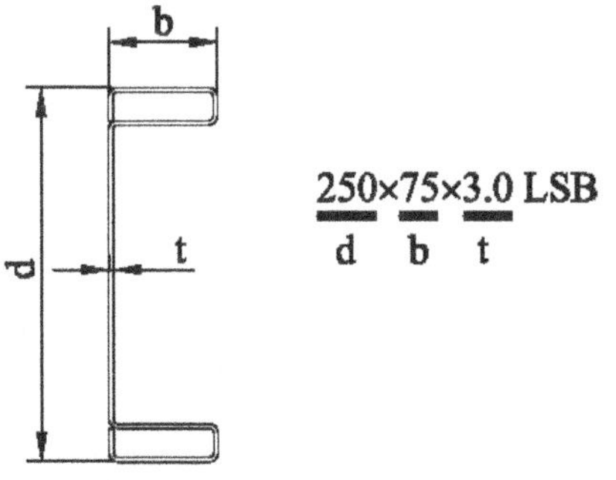

Figure 1. LSB section.

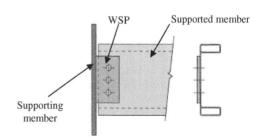

Figure 2. Typical WSP connection (provided by SSTM).

Wheeler et al. (2001), Astaneh et al. (2002) and Aribert et al. (2004).

Generally, structural steel connections can be classified into three categories: rigid connections, semi-rigid connections and flexible connections. An ideal rigid connection does not change the angle between the supporting and supported member during load and transfers bending moment fully between the connected members. On the contrary, an idealised flexible connection acts as a pin and transfer only shear force without develop bending moment. In reality, all connections exhibit semi-rigid behaviour to some extent – they all have some rotational rigidity and can hence develop moment. However for simplicity of design, it is generally accepted that connections can be treated as ideas rigid or flexible connections if a connection has "significant stiffness" or "is rotationally flexible" respectively (Hogan & Thomas, (1994)). It is also obvious that the moment developed in a connection associates closely with the end rotation of the beam (supported member). Researchers have tried to quantify the above notion. For example Kulak et al (1987) recommended that a particular flexible connection developed 10–20% of fixed end moment and a typical rigid connection would develop 75% of the fixed end moment. In Astaneh (1989) classification, a rigid connection would develop 90% of more of the fixed end moment at 10% or less end rotation of that of a pin-supported beam. While a simple connection would only develop up to 20% of the fixed end moment when end rotation is at least 80% of that in a pin-supported beam.

Another issue to be considered in connection classification is the rigidity of the supporting member at the point of the connection (often referred to as the support condition). Support conditions, as in the connection itself, can be rigid or flexible, or in reality somewhere in between. In an ideal rigid support condition, all rotation is absorbed by the deformation of elements of the connection such as yielding of the side plate, bearing yielding of the holes on side plate and web of supported member and yielding of the connectors. While in ideal flexible support condition rotation of the beam (supported member) is fully absorbed by the supporting member alone and therefore no yielding as described above would occur as a result of the beam rotation. It is considered in this paper that rigid support condition results in more severe end moment and requirement to the connection elements and thus employed in the tests.

In terms of designing, WSP connections are generally considered to qualify for the simple construction terms of the requirements of AS 4100 (Hogan & Thomas (1994)) and therefore should comply with the following requirements given in AS 4100:

1. The connection shall be capable of transmitting the calculated design action effects.

2. It shall be capable of deforming to provide the required rotation at the connection.
3. The rotation capacity of the connection shall be provided by the detailing of the connection and shall have been demonstrated experimentally.

From the above requirement, it is clear that not only that it is important to ensure that the connection can transfer shear force, but also it is required to do so while under appropriate rotation. While as also pointed out in Wheeler et al (2001) that excessive rotation may fail the connection with very little shear force, it is evident that rotation has some impact on the connections shear capacity. It is thus important when designing a simple connection to study the effect of rotation and, where applicable, consider reduction of shear capacity in relation to the rotation under serviceability load.

The current LSB Connection Design Manual (SSTM (2004)) is largely based on the Hogan & Thomas (1994) design modal which is based on research mainly on open sections (typically I sections). However the LSB Connection Design Manual had modified the detailing of the WSP connection to suit the different section shape. This fact, combined with the unique shape and properties of the section, necessitates experimental research in WSP connections of LSB. The aim of this paper is to study the behaviour of such connections by experimental tests. While verifying the applicability of the Hogan & Thomas (1994) design model and the configuration details described in the SSTM Connection Design Manual (2004), particular interest is paid to the relationship between the moment, rotation and the shear capacity of the connections and possible new failure modes. The effects of span – section depth ratio, the rotational stiffness of the connections and the developments of the point of contraflexure are also of interest.

2 TEST PROGRAM

2.1 Testing method

The testing method was chosen after careful review of the previous research. The goal of the test rig design was to simulate connections of beams with reasonable spans subjected to uniformly distributed loads. The cantilever testing method, where a point load was applied to a cantilevered beam, was omitted first due to that the linear relationship of connection moment and shear force was deemed not to represent the true behaviour of simple connections. The propped cantilever testing method, which is more often used, was not chosen for two main reasons. Firstly the reaction force at the propped end would introduce complicity into the statics of the connection. Secondly the load path, or the shear – rotation relationship was unknown to this type of WSP connection thus prohibiting usage of propped cantilever method.

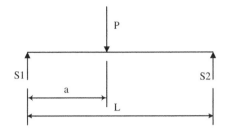

Figure 3. Loading position.

Full scale testing, in comparison with short beam testing, where tests were conducted on beams with spans that are likely to be used in, was as a result employed and the spans of the beams were chosen to maximise rotation while avoiding pre-mature failure of the beams. Also due to the asymmetrical shape of the LSB section, back-to-back tests were performed for simplicity of loading and data verification.

To simulate the uniformly distributed load (UDL) and to ensure that the failure of connection occurred at the desired end, a point load was applied offset to the centre of the beam as shown in Figure 3. The offset value was calculated as follows:

For simply supported beam subjected to a UDL w, the end rotation and the shear force are:

$$\theta = \frac{(wL)L^2}{24EI} \text{ and } S = \frac{wL}{2}$$

For simply supported beam subjected to an offset point load P as in Figure 3,

$$\theta = \frac{P(L-a)a^2}{2LEI} \text{ and } S1 = \frac{P(L-a)}{L}$$

To simulate the UDL case, let $S1 = S$ and $\theta = \theta$ and the result is:

$$a = \frac{L}{\sqrt{6}} \text{ and } w = 2\left(1 - \frac{1}{\sqrt{6}}\right)\frac{P}{L}$$

Therefore as far as the connection is concerned, ignoring the effect of the end moment, a concentrated load applied at a certain offset point can approximately simulate the loading conditions of a UDL load. Justification for ignoring the effect of the end moment partially rest upon the fact that the same rotation would cause same end moment at connections with the same configurations. It is realised that rotations at the two ends of the beam with off centre concentrated load are different. However the offset distance in the above calculations are less than 10% of the span of the beam and therefore the difference of the end rotations is ignored.

2.2 Test rig design and configuration

The testing rig consists of three frames, namely fixed end supporting frame, loading frame and movable end loading frame. Figure 4 shows the testing rig.

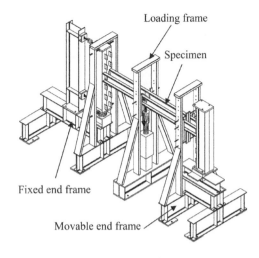

a. Isometric view

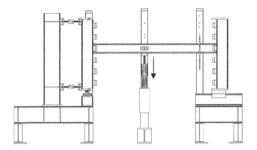

b. Front view (without lateral supporting frame)

c. Photo of Testing Rig

Figure 4. Overall view of testing rig.

Figure 5.   WSP connections.

Figure 6.   Roller support.

a. Flat load cell

b. S-shaped load cell

Figure 7.   Load cell.

The specimens were supported at both ends by WSP connections onto the supporting columns at both end frames (Fig. 5). At the movable end frame, the supporting column was positioned on a roller to allow for any axial shortening induced by the bending of the specimens (Fig. 6).

The supporting column of the fixed end frame sat on a load cell and was connected to its supporting frame through two S-shaped load cells (Fig. 7). This setup ensured that the statics of the supporting column can be obtained and calculated.

The loading frame, situated closer to the fixed end frame, contained a 250 kN capacity hydraulic actuator which applied load onto the specimens through a pinned connection (Fig. 8). The light frames made by hollow sections provided lateral support to the specimen and the fixed end supporting column to prevent pre-mature failure of the beams (Fig. 9).

A total number of six sets of side plates were welded onto each supporting column (Fig 10). This was because each set of side plates would become unusable after test and by welding six sets onto supporting column, six tests could be performed without significant interruption.

The tests were conducted on two sizes of LSB in different spans. Table 1 summarises the six tests.

All connection detailing were that of the SSTM Connection Design Manual. Side plates were 5 mm in thickness and all bolts were Grade 4.6 metric bolts. The side plates were cut from hot rolled flat bars manufactured to AS/NZS 3678 (1996) and all bolts, nuts and washers were to Australian standard AS 1111 (1996).

Welding of WSP onto supporting column was manual metal arc welding using Grade E48XX consumables to Australian/New Zealand standard AS/NZS 1553.1 (1995) – as specified in SSTM Connection Design Manual.

Table 2 outlines the predicted capacities and failure mode for each test. The calculation of connection capacities was based on the Hogan & Thomas (1994) design modal and the calculation of beam capacities was based on AS/NZS 4600 (1998) and SSTM Design Capacity Tables (2004).

Figure 8. Loading frame.

Figure 9. Lateral support.

Figure 10. Side plates.

Table 1. LSB WSP tests summary.

Test	Specimen	Bolting	Span (mm)
1	200 × 60 × 2.0 LSB	3 × M12	2560
2	200 × 45 × 1.6 LSB	2 × M16	2560
3	200 × 60 × 2.0 LSB	3 × M12	1680
4	200 × 45 × 1.6 LSB	2 × M16	1680
5	200 × 60 × 2.0 LSB	3 × M12	1280
6	200 × 45 × 1.6 LSB	2 × M16	1280

Table 2. Predicted capacities and failure modes.

Test	Predicted capacities		Predicted failure mode
	Connection shear (kN)	Beam section moment (kNm)	
1	21.3	22.3	Combined*
2	21.7	14.9	Beam
3	21.3	22.3	Bolt shear
4	21.7	14.9	Combined*
5	21.3	22.3	Bolt shear
6	21.7	14.9	LSB web shear

* – Connection failure and beam failure occurs at same load.

2.3 Testing procedure and instrumentation

The actuator was operated by a computerised controller in stroke control mode. Downwards load was applied onto the specimens through a pinned connection until the failure of the connection.

As shown in Figures 4b, 7 and 11a, the fixed end supporting column was supported by three load cells and hence the moment and shear force in the connection can be calculated by the statics of the supporting column. In addition, a number of strain gauges and transducers were used during the tests to obtain test data. Figure 11 shows the instrumentations of the specimen and WSP.

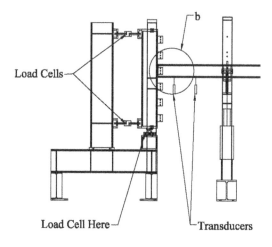

Load Cells

Load Cell Here

Transducers

b

a. Overall instrumentation

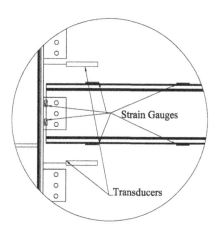

Strain Gauges

Transducers

b. Detailed instrumentation

Figure 11. Instrumentation of WSP tests.

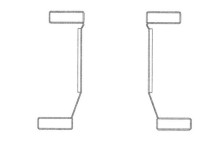

Figure 12. LSB web distortion.

Figure 13. Web yielding line.

Figure 14. Resting of top flange onto WSP.

3 TEST RESULTS

During the course of tests, a new failure mode was observed. In all tests, yielding was initiated by a tendency of twisting of the LSB section at the connection. The side plate prevented the twisting and resulted in a distortion of the LSB web which caused the top (compression) flange moved away from WSP while the bottom (tension) flange moved in the direction of WSP side. Figure 12 shows the web distortion in the LSB web.

After the initiation of the web distortion, in most tests the web distortion developed rapidly with little increase of the load. A web yield line starting from the top outside corner of the side plate and at approximately 45 degree towards the middle of the beam was subsequently developed as shown in Figure 13. Due to the small distance between the top flange of the LSB and the top of the WSP (nominally 9 mm, however due to that the holes were 2 mm larger in diameter than the bolts, the distance could be smaller), the top flange of LSB rested on top of the side plate at early stage of the development of the web yield line (Fig. 14). This in turn, combined with the web distortion, caused distortion of side plates and ultimately, the web of LSB tearing away from top flange as shown in Figure 15.

This failing mode was consistent throughout all six tests. There was no beam flexural failure observed in any test before web yielding and web tearing.

Figure 15. Web tearing.

a. M12 Bolt

b. M16 Bolt

Figure 16. Bolts after testing.

In term of the connectors, some yielding of washers was observed in the tests. In some bolts slight bearing marks were observed however there was no sign of bolt fracture (Fig 16).

Table 3 outlines the ultimate shear load that was achieved in each test and rotation at ultimate load. Figure 17 is the shear–rotation graph.

4 DISCUSSIONS

All specimens failed by the same failure mode – distortion in the LSB web at the connection. This unique failure mode is caused by the slender web of

Table 3. Testing results.

Test	Failure load P_u (kN)	Predicted connection capacity P_y (kN)	P_u/P_y	Rotation* (rad)
1	30.8	23.7	1.30	0.058
2	17.8	24.1	0.74	0.029
3	37.1	23.7	1.57	0.023
4	22.1	24.1	0.92	0.024
5	42.0	23.7	1.77	0.018
6	23.4	24.1	0.97	0.025
		Mean	1.21	
	Standard deviation		0.40	

* – Rotation at failure load.

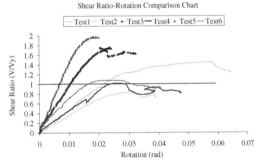

Figure 17. Shear–rotation graph.

the LSB section and the position of the weld (where the web tear occurred).

The development of the web yield line in the LSB web (Fig. 13), which is approximately 45 degree from the flanges, suggested a failure associated with bearing and shear caused by the contact at the side plate to the underside edge of the LSB top flange. Further investment should be carried out to establish the type of failure and perhaps a new design model is to be developed.

It is evident from Figure 17 that for connections with same configurations (Tests 1, 3, 5 and Tests 2, 4, 6) larger end rotation would result in smaller connection shear capacity. Deeper WSP connection (Tests 1, 3, 5 with three bolts compare to Tests 2, 4, 6 with two bolts) seems to be more sensitive to the beam rotation. It is to be noted that the span/section-depth (L/d) ratio of the test specimens in this paper is between 6.4 and 12.8 – which is much smaller then 25 recommended in Astaneh (2002). The recommendations in the SSTM Residential Construction Manual (2005) give L/d ratios up to 36 – almost 3 times as big as that in this paper. It is thus important that further investigation on the relationship of rotation and shear to be conducted.

It is also shown in Figure 17 that some specimens did not demonstrate the shear capacity calculated using Hogan & Thomas (1994) model. It is therefore essential to develop new design model for the WSP connections for LSB.

5 CONCLUSIONS

A total number of six tests were conducted on the WSP connections of LSB. The results showed that due to that section shapes and properties of the LSB, new failure mode was observed which was not covered in the current design codes. It is also evident that the joint rotation impacts the connection shear capacity – significantly in case of deep WSP. Further investigation of the WSP behaviour is recommended and new design model of LSB WSP connections should be developed.

A new version of the LSB Connection Design Manual (2005) was published and the recommendation of detailing of WSP connection was modified. Further experimental tests are expected to be carried out using the modified WSP detailing in later stages of this project.

Further investigation should be carried out on the relationship of the end moment and rotation and possible interaction of end moment and shear in the connections.

REFERENCES

Aribert J., Braham M. and Lachal A., (2004), "Testing of 'simple' joints and their characterisation for structural analysis", *Journal of Constructional Steel Research*, JCSR, Vol. 60, Issue 3–5, March–May 2004, pp 659–681.

Astaneh A., (1989), "Demand and Supply of Ductility in Steel Shear Connections", Journal of Constructional Steel Research, JCSR, Vol. 14, Issue 1, pp 1–19.

Astaneh-Asl A., Liu J. and McMullin K. M., (2002), "Behavior and design of single plate shear connections", Journal of Constructional Steel Research, JCSR, Vol. 58, Issue 5–8, May–August 2002, pp 1121–1141.

Hogan T.J. and Thomas I.R., (1994), *Design of Structural Connections*, 4th edition, Australian Institute of Steel Construction, Sydney.

Kulak J.L., Fisher J.W. and Struik J.H.A., (1987), *Guide to Design Criteria for Bolted and Riveted Joints*, second edition, John Wiley & Sons.

Richard R.M., Gillett P.E., Kriegh J.D. and Lewis B.A., (1980), "The Analysis and Design of Single Plate Framing Connections", *Engineering Journal*, AISC, Vol. 17, No.2, 1980, pp 38–52.

Smorgon Steel Tube Mills, (2004), *Connection Design Manual for Lite Steel Beams*, SSTM, Queensland, Australia.

Smorgon Steel Tube Mills, (2005), *Connection Design Manual for Lite Steel Beams*, Smorgon Steel LiteSteel Technologies, Queensland, Australia.

Smorgon Steel Tube Mills, (2004), *Design Capacity Tables for Lite Steel Beams*, SSTM, Queensland, Australia.

Smorgon Steel Tube Mills, (2005), *Residential Construction Manual for Lite Steel Beams*, SSTM, Queensland, Australia.

Standards Australia, (1996), *Australian Standard AS 1111 ISO metric hexagon commercial bolts and screws*, Standards Australia, Sydney.

Standards Australia, (1995), *Australian/New Zealand Standard AS/NZS 1553.1 Covered electrodes for welding – Part 1*, Standards Australia, Sydney.

Standards Australia, (1998), *Australian Standard AS 4100 Steel structures*, Standards Australia, Sydney.

Standards Australia, (1996), *Australian/New Zealand Standard AS/NZS 3678 Structural Steel – Hot-rolled plates, floor plates and slabs*, Standards Australia, Sydney.

Standards Australia, (1998), *Australian/New Zealand Standard AS/NZS 4600 Cold-Formed Steel structures*, Standards Australia, Sydney.

Wheeler A., Berry P. A. and Patrick M., (2001), "Improvements to the AISC Design Method for the Web-Side-Plate Steel Connection", Proceedings, The Australasian Structural Engineers Conference, Gold Coast, May 2001, pp 429–436.

Tubular Structures XI – Packer & Willibald (eds)
© 2006 Taylor & Francis Group, London, ISBN 0-415-40280-8

Tests on welded connections made of high strength steel up to 1100 MPa

M.H. Kolstein
Delft University of Technology, Faculty of Civil Engineering and Geosciences, Delft, The Netherlands

O.D. Dijkstra
TNO Building and Construction Research, Delft, The Netherlands

ABSTRACT: High strength steels with yield strength from 690 up to even 1100 MPa are very promising for high performance steel structures, e.g. bridges, cranes, transport, ships, offshore structures, etc. In spite of the base materials with excellent properties, today the design and fabrication of high strength steel structures are still facing difficulties. In the structural design the difficulties are in the assessment of the fatigue life in the joints, the buckling strength of the thin members and the deformation capacity of the (welded) connections. Several design as well as fabrication aspects of these steels have been investigated. This paper reports the experiments carried out covering some aspects related to the structural design of structures made of high strength steel.

1 INTRODUCTION

The interest for the application of steels with a very high yield and tensile stress, so called very high strength steel (VHSS), has been increased in recent years. In the test programme described in this paper we consider VHSS with a strength which lays a factor two to three higher than the maximum strength of the commonly used structural steels in the design codes. This increasing interest is based on economic benefits using VHSS. These benefits are due to weight reductions, which lead to reduce use of material and savings during fabrication (e.g. welding costs). Another aspect of this growing interest is the fact that using this type of VHSS in specific structures like very heavy lifting cranes, a real new scale of structure becomes possible. To achieve the same level of safety and comfort with the lightweight VHSS structures special attention has to be given to the fabrication and design of VHSS structures. In fabrication the welding and cold deformation of VHSS can lead to weld defects and material deterioration resulting in reduced safety of the structure. In design special attention has to be paid to the design of connections to avoid high stress concentrations and to achieve sufficient deformation capacity. The fatigue design of a VHSS structure needs more attention, as the fatigue strength is not proportional to the yield strength. Special attention has to be paid to the stiffness of a VHSS structure as the reduced dimensions may lead to more deflection and dynamic vibrations. The research project (IHSSS, 2002) Integrity of High Strength Steel Structures concerns the research into the design and fabrication of structures made of VHSS

with special emphasis on the structural safety. Due to lack of data, most existing codes for steel structures are restricting their scope to the steel grades up to yield strength of 355 MPa. The prEN 1993-1-1 (2003) – General Structural Rules for Steel Structures – deals with steel grades to S460. Additional rules for the extension of EN 1993 to steel grades S500 to S700 are given in prEN 1993-1-12 (2003). The rules for fabrication are given in ENV 1090 – 1 (1996) – Execution for Steel Structures. The sponsors of this project are steel suppliers, designers, fabricators and users of steel structures. The users are interested in better operational performance (lighter, faster, higher loaded) than conventional structures. Fabricators are interested in economic production methods, fulfilling all requirements. The activities are concentrated on two areas of interest, fabrication and structural design. TNO Institute of Industrial Technology carries out the fabrication work package. TNO Building and Construction Research and Delft University of Technology consider the structural design. This paper presents the experimental programme of the structural design work package.

2 DESIGN ASPECTS

2.1 General design philosophy

Depending on the application, a VHSS structure is built up as a space frame from (tubular) members (e.g. a crane jib) or as a box structure with plates (e.g. traffic bridges or ships). The design practice is mainly based on codes for conventional steel grades (prEN 1993-1-1, 2003 and prEN 1993-1-12, 2003). The research will concentrate on the applicability of these rules using the very high strength structural steels.

Currently using advanced finite element programs often carries out the general design of steel structures. Modelling of the structure and the interpretation of the numerical results is closely related to both the type of structure and the type of elements used and often needs specialist knowledge. The economical and technical advantages of using high strength steels can only be considered in the right way if at least the design predicts the behaviour of the structure under working conditions. In general, the use of VHSS is more attractive if strength is the governing factor in the design. Two stress states, deformation in the connections and fatigue design, are briefly discussed below.

2.2 *Deformation in the connections*

In a conventional steel structure the ratio of the tensile strength to the yield strength is relatively high and the weld is generally overmatched. The result is that before failure of the joints the connecting members or plates will yield, resulting in a large deformation capacity and a deformation tolerant structure. For VHSS the situation is more complex due to a higher yield to tensile strength ratio and under matched welds.

The lower ratio of the tensile strength to the yield strength means that the deformation capacity will be lower in connections and structural parts with (bolt) holes or other area reducing effects (e.g. fatigue cracks) and in spite of good toughness properties of the material (base and weld material) a low deformation failure can occur.

At the moment VHSS cannot be welded with an overmatched weld metal. In welded connections with under matched weld metal the deformation will be restricted to the weld metal, also resulting in a low deformation failure.

In the design of an VHSS structure the deformation capacity required at joints has to be determined from an overall analysis. A more detailed joint analysis should ascertain whether the required deformation capacity is available.

2.3 *Fatigue design*

The design stress in a VHSS structure is higher than in a structure made from conventional steel and the stress due to self weight will be lower. This results in absolute and relative higher stress variations due to the external load. As the fatigue strength is not proportional to the yield stress, fatigue is more often governing the design. For structures with a high number of cycles during the lifetime (e.g. bridges, ships) high cycle fatigue (HCF) is important. For structures with a low number of high stress cycles (e.g. cranes for heavy lifting operations) low cycle fatigue (LCF) has to be considered and for structures with a low number of very heavy loads in a survival condition (e.g. minesweepers) ratcheting may be important. It is expected that the HCF approach for conventional steel can be regarded as a lower approximation for VHSS structures. The LCF can be modelled with the cyclic stress-strain relation, an elastic stress concentration factor, Neuber's rule and strain controlled low-cycle-fatigue tests on small-scale specimens.

3 TEST PARAMETERS

3.1 *Strength of base material*

In the project it was decided to go up to material with yield strength of 1100 MPa. So, this will be the upper limit of the material used in this project. At the moment this is also the upper boundary of available material. For comparison and to make modelling possible, joints of material with lower yield strength have to be tested as well. The main purpose of testing material with lower yield strength is to have joints with overmatching welds. Material with yield strength of 690 MPa can be welded with overmatching welds. Therefore this level was chosen as the minimum value to be used in this project.

3.2 *Strength of weld material*

The strength of the weld material is an important parameter in the behaviour of a welded joint. Especially the strength relative to the base material is important. In steel with normal strength a weld metal will be chosen with a higher strength than the base material (overmatching welds). When joints with overmatching (OM) welds are loaded up to failure large strains will occur in the adjacent parent material and not in the welds. This means that the deformation capacity of the joint is large and especially a redundant structure has the possibility of load redistribution. This enhances the safety of a structure. In very high strength steel the weld metal will be less strong than the base material (under matching welds). When joints with under matching (UM) welds are loaded up to failure large strains will occur in the welds metal and not in the adjacent parent material. Due to small dimensions of the weld the deformation capacity of those joint can be limited, even when the weld metal itself has a good deformation capacity. This means that the possibility of load redistribution is also limited, as the joint will fail before the load is redistributed. This reduces the safety of a structure. So, the design of a joint in high strength steel with under matched welds needs special attention with respect to the deformation capacity.

3.3 *Global stress concentration factor*

The geometry of a welded joint is often responsible for a variation in stiffness in the joint. This will cause an unequal stress/strain distribution, resulting in a stress concentration factor (SCF). The SCF can be the result of a global variation in stiffness or of a more local variation in stiffness (e.g. due to the weld geometry). The latter will be dealt with in the next section. In joints with high global SCF the stress and strain distribution will be more unequal than in joints with low SCF. In joints with overmatched welds the unequal strain distribution can be compensated by the large deformation capacity in the parent material (see section above). However in joints with under matched welds the unequal strain distribution will be localized in the weld metal. Very high SCFs can result in local failure of the weld due to the high strains at the SCF location. This can reduce the deformation capacity and with this the safety of the structure. So, the design of high strength steel joints with under matched welds needs special attention with respect to the SCF.

3.4 *Weld type*

The incoming members at a joint can be welded together in different ways. In general the following weld types are used:

* Full penetration weld,
* Partly penetration weld,
* Fillet weld.

It is clear that the full penetration weld is the weld with the lowest local SCF as there is a complete connection. The fillet weld has a high local SCF due to the incomplete penetration. The local SCF of the partly penetration weld is depending on the amount of partly penetration.

3.5 *Material thickness*

The material thickness is a parameter in the high cycle fatigue strength of a welded joint. The thicker the plates the lower the fatigue under the same stress ranges. It is likely that the low cycle fatigue strength has also a thickness effect. Therefore tests at two thicknesses are foreseen.

3.6 *Loading mode*

A welded connection can be loaded in tension, compression, bending or shear. In heavily loaded joints tension and/or shear will be the most likely loading mode.

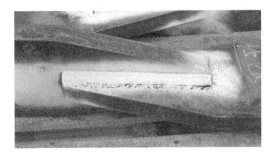

Figure 1.   Cross plate connection (Specimen type A).

Figure 2.   X-joint with low SCF (Specimen type B).

4   TEST PROGRAMME

4.1 *Joint geometry*

The joint geometry to be used in the design part must be practical on one side and serve as a basis for the modelling. Therefore joints with and without an SCF are included in the programme.

The following geometry types are chosen (see Figure 1 to Figure 3):

* A: Cross plate connection with fillet welds loaded in shear
* B: X-joint with load carrying full penetration welds with low SCF
* C: X-joint with load carrying full penetration welds with high SCF

In type A the weld is loaded in shear. There will be unequal stress distribution along the weld and at the end of the plates there is an SCF in the plate. In type B there are load carrying full penetration welds. There is no SCF along the weld. In type C there are also load carrying full penetration welds. Due to the rotated connection of the transverse plate on each side of the connecting plate there is a high SCF along the weld.

The width of the plates is 5 times the thickness. These geometries are a good starting point for the modelling of the behaviour of joints in high strength steel.

4.2 Material properties

Table 1 gives the chemical composition of the different steel grades used for the fabrication of the specimens. Table 2 gives results of the mechanical properties. Re and Rm stand for yield strength and ultimate strength respectively. The data in both tables are taken from the certificates of the material.

4.3 Welding parameters

The dimensions of the welds are as practical as possible. For the type B and C specimens this means that the weld will be carried out as a full penetration weld with the weld base at the intermediate plate as close to the plate thickness as possible. Doing so the plate is fully connected and in an overmatching situation the connection is stronger than the connecting plate. In the under matched situation the possibility that the weld will fail before the plate is present.

For the type A specimens the weld design is more complicated and for the test specimens the design depends on the aim of the tests. An important item is the relative strength of the welded connection to the plate. It was decided that the weld dimension should be such that the weld will fail before the plate in the overmatched situation.

Table 3 gives some data of the welding processes and welding consumables provided by the fabricator of the specimens. As shown the S690 specimens are welded using Flux Cord Arc Welding (FCAW) and the S1100 specimens using Shielded Metal Arc Welding (SMAW). The mechanical properties of the filler metal are theoretical values. In the fabrication part of the research program these values will be verified by measurements.

Figure 3. X-joint with high SCF (Specimen type C).

Table 2. Mechanical properties of the different steel Grades – data from certificates.

Steel grade	Plate (mm)	Re (MPa)	Rm (MPa)	A (%)	Re/Rm	Charpy-V (J) −20°C	Charpy-V (J) −40°C
S690	12	811	842	15	0.96	251	
S690	40	792	835	16.1	0.95		204
S1100	10	1197	1432	11	0.81		29
S1100	40	1106	1325	10.7	0.83		39

Table 3. Welding procedure specification data.

Steel grade	Plate (mm)	Specimen type	Weld process	OM weld	UM weld
S690	12/40	A/B/C	FCAW	S830	S550
S1100	10/40	A/B/C	SMAW		S900

Table 1. Chemical composition.

Steel grade	Plate (mm)	C	Si	Mn	P	S	Cr	Ni	Mo
S690	12	.16	.19	.87	.012	.002	.33	.06	.22
S690	40	.171	.324	1.33	.014	.012	.319	.106	.355
S1100	10	.16	.23	.86	.007	.002	.6	1.89	.586
S1100	40	.147	.18	.28	.007	.0007	1.48	2.62	.434

Steel grade	Plate (mm)	V	Ti	Cu	Al	Nb	B	N	CE
S690	12		.004	.03	.085	.026	.002	.0038	.42
S690	40	.002			.065	.03	.0016		
S1100	10	.029	.004	.04	.066	.02	.002	.005	
S1100	40	.004	.002		.041	.001	.0001		

4.4 *Parameters in the test programme*

The parameters to be investigated in the test programme are:

- Specimen type (A, B and C)
- Test type (static, fatigue and ratcheting)
- Material thickness (10/12 and 40 mm)
- Base material strength (690 and 1100 MPa)
- Relative weld strength (overmatched, only for 690 base material) and under matched

A review of the test programme including 48 tests is shown in Table 4. Each test series includes specimens of type A, B and C.

5 TEST PROCEDURES

5.1 *Static tests*

In the static tests the specimens will be loaded until failure occurs. To get a stable curve to failure the test will be deflection controlled. The main result will be a load deflection curve (Figure 4). For detailed information strain gauges will be fixed on the specimen at specific locations. As given in Figure 4 it is expected that for an over matched weld the failure load and the deformation capacity will be higher.

Table 4. Test plan.

Test series	Number of tests	Re base (MPa)	Re weld (MPa)	Plate (mm)	Parameter
I	12	690	900	12	Reference
II	9	690	900	40	Thickness
III	12	690	490	12	Under matched
IV	9	1100	900	10	High strength
V	6	1100	900	40	Thickness/ High strength

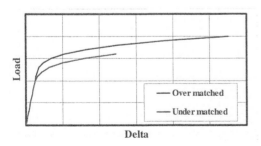

Figure 4. Static test procedures.

5.2 *Ratcheting tests*

The ratcheting tests will be carried in a load-controlled situation. In the ratcheting tests the specimens will be loaded until a certain deformation is reached. Than the specimen will be unloaded and subsequently loaded again to a larger amount of deformation (Figure 5). Based on the static test experience the deformation steps will be chosen in such a way that failure is expected to occur in approximately 10 loading steps. The deflection curve (Figure 5) will be recorded during the test and strain measurements will be taken as well.

5.3 *Low cycle fatigue tests*

The low cycle fatigue tests will be carried out in a load-controlled situation. The load range will be chosen in such a way that failure is expected between 1000 and 10000 cycles. The load deflection curve (Figure 6) will be recorded during the test and strain measurements will be taken as well.

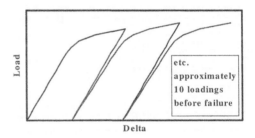

Figure 5. Ratcheting test procedure.

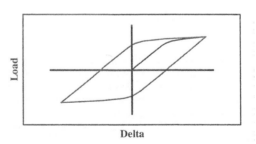

Figure 6. Fatigue test procedure.

Table 5. Test procedures.

Test series	Static	Ratcheting	Fatigue
I	X	X	X
II	X	X	X
III	X	X	X
IV	X		X
V	X		X

Figure 7a. 10.000 kN Test rig.

Figure 7b. 600 kN Test rig.

6 TEST RIG AND MEASUREMENTS

6.1 *Test rig*

Depending on the expected load level needed to obtain failure of the test specimen, two servo hydraulic test rigs are available. The capacity of these test rigs are 10.000 kN (Figure 7a) respectively 600 kN (Figure 7b).

6.2 *Measurements*

All test specimens have been instrumented with displacement transducers measuring locally the deformation of the weld or the deformation of the whole connection. A typical example of an instrumented specimen is shown in Figure 8.

Locally means that the measuring length includes the weld and the transverse plate (LVDT_1 and LVDT_2). The measuring length for the overall measurements is about 10 times the plate thickness (LVDT_3 and LVDT_4). Single strain gauges are used to measure the nominal stress in the connection plates. Strip strain gauges are used to obtain information about the strain development at the hot spot stress location of the welded connection.

7 TEST RESULTS

In this paper the results (Kolstein, 2006 and Dijkstra, 2006) of the X-joint specimens with load carrying full

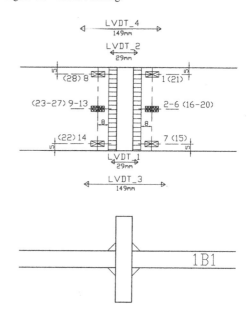

Figure 8. Instrumentation.

penetration welds with low SCF (Specimen type B) will be presented only.

7.1 *Static strength*

Typical results obtained are gathered in Table 6. The relative deformations in this table are the values measured at the maximum test load. In some cases

Table 6. Deformation capacity specimens type B.

Steel grade	Plate (mm)	Weld	Fmax (kN)	Failure mode	Deformation (%)	
					~10 T	local
S690	12	OM	545	plate	>2	~3.5
S690	40	OM	6509	plate	~6	~6
S690	12	UM	580	plate (weld toe)	>3	~6
S1100	10	UM	570	weld (HAZ)	~1.5	~1.5
S1100	40	UM	>9600		>1.5	>1.5

Table 7. Static strength specimens type B.

Steel grade	Plate (mm)	Weld	Fmax (kN)	Failure mode	Predicted/Test F_r/F_u
S690	12	OM	545	plate	1.11
S690	40	OM	6509	plate	1.03
S690	12	UM	580	plate (weld toe)	1.04
S1100	10	UM	570	weld (HAZ)	1.25
S1100	40	UM	>9600		>1.10

Table 8. Ratcheting results specimens type B.

Steel grade	Plate (mm)	Weld	Ratcheting Fmax (kN)	Static Fmax (kN)
S690	12	OM	570 plate	545 plate
S690	40	OM	6602 plate	6509 plate
S690	12	UM	590 weld	580 plate (weld toe)

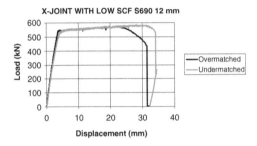

Figure 9. Overmatched versus under matched welding.

the LVDT's dropped down from the specimens too early. To compare the results in terms of overmatched welding versus under matched welding the piston displacement of the hydraulic actuator (S01) has been used. From the results as shown in Figure 9 it can be concluded that difference between the overmatched and under matched test specimen is very small.

Test results and predicted values for the X-joints with low SCF following the design rules in Eurocode 3 are gathered in Table 7. For these specimens the predicted values of the ultimate strength are based on the actual material properties. Results show the following:

In all cases the predicted ultimate strength F_r is higher than the maximum test load F_u. As expected the overmatched specimens failed all in the base material. Rather unexpected is the plate failure in the under matched specimen 3B1 (S690 12 mm thick). The crack however is located very close to the weld, which probably influenced here the contraction capacity of the base material. On the other hand the weld reinforcement for this type of specimen was relatively large. So, the under matched weld metal is compensated by the larger weld. Specimen 5B (S1100 40 mm thick) did not fail due to the capacity of the test rig.

7.2 Ratcheting behaviour

A review of the results obtained from the ratcheting tests is gathered in Table 8 and Figure 10 to 12. It can be concluded that repeated loading until the plastic zone

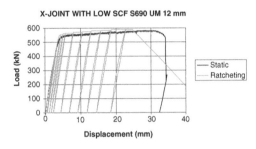

Figure 10. Static versus Ratcheting behaviour (1).

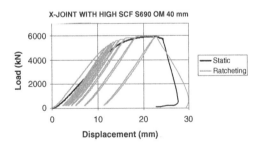

Figure 11. Static versus Ratcheting behaviour (2).

hardly influences the ultimate load or the deformation capacity of these welded joints.

7.3 Fatigue behaviour

The results of the low cycle fatigue tests are gathered in Table 9 and Figure 13. Despite the small number of test results it can be concluded that the fatigue endurance of the S1100 specimens are considerable longer than the S690 specimens.

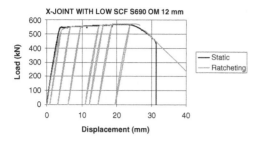

Figure 12. Static versus Ratcheting behaviour (3).

Table 9. Fatigue test results specimens type B.

| Steel grade | Plate (mm) | Weld | Number of cycles | |
			666 MPa	750 MPa
S690	12	OM	1535	1308
S690	12	UM	2933	5389
S1100	10	UM	32285	24150

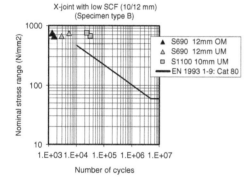

Figure 13. Fatigue test results specimens type B.

Next to this way of presenting the obtained results are analysed on a LCF approach based on notch strains. These strains are extrapolated values of strains measured at several locations perpendicular to the weld toe. The results compared to the S690 curves presented by Boller en Seeger (1987) are not reported in this paper.

8 CONCLUDING REMARKS

Overmatched welded joints in S690 have a good strength and deformation capacity. Under matched weld metal can be compensated by weld reinforcement. In general the ratio weld area at the vertical plate – plate thickness should be at least inversely proportional to the relative weld strength ($A_{weld} \geq (Rm_{plate}/Rm_{weld}) A_{plate}$. Under matched weld metal and incomplete penetration in a small scale S1100 X-joint lead to low deformation failure. Ratcheting loads give

Figure 14. Plate failure S690 40 mm thick.

nearly the same results with respect to strength and deformation capacity as static loads. The fatigue tests show a longer life for the small S1100 specimens.

ACKNOWLEDGEMENTS

The steering committee of this research project is gratefully acknowledged for their permission to publish this paper.

REFERENCES

IHSS project 'Integrity High Strength Steel Structures', Project Office for Research on Materials and Production technology (PMP), Apeldoorn, The Netherlands (2002).

prEN 1993-1-1 – Eurocode 3: Design of steel structures – Part 1-1: General rules and rules for buildings, Final draft, European Committee for Standardization, Brussels (2003).

ENV 1090 – 1 – Execution of steel structures – Part 1: General rules and rules for buildings, European Committee for Standardization, Brussels (1996).

prEN 1993-1-12 – Eurocode 3: Design of steel structures – Part 1-12: Additional rules for the extension of EN 1993 to steel grades S 500 to S 690, 2nd draft, European Committee for Standardization, Brussels (2003).

Boller, Chr., Seeger, T. Materials data for cyclic loading, Elsevier (1978).

Kolstein M.H., Dijkstra, O.D. Integrity High Strength Steel Structures – Experimental results, Stevin report 6-05-7 (2006).

Dijkstra, O.D., Van Wortel, J., Kolstein, M.H. Integrity High Strength Steel Structures – Summary report, TNO report 2005-BCS-R0470 (2006).

Tubular Structures XI – Packer & Willibald (eds)
© 2006 Taylor & Francis Group, London, ISBN 0-415-40280-8

Elasto-plastic behavior of high-strength bolted friction joints for steel beam-to-diaphragm connections

N. Tanaka
Kajima Technical Research Institute, Kajima Corporation, Japan

A. Tomita
ARTES Corporation, Japan

ABSTRACT: Bolted beam-to-diaphragm connections are often critical sections to under earthquake stresses. The Architectural Institute of Japan (AIJ) has recommended a coefficient for joints as a safety factor: $\alpha_y = 1.0$ to allow beam yield and $\alpha_u = 1.10 \sim 1.25$ to prevent rupture. However, the target for these coefficients is beam-to-beam joints subject to comparatively small stresses in welded diaphragm connections. In this study, eight cantilever-beam-type specimens incorporating bolted friction joints were cyclically loaded to determine a suitable joint coefficient to meet the AIJ recommendation. Test results and calculations led to conclusion that for grade of SS400 (mild steel), α_y needs to be greater than ζ (bending moment gradient), and α_u needs to be over 1.35.

1 INTRODUCTION

Three types of beam-to-column connections have been mainly used in Japan. These are outer, through, and inner diaphragms, as shown in Figure 1a to c, respectively. Since the 1995 Hyogoken-Nanbu Earthquake, mixed connection types such as the improved through diaphragm (shown in Figure 1d), proposed by Kurobane & Azuma (2003), have been also used to increase structural performance and construction speed. The through and inner diaphragms are welded to the column in the section that is critical under earthquake loads. Therefore, the beam-to-beam joints, which employ exclusively high-strength bolts (HSBs) utilizing friction, suffer no severe load. However, the diaphragms are jointed to the beam directly by HSBs in the outer and improved diaphragm, leading to concern that the critical section under earthquake load tends to be in the middle of the bolted joints.

Generally, two design criteria have been employed for the bolted friction joint in order to ensure the required strength and deformability. One is that no bolt slippage occurs before the beam yields, satisfying Equation 1, and the other is that no fracture occurs before the beam reaches its full plastic capacity, satisfying Equation 2.

$$_jM_y \geq a_y \cdot M_y \tag{1}$$

$$_jM_u \geq a_u \cdot M_p \tag{2}$$

where $_jM_y$, $_jM_u$ are yield and maximum bending moment of beam joint, respectively. M_y, M_p are yield and full plastic bending moment of beam end, as indicated in Figure 1. α_y, α_u are coefficients for yield and rupture, respectively.

For the through and inner diaphragms, so-called welded diaphragms, the Architectural Institute of Japan (AIJ) has recommended a joint coefficient as a safety factor, as shown in Table 1, in which a coefficient for beam-to-column connection is also indicated for reference. Therefore, suitable coefficients need to be determined for the outer and the improved diaphragms to ensure the same performance as for the welded diaphragm.

To investigate the elasto-plastic behavior and the required coefficient, the authors (Tanaka et al. 2003) have already carried out loading tests on eight cantilever-beam-type specimens consisting of outer diaphragm-to-beam bolted friction joints. This paper mainly reports the joint coefficients.

2 EXPERIMENTS

2.1 Specimens

Eight mock-up cantilever-beam-type specimens incorporating bolted friction joints, shown in Figure 2, were used. An H-shaped column was used instead of a hollow column for convenience of the specimen's setting and its irrelevance to the joint's behavior. Although specimens were of a typical outer diaphragm, the test results can also be utilized for the improved diaphragm-to-beam joints, as shown in Figure 1d.

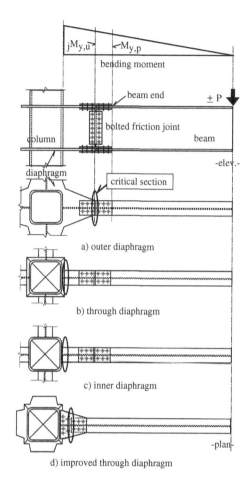

a) outer diaphragm

b) through diaphragm

c) inner diaphragm

d) improved through diaphragm

Figure 1. Beam-to-column connections used in Japan.

Table 1. Coefficient.

| Steel grade | α_y | Coefficient | |
| | | α_u | |
		Joint (beam-to-beam)	Connection (beam-to-column)
SS400	1.00	1.25	1.40
SM490		1.20	1.35
SN400		1.15	1.30
SN490		1.10	1.25

(Note) SS400: mild steel, SM490: low alloy high-strength steel, SN400: mild steel used exclusively for building structure, SN490: low alloy high-strength steel used exclusively for building structure.

No.1 to No.4 had similar beam sections (small-width beam series) while No.5 to No.8 had larger beam sections (middle-width beam series). The main parameters were α_y and α_u as shown in Table 2. The α_y

and α_u values were calculated from Equations 1 to 9. Each beam series had four α_y values and almost constant α_u. The α_y values were 1.12 to 1.35 and α_u was around 1.19 for the small-width beam series, while the α_y values were 1.08 to 1.24 and α_u was 1.21 for the middle-width beam series. A larger coefficient is less vulnerable to slip or fracture. Therefore, No.1 or No.5 were most vulnerable to slip in each series. Each α_u value was a little smaller than the recommended value of AIJ, $\alpha_u = 1.25$. The arrangement and number of HSBs and the thickness of the splice-plate were varied to ensure the required α_y and α_u. The grade of the steel and the HSB of the specimen were SS400 (mild steel) and F10T (tensile strength 1,000∼1,200 N/mm²), respectively, and the mechanical properties are listed in Table 3. The slip factor of the HSB friction joint was 0.59 as the average of three slip-test results.

2.2 Loading procedure

All specimens were loaded cyclically up to maximum strength. The load was applied at the tip of the beam by a hydraulic jack with the deflection controlled at $0.5\,\delta_p$, $2.0\,\delta_p$, $4.0\,\delta_p$, and $6.0\,\delta_p$. The δ_p value was a calculated deformation obtained from M_p/K. M_p was the full plastic bending moment of the beam and K was the initial stiffness obtained from the test, as shown in column (1) of Table 4. To prevent out-of-plane deflection of the beam, constraining rigs were set up on both sides.

The load, displacements, and strains of the specimens were measured during the test.

3 RESULTS

3.1 Load-deflection relationships

The load-deflection curves and the sequence of each loop are shown in Figures 3 and 4 where the size of each loop is drawn in slightly reduced. The white circle and white square indicate yielding of the beam and first slippage of the HSBs, respectively, while the jag indicates slippage of the HSBs. The figure shows that the beam flange yielded first in all specimens except No.5, in which slippage preceded yielding, and each slippage occurred at different times. That is to say, the bolt of No. 1 or No. 6 slipped immediately after beam yielding, while the slippage of No. 4 or No. 8 took more than a few loading cycles. This confirms that the larger α_y ensures slippage delay and inflation of the hysteresis loop, which implies an increase in energy absorption. Furthermore, α_y, although it is based on the nominal value, needs to be greater than 1.08 to ensure yield precedence according to No.5.

3.2 Yield strength

Table 4 shows the initial stiffness, the yield and the slipping loads as well as yielding strength calculations.

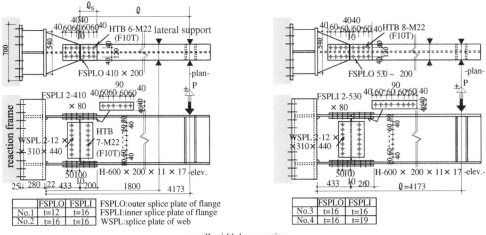

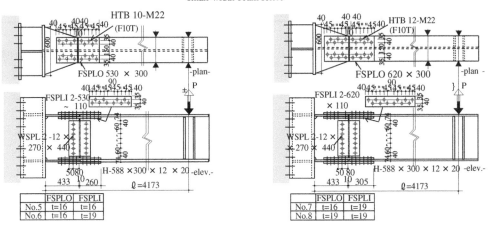

Figure 2. Test specimens.

Table 2. List of specimens.

Specimen	Series	Beam	Joint coefficient (designed)		Bending moment gradient	Thickness of splice plate of beam flange (mm)		High strength bolt	
			α_y	α_u	ζ	inside	outside	flange	web
No.1	Small-width beam	H-600 × 200 ×	1.12	1.19	1.06	12	16	6-M22	7-M22
No.2		11 × 17	1.17	1.19		16			
No.3		(SS400)	1.27	1.18	1.07			8-M22	
No.4			1.35	1.18			19		
No.5	Middle-width beam	H-588 × 300 ×	1.08	1.21			16	10-M22	
No.6		12 × 20	1.14	1.21			19		
No.7		(SS400)	1.14	1.21	1.09			12-M22	
No.8			1.24	1.21		19			

(Note) $\sigma_y = 235$, $\sigma_u = 400 \, \text{N/mm}^2$ are used in the design.

Table 3. Mechanical properties of the material.

Thickness or Diameter (mm)	Yield strength (N/mm²)	Tensile strength (N/mm²)	Yield strength ratio (%)	Elongation (%)	Remarks
16.2	292	443	66	31	for flange of No.1–No.4
10.6	335	470	71	25	for web of No.1–No.4
19.4	304	443	67	32	for flange of No.5–No.8
11.8	340	470	72	25	for web of No.5–No.8
11.5	407	492	83	28	for splice plate of 12 mm
15.5	263	434	61	32	for splice plate of 16 mm
18.6	273	454	60	28	for splice plate of 19 mm
M22	1030	1086	95	18	for HSB

(Note) Each value for plate is the average of three coupon specimens (JIS Z 2201-A) Value for HSB is determined by a tensile test of bolt itself.

Table 4. Test results of the initial stiffness, yielding and slipping load and the corresponding calculations.

Specimen	Initial stiffness (kN/mm) Test K (1)	Yielding load (kN) Beam Test (2)	Yielding load (kN) Beam Calculation $_bP_y$ (3)	(2)/(3)	Slipping load (kN) Joint Test (4)	Slipping load (kN) Joint Calculation $_jP_y$ (5)	(4)/(5)	Coefficient for yielding α_y Nominal (6)	Coefficient for yielding α_{ry} Real (5)/(3)
No.1	6.89	203	195	1.04	211	241	0.87	1.12	1.23
No.2	7.08	194	196	0.99	−215	235	0.92	1.17	1.20
No.3	7.06	209	199	1.05	−251	240	1.05	1.27	1.20
No.4	7.49	211	201	1.05	−249	259	0.96	1.35	1.29
No.5	9.99	291	326	0.89	−272	326	0.83	1.08	1.00
No.6	10.1	295	327	0.90	−316	353	0.89	1.14	1.08
No.7	10.3	307	329	0.93	350	345	1.01	1.14	1.05
No.8	10.4	329	333	0.99	−371	382	0.97	1.24	1.15

The test result in column (2) of Table 4 is the average of the values obtained by the strain gauges at the beam end and General Yield Point (GYP) method with the outline shown in Figure 5. The calculation, $_bP_y$ in column (3), which was the load when the beam reached the yield bending moment at the beam end (refer to Figure 1), shows good correspondence with majority of the test results, as shown in the column (2)/(3) of the Table. The disagreement in No.5 is due to the precedence of HSB slippage. For the joint, the slip load is indicated in column (4), where the minus sign shows where slippage occurred during negative loading. The calculation, $_jP_y$ in column (5), was based on Equations 3 to 5 proposed in the recommendation for design of connections by AIJ 2001. Figures 6 and 7 outline the calculation method.

$$_j M_y = {}_j M_{fy} + {}_j M_{wy} \tag{3}$$

$$_j M_{fy} = \min\{{}_j M_{fy1}, {}_j M_{fy2}, {}_j M_{fy3}\} \tag{4}$$

$$_j M_{wf} = \min\{{}_j M_{wy1}, {}_j M_{wy2}\} \tag{5}$$

where $_jM_{fy}$ is the yield bending strength of the flange determined as the minimum of three bending strengths shown in Figure 7: $_jM_{fy1}$ (slip strength of HSBs), $_jM_{fy2}$ (yield strength of splice plates after bolt holes reduction, added 33% of the slip strength of the first line of the HSBs), and $_jM_{fy3}$ (yield strength of splice plates). $_jM_{wy}$ is the yield bending strength of the web determined as the minimum of two bending strengths also shown in Figure 7: $_jM_{wy1}$ (slip strength of HSBs) and $_jM_{wy2}$ (yield strength of splice plate after bolt holes reduction). In the calculation using real material properties, $_jM_{fy3}$ and $_jM_{wy1}$ were minimum for all specimens except No.1, whose strength was $_jM_{fy1}$ of the flange and $_jM_{wy2}$ of the web. The load P can be obtained from the calculated M divided by the length between the load application point and the joint center, ℓ, as shown in Figure 2.

The calculation overestimates most of the test results, as shown in the column (4)/(5) of Table 4. The main reason for this is that the slip factor decreased due to beam yield. Based on Equations 1, the real

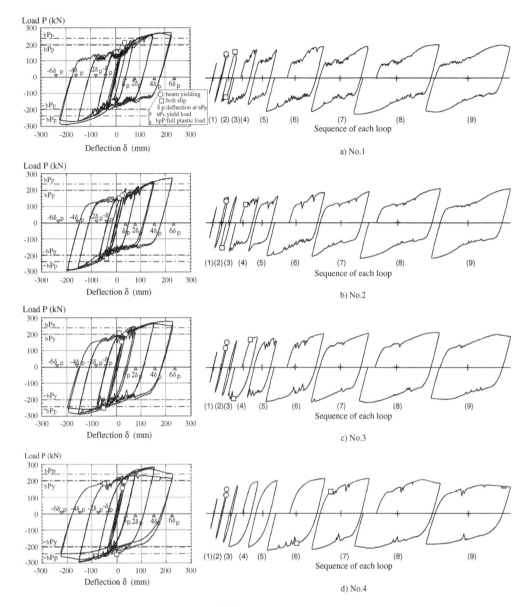

Figure 3.   Load-deflection curves for the small-width beam series.

joint coefficient α_{ry} for yielding can be calculated by $_bP_y$ and $_jP_y$. Each α_{ry} is shown in the column (5)/(3) along with the nominal coefficient α_y in column (6) of Table 4. Precious indication is that the slippage of the HSB preceded the beam yielding in specimen No.5 even if α_{ry} equals 1.0, as recommended by AIJ 2001, shown in Table 1. This is mainly due to the difference between the bending moments at beam-end and joint center. As the difference between the bending moments, bending moment gradient, ζ, is defined by Equation 6 and shown in Table 2. For the rest of the

specimens in which the beam yielding preceded, the minimum value of α_{ey} was 1.05 for No.7. This is nearly equal to the bending moment gradient, 1.07. Therefore, it is recommended that the coefficient for yielding be greater than the bending moment gradient of the joint.

$$\zeta = \frac{\ell}{\ell - \ell_s} \qquad (6)$$

where ℓ is beam length from joint center to beam tip, ℓ_s is half the joint, as shown in Figure 2.

579

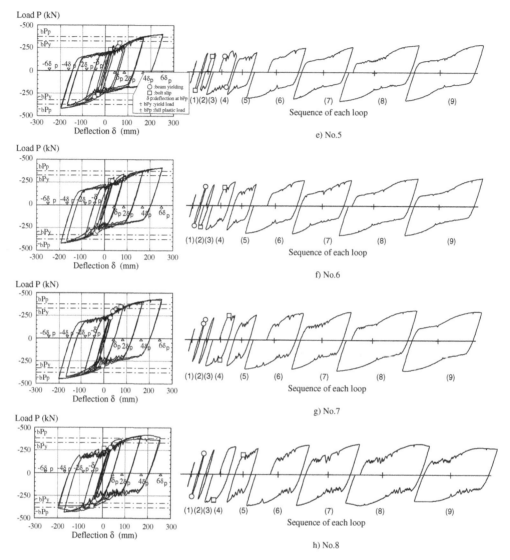

Figure 4. Load-deflection curves for the middle-width beam series.

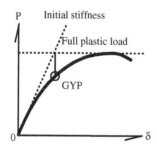

Figure 5. Method to obtain GYP.

3.3 *Maximum strength*

The test results and calculations for the maximum loads and the joint coefficient are shown in Table 5. Local buckling of the flange or web decided the maximum strength of all specimens. Three calculations, based on the maximum connection strength, the full plastic strength, and the local buckling strength, are compared with the test results. $_jP_u$, which represents the maximum connection strength, is given by Equations 7 to 9 described in the recommendation of AIJ 2001. Figures 6 and 8 outline the calculation method.

$$_j M_u = _j M_{fu} + _j M_{wu} \qquad (7)$$

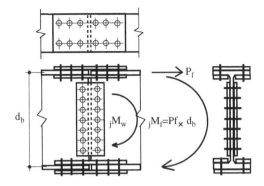

Figure 6. Calculation for joint strength.

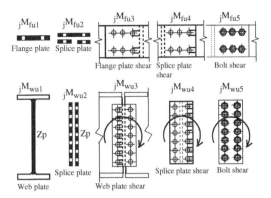

Figure 8. Maximum strength of joint (σ_u is used for plate).

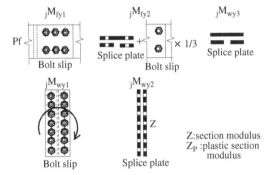

Figure 7. Yield strength of joint (σ_y is used for plate).

$$_j M_{fy} = \min\left\{_j M_{fu1}, _j M_{fu2}, _j M_{fu3}, _j M_{fu4}, _j M_{fu5}\right\} \quad (8)$$

$$_j M_{wy} = \min\left\{_j M_{wu1}, _j M_{wu2}, _j M_{wu3}, _j M_{wu4}, _j M_{wu5}\right\} \quad (9)$$

where $_j M_{fu}$ is the maximum bending strength of the flange, determined as the minimum of five maximum bending strengths shown in Figure 8: $_j M_{fu1}$ (strength of flange with bolt holes deducted), $_j M_{fu2}$ (strength of splice plates with bolt holes deducted), $_j M_{fu3}$ (strength due to shear rupture originating at the bolt hole in the end of the flange), $_j M_{fu4}$ (strength due to shear rupture originating at the bolt hole in the end of the splice plate), and $_j M_{fu5}$ (shear strength of the HSBs). $_j M_{wu}$ is the maximum bending strength of the web determined as the minimum of five bending strengths also shown in Figure 8: $_j M_{wu1}$ (strength of web), $_j M_{wu2}$ (strength of splice plates with bolt holes deducted) $_j M_{wu3}$ (strength due to shear rupture originating at the bolt hole in the end of the web), $_j M_{wu4}$ (strength due to shear rupture originating at the bolt hole in the end of the splice plate), and $_j M_{wu5}$ (shear strength of the HSBs).

In the calculation using real material properties, $_j M_{fu1}$ and $_j M_{wu2}$ were minimum for all specimens. Next, $_b P_p$ is the strength when the beam reaches the full plastic bending moment $_b M_p$. The local buckling strength, $_b P_{lbc}$, is calculated by Equation 10 based on the studies of Kato & Nakao (1994), Kato (2000a,b).

$$_b M_{lbc} = S' \cdot _b M_p \quad (10)$$

where $_b M_{lbc}$ is local buckling moment and S' is a coefficient of local buckling strength obtained mainly from the width-thickness ratio of the beam member.

The maximum connection strength, $_j P_u$, is smaller than the test results shown in column (7)/(8) of Table 5, while the local buckling strength, $_b P_{lbc}$, shows good correspondence as shown in column (7)/(10) of the same table. It is natural that the connection strength, $_j P_u$, should be larger than the test results or the local buckling strength because each specimen was not fractured at the joint but buckled in the flange or web. One of the reasons why $_j P_u$ is smaller than the test result could be residual frictional resistance of the joint after HSB slippage.

For the coefficient for rupture, the real design value, α_{ru}, is in column (8)/(9) of the table. This is much smaller than the nominal α_u used for the design of the specimen in column (11) of the table. This is because of the difference between the minimum strengths in Equations 8 and 9, due to the difference between the nominal and real strengths of the material. This indicates the importance of taking account of the dispersion of the real value from the nominal value of the material strength into the joint design. Generally, the dispersion is expressed as the ratio of real to nominal value, β, and $\beta = 1.15$ is recommended for SS400 (mild steel) in Japan. α_{eu} in column (7)/(9) of the table represents the real coefficient, being test results divided by the full plastic load, $_b P_p$. If α_{eu} ensures the precedence of the local buckling of the beam plates, this leads to the conclusion that the required coefficient for rupture design is given by multiplying α_{eu} by β. The results are shown as α_{ureq} in column (12) of Table 5. α_{ureq} is 1.35 to 1.36 for the small-width beam series and 1.26 to 1.34 for the middle-width beam series, averaging 1.32. This value nearly equals the average

Table 5. Test results of the maximum load and the corresponding calculations.

| Specimen | Maximum load (kN) | | | | | | Coefficient for maximum strength | | | |
| | Test (7) | Calculation | | | | | | | | |
		Connection strength $_jP_u$ (8)	Full plastic $_bP_p$ (9)	Local buckling $_bP_{lbc}$ (10)	(7)/(8)	(7)/(10)	α_u Nominal (11)	α_{ru} Real (8)/(9)	α_{eu} (7)/(9)	α_{ureq} (12)
No.1	279	244	236	280	1.14	1.00	1.19	1.04	1.19	1.36
No.2	281	246	237	282	1.14	1.00	1.19	1.04	1.19	1.36
No.3	283	249	240	286	1.14	0.99	1.18	1.04	1.18	1.35
No.4	288	251	242	289	1.15	1.00	1.18	1.04	1.19	1.37
No.5	410	382	374	442	1.07	0.93	1.21	1.02	1.10	1.26
No.6	417	383	375	443	1.09	0.94	1.21	1.02	1.11	1.28
No.7	441	387	378	447	1.14	0.99	1.21	1.02	1.17	1.34
No.8	427	390	382	451	1.10	0.95	1.21	1.02	1.12	1.29

Table 6. Joint coefficient proposed in this research.

| Steel grade | Joint coefficient | | |
	α_y	α_u	
SS400	ζ 1.06–1.09	1.35	$_jM_y \geqq \alpha_y M_y$ $_jM_u \geqq \alpha_u M_P$

of 1.25 (required for joint) and 1.40 (required for connection) for steel grade SS400, as shown in Table 1. It can be interpreted that the beam joint in the critical region should be considered more severely than that of an ordinary beam joint, although it is not to the level of beam-to-column connection. The authors propose the joint coefficients, $\alpha_y = \zeta$ and $\alpha_u = 1.35$, as shown in Table 6. Coefficients for steel grades other than SS400 need further experimental investigation.

4 CONCLUDING REMARKS

Mock-up tests and calculations were carried out to investigate the elasto-plastic behavior and required joint coefficient of outer diaphragm-to-beam bolted friction joints. The conclusions are the followings.

(a) Load-deflection relation is greatly affected by the joint coefficients, α_y and α_u, given in design.
(b) The coefficients recommended by the AIJ, which is targeted to beam-to-beam joints subjected to moderate stress, need to be increased for application to diaphragm-to-beam joints subjected to severe stress in the outer diaphragm. Therefore, new coefficients are proposed here for SS400 (mild steel). For precedence of beam yielding, α_y should be greater than ζ, which denotes the bending moment gradient of the joint. To prevent rupture of the beam and splice plates, α_u needs to be more than 1.35.

ACKNOWLEDGEMENTS

The authors would like to thanks Mr. K. Suzuki of Kajima Technical Research Institute for his help in the experimental works and Mr. K. Kawamoto of ARTES Corporation for his assistance in design of specimens.

REFERENCES

AIJ 2001. Recommendation for design of connection in steel structures (in Japanese), Architectural Institute of Japan.
AIJ 1997. Full-scale test on plastic rotation capacity of steel wide-flange beams connected with square tube steel(in Japanese), The Kinki Branch of the Architectural Institute of Japan
Kato B & Nakao M. 1994: Strength and deformation capacity of H-shaped steel members governed by local buckling (in Japanese). *Journal of structural and construction engineering*. Architectural Institute of Japan. No. 458: 127–136.
Kato B. 2000. Prediction of the vulnerability of premature rupture of beam-to-column connection (in Japanese). *Journal of structural and construction engineering*. Architectural Institute of Japan. No. 527: 155–160.
Kato B. 2000. Prediction of the vulnerability of premature rupture of beam-to-column connection – welded flanges and bolted web type – (in Japanese), *Journal of structural and construction engineering*. Architectural Institute of Japan. No. 529: 175–178.
Kato B. 2003. Seismic design of moment resisting connection, *International journal of steel structures* Vol. 3, No. 3: 163–170.
Kurobane Y. & Azuma K. 2003. Fully restrained beam-to-RHS column connections with improved details, Tubular structures X, ISBN 90 5809 552 5: 439–446.
Tanaka N., Tomita A., Kawamoto K. & Suzuki K. 2003. Elasto-plastic behavior of high strength bolted friction joints at steel beam (in Japanese), *Journal of structural and construction engineering*. Architectural Institute of Japan.

Parallel session 10: Fatigue & fracture

Tubular Structures XI – Packer & Willibald (eds)
© *2006 Taylor & Francis Group, London, ISBN 0-415-40280-8*

Selection of butt-welded connections for joints between tubulars and cast steel nodes under fatigue loading

M. Veselcic, S. Herion & R. Puthli
Research Centre for Steel, Timber and Masonry, University of Karlsruhe, Germany

ABSTRACT: Research work carried out at the University of Karlsruhe on fabrication of welded connections between hot-rolled tubes and cast sections is presented in this paper. The relative advantages of different layouts of the weld seam, weld preparation and non-destructive testing are demonstrated. This also includes the influence of different backing materials and influence of cast defects. A theoretical evaluation will be carried out later in the project, in phase with the experimental study, part of which is described here. These investigations should lead towards the development of a concept for improving the fatigue life for bridges using such connections.

1 INTRODUCTION

In the long history of steel bridge building, bridges made of hollow sections are an interesting alternative for engineers and architects. Especially, the usage of cast steel joints together with hollow sections leads to aesthetic improvements in structures. For example, the tree structures for columns of bridges, which leads to slender and optically aesthetic design, is more frequently being used. Complicated joint types are easily manufactured and the shape can be optimized for the distribution of forces. The stress concentration is reduced, which is very important for fatigue behaviour. Insights in the usage of cast steel in bridges were presented in a previous paper at the 10th International Symposium on Tubular Structures in Madrid (2003).

However, due to the rare use of cast steel, the special properties of the various grades of cast steel are not widely known. The advantages are unknown to a majority of civil engineers and the existing rules and design guides are seldom sufficient. Especially the assessment of when a cast connection or a welded connection should be selected has to be discussed and more experience on this field is required.

Normally, in areas subjected to high local stresses or shear, thick-walled chords are required. However, in order to reduce weight, thinner walled sections are desirable in other areas. The problem here is the transition between the two chord members, which could be fabricated in different variations, such as with same inner diameter, same outer diameter, with or without backing plate, misalignment tolerances etc. There is no known research on the fatigue resistance of such connections. Traditionally, for plates with different wall-thicknesses, a smooth transition of 1:4 is recommended (EN 1993-1-9).

In addition to butt-welded connections between hollow section chords, this problem also occurs for connections of hollow sections with joints made of cast steel.

To cover some of these aspects, two research projects are carried out. They are funded by CIDECT (Project 7W – Fatigue of end-to-end connections) and FOSTA (Project P591 – Economic use of structural hollow sections for highway and railway bridges), a German steel research association.

2 FABRICATION OF THE BUTT WELDS

2.1 *Steel grades and quality*

For the research project, high-strength fine-grained steels are considered for the tests. Substantial advantages lie in the simplified weldability compared to conventional steel (pre-heating can be omitted), which entails a significant cost reduction regarding the manufacturing process. A potential reduction of the member thickness due to the usage of high-strength steel is also regarded as substantial.

The majority of the test specimens are produced in steel grade S355. However, some of the tests will be carried out with specimens of fine-grained steel with a yield strength of $f_y = 460\,\text{N/mm}^2$. This steel grade does not represent a problem for either the hollow section manufacturer or the foundry (Table 1).

The quality levels of the cast steel members have to be chosen depending on the inspection method and welding procedures, where the relevant regulations

Table 1. Steel grades for the hollow sections and corresponding cast steels.

Steel grades	Standard
Hollow sections	
S355J2H	EN 10210
S460NH	EN 10210
Cast steel	
GS20Mn5(V)	EN 10293
G12MnMo 7.4	SEW 520

Table 2. Cast steel quality levels according to EN 12680-1.

Welding area	Quality level 1	V1 S1
Surface:	Quality level 2	V2 S2
Within cast steel	Quality level 3	up to V3

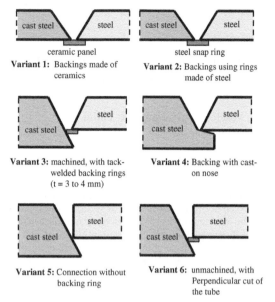

Variant 1: Backings made of ceramics

Variant 2: Backings using rings made of steel

Variant 3: machined, with tack-welded backing rings (t = 3 to 4 mm)

Variant 4: Backing with cast-on nose

Variant 5: Connection without backing ring

Variant 6: unmachined, with Perpendicular cut of the tube

Figure 1. Tested variants of butt welds (schematic sketches).

have to be considered (such as EN 12680-1 in case of ultrasonic testing methods). The quality levels of the cast joint are specified in Table 2. It has to be pointed out, that the size of the cast defects in the welding area should be smaller than the size of the probe for ultrasonic inspection, to get reasonable results.

The possibility of using cast steel grades with lower quality is considered. If a lower quality level than 1 is ascertained at the weld ends during testing, this is to be precisely documented for the tests. From this, conclusions can be drawn about the development of cracks and data can be provided on how far the grade quality influences the fatigue life of the component. The cast steels delivered until now have been inspected and a good quality observed.

2.2 Butt weld geometry

For the planned butt weld details, a total of six different design variants have been taken into account (Figure 1).

Variant 1: Backings made of ceramics
This variant is found in practice to be the optimum design for a butt joint between steel and cast steel. In spite of the high costs, the most advantageous solution with the best possible quality is normally achieved. In the course of the research project, these expensive specimens can be taken as the best quality that can be achieved. However, this is not used as a reference, because of the high costs involved.

In general, the experience with backing strips made of ceramics have been quite positive. However, some further aspects have to be considered. For smaller diameters of the hollow sections, a nose can occur between the individual strips that are arranged on a flexible adhesive strip. Whether and if so, to what extent the notch effect of these noses influences the

fatigue behaviour of the connection is not known at present. Also to be clarified is the question, what effects these noses have under fracture-mechanics criteria.

An alternative solution is to use a complete ring made of ceramics instead of using a flexible strip with single plates as discussed before. With this, the noses can be avoided. However, the usage of these complete rings in the test set-up lead to another problem. As a result of the tolerances between the ceramic ring, with tolerances of ±0.5 mm, the steel part and the cast steel part, with tolerances according to the given standards, in some cases a gap of up to 2 mm occurred. The tests so far have not shown any significant influence to the fatigue behaviour.

With this variant a weld preparation (machining/grinding) in the longitudinal direction of the sections becomes necessary. This is to be done inside and outside within the range of approximately two to four times the wall-thickness. It is recommended to prepare the outside so that it runs out with an angle of 2° maximum, in order to allow verification. These details are required by the ultrasonic examiner, in order to get a reliable inspection of the weld.

Variant 2: Backings using rings made of steel
Compared to variant 1, this variant will examine the influence of the fatigue behaviour of weld pool backing with a steel base. It is planned to take this variant as a reference for all other variants, therefore more test specimen are fabricated.

As self-fitting aid, a chamfered edge to the snap ring is helpful.

Variant 3: Backing rings made of steel on the front surface and chamfered tube

With this variant the butt jointed members have different wall thicknesses. The backing is partially welded to the thicker member. In addition, through rotation of an out of round hollow section specimen the tolerances can be minimized. This variant is considered to be the best possible way of a connection with different wall-thicknesses. The steel ring provides a good backing and the chamfered hollow section a good accessibility.

Variant 4: Backing with cast-on nose

In this variant, the backing is part of the cast member. To cast on an additional nose usually raises no problem, if the wall-thickness is not too small. With many surfaces on the cast member, the post-processing costs are, however, very high, since a post-treatment is very elaborate and very large processing machines are needed.

The advantage of this variant is in the good assemblage possibilities due to the cast-on nose, particularly with large and heavy structural members.

Variant 5: Connection without backing ring

This variant is designed to be as practical as possible and can be produced most easily. The production costs for this butt joint is the lowest and a predominant number of connections in practice are produced in this way. Here, the ends do not have to be treated, and the steel tube is not bevelled (cost savings).

For this variant, the dimensions of the root size should be investigated in order to be able to make any assessment about this influence. Deviations from the standard root size (2 mm) are documented. This should also clarify how the bevelled cast tube functions as a substitute for the backing.

Variant 6: Backing rings made of steel on the front surface with vertical cut of the steel tube

This variant was taken into account as a comparison to variant 3. Most parameters are the same as in variant 3. With this variation, a cost-saving possibility compared to variant 3, can be realized, because only the cast steel part will be fabricated with a chamfer as proposed and the hollow section is welded to it with a straight cut of the tube without bevelling and therefore does not need to be machined.

Altogether, this variant is error-prone and should be avoided in critical areas. However, with uncritical connections this method is interesting, provided sufficient qualification of the welders and quality control during the production is present.

2.3 Welding procedures

The discussions on the different welding methods resulted in the observation that the more practically

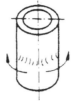 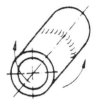

PC tube: fixed   PF tube: fixed
     axis: vertical   axis: horizontal
     welding: transversely   welding: upwards

Figure 2.   Weld positions for tubes according to EN 287.

oriented manual metal arc welding process should be used to be on the safe side concerning the results. Dependent on the qualification of the welder, the position of the weld and the size of the component, more defects may occur compared to the application of the metal active gas welding process or inner shield welding process. This must be considered for the assessment of the test results. A design in metal active gas welding or inner shield, can have a positive effect for critical welds with welding in the fabrication yard.

Preheating

Preheating on the building site is accompanied by considerable costs and accordingly has an enormous influence when considering profitability. A reduction in costs of up to 15% is possible without preheating.

According to existing rules and regulations, preheating is necessary. In EN 17205, preheating temperatures are specified. A temperature from 22°C up to 150°C is recommended, depending on the chemical analysis.

The ambient temperature is also important for preheating. Welding should not be carried out below a temperature of 3°C. In addition, preheating is appropriate for drying cast components. In general, it is known that hydrogen must be avoided during welding. When welding at room temperature, humidity has to be considered in particular. If humidity is still present on and in the components to be welded, this may result in pore formation during welding. Therefore, the development of condensation must be strictly avoided.

The shrinkage during welding is also measured with markings for further evaluation.

Welding position

The welding positions (Figure 2) horizontal-vertical PC and vertical upwards PF according to EN 287 were selected for the investigation. Test specimens are fabricated in order to investigate both welding positions, which are the most difficult for welding. The conditions on the building-site should be simulated in this way to get reliable results, compared to welding in a fabrication yard. With regard to the reproducibility, position PC should be preferred. With this,

Table 3. Shape of the weld.

Wall thickness	Weld type	Geometry
20/30	V-weld	$\alpha = 2*20°$
55	single V-butt joint	$\beta = 2*15°$
		gap b = 4–8 mm

it would be possible to turn the specimen, repair it and perform a new test if a fatigue crack occurs. This would considerably reduce the testing cost. Based on the comparatively higher fluctuations with regard to weld quality for designs in position PF, a multiple use of the same specimen would be less appropriate.

Weld preparation
For the weld preparation, several aspects should be taken into account. Some problems could occur with the sizes of possible defects at the weld-on ends. Although they are in the permissible range for cast joints at the weld-on ends (quality level 1), casting defects can lead to problems. This is the case if they lie in the heat affected zone, since, according to the relevant welding regulations, only smaller defect sizes are permissible there.

These defects, however, can be recognized by prior magnetic particle inspection or dye penetrant inspection and can be repaired if necessary. Particularly with critical weld seams, this procedure is recommended in advance.

The gap width for the root should be 2 mm. Such a gap, however, cannot be realized in practice. Experience shows that gap widths from 0 to 8 mm occur depending on the dimensions. It is attempted, however, to realize gap widths of 2 mm. Deviations are acceptable but must be recorded.

The weld preparation is based on EN ISO 9692. As a standard value it is decided that for the specimens with the smallest possible weld (V-weld), bevels with the smallest aperture angle (α, β) should be used. For the larger wall thicknesses, a single steep-flanked V-butt joint, however, will be applied in order to reduce the volume to be welded and thus, decrease costs.

It has been identified for variant 4 that the tolerances become larger, also due to the non-centric hole in the hollow sections. Therefore, an adjustment of the ends of the cast components has to be carried out. The optimum shape of the cast nose has been determined by the foundries involved with their knowledge based on previous projects. As a whole, the shape of the weld is arranged according to Table 3.

3 INSPECTION OF THE JOINTS

3.1 *Inspection procedures*

Different individual tests on cast steel components as well as on welded test specimens were carried out on

the specimens das delivered. All relevant regulations concerning the testing were adhered to.

For all delivered steel and cast steel members material testing has been carried out to determine the mechanical properties. Furthermore, a chemical analysis was performed. All obtained values were satisfactory concerning their normative regulations.

Also, non-destructive tests were carried out by the cast steel manufacturers on the individual members. All parts were tested with the following testing methods:

– Radiographic inspection
– Ultrasonic inspection
– Magnetic particle inspection

Later, the delivered cast steel was tested in Karlsruhe with a dye penetration test on the ground front surfaces. In all examined test specimens, the necessary cast quality was achieved.

After welding the test specimens another ultrasonic inspection was carried out. For this inspection the following points should be taken into account for the different variants.

The ultrasonic inspection for variant 2 to check for an error free and correct welding cannot be accomplished with the angle transmitter-receiver probe due to potentially the same sonic distances in case of insufficient penetration or nose formation. To get clear results, the weld seam must be ground off and inspected with a perpendicular transmitter-receiver probe.

A reasonable inspection for variant 3 is only possible by positioning the angle transmitter-receiver probe on the thinner member. The same procedure starting from the thicker member does not give reliable results. Furthermore, grinding of the welding seam is necessary, to find adhesive defects on the thicker member. The lower 2 to 3 mm of the weld are only verifiable with difficulty. Clear results can be obtained only by a perpendicular inspection with a ground weld seam.

For variant 4 the ultrasonic testing takes place perpendicular to the surface with ground welding seam.

An ultrasonic test of variant 5 is possible, since no backing is used. In order to be able to make a precise evaluation, grinding of the weld is necessary. Root failure is difficult or impossible to detect with X-Ray testing.

However, there are doubts concerning the inspection for variant 6. Defects in the vertical direction cannot always be recognized. Also, the sufficient fusion of the base metal, depending upon geometry, can be difficult and an adequate opening angle and a correct blowpipe guidance are necessary.

An optimal testing (NDT) of the connection is ensured by grinding off the weld seam and using a perpendicular transmitter-receiver probe as already mentioned. As an alternative to this, unmachined areas are also tested, using an angle transmitter-receiver probe to find out whether time and costs can be saved and equally good results obtained. It can be confirmed that this solution is very dependent on the weld-inspector. With high experience on ultrasonic inspection, the grinding off can be omitted. However, it is clearly evident that this is not so in every case.

Finally, it should be pointed out for the practical application of the testing methods, that with a multitude of regulations concerning ultrasonic inspection the necessary requirements should at an early stage of the project between the buyer and the contractor to avoid subsequent complications. Without a detailed consultation, as often occurs in practice, quality standards can not be maintained.

3.2 Inspection results

To date all inspections have been carried out for the first stage of this research project (see section 5). There are three different set of tests at this stage. Fatigue tests under tension, bending and also tests on small specimens. All welding was performed without any special instructions to the welder. In fact, the welder received no information other than that specified in the welding procedure sheets and was not informed that the welding was to be tested with an ultrasonically later on.

For all examined welds, an alternative error quality of 3 according to AD HP 5/3 was taken according to the experience of the ultrasonic examiner with such connections. In fact, this is the result of the required quality level 1 for cast steel and the necessary quality requirements for the weld depending on the wall-thickness. The detailed specifications for ultrasonic testing are also based on this regulation with details concerning the allowable ultrasonic display for longitudinal imperfections in the tested specimen. Whether the inspection is satisfactory or not depends on these regulations.

All specimens for the tension tests and small specimen tests were welded in welding position PC. Welding in this position is rather difficult, but it is very common, for example in columns of bridges at the building site. An example of this weld is given in Figure 3. The weld was not ground smooth, but this did not affect the results of the fatigue tests because all failures in the welds up to this point started on the backing. It can be also seen that the cast steel has many irregularities. These also did not lead to a failure in the cast steel.

The difference in the results in welding the tension specimens and small test specimens lie in the accessibility for the welder. The small test specimens could be welded in a much better position on a table, therefore providing better results in ultrasonic testing.

The welding in all different variants showed that the variants 1 and 2 were the best for verifiability and quality of the welding. The variants 3 and 4 provided moderate results concerning the weld quality. A special problem was the welding of the square edge in variants

Figure 3. Joint welded in position PC.

Figure 4. Weld welded in position PF.

5 and 6. Especially here was a lack of fusion observed. During ultrasonic inspection, every weld pass could be recognized.

All specimens for the bending tests were welded in the welding position PF. This welding position is easier to manufacture therefore leads to better results in the ultrasonic tests. A practical example for these welds is for joints in chords of bridges. An example of this weld is given in Figure 4. The weld was likewise not ground smooth, but this time the structure of the weld is much more even. Similarly to the previous specimens all failures in the welds up to this point started at the backing.

However, the results of the ultrasonic inspection for these welds are much better than those for welding position PC. The error signals showed only marginal errors in some cases for variants 1 and 2. The variants 3 and 4 gave no errors for the weld quality. The problem with the welding of the square edge in variants 5 and 6 was present again, but this time to a much lesser degree.

4 TESTING PROGRAM

4.1 Dimensions of the test specimens

The outer diameters for the hollow sections to be investigated are selected according to Table 4. They

Table 4. Testing program.

| Stage | Diameter [mm] | Wall thickness [mm] | |
		Steel hollow section	Cast steel part
1	193.7	20	20/30*
2	298.5	30	30/40*
3	508.0	55	55/65*

* depending on variant.

have been chosen based on the experiences of the participants in different working groups of the research project. The same dimensions are being used for a planned motorways bridge in Germany, the Killian bridge on the A73 near Suhl, a bridge where insights from this project are made use of and vice versa. In this way, costs can be minimized and a direct practical application of the results of this investigation is possible. The cast steel wall thicknesses are equally chosen based on the hollow sections thickness with 10 mm added for the variants with wall thickness discrepancy.

For dimensions t > 30 mm, a weld bending test is required in Germany for hollow section joints. This would then also apply to the planned test program. However, according to the manufacturers of hollow sections, it is possible to substitute these expensive weld bending tests with a combination of establishing the purity grade of the fused material and a tensile test in the thickness direction. This is possible for hot-rolled hollow sections, since the core with the impurities is rejected und thus, no lamellar tearing is possible. Therefore these tests are omitted to keep the costs down.

4.2 Structure of the tests

In the research project, the experimental work is aimed at fatigue testing of end-to end connections between cast steel members and rolled hollow sections. Three different test rigs are used.

In a first test rig, tension tests on specimens with two butt joints (Figure 5) are carried out. With this structure, a constant load along the section is possible. However, the drawback of this test set-up is the costly installation and the high loads necessary for the fatigue testing. Therefore, only the tests in stage 1 with the smallest wall thicknesses are investigated here. The variants investigated are variants 2, 3, 5 and 6, each with three specimens.

To reduce the necessary test loads, a second test set-up is established. Here the loads are applied in 4-point bending (Figure 6). This allows a significant reduction of the maximum load. It is clear that since only a small part of the butt weld will have the highest stress, so that the expected fatigue life would be higher than in the

Figure 5. Test rig for tension tests.

Figure 6. Test rig for bending tests.

tension tests if this section of the weld is free of any errors. These tests are carried out for all three stages of the research project with all variants in the testing program, where the test specimens are all larger than those in the tension tests.

The advantage of carrying out the tension tests and the bending tests with two butt welds is that two welds can be examined at the same time. The length of the specimens should normally at least be two to three times the diameter of the steel tube. Based on size, weight and installation into the testing machine, it is not possible to keep to this criterion for sections with large diameters. For these cases, the distance from weld to the other extremity is shortened to 1.5 d.

Figure 7.   Test rig for small specimens.

Table 5.   Load cycles for first bending series with constant amplitude.

Variant	R	S_R [N/mm^2]	Load cycles N
1	+0.2	170	4.947.713
2	+0.2	170	3.011.366
3	+0.2	170	1.193.693
4	+0.2	170	1.575.475
5	+0.2	170	1.838.046
6	+0.2	170	434.024

The last test set-up is for small tension test specimens cut out of hollow sections to cast steel butt weld joints (Figure 7).

Within the three stages, assessments from available results are directly used for the subsequent tests. In this way, it is possible to modify the plan where necessary. A reduction of the number of larger tests or more tests on small-size specimens is also possible, depending on unforeseen costs (delivery, fabrication, welding, testing).

All test specimens with a diameter of 193.7 mm for the first stage have been fabricated so far and are currently tested. For the following stages, the planning of the test set-up is completed.

4.3   Evaluation of the tests

In contrast to the small test specimens with only one weld being tested, the tension and bending tests are arranged so that two welds can be simultaneously tested. The cycles to failure of a specimen is chosen for the first weld to show a crack. Therefore there would always be two test results, with the weaker joint giving the first results. Subsequently, repair welding on the failed joint is carried out to provide data for the stronger joint or give insights on the durability of repair welds. It is obvious that sufficient data is necessary for the various series of tests in order to allow a statistical evaluation. Until now only a part of the testing program has been evaluated, therefore providing a first insight into the research work.

The first bending series (Table 5) was tested with a nominal calculated stress range of $S_R = 170$ N/mm^2

in the outer fiber of the steel part which corresponded to the measurements on tested specimens.

As mentioned in the previous section the quality of the investigated welds according to the planned testing structure with variants 1 and 2 is the best and variant 6 the worst case. However, a cast-on nose in variant 4 as planned in the program was not possible in this case due to the small wall thickness. Instead in this case this variant was welded in the same manner as variant 5 but with a special careful treatment for the root weld to provide good results. The results of variant 5 are better than for 3 and 4, which is attributed to the better weld quality.

In the tests carried out for variants 1 and 2 to compare the behaviour of ceramic panels versus standard steel ring. It was found that use of ceramic panel gave the same or better fatigue strength than the standard steel ring solution.

The results for the small test specimens established so far (see Figure 8) also indicate that the results from the bending tests are also applicable for the small specimens.

4.4   Further testing

The tests described above are not completed and still in progress. After the tests repair of the failed test specimens under similar conditions to those on the building will be attempted. The influence of normal fabrication and repair welding will therefore be considered where possible. Defects in the repairs that occur are also to be investigated

In a theoretical FE analysis the basic hot spot stresses for each variant are to be determined. In further investigations, effects concerning the practical design of the joints will be taken into account. The given material tolerances and the influence of wall-thickness discrepancies will particularly be modelled, to ascertain this influence. Therefore, special importance is attached to the measurements of the test specimens. For the later assessment of results and further FE-analyses, a precise documentation will be made. In a second step, a fracture mechanics approach to the development of cracks in selected areas will be taken into account.

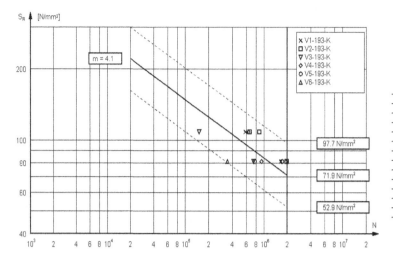

	Tension [N/mm²]	Load cycles
V1	90	1.986.325
	90	1.651.800
	120	581.602
V2	90	1959.602
	120	660.242
	120	873.730
V3	90	735.465
	90	742.455
	120	151.742
V4	90	936.431
	120	635.667
V5	90	770.255
	90	1.718.636
V6	90	341.972

Figure 8. Fatigue results of the small specimens tests.

5 CONCLUSIONS

It is clearly evident that the properties of cast steel from reputable foundries are nowadays only slightly different compared to steel. Particularly the weldability and the toughness are comparable to rolled steel products. For the tested specimens no failures different to those of steel are found. No influence of cast defects is observed. The provided cast steel was manufactured with the necessary quality level as desired in practice.

Preliminary experimental and numerical studies were initiated with the aim of establishing solutions to improve fatigue behaviour in bridges with respect to costs and desired quality levels. The preliminary results are presented in this paper. The research work is aimed towards an economic design of butt welds in hollow sections under fatigue loading.

ACKNOWLEDGEMENTS

The authors would like to thank FOSTA (Forschungsvereinigung Stahlanwendung e.V.) and CIDECT for financial support to carry out the investigations. The authors also like to thank the members of the working group of the research project P591 "Wirtschaftliches Bauen von Straßen-und Eisenbahnbrücken aus Stahlhohlprofilen" and of the research project CIDECT 7W "Fatigue of End-To-End CHS Connections".

REFERENCES

F. Mang; S. Herion: Guss im Bauwesen, Sonderdruck aus Stahlbau Kalender, Ernst & Sohn, 2001

M. Veselcic, S. Herion, R. Puthli: Cast steel in tubular bridges – New applications and technologies, Proceedings of the 10th International Symposium on Tubular Structures; Swets & Zeitlinger, Lisse, Netherlands 2003

CIDECT Report – Project 7W: Fatigue of End-To-End CHS Connections, Fourth Interim Report, University of Karlsruhe, July 2005

EN 287: Qualification test of welders – Fusion welding – Part 1: Steels, CEN, January 2004

EN 1369 : Founding, Magnetic particle Inspection, CEN, February 1997

EN 1370: Founding, Surface roughness inspection by visu-altactile comparators, CEN, February 1997

EN 1371-1: Founding, Liquid penetrant inspection – Part 1: Sand, gravity die and low pressure castings, CEN, October 1997

EN 1559: Founding, Technical conditions of delivery, CEN, August 1997

EN 1993-1-9: Design of steel structures – Part 1–9: Fatigue; CEN, July 2005

EN 10210: Hot finished structural hollow sections of non-alloy and fine grain structural steels – CEN, September 1994

EN 10293: Steel castings for general engineering uses, CEN, June 2005

EN 12680: Founding – Ultrasonic inspection – Part 1: Steel castings for general purposes, CEN June 2003

EN 12681: Founding – Radiographic inspection, CEN, June 2003

EN 17205: Quenched and tempered steel castings for general purposes, CEN, April 1992

EN ISO 5817: Welding – Fusion-welded joints in steel, nickel, titanium and their alloys (beam welding excluded) – Quality levels for imperfections, CEN, December 2003

EN ISO 9692 Welding and allied processes – Recommendations for joint preparation, CEN, Mai 2004

AD HP 5/3 and App.1; Edition: 2002–01; Manufacture and testing of joints – Non-destructive testing of welded joints

SEW 520 Hochfester Stahlguß mit guter Schweißeignung; Technische Lieferbedingungen, Stahl-Eisen, Sept. 1996.

Tubular Structures XI – Packer & Willibald (eds)
© 2006 Taylor & Francis Group, London, ISBN 0-415-40280-8

Fatigue assessment of weld root failure of hollow section joints by structural and notch stress approaches

W. Fricke & A. Kahl

Hamburg University of Technology, Hamburg-Harburg, Germany

ABSTRACT: The fillet-welded joints of hollow section members are usually made from the outer side only so that the weld root cannot be controlled. At partial- or non-penetrating fillet welds, non-fused root faces are present which may act as initial cracks in the case of cyclic loading. Therefore, a fatigue assessment is necessary which not only considers the possibility of cracks initiating at the weld toe, but also at the weld root. The latter can be well assessed using the crack propagation approach, however, simpler approaches are called for in practical design. In the paper, the effective notch stress approach is described which has been developed for the fatigue strength assessment of weld toes as well as weld roots. The numerical stress analysis can be performed using the submodel technique. In addition, structural stress approaches are presented which have recently been proposed for the assessment of root failures of welds being mainly subjected to throat bending. A relatively coarse finite element mesh can be used. The requirements for meshing are discussed in detail. The procedures are illustrated by the example of a fillet-welded RHS joint subjected to axial and bending loads. Fatigue tests have also been performed so that calculated fatigue lives can be compared with those observed in the tests, showing good agreement.

1 INTRODUCTION

The welded joints of hollow sections are mostly performed from one side only using full-penetration welds or fillet welds having no or partial penetration, Fig. 1. In these cases, fatigue cracks can start from the weld root as crack-like defects or the non-fused root faces may act as an initial crack.

Such a situation is well-known from cruciform joints with fillet welds having no or partial penetration, see Fig. 2. Fatigue cracks can initiate from the weld toe as well as from the root. The fatigue life of the latter has been investigated on the basis of nominal stresses (among others by Gurney and Maddox, 1973; Olivier and Ritter, 1979; Hobbacher, 1996), notch stresses (Anthes et al., 1993; Radaj and Sonsino, 1998) and particularly of fracture mechanics (Usami and Kusumoto, 1978; Frank and Fisher, 1979; Maddox, 1991).

In practice, the nominal stress approach is frequently applied, using a 'nominal weld stress' in the throat section. This is simply calculated by relating the force in the loaded plate to the weld throat area. This approach assumes constant normal and shear stresses in the throat section, resulting in the nominal stress as a vector sum of these two components. The corresponding design S-N curves or fatigue classes are for steel between FAT 36 and FAT 45, depending on the code considered. The FAT values correspond to

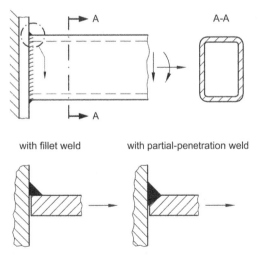

with fillet weld   with partial-penetration weld

Figure 1. Fillet-welded Rectangular Hollow Section (RHS) with no and partial penetration.

the characteristic fatigue strength (survival probability $P_S = 97.7\%$) in [MPa] at $2 \cdot 10^6$ cycles.

However, the nominal stress approach has shortcomings with respect to tubular joints as it is difficult to consider the effects of the weld eccentricity and bending on the stresses.

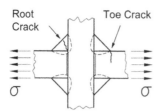

Figure 2. Possible fatigue cracks in fillet-welded cruciform joints.

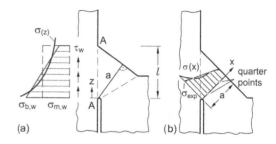

Figure 3. Definitions of structural weld stresses, (a) in the leg section A-A and (b) in the throat section, using quarter points.

Particularly in the case of non- or partial-penetrating welds, the crack propagation approach is much better suited as the initial crack length can be defined by the length of the non-welded root faces. The complex geometry of tubular joints may, however, require an extensive finite element analysis with a very fine mesh.

For this reason, simpler approaches are looked for in practice which consider the complex geometry, but can also be used for 'scanning' of a structure to identify fatigue-prone areas. Particularly the structural stress approach, taking into account the bending component of the relevant stress, as well as the effective notch stress approach offer possibilities for an appropriate fatigue analysis. These two approaches are described in the following and finally applied to a fillet-welded RHS-joint as illustrated in Fig. 1.

2 STRUCTURAL STRESS APPROACHES FOR WELD ROOT FAILURE

2.1 Description of approaches available

The structural stress is frequently defined as the sum of a membrane and a bending stress, e.g. in the plate or shell adjacent to a weld toe (Niemi, 1995). In this way, the local nonlinear stress peak at the weld toe, which is caused by the weld geometry, is excluded, but will be considered by the S-N curve. Alternatively, the structural stress can be obtained from stress extrapolation over two or three selected reference points, leading to the well-known hot-spot stress approach, which was at first successfully applied in the fatigue design of tubular structures in the 1970's (van Wingerde et al., 1995; Radaj and Sonsino, 1998).

The application of the structural hot-spot stress approach, as it has been designated in recent IIW recommendations (Niemi et al., 2004), is restricted to weld toe cracks. Based on a study by Zenner and Grzesiuk (2004) on axially loaded plates with one-sided welds having partial penetration, Fricke et al. (2005) proposed an approach for the fatigue assessment of fillet welds predominantly subjected to throat bending. They used the linearised stress in the leg plane in the extension of the root face (Section A-A

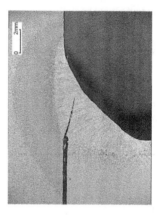

Figure 4. Observed root crack path in a one-sided fillet weld of the joint shown in Fig. 1.

in Fig. 3a), because fatigue tests as well as crack propagation analyses have shown that the crack path is closer to this plane than to the weld throat section, which one might expect as crack path at a first glance (Fig. 4). Furthermore, the stress in the leg section is – in contrast to the throat section – mainly characterised by the axial and bending stress, $\sigma_{m,w}$ and $\sigma_{b,w}$ respectively, and less by shear stresses τ_w in several cases. The linearised stresses in one-sided welds can even be evaluated from the loads in the attached plate, using equilibrium equations.

If mainly axial and bending stresses are acting and the shear stress is small, fatigue class FAT 80 has been found from different fatigue test series to be appropriate for steel. It should be noted, that larger shear stresses τ_w change the direction of the principal stress and, hence, the crack path. In case of doubts, the amount of τ_w should be checked (it should be, e.g. less than 20% of the normal stress, changing the principal stress direction by max. 10°).

An alternative approach has been derived by Sørensen et al. (2005) for fillet-welded doubler plates under out-of-plane loads. They use the stresses in the throat section, which are determined in a finite element analysis and extrapolated over the quarter points to the weld root, Fig. 3b. The authors recommend to extrapolate the individual stress components to the weld root and compute afterwards the principal stress. From fatigue tests of various doubler plates, a fatigue class corresponding to FAT 60 has been derived.

2.2 Computation of structural stresses

In the procedure proposed by Fricke et al. (2005), the structural stresses are usually derived from a finite element analysis. The stresses or forces in the leg section can be directly linearised to obtain the membrane and bending portion of the stress:

$$\sigma_{m,w} = \frac{1}{\ell}\int_0^\ell \sigma(z)\,dz \tag{1}$$

$$\sigma_{b,w} = \frac{6}{\ell^2}\int_0^\ell \sigma(z)\left(\frac{\ell}{2}-z\right)dz \tag{2}$$

$$\sigma_{s,w} = \sigma_{m,w} + \sigma_{b,w} \tag{3}$$

where $\sigma(z)$ is the stress normal to leg section, z the coordinate along weld leg, ℓ the leg length, $\sigma_{s,w}$ the structural stress in weld leg section, $\sigma_{m,w}$ the membrane portion and $\sigma_{b,w}$ the bending portion of structural stress. In the same way, also the shear stresses τ_w can be linearised.

However, the determination of structural weld stresses from finite element models can be problematic due to the influence of the notch singularity at the weld root. The element stresses in the vicinity of the weld root are highly unreliable (Fricke et al., 2003). Therefore it is better to use internal nodal forces instead of element stresses for the integration according to eqs (1) and (2), as these naturally satisfy equilibrium conditions.

The problem is illustrated by the example shown in the left part of Table 1, a simple T-joint with a one-sided fillet weld subjected to axial unit stress $\sigma_n = 1$ in the 6 mm thick branch. The analytical result for this statically determined case gives a structural weld stress $\sigma_{s,w} = 7.0$. The same result is only obtained from nodal forces for both coarse and fine meshes. The integration of the element stresses underestimates the stress. The accuracy is not better, if the elements at the root face are included, which show normal stresses in spite of the free element surface.

If a one-sided weld is present as shown in Fig. 1, the structural weld stress can generally be derived also from the membrane and bending part σ_m and σ_b of the structural stress in the attached tube, Fig. 5. This means that a relatively coarse mesh established for the analysis of structural hot-spot stresses

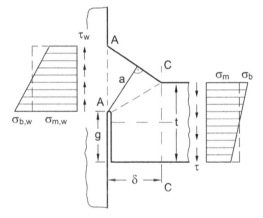

Table 1. Computed structural stress $\sigma_{s,w}$ in the weld leg section of a fillet-welded T-joint for unit loading (analytical result: 7.0).

Evaluated from:	Coarse mesh	Fine mesh	Further refined mesh
Nodal forces in section A-A:	7.0	7.0	7.0
Element stresses in section A-A:	2.8	5.1	5.9
Same as above + two add. elements·	–	8.1	8.1

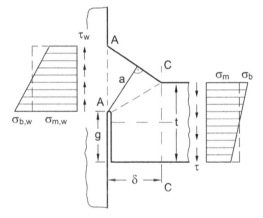

Figure 5. Calculation of linearised stress distribution in the weld leg section from stresses in attached tube wall.

(Niemi et al., 2004) can be used. The equilibrium conditions between sections A-A and C-C yield:

$$\sigma_{m,w} = \sigma_m \frac{t}{\ell} \tag{4}$$

$$\sigma_{b,w} = \sigma_m\left[3\frac{t}{\ell} + 6\frac{t}{\ell^2}\left(g-\frac{t}{2}\right)\right] - \sigma_b\frac{t^2}{\ell^2} - 6\tau\frac{t\cdot\delta}{\ell^2} \tag{5}$$

$$\tau_w = \tau\frac{t}{\ell} \tag{6}$$

where t is the thickness of attached plate, g the gap length and δ the distance between leg plane and Section C-C ref. to plate stresses (see Fig. 4). The stresses σ_m, σ_b and τ refer to Section C-C.

The procedure proposed by Sørensen et al. (2005) requires a finer finite element mesh in order to derive the stresses at the quarter points with sufficient accuracy. At least two higher-order elements with midside nodes should be arranged over the throat section, where the midside nodes are located at the quarter points.

The approach has been applied to the example shown in Table 1, yielding a principal stress of 5.7 after extrapolation of the stress components to the weld root gap. The smaller stress compared to that one used in the previous approach fits quite well with the difference in the FAT classes, so that also the fatigue assessments will be comparable.

3 EFFECTIVE NOTCH STRESS APPROACH

3.1 Description of approach

The notch stress approach takes into account the local stress increase due to the weld shape. At sharp corners, the theoretical elastic stress is infinite and, therefore, no more relevant for fatigue strength assessment. Based on the micro-structural support effect of the material postulated by Neuber, an effective notch stress approach has been developed by Radaj (1990), where a fictitious notch radius is assumed at the weld toe amounting to $r_f = 1$ mm for welds at steel. This fictitious notch radius has to be added to the actual notch radius, which is usually assumed to be zero in a conservative way. Therefore it is recommended to assume generally $r_f = 1$ mm for design purposes. The result is an effective notch stress or, if related to the nominal stress, a fatigue notch factor K_f. This approach together with a design S-N curve according to FAT 225 for steel has been included in the recommendations of the International Institute of Welding (Hobbacher, 1996).

The approach is not only restricted to weld toes, but can also be applied to weld roots. The fictitious radius may be arranged as shown in Fig. 6, leading to a keyhole shape. This can result in a stress over-estimation in case of stresses parallel to the slit. Therefore, oval shapes are sometimes preferred. Alternative approaches using a smaller notch radius have been developed particularly for thin-walled structures used in the automotive industry, e.g. 0.05 mm (Eibl et al., 2003).

3.2 Computation of notch stresses

The computation of the notch stresses at the fictitiously rounded weld root requires a relatively fine mesh. It is recommended to arrange at least 20 elements along the circumference of the whole circle, preferably with higher-order elements having midside nodes. The element subdivision normal to the notch surface should also be fine due to the high stress gradient in this direction.

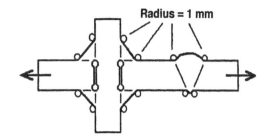

Figure 6. Fictitious rounding of weld toes and roots (Hobbacher, 1996).

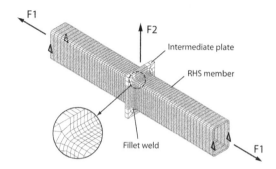

Figure 7. Finite element model and loading of the RHS-joint.

For some standard cases such as butt and cruciform joints, parametric investigations have been performed, illustrating the effect of geometric parameters such as plate and weld thickness on the notch stress concentration factor (Anthes et al., 1993).

4 APPLICATION TO A FILLET-WELDED RHS-JOINT

4.1 Description of the joint

The approaches described are applied to a fillet-welded rectangular hollow section (RHS) joint, for which also fatigue test results are available. The joint is principally shown in Fig. 1. By welding two RHS-members to an intermediate plate (see also Fig. 7), a symmetrical specimen was created, which can be subjected to tensile (F1) as well as to bending loads (F2), the latter by 3-point bending. The lever between the load points was 400 mm.

The chosen profile 120 × 80 × 6 was made of mild steel (S235 JR G2). The thickness of the intermediate plate was 15 mm. Two types of welding process were applied: manual metal-arc welding with covered electrode and the MAG process with flux-cored wires. The mean values of the realised throat thicknesses a are given in Table 2.

Table 2. Mean values of actual throat thickness a, loading mode and computed structural weld stress concentration factors $K_{s,w}$.

Specimen	a [mm]	Loading	$K_{s,w}$
RHS joint, man. arc-welded	4.3	tensile	4.25
RHS joint, man. arc-welded	4.3	bending	3.37
RHS-joint, MAG welded	5.1	bending	2.78

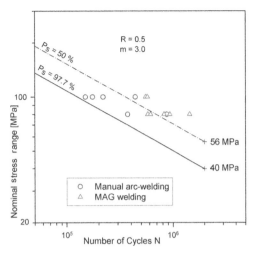

Figure 9. Fatigue lives obtained for bending load.

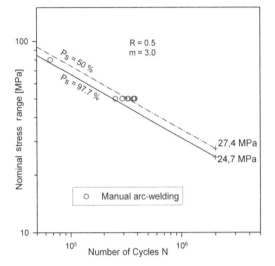

Figure 8. Fatigue lives obtained for tensile load.

4.2 Fatigue tests

The fatigue tests were performed in a testing machine with constant load amplitudes and a frequency of approx. 30 Hz. A high stress ratio R = 0.5 was conservatively chosen according to the IIW fatigue design recommendations (Hobbacher, 1996). One or two load levels were selected for the tests series of specimens. All specimens failed due to root cracking. The failure criterion for terminating the tests was a fatigue crack penetrating through the weld surface. After this, the remaining life of the specimens was very short.

The fatigue lives are shown in Fig. 8 for tensile load and in Fig. 9 for bending load using the nominal stress range in the RHS-member at the joint. A slope exponent m = 3 of the S-N curve was assumed to derive the characteristic fatigue strength at $2 \cdot 10^6$ cycles.

The results for the bending load show that the welding process affects the fatigue life, which is probably due to differences in the sharpness of the weld root.

When the characteristic fatigue strength is compared with that given in codes and recommendations such as Hobbacher (1996), it should be kept in mind that the nominal stress for root failure is referred to the weld throat area. Using this stress, the tests under tensile load show a fatigue class of just below FAT 36,

while the bending tests yield a fatigue class of over FAT 45. The difference can be explained by the effect of local weld throat bending. Under tensile load, the fatigue cracks appear at the long side of the RHS-joint, which is less restrained to local bending, while under bending load they appear at the short upper side of the joint, which is the most highly stressed one and where local bending is more restrained.

4.3 Computation and evaluation of structural stresses

The numerical analysis was performed with the finite element model shown in Fig. 7, using 20-node solid elements. Only one element was arranged over the wall and weld thickness.

The structural weld stress was analysed in the weld leg section according to Fig. 3a using the nodal forces. The last column of Table 2 shows the computed stress concentration factors $K_{s,w}$, i.e. the ratio between structural weld stress and nominal stress in the RHS-member. The stress concentration factors clearly show the effect of increased local bending under tensile loads mentioned above. Furthermore it can be seen that the weld throat thickness influences also the stress concentration factor, which explains the longer fatigue lives of the MAG-welded specimens as plotted in Fig. 9.

Figure 10 shows the test results in an S-N diagram using the structural stress range together with the test results for two other details (Fricke et al., 2005). All results are within a scatter band which is not unusually large (standard deviation of log. life = 0.273). The lower boundary of the scatter band justifies the design

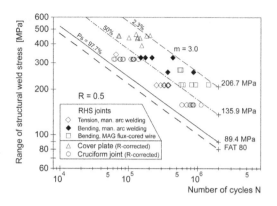

Figure 10.   S-N results of the fatigue tests of different welded joints using the computed range of structural weld stress.

S-N curve FAT 80 based on structural stress ranges mentioned earlier.

It can be seen from Fig. 10 that the results for tensile load are in the lower part of the scatter band, while those for bending load are in the upper part. This might be due to the more localised stress concentration under bending, resulting in semi-elliptical cracks showing a slower crack propagation.

4.4   Computation and evaluation of notch stresses

The computation of the notch stresses was performed using the submodel technique. The coarse model shown in Fig. 7 was slightly refined in order to obtain improved prescribed displacements for the submodels. These were created with very fine meshes and subjected to prescribed deformations at the boundaries taken from the coarse model. Fig. 11 shows typical models.

The weld root was modelled with a keyhole shape and a fictitious notch radius $r_f = 1$ mm. 8-noded solid elements were chosen, having a length of about 0.1 mm at the notch.

According to the actual throat thicknesses given in Table 2, two models have been created. The computed fatigue notch factor K_f along the weld root line, related to the nominal stress in the RHS, is shown in Fig. 12 with a max. value $K_f = 9.03$ for the tensile load case and $K_f = 6.50$ for the bending load case. The latter is decreased to $K_f = 5.97$ if the throat thickness is increased to $a = 5.1$ mm for the MAG welded joints.

Based on these factors, the fatigue lives are plotted in Fig. 13 in relation to the effective notch stress range. It can be seen that all test results are located above the S-N curve for the fatigue class FAT 225 recommended by Hobbacher (1996). Again, the values for bending are above those for axial loading as discussed in the previous sub-chapters.

The specimen has also been investigated numerically in a round robin, where six analysts computed the

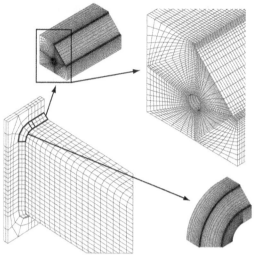

Figure 11.   Coarse finite element model and submodels.

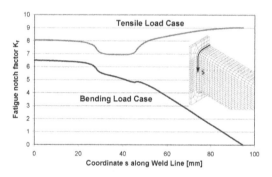

Figure 12.   Distribution of the fatigue notch factor along the weld root line for a weld throat thickness $a = 4.3$ mm.

notch stresses for the joint with nominal throat thickness $a = 4.5$ mm. The resulting notch fatigue factors are fairly close together, if the results of one rather coarse model are excluded, ranging from 8.46 to 9.13 for the tensile load case and from 6.01 to 6.50 for the bending load case. The results presented above are within the scatter band.

5   CONCLUSIONS

The fatigue assessment of weld root failure of hollow section joints, which are usually welded from one side only, can be performed by different approaches. In addition to the crack propagation approach, where the non-welded root faces are considered as initial crack, also local stress approaches are applicable, which take into account the local bending of the weld throat.

The first is the structural stress approach, based on the linearised stress in the weld. Two versions have

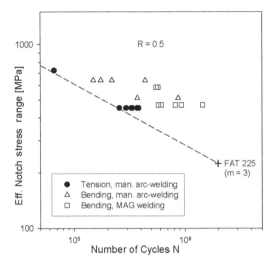

Figure 13. S-N results of the fatigue tests of the RHS-joints using the computed range of effective notch stress.

been presented, the first using the membrane and bending portion of the stress in the leg plane in extension of the non-welded root faces, while the second is based on an extrapolation of the stress components over the quarter points in the weld throat plane and on the computation of the principal stress at the weld root. Both versions use different fatigue classes.

The second approach is the effective notch stress approach, which is based on the notch stress in a fictitiously rounded weld root with a radius of 1 mm, requiring a very fine finite element mesh.

Both approaches have been applied to a fillet-welded joint of a rectangular hollow section, which was fatigue-tested under tensile and bending loads. The structural stresses were determined with a relatively coarse finite element model, while fine-meshed submodels were used for the computation of the notch stresses. The results show the applicability of the approaches and the appropriateness of the design S-N curves proposed. The differences in local bending between the two load cases were considered in an appropriate way.

ACKNOWLEDGEMENT

The major part of this work has been performed within the project 'Network of Excellence on Marine Structures' (MARSTRUCT). The participants are Instituto Superior Técnico, Universities of Glasgow & Strathclyde, Université de Liege, Technical University of Varna, Technical University of Denmark, Helsinki University of Technology, VTT Industrial Systems, Kvaerner Masa-Yards, Bureau Veritas, Principia Marine, Sirehna, Germanischer Lloyd, Hamburg University of Technology, Flensburger Schiffbau-Gesellschaft mbH & Co KG, CMT – Center of Maritime Technologies e.V., National Technical University of Athens, CETENA – Centro Tecnico Navale, Università di Genova, TNO – Netherlands Institute for Applied Scientific Research, Schelde Naval Shipbuilding, Norwegian University of Science and Technology, Det Norske Veritas, Technical University of Szczecin, CTO – Centrum Techniki Okretowej, Lisnave Estaleiros Navais SA, Estaleiros Navais de Viana do Castelo, University 'Dunarea de Jos' of Galati, IZAR Construcciones Navales S.A., Chalmers University of Technology, Technical University of Istanbul, University of Newcastle, University of Southampton and The Welding Institute. The work has been partially funded by the European Union through the Growth programme under contract TNE3-CT-2003-506141.

REFERENCES

Anthes, R.J., Köttgen, V.B. and Seeger, T. 1993. Kerbformzahlen von Stumpfstößen und Doppel-T-Stößen. Schweißen und Schneiden 45(12): 685–688.

Eibl, M., Sonsino, C.M., Kaufmann, H. and Zhang, G. (2003): Fatigue assessment of laser welded thin sheet aluminium. Int. J. of Fatigue 25, pp. 719–731.

Frank, K.H. and Fisher, J.W. 1979. Fatigue strength of fillet welded cruciform joints. J. of Structural Division, ASCE, 105: 1727–1740.

Fricke, W., Mertens, M. and Weißenborn, C. 2003. Finite element calculation and assessment of static stresses in load-carrying fillet welds. IIW-Doc. XV-1151-03, International Institute of Welding.

Fricke, W., Kahl, A. and Paetzold, H. 2005. Fatigue strength assessment of fillet welds predominantly subjected to throat bending. In: Maritime Transportation and Exploitation of Ocean and Coastal Resources (Ed. C. Guedes Soares, Y. Garbatov and N. Fonseca), Vol. 1, pp. 405–412, Tylor & Francis, London.

Gurney, T.R. and Maddox, S. 1973. A Re-Analysis of Fatigue Data for Welded Joints in Steel. Welding Res. Int., 3(4): 1–54.

Hobbacher, A., Ed. 1996. Fatigue Design of Welded Joints and Components. Cambridge (UK): Abington Publishing.

Maddox, S.J. 1991. Fatigue Strength of Welded Structures. Cambridge (UK): Abington Publishing.

Niemi, E., Ed. 1995. Recommendations concerning stress determination for fatigue analysis of welded components. Abington Publ., Cambridge.

Niemi, E., Fricke, W. and Maddox, S. 2004: Structural Hot-Spot Stress Approach to Fatigue Analysis of Welded Components – Designer's Guide. IIW-Doc. XIII-1819-00/XV-1090-01, International Institute of Welding.

Olivier, R. and Ritter, W. 1979. Catalogue of S-N Curves of Welded Joints in Structural Steels – Vol. 1-5. Report 56, Düsseldorf: Deutscher Verband für Schweißtechnik (DVS).

Radaj, D. 1990. Design and Analysis of Fatigue Resistant Structures. Abington Publishing, Cambridge (UK).

Radaj, D. and Sonsino, C.M. 1998. *Fatigue assessment of welded joints by local approaches*. Cambridge (UK): Abington Publ.

Sørensen, J.D., Tychsen, J., Andersen, J.U. and Brandstrup, R.D. 2005. Fatigue analysis of load-carrying fillet welds. Proc. JCSS workshop 23–25 August 2004 on 'Fatigue and Fracture', pp. 165–186, DTU, Copenhagen.

Usami, S. and Kusumoto, S. 1978. Fatigue strength at roots of cruciform, tee and lap joints. *Trans. Japan Welding Soc.*, 9(1): 3–10.

van Wingerde, A.M., Packer, J.A. and Wardenier, J. 1995. Criteria for the fatigue assessment of hollow structural section connections. *J. Construct. Steel Res.*, 35, pp. 71–115.

Zenner, H. and Grzesiuk, J. 2004. Einfluss der Nahtvorbereitung und Nahtausführung auf die Schwingfestigkeit hochwertiger Aluminiumkonstruktionen. *Schweißen und Schneiden*, 56(2): 58–62.

Tubular Structures XI – Packer & Willibald (eds)
© 2006 Taylor & Francis Group, London, ISBN 0-415-40280-8

Assessment of risk of brittle fracture for beam-to-column connections with weld defects

T. Iwashita
Ariake National College of Technology, Fukuoka, Japan

Y. Kurobane
Kumamoto University, Kumamoto, Japan

K. Azuma
Sojo University, Kumamoto, Japan

ABSTRACT: A number of beam-to-RHS column connections with various weld defects are modeled numerically. The weld defects are the same as those observed during the Kobe Earthquake and in the past full-scale tests. Focusing on fractures from weld defects, a procedure to evaluate fracture moments of beam-to-column connections is proposed based on a series of numerical analyses. Matrices showing the fracture moments as a function of the size and location of weld defects and the material toughness are presented. In addition, the rotation capacity of the beams in the beam-to-column assemblies is presented based on the past extensive test results and FE analyses.

1 INTRODUCTION

During the 1995 Kobe Earthquake many steel building frames experienced brittle fractures starting from various weld defects in beam-column joints. In instances where weld defects are found in steel structures, engineers have to decide whether repairs are necessary. In deciding on whether to repair or not, it is important to use a highly reliable assessment method to evaluate the possibility of brittle fracture from weld defects.

Our recent experimental and analytical investigations (Iwashita et al. 2003, 2005) indicated that occurrences of brittle fractures from weld defects could be predicted rather accurately. These investigations took into account the effects of a loss of plastic constraint at the crack tips. These effects are important because many of the weld defects in beam-to-column connections were surface cracks and sustained brittle fractures after extensive yielding of cracked area under strong earthquakes. Crack tips of these weld defects show much greater fracture toughness than the material toughness determined by conventional testing procedures like a single edge notched bend (SENB) test in which the notch root is under high plastic constraint (See Figures 1, 2, 3).

The present study follows the methodology developed earlier for predicting applied loads when brittle fractures initiate; the enhanced toughness due to the

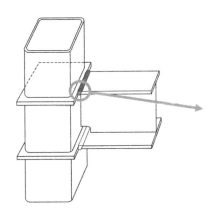

Figure 1. Beam-to-column connection.

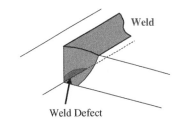

Figure 2. Weld defect in connection.

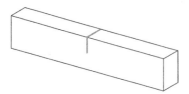

Figure 3. Notch in fracture toughness test specimen.

loss of plastic constraint at the crack tip is considered. However, detailed analyses of notch tip stress states are required for both beam-to-column connections with weld defects and SENB specimens.

A number of beam-to-RHS column connections with various weld defects are modeled numerically. The weld defects are the same as those observed following the Kobe Earthquake and in the past full-scale tests (AIJ Kinki 1997), which were caused by lack of fusion, lack of penetration, slag inclusions, etc. in the welded joints between the beam flanges and continuity plates. Representative locations of these defects are at the terminations of the welded joints where weld tabs exist and at the roots of the welds where backup bars exist. Especially, an edge crack at the beam flange edge groove welds, classified as a through crack later, creates high plastic constraints at the crack tips and induces a brittle fracture at a low level of the crack driving force (e.g. J-integral). Based on a series of numerical analyses it is possible to construct matrices showing the critical loads as a function of the size and location of weld defects and the material toughness. From the critical loads it is possible to infer the plastic rotation capacity of beam-to-column assemblies based on the past extensive test results.

2 EVALUATION OF FRACTURE MOMENT

2.1 Effects of plastic constraint

Under small scale yielding conditions (SSY), a single parameter, the J-integral, characterizes crack tip conditions and can be used as a geometry independent fracture criterion. Single parameter fracture mechanics breaks down in the presence of excessive plasticity and fracture toughness is dependent on the size and geometry of the test specimen. Anderson and Dodds (1991) found using numerical analyses that the apparent toughness of a material was highly dependent on the constraint conditions at the crack tip of the test specimens. In particular they examined the effects of crack depth on SENB specimens.

Figure 4 shows two SENB specimens with different notch depths (deep notch and shallow notch). This figure also shows the stress state at the crack tips when the same J value is applied to both specimens. The stress state shows a plane section of the volume

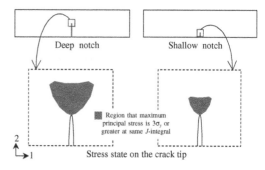

Figure 4. SENB specimens with different notch depth.

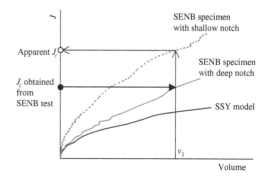

Figure 5. J-integral vs. volume curves.

surrounded by a contour of a certain value of the maximum principal stress, σ_1. The volume of the specimen with shallow notch is smaller than that of the specimen with deep notch, although the same J value is applied to each specimen. This is an example of the fracture toughness depending on the size and geometry of the test specimens. Anderson and Dodds proposed that the probability of brittle fracture is equal for the two specimens when the volumes are equal. Figure 5 is a chart containing J-volume curves as an example. The vertical axis is the applied J-integral while the horizontal axis is the volume surrounded by a contour of a certain value of σ_1 at the crack tip. The maximum principal stress is assumed as $\sigma_1 = 3\sigma_y$. From Figure 5, when the critical J value, J_c, is obtained through SENB testing of a deep notched specimen, the value of J_c can be corrected by determining the J value at which the crack tip volume of the shallow notched specimen equals that (v_1) of the deep notched specimen at J_c. The corrected J_c is called the apparent J_c. Using the apparent J_c, it is possible to evaluate an occurrence of brittle fracture even if plastic constraint is at a low level like the defect in Figure 2. We call this procedure as "the Anderson method" in this paper.

Matos et al. (2001) used the Weibull stress to assess the risk of brittle fracture for beam-column connections. However, the Weibull stress model necessitates

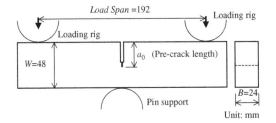

Figure 6. Specimen geometry and loading scheme.

Table 1. Results of SENB tests.

Specimen	a_0 (mm)	Δa (mm)	P_{max} (kN)	J_c (N/mm)	J_c (N/mm)
D1	25.6	0.06	32.1	58	Apparent
D2	25.1	0.04	33.5	46	154
D3	24.9	0.06	33.3	39	
S1	3.7	0.08	120.0	133	Average
S2	3.1	0.09	122.2	131	140
S3	3.6	0.07	120.4	155	

D = deep notch; S = shallow notch; a_0 = pre-crack length; Δa = ductile crack length; P_{max} = fracture load.

calibration of micromechanical parameters, which is not easy. The Anderson method employs directly the J_c value measured by conventional fracture specimens, with the only assumption that crack fronts of connections attain equivalent stressed volumes. The Anderson method is much simpler and was found to predict test results for connections with various defects very well, even if the threshold stress was varied between 2.8–$3.2\sigma_y$ (Iwashita et al. 2003).

2.2 Test of SENB specimen with different notch depth

SENB tests are conducted to confirm that the Anderson method is effective for materials used in steel building structures. Specimens have deep notches or shallow notches like Anderson's investigations and are made based on BS 7448 (BSI 1991). The material of the specimens has an absorbed energy of about 25 J at the test temperature. The specimen geometry and loading scheme are shown in Figure 6. The load, loading rig displacement and crack opening displacement (COD) are measured during the tests. Brittle fracture occurred in all of the specimens. Test results are summarized in Table 1. J_c is defined as J-integral at fracture and calculated based on BS 7448.

Figure 7 shows a comparison between the test and FE analysis results for load vs. COD curves. The analysis results are comparable to the test results for deep notch specimens. The analysis results for the shallow notch specimens also follow the test results. Figure 8

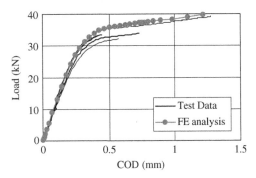

Figure 7. Load vs. COD curves for deep notch specimen.

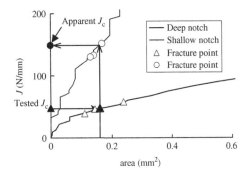

Figure 8. J-integral vs. area curves.

shows J-integral vs. area curves for deep and shallow notch specimens. The area in this case is the plane area (1–2 plane in Figure 4). The volume can be replaced with the area. Plotted marks on the curves for deep notch specimens represent fracture initiations in the tests that indicate J_c. Solid marks represent the average J_c. The apparent J_c for shallow notch specimens is obtained from the average J_c for deep notch specimens through the area surrounded by a contour of $\sigma_1 = 3\sigma_y$ at the crack tip as shown in Figure 8. The apparent J_c coincides well with the J_c for shallow notch specimens (See Table 1). Therefore it is reasonable to use the Anderson method to consider the effects of plastic constraint on fracture toughness. However, the test specimens showed small ductile cracks. If brittle fracture occurs with large ductile crack growth, another consideration is needed to predict occurrence of brittle fracture. The effects of ductile crack growth are ignored in this paper.

2.3 Evaluation of fracture moment by apparent J_c

The fracture toughness J_c is dependent on the size and geometry of the test specimen. Weld defects have variable profiles and are smaller as compared with notches of fracture toughness test specimens. It is therefore important to use the apparent J_c considering

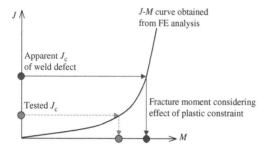

Figure 9. Evaluation of fracture moment by apparent J_c.

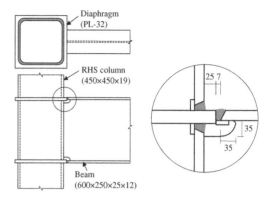

Figure 10. Connection design and detail of weld region.

the effects of plastic constraint to predict an occurrence of brittle fracture from weld defects. Figure 9 shows a procedure for the evaluation of the fracture moment of beam-to-column connections. For a conventional method, the fracture moment of connections is obtained from tested J_c and J-M curve of beam-to-column connections calculated by an FE analysis like the dashed line in Figure 9. For a method considering the effects of plastic constraint on the fracture toughness, the fracture moment is obtained from the J-M curve and apparent J_c of the weld defects in beam-to-column connections. The fracture moment of the beam-to-column connections evaluated by the apparent J_c is larger than that evaluated by the tested J_c as shown in Figure 9 due to the effects of plastic constraint. The fracture moment could be predicted rather accurately. In addition, the beam rotation at the fracture as well as fracture moment is obtained from an FE analysis to convert the moment to the rotation of the beam-to-column connection. Then, we can discuss the rotation capacity of connections.

3 BEAM-TO-COLUMN CONNECTION MODEL

3.1 Geometry

Geometry of beam-to-column connection models is based on the specimens in the past full-scale tests (AIJ Kinki 1997). The beam-to-column connections have RHS columns of $450 \times 450 \times 19$ mm in dimension, wide flange beams of $600 \times 250 \times 25 \times 12$ mm in dimension and 32 mm thick through diaphragms (continuity plates) at the beam flange positions as shown in Figure 10. The connections have complete joint penetration (CJP) groove welds at the end of the beam flanges. Weld tabs are flux tabs that are used as weld dams. These tabs are commonly used in Japan and are removed after welding. Figure 10 shows connection design and details.

3.2 Material properties

Table 2 shows material properties of the beam-to-column connection models. All of the members have

Table 2. Material properties.

Steel grade	Yield stress (Mpa)	Tensile strength (Mpa)	Maximum uniform (%)	Young's modulus (GPa)
SN490B	370	566	15.0	215

the same material properties in this FE analysis. The material grade is corresponding to grade SN490B according to Japanese Industrial Standard (JSA 2005). We assumed the weld metal has the same material properties as the base metal. Then, the mismatch between the strength of weld metal and that of base plate is ignored in the FE analysis. The stress-strain relationship is based on the previous tensile coupon tests (Iwashita et al. 2005). The fracture toughness J_c is very important to predict an occurrence of brittle fracture. J_c is defined as 50 N/mm or 200 N/mm in this paper. $J_c = 50$ N/mm indicates fracture toughness of a low level and $J_c = 200$ N/mm indicates fracture toughness of a medium level. Incidentally, a material of $J_c =$ about 50 N/mm has an absorbed energy of about 25 J in Charpy impact testing as already noted in Paragraph 2.2.

3.3 Weld defects

Three types of weld defect observed during the Kobe Earthquake are prepared at the termination of the weld as shown in Figure 11. A through crack (a) is introduced at the weld toes on the diaphragm plate side through the plate thickness. A surface crack (b) is introduced at the weld toes on the diaphragm plate side. The crack did not penetrate through the thickness. An internal crack (c) is introduced along the weld root on the diaphragm plate side. These weld defects (by lack of fusion, lack of penetration etc.) can occur easily around the ends of CJP groove welds.

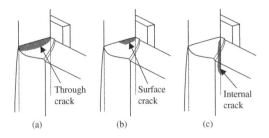

Figure 11. Type of weld defect.

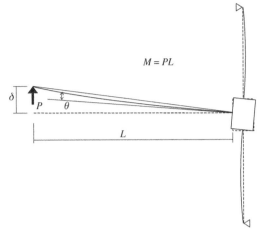

Figure 13. Definition of beam rotation and moment at column face.

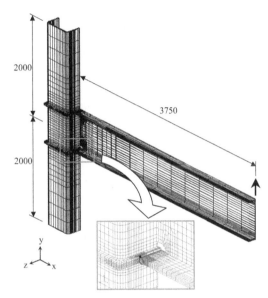

Figure 12. Finite element mesh for half-symmetric model for the beam-to-column connections.

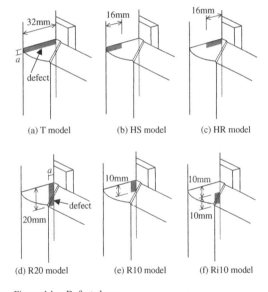

Figure 14. Defect shape.

3.4 *Model of FE analysis*

Finite element models were created and non-linear analyses performed using the finite element program ABAQUS (HKS 2001). Figure 12 shows the 3-D FE meshes for a typical beam-to-column connection with a through crack. Symmetry about the $z = 0$ plane permits the use of a half-model with a number of nodes about 18,000. The FE models employ 8-noded brick elements with reduced integration technique. The crack front consists of ten rings with 16 elements in each ring for J calculations. The minimum element dimension is 0.05 mm in the crack tip region. The plasticity of materials was defined by the Von Mises yield criteria. The material stress-strain data used for FE analyses was determined by tensile coupon tests of a material in the previous research. Loading occurs through an axial displacement imposed at the beam end, with the final displacement of $\delta = 400$ mm (See Figure 13).

Weld defects exist in the CJP groove welded joints to the bottom beam flange, lying on the edges of the diaphragms, on which weld metal is to be deposited. The profile of real defects as shown in Figure 11 is simplified for modeling. Then the profile of weld defects in FE analysis is rectangular as shown in Figure 14. "a" in Figure 14 means the crack depth and two crack depths ($a = 3$ mm or 5 mm) are prepared for the FE analysis. We refer to (a) ~ (c) defects as "weld end defects" and to (d) ~ (f) defects as "root defects" in this paper.

Inspections following the Kobe Earthquake showed that fractured surfaces at the beam ends invariably

605

accompanied ductile crack growth preceding initiations of brittle fractures. Since the evolution of plastic constraint at a crack tip with ductile crack growth is not negligible (Xia et al. 1996), the crack depth assumed in this analysis should be considered to include cracks due to ductile growth.

4 RESULTS AND DISCUSSION

4.1 *Results of FE analysis*

Figure 15 shows a normalized moment (M/M_p, M_p being the plastic moment) vs. rotation curve for FE analysis. This curve is similar to those observed in the past full-scale test results (AIJ Kinki 1997) although a comparison between the analysis and test curves is not presented herein. Figure 16 shows a contour plot of maximum principal stresses around the tip of a weld defect for a T model and stresses at the tip of the weld defect are very high.

Figure 17 shows J-integral vs. moment curves for each model with 3 mm crack depth. The J-integral values at the same value of moment are greater for the

weld end defect models than for the root defect models. These J-integral values for each model do not include constraint effects. Plastic constraint can be related to the stress triaxiality. Figure 18 shows stress triaxiality vs. J-integral curves for each model with 3 mm crack depth. The stress triaxiality, T_s is calculated as follows:

$$T_s = \sigma_h / \sigma_{eq} \qquad (1)$$

where σ_h is the hydrostatic pressure and σ_{eq} is the Von Mises equivalent stress. Stresses were calculated for elements forming the crack tip, being 1 mm apart from the crack tip itself, at the location where possibility of brittle fracture is high. The weld end defect models are subjected to high stress triaxiality as compared with that of the root defect models. The results indicate the risk of brittle fracture from weld end defects is higher than from root defects. The level of stress triaxiality varies with the defect type and, of course, the level of stress triaxiality varies with each model as shown in Figure 18. It is therefore important to use the apparent J_c considering the effects of plastic constraint for the prediction of brittle fracture.

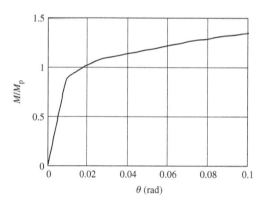

Figure 15. Moment vs. rotation curve.

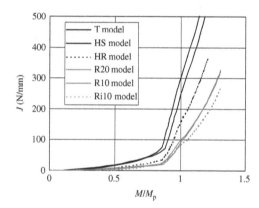

Figure 17. J-integral vs. moment curves.

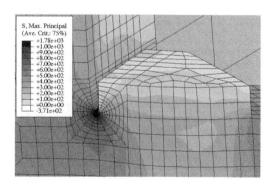

Figure 16. Contour plot of maximum principal stresses around the tip of a weld defect for a T model.

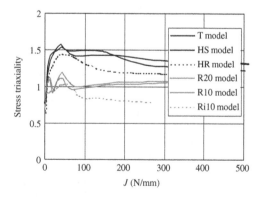

Figure 18. Stress triaxiality vs. J-integral curves.

Figure 19 shows J-integral vs. area curves for the models with 3 mm crack depth and evaluates the apparent J_c when the tested $J_c = 50$ N/mm. The apparent J_c of T model and HS model is small due to high plastic constraint. In fact, T model and HS model are subjected to high stress triaxiality as shown in Figure 18. The root defect models show large apparent J_c because the stress triaxiality of these models is low. In addition, the apparent J_c of Ri10 model does not exist in this FE analysis because of low stress triaxiality. This means that brittle fracture will not occur for Ri10 model within the load range conducted in the FE analysis.

4.2   Effects of defect condition on brittle fracture

The apparent J_c for all of the cases is calculated and Table 3 shows normalized fracture moments (M_f/M_p, M_f being the fracture moment) obtained using the procedure shown in Paragraph 2.3. Parameters are the defects profile, defect depth and fracture toughness J_c. The fracture moments for the models of the weld end defect type, except HR model with $J_c = 200$ N/mm,

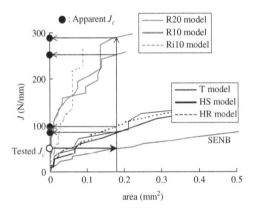

Figure 19. J-integral vs. area curves and evaluation of the apparent J_c.

did not reach 1.3. The fracture moments for the models of the root defect type except R20 model and R10 model with 5 mm crack depth and $J_c = 50$ N/mm reached values greater than 1.3. The main features found in these results are described as follows:

(1) The fracture moment of the weld end defect models (a) ~ (c) is small as compared with that of the root defect models (e) ~ (f). Especially, the fracture moment is very small for the through crack type (a) and for the surface crack type (b) in which the defects exist on the outside (remote from the root) of the welded joints, even if defect depth is small and the fracture toughness is on the medium level. The fracture moment of the HR model (c) belonging to the weld end defect models, however, is very large when $J_c = 200$ N/mm but is very small when $J_c = 50$ N/mm. Probability of brittle fracture is very small for the Ri10 model (f) with an internal crack even if $J_c = 50$ N/mm.

(2) The defect depth does not influence the fracture moment of the weld end defect models while the defect depth influences the fracture moment of the root defect models. The fracture moment of the weld end defect models is small regardless of the defect depth, except the HR model with higher fracture toughness (200 N/mm).

In addition, it is possible to infer the beam rotation capacity of beam-to-column assemblies from the present FE analysis and the past extensive test results. A procedure for the inference is described below.

The beam rotations at brittle fracture can be determined on the fracture moment vs. rotation curves given by FE analyses from the fracture moments shown in Figure 9. The moment vs. rotation curve obtained by a monotonic loading test is known to be approximately equal to the moment vs. rotation skeleton curve (See Figure 20) constructed from the moment vs. rotation curve observed in a cyclic loading test. Thus, the beam rotations according to FE analyses roughly correspond to beam-rotation skeleton curves in cyclic loading test. The beam rotation capacity is generally defined as the maximum rotation capacity θ_{max} achieved during

Table 3. M_f/M_p for each defect model.

J_c (N/mm²)	a (mm)	(a) T model	(b) HS model	(c) HR model	(d) R20 model	(e) R10 model	(f) Ri10 model
50	3	0.87	0.87	0.96	1.35	1.37	>1.40
50	5	0.85	0.86	0.95	0.96	1.05	>1.40
200	3	1.12	1.09	>1.40	>1.40	>1.40	>1.40
200	5	1.04	1.05	>1.40	1.31	>1.40	>1.40

cyclic loading tests. However, the rotation capacity on the skeleton curve $\theta_{s\,max}$ can be converted to the rotation capacity observed in a cyclic loading test by utilizing a correlation between θ_{max} and $\theta_{s\,max}$ as shown in Figure 21. Plotted data in Figure 21 are obtained from the previous test results (Nakashima et al. 1997, Akiyama et al. 1999, Iwashita et al. 2002). Then, it is possible to infer θ_{max} from $\theta_{s\,max}$. The beam

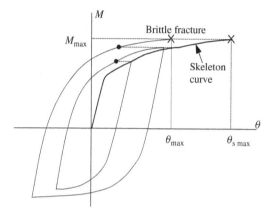

Figure 20. Definition of skeleton curve and relationship θ_{max} and $\theta_{s\,max}$.

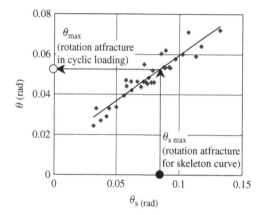

Figure 21. Estimation of θ_{max} in cyclic loading.

rotation capacity θ_{max} could be obtained from beam rotation at fracture according to the FE analyses.

Table 4 shows the plastic beam rotation capacity, which is a plastic component of θ_{max}, for each model. The FEMA Criteria (FEMA, 2000) define connections which failed at interstorey drift angle values less than 0.04 radian as having insufficient rotational capacity for the special moment frame (SMF). The elastic drift angle of typical moment frames is roughly in the range of 0.01 radian. Thus, the plastic rotation capacity of the beam is required to be greater than 0.03 radian. Shaded region in Table 4 shows plastic rotation capacity is less than 0.03 radian. This means that these models do not meet the plastic beam rotation demand on SMF. Especially, Table 4 indicates cases (a) and (b) are the most critical.

5 CONCLUSIONS

The Anderson method is effective for the materials used in steel building structures and provides the apparent J_c considering the effects of plastic constraint on fracture toughness.

A number of beam-to-RHS column connections with various weld defects are modeled numerically and the procedure to evaluate the fracture moment of beam-to-column connections is proposed based on a series of numerical analyses. The results of numerical analyses show the risk of brittle fracture is high for the weld end defect type as compared with the root defect type. Especially, the fracture moment of through crack type and surface crack type with defects existing on the outside of the beam flange to diaphragm welded joints (HS type) is very low. In contrast, probability of brittle fracture is very small for the Ri10 model with internal crack even if $J_c = 50\,\text{N/mm}$. It should be noted that the FE analysis have not reproduced ductile crack growth prior to an initiation of brittle fracture. Further research in this area is to consider effects of crack growth in an FE analysis.

The procedure to evaluate plastic rotation capacity at brittle fracture from weld defects in beam-to-column connections is proposed based on the past extensive test results and FE analysis. It is necessary to verify applicability of this procedure based on more experimental and numerical evidences for the future.

Table 4. Plastic rotation capacities for each defect model.

J_c	a	(a) T model	(b) HS model	(c) HR model	(d) R20 model	(e) R10 model	(f) Ri10 model
50	3	0.010	0.010	0.012	0.039	0.042	>0.060
50	5	0.009	0.010	0.012	0.012	0.015	>0.060
200	3	0.023	0.018	>0.060	>0.060	>0.060	>0.060
200	5	0.015	0.015	>0.060	0.046	>0.060	>0.060

REFERENCES

AIJ Kinki. 1997. *Full-scale test on plastic rotation capacity of steel wide-flange beams connected with square tube steel columns*. Committee on Steel Building Structures, The Kinki Branch of the Architectural Institute of Japan, Osaka, Japan. (in Japanese)

Akiyama, H., et al. 1999. Transition form ductile fracture to brittle fracture of full scale beam-to-column connections caused by temperature, *Journal of structural and construction engineering* 522: 105–112. (in Japanese)

Anderson, T.L. & Dodds, R.H. 1991. Specimen size requirements for fracture toughness testing in the ductile-brittle transition region. *Journal of Testing and Evaluation* 19 (2): 123–134.

British Standard Institution. 1991. *BS 7448-1. Fracture mechanics toughness tests – Part 1: Method for determination of K_{Ic}, critical CTOD and critical J values of metallic materials*. London.

FEMA. 2000. *FEMA-350 Recommended seismic design criteria for new steel moment-frame buildings*. Federal Emergency Management Agency, Washington D.C.

Hibbit, Karlsson & Sorensen. 2001. *ABAQUS Version 6.2 User Manuals*. Pawtucket: Hibbit, Karlsson and Sorensen, Inc.

Iwashita, T., et al. 2003. Prediction of brittle fracture initiating at ends of CJP groove welded joints with defects: study into applicability of failure assessment diagram approach. *Engineering Structures* 25 (14): 1815–1826.

Iwashita, T., et al. 2005. Effect of Crack Tip Constraint on Brittle Fracture Initiation at Weld Defects. *Proceedings of 3rd International Symposium on Steel Structures* 2: 732–743.

Iwashita T., et al. 2002. Brittle fracture from ends of CJP groove welded joints. *Proceedings of the 12th International Offshore and Polar Engineering Conference* 4: 117–124.

JSA. 2005. *JIS G 3136 Rolled steels for building structure*. Japanese Standards Association. (in Japanese)

Matos, C.G. & Dodds Jr. R.H. 2001. Probabilistic modeling of weld fracture in steel frame connections, part I: quasi-static loading. *Engineering Structures* 23: 1011–1030.

Nakashima, M., et al. 1997. Full-scale test on plastic rotation capacity of steel wide-flange beams connected with square tube steel columns. *Steel Construction Engineering* 4 (16): 59–74. (in Japanese)

Xia, L. & Shih, C.F. 1996. Ductile crack growth—III. transition to cleavage fracture incorporating statistics. *J. Mech. Phys. Solids* 44 (4): 603–639.

Tubular Structures XI – Packer & Willibald (eds)
© 2006 Taylor & Francis Group, London, ISBN 0-415-40280-8

Applicability of partial joint penetration groove welded joints to beam-to-RHS column connections and assessment of safety from brittle fracture

K. Azuma
Sojo University, Kumamoto, Japan

Y. Kurobane
Kumamoto University, Kumamoto, Japan

T. Iwashita
Ariake National College of Technology, Fukuoka, Japan

K. Dale
formerly Sojo University, Kumamoto, Japan

ABSTRACT: This paper concerns the applicability of partial joint penetration (PJP) groove welds to beam-to-column connections. Two full-sized beam-to-column connections, with rectangular hollow section columns and PJP groove welds at the ends of the beam flanges, were tested under cyclic loads. All the specimens were made of one-sided connections. When the unfused regions created by PJP groove welds were reinforced by fillet welds so that the welded joints have a sufficient cross-sectional area, the connections showed sufficient deformation capacity, although ductile cracks initiated at weld toes and grew stably. Test results were reproduced well by non-linear FE analyses. Strains sustained at points around internal discontinuities were found to be low because of greater cross-sectional areas of welded joints compared with the cross-sectional area of the beam flanges. Fracture toughness properties of numerically modeled connections were evaluated by using a failure assessment diagram approach, which was modified by considering the effect of enhanced apparent toughness of material due to the loss of crack tip constraint. Both the test results and the fracture mechanics-based assessment demonstrated that it is unlikely to initiate brittle fracture at these discontinuities when materials used in the connections have sufficient fracture toughness.

1 INTRODUCTION

The 1995 Kobe Earthquake revealed that beam-to-column connections in steel moment frames are susceptible to brittle fracture under the influence of strong earthquake. Many of the failures were caused by cracks growing from the corner of cope holes, as was predicted prior to the earthquake, or weld tab regions in the beam bottom flange groove welds. Post-earthquake inspections showed that many of these fractures occurred at or around beam flange groove welds starting from various weld defects and from tips of ductile cracks that grew from geometrical discontinuities. Majority of these fractures occurred after beam flanges sustained extensive yielding and/or local buckling. A large-scale post-earthquake research project on conventional and improved beam-column connections (AIJ Kinki. 1997) reproduced well the fracture behavior observed following the Kobe Earthquake. One of the post-earthquake proposals (AIJ. 1996) is to use improved profiles of cope holes. These new details, however, could introduce other weld defects. One issue raised by the earthquake concerns improvements of details of welded joints so that connections have not only sufficient deformation capacity but also are less costly. Previous testing of full-scale beam-to-column connections with partial joint penetration (PJP) groove welded joints without using any cope hole and back-up bar showed sufficient deformation capacity (Azuma et al. 2000, 2003). An assessment of unfused regions based on CTOD design curve predicted that brittle fracture would not occur.

This paper concerns the applicability of partial joint penetration (PJP) groove welds to beam-to-column connections. Two full-sized beam-to-rectangular hollow section (RHS) column connections with PJP groove welds at the ends of the beam flanges were tested under cyclic loads. Both specimens sustained local buckling of the beam flanges and webs accompanying a ductile crack extension. A non-linear finite element analysis was conducted, which showed that analysis results reproduce well not only overall deformations but also local strains observed in the tests. The modified fracture mechanics approach based on failure assessment diagram (FAD) approach (BS 7910, 1999), which was modified by considering the effect of enhanced apparent toughness of material due to the loss of crack tip constraint, was applied to the numerically modeled connections. The results of assessment were found to be consistent with the test results.

NOMENCLATURE

A_w	cross-sectional area of beam web
B	brittle surface ratio
b	material constant
E	Young's modulus
$E.L.$	elongation
J	J-integral obtained by elastic-plastic analysis
J_c	critical J-integral of material obtained by CT testing
J_e	J-integral obtained by elastic analysis
J_{ef}	J-integral at the fracture point obtained by elastic analysis
G	shear modulus of elasticity
H_b	Height of beam
L	length between loading point and column face
M	beam moment at column face
M_{max}	maximum moment of the beam
M_p	full plastic moment of the beam
t_f	thickness of beam flange
u_1, u_2	horizontal displacement
v_1, v_2	vertical displacement
vE	Charpy absorbed energy
vE_{shelf}	shelf energy obtained from Charpy tests
$v.E(T)$	Charpy absorbed energy at T (°C)
vTE	transition temperature
vT_{re}	energy transition temperature
η_s	cumulative total plastic rotation factor
θ	rotation of beam segment between loading point and column face
θ_p	beam rotation at M_p obtained from elastic stiffness
$n\theta_p$	nominal beam rotation at M_p
σ_h	hydrostatic stress
σ_{eq}	von Mises's equivalent stress
σ_u	tensile strength obtained from coupon tests
σ_y	yield stress of virgin steel
σ_{1-3}	principal stress

2 CYCLIC TESTING OF BEAM-TO-COLUMN CONNECTIONS

2.1 Specimens and loading procedures

Two full-size beam-to-column connections, with RHS columns with $400 \times 400 \times 12$ (height × width × thickness of flange) in grade BCR295 (nominal yield strength = 295 MPa), designated as RHS-H and RHS-Q, were tested. All the specimens were made of one-sided connections with wide flange beams with $500 \times 200 \times 10 \times 16$ in grade SN490B (nominal yield strength = 325 MPa). Specimens had PJP groove welds at the ends of the beam flanges and had 19 mm thick internal diaphragms. Weld metal was produced by electrodes designated as YGW-18. The roots of welds were reinforced by additional fillet welds after removing weld tabs. The beam webs were fillet welded directly to the column face. Stringer passes were used for all welding and heat inputs during welding were approximately 20 kJ/cm for each layer.

The configuration of the specimens and details of welded joints are shown in Figure 1. Specimen RHS-H had cross-sectional area of unfused region equal to the half of the beam flange area, while RHS-Q had the area equal to the three quarter of the beam flange area. Cyclic loads in the horizontal direction were applied to the end of the beams, while the both ends of column were fixed. Figure 2 shows the position of load application and the displacement measurements. The rotation of the beam θ was calculated by the following equation.

$$\theta = \frac{u_1 - u_2}{L} - \frac{v_1 - v_2}{H_b - t_f} \qquad (1)$$

The amplitude of the beam rotation was increased in increments of $2_n\theta_p$ up to failure. $_n\theta_p$ was defined as the elastic component of beam rotation when the beam moment at the column face reaches the full plastic moment M_p and can be calculated by the following equation:

$$_n\theta_p = \frac{L}{3EI}M_p + \frac{1}{GA_w}\frac{M_p}{L} \qquad (2)$$

The full plastic moment was calculated using measured yield strengths of materials. Two cycles of loading were applied at each displacement increment.

2.2 Material properties

The material properties, in terms of engineering stress-strain, were obtained by tensile coupon tests for the beam, column, internal diaphragms, stiffeners and weld materials, which are summarized in Table 1.

The fracture toughness was obtained by Charpy impact tests. Test pieces were taken from plates welded

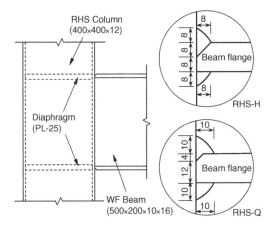

Figure 1. Specimen configuration.

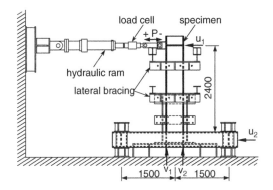

Figure 2. Positions of load application and displacement measurements.

Table 1. Results of tensile coupon tests.

Location	σ_y (MPa)	σ_u (MPa)	E.L. (%)	E (GPa)
Beam flange	365.7	521.4	21.9	213.1
Beam web	398.7	536.2	18.7	209.1
Column	402.3	469.8	21.1	208.4
Diaphragm Stiffener	330.0	510.6	21.0	212.2
Weld metal	492.5	619.9	24.9	215.2

under the same welding conditions as those for the specimens. The positions of notch roots were base metal of beam flange, HAZ (heat affected zone) and DEPO (deposited weld metal). Test pieces were cooled to temperatures between −150°C and 10°C by using liquid nitrogen or dry ice and alcohol. Three test pieces were tested at each temperature. The results of Charpy

Table 2. Results of Charpy impact test.

	$vE(0)$ (J)	vE_{shelf} (J)	vT_{re} (°C)
Base metal	214	239	−47
HAZ	242	246	−95
DEPO	125	179	−32

impact test are shown in Table 2. The energy transition curve was obtained by fitting test results into the follow equation:

$$vE(T) = \frac{v E_{shelf}}{e^{-b(T-vTE)} + 1} \tag{3}$$

2.3 Inspection of penetration conditions

The specimens of the previous testing (Azuma et al. 2003), ductile cracks grew from tips of incomplete penetration through the weld metal or HAZ. The premature failure, and subsequent lack of strength and rotation capacity, was due to incomplete penetration of the weld metal at the beam top flange.

Before the fabrication of the present specimens, the welding positions were decided by investigating the penetration status. Figure 3 shows the penetration of weld metal on cross sections of welded joints. Using the flat welding created an incomplete penetration at the root of PJP welding, while using the horizontal welding realized a sufficient penetration. This is because arcs could not reach the roots but only the groove surfaces when flat welding was used. Therefore, horizontal welding was adopted for the fabrication of specimens.

2.4 Deformation capacity and failure modes

Figure 4 shows hysteresis loops for the two specimens. The vertical axis is the moment at the column face and is non-dimensionalized by dividing it by the full-plastic moment of the beam. The moment is herein defined as the positive moment when the bottom flange is in tension. Figure 5 shows moment vs. rotation skeleton curves for both specimens. Table 3 summarized the maximum moments with the corresponding rotations and the cumulative plastic rotation factors η_s. η_s was obtained from the skeleton curves for each specimen. The beam rotations at full plastic moment θ_p, namely M_p divided by the elastic stiffness of the beam, were calculated. The elastic stiffness was determined by using slopes at unloading portions of hysteresis loops.

Both specimens failed owing to combined local and lateral buckling of the beams. Ductile cracks initiated at the weld toes at the edge of the beam flanges and extended stably along the toes of welds on the beam

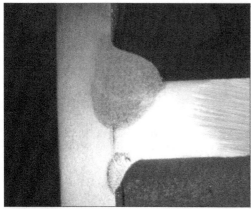

(a) Flat welding

(b) Horizontal welding

Figure 3.   Macro sections of PJP welds.

flange side, until the rotation of the beams reached 0.06 radians. The specimen RHS-Q after failure is shown in Figure 6.

2.5   FE analysis

A finite element analysis of the two specimens was carried out using the ABAQUS (2005) general-purpose finite element package. The models were constructed from 8-noded linear 3D elements. This element is nonconforming and isoparametric and employs the reduced integration technique with hourglass control. The plasticity of the material was defined by the von Mises yield criterion. The isoparametric hardening law was used for this analysis. The ABAQUS program requires the stress-strain data to be input in the form of true stress and logarithmic strain, and the stress-strain curves were transformed accordingly. The material data in the analysis were calculated from tensile coupon test results. Mesh models were generated for a half of the specimens because of symmetry in configuration.

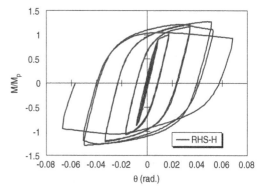

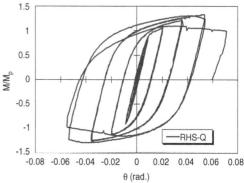

Figure 4.   Hysteresis loops.

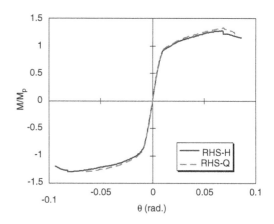

Figure 5.   Moment vs. rotation skeleton curves.

The unfused regions were generated by the nodes in the defect area on the contact surfaces between the beam flange and the column flange. These surfaces were separated in each element as double nodes. A monotonic load was applied to the beam end. The moment-rotation curves are compared with the skeleton curves that were obtained from experimental results for each specimen.

614

Table 3. Cumulative plastic deformation factors.

Specimen	M_{max} kNm	θ_{max} 10^{-2} rad.	M_{max}/M_p	η_s^+	η_s^-
RHS-H	1016	8.32	1.29	8.54	9.21
RHS-Q	1046	6.86	1.33	7.48	6.14

Table 4. Ultimate strains obtained from strain gage measurement and FE analysis results.

Specimen	Test results $\times 10^4 \mu$	FE analysis results $\times 10^4 \mu$
RHS-H	8.30	8.26
RHS-Q	9.48	9.00

Figure 6. RHS-Q specimen after failure.

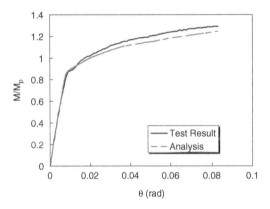

Figure 7. Moment vs. rotation skeleton curves obtained from FE analysis.

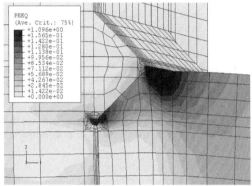

Figure 8(a). Contour plot of equivalent plastic strain in cross section on plane of symmetry.

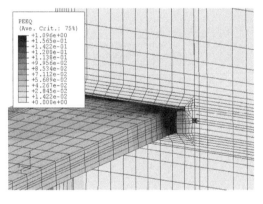

Figure 8(b). Contour plot of equivalent plastic strain around the weld toes.

Though ductile cracks (shallow cracks) grew stably from the weld toes at the edges of the beam flanges, the depth of cracks could not be measured. It was assumed that the depth of flaw size was 0.5 mm.

The analysis results coincide well with the test results as seen in Figure 7. Strains in the side of the beam flange were measured by a strain gage at a position 25 mm away from the root face on the column side in each specimen. Ultimate strains in the direction of the beam axis were obtained from FE models. The strains according to FE analysis are comparable to skeleton strains obtained from strain gage measurements, as shown in Table 4.

Figures 8(a) and 8(b) show the contour plot of equivalent plastic strains in cross section on plane of symmetry and around the weld toes when the deformation reached the final failure stage in RHS-H specimen. Strains at the tip of the unfused regions are not high and even smaller than remote strains. The cross-sectional area on the plane where discontinuities exist is much larger than the cross-sectional area of the beam flange, due to the existence of reinforcing fillet welds. This may account for a low level of strain concentration at discontinuities. The greatest strain concentration was found at the weld toes at both ends of the beam flanges.

Figure 9 shows the stress triaxiality, T_s, vs. J-integral curve. The stress triaxiality, which is related to plastic constraint, was defined using the following equations:

$$T_s = \frac{\sigma_h}{\sigma_{eq}} \qquad (4)$$

T_s was taken as a peak value found below the blunted crack tips in FE analysis models. Plotted marks on each curve represent the failure point. The failure point is defined in this report as the instant when the beam rotation reached 0.06 radians. Test results for CT specimens are also plotted in this figure for reference. The fracture points show a tendency for J at fracture to be large while the stress triaxiality is low, which is in contrast to with CT specimens. This may suggests that the risk of brittle fracture from the tips of discontinuities and ductile cracks is low.

2.6 Parametric study varying size of beams and columns

For a parametric study, an additional analysis of six numerical models was conducted to find the influence of specimen sizes. The size of beams, columns and

additional fillet welds of each specimen was specified in Table 5, while specimen configuration was shown in Figure 10. Each specimen has a cross-sectional area of unfused region created by PJP groove welds being equal to the half of the beam flange area. Additional fillet weld sizes were decided so that over-strength factor α of each specimen was about 1.25. The element types and material data in the analysis were assumed as the same as those for RHS-H and RHS-Q models. The failure assessment of brittle fracture was conducted following the procedures described in Section 3.4.

3 ASSESSMENT OF WELD DEFECTS

3.1 Assessment procedure

Fracture toughness properties of the simplified beam-to-column connection specimens, which sustained brittle fractures, were assessed by using a modified FAD approach (Iwashita et al. 2003). Ductile crack growth is ignored in the calculations, but plastic constraint is taken into account. The same approach was applied to the numerically modeled connections with PJP groove welded joints to evaluate the fracture toughness properties of the two specimens.

BS 7910 gives guidance on fracture mechanics based method for assessing the acceptability of defects in structures. Assessment is generally made

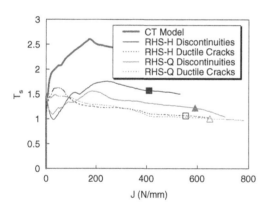

Figure 9. Stress triaxiality vs. J-integral curve.

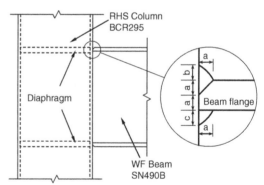

Figure 10. Specimen configuration.

Table 5. The sizes of column, beam and additional fillet welds.

FE Model	Column	Beam	a (mm)	b (mm)	c (mm)
RHS-1	$500 \times 500 \times 14$	$500 \times 200 \times 12 \times 19$	9.5	6	7
RHS-2	$500 \times 500 \times 14$	$500 \times 200 \times 12 \times 25$	12.5	7	8
RHS-3	$500 \times 500 \times 16$	$600 \times 200 \times 12 \times 19$	9.5	7	7
RHS-4	$500 \times 500 \times 16$	$600 \times 200 \times 12 \times 25$	12.5	8	8
RHS-5	$550 \times 550 \times 16$	$700 \times 300 \times 12 \times 19$	9.5	6	7
RHS-6	$550 \times 550 \times 16$	$700 \times 300 \times 12 \times 25$	12.5	7	8

by means of a FAD based on the principles of fracture mechanics. The defect is assessed by evaluating the fracture and plastic collapse parameters and plotting the corresponding point on the FAD. The vertical axis of the FAD is the ratio of the applied fracture toughness to the required fracture toughness. The horizontal axis is the ratio of the applied load to that required to cause plastic collapse. An assessment lines is plotted on the diagram. Calculations for a flaw provide the co-ordinates of an assessment point. The location of the point is compared with the assessment line to determine the acceptability of the flaw.

The assessment procedures are given as follows:

1. Determination of assessment curve
2. Calculation of plastic collapse parameters (L_r) using the ratio of moment
3. Calculation of fracture parameters (K_r) taking into account the effect of plastic constraint.

3.2 Application of FAD

The assessment curve of Level 3C in the BS 7910 is applied to the FAD. L_r for the assessment curve is calculated from the following equation:

$$L_r = \frac{M}{M_p} \qquad (L_r \leq L_{r,max}) \qquad (5)$$

$L_{r,max}$ is determined using the following equation:

$$L_{r,max} = \frac{\sigma_y + \sigma_u}{2\sigma_y} \qquad (6)$$

K_r is calculated from the following equation:

$$K_r = \sqrt{\frac{J_e}{J}} \qquad (7)$$

where J and J_e are the values at same applied load and obtained from FE analysis.

K_r and L_r at the fracture point of each specimen were determined by using the following equations:

$$L_r = \frac{M_{max}}{M_p} \qquad (8)$$

$$K_r = \sqrt{\frac{J_{ef}}{J_c}} \qquad (9)$$

3.3 Plastic constraint effects

The fracture toughness of a material is frequently measured by a three points bending test using SENB (single edge notched bend) specimens. SENB specimens may be subjected to much greater plastic constraint at the crack tips as compared with tips of surface cracks in wide plate specimens. Therefore, critical fracture toughness for a wide plate under tensile loads is possibly under-estimated. The beam bottom flanges are close to a wide plate under tensile load in stress state and correspond to this case.

The stress state at crack tips resembles that of a notched bar with a volume of material surrounded by a contour of a certain value of the maximum principal stress, σ_1. The volume for the specimen with shallow notch is smaller than that for the specimen with deep notch, even if the same J value is applied to each specimen. This is an example of the fracture toughness depending on the size and geometry of the specimens. Anderson and Dodds (1991) proposed that the probability of brittle fracture may be equal for two specimens when the volumes are equal. The volume could be replaced with the area. The area in this case is the plane area at the crack tip surrounded by the contour of a certain value of σ_1. The maximum principal stress is assumed as $\sigma_1 = 3\sigma_y$. Figure 11 shows J-integral vs. area curves obtained from FE analysis. The vertical axis is the applied J-integral while the horizontal axis is the area surrounded by a contour of a certain value of σ_1 at the crack tip. When the critical J value, J_c, is obtained from SENB testing of a deeply notched specimen, the value of J_c can be corrected by determining the J value at which the crack tip area of the shallowly notched specimen equals that of the deeply notched specimen at J_c. The corrected J_c is called the apparent J_c. Table 6 shows apparent J_c values at the assessed point.

3.4 Assessment

Following the procedures discussed in previous sections, the assessment of susceptibility to brittle fracture is made for numerically modeled joints with discontinuities and ductile cracks. Figures 12 and 13 shows

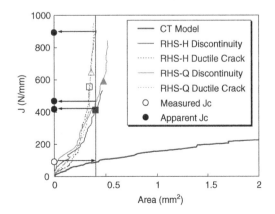

Figure 11. J-integral vs. area curves.

Table 6. Apparent J_c.

Specimen	Position	Apparent J_c (N/mm)
RHS-H	Discontinuities	418
	Ductile Cracks	894
RHS-Q	Discontinuities	467
	Ductile Cracks	951
RHS-1	Discontinuities	514
	Ductile Cracks	99
RHS-2	Discontinuities	334
	Ductile Cracks	132
RHS-3	Discontinuities	435
	Ductile Cracks	102
RHS-4	Discontinuities	137
	Ductile Cracks	81
RHS-5	Discontinuities	445
	Ductile Cracks	124
RHS-6	Discontinuities	226
	Ductile Cracks	155

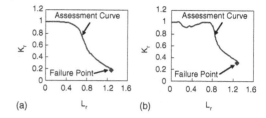

Figure 13. Example of FAD for a parametric study.

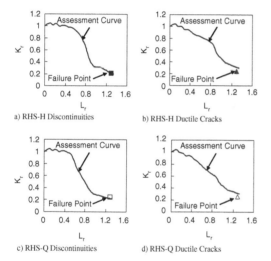

a) RHS-H Discontinuities
b) RHS-H Ductile Cracks
c) RHS-Q Discontinuities
d) RHS-Q Ductile Cracks

Figure 12. Failure assessment diagram.

the modified FAD plotting the failure point that is obtained by using the apparent J_c in the calculation of K_r. Although notch toughness of base metal of specimens is not measured, it was conservatively assumed that J_c was 90 N/mm, which was obtained from CT testing (See reference, Iwashita et al. 2003).

Since the partial joint penetration groove welds in specimens created discontinuities at the root of the welds. These discontinuities formed internal defects because the roots of the welds were reinforced by additional fillet welds. Plastic constraint at tips of discontinuities is greater than that of surface flaws. For the unfused region, apparent J_c could not be obtained because the area at crack tips surrounded by the

contour of maximum principal stress did not reached that of the CT model, as shown in Figure 11. Therefore, the FE analysis for RHS-H and RHS-Q specimens was continued until the area could be obtained beyond the loads at which actual specimens failed. The failure point plotted close to the assessment curves. Note that these results were obtained from rotations of the beams reached much larger than the rotation capacity of the beams in the testing. Therefore, it is unlikely to have a brittle fracture starting from the roots of the PJP welds.

Strains concentrated at weld toes at the edges of the beam flanges rather than at internal defects, ductile cracks grew stably from the weld toes at the edges of the beam flanges. The failure point plotted inside of the assessment curves. These results show that brittle fractures would not occur from the tips of the ductile cracks according to the present assessment. The test results supported the prediction; ductile cracks grew only slowly from these defects under cyclic loading.

The analysis results for 6 additional connections (Table 5) are shown in Figure 13 and Table 6. These results are found to be equal to those observed for the two specimens mentioned previously.

4 CONCLUSIONS

The applicability of PJP groove welds to beam-to-column connections was examined. Strain and deflection measurements as well as FE analysis results showed that strains sustained at the end portions of the beam flanges before cracks extended significantly were about equal in magnitude irrespectively of the specimen type.

Partial joint penetration groove welds formed internal defects because the roots of the welds were reinforced by additional fillet welds. Strains sustained at points around these defects were found to be low because of greater cross-sectional areas of welded joints compared with the cross-sectional area of beam flanges. Both the test results and the fracture mechanics-based assessment demonstrated that it is unlikely to initiate brittle fracture at these defects. Namely, the PJP welds were strong enough, when effective throat thickness was large enough. One of the important subjects requiring further basic study concerns the non-destructive inspection of incomplete penetration at the root of welding.

A modified fracture mechanics approach was examined on two full-scale connections using strains obtained by FE analyses. The method was found to be applicable; the predictions happened to exactly coincide with the test results. However, some difficulties found in the proposed approach lies in how to evaluate the effect of ductile crack growth. To appraise this approach, the ductile tearing analysis had to be included into FAD approach. Connections and loading conditions had to be represented by reproducible numerical models. Further experimental verifications to evaluate the fracture toughness of various joints with weld defects are required to make the proposed assessment method more reliable.

ACKNOWLEDGMENT

The proposals made here are based on experimental investigations conducted as joint research projects between Kumamoto and Sojo Universities. The authors wish to thank H. Shinde, T. Nakayama and I. Tanaka, M.Sc. students for hard work in laboratories. This work was partly supported by the Japanese Society for the Promotion of Science Grant-in-Aid for Scientific Research under the number 13650645.

REFERENCES

AIJ Kinki. 1997. Full-scale test on plastic rotation capacity of steel wide-flange beams connected with square tube steel columns. Committee on Steel Building Structures, Kinki Branch of Architectural Institute of Japan, Osaka, Japan. (in Japanese)

AIJ. 1996. Technical recommendations for steel construction for buildings. Part 1 guide to steel-rib fabrications. Architectural Institute of Japan, Tokyo, Japan. (in Japanese)

Azuma, K., Kurobane, Y. and Makino, Y. 2000. Cyclic testing of beam-to-column connections with weld defects and assessment of safety of numerically modeled connections from brittle fracture. Engineering Structures, Vol. 22, No. 12, Elsevier Science Ltd., pp. 1596-1608.

Azuma, K., Kurobane, Y. and Makino, Y. 2003. Full-scale testing of beam-to-column connections with partial joint penetration groove welded joints. Proceedings of the 10th International Symposium on Tubular Structures, pp. 419–427.

BSI. 1999. Guidance on methods for assessing the acceptability of flaws in metallic structures. BS 7910.

ABAQUS 2005. ABAQUS v6.5 Manuals (User's Manuals I, IIand III), Hibbitt, Karlsson and Sorensen, Inc.

Iwashita, T., Kurobane, Y., Azuma, K. and Makino, Y. 2003. Prediction of brittle fracture initiating at ends of CJP groove welded joints with defects: study into applicability of failure assessment diagram approach. Engineering Structures, Vol. 25, Issue 14, Elsevier Science Ltd., pp. 1815–1826.

Anderson, T.L. and Dodds, R.H., Jr. 1991. Specimen size requirements for fracture toughness testing in the ductile-brittle transition region. Journal of Testing and Evaluation, Vol. 19(2), pp. 123–134.

Tubular Structures XI – Packer & Willibald (eds)
© 2006 Taylor & Francis Group, London, ISBN 0-415-40280-8

Crack surface under weld toe in square hollow section T-joints

S.P. Chiew, S.T. Lie, C.K. Lee & H.L. Ji
School of Civil and Environmental Engineering, Nanyang Technological University, Singapore

ABSTRACT: Large-scale fatigue tests have shown that the crack propagation through the chord wall in square hollow section T-joints is very different from those observed in circular hollow section T-joints, especially the fatigue crack surface under the weld toe. This paper details an experimental investigation on the crack surface under the weld toe in four (4) large-scale square hollow section T-joint specimens. The test results revealed that the crack has a twisted crack surface. The crack not only curved under the weld toe but also changed significantly with the crack location around the weld toe. Further numerical investigation is conducted to study the significance of the crack surface on the distributions of the stress intensity factors around the crack front. A general formula is proposed to define the crack surface for fracture mechanics analysis to predict the residual life of cracked square hollow section T-joints.

1 INTRODUCTION

Many researchers have been involved in the modelling of the circular hollow section (CHS) joints with cracks (Rhee et al, 1991, Cao et al, 1997, Lee et al, 1999, and Lie et al, 2000) but most of them simplified the crack surface with a straight line perpendicular to the chord surface (Rhee et al, 1991 and Cao et al, 1997) or a straight line going through the intersection point of the chord and brace's midlines (Lee et al, 999 and Lie et al, 2000). This is because of the extreme difficulty in modelling the actual situation involving a crack curving around the hollow section joint intersection as well as under the weld toe. Huang (2002) and Shao (2004) studied the crack surface in CHS joints by experimental investigation and found that the crack surface did not curve mostly before the crack depth is less than 0.8T (T is the chord thickness). Bowness and Lee (1998) also investigated the fatigue crack surface under the weld toe in the CHS joint and found that the stress intensity factors derived from a straight perpendicular crack is not significantly different from that for a crack with the correct crack surface. It means that the crack surface under the weld toe can be simplified to a straight perpendicular line in CHS joints. However, these experiences cannot be used in the square hollow section (SHS) joints directly due to the geometric difference. Hence, experimental and finite element analysis is carried out to investigate the crack surface in SHS joints.

2 EXPERIMENTAL INVESTIGATION

2.1 Test rig and specimen details

The test rig designed for uniplanar hollow section joints which can apply axial load (AX), in-plane bending (IPB) and out-of-plane bending (OPB), has been introduced previously by Chiew et al (2005).

The brace and chord of the T-joint specimens are fabricated from hot-finished S355J2H square hollow sections to EN10210. The dimensions and geometrical ratio of the specimens are given in Figure 1. The mechanical properties of the specimens are listed in Table 1. The weld profile and the specimen preparation are carried out in accordance with AWS D1.1-2000 specifications. Ultrasonic technique is used to confirm the weld quality and compliance with standard specifications.

2.2 Test load and test procedure

The four specimens are applied to cyclic loads with different component of basic loads, AX, IPB and OPB. The actual nominal tensile stresses caused by peak cyclic loading applied to the specimens are calculated by the four midway brace member strain gauges readings and tabulated in Table 2.

During the tests, alternative current potential drop (ACPD) technique is used to record the crack propagation in chord thickness. The four specimens are tested till the crack depth measured by ACPD exceeds the

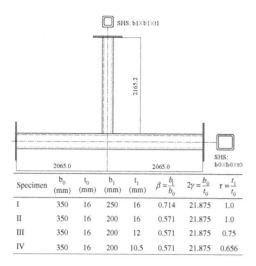

SHS: b1Xb1Xt1

2165.2

SHS:
b0Xb0Xt0

2065.0   2065.0

Specimen	b_0 (mm)	t_0 (mm)	b_1 (mm)	t_1 (mm)	$\beta = \frac{b_1}{b_0}$	$2\gamma = \frac{b_0}{t_0}$	$\tau = \frac{t_1}{t_0}$
I	350	16	250	16	0.714	21.875	1.0
II	350	16	200	16	0.571	21.875	1.0
III	350	16	200	12	0.571	21.875	0.75
IV	350	16	200	10.5	0.571	21.875	0.656

Figure 1. The dimensions and geometrical ratios of specimens.

Table 1. Mechanical properties of the four specimens.

Specimen	Yield strength (MPa)	Ultimate tensile strength (MPa)	Elongation (%)
I	366	511	32
II	370	518	32
III	370	518	32
IV	370	518	32

Table 2. Maximum and minimum nominal stresses (in MPa) caused by cyclic loads.

Load type		AX	IPB	OPB
Specimen I	max	8.9	7.83	0.0
	min	0.0	0.0	−6.63
Specimen II	max	6.35	0.0	0.0
	min	0.0	−10.8	−1.40
Specimen III	max	0.0	0.0	0.0
	min	0.0	−19.0	−16.0
Specimen IV	max	9.1	5.86	0.0
	min	0.0	0.0	−6.56

Note: Positive loads are tension loads, and negative loads are compression loads applied by actuators. AX – axial, IPB – in plane bending, OPB – out of plane bending.

chord thickness. Then the cracked joints are cut out from the specimens and split into two parts, namely, the chord face part and the brace wall part.

2.3 Test results and measurement

Plane and sections views of the crack surfaces exposed are shown in Figure 2 and 3 respectively. Figure 2

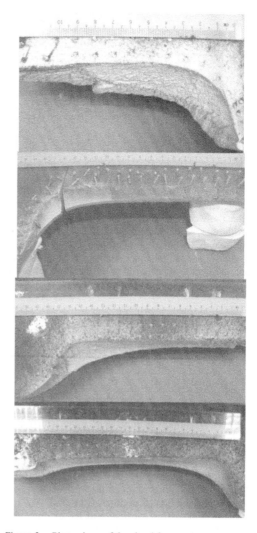

Figure 2. Plane views of the chord face parts.

shows that the geometry of the crack surface changes along brace-chord intersection and is highly unsymmetrical with respect to the deepest point. Figure 3 shows that for a given section, the crack surface curves as the crack penetration into the chord wall. In order to gather further details for the development of an appropriate numerical model, a simple clay molding procedure is employed to measure the section profile (shape of curve **O'd'** in Figure 4). In order to define the shape of curve **O'd'**, the angel θ_s is introduced as shown in Figure 4.

This molding procedure consists of the following steps:

(1) After the crack surface is cleaned properly, soft fine clay is applied to the brace wall part (Figure 5 (a)).

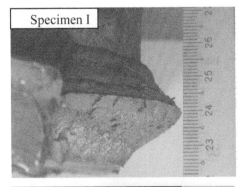

Specimen I

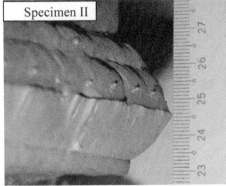

Specimen II

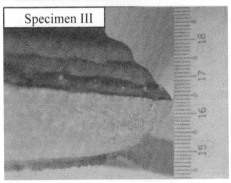

Specimen III

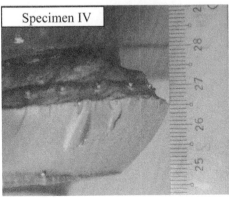

Specimen IV

Figure 3.   Section views of the brace wall parts.

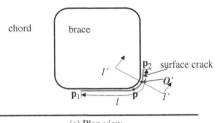

chord   brace

$1'$ p_2 surface crack

Q'

p_1 p $1'$

l

(a) Plan view

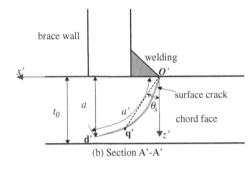

brace wall

welding

x' O'

surface crack

t_0 a a' θ_s

chord face

d' q' z'

(b) Section A'-A'

Figure 4.   Definitions of geometrical parameters of a surface crack.

(2) The clay mold is then removed when it is half-dry. As a result, the shape of the crack is imprinted onto the clay mold (Figures 5 (b) and (c)).

(3) The clay mold is allowed to dry completely. Slices of clay are trimmed off at 10 mm intervals in directions perpendicular to the weld toe (Figure 5 (d)).

(4) For each section, the profile (curve **O'd'** in Figure 4) is transferred to a tracing paper (Figure 5 (e)).

(5) The profiles are digitalized by scanning and processed by the AutoCAD software and the variation of θ_s with y' is measured (Figure 5 (f)).

According to the data measured from the experiments, the geometry of the surface crack will be devised from the following assumptions (as shown in Figure 6):

(1) The crack profile will be modeled as a curve for $0 \leq z' \leq z_c'$ and a straight line for $z_c' \leq z' \leq t_0$. The value z_c' is known as the *critical* depth for the crack and a value of $z_c' = 0.8t_0$ is adopted.

(2) For $0 \leq z' \leq z_c'$, $\theta_s(z')$ will increase linearly with z' from $\theta_s(0)$ to $\theta_s(z_c')$.

(3) For $z_c' \leq z' \leq t_0$, the crack profile will be a straight line with slope equal to the gradient of the curve at $z' = z_c'$.

By using the above assumptions and the data measured from experiment, equations of the crack surface profile with respect to the **x'-y'-z'** coordinate system

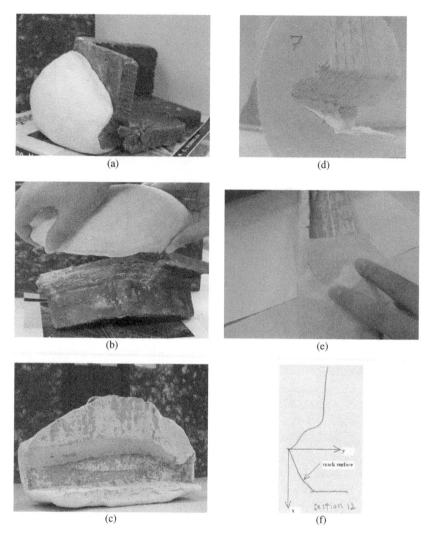

(a)

(d)

(b)

(e)

(c)

(f)

Figure 5.   Procedure to measure the geometry of a surface crack.

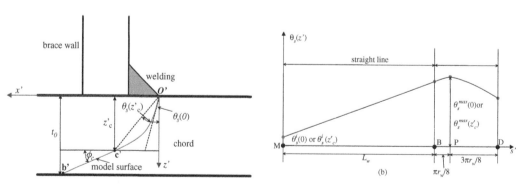

Figure 6.   Geometrical model for the crack surface.

Figure 7.   Assumed variation of $\theta_s(z')$ with s'.

(Figure 7) for each specimen can be obtained as follows,

For Specimen I,

For $0 \le z' \le z'_c$

$$
\begin{cases}
x' = z' \tan\left(\dfrac{12z' + 4t_0}{3t_0(16L_w + \pi r_w)}\pi s' + \dfrac{1}{12}\pi\right) \\
\qquad\qquad\qquad\qquad (0 \le s' \le L_w) \\
x' = z' \tan\left(-\dfrac{48z' + 16t_0}{3t_0 r_w(16L_w + \pi r_w)}(s' - L_w - \dfrac{\pi r_w}{8})^2 \right. \\
\left. + (\dfrac{z'}{4t_0} + \dfrac{1}{6})\pi\right) \quad (L_w \le s' \le L_w + \pi r_w/2)
\end{cases}
$$
(1a)

For $z'_c \le z' \le t_0$,

$$
x' = \left[\tan(\theta_s(z'_c, s')) + \dfrac{(\theta_s(z'_c, s') - \theta_s(0, s'))}{\cos^2(\theta_s(z'_c, s'))}\right](z' - 0.8t_0) + 0.8t_0 \tan(\theta_s(z'_c, s'))
$$
(1b)

where,

$$
\begin{cases}
\theta_s(0, s') = \dfrac{4}{3(16L_w + \pi r_w)}\pi s' + \dfrac{1}{12}\pi \\
\theta_s(z'_c, s') = \dfrac{68}{15(16L_w + \pi r_w)}\pi s' + \dfrac{1}{12}\pi
\end{cases}
$$

$$(0 \le s' \le L_w)$$

$$
\begin{cases}
\theta_s(0, s') = -\dfrac{16}{3r_w(16L_w + \pi r_w)}(s' - L_w \\
\qquad - \dfrac{\pi r_w}{8})^2 + \dfrac{1}{6}\pi \\
\theta_s(z'_c, s') = -\dfrac{208}{15r_w(16L_w + \pi r_w)}(s' - \\
\qquad L_w - \dfrac{\pi r_w}{8})^2 + \dfrac{11}{30}\pi
\end{cases}
$$

$$(L_w \le s' \le L_w + \pi r_w/2)$$

For Specimen II,

For $0 \le z' \le z'_c$

$$
\begin{cases}
x' = z' \tan\left(\dfrac{20z' + 8t_0}{9t_0(16L_w + \pi r_w)}\pi s' + \dfrac{1}{12}\pi\right) \\
\qquad\qquad\qquad\qquad (0 \le s' \le L_w) \\
x' = z' \tan\left(-\dfrac{80z' + 32t_0}{9t_0 r_w(16L_w + \pi r_w)}(s' - L_w - \dfrac{\pi r_w}{8})^2 \right. \\
\left. + (\dfrac{z'}{t_0} + 1)\dfrac{5\pi}{36}\right) \quad (L_w \le s' \le L_w + \pi r_w/2)
\end{cases}
$$
(2a)

For $z'_c \le z' \le t_0$,

$$
x' = \left[\tan(\theta_s(z'_c, s')) + \dfrac{(\theta_s(z'_c, s') - \theta_s(0, s'))}{\cos^2(\theta_s(z'_c, s'))}\right](z' - 0.8t_0) + 0.8t_0 \tan(\theta_s(z'_c, s'))
$$
(2b)

where,

$$
\begin{cases}
\theta_s(0, s') = \dfrac{8}{9(16L_w + \pi r_w)}\pi s' + \dfrac{1}{12}\pi \\
\theta_s(z'_c, s') = \dfrac{8}{3(16L_w + \pi r_w)}\pi s' + \dfrac{1}{12}\pi
\end{cases}
$$

$$(0 \le s' \le L_w)$$

$$
\begin{cases}
\theta_s(0, s') = -\dfrac{32}{9r_w(16L_w + \pi r_w)}(s' - L_w \\
\qquad - \dfrac{\pi r_w}{8})^2 + \dfrac{5}{36}\pi \\
\theta_s(z'_c, s') = -\dfrac{32}{3r_w(16L_w + \pi r_w)}(s' - L_w \\
\qquad - \dfrac{\pi r_w}{8})^2 + \dfrac{1}{4}\pi
\end{cases}
$$

$$(L_w \le s' \le L_w + \pi r_w/2)$$

For Specimen III,

For $0 \le z' \le z'_c$

$$
\begin{cases}
x' = z' \tan\left(\dfrac{430z' + 104t_0}{90t_0(16L_w + \pi r_w)}\pi s' + \dfrac{1}{12}\pi\right) \\
\qquad\qquad\qquad\qquad (0 \le s' \le L_w) \\
x' = z' \tan\left(-\dfrac{860z' + 208t_0}{45t_0 r_w(16L_w + \pi r_w)}(s' - L_w - \dfrac{\pi r_w}{8})^2 \right. \\
\left. + (\dfrac{43z'}{144t_0} + \dfrac{5}{36})\pi\right) \quad (L_w \le s' \le L_w + \pi r_w/2)
\end{cases}
$$
(3a)

For $z'_c \le z' \le t_0$,

$$
x' = \left[\tan(\theta_s(z'_c, s')) + \dfrac{(\theta_s(z'_c, s') - \theta_s(0, s'))}{\cos^2(\theta_s(z'_c, s'))}\right](z' - 0.8t_0) + 0.8t_0 \tan(\theta_s(z'_c, s'))
$$
(3b)

where,

$$
\begin{cases}
\theta_s(0, s') = \dfrac{52}{45(16L_w + \pi r_w)}\pi s' + \dfrac{1}{12}\pi \\
\theta_s(z'_c, s') = \dfrac{224}{45(16L_w + \pi r_w)}\pi s' + \dfrac{1}{12}\pi
\end{cases}
$$

$$(0 \le s' \le L_w)$$

625

$$\begin{cases} \theta_s(0,s') = -\dfrac{208}{45r_w(16L_w + \pi r_w)}(s' - L_w \\ \quad -\dfrac{\pi r_w}{8})^2 + \dfrac{5}{36}\pi \\ \theta_s(z_c',s') = -\dfrac{896}{45r_w(16L_w + \pi r_w)}(s' - L_w \\ \quad -\dfrac{\pi r_w}{8})^2 + \dfrac{17}{45}\pi \end{cases}$$

$$(L_w \le s' \le L_w + \pi r_w/2)$$

For Specimen IV,

For $0 \le z' \le z_c'$

$$\begin{cases} x' = z'\tan(\dfrac{175z' + 48t_0}{45t_0(16L_w + \pi r_w)}\pi s' + (\dfrac{5z'}{72t_0} \\ \quad + \dfrac{2}{45})\pi) \qquad (0 \le s' \le L_w) \\ x' = z'\tan(-\dfrac{700z' + 192t_0}{45t_0 r_w(16L_w + \pi r_w)}(s' - L_w - \dfrac{\pi r_w}{8})^2 \\ \quad + (\dfrac{5z'}{16t_0} + \dfrac{1}{9})\pi) \qquad (L_w \le s' \le L_w + \pi r_w/2) \end{cases} \quad (4a)$$

For $z_c' \le z' \le t_0$,

$$x' = \left[\tan(\theta_s(z_c',s')) + \dfrac{(\theta_s(z_c',s') - \theta_s(0,s'))}{\cos^2(\theta_s(z_c',s'))} \right](z' \\ - 0.8t_0) + 0.8t_0\tan(\theta_s(z_c',s')) \quad (4b)$$

where,

$$\begin{cases} \theta_s(0,s') = \dfrac{16}{15(16L_w + \pi r_w)}\pi s' + \dfrac{2}{45}\pi \\ \theta_s(z_c',s') = \dfrac{188}{45(16L_w + \pi r_w)}\pi s' + \dfrac{1}{10}\pi \end{cases}$$

$$(0 \le s' \le L_w)$$

$$\begin{cases} \theta_s(0,s') = -\dfrac{192}{45r_w(16L_w + \pi r_w)}(s' - L_w - \\ \quad \dfrac{\pi r_w}{8})^2 + \dfrac{1}{9}\pi \\ \theta_s(z_c',s') = -\dfrac{752}{45r_w(16L_w + \pi r_w)}(s' - L_w - \\ \quad \dfrac{\pi r_w}{8})^2 + \dfrac{13}{36}\pi \end{cases}$$

$$(L_w \le s' \le L_w + \pi r_w/2)$$

3 NUMERICAL INVESTIGATION

One numerical model is developed by the authors. The details of the model and mesh generation have been reported by Chiew (2005). The model is used to investigate the influence of the crack surface on the SIF's calculation.

The dimensions of the models are the same as the specimens. The load is the same as the actual fatigue load applied to the specimens (please refer to Table 1). The crack location and crack shape are all from the measurement of the four tests. Straight line crack surfaces, inclined at five different angles, i.e. $\theta_s(z') = 0$, $\pi/12$, $\pi/6$, $\pi/4$ and $\pi/3$ are studied.

The SIF values at the deepest point for different crack surface are presented in Figures 8 to 11. Furthermore, the SIF at the deepest point from the test results are also shown in these figures as benchmark. From the figures, the SIF values at the deepest point diverge largely with the experimental results and the SIF values from these different crack surfaces are quite different. That means the crack surface has significant effect on the SIF values. In the figures, when the crack depth is quite small or quite big, especially when the crack depth ratio $a'/t_0 \ge 0.6$, the influence is notable. The models with surface inclined at small values of $\theta_s(z')$ $(0, \pi/12$ and $\pi/6)$ result in overestimation SIF values when a'/t_0 is small but severe underestimation SIF values when $a'/t_0 \ge 0.6$. However, the SIF values in the models with crack surface inclined at $\theta_s(z') = \pi/4$ and $\theta_s(z') = \pi/3$ are severe higher than the experimental results when $a'/t_0 \ge 0.6$. These comparisons show that the straight lines cannot represent the crack surface in SHS T-joints. This also indicates that the crack surface is one of the factors that determine the crack growth in SHS joints and the models with constant $\theta_s(z')$ cannot lead to satisfactory prediction of SIFs. Hence, accurate FE modeling of the crack surface is necessary in the calculation of SIFs.

A general crack surface equation which can be used for all the load cases and joint dimensions is proposed based on the test results. This equation may not be expected to give very accurate SIF values for all different cases since it is nearly impossible to be achieved, but at least it must be able to generate an acceptable SIF value for each case within certain error limits. The general equation can be expressed as follows,

For $0 \le z' \le z_c'$

$$\begin{cases} x' = z'\tan(\dfrac{10z' + 4t_0}{3t_0(16L_w + \pi r_w)}\pi s' + \dfrac{1}{12}\pi) \\ \qquad (0 \le s' \le L_w) \\ x' = z'\tan(-\dfrac{40z' + 16t_0}{3t_0 r_w(16L_w + \pi r_w)}(s' - L_w - \dfrac{\pi r_w}{8})^2 \\ \quad + (\dfrac{5z'}{36t_0} + \dfrac{5}{6})\pi) \qquad (L_w \le s' \le L_w + \dfrac{\pi r_w}{2}) \end{cases} \quad (5a)$$

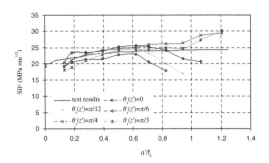

Figure 8. SIF of the deepest point from the model with different straight line crack surface in Specimen I.

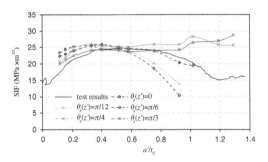

Figure 9. SIF of the deepest point from the model with different straight line crack surface in Specimen II.

For $z'_c \le z' \le t_0$,

$$x' = \left[\tan(\theta_s(z'_c, s')) + \frac{(\theta_s(z'_c, s') - \theta_s(0, s'))}{\cos^2(\theta_s(z'_c, s'))} \right](z' \quad (5b)$$
$$- 0.8t_0) + 0.8t_0 \tan(\theta_s(z'_c, s'))$$

where,

$$\begin{cases} \theta_s(0, s') = \frac{4}{3(16L_w + \pi r_w)} \pi s' + \frac{1}{12}\pi \\ \theta_s(z'_c, s') = \frac{4}{(16L_w + \pi r_w)} \pi s' + \frac{1}{12}\pi \end{cases}$$

$$(0 \le s' \le L_w)$$

$$\begin{cases} \theta_s(0, s') = -\frac{16}{3r_w(16L_w + \pi r_w)}(s' - L_w) \\ \qquad - \left(\frac{\pi r_w}{8}\right)^2 + \frac{5}{6}\pi \\ \theta_s(z'_c, s') = -\frac{16}{r_w(16L_w + \pi r_w)}(s' - L_w) \\ \qquad - \left(\frac{\pi r_w}{8}\right)^2 + \frac{17}{18}\pi \end{cases}$$

$$(L_w \le s' \le L_w + \frac{\pi r_w}{2})$$

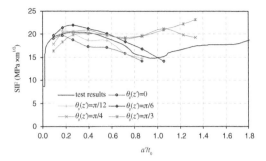

Figure 10. SIF of the deepest point from the model with different straight line crack surface in Specimen III.

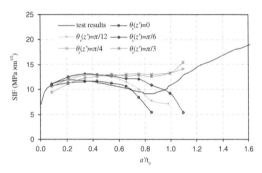

Figure 11. SIF of the deepest point from the model with different straight line crack surface in Specimen IV.

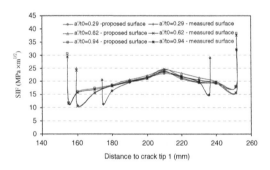

Figure 12. SIF comparisons along crack front of the model with proposed surface and measured surface in Specimen I.

In order to validate the proposed crack surface equation, the SIF distributions along the crack front from the models with the proposed crack surface and measured crack surface are compared in Figures 12 to 15. The figures show that the SIF distribution from the models with proposed crack surface and measured one match well and the divergence is less than 5%. This proves that the proposed crack surface equation can be used as a general representation for crack surface in SHS T-joints.

627

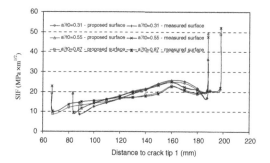

Figure 13. SIF comparisons along crack front of the model with proposed surface and measured surface in Specimen II.

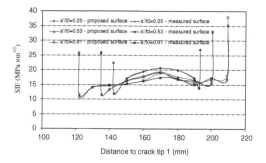

Figure 14. SIF comparisons along crack front of the model with proposed surface and measured surface in Specimen III.

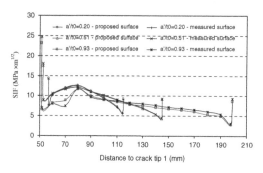

Figure 15. SIF comparisons along crack front of the model with proposed surface and measured surface in Specimen IV.

4 CONCLUSION

Both experimental and numerical investigations on crack surface in square hollow section T-joints are carried out. The experimental results show that the crack surface is twisted since it not only curved under weld toe but also changed with the crack location at the intersection. Detailed measurements are carried out, and crack surface equations are given for the four specimens. From the numerical studies, it is known that the crack surface has significant effect on the SIF values at the deepest point. The crack surface cannot be represented by a straight line or a single-variable equation. One general crack profile equation is suggested based on the full-scale test results and comparison with numerical results. The SIF along the crack front are compared between the models with actual measured crack surface and proposed crack surface. The good agreements show that the proposed crack surface equation is good enough for SHS fatigue analysis.

REFERENCES

Bowness, D. & Lee, M.M.K. 1998. Fatigue Crack Curvature under the Weld Toe in a Offshore Tubular Joint. *International Journal of Fatigue* vol. 20, No.6: 481–490.

Cao, J.J. et al. 1997. FE mesh Generation for Circular Tubular Joints with or without Cracks. *Proceedings of the Seventh International Offshore and Polar Engineering Conference* Vol. IV: 98–105.

Chiew, S.P. et al. 2005. Numerical Modeling of Square Tubular T-Joints with Surface Crack at Corner. *Proc. of the Fourth International Conference on Advances in Steel Structures* Vol 1: 1159–1164.

Huang, Z.W. 2002. Stress Intensity Factor of Steel Tubular T and Y-joints under Combined Loads. *PhD Thesis,* Nanyang Technological University, Singapore.

Lee, C.K. et al. 1999. Model and Mesh Generation of Cracked Tubular Joints. *Proceedings of the European Conference on Computational Mechanics:* 182–183.

Lie, S.T. et al. 2000. Modelling Arbitrary Through-thickness Crack in a Tubular T-joint. *Proceedings of the 10th International Offshore and Polar Engineering Conference (ISOPE 2000)* Vol. IV: 53–58.

Rhee, C.H. et al. 1991. Reliability of Solution Method and Empirical Formulas of Stress Intensity Factors for Weld Toe Cracks of Tubular Joints. *Proc. 10th Offshore Mechanics and Arctic Engineering Conference* ASME, Vol. 3-B: 441–452.

Shao, Y.B. 2004. Fatigue Behaviour of Uniplanar CHS Gap K-joints under Axial and In-plane Bending Loads. *PhD Thesis,* Nanyang Technological University, Singapore.

Tubular Structures XI – Packer & Willibald (eds)
© 2006 Taylor & Francis Group, London, ISBN 0-415-40280-8

J-integral assessment of low temperature fracture of a welded K-joint

T. Björk, S. Heinilä and G. Marquis

Lappeenranta University of Technology, Lappeenranta, Finland

ABSTRACT: The deformation capacity and ultimate load capacity of K-joints fabricated from cold-formed rectangular hollow sections at −40°C has been studied both experimentally and numerically. The predicted load-deformation behaviour of a K-joint based on non-linear finite element analysis compared well with the experimentally measured behaviour. Such analysis, however, was not able to provide information on the ultimate strength or ultimate deformation capacity of a joint. An extensive experimental program on welded K-joints at low temperatures revealed that the primary failure mode was ductile tearing of the chord flange at the weld toe of the tension brace flange. For this reason it was hypothesized that *J*-integral evaluation of an advancing crack could be combined with measured materials data to assess the ultimate strength of a joint. *J*-Δ*a* material testing was performed at −40°C on material taken from the wall of a welded structural hollow section. This data has been combined with detailed FE-based *J*-integral analysis of a crack growing from the weld toe. This assessment revealed that crack advance, once initiated, would be expected to continue through the wall thickness of the chord member with very little change in load. This ductile fracture assessment method was found to be effective in modelling the ultimate ductility of the joint.

1 INTRODUCTION

Rectangular hollow sections (RHS) are widely used in load-carrying structures due to their good load transfer behaviour and aesthetic form. Experimental tests and numerical analyses on the capacity and fracture of structural hollow section joints are regularly reported, however, most of these investigations involved joints fabricated using hot-formed structural hollow section members or testing has been at room or elevated temperatures [Wardenier 1982 and 2002, Packer et al. 1992, Packer & Henderson 1997, Puthli 1998, Vainio 2000, Zhao et al. 2005]. Only limited test data is available for cold-formed structural hollow section joints at sub-zero temperatures [Niemi 1990, Niemi 1996]. One reason for this is that sub-zero temperature tests are complex and relatively expensive, especially if a significant number of joint geometries and section profiles are of interest.

Finite element (FE) modelling can be an effective tool for studying the nonlinear load deformation behaviour of RHS joints. For cold formed structural hollow sections (CFSHS), the yield strength of the corner is typically higher than that of the adjacent flat faces because of the higher cold forming. True stress-strain curves determined by material tests from both the flat face and the corner region can be used to develop material models for the FE analyses. It was found that the load-deformation behaviour predicted

by FE analysis was in good agreement with that measured in the laboratory experiments, However, the ultimate strength or ultimate deformation capacity of a joint based on FE analysis is strongly dependent on the choice of the failure mode. Previous work showed that the simple von Mises theory over predicted the deformation at failure [Björk 2005, Björk et al. 2003a, Björk et al. 2005].

In practice, sufficient load carrying capacity means that the joint can withstand the required design load with a sufficiently low failure probability. The design load can be assessed using design guidance documents [EN 2003], where the capacity is based on the nominal yield strength of the parent metal. Even though load carrying capacity is the most significant practical design attribute, deformation capacity is an extremely relevant parameter when considering structural safety. In order to fulfil the requirements for the deformation capacity, the joint must have sufficient plastic deformation before final failure. The concept of "sufficient deformation capacity" of the RHS-joint is not standardized, but a widely adopted practice is to ensure that the plastic deformation/chord width-ratio $\delta_p/b_0 \geq 0.005$ for joints where the widths of the brace and chord members are equal, $\beta = 1.0$, and $\delta_p/b_0 \geq 0.01$ for joints with $\beta < 1.0$ [Wardenier 1982].

The current study focuses on the use of nonlinear FE analysis of a K-joint fabricated from cold formed rectangular sections. Elastic-plastic analysis was used to model the load-deformation behaviour of the joint. Additionally, J-integral values at several crack depths were computed. J-Δa crack resistance curves measured for the material near the weld in the through thickness direction at $-40°$C were experimentally measured using standard procedures. J-integral values were used in conjunction with the J-Δa curve in order to asses both the ultimate load carrying capacity of the joint as well as the ultimate deformation capacity.

2 EXPERIMENTS

2.1 *Full-scale tests*

Quasi-static tests of full-scale symmetric K-joints at ambient and sub-zero temperatures have been performed. The gap type K-joint was used because it is very common joint type in truss and other types of structures. The joint geometries were chosen so that the minimum deformation capacity will be found. Thus, in order to restricting the deformation capacity of the joint the gap between the brace members was chosen to be as small as the design recommendations allow. The material of test tubes had a nominal yield strength $f_y = 355$ MPa. A schematic of the test specimen is shown in Figure 1 and a typical test specimen is shown in Figure 2.

The joint chosen for detailed analysis had chord dimensions $100 \times 100 \times 6$ ($b_0 \times h_0 \times t_0$ [mm]), brace dimension $50 \times 50 \times 5$, a gap of 8 mm and an angle between chord and braces of $50°$. The test temperature was $-40°$C and the pretension load in the chord member was 460 kN.

The specimen chord was fixed in a frame and preloaded in tension by two jacks. Pinned boundary conditions were used for the brace ends. A schematic of the test frame is given in Figure 3. More full details of the test program are reported elsewhere [Björk 2005, Björk et al. 2003a].

The experimental ultimate load was 365 kN which was in good agreement to the design code value [EN 2003]. Measured joint deformation at fracture, δ_p/b_0, was 1.5%.

2.2 *Fracture surface evidence*

The experimental program revealed that the primary failure mode for many joints tested at sub-zero temperatures was ductile tearing of the chord flange followed by brittle fracture. Failure started from the weld of the tension brace at the chord flange side of the weld toe. The initiation point was located where, according to FE analysis, the stresses of the gap were greatest. The failure surface in the chord flange was generally perpendicular to the tension brace, i.e., at an angle of approximately $50°$ with respect to the longitudinal chord axis.

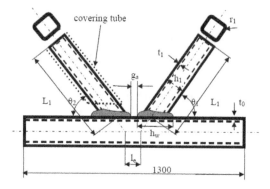

Figure 1.   Schematic of K-joint test specimen.

Figure 2.   A typical K-joint test specimen.

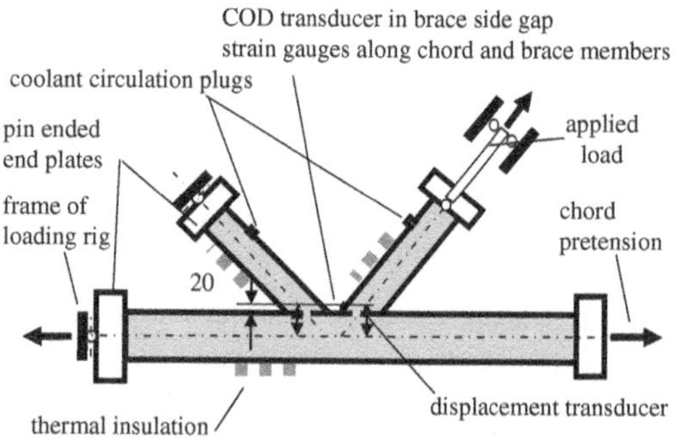

Figure 3.   Schematic illustration of K-joint test arrangement.

The fracture initiated by ductile tearing and then, as the crack size increased, the fracture type changed to brittle. This is seen in Figure 4. In the ductile zone the surface was rougher and appears as the darker region in the photograph. The brittle surface has a lighter shade and a typical chevron pattern. SEM scanning was carried out in order to confirm the failure mode. The failure types find by SEM agrees well with the macro-level observation about the change of failure mechanism.

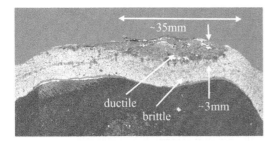

Figure 4. Failure surface of a K-joint at −40°C.

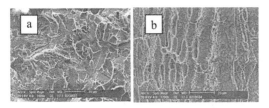

Figure 5. Ductile (a) and brittle (b) failures defined by SEM.

The failure started as ductile tearing until the crack depth was approximately 3 mm through the wall thickness of the chord. The surface crack width at this point was about 35 mm. After this, the fracture type changed to brittle. Once initiated, brittle fracture continued until complete separation of the chord member.

Similar types of failures were observed in other K-joints. The crack length at which the fracture behaviour changed from ductile to brittle varied from one joint to another but the mechanism remained constant. In some cases the critical crack depth was only a fraction of the chord wall thickness while in other cases a full through thickness ductile crack was observed. In those cases where the critical crack was through thickness, the surface crack length was approximately equal to the width of the brace member. Comparable failure cases have also been observed in the X-joints [Björk 2005].

2.3 J-Δa tests

Crack resistance test specimens $(5 \times 10 \times 40\,\text{mm})$ seen in Figure 6 were cut from the same RHS from which the chord of the K-joint was taken. In order to simulate the weld, a cladding operation was performed on the specimen after to removing the test specimens. Cladding was performed for some specimens using a one-pass MAG process. The same weld parameters were used as those used during K-joint fabrication. The cladding was applied to the cut surface of the specimen. After cladding, a notch was machined for its final dimensions and fatigue pre-cracking was used to achieve a sharp crack.

The tip of the fatigue pre-crack was about 1 mm from the fusion boundary as seen in Figure 7. The

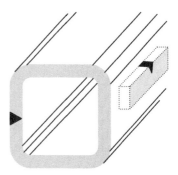

Figure 6. Specimen for J-integral test.

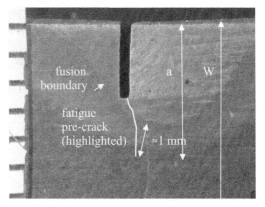

Figure 7. An example of a re-cracked J-Δa test specimen.

fusion boundary was located near the specimen surface at a depth similar to that observed in the welded K-joint. Specimens were pre-cracked at ambient temperature and tested at −40°C according to ASTM E 1921-03 [ASTM 2002].

3 THE J-INTEGRAL

Because the experimental program revealed that ductile tearing followed by brittle failure was observed for many welded K-joints at sub-zero temperatures, it was decided to include ductile crack growth in the analysis model using a J-integral fracture mechanics approach. The J-integral as developed by Rice for elastic materials generally describes the flow of energy into the crack tip region [Rice & Tracey 1969]. The theoretical load carrying capacity is based on J-integral value defined as

$$J = \int_0^\varepsilon W\,dy + \int_s \left[t_x \frac{du_x}{dx} + t_y \frac{du_y}{dx} \right] ds \qquad (1)$$

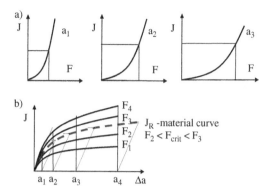

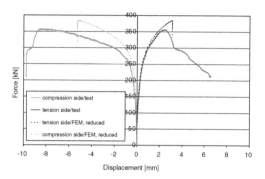

Figure 9. Measured and numerically simulated load displacement behaviour of a K-joint.

Figure 8. a) Numerically defined F and J-integral dependence for a structure and b) critical load F_{crit} assessment procedure.

where u_i are the nodal displacements, s is the path length and W is the strain energy density which can be divided into elastic and plastic component

$$W = W_e + W_{pl} = \int_0^{\varepsilon_{ij}} \sigma_{ij} d\varepsilon_{ij} + W_{pl} \qquad (2)$$

In the case of FEA, the plastic strain energy density W_{pl} can be obtained directly within the FE software. The tractions t_x and t_y are

$$t_x = \sigma_x \overline{n}_x + \sigma_{xy} \overline{n}_y \qquad (3)$$

$$t_y = \sigma_{xy} \overline{n}_x + \sigma_y \overline{n}_y \qquad (4)$$

where $\overline{n}_x$ and $\overline{n}_y$ are the normal vector x- and y-components.

The procedure for defining the critical load, F_{crit}, and subsequently the critical deformation capacity is based on the EPRI approach [Kumar et al. 1984]. In this method the calculated elastic-plastic J-integral is compared to J_R – values obtained from material test results. The procedure is illustrated schematically in Figure 8.

The applied force F and J-integral dependence of a joint is defined using FE analysis for several crack sizes, a_1, a_2, a_3, etc., see Figure 8a. From these curves the J-Δa relationships for alternate force levels, F_1, F_2, F_3, F_4, etc., can be defined as shown in Figure 8b. By plotting the J_R curve from material tests on the same diagram, the critical load F_{crit} can be estimated.

It should be noted that the shape of the constant force J-Δa curves for a structure is highly dependent on the structure itself. A structure for which the crack growth is predominantly load-controlled will have a generally positive curvature. While the curve for a structure in which the crack growth is displacement controlled will have a negative curvature or even a negative slope.

4 FINITE ELEMENT CALCULATIONS

4.1 Load-deformation evaluation

The program MSC MARC was chosen for the FE load-deformation analysis. Eight node linear solid elements and 4-node linear tetra elements were used. A solid model was used in order to take into account the real joint geometry with the welds. A Ramberg-Osgood type true $\sigma - \varepsilon$ material model was for the analysis. Details of this analysis have been reported previously [Björk 2005, Björk et al. 2003b]. A comparison of the experimentally and the numerically predicted load-deformation behaviour is shown in Figure 9.

The FE analysis provided good estimates of the elastic–plastic joint behaviour of the analysed joints. However, FE analysis tended to overestimate the ultimate deformation and load carrying capacity. This is mainly due to the von Mises failure hypotheses used in the analysis. This failure hypothesis does not take into consideration the void nucleation or crack growth that lead to ductile failure in a joint. Experimentally, joints failed before the local strains reached the measured critical strain value obtained from the coupon tests.

4.2 J-integral evaluation

Quasi-static tests of full-scale symmetric K-joints at ambient and sub-zero temperatures were simulated using ANSYS. Measured geometrical data from the laboratory tested K-joint was used in the model. The finite element model of a K-joint was solved for several different crack depths and the J-integral was computed at numerous load steps for each crack depth. Identical FE models were used except that the crack depth varied from 0.2 mm to 3 mm. Material properties for the brace and the chords were based on the measured material data. Elastic linear hardening material models were assumed.

Due to the geometric symmetric and the loading condition, ½ of the joint was modelled. The boundary

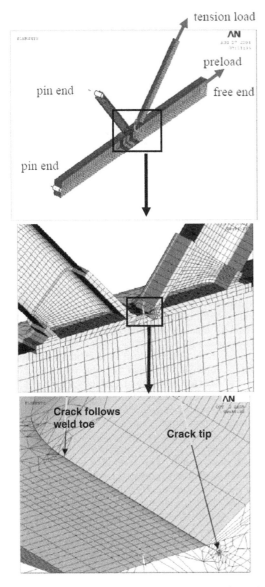

tension load

AN

preload

pin end

free end

pin end

Crack follows
weld toe

Crack tip

AN

Figure 10. The FE model used for assessing the J-integral in a K-joint.

Regions near the crack were meshed with 20-node solid elements. Regions distant from the crack, but that were expected to experience nonlinear deformation, were modelled with 8-node solid elements. Pyramid or tetrahedral versions of the 20-node solid elements were used in some difficult transition volumes near the crack. Connections between the parabolic and linear solids were done by reducing the mid-side nodes. Other regions of the structure, such as chords and the ends of the brace, were modelled with linear or parabolic shell elements depending on the connection to the solid elements.

The crack tip mesh consisted of two rows of crack tip elements. Only a limited number of crack tip elements around the crack tip were used because of the difficult joint geometry. The crack tip elements were also modelled as elastic crack tip elements having middle nodes ¼ distance from the crack tip and the nodes at the tip connected to each other. Each of the FE models had approximately 35 000 elements and 462 000 degrees of freedom.

The J-integral along the weld toe between the corner of the brace member and the centreline of the joint was assessed. For smaller crack depths and at lower loads, the J-integral reached a maximum value at a distance approximately 3.2 mm from the brace corner. This corresponds reasonably well with the location of the ductile crack from the fracture surface (see Fig 4.).

5 RESULTS AND DISCUSSION

The J-Δa crack resistance test results are summarized in Figure 11. These were generated using ASTM test procedures at −40°C. The symbols LTY2, LTY8 and LTY12 refer to test specimen numbers. Specimen LTY2 was without cladding while specimens LTY8 and LTY12 included a cladding operation and the ductile crack growth was initiated in the HAZ approximately 1 mm from the fusion line (refer to Figure 7). In all cases the crack propagation direction represented a through-wall ductile crack extension.

The J-integral calculation results for the K-joint of interest are shown in Figure 12 for crack depths of 0.1, 0.2, 0.5, 1.0 and 2.0 mm. The J-integral is calculated for a path situated 3.2 mm from the corner area of the brace where the highest elastic stresses were computed. This location corresponds well to the location where ductile fracture was observed to occur based on the fracture surface analyses.

From the curves in Figure 12, the constant force curves ($F = 200$, 250, 300 and 350 kN) were derived. These are plotted in Figure 13. Experimental J-Δa crack resistance curves of the chord material from Figure 8 are shown in this same figure. The upper curve is for a crack initiating in the HAZ of the welded material while the lower curve is for a crack initiating

conditions, applied loads and the element model used to define the J-integral in the critical region of the K-joint of interest are shown in Figure 10. Boundary conditions were assumed to represent the laboratory test rig boundary conditions. Modelling of the crack and crack tip required a very fine mesh so the FE models were large. For this reason only the regions near the crack were modelled using solid elements while shell elements were used elsewhere. The boundary conditions required some beam elements and spider web meshing on the end of one chord.

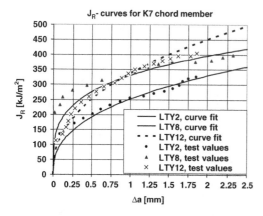

Figure 11. The measured crack resistance curves for specimens taken from RHS.

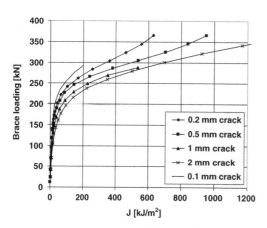

Figure 12. Computed J-integral values for the K-joint at a path 3.2 mm from the brace corner.

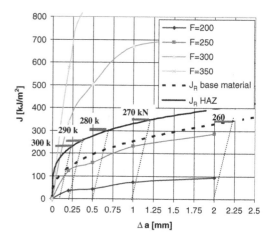

Figure 13. Assessment of the critical brace load to initiate ductile crack extension.

and propagating in the base material. The upper crack resistance curve was assumed to be valid for crack depths up to 1 mm. For cracks deeper than this the base metal curve was considered more representative.

From Figure 13 it can be immediately seen that the 0.1 mm deep continuous crack-like defect along the weld toe would begin to grow by ductile fracture when the force in the brace member attains a force level between 250–300 kN. Further numerical analysis revealed that this value was approximately 300 kN. Once ductile fracture is initiated, this assessment shows that the crack would continue to propagate even at slightly lower force levels. For example, a crack which has extended to a depth of 1 mm would continue to advance even if the load on the brace member is reduced to 270 kN.

When the crack has passed through the HAZ area, it was estimated that the force required for continued crack extension drops still further, i.e., about

260 kN for a 2 mm deep crack. The ductile crack extension for this K-joint is therefore combines both load- and displacement controlled mechanisms. Consequently, crack propagation does not become completely unstable, but the crack is predicted to advance at a relatively constant load. The estimated critical force is somewhat less than the value obtained from experimental test of this joint geometry, $F_u = 365$ kN. The numerical procedure, however, has included some assumptions that may need some refinement. For example, initial crack shapes other then the assumed continuous toe crack should be studied. The method does, however, provide a promising starting point for further studies. The method was a significant improvement when compared to estimating the ultimate deformation capacity using a more conventional von Mises type of failure criterion that tended to give overly optimistic assessment of joint ductility.

Once the FE model is constructed, it is possible to assess the temperature at which a joint will not exhibit sufficient ductility. This can be accomplished by substituting alternate J-Δa crack resistance curves measured at different temperatures. These small-specimen material tests are far less expensive that full joint tests. Also the influence of alternate welding processes involving different heat inputs can be assessed in this fashion.

6 CONCLUSIONS

The deformation capacity and ultimate strength of K-joints fabricated from cold-formed rectangular hollow sections at sub-zero temperatures has been studied both experimentally and numerically. Non-linear elastic-plastic finite element analysis was used to model the load-deformation behaviour of a K-joint.

The true $\sigma - \varepsilon$ material model used in the FE analysis was based on actual material tensile tests performed at $-40°C$. The FE model geometry corresponded to the measured geometry of a joint tested in the laboratory, also at $-40°C$.

The nonlinear FE analysis with solid elements could be used to accurately estimate the elastic-plastic load-displacement behaviour of K-joints, but did not provide an estimate of the deformation capacity or ultimate strength of the joint. The conventional von Mises type of failure criterion gave overly optimistic assessments of joint ductility and ultimate load-carrying capacity.

Elastic-plastic fracture mechanics analyses including the J-integral for one K-joint were combined with J-Δa crack resistance material data obtained using standardized test procedure. The ultimate load predicted using this procedure was slightly lower than the experimentally measured capacity of the joint. The method did, however, give an accurate representation of the location from which ductile fracture was expected to initiate. The method is promising and some the assumptions used in the analysis should be refined. Specifically, smaller crack sizes and initially semi-elliptical crack shapes should be investigated.

REFERENCES

ASTM E 1921–03. Standard test method for determination of reference temperature, T_0, for ferritic steels in the transition range. ASTM International, USA, 2002.

Björk, T., Ductility and ultimate strength of cold-formed rectangular hollow section joints at sub-zero temperatures, D. Sc. Thesis Lappeenranta University of Technology, *Acta Universitatis Lappeenrantaensis 233*, 2005

Björk, T., Kemppainen, R., Ilvonen, R. & Marquis, G., The capacity of cold-formed rectangular hollow section K-gap joints. *Proc. 10th symposium on tubular structures*, Edited by Jaurrieta, Alonso and Chica. ISBN 90 5809 552 5. Swets and Zeitlinger, Lisse, 2003.

Björk, T., Kemppainen, R., Ilvonen, R. & Marquis, G., On the definition of a material model for finite element analysis of cold-formed structural hollow section K-joints, *Proc.*

10th symposium on tubular structures, Edited by Jaurrieta, Alonso and Chica. ISBN 90 5809 552 5. Swets and Zeitlinger, Lisse, 2003.

Björk, T., Pellikka, V., Marquis, G. & Ilvonen, R., An experimental and numerical study on the fracture strength of welded structural hollow section X-joints. *Proc. 5th international ASTM/ESIS symposium of fatigue and fracture mechanics*. Reno, 2005.

Kumar, V. et al., *Advances in elastic-plastic fracture analysis*. *EPRI NP-3607*. Electric Power Research Institute, Palo Alto 1984.

Niemi, E., Behaviour of rectangular hollow section K-joints at low temperatures. *Proc. 3rd International symposium on tubular structures*, Edited by E. Niemi and P. Mäkeläinen, Elsevier Applied Science, London, 1990.

Niemi, E. J., Use of cold-formed rectangular hollow sections in welded structures subjected to low ambient temperatures. CIDECT project 5AQ/2, *Cold-formed RHS in artic steel structures, final report 5AQ-5-96*, 1996.

Packer, J. A., Wardenier, J., Kurobane, Y., Dutta, D. & Yeomans, N., *Design Guide for rectangular hollow section (RHS) joints under predominately static loading*. CIDECT. ISBN 3-8249-0089-0 Verlag TÜV Rheinland GmbH, Köln 1992.

Packer, J. A. & Henderson, J. E., *Hollow structural section connections and trusses – design guide 2nd ed.*, ISBN 0-888111-086-3. Canadian Institute of Steel Construction, Universal Offset Limited, Allison, Canada, 1997.

Puthli, R., Hohlprofilkonstruktionen aus Stahl nach DIN ENV 1993 (EC3) und DIN 18800. ISBN 3-8041-2975-5, Verner Verlag, Düsseldorf, 1998.

Rice, J. R. & Tracey, D. M., On the ductile enlargement of voids in triaxial tress fields, *Journal of Mechanics and Physics of Solids*, Vol. 17, pp. 201–217, 1969.

Vainio, H., *Design Handbook for Rautaruukki structural hollow sections*, Rautaruukki Metform. ISBN 952-5010-47-3, Otava Book Printing Ldt, Keuruu 2000.

Wardenier, J., *Hollow sections joints*. ISBN 9062750842. Delft University Press, Delft 1982.

Wardenier, J., *Hollow sections in structural applications*, Delft University of Technology, Bouwen met Staal, ISBN 90-72830-39-3, 2002.

Zhao, X. L., Wilkinson, T. & Hancock, G. J., *Cold-Formed Tubular Members and Connections. Structural behaviour and Design*. ISBN 0-080-4410-17, Elsevier, Oxford 2005.

EN1993-1-8. Design of steel structures, Part 1-8: Design of joints, European Committee for Standardization, 2003.

Plenary session C: Seismic

Tubular Structures XI – Packer & Willibald (eds)
© 2006 Taylor & Francis Group, London, ISBN 0-415-40280-8

Influence of connection design on the inelastic seismic response of HSS steel bracing members

M. Haddad & R. Tremblay

Department of Civil, Geological and Mining Engineering, Ecole Polytechnique, Montreal, Canada

ABSTRACT: The design of brace connections for concentrically braced steel frames detailed for ductile inelastic seismic response is examined. The study focuses on bracing members made of square hollow structural steel shapes with welded slotted connections. Different design strategies are investigated to prevent net section fracture and develop the full tensile capacity of the braces. Connections designs for out-of-plane and in-plane buckling of the braces are also proposed and discussed. The response of a brace sample subjected to monotonic loading as well as under quasi-static cyclic displacement history is examined through detailed finite element analysis. Un-reinforced brace connections are found to be prone to premature fracture under tension, especially when defects are present along the slots cut in the tube walls. Connections designed for out-of-plane and in-plane brace buckling resulted in nearly identical cyclic inelastic brace response. For the out-of-plane buckling design, satisfactory brace end rotation response was achieved without the free length gusset plate detail proposed in codes.

1 INTRODUCTION

Concentrically braced steel frames are commonly used to provide lateral stiffness and strength against wind and earthquake loads. Under strong seismic ground motion, the bracing members are expected to dissipate input energy through yielding in tension and inelastic buckling. Due to their high efficiency in compression, hollow structural section shapes (HSS) are often selected for the bracing members. For adequate seismic performance, the brace connections must be designed and detailed to develop the expected yield tensile strength of the braces and field welded slotted connections as shown in Figure 1a are typically used to achieve this performance.

Past test programs (e.g., Lee and Goel 1987, Tremblay et al. 2003, Haddad 2004, Yang and Mahin 2005, Fell et al. 2006) revealed that net section fracture is likely for rectangular HSS tubular steel braces with

welded slotted connections subjected to reversals of inelastic cyclic loading, even when welds are carefully done and/or end returns are used (Fig. 1b). Adequate performance was obtained by Yang and Mahin for brace specimens with a reinforced net section area, but limited guidance is available on how to adequately address shear lag effects when sizing the reinforcing plates.

For out-of-plane buckling system, Astaneh et al. (1985) proposed to include a free length equal to twice the gusset plate thickness, t_g, between the end of the brace and a line perpendicular to the brace axis about which the gusset plate may bend unrestrained by the beam and column to accommodate the inelastic rotation associated with buckling of the brace (Fig. 2a). This detail has been recommended in seismic design provisions (e.g., AISC 2005a) but has been found to lead to large gusset plate dimensions. Herman et al. 2006 showed that gusset plates without the free length could perform well when properly welded to the beams and columns. Research to date on the inelastic cyclic response of steel bracing members has mainly dealt with out-of-plane buckling systems. In-plane buckling design as shown in Figure 2b could represent an attractive design alternative as it would prevent damage to nearby wall partitions or other non-structural components.

A test program has been undertaken at Ecole Polytechnique of Montreal to examine the seismic performance of large size HSS braces with particular emphasis on the low-cycle fracture life of braces subjected to quasi-static cyclic loading histories simulating seismic

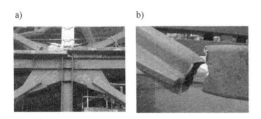

a)   b)

Figure 1. a) Typical HSS brace connection for out-of-plane buckling response; b) Net section fracture of HSS member under cyclic inelastic loading.

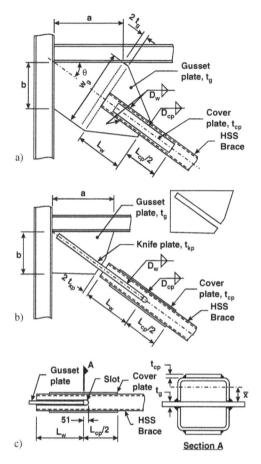

a)

b)

c)

Figure 2. Brace connection details: a) Out-of-plane design; b) In-plane design; and c) Detail of the net section (out-of-plane design).

loading. Connections that allow both out-of-plane and in-plane buckling of the braces will be considered in this experimental study. In this paper, several aspects related to the design of the connection of the brace specimens are presented and discussed. AISC seismic provisions are used as a reference in the study. Different approaches are investigated to prevent net section fracture and develop the full tensile capacity of the braces by adding reinforcing plates at the HSS connection net area. Block shear failure modes and gusset stability are also considered. The resulting weld length and reinforcement plate sizes are compared for a typical HSS brace size. The influence of the values adopted for the resistance factor is examined. Connection designs for out-of-plane and in-plane brace buckling are then proposed and discussed. For the out-of-plane buckling bracing system, the need for a free length at the end of the braces is examined. The response of the brace sample under monotonic

tension and compression loading as well as under quasi-static cyclic displacement history is examined through detailed finite element analysis.

2 TEST PROGRAM AND SAMPLE BRACE

The test program undertaken at Ecole Polytechnique mainly aims at studying the seismic inelastic cyclic response of large size, stocky steel bracing members with particular interest in the low-cycle fatigue resistance. The test matrix includes specimens made of square shaped structural tubing ranging from RHS 152 × 152 × 9.53 to RHS 305 × 305 × 16 and circular structural tubing ranging between CHS 245 × 6.4 and CHS 324 × 13. Braces made of ASTM A992 W shapes up to W310 × 179 will also be investigated. Two brace slenderness ratios are studied: 40 and 60. The influence of varying the width-to-thickness ratios of the flat elements of the brace cross-sections is also examined.

In this paper, the connection design for an RHS 254 × 254 × 13 brace specimen is examined (gross area, $A_g = 11100\,\text{mm}^2$; radius of gyration, $r = 98.1\,\text{mm}$). The brace is assumed to be connected to an W460 × 144 beam and W360 × 120 column, with an inclination angle with respect to the horizontal, θ, of 35° (Fig. 2a). An effective brace slenderness ratio, $KL/r = 40$, was also assumed in the calculations. The brace is made of ASTM A500, grade C, material with a specified minimum yield strength, F_y, of 345 MPa and a minimum tensile strength, F_u, of 427 MPa. The expected yield stress, R_yF_y, and the expected tensile strength, R_tF_u, for this material are 483 MPa and 555 MPa, respectively (AISC 2005a). The gusset plates were made of ASTM A572 steel ($F_y = 345\,\text{MPa}$, $F_u = 448\,\text{MPa}$) and E490XX electrodes ($X_u = 490\,\text{MPa}$) were used. In the test program, the connections were designed to sustain the expected yield tensile capacity, including 10% increase for strain hardening (Tremblay 2002):

$$T_u = 1.1R_yF_yA_g = 5.9\ \text{MN} \qquad (1)$$

Similarly, the brace connections were sized for a maximum brace axial compression load given by:

$$P_u = 1.1F_{cr}A_g = 5.0\ \text{MN} \qquad (2)$$

where:

$$F_{cr} = (0.658)^{\lambda^2} R_yF_y,\ if\ \lambda \leq 1.5$$
$$F_{cr} = (0.877/\lambda^2)R_yF_y,\ if\ \lambda > 1.5$$

$$\lambda = \frac{KL}{r}\sqrt{\frac{R_yF_y}{\pi^2 E}}$$

640

3 OUT-OF-PLANE BUCKLING DESIGN

3.1 Net section fracture

According to the AISC specification (AISC 2005b), the factored tension resistance for net section fracture including shear lag effects is:

$$\phi_t F_u A_e \geq T_u \qquad (3)$$

with: $A_e = A_n U$, where $U = 1 - \bar{x}/L_w$

In these expressions, ϕ_t is the resistance factor, A_e is the effective net section area, A_n is the net section area, and U is the shear lag reduction factor. For slotted HSS members, the critical section is at the end of the gusset plate (section A in Fig. 2c). Hence, L_w is the length of the longitudinal welds connecting the HSS to the gusset plate and $\bar{x}$ is the position of the center of gravity of the portion of the net section area located on either side of the gusset plate, including the reinforcing cover plate, taken with respect to the surface of the gusset plate. The net area is the sum of the cover plate area, A_{cp}, and the HSS net area, $A_{n,HSS}$. It is noted that the width of the slot is set 3.2 mm wider than the gusset plate thickness and that the slot is extended 51 mm beyond the end of the gusset plate to provide for fabrication and erection tolerances. When applying Equation 3 in the context of capacity design, different options can be considered in terms of the resistance factor and material strength. Five of these options are examined herein.

Following the AISC requirements (AISC 2005a, 2005b), the factored tension resistance for net section fracture including shear lag effects is:

$$\left(\phi_t F_{u,cp}(2A_{cp}) + \phi_t R_t F_{u,HSS} A_{n,HSS}\right)U \geq T_u \qquad (4)$$

In this expression, the expected tensile strength is used for the HSS ($R_t F_{u,HSS}$) as the rupture takes place in the same member for which the required strength is determined. For both the contributions of the cover plates and the HSS, the resistance factor is taken equal to 0.75. In the AISC seismic provisions, Equation 4 does not need to be satisfied if the effective net area is equal to or exceeds the brace gross area, i.e.:

$$\left(2A_{cp} + A_{n,HSS}\right)U \geq A_g \qquad (5)$$

The third approach is the one proposed by Yang and Mahin (2005) in which the ϕ_t and R_t factors are set equal to unity in Equation 4. In the fourth approach, referred to herein as Proposal 1, different resistance factors are applied to the cover plate and the HSS contributions in Equation 4: 0.75 and 0.9, respectively. The rationale for increasing the resistance factor for the HSS portion is that the demand and the supply are both associated to the same material, which is consistent with the use of $R_t F_{u,HSS}$ with $A_{n,HSS}$. In the last method

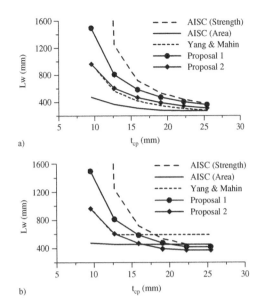

Figure 3. Influence of the cover plate thickness on the length of the weld connecting the gusset plate to the HSS considering: a) Net section fracture; b) All failure modes.

(Proposal 2), the resistance factor is set equal to 0.9 for both the cover plates and the HSS, as suggested for non-ductile limit states in the AISC 358-05 standard for the design of pre-qualified connections for steel moment frames for seismic applications (AISC 2005c).

Figure 3a shows the variation of the required length L_w as a function of the cover plate thickness for the RHS 254 × 254 × 13 brace sample, assuming a 32 mm thick gusset plate. As shown, the length of the weld is strongly influenced by the assumption made in design. For the whole range of cover plate thicknesses, the AISC (Strength) and AISC (Area) methods require the maximum and minimum weld length, respectively. As anticipated, Proposal 1 is more conservative than Proposal 2. However, the latter produces similar results compared to the approach suggested by Yang and Mahin.

3.2 Other connection failure modes

In the design procedure adopted in the study, the gusset plate dimensions a and b and the gusset plate thickness t_g are determined through an iterative procedure. For a given thickness, a and b are obtained following the modified Uniform Force Method, with the working point being located at the intersection of the beam bottom flange and the adjacent column inner flange to minimize the gusset plate dimensions (Sabelli 2005). Knowing the dimensions a and b, the width of the gusset, w_g, can be determined and the assumed thickness

t_g is then checked to satisfy the required buckling capacity of the gusset plate:

$$\phi_y F_{cr} \min\left[w_g, \left(b_{HSS} + 2L_w \tan(30°)\right)\right] t_g \geq P_u \qquad (6)$$

In this equation, the critical stress is computed as described in Equation 2 with an effective length equal to the central length multiplied by a K factor of 1.2, based on past test results and recommendations (Brown 1988, Astaneh 1998). Lin et al. (2005) recently suggested the use of $K = 2.0$ for gusset plate design. This value was not retained for this study because it was proposed for gusset plates with long unsupported edges, which was not the case herein. If required by Equation 6, the thickness t_g is modified.

The other failure modes that are then verified include tear-out failure in the brace and in the gusset plate, tension yielding on the Whitmore section of the gusset plate, and failure of the welds. Resistances associated to the first three of these failures modes are, respectively:

$$\phi_t 0.6 R_t F_{u,HSS} (2 \times 2 L_w t_{HSS}) \geq T_u \qquad (7)$$

$$\left(\phi_t 0.6 F_{u,g} (2 L_w) + \phi_y F_{y,g} b_{HSS}\right) t_g \geq T_u \qquad (8)$$

$$\phi_y F_{y,g} t_g \min\left[w_g, \left(b_{HSS} + 2 L_w \tan(30°)\right)\right] \geq T_u \qquad (9)$$

When applying the Uniform Force Method, a resistance factor of 0.9 was used for all design approaches except for Proposal 2 in which case a value of 1.0 was adopted. For the two AISC requirements, the design proposal by Yang and Mahin, and for Proposal 1, the resistance factors ϕ_t and ϕ_y in Equations 6 to 9 were taken equal to 0.75 and 0.9, respectively. In Proposal 2, ϕ_y in Equation 6 was kept equal to 0.9 as gusset plate buckling was not considered as a ductile limit state. The resistance factors ϕ_t and ϕ_y were however increased respectively to 0.9 and 1.0 in Equations 7 to 9. In Equation 7, the expected tensile strength of the HSS material ($R_t F_u$) was used in all cases except when following the approach by Yang and Mahin in which case $R_t = 1.0$ was used. Welds were checked according to the AISC specification with a resistance factor of 0.75 except for Proposal 2 where a ϕ factor of 0.9 was applied.

3.3 Optimum out-of-plane buckling design

In the iterative procedure, thick gusset plate may lead to less steel tonnage. However, the proper choice for seismic resistance should be the minimum thickness that provides sufficient compression capacity such that rotation upon brace buckling can develop more easily in the gusset plate, which will result in less inelastic demand into the brace (Haddad 2004). This approach was used herein and the thickness t_g was kept to the minimum value satisfying Equation 6. A cover plate thickness was selected to provide a reasonable

length for the welds connecting the gusset plate to the HSS. In cases where the Whitmore check of Equation 9 required a gusset plate thicker than required for compression resistance, the weld length L_w was increased to maintain the minimum thickness. Sheng et al. (2002) recommended the 30 degree angle to reduce the size of the gusset plate.

The influence of incorporating these additional design requirements and constraints on the L_w values is illustrated in Figure 3b. When compared to Figure 3a, the use of an R_t factor of 1.0 and a resistance factor $\phi_t = 0.75$ for the brace tear-out check in the design according to the proposal by Yang and Mahin resulted in a significant increase of the weld length for t_{cp} greater than 13 mm. A similar increase is found when following the AISC (Area) requirement. For the particular brace and gusset plate thickness examined herein, Proposals 1 and 2 generally resulted in the less conservative designs. For thick cover plates, the figure shows that both AISC methods and the two proposals give similar weld lengths.

Proposal 2 was selected in the design of the brace specimens. For the brace examples studied herein, the resulting final gusset plate dimensions were: $a = 629$ mm, $b = 425$ mm, $w_g = 709$ mm, and $t_g = 32$ mm. A reinforcing cover plate thickness, t_{cp}, of 15.9 mm was selected, which resulted in a weld length $L_w = 475$ mm between the gusset plate and the HSS brace (Fig. 3b). The width of the cover plate was set equal to the flat width of the HSS, i.e $w_{cp} = 176$ mm. The cover plates are welded to the HSS only along their two sides parallel to the brace to avoid less ductile welds normal to the axial load. The length of the cover plates, $L_{cp} = 555$ mm, was then governed by the weld length needed to develop the tensile strength of the cover plates assumed in design ($F_{u,cp} A_{cp}$).

4 IN-PLANE BUCKLING DESIGN

The philosophy of in-plane buckling has been introduced only recently in practice (Fig. 2b). However, the literature is absent from any in-plane buckling tests to quantify the seismic performance of brace including the end connections. In the system, a rectangular knife plate is inserted into the slot of the HSS brace and into a slot fabricated into the gusset plate, as shown in Figure 2b. As was the case for the out-of-plane design, the gusset plates are welded to the beam and column flanges. Locating the slot in the gusset plate, rather than in the knife plate, was selected as it results in thinner knife plates. The design requirements for the knife plates are similar to those used for the gusset plates in the out-of-plane buckling scenario. For practical considerations, the width of the knife plate, b_{kp}, would however be limited by the width of the beam and column flanges. As shown in Figure 2b, a free length equal to 2 t_{kp} is left between the end of the HSS and the gusset plate to accommodate the rotations anticipated upon brace buckling. Proposal 2 was also adopted for this connection configuration. The same assumptions were made regarding the resistance factors and the material properties except that $\phi = 0.9$ was used when verifying the tension yielding limit state of the knife plate near the gusset plate. At this location, the plate is subjected to nearly uniformly distributed tension stresses and using a resistance factor or 1.0 would likely result in significant inelastic deformations when the brace reaches its expected tensile strength. For the brace sample, the final knife plate dimensions were $t_{kp} = 44$ mm and $b_{kp} = 445$ mm. The gusset plate dimensions were: $a = 629$ mm, $b = 425$ mm, and $t_g = 38$ mm.

5 FINITE ELEMENT ANALYSIS

A three dimensional nonlinear finite element model was developed using Abaqus 6.6 to predict the cyclic behavior of the bracing members. Eight-node brick elements were used in the analysis, and five brick elements were used through the thickness of the HSS member, the gusset plates and the cover plates. The total number of elements of the reinforced and the un-reinforced braces were 117,175 and 76,231, respectively. Stress-strain curves based on tensile tests performed on a flat plate specimen and on a specimen extracted from the flange of an HSS were used for the plates (gusset plates and reinforcing plates) and the brace, respectively. The measured stress-strain curves for the plates and the HSS were then modified to fit the nominal and expected material strengths, respectively. Isotropic hardening material model was adopted in all cases except that the combined isotropic-kinematic model was considered for some of the cyclic loading cases.

Two different loading protocols were applied: monotonically increasing loading, either in tension or in compression, and the quasi-static cyclic displacement history specified in the AISC 2005 seismic provisions for Buckling Restrained Braces. The yield deformation of the brace, δ_y, is 7.5 mm and the design storey drift was set equal 3.75 δ_y. Prior to applying compression or cyclic loading, a small transverse displacement was imposed at mid-length of the brace to reproduce the initial crookedness of the HSS. In the analysis, equilibrium was satisfied at each increment within the cycles using the Newton-Raphson's algorithm. Eigenvalue analysis was also performed to determine the critical load and effective buckling length.

5.1 Behaviour under tension

The results of the finite element analysis under monotonic tension loading are shown in Figure 4. The

brace with the 15.9 mm cover plates exhibited high ductility with a tensile capacity equal to that predicted by Equation 1 (5.9 MN = $1.54F_yA_g$). Figure 5a shows that yielding developed uniformly over the length of the brace, as desired. A second analysis was performed with an increased 19.1 mm cover plate thickness, without changing the other design parameters. The behaviour was identical to that obtained with $t_{cp} = 15.9$ mm, as shown in Figure 4a.

In a third analysis, the cover plate was removed ($t_{cp} = 0$). The limit load of the brace was reached at an axial displacement of 83.3 mm (11.1 δ_y). As shown in Figure 5b, stress demand concentrated in the slot region. Figure 6 compares the equivalent plastic strain (PEEQ) for the cases $t_{cp} = 15.9$ mm and $t_{cp} = 0$ mm. The figure clearly shows that the addition of cover plates pushed the inelastic demand into the bracing member, away from the connections. Figure 4b shows the increase in PEEQ values computed both at the end of the slot and at mid-length of the brace for both cover plate conditions. For $t_{cp} = 0$, the PEEQ at the slot increased at a constant and much higher rate compared to the values monitored in the brace. Conversely, the growth of the equivalent plastic strain near the slot for the brace with the cover plate decayed rapidly while plastic strains developed in the core of the brace. In a fourth analysis, a 10 mm deep triangular notch was

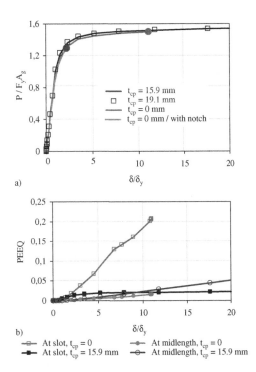

Figure 4. Tensile capacities of the brace for different cover plate thicknesses.

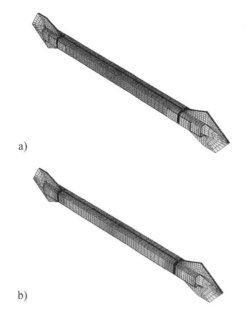

a)

b)

Figure 5. Distribution of von Mises stress under tension loading at a ductility of 7.0: a) $t_{cp} = 15.9$ mm (darkest = 557 MPa); b) $t_{cp} = 0$ (darkest = 649 MPa).

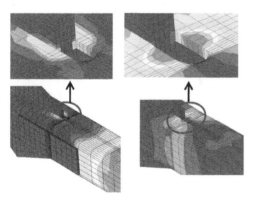

Figure 6. Distribution of the PEEQ for $t_{cp} = 15.9$ mm (left) and $t_{cp} = 0$ mm (right) under tension loading at a ductility of 11.1.

introduced at the slot end, as shown in Figure 7. At a ductility of 2.2, the PEEQ in the vicinity of the notch reached a value of 0.8, which has been suggested as sufficient to initiate fracture in structural steel tubing (Cheng et al. 1998). This illustrates that the performance of un-reinforced connections is sensitive to fabrication defects.

The analysis under monotonic tension loading was also performed for the in-plane design with the reinforced brace. Similar response was observed in both cases. In Figure 8, the stress distribution obtained for the two designs is compared.

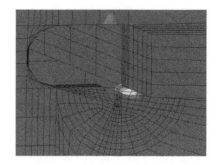

Figure 7. Equivalent plastic strains in the HSS wall at the notch location.

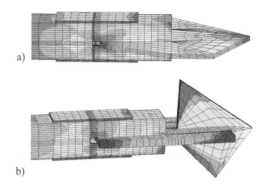

a)

b)

Figure 8. Distribution of von Mises stresses under tension loading for: a) Out-of-plane buckling design; b) In-plane buckling design.

5.2 Behaviour under compression and cyclic loading

For the out-of-plane and in-plane designs, the total length of the brace including the free length at both ends was equal to 4360 mm in the models. The same length was assigned to the HSS member for the out-of-plane design without the end free length hinge zones. For these three cases, the effective length factors obtained from eigenvalue analysis were respectively 0.94, 0.89, and 0.96, which agrees well with the K factor assumed in design (0.9).

Figure 9 presents the inelastic buckled shape under monotonic compression loading for the two buckling systems. As shown, both designs behaved as intended. Figure 10 shows detailed views of the connection regions. Stress concentrations developed in the HSS near the end of the welds as well as in the gusset plates. Figure 11 shows the location of high PEEQ values at three different loading stages for the out-of-plane design without the free length detail. At buckling, higher strain demand developed near the end of the slot location. In the post-buckling range, the contours of the accumulative plastic strains appeared in the gusset plate while maintaining the contours around the slot.

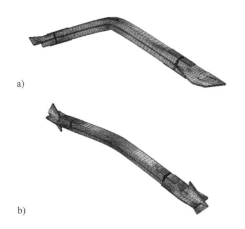

a)

b)

Figure 9. Von Mises stresses under brace buckling: a) Out-of-plane buckling design; b) In-plane buckling design.

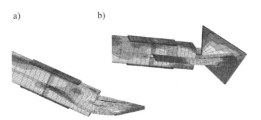

a)   b)

Figure 10. Von Mises stresses in the connection region under brace buckling: a) Out-of-plane design; b) In-plane design.

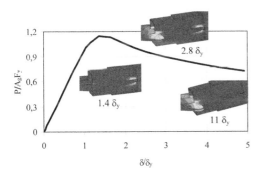

Figure 11. Axial load-deformation response under monotonic compression loading for the out-of-plane buckling design.

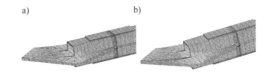

a)   b)

Figure 12. Active yielding under out-of-plane buckling: a) With the free length; b) Without the free length.

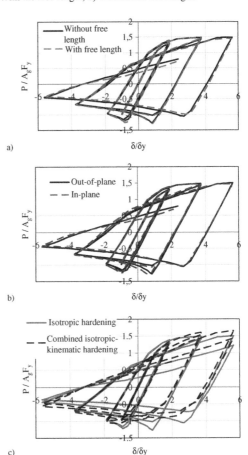

a)

b)

c)

Figure 13. Hysteretic brace response under cyclic loading: a) Effect of the free length for the out-of-plane buckling design; b) Comparison between out-of-plane and in-plane designs; and c) Influence of the material strain hardening model.

Under larger deformations, the contours disappeared around the slot.

Figure 12 shows the active yielding pattern for the out-of-plane design for the cases with and without the free length detail. As shown, yielding propagates more on both sides of the bending line when no free length is provided. In Figure 13a, the overall behaviour under cyclic loading is however nearly identical for both cases. This result suggests that there would be no need for the free length left for the out-of-plane buckling, as was observed in tests by Hermann et al. (2006).

The comparison of the hysteretic response obtained with the out-of-plane and in-plane designs is illustrated in Figure 13b. Again, the response is very similar. For the in-plane design, the knife plates would typically be thicker for large brace sizes, which can result in higher rotation demand imposed on the bracing member and, thereby, a reduced fracture life, as observed by

Haddad (2004). The responses in Figures 13a & b were obtained using the isotropic strain hardening model. Figure 13c shows that the combined isotropic-kinematic hardening model results in lower compressive capacity and higher tensile strength, which is due to the shift in the yield surface in stress space.

6 CONCLUSIONS

The design of the welded slotted connection for an RHS $254 \times 254 \times 13$ tubular bracing member was examined and finite element analysis was performed to evaluate the performance of the brace under simulated earthquake loading. The main conclusions of this study can be summarized as follows.

Reinforcement is needed at the net section area of brace end connections to prevent premature fracture and allow uniform yielding to develop along the entire length of the brace under tension loading. Increasing the thickness of the reinforcing cover plates reduces the length of welds required to account for shear lag effects. Both the cover plate dimensions and the length of the welds are influenced by the values adopted in design for the resistance factors.

According to the finite element analysis, the behaviour of the brace with and without the 2 t_g free length is similar, which is in accordance with recent test results indicating that the free length may not be required to achieve satisfactory seismic response.

For the studied brace specimen, the in-plane buckling design resulted in nearly identical cyclic inelastic brace behaviour compared to the more common out-of-plane design. Further studies are needed to investigate the need to stiffen the gusset plate to eliminate the possibility of local gusset plate instability and out-of-plane brace buckling. For larger brace sizes, thick knife plates may be needed to meet geometric constraints on the width of the knife plate, which may impact the gusset and brace behaviour.

According to the finite element analysis, the combination of resistance factors and strength properties that were used in this study resulted in satisfactory inelastic brace performance. However, it must be recognized that these factors must also account for the variability and the uncertainty in strength evaluation, which was not taken into account in this analytical study. Therefore, these factors should also be adjusted to achieve the level of confidence implicitly assumed in seismic design codes.

ACKNOWLEDGMENT

The authors gratefully acknowledge the financial support of the Natural Science and Engineering Research Council of Canada.

REFERENCES

Abaqus. 2006, User's Manual, Version, 6.6, Hibbitt, Karlsson, and Sorensen, Inc., Providence, RI.

AISC. 2005a. ANSI/AISC 341-05, Seismic Provisions for Structural Steel Buildings. American Institute of Steel Construction, Chicago, IL.

AISC. 2005b. ANSI/AISC 360-05, Specification for Structural Steel Buildings. American Institute of Steel Construction, Chicago, IL.

AISC. 2005c. ANSI/AISC 358-05, Prequalified Connections for Special and Intermediate Steel Moment Frames for Seismic Applications. American Institute of Steel Construction, Chicago, IL.

Astaneh-Asl, A. 1998. Seismic behavior and design of gusset plates, Steel Tips Report. Structural Steel Education Educational Council, Moraga, CA. www.steel.tips.org

Astaneh-Asl, A., Goel, S.C. & Hanson, R.D. 1985. Cyclic Out-of-Plane Buckling of Double-Angle Bracing. J. of Struct. Eng., ASCE, 111, 1135–1153.

Brown, V.L.S. 1988. Stability of gusset connections in steel structures. Doctoral Dissertation. Dept. of Civil Eng. University of Delaware, Newark, Delaware.

Cheng, J.J.R., Kulak, G. & Khoo, H.-A. 1998. Strength of slotted tubular tension members. Can. J. of Civ. Eng. 25, 982–991.

Fell, B.V., Myers, A.T., Deierlein, G.G. & Kanvinde, A.M. 2006. Testing and simulation of ultra-low cycle fatigue and fracture in steel braces. Proc. 8th US Nat. Conf. on Earthquake Eng., San Francisco, CA, Paper No. 587.

Haddad, M.A. 2004. Design of concentrically braced steel frames for earthquakes. PhD Dissertation. Civil Engineering Department. University of Calgary, Calgary, Alberta, Canada.

Herman, D., Johnson, S., Lehman, D. & Roeder, C. 2006. Improved seismic design of special concentrically braced frames. Proc. 8th US Nat. Conf. on Earthquake Eng., San Francisco, CA, Paper No. 1356.

Lee, S. & Goel, S.C. 1987. Seismic behavior of hollow and concrete-filled square tubular bracing members. Research Rep. No. UMCE 87–11, Univ. of Michigan, Ann Arbor, MI.

Sabelli, R. 2005. Seismic braced frames – Design concepts and connections. An AISC Short Course. North American Steel Construction Conference. Montreal, QC.

Sheng, N., Yam, C.H. & Lu, V.P. 2002. Analytical investigation and the design of the compressive strength of steel gusset plate connections. Journal of Constructional Steel Research. 58: 1473–1493.

Tremblay, R., Archambault, M.H. & Filiatrault, A. 2003. Seismic Performance of Concentrically Braced Steel Frames made with Rectangular Hollow Bracing Members. J. of Struct. Eng., ASCE, 129 (12), 1626–1636.

Tremblay, R. 2002. Inelastic Seismic Response of Steel Bracing Members. J. of Const. Steel Research, 58, 665–701.

Yang, F. & Mahin, S. 2005. Limiting net section fracture in slotted tube braces. Steel Tips Report, Structural Steel Education Educational Council, Moraga, CA. www.steel.tips.org

Tubular Structures XI – Packer & Willibald (eds)
© 2006 Taylor & Francis Group, London, ISBN 0-415-40280-8

Experimental study on behavior of tubular joints under cyclic loading

W. Wang & Y.Y. Chen

College of Civil Engineering, Tongji University, Shanghai, China

ABSTRACT: This paper examines the cyclic performance of CHS joints used in steel tubular structures. Quasi-static experimental study into the response of eight T-joint specimens is described. Four of them are subjected to cyclic axial load, and the other four are subjected to cyclic in-plane bending. The general test arrangement, specimen details, and most relevant results (failure modes and load-relative deformation hysteretic curves) are presented. Some indexes to assess the seismic performance of tubular joints, including strength, ductility and energy dissipation, are synthetically analyzed and compared. Test results show that failure modes of axially loaded joints mainly contain weld cracking in tension and chord plastification in compression. But for joints under cyclic in-plane bending, punching shear may become a regular failure mode accompanied by ductile fracture of the welds. Hysteretic curves behave stably in general. Ultimate strengths of joints are also compared with equation values for monotonic loading from various design codes. Results indicate existing codes can be used to check the ultimate capacity of tubular T-joints under cyclic loading to a certain extent. It is also found that there is a significant distinction in the energy dissipation mechanism for tubular joints under different loading conditions. Additionally, it is deserved to note that weld cracking sometimes happened under lower cyclic load level compared with design resistance specified by the codes.

1 INTRODUCTION

Lattice girders are stronger and lighter than I-section girders in general. Loads other than seismic action usually govern the design of large-span trusses. However, there exist trusses seriously affected by earthquakes. High-rise apartment buildings with trussed frames damaged during the Kobe Earthquake are one outstanding example. Trussed structures are more difficult than moment resisting frames to design for earthquakes owing to more scant research on the seismic performance. The hysteretic behavior of tubular connections forms a basis for a rational aseismic design of such tubular structures as large-span lattice girders, space frames and towers.

Existing research focused on ultimate capacity of tubular connections under axial loads as their main target. Rare studies on cyclic behavior of tubular connections include investigations into the behavior of a completely overlapped tubular joint (Qin et al. 2001) and interacting behavior of connections and frames (Kurobane 1998). Because of the lack of test evidence, the current design approach for tubular connections is limited to static and fatigue strength. Hence an experimental program has been set up to investigate the hysteretic behavior of CHS T-joints under both axial load and in-plane bending. A total of 8 tests have been conducted, with the results being reported in this paper.

2 EXPERIMENTAL PROGRAM

2.1 Test specimen

In total eight specimens of CHS T-joints were tested. Four of them are subjected to cyclic axial load on brace, and the other four are subjected to cyclic bending moment on brace. The schematic view of the T-joint specimen is shown in Figure 1. The geometrical characteristics and details of all the specimens are listed in Table 1. The chord length between supports is $l_c = 1500$ mm (with $\alpha = 2l_c/D = 12.2$) while the brace length l_b is approximately 5 times the brace diameter d. The chosen geometrical parameters (with $\gamma = D/(2T) = 10.2$ and 15.3; $\beta = d/D = 0.49$ and 0.79; $\tau = t/D = 0.75$) correspond to typical values for T-joints in steel tubular structures. The chord and brace members are jointed together by fillet welds, the throat thickness of which, h_f is shown in Table 1. Table 2 summarizes the measured material properties of the chord and brace.

2.2 Test setup

Figure 1 shows the general arrangement for the T-joint tests. The chord was placed horizontally with brace placed in an upright position. The two ends of the chord were bolted to the supports that extended to the base of

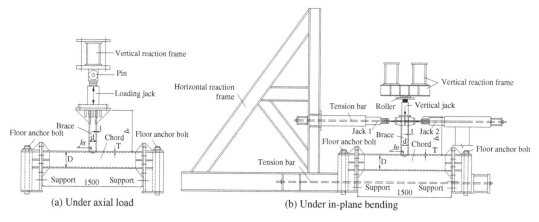

| | (a) Under axial load | | (b) Under in-plane bending |

Figure 1.   Joint specimen and loading rig.

Table 1.   Geometrical characteristics of specimens.

No.	D × T (mm × mm)	d × t (mm × mm)	l_b (mm)	h_f (mm)	β	γ	τ	Load on brace
A1	245 × 8	121 × 6	600	12	0.49	15.3	0.75	Cyclic axial loading, with a larger magnitude for tension than compression in each displacement amplitude.
A2	245 × 8	121 × 6	600	12	0.49	15.3	0.75	Cyclic axial loading, with a smaller magnitude for tension than compression in each displacement amplitude.
A3	245 × 12	121 × 8	600	12	0.49	10.2	0.75	Cyclic axial loading, with a smaller magnitude for tension than compression in each displacement amplitude.
A4	245 × 8	194 × 6	1000	8	0.79	15.3	0.75	Cyclic axial loading, with a smaller magnitude for tension than compression in each displacement amplitude.
B1	245 × 8	121 × 6	600	12	0.49	15.3	0.75	Cyclic in-plane bending, with no axial load.
B2	245 × 8	121 × 6	600	12	0.49	15.3	0.75	Cyclic in-plane bending, keeping the brace member with 20% of plastic axial resistance.
B3	245 × 12	121 × 8	600	12	0.49	10.2	0.75	Cyclic in-plane bending, with no axial load.
B4	245 × 8	194 × 6	1000	8	0.79	15.3	0.75	Cyclic in-plane bending, with no axial load.

Table 2.   Material properties of specimens.

Section Size (mm)	f_y (MPa)	f_u (MPa)	ζ (%)	f_y/f_u
121 × 6 CHS	345	485	26	0.71
194 × 6 CHS	344	482	27	0.71
121 × 8 CHS	392	601	25	0.65
245 × 8 CHS	398	564	28	0.71
245 × 12 CHS	356	583	26	0.61

the laboratory floor. For axially loaded joints, the 100 ton actuator was bolted to the free end of the brace to apply the force. For joints loaded by in-plane bending, the vertical jack was first used to apply compressive load on the brace and this load was maintained constant, in-plane bending moment was then produced by two horizontal jacks to apply cyclic lateral forces to the brace end.

2.3   Instrumentation arrangement

For each specimen, the instrumentation includes strain gauging around the crown and saddle points to measure the strain distributions at the hot spots, and transducers to measure the displacements at selected points. The general arrangement of displacement transducers and strain gauges is shown in Figure 2. The local deformation of the chord wall can be obtained by measuring the brace displacement around the brace-chord intersection and the mid-span deflection of the chord as a beam element.

2.4   Loading history

Axially loaded joints were tested by applying cycles of alternated load with tip vertical displacement Δ_v as shown in Table 3. For joints loaded by in-plane bending, a typical displacement-controlled loading history is also shown in Figure 3.

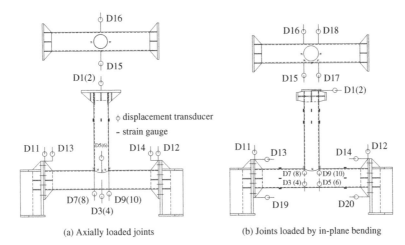

(a) Axially loaded joints   (b) Joints loaded by in-plane bending

Figure 2. Arrangement of measuring instruments.

Table 3. Loading history for axially loaded joints.

Specimen A1 Δ_v(mm)	Specimen A2 Δ_v(mm)	Specimen A3 Δ_v(mm)	Specimen A4 Δ_v(mm)	Number of cycles
+6.6, −1.8	+1.0, −2.0	+1.6, −1.4	+1.1, −1.5	1
+14.9, −2.8	+2.0, −3.6	+6.7, −5.2	+2.0, −3.6	1
+19.3, −3.5	+6.1, −8.1	+9.0, −8.2	+4.4, −8.1	3
+22.6, −5.7	+8.1, −10.5	+15.3, −11.3	+7.2, −10.2	3
+30.3, −8.2	+9.5, −12.6	+26.9, −14.0	+11.8, −13.0	3
–	+10.2, −14.3	–	+12.2, −16.7	3

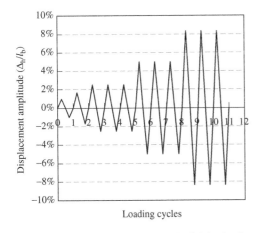

Loading cycles

Figure 3. Loading history for joints loaded by in-plane bending.

Before the actual test, one complete cyclic load of ±2 mm was applied to check the working conditions of the strain gauges and other equipment and to loosen the friction and remove any lock-in forces within the test setup.

3 EXPERIMENTAL RESULTS

3.1 Observations and failure modes

Under axial loads, for the test of specimens A1, it was observed that at the tension phase of the second loading cycle, a small crack initiated at the weld toe between the chord and the brace. In the next cycles, the crack extended and expanded gradually until the applied load dropped quickly with the displacement greatly increased at the tension phase of the ninth loading cycle. The large crack almost cut through the chord wall near the saddle point. No obvious plastic deformation of the chord wall was found during the whole process of the test. The failure mode of A1 is shown in Figure 4a.

For the test of specimen A2, A3 and A4, cracking of the weld under tension happened after the maximum Von Mises stress at the intersection area reached the yield stress of the steel. When reverse load was applied, the crack closed up and plastic deformation developed continuously. After a number of cycles like this, the peak compressive load was reached and obvious chord ovalization and concave deformation was noticed at the end of the test. Figure 4b shows the failure mode of specimen A3.

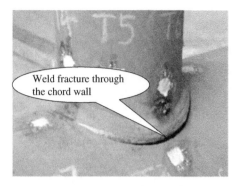

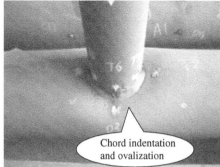

(a) Specimen A1   (b) Specimen A3

Figure 4.   Typical failure modes of axially loaded joints.

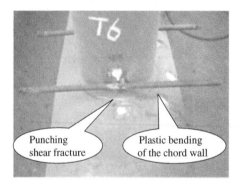

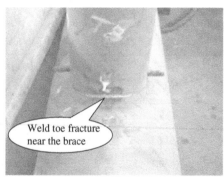

(a) Specimen B2   (b) Specimen B3

Figure 5.   Typical failure modes of joints loaded by in-plane bending.

Under in-plane bending loads, for all specimens, the following observations were made: (1) In the first two displacement amplitude, there was no weld cracking. But after intersection area between the chord and the brace stepped into plastic phase, small cracks formed at the tension side of the weld toe along the chord. (2) Weld cracking did not occur in the first loading cycle of certain displacement amplitude but in the second or third cycle. (3) In the following testing process, small cracks expanded and extended with the chord wall sustaining plastic deflection. Two typical failure modes of joints loaded by bending are graphically illustrated in Figure 5. Among them, B2 and B4 can be attributed to the combination of punching shear fracture and plastic bending of the chord wall (Figure 5a), while B1 and B3 belong to ductile fracture of welds or heat-affected zone (Figure 5b).

3.2 Hysteretic curves

The load-displacement hysteretic loops obtained for axially loaded joint tests are presented in Figure 6. The non-dimensionalized axial load N/N_{bp} has been

plotted against the intersection line displacement δ, which is defined as the average displacement of the crown points and saddle points minus that of the chord centerline, thus effectively excluding the displacement due to chord beam bending. N_{bp} is the plastic axial resistance of the brace. Tensile forces are represented in positive signs with compressive forces in negative signs. Yura's deformation limit (Yura et al. 1981) has been shown in these figures with dashed lines.

The moment-rotation hysteretic loops obtained for joint tests loaded by in-plane bending are presented in Figure 7. The non-dimensionalized moment M/M_{bp} has been plotted against the rotation of the brace θ, which is acquired by dividing the sum of relative concave and convex at the crown points by brace diameter. The moment are given at the chord surface at the crown point. M_{bp} is the plastic moment capacity of the brace member.

In addition, the first weld cracking positions are marked in these Figures.

For most joints, hysteretic curves behave stably without exhibiting "pinching" at higher load level. Pinching is characterized by an increase in

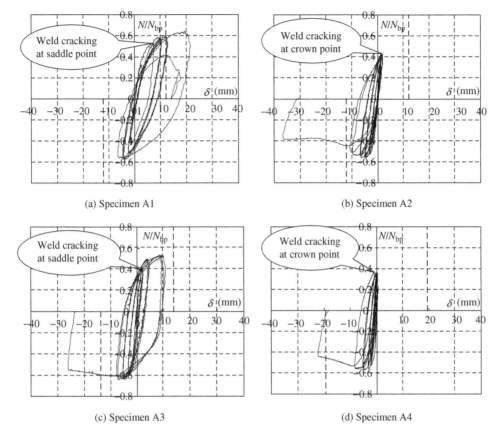

(a) Specimen A1	(b) Specimen A2
(c) Specimen A3	(d) Specimen A4

Figure 6. Load-displacement curves of axially loaded joints.

deformation without a significant increase in load, thus resulting in a loss in stiffness of the connection. The hysteretic plots were observed to become plump as the load was enhanced.

From the results for specimens A1 and A3, it can be found that after small cracks formed in weld toe or weld heat-affected zone, the joint can sustain a further increase in load with stiffness descending gradually in the next cycles until the joint failed. The displacement values of hysteretic curves under tension seem very large by the end of the test, because they contained displacements due to crack expansion. From the results for specimens A2, A3 and A4, peak compressive force is clearly shown, accompanied by deterioration in capacity and stiffness at the compressive phase near the end of test.

For specimens B1 and B2 with the same geometry, it would be of interest to investigate the effect of the brace axial load on cyclic joint behavior under in-plane bending. From the hysteretic curves, initial elastic rigidities are very close. The first cracking moment of B2 is a little higher than that of B1. However, a significant difference can be noted between them: After first

cracking the bending load could still be increased for B1, whereas there appeared significant strength and stiffness deteriorations for B2. This may be caused by enhanced plastic bending of the chord wall under brace axial load due to punching shear crack initiation. For specimens B3 and B4, the first crack did not emerge until the maximum loads had been achieved. Afterwards, the strength and the stiffness began to decrease. Moreover, a little pinching was observed to occur for B2 and B4.

3.3 Load-carrying capacity and comparison with code estimation

One way of presenting extreme values of connection load-displacement curves is to plot the skeleton curves. Skeleton curves of the cyclic response of connections have been shown experimentally to be a reasonable estimate of the probable monotonic response. For specimens A2, A3 and A4, they all exhibited a peak compressive load. Therefore, the maximum load on the skeleton curve is defined as the ultimate strength of the joint (N_u). But for specimen A1, no peak load is shown.

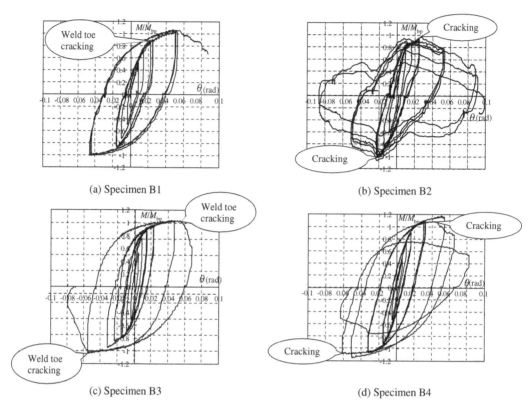

(a) Specimen B1

(b) Specimen B2

(c) Specimen B3

(d) Specimen B4

Figure 7. Moment-rotation curves of joints loaded by in-plane bending.

Table 4. Comparison between test resistance of axially loaded joints and the predicted strengths.

	Joint test resistance(kN)			Joint strength predictions(kN)*			Brace strength(kN)			
No.	N_y	N_{wcr}	N_u	$N_{uc,GB50017}^{pj}$	$N_{ut,GB50017}^{pj}$	$N_{u,EC3}^{sj}$	N_{bp}	$N_{wcr}/N_{ut,GB50017}^{pj}$	$N_u/N_{u,GB50017}^{pj**}$	N_u/N_{bp}
A1	330	400	447	−305	427	699	745	0.94	1.05	0.60
A2	−396	327	−418	−305	427	−699	−745	0.77	1.37	0.56
A3	−670	480	−707	−566	792	−938	−1105	0.61	1.25	0.64
A4	−520	438	−693	−523	633	−1120	−1216	0.69	1.33	0.57

* Superscript pj or sj shows failure mode for the joint is chord plastification or punching shear; Subscript uc or tc denotes compressive or tensile resistance.
** For A1, $N_{u,GB50017}^{pj}$ denotes $N_{ut,GB50017}^{pj}$; For A2, A3 and A4, $N_{u,GB50017}^{pj}$ denotes $N_{uc,GB50017}^{pj}$.

So Yura's deformation limit (Yura et al. 1981) is used to define the ultimate tensile resistance of A1, although it should be based on the local deformation of the chord face at the intersection under monotonic loads. Following the same principle, the ultimate bending strength (M_u) of specimens loaded by in-plane bending can be determined. The yield strength observed on load-deformation curves can be used as another measure for the capacities of the joints. The yield strength (N_y or M_y) defined by Kurobane et al. (1984) is determined by secant modulus of $0.779K_N$ or $0.779K_M$. Where, K_N or K_M refers to initial joint rigidity under axial force or bending moment.

Test values of N_y, N_u, M_y, M_u are presented in Tables 4 and 5 respectively. Furthermore, the first weld cracking load (N_{wcr} or M_{wcr}) for these joints are also listed in the tables. They are compared with unfactored design strength calculated from Chinese ($N_{uc,GB50017}^{pj}$, $N_{ut,GB50017}^{pj}$) (2003) and European ($N_{u,EC3}^{sj}$, $M_{u,EC3}^{pj}$, $M_{u,EC3}^{sj}$) (2005) structural steelwork design specifications. Predicted values of strengths were

Table 5. Comparison between test capacities of in-plane bending loaded joints and the predicted strengths.

| No. | Joint test resistance (kN-m) | | | Joint strength predictions (kN-m)* | | Brace strength (kN-m) | | | |
	M_y	M_{wcr}	M_u	$M_{u,EC3}^{pj}$	$M_{u,EC3}^{sj}$	M_{bp}	$M_{wcr}/M_{u,EC3}^{sj}$	$M_u/M_{u,EC3}^{sj}$	M_u/M_{bp}
B1	21.0	23.4	27.7	28.65	26.91	26.04	0.87	1.03	1.06
B2	23.0	26.0	26.5	28.65	26.91	26.04	0.97	0.98	1.02
B3	31.8	38.6	39.0	47.08	36.11	37.47	1.07	1.08	1.04
B4	67.5	76.1	81.7	74.06	69.19	70.74	1.10	1.18	1.15

* Superscript pj or sj shows failure mode for the joint is chord plastification or punching shear.

calculated using material properties obtained from the tensile coupon tests described previously. In Table 4, tensile loads are represented in positive signs with compressive loads in negative signs.

For axially loaded joints, strength predictions based on failure modes of chord plastification in general underestimate the actual joint resistances under cyclic loading conditions. It also can be found that the brace member efficiency (N_u/N_{bp}) is less than one for these joint specimens. This means they dissipated energy mainly by plastic deformation of chord wall.

For joints loaded by in-plane bending, it can be noticed that predicted strengths based on failure modes of chord plastification are higher than that based on failure modes of punching shear, which tend to agree with the actual joint resistances under cyclic loading conditions reasonably. The brace member efficiency (M_u/M_{bp}) is larger than one for these joint specimens. This feature is different from the joints under axial load, meaning they have enough strength to make plastic hinge formed in the connecting brace.

Whether for axially loaded joints or for in-plane bending loaded joints, weld cracking sometimes happened under lower load level compared with design resistance specified by the codes. Early development of cracks may be caused by local material deterioration due to reversals of inelastic strain and triaxial tensile stress distribution around intersection lines.

3.4 Ductility ratio

The ductility ratio is an important index to assess deformability. The ductility ratio is defined as $\mu = \delta_u/\delta_y$ for axially loaded joints and θ_u/θ_y for joints loaded by in-plane bending, where δ_u or θ_u is assumed to be local deformation of the chord face corresponding to ultimate strength and δ_y or θ_y is the yield deformation. From the hysteretic curve shown in Figures 6 and 7, the ductility ratios under axial load and under in-plane bending are determined and listed in Tables 6 and 7 respectively. From Table 7, θ_u of B2 is obviously smaller than that of B1 due to the compressive preload on the brace.

It is important to consider if cyclic loadings would deteriorate the ductility of unstiffened tubular joints.

Table 6. Ductility ratio of axially loaded joints.

No.	δ_y (mm)	δ_u (mm)	$\mu = \delta_u/\delta_y$
A1	3.80	12.16	3.2
A2	−3.32	−4.40	1.4
A3	−4.20	−7.42	1.8
A4	−1.00	−4.66	4.7

Table 7. Ductility ratio of joints loaded by in-plane bending.

No.	θ_y (rad)	θ_u (rad)	$\mu = \theta_u/\theta_y$
B1	0.017	0.048	2.8
B2	0.016	0.027	1.7
B3	0.012	0.047	3.9
B4	0.017	0.051	3.0

According to test results of four multi-planar KK tubular joints (Chen et al. 2003), the ductility ratio under cyclic axial loadings would be smaller than that under monotonic loadings. Weld cracking or punching shear cracking of tubular T-joints, which happened at certain load level in this study, may lead to the same result.

3.5 Energy dissipation

The capacity of structural connections to dissipate energy when subjected to seismic loads is as important as their strength or stiffness in the evaluation process. The dissipated energy is calculated by evaluating the area enclosed by the load-displacement curves. The accumulative energy dissipation ratio is used to measure the energy dissipated. The ratio is defined as $\eta_a = \sum_{i=1}^{n} (E_i^+ + E_i^-)/E_y$, where E_y is the energy absorbed at the first yield displacement δ_y and is defined as $E_y = P_y\delta_y/2$. E_i^+ and E_i^- are the energy dissipation in the tension half-cycle and in the compression half-cycle, respectively. This ratio values of the specimens are listed in Table 8. They can be found to be higher than the value of 4 required by the API (1993) provision.

Table 8. Accumulative energy dissipation ratio of joints.

No.	$\sum_{i=1}^{n} (E_i^+ + E_i^-)$(kN-m)	E_y(kN-m)	η_a
A1	49.32	0.627	78.7
A2	29.06	0.657	44.2
A3	74.61	1.407	53.0
A4	26.62	0.260	102.4
B1	16.94	0.179	94.6
B2	22.14	0.184	120.3
B3	18.69	0.191	97.9
B4	48.03	0.608	79.0

4 CONCLUSIONS

The aim of this paper was to provide information regarding the cyclic performance of unstiffened tubular joints. Based upon the analysis on the test results of eight tubular T-joint specimens, the following conclusions can be drawn:

(1) From the overall deformed shapes of the joint specimens, the T-joints under cyclic axial loading primarily failed by weld cracking in tension and excessive plastic deformation of the chord wall in compression. But for the T-joints under cyclic in-plane bending, punching shear may become a regular failure mode accompanied by ductile fracture of the welds.
(2) To a certain extent existing codes including GB50017 (2003) and Eurocode 3 (2005) can be used to check the ultimate capacity of tubular T-joints under cyclic loading.
(3) Hysteretic curves behave stably. The energy analysis showed that the joints have good energy dissipating capacity. However, compared with monotonic loads, cyclic reverse loads may reduce the ductility of unstiffened tubular joints.
(4) For tubular joints under differing loading conditions, there is a significant distinction in the energy dissipation mechanism. The axially loaded joints dissipate energy mainly by plastic deformation of the chord wall with brace member efficiency smaller than one. The in-plane bending loaded joints dissipate energy mainly by plastic deflection of the brace with brace member efficiency larger than one.
(5) In cyclic loading conditions, weld cracking sometimes happened under lower load level compared with design resistance specified by the codes. This may be result from material deterioration owing to repeated cold-working and triaxial tensile stress distribution around intersection lines.

ACKNOWLEDGEMENTS

The reported work is supported by National Natural Science Foundation of China (50578117). The specimens were contributed by Jiangsu Huning Steel Mechanism Company Ltd. (China). The experimental investigation was conducted in the Structural Engineering Laboratory, Department of Building Engineering, Tongji University.

NOMENCLATURE

D   outer diameter of the chord
d   outer diameter of the brace
T   wall thickness of chord
t   wall thickness of brace
f_y yield stress
f_u ultimate tensile stress
ζ the prolongation of tensile coupons
E   Young's modulus
α chord length parameter $2l_c/D$
β the brace-to-chord diameter ratio
γ the ratio of chord diameter to twice thickness
τ the brace-to-chord thickness ratio

REFERENCES

American Petroleum Institute (API). 1993. Recommended practice for planning, designing, and constructing fixed offshore platforms. RP2a. Washington, D.C.

Chen, Y. Y., Shen, Z. Y. & Zhai, H. 2003. Experimental research on the hysterical property of unstiffened space tubular Joints. Building Structure 24(6): 57–62 (in Chinese).

EN1993-1-1, Eurocode 3. 2005. Design of steel structures, Part 1–1: General rules and rules for buildings. CEN.

EN1993-1-8, Eurocode 3. 2005. Design of steel structures, Part 1–8: Design of joints. CEN.

GB50017-2003. 2003. Chinese code for Design of Steel Structures.

Kurobane, Y., Makino, Y. & Ochi, K. 1984. Ultimate resistance of unstiffened tubular joints. Journal of Structural Engineering 110(2): 385–400.

Kurobane, Y. 1998. Static behavior and earthquake resistant design of welded tubular structures. In Jármai, K. and Farkas, J. (ed.), Mechanics and Design of Tubular Structures: 53–116. Wien, Austria: Springer-Velag.

Qin, F., Fung, T. C. & Soh, C. K. 2001. Hysteretic behavior of completely overlap tubular joints. Journal of Constructional Steel Research 57: 811–829.

Yura, J. A., Zettlemoyer, N. & Edwards, I. F. 1981. Ultimate capacity of circular tubular joints. Journal of Structural Engineering 107(10): 1965–1982.

Tubular Structures XI – Packer & Willibald (eds)
© *2006 Taylor & Francis Group, London, ISBN 0-415-40280-8*

Concrete filled hollow sections under earthquake loading: Investigation and comparison of structural response

M. Angelides & G. Papanikas
AMTE Consulting Engineers, Athens, Greece

ABSTRACT: The basic principles of design of braced buildings under high earthquake loading are illustrated at first, followed by the parametric investigation of the response of concrete filled hollow sections as vertical elements in braced bays and their comparison to equivalent steel braces and reinforced concrete shear walls. The investigation is extended to high industrial buildings under significant vertical and seismic loading, through the analysis of the response of 'super columns', i.e. combination of concrete filled large hollow sections with reinforced concrete shear walls for the optimization of earthquake response.

1 INTRODUCTION

The use of concrete filled hollow sections is customarily adopted for low cost fire protection. The advantages of the concrete core contribution to the resistance of the vertical elements are obvious but seldom utilized in earthquake resistant design due to uncertainties in the connection design and the corresponding load transfer mechanism. This however primarily applies to moment resisting frames, which are gradually replaced by braced bays as the preferred lateral load resisting elements in areas of high seismicity, such like the ones increasingly defined by modern revised codes.

The revisions of the seismic design codes have typically resulted from measured higher ground acceleration values and from observed weaknesses in standard (non seismic) structural detailing. In recent years, the observation of typical failures along with the need to incorporate severe shocks into the design of ordinary structures, have demonstrated the limitations of moment resisting frames. A moment resisting frame cannot efficiently provide the high stiffness required for controlling excessive drift, nor can it mobilize member strengths the same way that an axially loaded braced arrangement does, particularly in steel structures. The use of hollow steel sections provides the additional advantage of the possibility of filling the core with reinforced concrete, thus significantly increasing axial stiffness and strength. Basic theory and design guidelines for this construction method are available in standard texts (for example, Bode 1998), as well as in specialized associations' editions (for example, Bergmann et al. 1995).

In the present paper, the application of this method to the earthquake response of commercial and industrial structures is investigated. The results of a comparative analysis of different structural systems for a 4-storey office building are presented first. The structural response of the basic lateral load resisting element is then parametrically investigated for regular steel bracings, concrete filled sections and reinforced concrete shear walls. Design calculations have been based on the European Standards (Eurocodes 2, 3, 4) and the Greek Code for earthquake resistant design.

2 COMPARATIVE DESIGN OF A 4-STOREY OFFICE BUILDING

The work was carried out on behalf of the Union of Cement Producers of Greece, in order to investigate cost differences for the construction of a typical 4-storey office building using different structural systems. Three different systems were investigated: Reinforced concrete structure with shear walls and cores as lateral load resisting elements, structural steel system with braced bays using concrete filled hollow sections as lateral load resisting elements and a structural steel system using reinforced concrete shear walls and cores as lateral load resisting elements.

The use of reinforced concrete as bearing structure resulted in a significantly heavier structure. The total weight above ground (including finishings and live load participation) in the seismic combination was approximately double the one in the structural steel solution. This resulted in correspondingly higher seismic forces for the case of the reinforced concrete structure and higher foundation loads. In the specific case study however, these differences did not offset the advantage of low unit price for reinforced concrete construction in Greece and, given also the existence of two reinforced concrete underground levels, the comparison yielded the result that the steel solution is approximately 15% to 18% more expensive than the reinforced concrete solution. In the course of the investigations that were carried out in order to optimize the main structural elements in the different models, it became apparent that the use of concrete filling in hollow steel columns (initially adopted for fire protection) resulted in improvement of the structural response under earthquake loading.

In order to analyze and quantify this impact, a series of parametric analyses were carried out at the end of the comparative study. The results are presented in the following Chapter.

3 INVESTIGATION OF ECCENTRICALLY BRACED BAY UNDER EARTHQUAKE LOADING

3.1 The eccentrically braced bay

Several arrangements of bracings are available and are primarily categorized into concentric and eccentric bracings. The most common types of concentric bracings are: the single diagonal brace, the double diagonal (X) brace, the V brace and the inverted V (Λ) brace. The single diagonal is generally avoided in seismic applications due to lack of symmetry and development of resistance through alternating compression and tension at alternating seismic directions. The double diagonal (X) brace places significant constraints in architectural (doors, windows) and other operational requirements and is also generally avoided in seismic applications. Similarly, the V brace presents difficulties in arranging doors and passageways. This leads to the inverted V (Λ) brace as the most popular arrangement for vertical bracings from the point of view of least operational hindrances. The eccentric arrangement of the inverted V brace is adopted in the present work, since, apart from the obvious improved operational and architectural advantages, it also provides a means for seismic energy dissipation through the bending of the girder, thus leading to possibilities of seismic force reduction or absorption of higher seismic forces than the design earthquake. As far as choice of section types is concerned, hollow sections are customarily adopted for diagonals and columns due to symmetric axial properties (with respect to buckling resistance), while double flange sections (HE-A, HE-B) are usually selected for girders as part of the general slab structural system. At higher load intensities however, hollow welded sections are also more advantageous as bracing girders.

The use of concrete filling of hollow sections is a popular measure for fire protection and it is obvious that it provides an increase in the axial stiffness and

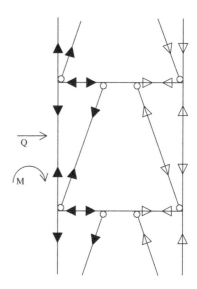

Figure 1. Force flow in eccentric inverted V brace.

strength of the member involved. The following investigations attempt to quantify its impact on the total system response, where concrete filling is considered only for the columns. This assumption is based on the fact that concrete filling of diagonal members induces bending moments which, depending on inclination and size of member, may lead to significant stressing of the member (primary bending and bending due to load eccentricity).

The response of the eccentric inverted V brace is well documented (see for example Stafford Smith & Coull 1991). The resistance to vertical (service) loading is provided by the columns, since the contribution of the diagonals is conservatively ignored. The resistance to horizontal loading is provided by both the columns and the diagonals, while the central part of the girder between the diagonals is subjected to shear and bending.

The lateral stiffness of the system is a combination of flexural stiffness provided by the columns' axial stiffness and shear stiffness provided by the diagonals' axial stiffness and the girder's axial and flexural stiffness. The contribution of each of the parameters (column section area, diagonal section area, girder section area and inertia) to the system's stiffness and strength depends on the bracing's width, height and loading distribution and magnitude. For example, in a 4 m wide and 16 m high brace under uniform horizontal loading, if all elements involved had similar section properties, then the contribution of the columns to the total system lateral flexibility would be in the order of 63% and the contribution of the diagonals would be in the order of 27%. In a wider bay (8 m width) under the same conditions concerning height and loading, the columns' contribution to total lateral flexibility would be reduced to approximately 35%. It follows from these simple observations that the impact of concrete filling on total system response depends on the geometry, loading and section properties.

Earthquake loading however is rarely applied uniformly on a structure. Most modern codes require either an inverted triangular distribution of seismic shear or prescribe dynamic analysis, which for most predominantly cantilever structures yields horizontal load distributions approximating an inverted triangular distribution (resulting from the high contribution of the first mode of vibration). The contributions of the individual elements to total lateral system stiffness are thus modified under inverted triangular loading as follows: For the 4 m wide bay of the previous example, the columns contribute approximately 66% of the total stiffness and the diagonals 25%, while in a 8 m wide bay the contribution of the columns is 36% and the diagonals contribute 24%.

The following section presents the results of a number of analyses carried out on eccentric inverted V bracings of different widths and heights under varying seismic loading using concrete filled and standard hollow sections.

3.2 *Investigation of response under earthquake loading*

The levels of seismic loading have been based on the requirements of the Greek Code for the two major earthquake zones of the country. Zone 1 involves a 0.16 g basic ground acceleration, while Zone 2 involves 0.24 g. The final system acceleration depends on the natural frequency of the system and the behaviour and importance factors. In the present work, no reduction of seismic forces due to dynamic response is used, assuming relatively stiff structures. The importance factor is taken equal to 1.0 and the behaviour factor is assumed equal to 1.5 for all arrangements investigated. The use of a low behaviour factor leads to a practically elastic design. The adoption of a higher behaviour factor could have been justified by the eccentric arrangement of the diagonals and the possibility thus offered for dissipating seismic energy through girder bending. However, due to the inherent uncertainties related to capacity design, i.e. the quantification of the energy dissipated and the behaviour of the plasticized structure in case of aftershocks, a conservative value has been selected for the behaviour factor. The resulting system design accelerations for the two zones are 0.27 g and 0.40 g respectively. These values are derived from the basic ground accelerations (0.16 g and 0.24 g) multiplied by a spectral amplification factor of 2.5 and divided by a behaviour factor of 1.5.

Two types of bracings have been investigated: a 4 m wide brace and an 8 m wide brace. Heights vary

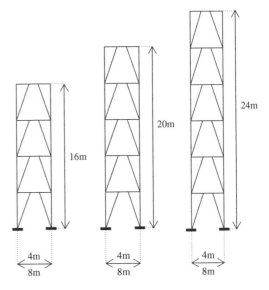

Figure 2. Layout of bracings investigated.

from 16 m to 24 m, while a standard storey height o 4 m has been used. For each case, hollow sections and concrete filled hollow sections have been used. Equivalent reinforced concrete shear walls have also been analyzed for comparison. All braces were loaded with lateral forces resulting from seismic accelerations of 0.27 g and 0.40 g. Vertical loading corresponds to structural self weights (including concrete topping of 12 cm), flooring weight of $1.5 \, kN/m^2$ and live load of $3.0 \, kN/m^2$. An 8 m × 8 m tributary area has been assumed for the vertical load distribution and 30% of the live loads have been taken into account at the seismic combination.

Comparisons have been based on structural equivalence, i.e. satisfaction of strength criteria for the sections used and equal maximum deflections. Additionally, such detailed design criteria as minimum hollow section dimensions for concrete filling and compatibility of column width with girder flanges to facilitate connections have also been used. The results are summarized in tables 1,2,3 & 4.

The correspondence between steel reduction and actual construction cost impact naturally depends on local market values for the materials and labour involved. In the present case study, the cost comparison is based on Greek market prices, where a kg of erected steel costs in the order of 1.5 Euros, while a cubic meter of poured reinforced concrete costs approximately 180 Euros. The reinforced concrete price quoted is based on the absence of formwork and the associated labour cost and on a reinforcement content of up to 1%. Using these figures and the results tabulated above, it follows that the ratio of the additional cost of concrete filling to the cost savings due to steel reduction ranges from

Table 1. Design sizes of bottom members.

| | Size of bottom members | | |
	Columns	Diagonals	Girder
8 m width, 16 m height, 0.27 g			
hollow sections	SHS 400/12.5	SHS 200/16	HE-B450
concrete filling	SHS 300/6	SHS 200/16	HE-B450
8 m width, 16 m height, 0.40 g			
hollow sections	SHS 400/12.5	SHS 250/16	HE-B600
concrete filling	SHS 300/6	SHS 250/16	HE-B600
8 m width, 20 m height, 0.27 g			
hollow sections	SHS 400/12.5	SHS 250/16	HE-B550
concrete filling	SHS 300/6	SHS 250/16	HE-B550
8 m width, 20 m height, 0.40 g			
hollow sections	SHS 400/16	SHS 300/16	HE-B800
concrete filling	SHS 300/12.5	SHS 300/16	HE-B800
8 m width, 24 m height, 0.27 g			
hollow sections	SHS 400/16	SHS 300/16	HE-B800
concrete filling	SHS 300/12.5	SHS 300/16	HE-B800
8 m width, 24 m height, 0.40 g			
hollow sections	SHS 500/25	SHS 350/16	HE-B1000
concrete filling	SHS 400/16	SHS 350/16	HE-B1000
4 m width, 16 m height, 0.27 g			
hollow sections	SHS 400/16	SHS 300/16	HE-B600
concrete filling	SHS 300/12.5	SHS 300/16	HE-B600
4 m width, 16 m height, 0.40 g			
hollow sections	SHS 500/25	SHS 350/16	HE-B650
concrete filling	SHS 400/16	SHS 350/16	HE-B650
4 m width, 20 m height, 0.27 g			
hollow sections	SHS 700/30	SHS 350/16	HE-B600
concrete filling	SHS 500/20	SHS 350/16	HE-B600
4 m width, 20 m height, 0.40 g			
hollow sections	SHS 1000/35	SHS 400/20	HE-B800
concrete filling	SHS 700/25	SHS 400/20	HE-B800
4 m width, 24 m height, 0.27 g			
hollow sections	SHS 1000/40	SHS 400/16	HE-B650
concrete filling	SHS 700/30	SHS 400/16	HE-B650
4 m width, 24 m height, 0.40 g			
hollow sections	SHS 1000/60	SHS 500/20	HE-B900
concrete filling	SHS 900/40	SHS 500/20	HE-B900

Table 2. Weight reduction in structural steel due to concrete filling impact.

| | Height of brace | | |
Brace width, Seismic acceleration	16 m (%)	20 m (%)	24 m (%)
8 m width, 0.27 g	18	15	16
8 m width, 0.40 g	15	17	14
4 m width, 0.27 g	19	21	24
4 m width, 0.40 g	17	30	32

Table 3. Weight participation of the structural elements in the brace (without concrete filling).

| | Brace elements | | |
	Columns (%)	Diagonals (%)	Girders (%)
8 m wide braces			
16 m height, 0.27 g	35%	25	40
16 m height, 0.40 g	29	29	42
20 m height, 0.27 g	29	29	42
20 m height, 0.40 g	34	27	39
24 m height, 0.27 g	35	26	38
24 m height, 0.40 g	39	25	36
4 m wide braces			
16 m height, 0.27 g	46	33	21
16 m height, 0.40 g	51	32	17
20 m height, 0.27 g	68	19	13
20 m height, 0.40 g	77	14	9
24 m height, 0.27 g	69	23	8
24 m height, 0.40 g	88	7	5

Table 4. Weight increase due to increased seismic acceleration level.

| | Weight increase due to design acceleration increase from 0.27 g to 0.40 g | |
Brace width, Brace height	Plain steel (%)	Concrete filled (%)
8 m width, 16 m height	18	22
8 m width, 20 m height	40	37
8 m width, 24 m height	29	33
4 m width, 16 m height	36	39
4 m width, 20 m height	74	55
4 m width, 24 m height	80	60

11% to 18%. In all cases, therefore, the reduction in steel quantities achieved through concrete filling leads to significant reduction in construction cost.

It is evident from the above data that in narrow bracings the participation of the columns to total system response and weight is more significant than in wider bracings. Consequently, concrete filling of columns in narrow bracings tends to be more effective. This observation led to the development and investigation of a 'super column' arrangement described in Chapter 5.

A direct equivalence to reinforced concrete shear walls could not be made for all cases, because of the larger shear wall widths required to develop the seismic resistance. Typically, a 4 m wide shear wall could only accommodate seismic moments resulting from the 16 m high structure under 0.27 g acceleration. For all other heights and seismic accelerations, shear walls wider than 4.0 m would be required. On the other hand, a shear wall of 8 m width would provide

Table 5. Industrial structure: design sizes of bottom members.

	Columns	Diagonals	Girder
Hollow	SHS 1500/30	SHS 1000/30	SHS 1000/30
Filling	SHS 1000/10	SHS 1000/30	SHS 1000/30

sufficient stiffness and strength to carry the seismic forces from all height and acceleration cases considered. Such a wall would however not be utilized in an actual design due to the significant constraints it would place on operation (windows, passageways, etc) and aesthetics.

4 INDUSTRIAL STRUCTURE

Industrial structures usually involve higher service loads and little, if any, reduction of loads in the seismic combination. Consequently, the bracings tend to be higher stressed than in commercial or residential applications. The case study presented here corresponds to a 61 m high preheating tower for a cement producing plant. The structure is 20 m × 20 m in plan and comprises 6 levels. The operational requirements of the structure did not allow any intermediate columns, so the structure was laterally stabilized through eccentric inverted V bracings of 20 m width. The structure is to be built in a Zone 2 seismic area (design acceleration: 0.40 g) and will carry significant service loadings from equipment and heavy refractory lining. The total horizontal seismic shear on each bracing was 25,290 kN. Again, alternative arrangement with hollow sections and concrete filled sections were investigated for the columns.

The results from the design calculations and detailing showed that using concrete filling in the corner columns resulted in a total steel weight reduction of 27% per bracing (i.e. per building side).

5 SUPER COLUMNS

On the basis of the observation that in narrow bracings the column participation to total system stiffness and strength is significant, the following composite arrangement is investigated: Concrete filled large welded hollow sections for columns connected through a reinforced concrete shear wall. The arrangement was investigated for a 24 m high and 4 m wide bay under 0.40 g design acceleration, with service loads as in the corresponding bracing in Chapter 3.

The columns provide approximately 85% of the total system stiffness, while the shear wall contributes the remaining (shear) stiffness. This system is compared to a regular steel structure, a reinforced concrete

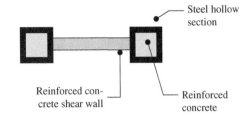

Figure 3. Section of 'super column'.

shear wall and a steel structure with concrete filled sections.

In the case of the reinforced concrete shear wall, the section is not sufficient to carry the seismic moment, so that end columns need to be arranged in order to increase moment capacity. The concrete filled arrangement yields a weight reduction in the order of 45% with respect to the simple steel structure. The combination of the concrete filled hollow sections with a reinforced concrete shear walls leads to a weight reduction in the order of 70% with respect to the simple steel structure and 40% with respect to the concrete filled case. When the cost of the reinforced concrete is weighted in, the final cost difference between simple steel structure and 'super column' is in the order of 45%. The reinforced concrete cost is again based on Greek market prices and is assigned a value of 300 Euros per cubic meter. This value includes the formwork and reinforcement for the shear wall and the reinforcements in the super columns.

Further investigation into this system will be carried out, particularly in the area of shear wall connection and the use of precast elements. It appears however that in tall structures under heavy seismic loading a 'super column' arrangement, i.e. the combination of concrete filled hollow sections with a reinforced concrete shear wall, offers a structural element with high moment capacity and significant lateral stiffness occupying a relatively narrow width.

6 CONCLUSIONS

The use of concrete filled hollow sections offers the advantage of reduction of structural steel weight in bracing arrangements subjected to earthquake loading. The magnitude of weight reduction was found to range in the order of 15% to 30% for ordinary heights and loads. The reduction is more effective in narrower bays as well as in higher structures with significant loading.

For cases of high structures with heavy loads under significant seismic accelerations a 'super column' arrangement is suggested consisting of concrete filled hollow sections coupled to a reinforced concrete shear wall. This arrangement offers the advantage of high moment capacity and lateral stiffness with relatively low construction cost.

REFERENCES

Bergmann, R., Matsui, C., Meinsma, C. & Dutta, D. 1995. Design guide for concrete filled hollow section columns under static and seismic loading. In Comité International pour le Développement et l'Etude de la Construction Tubulaire (ed.), *Construction with hollow sections*. Köln: Verlag TÜV Rheinland GmbH.

Bode, H. 1998. *Euro-Verbundbau. Konstruktion und Berechnung.* Düsseldorf: Werner Verlag.

CEN, 1992. ENV 1992. Eurocode 2. Design of reinforced concrete structures.

CEN, 1993. ENV 1993. Eurocode 3. Design and construction of steel structures.

CEN, 1994. ENV 1994. Eurocode 4. Design and construction of steel – concrete composite structures.

Government of Republic of Greece. Ministry of Public Works. 2000. *Greek code for seismic resistant structures.*

Stafford Smith, B. & Coull, A. 1991. *Tall building structures: Analysis and design.* New York: John Wiley & Sons.

Tubular Structures XI – Packer & Willibald (eds)
© 2006 Taylor & Francis Group, London, ISBN 0-415-40280-8

Low-cycle fatigue behavior of circular CFT columns

G.W. Zhang
Center for Integrated Protection Research of Engineering Structures (CIPRES),
Hunan University, P.R. China

Y. Xiao
Department of Civil Engineering, University of Southern California, USA;
Cheung Kong Scholar, Hunan University, P.R. China

Sashi K. Kunnath
Civil & Environmental Engineering, University of California at Davis, USA

ABSTRACT: Under seismic actions, the induced damage in critical regions of a CFT column could be attributed to cumulative damage caused by repeated loading with different displacement levels after yielding. To develop a more rational damage criterion for circular CFT columns subjected to a series of earthquake excitations, an experimental study is undertaken to investigate the cumulative damage and relationship between low cycle fatigue life and displacement amplitude. The results of the testing containing both standard cyclic loads and constant amplitude tests are summarized in this paper. The low-cycle fatigue characteristics of typical flexural CFT columns are discussed. For comparison, a confined CFT column (CCFT) confined with steel collar was also tested, which showed improved fatigue life. A fatigue life expression is developed that would be used in damage-based seismic design of circular CFT columns.

1 INTRODUCTION

During the past decade, CFT structures have been accepted as a new structural component and used in tall buildings as well as arch bridges. According to the previous work carried out by several researchers, CFT columns have the following advantages: (1) the steel tube provides efficient confinement to the concrete core, developing full composite capacity of the column. (2) Concrete core delays the steel tube from local bulking and concrete spalling is also prevented by the confined steel tube, contributing to the strength and ductility and improving the seismic energy dissipation. (3) The potential economical advantages of CFT columns in tall buildings have long been recognized (Tarsics 1972) and the concept could lead to 60% total savings of steel in comparison to a steel system (Zhong, S.T. 1989).

Under earthquake loading, the components of a structure resist cyclic loading with different displacement levels, undergoing yielding to dissipate energy and dampen the seismic effects. To investigate the seismic behavior of CFT members, numerous cyclic tests had been carried out in the past decade. All of them were focused on standard loading to obtain the cyclic capacity without considering the effects of amplitude

and number of cycles on damage accumulation. In fact, many structures and bridges have collapsed after peak acceleration with few inelastic cycles suggesting that there is a need to develop fatigue-life relationships. For this purpose, low-cycle fatigue experiments were carried out on two types of CFT columns subjected to constant amplitude.

A new CFT column system referred to as confined CFT columns (CCFT) conceived by the second author (Xiao et al. 2003) following extensive previous research on CFT columns and tubed columns is also included in this study. For CCFT columns, additional transverse reinforcement is designed at the potential plastic hinge regions to achieve additional seismic performance. Based on fundamental mechanics, the design concept is aimed at controlling the local buckling of the steel tube and more efficiently confining concrete in the critical regions of a CFT column. A CCFT column confined with steel collar was also tested with constant amplitude for comparison.

In order to meet the needs of elastic-plastic analysis for earthquake resistant design, many hysteretic models for CFT columns have been established. However, none of these models explicitly consider the effects of cumulative damage on component behavior under cyclic loading. Characterizing damage based on changes in the effective elastic modulus and strength degradation during loading and unloading process is usually more apparent only at advanced damage states. Based on the cumulative damage characteristics of CFT columns through low-cyclic fatigue test, a damage criterion was developed.

2 DETAILS OF EXPERIMENTAL PROGRAM

2.1 Design of test specimens

To ensure the characteristics and dimensions of the tested specimens represent real prototypes, the model CFT columns were designed in full scale. Two types of circular CFT columns with a total of nine specimens were considered to simulate typical columns in multi-story buildings in seismic regions.

Type I: The diameter and thickness of steel tube were 325.0 and 5.78 mm with a D/t ratio of 56. Type II: The diameter and tube thickness were 336.0 mm and 3.0 mm with a D/t ratio of 112. The height of the columns from the point of lateral loading to the top of the footing was 1,500 mm. All specimens were designed and constructed with a stiff stub footing of 2000 mm × 700 mm × 420 mm. Stub footings were heavily reinforced to eliminate any premature failure during testing. For Type I, the tubes were first welded to an end plate and stiffeners for the bottom to ensure a strong connection with the stub footing. For Type II, a local manufacturer rolled the circular steel tubes from steel plates into cylindrical shells and then the seams were welded. The tubes were first welded to an end plate and four stiffeners at the bottom to ensure a strong connection with the stub footing. The tubes were then shipped to the laboratory for preparing the steel caging for the footing and top stub as well as casting the concrete. The basic detail of the CFT columns is illustrated in Figure 1.

As shown in Figure 2, three circular steel collars were fixed around the steel tube (labeled as SC-1 to SC-3) along the column from concrete foundation. The thickness of these collars is 7 mm with a height of 100 mm. There was 10 mm space between two adjacent collars. The gap in the steel tube and the collars was about 1 mm to yield a cushion effect based on basic mechanical testing studies.

2.2 Material properties

The compressive strength of concrete, f'_c, in the steel tube was 24.8 MPa for Type I and 39.1 MPa for Type II. This is based on compression tests on 150 mm cubic specimens, however converted into cylinder strength properties by applying a modification factor of 0.8. The mixture proportions per cubic meter of concrete for Type I were 205 kg water, 394 kg Portland cement, 1125 kg coarse aggregates, and 633 kg fine aggregates,

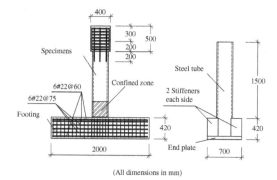

(All dimensions in mm)

Figure 1. CFT specimen details.

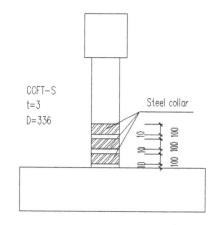

Figure 2. Steel collars confined CFT column.

and for Type II consisted of 190 kg water, 425 kg Portland cement, 1211 kg coarse aggregates, and 570 kg fine aggregates. The water-to-cement ratio for Type I was 0.52 and Type II was 0.45 with a maximum dimension of the coarse aggregate approximately 20 mm. The steel tube for Type I specimens had an average yield strength of 272 MPa with an elongation of 24.1%, and for Type II the yielding strength and elongation of the steel tube and the steel collar for the CCFT specimens were all 248.9 MPa and 31.4%. The longitudinal reinforcement in the column head and footing was Grade III reinforcing bars with average yield strength of 361 MPa and the transverse reinforcement was Grade II with average yield strength of 243 MPa.

2.3 Test setup

The tests were performed in the structural testing facility at the College of Civil Engineering in Hunan University. All the model columns were tested using the test setup shown in Figure 3. Lateral force was applied to the top of the model column to simulate a vertical cantilever through a 630 kN capacity

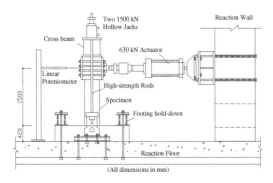

Figure 3. Test setup.

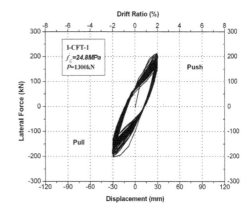

Figure 4. Response of I-CFT-1 at 2% drift.

pseudo controlled hydraulic actuator. Constant axial load could be applied to the column through post-tensioning two 50 mm diameter high-strength steel rods using two 1,500 kN capacity hydraulic hollow jacks. The forces of the rods were transferred to the model column by a cross beam mounted on top of the load stub. In order to eliminate the bending of the high-strength rods, a specially designed pin device was connected to the lower end of each rod. The imposed lateral displacement was measured by both the displacement transducer of the actuator and a linearly variable differential transformer (LVDT). The built-in load cell of the actuator recorded the corresponding lateral force. Forty-seven electrical resistance strain gauges were affixed on the surface of the steel tube to allow assessment of the moment, the shear force and moment transferred from the steel tube to the top and bottom plates, and flexure and axial forces at various locations in the steel tube and in the foundation.

3 RESULTS OF EXPERIMENTAL TESTING

3.1 Specimen configuration and details

For Type I, the five tested specimens were labeled as I-CFT-1 to I-CFT-5 and the remaining four specimens of Type II were labeled as II-CFT-1 to II-CFT-3 for ordinary CFT columns and CCFT-S for the steel collar confined CFT column. For all the tests, the axial load of 1,300 kN was first applied to the column and held constant for the duration of the test. I-CFT-1, I-CFT-2, I-CFT-3, II -CFT-1 and II-CFT-2 were tested under constant amplitude loading. For calibrating fatigue-based damage models of CFT columns, these tests provided the basis for developing a fatigue-life expression for the specimen. I-CFT-4 and I-CFT-5 were subjected to different amplitudes for comparison and measurement, considering the effects of loading path on cumulative damage. II-CFT-1 was subjected to standard loading. CCFT-S was tested under constant amplitude for comparing its performance with the response of CFT columns.

3.2 Low-cycle fatigue test of CFT columns

Figure 4 shows the hysteretic lateral force-deflection curves of specimen I-CFT-1 subjected to constant amplitude cycles of ±30 mm corresponding to 2 percent lateral drift. In the first cycle yielding occurred as the deflection increased beyond 1 percent drift and the force–deflection curve began to deviate from the straight elastic line. The data from some of the strain gauges at about 500 mm height from the stub footing on the steel tube, showed the steel tube yield in both tension and compression. Slight buckling at about 70 mm height from the footing of column on the compression side of the column was observed when the deflection reached 2% drift and on reverse excursion, it was also seen on the compression side. The maximum applied horizontal force was reached in this cycle. With additional cycling, the buckling progressed on both sides and the plastic hinge area formed at about 60 mm height from the base of concrete foundation with the lateral strength dropping slowly and the steel remaining ductile in all subsequent cycles. Pinching in the hysteretic curve was not observed before the steel tube cracked. At the 44th cycle, necking started to develop through the plastic hinge area of the steel shell and cracking was observed on both sides at the 45th cycle without a sudden drop in strength. The crack appeared to have penetrated through the tube as the testing went on. Pulverized concrete spilled out through the crack in the buckled region as the crack width increased.

Specimen I-CFT-2 was tested under repeated loading at a constant amplitude of ±60 mm, corresponding to a drift of 4 percent. The force-deflection result obtained during testing is shown in Figure 5. In the first cycle, lateral force reached the maximum load at a deflection of 2.5 percent drift, and then the force became steady as deflection increased. Data from strain gauges showed that the steel tube at about 700 mm height from the base of concrete foundation yielded. Obvious buckling was visible on both sides

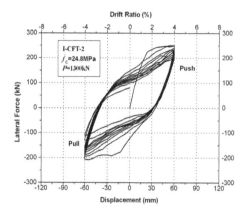

Figure 5. Response of I-CFT-2 at constant amplitude 4% drift.

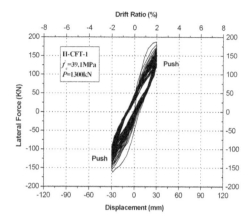

Figure 7. Low-cycle fatigue test of II-CFT-1 at 2% drift.

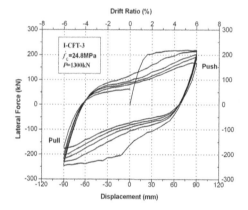

Figure 6. Repeated cyclic response of I-CFT-3 at 6% drift.

and the plastic hinge area formed at about 310 mm height from the base of the concrete foundation. After the second cycle, buckling continued to progress on both sides with increasing cycles and cracks in the steel tube surface were obvious. Pinching in the hysteretic curve became more and more pronounced. Necking developed through the plastic hinge area on the steel shell and significant cracking was observed on both sides at the 10th cycle with a sudden drop in strength.

Specimen I-CFT-3 was subjected to repeated loading at a constant amplitude of ±90 mm, corresponding to 6 percent drift. The experimentally obtained hysteretic force-deflection behavior is shown in Figure 6. Similar to the second specimen, lateral force increased very slowly as the deflection exceeded 3 percent drift in the first cycle. The hysteretic curve had slight pinching. Data from strain gauges at about 700 mm height from the base of concrete foundation showed yielding of the steel shell. Serious buckling was most obvious in the plastic hinge area on both sides at about 280 mm height. With continued loading, the buckling

progressed on both sides and paint on steel tube surface around plastic hinge area started to peel. Pinching in the hysteretic curve became more obvious. At the 5th cycle, necking developed through the plastic hinge area in the steel tube at about 4 percent drift and cracking was visible on both sides with a sudden drop in strength as the deformation increased.

Figure 7 shows the hysteretic lateral force-deformation curves and buckling of specimen II-CFT-1 subjected constant amplitude cycles of ±30 mm corresponding to 2 percent lateral drift. In the first cycle yielding occurred as the deflection increased beyond 1 percent drift and the force–deflection curve began to deviate from the straight elastic line. Data from some of the strain gauges on the steel tube showed that the steel tube yielded in both tension and compression. Slight buckling at about 80 mm height from the footing of column on the compression side of the column was observed when the deflection reached 2% drift and on the reverse excursion, it was also seen on the opposite side. The maximum applied horizontal force was reached at the peak deformation in this cycle. Pinching in the hysteretic curve was visible. With continued loading, buckling grew on both sides and the plastic hinge area formed at about 70 mm height from the base of concrete foundation with the lateral strength dropping slowly and the steel remaining ductile in all subsequent cycles. Pinching in the hysteretic curve was significant before the steel tube cracked. At the 41st cycle, necking started to develop through the plastic hinge area of the steel shell and then a visible crack appeared along the necking line on both sides in the 42nd cycle with an obvious drop in strength. Pulverized concrete spilled out through the crack in the buckled zone as it opened wider at the peak drift.

Specimen II-CFT-2 was tested under repeated loading at a constant amplitude of ±60 mm, corresponding to a drift of 4 percent. The force-deflection response

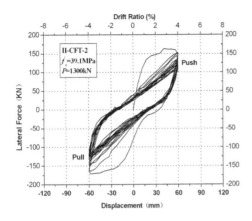

Figure 8. Low-cycle fatigue test of II-CFT-2 at 4% drift.

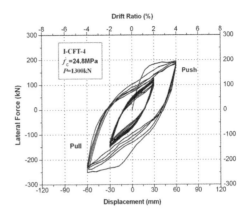

Figure 9. Response of I-CFT-4 at 4% and 2% drift.

obtained during testing is shown in Figure 8. In the first cycle, the lateral force reached the maximum load at the peak deformation. Data from strain gauges showed that the steel tube yielded about 350 mm from the base of concrete foundation. Buckling of the steel tube was visible on both sides and the plastic hinge area formed around that area. There was significant stiffness degradation in the second cycle compared to the first cycle. Buckling increased progressively with increasing cycles and cracks in the paint on the steel tube surface were obvious. Hysteretic curve pinched seriously at entirely test. Necking of the tube was observed in the plastic hinge area during the 15th cycle and at peak deformation, the crack in the steel tube opened suddenly. When the peak drift approached 4%, the load capacity began to drop significantly and the testing was stopped.

3.3 Load path effects

In order to consider the effect of load path on column capacity, specimen I-CFT-4 was first subjected to cyclic loading at a constant drift of 4 percent and then after the lateral force capacity decreased by 10 percent, it was tested under 2 percent drift till failure. The hysteretic curve obtained during testing is shown in Figure 9. Altogether the specimen was subjected to 5 cycles at 4 percent drift and 25 cycles at 2 percent drift. Test results showed that in the first five cycles the behavior of the specimen was similar to I-CFT-2. Buckling was visible on both sides of the steel tube at about 70 mm height from the base. When testing resumed at constant amplitude cycles of 2 percent drift, the strength and stiffness degradation of the column was more significant at this amplitude of ±30 mm compared to column I-CFT-1. As the testing progressed, buckling did not progress much further at the reduced amplitude and no pinching of the hysteretic response was observed. The height of plastic

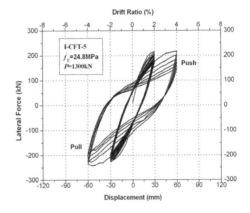

Figure 10. Response of I-CFT-5 at 2% and 4% drift.

hinge length at the foot of tested column was about 220 mm.

For comparison, specimen I-CFT-5 was first subjected to constant amplitude cycles of 2 percent drift. After 25 complete cycles at ±30 mm amplitude, the drift was exceeded to ±60 mm and tested till failure, which consisted of 6 complete cycles, as shown in Figure 10. In the 7th cycle at 2 percent drift, minor cracking from the brazing seam was observed at the bottom of the column near the Reaction Wall.

An important observation in the hysteretic response was that there was an obvious drop in strength in the push direction. As the test went on, the strength dropped slowly in both directions with imperceptible crack growth. In the 25th cycle, the buckling was visible on both sides without any pinching in the force-deformation curve and the plastic hinge area extended to about 50 mm to 70 mm height. When the imposed drift amplitude exceeded 4 percent, the buckled profile grew on both sides and cracking extended along the seam to more than 20 mm, resulting in increasing strength loss in the push direction compared to

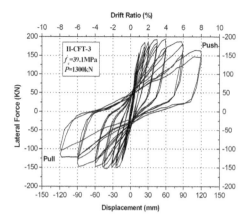

Figure 11. Standard cyclic loading of II-CFT-3.

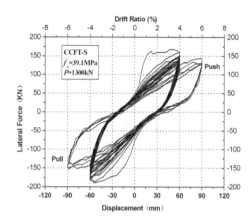

Figure 12. Low-cycle fatigue test of CCFT-S at 4% drift.

the corresponding strength loss in the pull direction. Pinching in the hysteretic curve was more pronounced than the previous case. In the 6th cycle at 4 percent drift, the crack length grew to 42 mm with pulverized concrete spilling out of the crack and necking developed through the plastic hinge area on the opposite side at about 38 mm height from the stub footing.

During the standard cyclic loading, three single cycles were initially applied to specimen II-CFT-3 corresponding to an increment of 0.25% peak drift, and then three repetitive loading cycles were applied at respective drifts of 1%, 1.5%, 2%, 3%, 4%, 6%, and 8%. This loading procedure was continued till failure of the specimen which consisted of the splitting of the steel tube accompanied with an obvious drop in strength. Figure 11 shows the observed hysteretic lateral force-deflection curves of the specimen. During testing, no evidence of yielding or damage could be seen until the third positive cycle at 0.75 percent drift at a peak deformation of 12 mm. During this cycle, the strain gauges on the steel tube had reached yield, and the force-deflection plot began to deviate from the straight elastic line. For the next three cycles with peak deformation of 15 mm corresponding to 1% drift, most of the strain gauges on the steel tube, near the concrete foundation, reached yield in both tension and compression and slight buckling on both sides of the tested column at about 60 mm height from the foundation was observed during the third cycle. With increased deformations, the buckling progressed more aggressively and was clearly discernable on both sides around the plastic hinge region.

At 3% drift, the maximum lateral force was reached and pinching in the hysteretic curve was obvious. But the strength didn't drop until the peak deformation reached 6% drift. In the second cycle at a peak deformation of 90 mm, necking of the steel shell was observed in the plastic hinge area along with visible cracking on both sides at about 25 mm height from

the concrete foundation supplemented by a drop in strength. The crack began to penetrate the tube as the test went on. At 8% drift, pulverized concrete spilled out through the crack in the buckle.

3.4 Low-cycle fatigue tests of CCFT column

Specimen CCFT-S was subjected to repeated loading at a constant amplitude drift of 4 percent followed later by a few cycles at 6%. The hysteretic curve obtained during testing is shown in Figure 12.

During the very first cycle buckling of the steel tube was visible and the gap between steel tube and steel collar SC-1 was closed. Data in the strain gauges showed that SC-1 was in tension. As cycles were added, the buckled profile extended up the column height and the remaining two collars were put into action. At the peak deformation during the 18th cycle, a 20 mm long crack appeared along the space between SC-1 and SC-2 without visible drop in the loading capacity. Compared to the performance of specimen II-CFT-2, three more cycles were gained and there was no obvious strength and stiffness degradation in this specimen (CCFT-S). The column was then subjected to constant amplitude drift of 6 percent drift; the crack began to penetrate through the tube along the space between the steel collars with gradual loss of loading capacity. After three cycles at 6 percent drift ratio, the crack on "pull" side of the specimen was more than 45 mm long, and the tested was stopped.

4 FATIGUE LIFE RELATIONSHIPS

The quantification of damage on structures and their components is an important aspect of seismic evaluation and design. Since the numerous factors associated with seismic damage are complex and interrelated, different damage index expressions

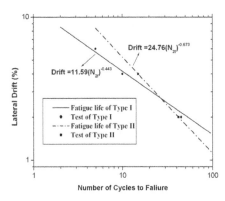

Figure 13. Fatigue life relationship for CFT columns.

Table 1. The number of complete cycles to failure.

Drift ratio	1%	1.5%	2%	3%	4%	6%	8%
N_{2f} of Type I	252	101	45	21	10	4.5	2.5
N_{2f} of Type II	118	65	42	23	15	8	5

Table 2. Damage of tested specimens using Miner's hypothesis.

No.	Specimen label	D
1	I-CFT-1	1.0
2	I-CFT-2	1.0
3	I-CFT-3	1.0
4	I-CFT-4	1.06
5	I-CFT-5	1.16
6	II-CFT-1	1.0
7	II-CFT-2	1.0
8	II-CFT-3	1.05
9	CCFT-S	1.2

and damage modeling schemes have been proposed. Among them, a common yet simple approach is the Coffin-Manson relationship and Miner's rule for damage accumulation. The Coffin-Manson hypothesis is given by an expression of the following form:

$$\varepsilon_p = \varepsilon_f{}'(N_{2f})^c \qquad (1)$$

where ε_p is the plastic strain amplitude, c and $\varepsilon_f{}'$ are material constants to be determined from fatigue testing and N_{2f} is the number of complete cycles to failure. A similar formula more suitable for developing damage in structural components was suggested by Krawinkler et al.:

$$N_{2f} = C^{-1}(\Delta\delta_p)^c \qquad (2)$$

where $\Delta\delta_p$ is the plastic deformation and c and C are parameters to be determined experimentally.

The results of the constant amplitude tests were utilized in a curve-fitting exercise using a format similar to Equation (1) that yielded the following relationships:

$$\delta, \text{percent} = 11.59(N_{2f})^{-0.443} \qquad (3)$$

$$\delta, \text{percent} = 24.76(N_{2f})^{-0.673} \qquad (4)$$

where δ is the lateral drift. Equation (3) is for CFT Type I specimens and Equation (4) is for Type II columns. Figure 13 shows the test results of the two column types and fatigue life expressions for the CFT columns with and without added confinement.

The two fatigue-life equations were utilized to predict the damage of the two types of CFT columns subjected to all the different loading scenarios. For all tests, significant departure from the elastic curve occurred at approximately 1% drift, and data in the strain gauges at the plastic region also showed yielding of the steel tube at this stage. So the yield deformation

was taken as 1% drift. It is assumed that deformation below 1% drift induced no damage to the specimen. The calculated fatigue lives of the CFT columns based on this assumption and the fatigue life expressions (Equations 3 and 4) are illustrated in Table 1.

The second hypothesis is Miner's rule of linear damage accumulation, which postulates that the damage per excursion is $1/N_f$, and that the damage from excursion with different plastic deformation ranges could be combined linearly. It is computed as follows:

$$D = \sum D_i = \sum n_i/N_i = 1 \qquad (5)$$

where D is the damage index, D_i is the damage index at a certain drift, N_i and n_i are the number of complete cycles to failure and the number of complete cycles contained at the drift during the complete loading process respectively.

With this formula, the damage indices of the nine tested specimens were computed as illustrated in Table 2. For Type I, the fatigue life expressions and the application of Miner's Rule produced very accurate results with the exception of specimen I-CFT-5. This is because this specimen sustained one additional cycle at 4 percent drift than specimen I-CFT-4. The reason is that the cracking of the brazing seam decreased the loading capacity and energy absorption of CFT columns during the cyclic loading, delaying its failure. For Type II, the formula produced reasonably accurate results. The fatigue life of specimen CCFT-S showed that confinement to CFT columns around the plastic region could improve the ductility of CFT columns and delay cracking of the steel tube, prolonging their fatigue lives. Additional testing is underway to extend the findings and applicability of this research.

5 CONCLUSIONS

Based on low-cycle fatigue testing, fatigue life expressions were obtained for CFT columns. Compared with test results, the expressions were reasonably accurate and there is good reason to believe that it will be applicable to other components of structures and bridges. Compared with other damage models, it has the advantage of being simple to apply and also has the great advantage that it permits performance assessment for any arbitrary loading history. Test result also showed that CCFT columns improved the seismic performance of normal CFT columns.

ACKNOWLEDGEMENTS

The research reported in this paper was sponsored by the Center for Integrated Protection Research of Engineering Structures (CIPRES) of the Hunan University, under the Project 985-CIPRES and the Cheung Kong Scholarship.

REFERENCES

Coffin, L.F., Jr. 1954. A study of effects of cyclic thermal stresses on ductile metal. *Transactions of the American Society of Mechanical Engineers*, New York, 75, 1954, pp. 931–950.

El-Bahy A., Kunnath S.K. & Stone W. et al. 1999. Cumulative seismic damage of circular bridge columns: Benchmark and low-cycle fatigue tests. *ACI Struct. J.*, Vol. 96(4), 633–641.

El-Bahy A., Kunnath S.K. & Stone, W. et al. 1999. Cumulative seismic damage of circular bridge columns: Variable amplitude tests. *ACI Struct. J.* Vol. 96(5), 711–719.

Furlong, R.W. (1968). Design of steel-encased concrete beam-columns. *J. Struct. Div.* ASCE, 94(1), 267–281.

Krawinkler H. et al. 1983. Recommendations for experimental studies on seismic behavior of steel components and materials. *John Blume Center Report* No. 61, Stanford University, 1983.

Manson, S.S. 1953. Behavior of materials under conditions of thermal stress. Ann Arbor, Mich., *Engng. Frac. Mech*, 9–75.

Miner, M.A. 1945. Cumulative damage in fatigue. *J. Appl. Mech.*, 12 (3): A159–A164.

Tarics, A.G. 1972. Concrete-filled steel columns for multi-story construction. *Modern Steel Constr.*, 12: 12–15.

Xiao Y., He W.H. & Mao X.Y. et al. 2003. Confinement Design of CFT Columns for Improved Seismic Performance. *Proceedings of the International Workshop on Steel and Concrete Composite Construction (IWSCCC-2003)*, Taipei, 217–226.

Xiao Y., He W.H. & Mao, X.Y. 2004. Development of Confined Concrete Filled Tubular (CCFT) Columns. *Journal of Building Structures*, 25(6), 59–66.

Xiao Y., He W.H. & Choi, K.K. 2005. Confined Concrete-Filled Tubular Columns. *ASCE Journal of Structural Engineering*, 131(3), 488–497.

Zhong S.T. 1989. The development of concrete filled tubular structures in China. *Proceeding of the International Conference on Concrete Filled Steel Tubular Structures*, China.

Tubular Structures XI – Packer & Willibald (eds)
© 2006 Taylor & Francis Group, London, ISBN 0-415-40280-8

Seismic performance of Confined Concrete Filled Tubular (CCFT) columns

W.H. He
Center for Integrated Protection Research of Engineering Structures (CIPRES), Hunan University,
Changsha, Hunan Province, P. R. China

Y. Xiao
Department of Civil Engineering, University of Southern California, Los Angeles, California, USA;
Cheung Kong Scholar, Hunan University, Changsha, Hunan Province, P. R. China

ABSTRACT: This paper summarizes experimental results of an innovative concrete filled steel tubular (CFT) column system, named as confined CFT or CCFT, for improved seismic design of steel and concrete composite structures. Based on fundamental mechanics, the design concept is aimed at controlling the local buckling of the steel tube and confining the concrete in the potential plastic hinge regions of a CFT column. All the specimens tested can be considered as full-scale to two-third scale models of the structural components generally used in practice. The results showed that composite columns had very high ductility and capacity of energy absorption because of the interaction between the steel tube, concrete and the additional transverse confinement. Experimental tests validated the concept of the CCFT. As demonstrated from the results of seismic loading tests, the new type of CFT column system can provide excellent seismic performance.

1 INTRODUCTION

A new concrete filled tubular (CFT) column is conceived by the second author from extensive previous research on CFT column and the tubed column. The two different but related systems are conceptually shown in Fig. 1.

In a conventional CFT column system, concrete is filled in steel tubes which typically continue throughout several stories or the full-height of a building. The steel tube is expected to carry stresses in longitudinal direction caused by axial loading and moments, as well as transverse stresses caused by shear and the internal passive pressure due to concrete deformation, i.e. the confining stress (e.g. Zhong, S.T. 1987, Cai, S.H. 1989). Due to the fact that a steel tube is used as longitudinal reinforcement to resist axial force and moment, when steel tube yields under excessive longitudinal stresses due to moment or axial load, its transverse confinement (particularly in terms of stiffness) to internal concrete is drastically reduced. Besides, local buckling of steel tube upon cyclic loading also hinders the seismic behavior of a CFT column (Sakino, K. 1981).

The concept of using steel tube as primarily transverse reinforcement for reinforced concrete (RC) columns was first studied by a research group lead by Tomii. The terminology of "tubed column" first adopted by Tomii et al., refers to the function of the tube as that of the hoops in a hooped RC column. Thus, the composite action between the steel tube and concrete is primarily expected in transverse direction only for a tubed column, however, in both longitudinal and transverse directions for a conventional CFT column (Tomii, M., Sakino, K. Xiao, Y. etc. 1985–1987).

One of the key features of a tubed RC column is to properly detail the tube to avoid or reduce direct transfer of the longitudinal stresses into the tube, which is designed primarily as transverse reinforcement. This is achieved by providing a gap between the tube and the beam or footing at the ends of a column, as shown in Fig. 1(b). The tubed columns may only be used for the critical columns such as the short columns, wall boundaries or columns in lower stories of a structure. The concept of the tubed column was validated through testing model columns under constant axial load and cyclic shear in double-curvature condition.

Jacketing retrofit of existing deficient RC columns can also be considered as the application of tubed column concept (Priestley, M.J.N. 1994, Xiao, Y. 1997, 2000). For most cases, the jacket is used to provide additional transverse reinforcement to increase the capacity and to improve the ductility of an existing column. This is achieved by welding steel shells or wrapping FRPs to enclose an existing column to form a tubed system (Xiao, Y. 2000, 2001, 2003, Teng, J.G. 2000).

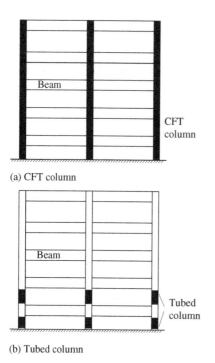

(a) CFT column

(b) Tubed column

Figure 1.   Two different tubular column systems.

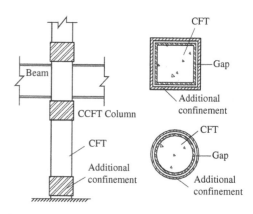

Figure 2.   Proposed CCFT column system.

2   PROPOSED CCFT COLUMN

A new CFT column system is conceived by the second author following extensive previous research studies on CFT columns and steel or FRP tubed columns (Xiao, Y. 2003, 2005). In the new CFT column, additional transverse reinforcement is designed for the potential plastic hinge regions, as illustrated in Fig. 2, to achieve improved seismic performance. Based on fundamental mechanics, the design concept is aimed at controlling the local buckling of the steel tube and more efficiently confining concrete in the critical regions of a CFT column. For this reason, the proposed new CFT column system can be named as confined CFT column system, and is referred to as CCFT hereafter. Though current research primarily concerns with building structure design, it is the authors' attention to apply the concept of CCFT in other structures, such as bridge piers and CFT arch bridges in near future.

The CCFT column is expected to overcome many disadvantages of the conventional CFT column and to provide the ideal choice for structural design of tall buildings or bridges, particularly for seismic regions.

In a CCFT column the functions of the through-tube (similar to the tube in a conventional CFT column) and the additional transverse reinforcement are separated, with the former mainly resists longitudinal stresses caused by axial load and moment as well as shear in the middle portion of the column, whereas the additional reinforcement mainly enhances the potential plastic hinge regions.

The additional transverse reinforcement can effectively prevent or delay the local buckling of the through-tube in the plastic hinge regions of a CFT column, thus improving its seismic performance with stable load carrying capacity and ductility.

The concrete in the column plastic hinge regions can be more efficiently confined by the additional transverse reinforcement, and as a consequence, the ductility of the column can be assured.

Due to the additional transverse confinement, the through-tube in the compression zone of the plastic hinge region is subjected to biaxial compressive stresses (strictly speaking, should be triaxial compression). This is a more efficient working state for steel tube as compared with the combination of axial compression and transverse tension, which is the working stress state of the tube in compression zones of a conventional CFT column.

In a conventional CFT column, in order to prevent the local buckling of the steel tube in the plastic hinge regions, relatively thicker steel tube is required, and typically such thickness is provided throughout the length of the column, particularly for columns with a rectangular section. On the other hand, in a CCFT column, the through-tube is designed mainly as longitudinal reinforcement to resist axial load and moment, and is enhanced transversely by the additional transverse reinforcement in the potential plastic hinge regions. The secondary function of the through-tube is to resist shear in the middle portion of the column, which can typically be achieved by using the same thickness of the through-tube. Thus, it is expected that even with the addition of the transverse reinforcement for the potential hinge regions, the total amount of steel usage in a CCFT column may be less than the identical CFT column.

Apparently, since the additional transverse reinforcement is only provided for the potential plastic hinge regions of the columns, a structural system using CCFT columns remains essentially the same as CFT structure. Thus, design details such as connections developed for conventional CFT structures are still applicable to the proposed CCFT system.

For the design of the additional transverse reinforcement, the followings are considered but not limited as potential options and are examined in the research program:

i. Additional steel shells or tubes are welded to the potential plastic hinge regions of the through-tube, schematically shown in Fig. 2;
ii. Angles, small tubes or pipes, etc., which have larger transverse stiffness and resistance can be used, similar as those proved to be effective to enhance the retrofitting efficiency of rectangular jacketing by the first author.
iii. Fiber-reinforced plastics (FRP) can be used to create a new composite system, as shown in this paper;
iv. Reinforced concrete shells that can also serve as the fireproof for the steel tube.

A joint research program designed to develop the design methodology of structures with CCFT columns is currently being conducted at the Hunan University and the University of Southern California. This paper describes the concept of CCFT and the experimental validation through simulated seismic loading tests. Prior to the seismic loading tests, the authors also conducted axial compressive loading test on CCFT cylinder specimens and validated the basic mechanical behavior of CCFT columns.

3 SIMULATED SEISMIC TESTING

3.1 Specimen design

Seven large-scale model CFT and CCFT columns were designed to simulate typical columns in multi-story buildings in seismic regions. The testing matrix is shown in Table 1 and the basic details of the CFT columns are illustrated in Fig. 3. All specimens were circular and five of them were with a diameter of 336 mm and two specimens were with a diameter of 325 mm. Height of the columns was 1500 mm from the point of lateral loading to the top of the footing. Thickness of steel tube was 3 mm for five those with larger diameter and 6 mm for the others. The specimens were designed and constructed with a stiff stub footing of 2000 mm × 700 mm × 420 mm. The stub footings were heavily reinforced to eliminate any premature failure during testing.

There was no any special confinement for model columns C1-CFT3 and C6-CFT6 for providing

Table 1. Seismic loading test matrix.

Specimen	Diameter (mm)	Steel tube D/t ratio	Confinement	Concrete strength f_c' (MPa)
C1-CFT3	336	112	–	39.1
C2-CCFT3			CFRP	39.1
C3-CFT3			GFRP	39.1
C4-CCFT3			Steel collar	39.1
C5-CCFT3			RC cover	39.1
C6-CFT6	325	54	–	28.0
C7-CCFT6			CFRP*	28.0

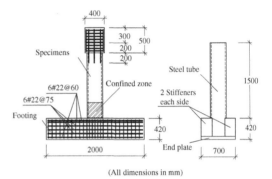

(All dimensions in mm)

Figure 3. Specimen details.

benchmark data of seismic behavior of conventional CFT columns. Four layers of CFRP wrapping and five layers GFRP wrapping were used to provide additional confinement to the potential plastic hinge regions of columns C2-CCFT3, C7-CCFT6 and C3-CCFT3, respectively. The length of the confined zone was 300 mm, close to the section diameter. Prior to applying the FRP wrapping, a layer of 1 mm thick foam tapes were affixed to the surface of the tube to yield a cushion effect based on the basic mechanical testing studies, shown in Fig. 4. Model column C4-CCFT3 was confined in the potential plastic hinge region by spot welded steel collars with a thickness of 8 mm Spiral hoops with a diameter of 8 mm were used to reinforce the additional concrete cover when confining model column C5-CCFT3.

3.2 Specimen construction

Material properties for all specimens are summarized in Table 1 and Table 2. The mixture proportions per cubic meter concrete were 190 kg water; 425 kg Portland cement; 1211 kg coarse aggregates; and 570 kg fine aggregates. The concrete compressive strength values shown in Table 1 are based on compression tests on 150 mm cubic specimens, however converted into the cylinder strength, fc', by a factor of 0.8. The Chinese standard Q235 grade steel with average yield

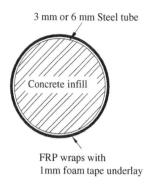

3 mm or 6 mm Steel tube

Concrete infill

FRP wraps with
1mm foam tape underlay

Figure 4.   Confinement detail of FRP confined specimens.

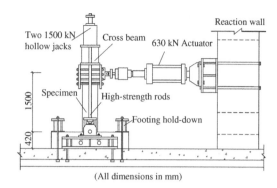

(All dimensions in mm)

Figure 5.   Test set-up.

Table 2.   Material performance of confinements.

Specimen	Confinement	Yield strength (MPa)	Tensile strength (MPa)	Young's modulus (MPa)/	Thickness (mm/ layer)
	Confinement element				
C2-CCFT3	CFRP	—	2500	2.1×10^5	0.22
C3-CCFT3	GFRP	—	469	3.3×10^4	0.8
C4-CCFT3	Steel collar	255	372	2.0×10^5	8
C5-CCFT3	RC*	241	317	2.0×10^5	8
C7-CCFT6	CFRP	—	2500	2.1×10^5	0.22

* Parameters in this line refer to the properties of spiral hoops. Properties of the concrete cover were similar with the concrete infilled.

strength of 235 MPa was used for steel tube in all the columns. The tensile strength and the modulus of the unidirectional CFRP sheets used for confining the C2-CCFT3 and C7-CCFT6 were 2500 MPa, and 210 GPa, respectively, based on 0.22 mm/layer thick flat coupon tests. The tensile strength and the modulus of the unidirectional GFRP sheets used for confining C3-CCFT3 were 469 MPa, and 33 GPa, respectively, based on 0.8 mm/layer thick flat coupon tests. The yield strength and the tensile strength of the steel collar used in confined C4-CCFT3 were 255 MPa and 372 MPa, respectively. The yield strength and the tensile strength of the hoops used in confined C5-CCFT3 were 241 MPa and 317 MPa, respectively.

3.3   Testing methods

All the model columns were tested using the test setup shown in Fig. 5. The test setup can apply lateral loading using a pseudo controlled hydraulic actuator to large-scale model column in a condition of vertical cantilever. A constant axial load of 2000 kN was applied to the column through post-tensioning two 50 mm diameter high-strength steel rods using two 1500 kN capacity hydraulic hollow jacks. The forces of the rods were transferred to the column by a cross beam mounted on top of the load stub. In order to eliminate the bending of the high-strength rods, a specially designed pin was connected to the lower end of each rod.

The axial load applied to the column was measured by a set of strain gauges affixed on the high-strength rods. The imposed lateral displacement was measured by both the displacement transducer of the actuator and a separate linear potentiometer. The corresponding lateral force was recorded by the built-in load cell of the actuator. Electrical resistance strain gauges were affixed on the surfaces of the steel tube and the additional confinements near column end.

During testing, the axial load was maintained constant by the hydraulic system, whereas the lateral force was cycled under lateral displacement control condition. Three single cycles were initially applied corresponding to an increment of 0.25% peak drift ratio, Δ/L, here Δ is the lateral displacement and L is the clear length of the model column measure between the application point of the lateral force and the top of footing. Then, three repetitive loading cycles were applied for each of the peak drift ratios, $\Delta/L = 1\%$, 1.5%, 2%, 3%, 4%, 6%, 8% and 10%. Such standard loading procedure was attempted until the stage where the model column under testing was judged as unsuitable for further loading.

3.4   Experimental results

3.4.1   Hysteretic responses
As shown in Fig. 6(a), model column C1-CFT with a D/t ratio of 112 had a stable behavior only until cycles corresponding to a peak drift ratio of 2% in the push direction whereas 1.5% in the pull direction. At these stages, the so-called "elephant foot" type local buckling of the steel tube was observed at the position about

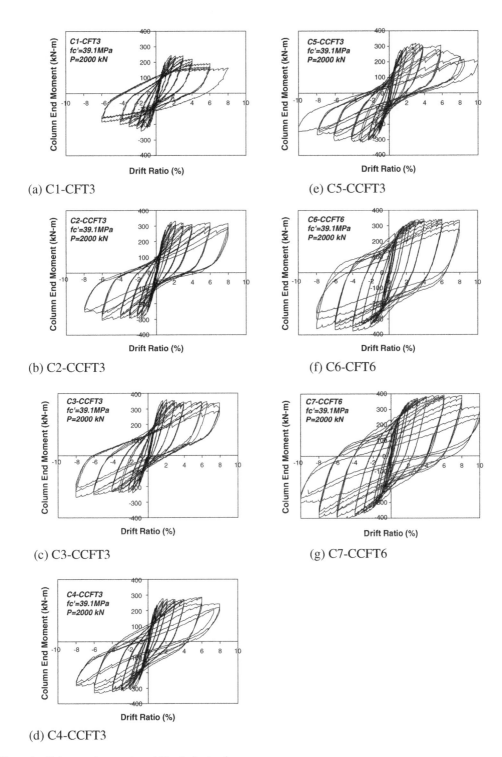

Figure 6.   Column end moment vs. drift ratio hysteretic curve.

20 mm from the bottom end of the column. In the subsequent loading cycles, the local buckling of the steel tube severed, forming several cripples in the column end region with a length approximately equal to the diameter. During the loading cycles corresponding to a peak drift ratio exceeding 3%, the column sections within the end region expanded drastically indicating the insufficient lateral confinement provided only by the thin steel tube. The moment carrying capacity degraded below 80% of the maximum value at 4% peak drift ratio in the pull loading direction. The loading was continued until the loss of the axial loading capacity during the first cycle at 8%.

As shown in Fig. 6(f), with a thicker tube, CFT model C6-CFT6 behaved satisfactorily until a drift ratio of 8%. Local buckling and low cycle fatigue caused rupture of the steel tube along the circular section at a height of about 50 mm.

Drastically improved behavior can be seen from Fig. 6(b) for CCFT model column C2-CCFT. The additional CFRP jacket wrapped in the potential plastic hinge region effectively restrained the local buckling of the steel tube and provided better confinement to the section. As the consequence, the CCFT column was able to develop a ductile and stable hysteretic behavior until a peak drift ratio of 8%, where the test was terminated due to rupture of steel tube. Rupture of the CFRP jacket near the column end was observed during the cycles corresponding to a drift ratio of 6%. The first rupture was initiated in the compression side of the column at a height of about 30 mm from the bottom end. The rupture of CFRP then formed a ring of about 20 mm wide where the steel tube was exposed and subsequently squeezed out due to the "elephant foot" type local buckling. Local buckling was also observed above the CFRP jacket. As shown in Fig. 8(b).

Comparing the hysteretic loops shown in Fig. 6, it is clear that other CCFT model columns also demonstrated improved behavior compared with their counterpart CFT model.

3.4.2 Skeleton curve

Figure 7 shows the influence of confinement affection on the lateral load versus lateral drift ratio envelope curves. Test results indicate that from initial loading stage till yielding, corresponding to a drift ratio of 1.5% or so, there are no significant differences between the counterpart model column and those confined ones. But after yielding, the curves are different due to the confinements: for counterpart specimen C1-CFT3, with a larger diameter to thickness ratio, load bearing capacity reached maximum rapidly and decreased in the subsequent loading cycles, while for the confined specimens, the loading capacity reached maximum after significant deformation and almost kept constant during the following loading cycles, as

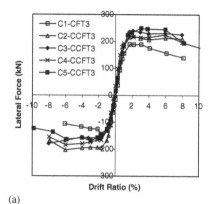

(a)

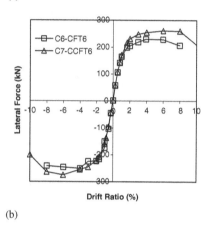

(b)

Figure 7. Skeleton curves comparison: (a) 3 mm thick steel tube; (b) 6 mm thick steel tube.

shown in Fig. 7(a). The differences of loading capacity and ductility between the two 6 mm thick steel tube specimens are illustrated in Fig. 7(b). It can be found that the additional transverse confinement obviously improves the ductility of specimens.

The final conditions of specimens are compared in Fig. 8. It is concluded that the delaying of local buckling of the steel tube and efficient confinement contributed to the improved seismic behavior of the CCFT column compared with the counterpart CFT specimen.

4 CONCLUSIONS

The paper presented a study on seismic behavior of a new CFT column system, named as CCFT, in which additional confinement is provided to improve seismic performance. The proposed CCFT columns combine the advantages of conventional CFT column and tubed RC column systems. Simulated seismic loading tests

674

(a) C1-CFT3 (b) C2-CCFT3

(c) C3-CCFT3 (d) C4-CCFT3

(e) C5-CCFT3 (f) C6-CFT6

(g) C7-CCFT6

Figure 8. Final conditions of specimens.

on large-scale specimens described in this paper successfully show superior seismic performance of the CCFT columns.

ACKNOWLEDGEMENTS

Main support for this study was provided by the Cheung Kong Scholarship awarded to the first author. Supports were also provided by the NSF China (50278032) and the Hunan Provincial Science and Technology Bureau (02JJY3046). The authors would like to thank the following gentlemen for their helps at various stages during the study: Prof. Yi, Weijian; Prof. Liu, Yijiang; Prof. Guo, Yurong; Mr. Shan, Bo; Mr. Zhang, Guowei; Mr. Mao, Weifeng. The research was also conducted as collaboration between the Hunan University and the University of Southern California, USA.

REFERENCES

CAI, S.H. 1989. Analysis and Application of Concrete Filled Steel Tubular Structures[M]. *Beijing: China Construction Press.*

Council on Tall Buildings and Urban Habitat. 1992. McGraw Hill: Cast-inPlace Concrete in Tall Building Design and Construction – 7.3 concrete filled steel tubes [C] 202–222.

Priestley, M.J.N., Seible, F., Xiao, Y. & Verma, R. 1994. Steel Jacket Retrofit of Square RC Bridge Columns for Enhanced Shear Strength – Part 1 – Theoretical Considerations and Test Design [J]. *American Concrete Institute ACI Structural Journal* 91(4): 394–405.

Priestley, M.J.N., Seible, F., Xiao, Y. & Verma, R. 1994. Steel Jacket Retrofit of Square RC Bridge Columns for Enhanced Shear Strength – Part 2 – Experimental Results [J]. *American Concrete Institute ACI Structural Journal* 91(5): 537–551.

Sakino, K. & Tomii, M. 1981. Hysteretic Behavior of Concrete Filed Square Steel Tubular Beam-Columns Failed in Flexure [J]. *Transactions of Japan Concrete Institute* 3: 65–72.

Teng, J.G., Chen, J.F., Smith, S.T. & Lam, T. 2000. RC Structures Strengthened with FRP Composites [R]. *Hong Kong: Research Centre for Advanced Technology in Structural Engineering, Department of Civil Engineering, Hong Kong Polytechnic University* 134.

Tomii, M. Sakino, K. Watanabe, K. & Xiao, Y. 1985. Lateral Load Capacity of Reinforced Concrete Short Columns Confined by Steel Tube [A]. *Harbin: Proceedings of the International Specialty Conference on Concrete Filled Steel Tubular Structures* [C]. 19–26.

Tomii, M. Sakino, K., Xiao, Y. & Watanabe, K. 1985. Earthquake Resisting Hysteretic Behavior of Reinforced Concrete Short Columns Confined by Steel Tube [A]. *Harbin: Proceedings of the International Specialty Conference on Concrete Filled Steel Tubular Structures* [C] 119–125.

Tomii, M. & Sakino, K. & Xiao, Y. 1987. Ultimate Moment of Reinforced Concrete Short Columns Confined in Steel Tube [A]. *New Zealand: Proceedings of Pacific Conference on Earthquake Engineering* [C]. 2: 11–22.

Xiao, Y., Tomii, M. & Sakino, K. 1986. Design Method to Prevent Shear Failure of Reinforced Concrete Short Circular Columns by Steel Tube Confinement (in Japanese) [A]. *Proceedings of the Annual Conference of Japan Concrete Institute* [C]. 517–520.

Xiao, Y., Tomii, M. & Sakino, K. 1986. Experimental Study on Design Method to Prevent Shear Failure of Reinforced Concrete Short Circular Columns by Confining in Steel Tube [J]. *Transactions of Japan Concrete Institute* 8: 535–542.

Xiao, Y. & Ma, R. 1997. Seismic Retrofit of RC Circular Columns Using Prefabricated Composite Jacketing [J]. *ASCE Journal of Structural Engineering* 123(10): 1357–1364.

Xiao, Y. & Mahin, S. 2000. Composite and Hybrid Structures [A]. Los Angeles: *Proceedings of 6th ASCCS International Conference* [C]. 1220. ISBN 0-9679749-0-9.

Xiao, Y. & Wu, H. 2000. Compressive Behavior of Concrete Stub Columns Confined by Fiber Composite Jackets [J]. *Journal of Materials, ASCE* 12(2): 139–146.

675

Xiao, Y. 2001. From Steel Tubed Columns to FRP Tubed Columns [A]. *Washington DC: Proceedings of ASCE Structural Congress* [C]. 2001.

Xiao, Y. & Wu, H. 2003. Retrofit of Reinforced Concrete Columns Using Partially Stiffened Steel Jackets [J]. *Journal of Structural Engineering, ASCE* 129(6): 725–732.

Xiao, Y. & He, W. 2003. Development of Confined CFST Columns (in Chinese) [J]. *Journal of Harbin Institute of Technology*. 8: 40–42, 47.

Xiao, Y., He, W., Mao, X., Choi, K. & Zhu P. 2006 Confinement Design of CFT Columns for Improved Seismic Performance [A], *Taipei, China: Proceedings of the International Workshop on Steel and Concrete Composite Construction (IWSCCC-2003)* [C]. 10: 217–226.

Xiao, Y., He, W. & Mao, X. 2004. Development of Confined Concrete Filled Tubular (CCFT) Columns [J]. *Journal of Building Structures* 25(6): 59–66.

Xiao, Y., He, W. & Choi, K. 2005. Confined Concrete Filled Tubular Columns [J]. *Journal of Structural Engineering, ASCE* 131(3): 488–497.

Zhong, S.T. 1987. Concrete Filled Steel Tubular Column Structures [M]. *Harbin: Heilongjiang Science Press.*

Tubular Structures XI – Packer & Willibald (eds)
© 2006 Taylor & Francis Group, London, ISBN 0-415-40280-8

Proposition of multi-column pier system with high seismic performance using steel tubes for elevated bridges

M. Matsumura & T. Kitada
Osaka City University, Osaka, Japan

S. Okashiro
Japan Bridge Engineering Center, Tokyo, Japan

ABSTRACT: Steel bridge piers of high seismic performance and economical advantage are required in Japan especially after the Hyogo-ken Nambu Earthquake. Presented in this paper are a design concept and a numerical calculation on a newly developing multi-column pier system with high seismic performance for elevated bridges using plural standardized steel tubes for structural purpose. Fundamental behavior and adoptability of the multi pier system and its design concept are verified through FEM analysis.

1 INTRODUCTION

Steel structures are more expensive than RC structure. On the other hand, the use of steel brings advantages of superior constructability at the site because fabricated structural members make construction period short and light weight makes the inertial force of superstructure small.

Steel bridge pier with a single column, supporting bridge superstructures in an elevated bridge, had been brought severe damage by cyclic seismic force at the Hyogo-ken Nambu Earthquake (Fukumoto et al. 1996). After the earthquake, existing steel bridge piers have been retrofitted to improve their ductility and deformation capacity according to the guideline on seismic retrofitting methods (HEPC, 1996). Steel bridge piers however have been constructed at the place where RC bridge piers are unfit, for instance, on a soft ground in bay areas and in downtown areas of tight spatial restrictions and so on even after the earthquake. Therefore steel bridge piers of high seismic performance and economical advantage are strongly required (JSCE, 1995).

Easiness in retrofitting work as well as clearness of design concept can be a key in future constructions of steel structures. Then the following items are to be considered; i) Use of standardized steel tubes for structural purpose, ii) Adoption of composite structure, iii) Easiness in repair and retrofitting work, iv) Simple anchoring system to the basement and so on.

Presented in this paper are a design concept and a numerical calculation example on a newly developing multi-column pier system for the elevated bridges, using plural numbers of steel tubes for structural purpose.

Firstly presented are the design concept of the multi-column pier system and the results of trial design against 1000 gal of the elastic response acceleration as seismic ground motion, which is highly probable to occur during service period of bridge and is defined in Japanese Specifications for Highway Bridges (JRA, 2002). Three dimensional elasto-plastic and finite displacement analysis is carried out in order to verify the adoptability of the design concept and to obtain fundamental behavior of the multi-column pier system.

2 MULTI-COLUMN PIER SYSTEM

The multi-column pier system consists of plural steel tubes, which are the standardized steel tubes for structural purpose in the Japanese Industrial Standard (JIS). The configuration of the multi-column pier system and the elevated bridge are illustrated in Figure 1.

In the multi-column pier system, a center column and 8 surrounding ones are arrayed and all the columns are connected by lateral connectors in the longitudinal direction. The dead load of the bridge superstructures are supported by the center column and the lateral force and shear force generated by seismic force are supported by the surrounding columns in the multi-column pier system. For the effective work of such design scheme for the multi-column pier system, the center column is constructed firstly and supports bridge superstructures. Both the surrounding columns

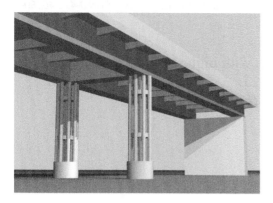

Figure 1. Configuration of multi-column pier system.

and the lateral connections are installed to right places not to deliver any initial load.

Characteristics of the multi-column pier system can be summarized as follows:

– Adoption of CFT and the standardized steel tube for structural purpose can reduce the total costs and manufacturing period in construction.
– Clear separation of structural function between the center column and the surrounding ones can bring easiness in structural design calculation.
– Rigid connections to the superstructure and the basement structure can reduce the stress concentrations, which cause fatigue problem.
– Plural numbers of the surrounding columns, which work against the inertial forces due to seismic motion, bring seismically advantageous because the sudden collapse of the bridge pier can be escaped.

3 TRIAL DESIGN OF MULTI-COLUMN PIER SYSTEM

The multi-column pier system can be designed simply according to the following design conditions and assumptions.

– Center column is designed as a column member subjected to an axial compression force to have enough cross-sectional area to support the dead load of the bridge superstructure.
– Diameter and plate thickness and setting location of the surrounding columns are decided so that the working stresses are within elastic against Level 1 earthquake, which is defined in Japanese Specifications for Highway Bridges (JRA, 2002).

In the following trial design calculations, the dead load of the bridge superstructures and the height of the pier are supposed 7155 kN and 19.2 m by referring to an existing 3 span continuous steel girder elevated bridges. All the steel tubes are made of STK

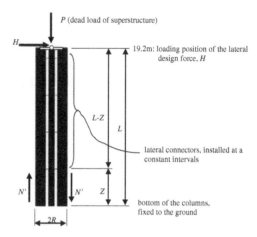

Figure 2. Design model of multi-column pier system for design calculation.

400 ($\sigma_Y = 235\,\text{N/mm}^2$) and are filled with concrete ($\sigma_{ck} = 30\,\text{N/mm}^2$). 1000 gal of the elastic response acceleration of the structure is considered as the design lateral force, which is subjected to the top of the pier. An appropriate value of the elastic response acceleration is to be investigated and determined in the future by considering the dynamic behavior of the pier system.

Illustrated in Figure 2 is a design model of the multi-column pier system in the design calculation. All the columns are rigidly connected at the top part of the columns and the lateral connectors are installed at a constant spacing in the longitudinal direction.

3.1 Design of center column

The center column supporting the dead load of bridge superstructure is designed according to Equation 1 as the column member subjected to the axial compression force.

$$\sigma_Y/\nu \times A_S + \sigma_{ck} \times A_C > P \qquad (1)$$

where ν = safety factor (=1.7); A_S and A_C = cross sectional area of steel tube and concrete inside; P = axial compression load of the dead load; and σ_{ck} = design strength of concrete (=30 N/mm^2).

3.2 Design of surrounding columns

The surrounding columns are designed as the column member subjected to both the axial force and the shear force. The surrounding columns are arranged on a circumference with radius R.

3.2.1 Spacing between lateral connectors

As bending damage at base of the pier predominates, the height of the lowest lateral connector is decided

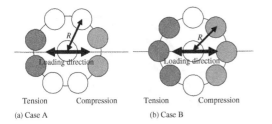

Tension   Compression   Tension   Compression

(a) Case A   (b) Case B

Figure 3. Loading direction considered in design calculation.

to satisfy Equation 2 so that the column buckling can be prevented, that is, the slenderness parameter of the lowest part is less than 0.2.

$$\lambda = \frac{1}{\pi}\sqrt{\frac{\sigma_y}{E}} \times \frac{l_e}{r} < 0.2 \qquad (2)$$

where l_e = effective buckling length of column member ($= 1.0 \times Z$); Z = column length in the lowest part; and r = radius of gyration of area in steel tube.

3.2.2 Design of cross sectional area of surrounding columns

Two of the design cases, Case A and Case B, are considered in the design calculation to decide the radius of the circumference locating the surrounding columns and the size of the surrounding columns as shown in Figure 3.

The differences between Case A and Case B are the loading direction of the lateral design force and the numbers of the surrounding columns, which distribute the axial force, N'. N' are axial compressive and tensile forces generated by the design lateral force as shown in Figure 2. It is supposed that the compressive and tensile forces are shared by two of the surrounding columns in Case A and by three of them in Case B, respectively. In both the cases, the shear force is distributed by all the column members. Moreover, it is assumed that steel and concrete can work together against axial compressive force and concrete subjected to axial tensile force can be neglected.

According to these assumptions, the dimensions of the surrounding columns are decided by Equations 3 and 4.

$$n' \times (\sigma_Y /\nu \times A_S + \sigma_{ck} A_C) > N' \qquad (3)$$

$$n' \times (\sigma_Y \times A_S) > N' \qquad (4)$$

$$H \times (L - Z) = 2 \times N' \times R \qquad (5)$$

where L = full length of the pier; H = lateral force; N' = design axial force distributed by surrounding

Table 1. Dimensions of steel tubes.

Case	D (mm)	t (mm)	R (mm)	t_{eq} (mm)
(a) Surrounding columns				
A	818.2	16	5600	36.03
B	818.2	16	4800	36.03
(b) Center column				
A	500	9		21.37
B	500	9		21.37

D, t, R and t_{eq} indicate diameter and plate thickness of steel tube, radius of the circumference locating the surrounding columns and equivalent plate thickness.

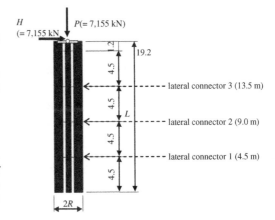

Figure 4. Results of design calculation (unit in m).

columns; and n' = numbers of surrounding columns distributing N'.

For more simplicity, CFT in compression side are modeled as the steel tube with an equivalent plate thickness t_{eq}, decided by Equation 6.

$$A' = A_S + A_C /10 \qquad (6)$$

where A' = cross sectional area of steel tube with t_{eq}.

3.3 Calculation results

The dimensions of all the steel tubes are selected among the prepared size of the standardized steel tube and are summarized in Table 1 and Figure 4.

The center column is diameter, $D = 500$ mm and plate thickness, $t = 9$ mm. Between Case A and Case B, radius R for the circumference locating the surrounding columns takes different values and Z does the same length. Spacing between the lateral connectors takes 4.5 m in the longitudinal direction. An equivalent plate thickness, t_{eq}, where concrete is replaced by steel tube with heavier wall, can be derived from Equation 6.

679

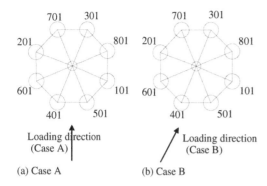

(a) Case A   (b) Case B

Figure 5.   Column IDs and loading direction.

Table 2.   Analytical cases.

ID	Explanation
A-0	lateral connectors are modeled by rigid element
A-1	lateral connectors are modeled by elastic element
A-2	lateral connector 1 is omitted from A-1
A-3	lateral connectors 1 & 2 are omitted from A-1
A-B	horizontal load is applied to design direction B to A-1
B-0	lateral connectors are modeled by rigid element
B-1	lateral connectors are modeled by elastic element
B-2	lateral connector 1 is omitted from B-1
B-3	lateral connectors 1 & 2 are omitted from B-1
B-A	horizontal load is applied to design direction A to B-1

4   ELASTO-PLASTIC AND FINITE DISPLACEMENT ANALYSIS

The loading capacity of the designed pier system is computed by the elasto-plastic and finite displacement analysis by the program EPASS/USSP (Kano et al. 2005). The tube members in the longitudinal direction are modeled by using elasto-plastic beam-column elements. The cross-sectional configurations of the lateral connectors but the top one are the steel tubes of $D = 600$ mm and $t = 12$ mm and the top one rigidly connect the center columns and the surrounding columns.

4.1   Analytical model and condition

Illustrated in Figure 5 are analytical cases corresponding to the two design cases. All the analytical cases are listed in Table 2. The numbers and/or stiffness of the lateral connectors are varied.

Analytical Cases A-2 and B-2, in which the lateral connectors of the lowest part are omitted from A-1 and B-2, take 0.35 of the slenderness parameter of the column members. Case A-B means that cross-sectional areas of the columns are designed by Case A and the lateral design force at the top of the pier system is

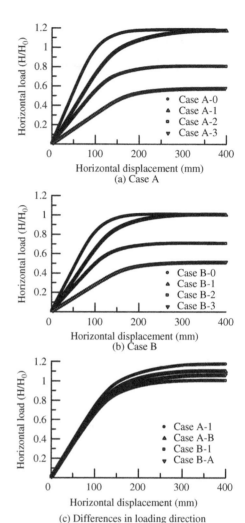

(a) Case A

(b) Case B

(c) Differences in loading direction

Figure 6.   Relationship between lateral load and lateral displacement.

subjected in the analysis to the direction considered in Case B.

The stress–strain relationship of steel material are modeled as a bi-linear relationship without considering the strain-hardening effect and the yield stress and the elastic modulus are 235 N/mm^2 and 2.0×10^5 N/mm^2, respectively.

In the analysis, the applied force due to the dead load of the superstructure is introduced to the center column as pre-stress and the lateral load H_0 is increased at the top of the pier.

4.2   Analytical results

Plotted in Figure 6 is the relationship between the applied lateral load, H divided by the designed lateral

Table 3. Axial forces of all the columns when applied tensile force reaches to 0.9 of yield state of surrounding columns (Analytical case A-1, unit in kN).

No. of column	1	301	201	101	401
		701	801	601	501
	101	−15,598	−1944	5442	8471

Table 4. Axial forces of all the columns when applied tensile force reaches to 0.9 of yield state of surrounding columns (Analytical case B-1, unit in kN).

No. of column	1	301	701	101	501	
			801	201	601	401
	−3502	−14,198	−8088	2317	6805	8471

load, H_0 and the lateral displacement at the top of the pier system. Tables 3 and 4 indicate the axial force of all the columns when the applied tensile force to surrounding columns reaches to 90% of their yield state in the analytical Cases A-1 and B-1.

As Case A and Case B show the similar tendency as shown in Figure 6, the loading directions supposed in the design do not influence on the loading capacity of the multi-column pier system when all the lateral connectors works effectively.

Tables 3 and 4 indicate that the axial forces predominantly are subjected to the surrounding columns and the neutral axis of the pier system is positioned in the compression side from the center column. The more accurate modeling for the analysis is needed but the bending moment of the columns are not so large compared with the axial forces as it is assumed in the design calculations.

5 CONCLUSION

In order to develop a new steel bridge pier of high seismic performance and of economical advantage, the multi-column pier system using plural numbers of concrete filled steel tubes for structural purpose is suggested in this paper. Further investigations are necessary, but the adoptability of the design concept and assumptions of the multi-column pier system are verified.

Effects of live load, dynamic response during an earthquake and calculation on anchoring technique to the basement structure are not included in this paper but will be investigated in near future.

ACKNOWLEDGEMENT

The authors would like to thank Dr. Katsuhiro Tanaka at JIP Techno Science Co., Ltd. for his advice on modeling of the multi-column pier system for FEM analysis.

REFERENCES

Fukumoto, Y., Watanabe, E., Kitada, T., Suzuki, I., Horie, Y. and Sakota, H. 1996. Reconstruction and repair of steel highway bridges damaged by the Great Hanshin Earthquake, *Proceedings of Bridge Management 3*, University of Surry, Guildford, UK: 8–16.

HEPC (Hanshin Expressway Public Corporation), 1998. *Guide-lines for seismic retrofitting methods for existing steel bridge piers (Draft)* (in Japanese).

JRA (Japan Road Association), 2002. *Japanese Specification for Highway Bridges, Part V. Seismic Design* (in Japanese).

JSCE (Japanese Society of Civil Engineers). 1996. *A proposal for seismic design of steel bridges,* Seismic Design WG. Committee on New Technology for Steel Structures (in Japanese).

Kano, M., Tanaka, K., Yamaguchi, T. and Kitada, T. 2005. Nonlinear static/dynamic FEM system for analyzing spatial bridge structures consisting of thin-walled steel and composite members, *Proceedings of 3rd International Conference on Structural Stability and Dynamics*, Florida, USA.

Tubular Structures XI – Packer & Willibald (eds)
© 2006 Taylor & Francis Group, London, ISBN 0-415-40280-8

Author index

Product Safety Concerns and Information please contact our EU
representative GPSR@taylorandfrancis.com Taylor & Francis Verlag GmbH,
Kingerstraße 24, 80331 München, Germany

ted and bound by CPI Group (UK) Ltd, Croydon, CR0 4YY